Names, Formulas, and Charges of Common Ions

Positive Ions (Cations)

1+	Ammonium	NH_4^+
	Copper(I) (Cuprous)	Cu^+
	Hydrogen	H^+
	Potassium	K^+
	Silver	Ag^+
	Sodium	Na^+
2+	Barium	Ba^{2+}
	Cadmium	Cd^{2+}
	Calcium	Ca^{2+}
	Cobalt(II)	Co^{2+}
	Copper(II) (Cupric)	Cu^{2+}
	Iron(II) (Ferrous)	Fe^{2+}
	Lead(II)	Pb^{2+}
	Magnesium	Mg^{2+}
	Manganese(II)	Mn^{2+}
	Mercury(II) (Mercuric)	Hg^{2+}
	Nickel(II)	Ni^{2+}
	Tin(II) (Stannous)	Sn^{2+}
	Zinc	Zn^{2+}
3+	Aluminum	Al^{3+}
	Antimony(III)	Sb^{3+}
	Arsenic(III)	As^{3+}
	Bismuth(III)	Bi^{3+}
	Chromium(III)	Cr^{3+}
	Iron(III) (Ferric)	Fe^{3+}
	Titanium(III) (Titanous)	Ti^{3+}
4+	Manganese(IV)	Mn^{4+}
	Tin(IV) (Stannic)	Sn^{4+}
	Titanium(IV) (Titanic)	Ti^{4+}
5+	Antimony(V)	Sb^{5+}
	Arsenic(V)	As^{5+}

Negative Ions (Anions)

1−	Acetate	$C_2H_3O_2^-$
	Bromate	BrO_3^-
	Bromide	Br^-
	Chlorate	ClO_3^-
	Chloride	Cl^-
	Chlorite	ClO_2^-
	Cyanide	CN^-
	Fluoride	F^-
	Hydride	H^-
	Hydrogen carbonate (Bicarbonate)	HCO_3^-
	Hydrogen sulfate (Bisulfate)	HSO_4^-
	Hydrogen sulfite (Bisulfite)	HSO_3^-
	Hydroxide	OH^-
	Hypochlorite	ClO^-
	Iodate	IO_3^-
	Iodide	I^-
	Nitrate	NO_3^-
	Nitrite	NO_2^-
	Perchlorate	ClO_4^-
	Permanganate	MnO_4^-
	Thiocyanate	SCN^-
2−	Carbonate	CO_3^{2-}
	Chromate	CrO_4^{2-}
	Dichromate	$Cr_2O_7^{2-}$
	Oxalate	$C_2O_4^{2-}$
	Oxide	O^{2-}
	Peroxide	O_2^{2-}
	Silicate	SiO_3^{2-}
	Sulfate	SO_4^{2-}
	Sulfide	S^{2-}
	Sulfite	SO_3^{2-}
3−	Arsenate	AsO_4^{3-}
	Borate	BO_3^{3-}
	Phosphate	PO_4^{3-}
	Phosphide	P^{3-}
	Phosphite	PO_3^{3-}

Atomic Masses of the Elements
Based on the 1995 IUPAC Table of Atomic Masses

Name	Symbol	Atomic Number	Atomic Mass	Name	Symbol	Atomic Number	Atomic Mass
Actinium*	Ac	89	227.0277	Mercury	Hg	80	200.59
Aluminum	Al	13	26.981538	Molybdenum	Mo	42	95.94
Americium*	Am	95	243.0614	Neodymium	Nd	60	144.24
Antimony	Sb	51	121.760	Neon	Ne	10	20.1797
Argon	Ar	18	39.948	Neptunium*	Np	93	237.0482
Arsenic	As	33	74.92160	Nickel	Ni	28	58.6934
Astatine*	At	85	209.9871	Niobium	Nb	41	92.90638
Barium	Ba	56	137.327	Nitrogen	N	7	14.00674
Berkelium*	Bk	97	247.0703	Nobelium*	No	102	259.1011
Beryllium	Be	4	9.012182	Osmium	Os	76	190.23
Bismuth	Bi	83	208.98038	Oxygen	O	8	15.9994
Bohrium	Bh	107	—	Palladium	Pd	46	106.42
Boron	B	5	10.811	Phosphorus	P	15	30.973762
Bromine	Br	35	79.904	Platinum	Pt	78	195.078
Cadmium	Cd	48	112.411	Plutonium*	Pu	94	244.0642
Calcium	Ca	20	40.078	Polonium*	Po	84	208.9824
Californium*	Cf	98	251.0796	Potassium	K	19	39.0983
Carbon	C	6	12.0107	Praseodymium	Pr	59	140.90765
Cerium	Ce	58	140.116	Promethium*	Pm	61	144.9127
Cesium	Cs	55	132.90545	Protactinium*	Pa	91	231.03588
Chlorine	Cl	17	35.4527	Radium*	Ra	88	226.0254
Chromium	Cr	24	51.9961	Radon*	Rn	86	222.0176
Cobalt	Co	27	58.933200	Rhenium	Re	75	186.207
Copper	Cu	29	63.546	Rhodium	Rh	45	102.90550
Curium*	Cm	96	247.0703	Rubidium	Rb	37	85.4678
Dubnium	Db	105	—	Ruthenium	Ru	44	101.07
Dysprosium	Dy	66	162.50	Rutherfordium	Rf	104	261.1089
Einsteinium*	Es	99	252.0830	Samarium	Sm	62	150.36
Erbium	Er	68	167.26	Scandium	Sc	21	44.955910
Europium	Eu	63	151.964	Seaborgium	Sg	106	—
Fermium*	Fm	100	257.0951	Selenium	Se	34	78.96
Fluorine	F	9	18.9984032	Silicon	Si	14	28.0855
Francium*	Fr	87	233.0197	Silver	Ag	47	107.8682
Gadolinium	Gd	64	157.25	Sodium	Na	11	22.989770
Gallium	Ga	31	69.723	Strontium	Sr	38	87.62
Germanium	Ge	32	72.61	Sulfur	S	16	32.066
Gold	Au	79	196.96655	Tantalum	Ta	73	180.9479
Hafnium	Hf	72	178.49	Technetium*	Tc	43	97.9072
Hassium	Hs	108	—	Tellurium	Te	52	127.60
Helium	He	2	4.002602	Terbium	Tb	65	158.92534
Holmium	Ho	67	164.93032	Thallium	Tl	81	204.3833
Hydrogen	H	1	1.00794	Thorium*	Th	90	232.0381
Indium	In	49	114.818	Thulium	Tm	69	168.93421
Iodine	I	53	126.90447	Tin	Sn	50	118.710
Iridium	Ir	77	192.217	Titanium	Ti	22	47.867
Iron	Fe	26	55.845	Tungsten	W	74	183.84
Krypton	Kr	36	83.80	Ununnilium	Uun	110	—
Lanthanum	La	57	138.9055	Unununium	Uuu	111	—
Lawrencium*	Lr	103	262.110	Ununbium	Uub	112	—
Lead	Pb	82	207.2	Uranium*	U	92	238.0289
Lithium	Li	3	6.941	Vanadium	V	23	50.9415
Lutetium	Lu	71	174.967	Xenon	Xe	54	131.29
Magnesium	Mg	12	24.3050	Ytterbium	Yb	70	173.04
Manganese	Mn	25	54.938049	Yttrium	Y	39	88.90585
Meitnerium	Mt	109	—	Zinc	Zn	30	65.39
Mendelevium*	Md	101	258.0984	Zirconium	Zr	40	91.224

*This element has no stable isotopes. The atomic mass given is that of the isotope with the longest known half-life.

FOUNDATIONS OF
COLLEGE CHEMISTRY

FOUNDATIONS OF COLLEGE CHEMISTRY

TENTH EDITION

Morris Hein
Mount San Antonio College

Susan Arena
University of Illinois, Urbana-Champaign

Brooks/Cole Publishing Company

 An International Thomson Publishing Company

Pacific Grove ▪ Albany ▪ Belmont ▪ Boston ▪ Cincinnati ▪ Johannesburg ▪ London ▪ Madrid
Melbourne ▪ Mexico City ▪ New York ▪ Scottsdale ▪ Singapore ▪ Tokyo ▪ Toronto

Sponsoring Editor: *Jennifer Huber*
Project Development Editor: *Beth Wilbur*
Marketing Team: *Steve Catalano,*
   Christine Davis, and Christina DeVeto
Editorial Associate: *Nancy Conti*
Production Service: *Ex Libris*
Production Editor: *Jamie Sue Brooks*
Manuscript Editor: *Luana Richards*
Text Design: *Nancy Benedict Breuer*

Cover Design: *Roy A. Neuhaus*
Cover Photo: *H. A. Roberts/*
   American Stock Photography
Photo Researcher: *Stuart A. Kenter*
Illustrations: *Lotus Art and Pat Rogondino*
Typesetting: *The Clarinda Company*
Cover Printing: *Phoenix Color Corp.*
Printing and Binding: *R. R. Donnelley*
   & Sons, Willard, Ohio

For more information, contact:

BROOKS/COLE PUBLISHING COMPANY
511 Forest Lodge Road
Pacific Grove, CA 93950
USA

International Thomson Publishing Europe
Berkshire House 168-173
High Holborn
London WC1V 7AA
England

Thomas Nelson Australia
102 Dodds Street
South Melbourne, 3205
Victoria, Australia
Nelson Canada

1120 Birchmount Road
Scarborough, Ontario
Canada M1K 5G4

International Thomson Editores
Seneca 53
Col. Polanco
11560 México, D. F., México

International Thomson Publishing GmbH
Königswinterer Strasse 418
53227 Bonn
Germany

International Thomson Publishing Asia
60 Albert Street
#15–01 Albert Complex
Singapore 189969

International Thomson Publishing Japan
Palaceside Building, 5F
1-1-1 Hitotsubashi
Chiyoda-ku, Tokyo 100-0003
Japan

Printed in the United States of America

10 9 8 7 6 5 4 3 2 1

Library of Congress Cataloging-in-Publication Data
Hein, Morris.
   Foundations of college chemistry / Morris Hein, Susan Arena.—
10th ed.
      p.   cm.
   Includes index.
   ISBN 0-534-35749-0 (acid-free, recycled paper)
   1. Chemistry.   I. Arena, Susan,   II. Title.
QD33.H45 1998b
540—dc21
                                            98-48677
                                            CIP

An Education isn't how much you have committed to memory, or even how much you know. It's being able to differentiate between what you do know and what you don't. It's knowing where to go to find out what you need to know; and it's knowing how to use the information once you get it.

William Feather

PREFACE

Foundations of College Chemistry was originally intended for students who had never taken a chemistry course or those who had a significant interruption in their studies but planned to continue with the general chemistry sequence. Since its inception this book has helped define the preparatory chemistry course and has since developed a much wider audience. In addition to preparatory chemistry, our text has been used extensively in one-semester general purpose courses (such as those for applied health fields) and in courses for nonscience majors. Our goal for this edition is to continue to make introductory chemistry accessible to all beginning students. The central focus is the same as it has been from the first edition: How can we make chemistry interesting and understandable to students and teach them the problem-solving skills they will need?

In preparing the Tenth Edition we considered the comments and suggestions of students and instructors to design a revision that builds on the strengths of previous editions and presents chemistry as a modern, vital subject. We have especially tried to relate chemistry to the real lives of our students as we develop the principles that form the foundation for the further study of chemistry.

Development of Problem-Solving Skills

We all want our students to develop real skills in solving problems. We believe that a key to the success of this text is the fact that our problem-solving approach works for students. It is a step-by-step process that teaches the use of units and shows the change from one unit to the next. Students learn concepts most easily in a step-by-step process. In this edition we continue to use examples to incorporate fundamental mathematical skills, scientific notation, and significant figures. Painstaking care has been taken to show each step in the problem-solving process *(see pp. 108, 140–141)* and to give *alternative methods for solution* where appropriate. These alternative methods give students flexibility in choosing the one that works best for them. We continue to use four significant figures for atomic and molar masses for consistency and for rounding off answers appropriately. We have been meticulous in providing answers, correctly rounded, for students who have difficulty with mathematics.

Fostering Student Skills *Attitude* plays a critical role in problem solving. We encourage students to learn that a systematic approach to solving problems is better than simple memorization. Throughout the book we encourage students to begin by writing down the facts or data given in a problem *(see p. 268)* and to think their way through the problem to an answer, which is then checked to see if it makes sense. Once we have laid the foundations of concepts, we highlight the steps in blue so students can locate them easily. Important rules and equations are highlighted for emphasis and ready reference.

Student Practice Practice problems follow the examples in the text, with answers provided at the end of the chapter. The end of each chapter begins with *Questions,* which help students review key terms and concepts, as well as material presented in tables and figures. This is followed by *Paired Exercises,* covering concepts and numerical exercises, where two similar exercises are presented side by side. The final section, *Additional Exercises,* includes further practice problems presented in a more random order. Challenging questions and exercises are denoted with an asterisk. Answers for *all* even-numbered questions and exercises appear in Appendix VI.

Emphasis on Real-World Aspects

We continue to emphasize the less theoretical aspects of chemistry early in the book, leaving the more abstract theory for later. This sequence seems especially appropriate in a course where students are encountering chemistry for the very first time. Atoms, molecules, and reactions are all an integral part of the chemical nature of matter. A sound understanding of these topics allows the student to develop a basic understanding of chemical properties and vocabulary.

Chapters 2 and 3 presents the basic mathematics and the language of chemistry, including an explanation of the metric system and significant figures. In Chapter 4 we present chemical properties—the ability of a substance to form new substances. Then, in Chapter 5, students encounter the history and language of basic atomic theory.

We continue to present new material at a level appropriate for the beginning student by emphasizing nomenclature, composition of compounds, and reactions in Chapters 6 through 9 before moving into the details of modern atomic theory. The Periodic Table is introduced and discussed in Chapters 10 and 11. Students gain confidence in their own ability to identify and work with chemicals in the laboratory before tackling the molecular models of matter. As practicing chemists we have little difficulty connecting molecular models and chemical properties. Students, especially those with no prior chemistry background, may not share this ability to connect the molecular models and the macroscopic properties of matter. Those instructors who feel it is essential to teach atomic theory and bonding early in the course can cover Chapters 10 and 11 immediately following Chapter 5. Finally, the entire text has been reexamined and the prose updated and rewritten to improve its clarity.

Learning Aids

To help the beginning student gain the confidence necessary to master technical material we have refined and enhanced a series of learning aids:

- Important **terms** are set off in bold type where they are defined, and are printed in blue in the margin. These terms are also defined in the glossary and printed in bold type in the index.
- Worked **examples** show students the how of problem solving before they are asked to tackle problems on their own.
- **Practice problems** permit immediate reinforcement of a skill shown in the example problems. Answers are provided at the end of the chapter to encourage students to check their problem solving immediately.
- **Marginal notations** help students understand basic concepts and problem-solving techniques. These are printed in magenta ink to clearly distinguish them from text and vocabulary terms.

Numerous worked examples
▼

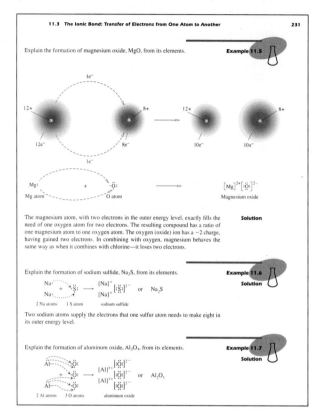

Explain the formation of magnesium oxide, MgO, from its elements.   **Example 11.5**

The magnesium atom, with two electrons in the outer energy level, exactly fills the   **Solution**
need of one oxygen atom for two electrons. The resulting compound has a ratio of
one magnesium atom to one oxygen atom. The oxygen (oxide) ion has a -2 charge,
having gained two electrons. In combining with oxygen, magnesium behaves the
same way as when it combines with chlorine—it loses two electrons.

Explain the formation of sodium sulfide, Na_2S, from its elements.   **Example 11.6**
                                                                         Solution

Two sodium atoms supply the electrons that one sulfur atom needs to make eight in
its outer energy level.

Explain the formation of aluminum oxide, Al_2O_3, from its elements.   **Example 11.7**
                                                                          Solution

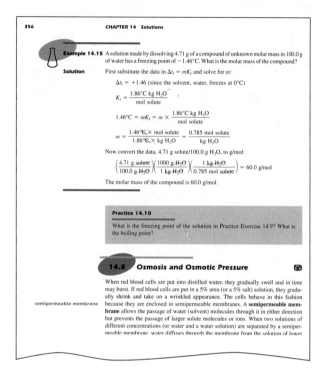

Example 14.15 A solution made by dissolving 4.71 g of a compound of unknown molar mass in 100.0 g
of water has a freezing point of $-1.46°C$. What is the molar mass of the compound?

Solution First substitute the data in $\Delta t_f = mK_f$ and solve for m:

$$\Delta t_f = +1.46 \text{ (since the solvent, water, freezes at } 0°C\text{)}$$

$$K_f = \frac{1.86°C \text{ kg } H_2O}{\text{mol solute}}$$

$$1.46°C = mK_f = m \times \frac{1.86°C \text{ kg } H_2O}{\text{mol solute}}$$

$$m = \frac{1.46°C \times \text{mol solute}}{1.86°C \times \text{kg } H_2O} = \frac{0.785 \text{ mol solute}}{\text{kg } H_2O}$$

Now convert the data, 4.71 g solute/100.0 g H_2O, to g/mol:

$$\left(\frac{4.71 \text{ g solute}}{100.0 \text{ g } H_2O}\right)\left(\frac{1000 \text{ g } H_2O}{1 \text{ kg } H_2O}\right)\left(\frac{1 \text{ kg } H_2O}{0.785 \text{ mol solute}}\right) = 60.0 \text{ g/mol}$$

The molar mass of the compound is 60.0 g/mol.

Practice 14.10

What is the freezing point of the solution in Practice Exercise 14.9? What is
the boiling point?

14.8   Osmosis and Osmotic Pressure

When red blood cells are put into distilled water, they gradually swell and in time
may burst. If red blood cells are put in a 5% urea (or a 5% salt) solution, they gradually shrink and take on a wrinkled appearance. The cells behave in this fashion
semipermeable membrane   because they are enclosed in semipermeable membranes. A **semipermeable mem-**
brane allows the passage of water (solvent) molecules through it in either direction
but prevents the passage of larger solute molecules or ions. When two solutions of
different concentrations (or water and a water solution) are separated by a semipermeable membrane, water diffuses through the membrane from the solution of lower

▲
Examples, practice problems,
and glossary terms

- **Each chapter opens** with a *color photograph* relating the chapter to our daily lives.
  A *chapter outline* assists students in viewing the topics covered in the chapter, and
  the *introductory paragraph* further connects the chapter topic to real life.
- Each chapter contains at least one **Chemistry in Action** section that shows
  the impact of chemistry in a variety of practical applications. These essays
  cover such topics as scuba diving, killer lakes, and gumballs. Other Chemistry
  in Action essays introduce experimental information on new chemical discoveries and applications.
- **Steps for solving problems** are printed in blue for easy reference.
- **Important statements,** equations, and laws are highlighted for emphasis.
- A list of **Concepts in Review** at the end of each chapter guides students in determining the most important concepts in the chapter.
- **Key terms** are listed alphabetically at the end of each chapter with section
  references to assist students in review of new vocabulary. A **glossary** is provided to help students review key terms. As a study aid, section numbers are
  given for each term to guide the student to the contextual definition. The glossary pages are color tinted to provide ready access. Glossary terms are also
  printed in bold type in the index.
- **End of chapter exercises** provide practice and review of the chapter material. *Paired exercises* present two parallel exercises, side by side, so the

student can use the same problem-solving skills with two sets of similar information. Answers to the even-numbered paired exercises are given in Appendix VI.

- **Additional exercises** are provided at the end of most chapters. They are arranged in a more random order, to encourage students to review the chapter material.
- A **Review of Mathematics** is provided in Appendix I.
- Units of measurement are shown in table format in Appendix IV and in the endpapers. (see p. A-18)
- **Answers** to even-numbered questions and exercises are given in Appendix VI.

New to This Edition

A number of changes have been made in this edition, but the level of the material remains the same. The entire text was reviewed and reworked to reflect the latest in chemistry education and to provide students with greater assistance in solving problems. Specific new features and changes include the following:

Chapter 14, "Solutions," has been updated to improve clarity for the students. Material on saturated solutions is now included in the discussion of solubility. Normality has been deleted from the chapter to allow students to focus on the role of the mole in concentrations of solutions.

Chapter 15, "Acids, Bases, and Salts," has been updated with new figures to clarify what is happening at the molecular level. The discussion of colloids has been modernized and streamlined to focus attention on properties and applications.

Chapter 16, "Chemical Equilibrium," has been completely updated to focus student attention on the changes occurring as a system approaches equilibrium.

A fresh, new design includes a modern art and photo program; illustrations for molecules and chemical reactions have been enhanced to improve students' ability to visualize.

- Eight new **Chemistry in Action** essays have been added, including topics that address current environmental concerns such as household pollution and fuel-efficient cars. Other Chemistry in Action sections have been revised and updated.
- We have added more information on the use of calculators, including a new appendix, **Using a Scientific Calculator** (Appendix II).
- **Putting It Together,** a new review section, summarizes every 3 to 4 chapters and includes additional conceptually oriented exercises for effective self-review.
- **Online and Multimedia Resources.** Web icons encourage students to explore chemistry topics on the World Wide Web. The icons are placed at section headings to indicate related Web material. *InfoTrac® College Edition* is a fully searchable online university library with full articles from many periodicals, including *Discover* and *Science World.* Students can print complete articles and join in online chat sessions.

Available in Two Versions

For the convenience of instructors and to accommodate the various lengths of academic terms, two versions of this book are available. *Foundations of College Chemistry,* 10th Edition, includes 20 chapters and is our main text. *Foundations of College Chemistry,* 7th Alternate Edition, provides a shorter, 17 chapter paperbound version with the same material, but without the nuclear, organic, and biochemistry chapters.

A Complete Ancillary Package

The following comprehensive teaching package accompanies these books.

For the student:

Media Resources

Brooks/Cole Chemistry Resource Center. This is Brooks/Cole's website for chemistry, which contains a homepage for *Foundations of College Chemistry.* All information is arranged according to the *Foundations'* table of contents. Students can access flash cards for all glossary terms, supplementary practice and conceptual problems, practice quizzes for every chapter, and hyperlinks that relate to each chapter's contents. In addition, students can interact with the Chemistry Hall of Fame tracing the accomplishments of past and present contributors to the field of chemistry in timeline format. Or, they can explore the chemistry of household items, food, medicinal and cosmetic items, plastics, and other common materials through our Molecule Media Guide which displays structures and describes the physical properties and practical applications of substances that have common use in everyday life.

   InfoTrac® College Edition is an online library available free with each copy of *Foundations of College Chemistry.* It gives students access to full-length articles—not simply abstracts—from more than 700 scholarly and popular periodicals, updated daily, and dating back as much as four years. Student subscribers receive a personalized account ID that gives them four months of unlimited Internet access—at any hour of the day—to readings from *Discover, Science World,* and *American Health* magazines.

Other Study Aids

Study Guide by Peter Scott of Linn-Benton Community College and Rachael Henriques Porter of LeMoyne College is a self-study guide for students. For each chapter, it includes a self-evaluation section, a recap section, one or more "challenge problems," and answers and solutions to all the exercises.

   Solutions Manual by Morris Hein and Susan Arena includes answers and solutions to all end-of-chapter questions and exercises.

   Foundations of Chemistry in the Laboratory, 10th Edition, by Morris Hein, Leo R. Best, Robert L. Miner, and Judith N. Peisen includes 28 experiments for a laboratory program that may accompany the lecture course. Featuring updated information on waste disposal and emphasizing safe laboratory procedures, the lab manual also includes study aids and exercises: Six new class-tested experiments have been added.

A Basic Math Approach to Concepts of Chemistry, 6th edition, by Leo Michels is a self-paced paperbound workbook that has proven an excellent resource for students who need help with mathematical aspects of chemistry. Evaluation tests are provided for each unit, and the test answers are given in the back of the book. A glossary is also included.

For the instructor:

For the Laboratory

Foundations of Chemistry in the Laboratory, 10th Edition, by Morris Hein, Leo Best, Robert Miner, and Judith Peisen, includes 28 experiments (six new to this edition), five study aids, and 19 exercises. It has been completely updated and revised to reflect the most current terminology and environmental standards. In addition, a new study aid that incorporates computer graphing has been added. Instructors can customize their own lab manual to meet the distinct needs of their laboratory by selecting from any of the 28 experiments, adding their own experiments or exercises.

Instructor's Manual for Foundations of Chemistry in the Laboratory, 10th Edition, includes information on the management of the lab, evaluation of experiments, notes for individual experiments, a list of reagents needed, and answer keys to each experiment's report form and to all exercises.

Assessment Tools and Materials *Instructor's Manual with Test Items for Foundations of College Chemistry,* 10th edition, by Morris Hein and Susan Arena. Includes a copy of the test questions provided electronically in Thomson World Class Learning Testing Tools, Review Exercise Worksheets, answers to the test item questions, answers to the Review Exercise Worksheets, and the solutions to the Putting It Together exercises from the main text.

Thomson World Class Learning™ Testing Tools is a fully integrated suite of test creation, delivery, and classroom management tools. This invaluable set of tools includes World Class *Test, Test Online,* and World Class *Manager* software. World Class *Test* allows instructors to create dynamic, algorithmic questions that regenerate the values of variables and calculations between multiple versions of the same test. Tests, practice tests, quizzes, and tutorials created in World Class *Test* can be delivered via paper, diskette or local hard drive, LAN (Local Area Network), or our Internet server. Both testing and tutorial results can then be integrated into a complete classroom management tool with scoring, gradebook, and reporting capabilities.

With World Class *Test* instructors can create a test from an existing bank of objective questions including multiple-choice, true/false, and matching questions; or instructors can easily edit existing questions and add their own questions and graphics. The online system can automatically score *objective* questions. *Subjective* essay and fill-in-the-blank questions that the instructor evaluates can also be added. Results can be scored, merged with final test results, and entered automatically into the gradebook.

Presentation Tools and Online Resources *Transparencies* in full color include illustrations from the text, enlarged for use in the classroom and lecture halls.

CNN Chemistry Video, produced by Turner Learning, can stimulate and engage your students by launching a lecture, sparking a discussion, or demonstrating an application. Each chemistry-related segment from recent CNN broadcasts clearly demonstrates the relevancy of chemistry to everyday life.

With Brooks/Cole's *ChemLink,* a cross-platform CD-ROM, creating lectures has never been easier. Using multi-tiered indexing, search capabilities, and a comprehensive resource bank that includes glossary, graphs, tables, illustrations, photographs, and animations, instructors can conduct a quick search to incorporate these materials into presentations and tests. And, any ChemLink file can be posted to the web for easy student reference.

Brooks/Cole Chemistry Resource Center is Brooks/Cole's website for chemistry, which contains a homepage for *Foundations of College Chemistry* and is described under student media resources.

InfoTrac® College Edition is an online library available free with each copy of *Foundations of College Chemistry.* See the description under student media resources.

Thomson World Class Learning™ Course. This online resource allows a professor to easily post course information, office hours, lesson information, assignments, sample tests, and links to rich web content, including review and enrichment material from Brooks/Cole. Updates are quick and easy and customer support is available 24 hours a day, seven days a week. For more information, go to **www.worldclasslearning.com** on the World Wide Web.

Acknowledgments

Books are the result of a collaborative effort of many talented and dedicated people. Among the friends and colleagues who have helped us, we appreciate the enthusiasm and energy of Connie Grosse who developed test items, and Rachael Henriques who created the Putting It Together sections and revised the study guide for this edition. We are grateful for the many helpful comments from colleagues and students who, over the years, have made this book possible. We hope they will continue to share their ideas for change with us, either directly or through our publisher.

We are especially thankful for the support, friendship, and constant encouragement of our spouses, Edna and Steve, who have endured many lost weekends and been patient and understanding through the long hours of this process. Their optimism and good humor have given us a sense of balance and emotional stability.

No textbook can be completed without the untiring efforts of many publishing professionals. Special thanks to the talented staff at Brooks/Cole, especially Jamie Sue Brooks, Production Services Manager, Beth Wilbur, Project Development Editor, Jennifer Huber, Chemistry Editor, and Melissa Henderson, Assistant Editor, who worked very hard to control the many aspects of this project. Much credit goes to Julie Kranhold of Ex Libris for her unfailing attention to detail and persistence in moving the book through the numerous stages of production. Julie is amazing in her ability to track each change and translate our ideas for art and photos into the reality of this edition. Our special thanks also to Nancy Benedict for the colorful and appealing design of the book. Our heartfelt thanks also to a great group of ITP/Brooks/Cole book "reps" for their interest and enthusiasm as they talk with instructors across the country.

Our sincere appreciation goes to the following reviewers who were kind enough to read and give their professional comments: Kathleen Ashworth, Yakima Valley Community College; Aaron Brown, Los Angeles Pierce College; Thomas Everton, Ventura College; Donna G. Friedman, St. Louis Community College at Florissant Valley; Robert Fremland, San Diego Mesa College; Allan A. Gahr, Gordon College; Robert Howell, University of Cincinnati—Raymond Walters College; E. Jerome Maas, Oakton Community College; Kathy Mitchell, St. Petersburg Junior College; Tracy Lynn Moore, Louisiana State University at Eunice; Judith N. Peisen, Hagerstown Community College; Jeffrey A. Schneider, State University of New York College at Oswego; and Tony Taylor, Santa Rosa Junior College.

Morris Hein and Susan Arena

About the Authors

Morris Hein is professor emeritus of chemistry at Mt. San Antonio College, where he regularly taught the preparatory chemistry course. His name has become synonymous with clarity, meticulous accuracy, and a step-by-step approach that students can follow. Over the years, more than three million students have learned chemistry using a text by Morris Hein. In addition to *Foundations of College Chemistry, Tenth Edition,* he is co-author of *Introduction to General, Organic, and Biochemistry, Sixth Edition,* and *Introduction to Organic and Biochemistry.* He is also co-author of *Foundations of Chemistry in the Laboratory, Tenth Edition,* and *College Chemistry in the Laboratory, Sixth Edition.*

Susan Arena currently teaches general chemistry and is director of the Merit Program for Emerging Scholars at the University of Illinois, Urbana-Champaign. She collaborated with Morris Hein on the seventh edition of *Foundations of College Chemistry,* and became co-author of the eighth edition. She is also co-author of *Introduction to General, Organic, and Biochemistry, Sixth Edition,* and *Introduction to Organic and Biochemistry.*

BRIEF CONTENTS

CONTENTS

FOUNDATIONS OF
COLLEGE CHEMISTRY

CHAPTER

1

An Introduction to Chemistry

CHAPTER 1 / OUTLINE

▲ The colors, fragrances, and textures of a rose garden are all the results of chemical reactions—examples of chemistry in our lives.

CHAPTER 1

Have you ever wondered what causes a rose to be fragrant? Perhaps you have been mesmerized by the flames in your fire on a romantic evening as they change color and form. And think of your relief when you dropped a container and found that it was plastic, not glass. These phenomena are the result of chemistry that occurs all around us, all the time. Chemical changes bring us beautiful colors, warmth, light, and products to make our lives function more smoothly. Understanding, explaining, and using the diversity of materials we find around us is what chemistry is all about.

1.1 Why Study Chemistry?

Chemistry is fascinating to many people. Learning about the composition of the world around us can lead to interesting and useful inventions and new technology. More than likely you are taking this chemistry course because someone has decided that it is an important part of your career goals. Chemistry is central to understanding many fields including agriculture, astronomy, animal science, geology, medicine, applied health technology, fire science, biology, molecular biology, and materials science. Even if you are not planning to work in any of these fields, chemistry is used by each of us every day in our struggle to cope with our technological world. Learning about the benefits and risks associated with chemicals will help you to be an informed citizen, able to make intelligent choices concerning the world around you. Studying chemistry teaches you to solve problems and communicate with others in an organized and logical manner. These skills will be helpful in college and throughout your career.

1.2 The Nature of Chemistry

chemistry

Key words are highlighted in bold and color in the margin to alert you to new terms defined in the text.

What is chemistry? One dictionary gives this definition: "**Chemistry** is the science of the composition, structure, properties, and reactions of matter, especially of atomic and molecular systems." Another, somewhat simpler definition is "Chemistry is the science dealing with the composition of *matter* and the changes in composition that matter undergoes." Neither of these definitions is entirely adequate. Chemistry, and physics, form a fundamental branch of knowledge. Chemistry is also closely related to biology, not only because living organisms are made of material substances but also because life itself is essentially a complicated system of interrelated chemical processes.

The scope of chemistry is extremely broad. It includes the whole universe and everything, animate and inanimate, in it. Chemistry is concerned with the composition and changes in the composition of matter, and also with the energy and energy changes associated with matter. Through chemistry we seek to learn and to understand the general principles that govern the behavior of all matter.

The chemist, like other scientists, observes nature and attempts to understand its secrets: What makes a rose red? Why is sugar sweet? What is occurring when iron rusts? Why is carbon monoxide poisonous? Problems such as these—some of which have been solved, some of which are still to be solved—are all part of what we call chemistry.

A chemist may interpret natural phenomena; devise experiments that reveal the composition and structure of complex substances; study methods for improving natural processes; or synthesize substances. Ultimately, the efforts of successful chemists advance the frontiers of knowledge and at the same time contribute to the well-being of humanity.

1.3 The Process of Chemistry

How does a chemist discover a new compound and study its behavior? A combination of hard work, accident, and luck produced "memory metal," an advanced substance now used to make toys, braces for teeth, and to treat problems such as blood clots and reattachment of tendons. However, none of these applications was the target of the original search for a new material. They all developed from the investigation of the properties of memory metal and the creativity of people associated with the work.

The "memory metal" discovery story began in Maryland at the Naval Ordnance Laboratory (NOL) where William J. Buehler had been given the task of finding a suitable substance for the nose cone of underwater missile launchers. Buehler was experiencing personal difficulties at this particular time in his life and so he became particularly immersed in his work. He eventually eliminated all but 12 compounds. He then began measuring their impact resistance. His test was simple but effective: He made buttons of the compound to be tested and hit them with a hammer. One substance, a 50-50 mixture of nickel and titanium, demonstrated greater impact resistance, elasticity, malleability, and resistance to fatigue than the others. Buehler named it *nitinol*, from *Ni*ckel *Ti*tanium *N*aval *O*rdnance *L*aboratory.

▲ **The fragrance of this beautiful rose and the rust on this car are two examples of the chemical reactions.**

As Buehler and his staff tested nitinol, they varied the percentages of nickel and titanium to determine the effect of composition on the properties of the compound. They made a series of bars in a furnace and, after cooling them, smoothed them on a shop grinder. Buehler accidentally dropped one of the bars and noticed that it made a dull thud—much like that of a bar of lead. This aroused his curiosity and he began dropping other bars. To his surprise, he found that the cooled bars made a dull sound whereas the warm ones made a bell-like tone. Fascinated, he then began reheating and cooling the bars, discovering that the sound changed consistently back and forth from dull (cool) to bell-like (warm). Why was this so intriguing? Because these variances indicated a change in the atomic structure of the metal!

During a project review at NOL, Buehler demonstrated the fatigue resistance of nitinol by bending a long strip of wire into accordion folds. He passed it around, letting the directors flex it and straighten it. One of them wanted to see what would happen when it was heated and held the pleated nitinol over a flame. To everyone's amazement it stretched out into a straight wire. Buehler recognized that this behavior related to the different sounds made when nitinol was heated or cooled.

Left: This nitinol filter can ▶ trap potentially fatal blood clots. When cooled below body temperature, it is collapsed into a straight bundle of wire. Then, with minor surgery, it is inserted into a large vein where it springs back into shape as it reaches body temperature.
Right: Glasses with frames that "remember" their shape are a product of nitinol technology.

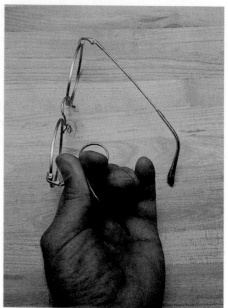

Buehler recruited Frederick Wang, a crystallographer, to define the "memory" properties of the metal. Wang determined the structural changes that give nitinol its unique characteristic of memory—changes that involve the rearrangement of the position of particles within the solid. Changes between solids and liquids or liquids and gases are well known (e.g., boiling water or melting ice), but the same sort of changes can also occur between two solids. In the case of nitinol, a *parent shape* (the return shape) has to be defined. To fix the parent shape of nitinol it must be heated. No changes are apparent, but when it cools, the atoms in the solid rearrange into a slightly different structure. Thus, whenever the nitinol is heated, the atoms rearrange themselves back to the structure that produces the parent shape.

Buehler and Wang then proceeded to move this "memory" metal from the experimental world of the laboratory to the commercial world of applications. By the late sixties, nitinol was being used in pipe couplings in the aircraft industry.

In 1968, George Andreason, a dentist, began experimenting with nitinol in his metalworking shop where he made jewelry as a hobby. He developed a fine wire that could be molded to fit a patient's mouth (parent shape), which when cooled could be bent to fit the misaligned teeth. When warmed to body temperature, it would exert a gentle constant pressure on the teeth. A major breakthrough in orthodontics, nitinol braces cut the treatment time in half from that of steel braces.

Dr. Wang left NOL in 1980 to become a supplier of nitinol to the numerous manufacturers who wanted to use this amazing substance for such things as eyeglass frames that can take extreme abuse (e.g., bending them, sitting on them, or twisting them) and return to their original shape; antiscalding devices for showerheads and faucets that automatically shut off if the water approaches 120°F; and toys such as blinking movie posters and tail-swishing dinosaurs.

The uses of nitinol are varied and growing. Today it has many applications in medicine, engineering, and safety; in housewares; and even in the lingerie business in underwire brassieres. Nitinol was the first of the "intelligent" materials—ones that respond to changes in the environment.

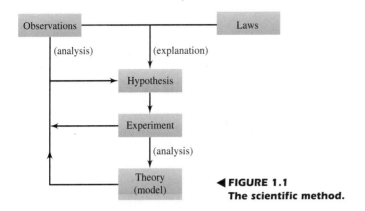

◀ **FIGURE 1.1**
The scientific method.

▲
Scientists employ the scientific method every day in their laboratory work.

scientific method

1.4 The Scientific Method

The nitinol story illustrates how chemists work with other scientists to solve problems. Chemists, metallurgists, physicists, engineers, and a wide variety of technicians were involved in the development of the memory metal. As scientists conduct studies they ask many questions, and their questions often lead in directions that are not part of the original problem. The amazing developments from chemistry and technology usually involve what we call the **scientific method,** which can generally be described as follows:

1. **Collect the facts or data** that are relevant to the problem or question at hand. This is usually done by planned experimentation. The data are then analyzed to find trends or regularities that are pertinent to the problem. In the nitinol story, Buehler compared the properties of various substances until he selected nitinol. He then investigated the properties of nitinol, including its ability to produce different sounds when warm and cool, and its unique ability to return to its original shape.
2. **Formulate a hypothesis** that will account for the data and that can be tested by further experimentation. Wang and Buehler proposed a hypothesis regarding the two phases of nitinol. Further tests supported their model.
3. **Plan and do additional experiments to test the hypothesis.** The nitinol group continued its testing process until the mechanism was well understood.
4. **Modify the hypothesis** as necessary so that it is compatible with all the pertinent data.

Confusion sometimes arises regarding the exact meanings of the words *hypothesis, theory,* and *law.* A **hypothesis** is a tentative explanation of certain facts that provides a basis for further experimentation. A well-established hypothesis is often called a **theory** or model. Thus a theory is an explanation of the general principles of certain phenomena with considerable evidence or facts to support it. Hypotheses and theories explain natural phenomena, whereas **scientific laws** are simple statements of natural phenomena to which no exceptions are known under the given conditions.

hypothesis

theory

scientific laws

These four steps are a broad outline of the general procedure that is followed in most scientific work, but they are not a "recipe" for doing chemistry or any other science (Figure 1.1). Chemistry is an experimental science, however, and much of its progress has been due to application of the scientific method through systematic research.

We study many theories and laws in chemistry; this makes our task as students easier because theories and laws summarize important aspects of the sciences. Certain theories and models advanced by great scientists in the past have since been substantially altered and modified. Such changes do not mean that the discoveries of the past are any less significant. Modification of existing theories and models in the light of new experimental evidence is essential to the growth and evolution of scientific knowledge. Science is dynamic.

1.5   Relationship of Chemistry to Other Sciences and Industry

Besides being a science in its own right, chemistry is the servant of other sciences and industry. Chemical principles contribute to the study of physics, biology, agriculture, engineering, medicine, space research, oceanography, and many other sciences. Chemistry and physics are overlapping sciences, because both are based on the properties and behavior of matter. Biological processes are chemical in nature. The metabolism of food to provide energy to living organisms is a chemical process. Knowledge of the molecular structure of proteins, hormones, enzymes, and the nucleic acids assists biologists in their investigations of the composition, development, and reproduction of living cells.

Chemistry plays an important role in our quest to alleviate the growing shortage of food in the world. Agricultural production has been increased with the use of chemical fertilizers, pesticides, and improved varieties of seeds. Chemical refrigerants make possible the frozen food industry, which preserves large amounts of food that might otherwise spoil. Chemistry is also producing synthetic nutrients, but much remains to be done as the world population increases relative to the land available for cultivation. Expanding energy needs have brought about difficult environmental problems in the form of air and water pollution. Chemists and other scientists are working diligently to alleviate these problems.

Advances in medicine and chemotherapy, through the development of new drugs, have prolonged life and relieved human suffering. More than 90% of the drugs and pharmaceuticals being used in the United States today have been developed commercially within the past 50 years. The plastics and polymer industry, unknown 60 years ago, has revolutionized the packaging and textile industries and continues to produce durable and useful construction materials. Energy derived from chemical processes is used for heating, lighting, and transportation. Virtually every industry is dependent on chemistry.

People outside of science usually have the perception that science is an intensely logical field. However, scientific discoveries are often the result of trial and error. In the nitinol story its memory nature was discovered only because someone wanted to see what would happen if a flame was applied to compressed nitinol. Buehler's creativity and insight enabled him to make a key connection. This is what we mean by serendipity in science. Buehler's search led him in a different direction than his assigned task because the results of his experiments were quite unexpected. The Navy found a new material for nose cones and at the same time discovered a material that has applications in medicine, engineering, and everyday life.

Household Pollution—
A Health Hazard??

How often are you exposed to potentially harmful substances? Scientists have developed highly sensitive instruments and portable monitors to judge where, when, and how people come in contact with potentially dangerous chemicals. Since 1980, Wayne Ott from Stanford University and Lance Wallace from the Environmental Protection Agency (EPA) have been studying our everyday exposure to toxic substances. They have analyzed exposure rates in 15 states and parts of Canada. To look for volatile organic compounds and pesticides they often have people carry monitors during their daily activities.

The results are surprising: Most people have the greatest contact with potentially toxic chemicals in places we generally consider to be free of pollution—our homes, offices, and cars. Places that are usually targets of environmental laws (such as factories in urban areas) are relatively small contributors to our toxic exposure. Does this mean our homes pose a greater health risk than industrial pollution? It appears that this is exactly the case. Let's consider three types of indoor pollution: (1) volatile organic compounds, (2) fine dust particles, and (3) pesticides.

We are regularly exposed to low levels of volatile organic chemicals in our daily routines. Freshly dry-cleaned clothes contain traces of tetrachloroethylene ("perc"), known to cause cancer in lab animals. Benzene, which is known to cause leukemia in people exposed to high concentrations, is found in gasoline and tobacco smoke. In 1985, Wallace determined that the majority of benzene exposure comes from homes and offices. We can most easily lower our exposure to these and other volatile organic compounds by avoiding products containing these pollutants and by improving ventilation.

Fine dust particles in the air contribute to such health problems as asthma and allergies. A study conducted in Riverside, California, found that exposure to these tiny particles is 60% greater indoors than outdoors. As we move through our homes and offices we constantly stir up our own "dust cloud." When this is combined with particles from combustion (cooking, candles, and fires), we find levels of dust that would be called hazardous if found outdoors. Carpets are effective reservoirs for these dust particles (as well as for bacteria and allergens). Even regularly vacuumed carpets contain high levels of these pollutants. What can you do to lower these levels? Use floor coverings that are smooth and easy to clean (such as wood, tile, or linoleum) and wipe your feet on a doormat before entering your home.

In the late 1980s, studies were completed on pesticide levels in indoor air. In both Florida and Massachusetts, indoor air contained a minimum of 5 times the pesticide concentration of outdoor air. Pesticide residues enter our homes in many ways. Some are directly applied, while others are tracked in on the bottom of shoes. Pesticides can last for years in carpets protected from sunlight. Simply removing your shoes before entering your home can lower indoor levels of pesticides by about 90%.

Although environmental laws governing air quality have helped lower levels of pollution outdoors, the same air pollutants are found at much higher levels indoors. You can reduce your exposure to these chemicals by making small changes in your daily routine and by carefully examining the products you use. Knowledge of chemistry gives you the ability to make choices that lead to a less polluted and healthier lifestyle.

▲ **Indoor plants and less carpeting can help curb the pollution in our homes.**

The nitinol story also illustrates the fact that scientists do not work alone. Buehler, an engineer, was joined by Wang, a crystallographer, and many other chemists, engineers, dentists, and physicians in developing the applications of nitinol. Each of them had a contribution to make to the body of knowledge surrounding the substance nitinol. In chemistry teamwork and cooperation are vital.

1.6   Risks and Benefits

▲
Chemists continue to explore the effects of nicotine on the smoker in order to help those who want to quit smoking.

Many problems that rely on science for an answer confront us today. Virtually every day we read or hear about stories on

- developing an AIDS vaccine
- banning the use of herbicides and pesticides
- analyzing DNA to determine genetic disease, biological parents, or to place a criminal at the scene of a crime
- removing asbestos from public buildings
- removing lead from drinking water
- the danger of radon in our homes
- global warming
- the hole in the ozone layer
- health risks associated with coffee, alcohol, margarine, saturated fats, and other foods
- burning of tropical rain forests and the effect on global ecology
- health risks from tobacco use

Which of these risks present true danger to us and which are less threatening? These problems will be around for many years, and new ones will be continually added to the list. Wherever we live and whatever our occupation, each of us is exposed to chemicals and chemical hazards every day. The question we must address is: Do the risks outweigh the benefits?

Risk assessment is a process that brings together professionals from the fields of chemistry, biology, toxicology, and statistics in order to determine the risk associated with exposure to a certain chemical. Assessment of risk involves determining both the probability of exposure and the severity of that exposure. Once this is done, an estimate of the overall risk is made. Studies have shown how people perceive various risks. The perception of risk depends on some rather interesting factors. Voluntary risks, such as smoking or flying, are much more easily accepted than involuntary ones, such as herbicides on apples or asbestos in buildings. People also often conclude that anything "synthetic" is bad, while anything "organic" is good. Risk assessment may provide information on the degree of risk but not on whether the chemical is "safe." Safety is a qualitative judgment based on many personal factors including beliefs, preferences, benefits, and costs.

Once a risk has been assessed, the next step is to manage it. This involves ethics, economics, and equity as well as government and politics. For example, some things perceived as low risk by scientists (such as asbestos in buildings) are classified as high risk by the general public. This inconsistency may result in the expenditure of millions of dollars to rid us of a perceived threat that may be much lower than the general public believes.

Risk management involves value judgments that integrate social, economic, and political issues. These risks must then be weighed against the benefits of new tech-

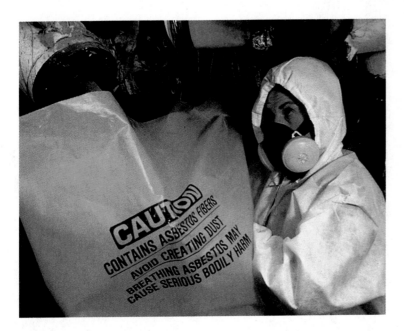

◀ *Asbestos, once a widely used building material, was banned by the EPA in 1986 because of its health hazards. Today, the removal of asbestos from our homes and our public buildings has become a big concern.*

nologies and products that will replace the old problem. We use both risk assessment and risk management to decide whether to buy a certain product (such as a pesticide), take a certain drug (such as a pain reliever), or eat certain foods (such as hot dogs). We must realize that all risks can never be completely eliminated. Our goal should be to minimize unnecessary risks and to make responsible decisions regarding necessary risks.

The theories and models used in risk assessment are based on assumptions and therefore contain uncertainties. Improve your understanding of the concepts of chemistry and you will be better able to understand the capabilities and limitations of science. You can then intelligently question the process of risk assessment and make decisions that will lead to a better understanding of our world and our responsibilities to each other.

Key Terms

The terms listed here are defined in the chapter. Section numbers are given in parentheses. More detailed definitions are given in the Glossary.

chemistry (1.2)   scientific method (1.4)
hypothesis (1.4)   theory (1.4)
scientific laws (1.4)

CHAPTER 2

Standards for Measurement

▲ Measuring instruments come in various shapes and sizes, all necessary to allow components to fit together, and permit us to quantify our world.

CHAPTER 2 / OUTLINE

Doing an experiment in chemistry is very much like cooking a meal in the kitchen. It's important to know the ingredients *and* the amounts of each in order to have a tasty product. Working on your car requires specific tools, in exact sizes. Buying new carpeting or draperies is an exercise in precise and accurate measurement for a good fit. A small difference in the concentration or amount of medication a pharmacist gives you may have significant effects on your well-being. In all of these cases making and properly using good measurements is key to a successful outcome. In chemistry, we begin by learning the metric system and the proper units for measuring mass, length, volume, and temperature.

2.1   Mass and Weight

Chemistry is an experimental science. The results of experiments are usually determined by making measurements. In elementary experiments the quantities that are

◀ **Mark Lee and Steven Smith work in the weightless environment of space to repair the Hubble Space telescope.**

commonly measured are mass, length, volume, pressure, temperature, and time. Measurements of electrical and optical quantities may also be needed in more sophisticated experimental work.

mass

Although we often use mass and weight interchangeably in our everyday lives, they have quite different meanings in chemistry. In science we define the **mass** of a body as the amount of matter in that body. The mass of an object is a fixed and unvarying quantity that is independent of the object's location. The mass of an object can be measured on a balance by comparison with other known masses.

weight

An example of a balance is shown in Figure 2.1. These balances were used by assayers during the gold rush to determine the mass of gold brought in by the prospectors. The gold was placed on one pan of the balance and known masses on the other side until the pans were level with each other. (Several modern balances are shown in Figure 2.4, page 27.)

Scientists define the **weight** of an object as the measure of the earth's gravitational attraction for that object. Weight is measured on an instrument called a scale, which measures force against a spring. Unlike mass, weight varies in relation to the position of an object on Earth or its distance from Earth.

Consider an astronaut of mass 70.0 kilograms (154 pounds) who is traveling to space. At the instant before blast-off the weight of the astronaut is also 70.0 kilograms. As his distance from Earth increases and the rocket turns into an orbiting course, the gravitational pull on the astronaut's body decreases until a state of weightlessness (zero weight) is attained. However, the mass of the astronaut's body has remained constant at 70.0 kilograms during the entire event.

▲
FIGURE 2.1
An assayer's balance is used for weighing gold.

2.2   Measurement and Significant Figures

To understand certain aspects of chemistry it is necessary to set up and solve problems. Problem solving requires an understanding of the mathematical operations used to manipulate numbers. Numerical values or data are obtained from measurements made in an experiment. Chemists use these data to calculate the extent of the physical and chemical changes occurring in the substances that are being studied. By appropriate calculations an experiment's results can be compared with those of other experiments and summarized in ways that are meaningful.

A measurement is expressed by a numerical value together with a unit of that measurement. For example,

$$\overbrace{70.0 \text{ kilograms}}^{\text{numerical value}} = \underbrace{154 \text{ pounds}}_{\text{unit}}$$

Numbers obtained from a measurement are never exact values. They always have some degree of uncertainty due to the limitations of the measuring instrument and the skill of the individual making the measurement. The numerical value recorded for a measurement should give some indication of its reliability (precision). To express maximum precision this number should contain all the digits that are known plus one digit that is estimated. This last estimated digit introduces some uncertainty. Because of this uncertainty every number that expresses a measurement can have only a limited number of digits. These digits, used to express a measured quantity, are known

significant figures   as **significant figures.**

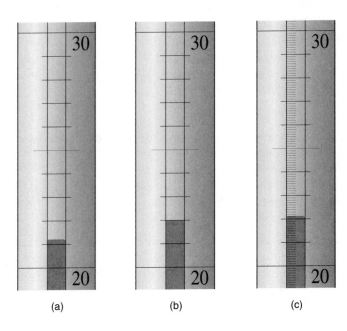

(a)   (b)   (c)

Suppose we measure temperature on a thermometer calibrated in degrees and observe that the mercury stops between 21 and 22 (see Figure 2.2a). We then know that the temperature is at least 21 degrees and is less than 22 degrees. To express the temperature with greater precision, we estimate that the mercury is about two-tenths the distance between 21 and 22. The temperature is therefore 21.2 degrees. The last digit (2) has some uncertainty because it is an estimated value. The recorded temperature, 21.2 degrees, is said to have three significant figures. If the mercury stopped exactly on the 22 (Figure 2.2b), the temperature would be recorded as 22.0 degrees. The zero is used to indicate that the temperature was estimated to a precision of one-tenth degree. Finally, look at Figure 2.2c. On this thermometer, the temperature is recorded as 22.11°C (four significant figures). Since the thermometer is calibrated to tenths of a degree, the first estimated digit is the hundredths.

Some numbers are exact and have an infinite number of significant figures. Exact numbers occur in simple counting operations; when you count 25 dollars, you have exactly 25 dollars. Defined numbers, such as 12 inches in 1 foot, 60 minutes in 1 hour, and 100 centimeters in 1 meter, are also considered to be exact numbers. Exact numbers have no uncertainty.

Evaluating Zero

In any measurement all nonzero numbers are significant. However, zeros may or may not be significant depending on their position in the number. Here are some rules for determining when zero is significant.

A zero is *significant* when it is:

1. between nonzero digits:
   205 has three significant figures
   2.05 has three significant figures
   61.09 has four significant figures

Rules for significant figures should be memorized for use throughout the text.

2. at the end of a number that includes a decimal point:
    0.500 has three significant figures (5, 0, 0)
    25.160 has five significant figures (2, 5, 1, 6, 0)
    3.00 has three significant figures (3, 0, 0)
    20. has two significant figures (2, 0)

A zero is *not significant* when it is:

1. before the first nonzero digit. These zeros are used to locate a decimal point:
    0.0025 has two significant figures (2, 5)
    0.0108 has three significant figures (1, 0, 8)
2. at the end of a number without a decimal point:
    1000 has one significant figure (1)
    590 has two significant figures (5, 9)

One way of indicating that these zeros are significant is to write the number using a decimal point and a power of 10. Thus if the value 1000 has been determined to four significant figures, it is written as 1.000×10^3. If 590 has only two significant figures, it is written as 5.9×10^2. We will learn more about writing numbers in this form in Section 2.4.

Answers to Practice Exercises are found at the end of each chapter.

Practice 2.1

How many significant figures are in each of these numbers?
(a) 4.5 inches   (e) 25.0 grams
(b) 3.025 feet   (f) 12.20 liters
(c) 125.0 meters   (g) 100,000 people
(d) 0.001 mile   (h) 205 birds

2.3   Rounding Off Numbers

When we do calculations we often obtain answers that have more digits than are justified. It is therefore necessary to drop the excess digits in order to express the answer with the proper number of significant figures. When digits are dropped from a number, the value of the last digit retained is determined by a process known as **rounding off numbers.** Two rules will be used in this book for rounding off numbers:

rounding off numbers

Not all schools use the same rules for rounding. Check with your instructor for variations in these rules.

Rule 1. When the first digit after those you want to retain is 4 or less, that digit and all others to its right are dropped. The last digit retained is not changed. The following examples are rounded off to four digits:

$74.693 = 74.69$ $1.00629 = 1.006$
        ↖ This digit is dropped.   ↖ These two digits are dropped.

Rule 2. When the first digit after those you want to retain is 5 or greater, that digit and all others to the right are dropped and the last digit retained is increased by one. These examples are rounded off to four digits:

1.026868 = 1.027

— These three digits are dropped.

— This digit is changed to 7.

18.02500 = 18.03

— These three digits are dropped.

— This digit is changed to 3.

12.899 = 12.90

— This digit is dropped.

— These two digits are changed to 90.

Practice 2.2

Round off these numbers to the number of significant figures indicated:
(a) 42.246 (four)   (d) 0.08965 (two)
(b) 88.015 (three)   (e) 225.3 (three)
(c) 0.08965 (three)   (f) 14.150 (three)

2.4 Scientific Notation of Numbers

Earth's age has been estimated to be about 4,500,000,000 (4.5 billion) years. Because this is an estimated value, say to the nearest 0.1 billion years, we are justified in using only two significant figures to express it. Thus we write it, using a power of 10, as 4.5×10^9 years.

Very large and very small numbers are often used in chemistry and can be simplified and conveniently written using a power of 10. Writing a number as a power of 10 is called **scientific notation.**

To write a number in scientific notation, move the decimal point in the original number so that it is located after the first nonzero digit. This new number is multiplied by 10 raised to the proper power (exponent). The power of 10 is equal to the number of places that the decimal point has been moved. If the decimal is moved to

scientific notation

 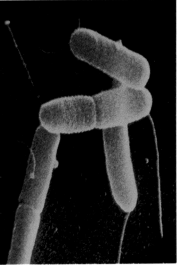

◀ **Scientific notation is a useful way to write very large numbers such as the distance between Earth and the moon, or very small numbers such as the length of these E. coli bacteria (shown here as a colored scanning electron micrograph × 30,000).**

the left, the power of 10 will be a positive number. If the decimal is moved to the right, the power of 10 will be a negative number.

Examples show you problem-solving techniques in a step-by-step form. Study each one and then try the Practice Exercises.

The scientific notation of a number is the number written as a factor between 1 and 10 multiplied by 10 raised to a power. For example,

$$2468 = 2.468 \times 10^3$$

number   scientific notation
of the number

Example 2.1   Write 5283 in scientific notation.

Solution

5283.   Place the decimal between the 5 and the 2. Since the decimal was moved
3   three places to the left, the power of 10 will be 3 and the number 5.283
      is multiplied by 10^3.

5.283×10^3

Example 2.2   Write 4,500,000,000 in scientific notation (two significant figures).

Solution

4 500 000 000.   Place the decimal between the 4 and the 5. Since the decimal
9   was moved nine places to the left, the power of 10 will be 9
          and the number 4.5 is multiplied by 10^9.

4.5×10^9

Example 2.3   Write 0.000123 in scientific notation.

Solution

0.000123   Place the decimal between the 1 and the 2. Since the decimal was
4   moved four places to the right, the power of 10 will be -4 and the
     number 1.23 is multiplied by 10^{-4}.

1.23×10^{-4}

Practice 2.3

Write the following numbers in scientific notation:
(a) 1200 (four digits)   (c) 0.0468
(b) 6,600,000 (two digits)   (d) 0.00003

2.5   Significant Figures in Calculations

The results of a calculation based on measurements cannot be more precise than the least precise measurement.

Multiplication or Division

In calculations involving multiplication or division, the answer must contain the same number of significant figures as in the measurement that has the least number of significant figures. Consider the following examples:

$(190.6)(2.3) = 438.38$

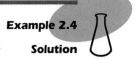

Example 2.4

Solution

The value 438.38 was obtained with a calculator. The answer should have two significant figures because 2.3, the number with the fewest significant figures, has only two significant figures.

Round off this digit to 4.

Drop these three digits.

438.38

Move the decimal 2 places to the left to express in scientific notation.

The correct answer is 440 or 4.4×10^2.

$$\frac{(13.59)(6.3)}{12} = 7.13475$$

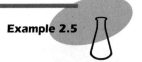

Example 2.5

Solution

The value 7.13475 was obtained with a calculator. The answer should contain two significant figures because 6.3 and 12 each have only two significant figures.

Drop these four digits.

7.13475

This digit remains the same.

The correct answer is 7.1.

Practice 2.4

(a) $(134 \text{ in.})(25 \text{ in.}) = ?$

(b) $\dfrac{213 \text{ miles}}{4.20 \text{ hours}} = ?$

(c) $\dfrac{(2.2)(273)}{760} = ?$

Addition or Subtraction

The results of an addition or a subtraction must be expressed to the same precision as the least precise measurement. This means the result must be rounded to the same number of decimal places as the value with the fewest decimal places (blue line in examples).

Example 2.6   Add 125.17, 129, and 52.2.

Solution
$$
\begin{array}{r}
125.\!|17 \\
129. \\
\underline{52.\!|2} \\
306.\!|37
\end{array}
$$

The number with the least precision is 129. Therefore the answer is rounded off to the nearest unit: 306.

Example 2.7   Subtract 14.1 from 132.56.

Solution
$$
\begin{array}{r}
132.5\!|6 \\
\underline{-14.1} \\
118.4\!|6
\end{array}
$$

14.1 is the number with the least precision. Therefore the answer is rounded off to the nearest tenth: 118.5.

Example 2.8   Subtract 120 from 1587.

Solution
$$
\begin{array}{r}
158\!|7 \\
\underline{-12\!|0} \\
146\!|7
\end{array}
$$

120 is the number with least precision. The zero is not considered significant; therefore the answer must be rounded to the nearest ten: 1470 or 1.47×10^3.

Example 2.9   Add 5672 and 0.00063.

Solution
$$
\begin{array}{r}
5672 \\
\underline{+0.\!|00063} \\
5672.\!|00063
\end{array}
$$

Note when a very small number is added to a large number the result is simply the original number.

The number with least precision is 5672. So the answer is rounded off to the nearest unit: 5672.

Example 2.10 $\dfrac{1.039 - 1.020}{1.039} = 0.018286814$

Solution   The value 0.018286814 was obtained with a calculator. When the subtraction in the numerator is done,

$$1.039 - 1.020 = 0.019$$

the number of significant figures changes from four to two. Therefore the answer should contain two significant figures after the division is carried out:

┌────── Drop these six digits.
│ ↓
┌──┐
0.018286814
↑
└──────────This digit remains the same.

The correct answer is 0.018, or 1.8×10^{-2}.

Practice 2.5

How many significant figures should the answer in each of these calculations contain?

(a) (14.0)(5.2)   (e) $119.1 - 3.44$

(b) (0.1682)(8.2)   (f) $\dfrac{94.5}{1.2}$

(c) $\dfrac{(160)(33)}{4}$ (g) $1200 + 6.34$

(d) $8.2 + 0.125$ (h) $1.6 + 23 - 0.005$

If you need to brush up on your math skills refer to the "Mathematical Review" in Appendix I.

metric system or SI

Additional material on mathematical operations is given in Appendix I, "Mathematical Review." Study any portions that are not familiar to you. You may need to do this at various times during the course when additional knowledge of mathematical operations is required.

2.6   The Metric System

The **metric system,** or **International System (SI,** from *Système International*), is a decimal system of units for measurements of mass, length, time, and other physical quantities. Built around a set of standard units, the metric system uses factors of 10 to express larger or smaller numbers of these units. To express quantities that are larger or smaller than the standard units, prefixes are added to the names of the units. These prefixes represent multiples of 10, making the metric system a decimal system of measurements. Table 2.1 shows the names, symbols, and numerical values of the prefixes. Some examples of the more commonly used prefixes are

1 *kilo*meter = 1000 meters
1 *kilo*gram = 1000 grams
1 *milli*meter = 0.001 meter
1 *micro*second = 0.000001 second

The seven standard units in the International System, their abbreviations, and the quantities they measure are given in Table 2.2. Other units are derived from these units. The metric system, or International System, is currently used by most of the countries in the world, not only in scientific and technical work but also in commerce and industry.

▲
Most products today list both systems of measurement on their labels.

TABLE 2.1	Prefixes and Numerical Values for SI Units		
Prefix	**Symbol**	**Numerical value**	**Power of 10 equivalent**
exa	E	1,000,000,000,000,000,000	10^{18}
peta	P	1,000,000,000,000,000	10^{15}
tera	T	1,000,000,000,000	10^{12}
giga	G	1,000,000,000	10^{9}
mega	M	1,000,000	10^{6}
kilo	k	1,000	10^{3}
hecto	h	100	10^{2}
deka	da	10	10^{1}
—	—	1	10^{0}
deci	d	0.1	10^{-1}
centi	c	0.01	10^{-2}
milli	m	0.001	10^{-3}
micro	μ	0.000001	10^{-6}
nano	n	0.000000001	10^{-9}
pico	p	0.000000000001	10^{-12}
femto	f	0.000000000000001	10^{-15}
atto	a	0.000000000000000001	10^{-18}

The prefixes most commonly used in chemistry are shown in bold.

TABLE 2.2	International System's Standard Units of Measurement	
Quantity	**Name of unit**	**Abbreviation**
Length	meter	m
Mass	kilogram	kg
Temperature	kelvin	K
Time	second	s
Amount of substance	mole	mol
Electric current	ampere	A
Luminous intensity	candela	cd

2.7   Measurement of Length

Standards for the measurement of length have a long history. The Old Testament mentions such units as the *cubit* (the distance from a man's elbow to the tip of his outstretched hand). In ancient Scotland the inch was once defined as a distance equal to the width of a man's thumb.

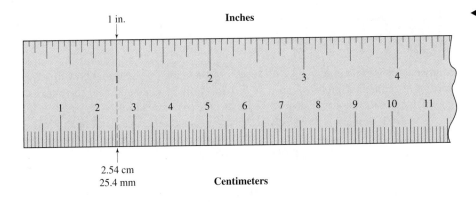

◀ **FIGURE 2.3**
**Comparison of the metric
and American systems of
length measurement:
2.54 cm = 1 in.**

Reference standards of measurements have undergone continuous improvements in precision. The standard unit of length in the metric system is the **meter (m)**. When the metric system was first introduced in the 1790s, the meter was defined as one ten-millionth of the distance from the equator to the North Pole, measured along the meridian passing through Dunkirk, France. In 1889, the meter was redefined as the distance between two engraved lines on a platinum–iridium alloy bar maintained at 0° Celsius. This international meter bar is stored in a vault at Sèvres near Paris. Duplicate meter bars have been made and are used as standards by many nations.

meter (m)

By the 1950s, length could be measured with such precision that a new standard was needed. Accordingly, the length of the meter was redefined in 1960 and again in 1983. The latest definition describes a meter as the distance that light travels in a vacuum during 1/299,792,458 of a second.

A meter is 39.37 inches, a little longer than 1 yard. One meter equals 10 decimeters, 100 centimeters, or 1000 millimeters (see Figure 2.3). A kilometer contains 1000 meters. Table 2.3 shows the relationships of these units.

Common length relationships:
$1 \text{ m} = 10^6 \text{ }\mu\text{m} = 10^{10} \text{ Å}$
$\phantom{1 \text{ m}} = 100 \text{ cm} = 1000 \text{ mm}$
$1 \text{ cm} = 10 \text{ mm} = 0.01 \text{ m}$
$1 \text{ in.} = 2.54 \text{ cm}$
$1 \text{ mile} = 1.609 \text{ km}$

The nanometer (10^{-9} m) is used extensively in expressing the wavelength of light as well as in atomic dimensions. See inside back cover for a complete table of common conversions.

See inside back cover for a table of conversions.

TABLE 2.3	Metric Units of Length		
Unit	Abbreviation	Meter equivalent	Exponential equivalent
kilometer	km	1000 m	10^3 m
meter	m	1 m	10^0 m
decimeter	dm	0.1 m	10^{-1} m
centimeter	cm	0.01 m	10^{-2} m
millimeter	mm	0.001 m	10^{-3} m
micrometer	μm	0.000001 m	10^{-6} m
nanometer	nm	0.000000001 m	10^{-9} m
angstrom	Å	0.0000000001 m	10^{-10} m

2.8   Problem Solving

See Appendix II for help in using a scientific calculator.

Many chemical principles can be illustrated mathematically. Learning how to set up and solve numerical problems in a systematic fashion is *essential* in the study of chemistry. A calculator will save you a lot of time in computation.

Usually a problem can be solved by several methods. But in all methods it is best, especially for beginners, to use a systematic, orderly approach. The *dimensional analysis method* is emphasized in this book because it

1. provides a systematic, straightforward way to set up problems.
2. gives a clear understanding of the principles involved.
3. trains you to organize and evaluate data.
4. helps to identify errors because unwanted units are not eliminated if the setup of the problem is incorrect.

Steps for solving problems are highlighted in color for easy reference.

The basic steps for solving problems are

1. Read the problem carefully. Determine what is to be solved for, and write it down.
2. Tabulate the data given in the problem. It is important to label *all* factors and measurements with the proper units.
3. Determine which principles are involved and which unit relationships are needed to solve the problem. You may need to refer to tables for needed data.
4. Set up the problem in a neat, organized, and logical fashion, making sure that unwanted units cancel. Use sample problems in the text as guides for setting up the problem.
5. Proceed with the necessary mathematical operations. Make certain that your answer contains the proper number of significant figures.
6. Check the answer to see if it is reasonable.

A few more words about problem solving: Don't allow any formal method of problem solving to limit your use of common sense and intuition. If a problem is clear to you and its solution seems simpler by another method, by all means use it. But in the long run you should be able to solve many otherwise difficult problems by using dimensional analysis.

Dimensional analysis converts one unit to another unit by using conversion factors.

Important equations are boxed or highlighted in color.

$$\text{unit}_1 \times \text{conversion factor} = \text{unit}_2$$

If you want to know how many millimeters are in 2.5 meters, you need to convert meters (m) to millimeters (mm). Start by writing

$$\text{m} \times \text{conversion factor} = \text{mm}$$

This conversion factor must accomplish two things: It must cancel (or eliminate) meters *and* it must introduce millimeters—the unit wanted in the answer. Such a conversion factor will be in fractional form and have meters in the denominator and millimeters in the numerator:

$$\cancel{m} \times \frac{mm}{\cancel{m}} = mm$$

We know that 1 m = 1000 mm. From this relationship we can write two conversion factors—by dividing both sides of the equation by the same quantity:

$$\frac{1\ m}{1000\ mm} = \frac{1000\ mm}{1000\ mm} \longrightarrow \frac{1\ m}{1000\ mm} = 1$$

or

$$\frac{1\ m}{1\ m} = \frac{1000\ mm}{1\ m} \longrightarrow 1 = \frac{1000\ mm}{1\ m}$$

Therefore the two conversion factors are

$$\frac{1\ m}{1000\ mm} \quad \text{and} \quad \frac{1000\ mm}{1\ m}$$

Choosing the conversion factor $\frac{1000\ mm}{1\ m}$, we can set up the calculation for the conversion of 2.5 m to millimeters:

$$(2.5\ \cancel{m})\left(\frac{1000\ mm}{1\ \cancel{m}}\right) = 2500\ mm \qquad \text{or} \qquad 2.5 \times 10^3\ mm$$

(two significant figures)

> We know that multiplying a measurement by 1 does not change its value. Since our conversion factors both equal 1 we can multiply the measurement by the appropriate one to convert units.

Note that, in making this calculation, units are treated as numbers; meters in the numerator are canceled by meters in the denominator.

Now suppose you need to change 215 centimeters to meters. We start with

cm × conversion factor = m

The conversion factor must have centimeters in the denominator and meters in the numerator:

$$\cancel{cm} \times \frac{m}{\cancel{cm}} = m$$

From the relationship 100 cm = 1 m, we can write a factor that will accomplish this conversion:

$$\frac{1\ m}{100\ cm}$$

Now set up the calculation using all the data given.

$$(215\ \cancel{cm})\left(\frac{1\ m}{100\ \cancel{cm}}\right) = 2.15\ m$$

Some problems require a series of conversions to reach the correct units in the answer. For example, suppose we want to know the number of seconds in 1 day. We need to convert from the unit of days to seconds in this manner:

day → hours → minutes → seconds

> Units are emphasized in problems by using color and flow diagrams to help visualize the steps in the process.

This series requires three conversion factors, one for each step. We convert days to hours (hr), hours to minutes (min), and minutes to seconds (s). The conversions can be done individually or in a continuous sequence:

$$\text{day} \times \frac{\text{hr}}{\text{day}} \longrightarrow \text{hr} \times \frac{\text{min}}{\text{hr}} \longrightarrow \text{min} \times \frac{\text{s}}{\text{min}} = \text{s}$$

$$\text{day} \times \frac{\text{hr}}{\text{day}} \times \frac{\text{min}}{\text{hr}} \times \frac{\text{s}}{\text{min}} = \text{s}$$

Inserting the proper factors we calculate the number of seconds in 1 day to be

$$(1 \text{ day})\left(\frac{24 \text{ hr}}{1 \text{ day}}\right)\left(\frac{60 \text{ min}}{1 \text{ hr}}\right)\left(\frac{60 \text{ s}}{1 \text{ min}}\right) = 86{,}400. \text{ s}$$

All five digits in 86,400 are significant, since all the factors in the calculation are exact numbers.

Label all factors with the proper units.

The dimensional analysis used in the preceding examples shows how unit conversion factors are derived and used in calculations. As you become more proficient, you can save steps by writing the factors directly in the calculation. The following examples show the conversion from American units to metric units.

Example 2.11   How many centimeters are in 2.00 ft?

Solution   The stepwise conversion of units from feet to centimeters may be done in this manner. Convert feet to inches; then convert inches to centimeters:

$$\text{ft} \rightarrow \text{in.} \rightarrow \text{cm}$$

The conversion factors needed are

$$\frac{12 \text{ in.}}{1 \text{ ft}} \quad \text{and} \quad \frac{2.54 \text{ cm}}{1 \text{ in.}}$$

$$(2.00 \text{ ft})\left(\frac{12 \text{ in.}}{1 \text{ ft}}\right) = 24.0 \text{ in.}$$

$$(24.0 \text{ in.})\left(\frac{2.54 \text{ cm}}{1 \text{ in.}}\right) = 61.0 \text{ cm}$$

Since 1 ft and 12 in. are exact numbers, the number of significant figures allowed in the answer is three, based on the number 2.00.

Example 2.12   How many meters are in a 100.-yd football field?

Solution   The stepwise conversion of units from yards to meters may be done in this manner, using the proper conversion factors:

$$\text{yd} \rightarrow \text{ft} \rightarrow \text{in.} \rightarrow \text{cm} \rightarrow \text{m}$$

$$(100. \text{ yd})\left(\frac{3 \text{ ft}}{1 \text{ yd}}\right) = 300. \text{ ft}$$

$$(300. \text{ ft})\left(\frac{12 \text{ in.}}{1 \text{ ft}}\right) = 3600 \text{ in.}$$

$$(3600 \text{ in.})\left(\frac{2.54 \text{ cm}}{1 \text{ in.}}\right) = 9144 \text{ cm}$$

$$(9144 \text{ cm})\left(\frac{1 \text{ m}}{100 \text{ cm}}\right) = 91.4 \text{ m} \qquad \text{(three significant figures)}$$

Examples 2.11 and 2.12 may be solved using a linear expression and writing down conversion factors in succession. This method often saves one or two calculation steps and allows numerical values to be reduced to simpler terms, leading to simpler calculations. The single linear expressions for Examples 2.11 and 2.12 are

$$(2.00 \text{ ft})\left(\frac{12 \text{ in.}}{1 \text{ ft}}\right)\left(\frac{2.54 \text{ cm}}{1 \text{ in.}}\right) = 61.0 \text{ cm}$$

$$(100. \text{ yd})\left(\frac{3 \text{ ft}}{1 \text{ yd}}\right)\left(\frac{12 \text{ in.}}{1 \text{ ft}}\right)\left(\frac{2.54 \text{ cm}}{1 \text{ in.}}\right)\left(\frac{1 \text{ m}}{100 \text{ cm}}\right) = 91.4 \text{ m}$$

If you need help in doing this calculation all at once on your calculator see Appendix II.

Using the units alone (Example 2.12), we see that the stepwise cancellation proceeds in succession until the desired unit is reached.

$$\text{yd} \times \frac{\text{ft}}{\text{yd}} \times \frac{\text{in.}}{\text{ft}} \times \frac{\text{cm}}{\text{in.}} \times \frac{\text{m}}{\text{cm}} = \text{m}$$

Practice 2.6

How many meters are in 10.5 miles?

How many cubic centimeters (cm^3) are in a box that measures 2.20 in. by 4.00 in. by 6.00 in.?   **Example 2.13**

First we need to determine the volume of the box in cubic inches ($in.^3$) by multiplying the length × width × height: **Solution**

$$(2.20 \text{ in.})(4.00 \text{ in.})(6.00 \text{ in.}) = 52.8 \text{ in.}^3$$

Now we convert $in.^3$ to cm^3 by using the inches and centimeters relationship three times:

$$\text{in.}^3 \times \frac{\text{cm}}{\text{in.}} \times \frac{\text{cm}}{\text{in.}} \times \frac{\text{cm}}{\text{in.}} = \text{cm}^3$$

$$(52.8 \text{ in.}^3)\left(\frac{2.54 \text{ cm}}{1 \text{ in.}}\right)\left(\frac{2.54 \text{ cm}}{1 \text{ in.}}\right)\left(\frac{2.54 \text{ cm}}{1 \text{ in.}}\right) = 865 \text{ cm}^3$$

Busy as a Bee

Monitoring pollutants in the air is expensive and time-consuming. What if you could employ a force of millions of free workers to collect samples for you? That's exactly what Jerry Bromenshenk and his colleagues at the University of Montana, Missoula, are doing. The scientists use honeybees to collect contaminants as they go about their daily business of collecting nectar. The bees ingest water, nectar, and pollen, and all sorts of particles and pollutants from the air, soil, water, and plants stick to their bodies. When the bees return to their hive, they release the pollutants as they fan the air to regulate the temperature. The air in the hives is then analyzed to identify the chemicals in the area sur-rounding the hive. Studies of the Puget Sound (on the coast of Washington) measuring metal contamination found that monitoring bees gives results that correspond closely to independent results done through soil analysis.

Now researchers have learned how to measure organic contaminants as well. Electronic hives, each with 7000–10,000 bees, have been installed at a variety of sites. Small copper tubes attached to the hives pump out air without disturbing the bees. The scientists then analyze the collected air to determine the chemical components. Millions of hives exist across the United States so a huge network of pollution monitoring stations could be established with each hive serving an area of about a mile in diameter. Instead of a reputation for producing a sting, bees may be changing reputation to monitors for pollution!

▲ **Honeybee on goldenrod plant.**

Example 2.14 A driver of a car is obeying the speed limit of 55 miles per hour. How fast is the car traveling in kilometers per second?

Solution Two conversions are needed to solve this problem:

$$mi \rightarrow km$$
$$hr \rightarrow min \rightarrow s$$

To convert mi → km,

$$\left(\frac{55 \text{ mi}}{hr}\right)\left(\frac{1.609 \text{ km}}{1 \text{ mi}}\right) = 88 \frac{km}{hr}$$

Next we must convert hr → min → s. Notice that hours is in the denominator of our quantity, so the conversion factor must have hours in the numerator:

$$\left(\frac{88 \text{ km}}{hr}\right)\left(\frac{1 \text{ hr}}{60 \text{ min}}\right)\left(\frac{1 \text{ min}}{60 \text{ s}}\right) = 0.024 \frac{km}{s}$$

Practice 2.7

How many cubic meters are in a room measuring 8 ft × 10 ft × 12 ft?

2.9 Measurement of Mass

The gram is a unit of mass measurement, but it is a tiny amount of mass; for instance, a U.S. nickel has a mass of about 5 grams. Therefore the *standard unit* of mass in the SI

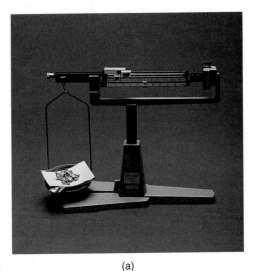
(a)

(b)

(c)

FIGURE 2.4
(a) A quadruple beam balance with a precision of 0.01 g; (b) digital electronic, top-loading balance with a precision of 0.001 g (1 mg); and (c) a digital electronic analytical balance with a precision of 0.0001 g (0.1 mg).

TABLE 2.4	Metric Units of Mass		
Unit	Abbreviation	Gram equivalent	Exponential equivalent
kilogram	kg	1000 g	10^3 g
gram	g	1 g	10^0 g
decigram	dg	0.1 g	10^{-1} g
centigram	cg	0.01 g	10^{-2} g
milligram	mg	0.001 g	10^{-3} g
microgram	µg	0.000001 g	10^{-6} g

system is the **kilogram** (equal to 1000 g). The amount of mass in a kilogram is defined by international agreement as exactly equal to the mass of a platinum–iridium cylinder (international prototype kilogram) kept in a vault at Sèvres, France. Comparing this unit of mass to 1 lb (16 oz), we find that 1 kg is equal to 2.205 lb. A pound is equal to 453.6 g (0.4536 kg). The same prefixes used in length measurement are used to indicate larger and smaller gram units (see Table 2.4).

kilogram

A balance is used to measure mass. Some balances can determine the mass of objects to the nearest microgram. The choice of balance depends on the precision required and the amount of material. Several balances are shown in Figure 2.4.

To change grams to milligrams, we use the conversion factor 1000 mg/g. The setup for converting 25 g to milligrams is

$$(25 \text{ g})\left(\frac{1000 \text{ mg}}{1 \text{ g}}\right) = 25,000 \text{ mg} \qquad (2.5 \times 10^4 \text{ mg})$$

Common mass relationships:
1 g = 1000 mg
1 kg = 1000 g
1 kg = 2.205 lb
1 lb = 453.6 g

Note that multiplying a number by 1000 is the same as multiplying the number by 10^3 and can be done simply by moving the decimal point three places to the right:

$$(6.428)(1000) = 6428 \qquad (6.428)$$
3

To change milligrams to grams, we use the conversion factor 1 g/1000 mg. For example, to convert 155 mg to grams:

$$(155 \text{ mg})\left(\frac{1 \text{ g}}{1000 \text{ mg}}\right) = 0.155 \text{ g}$$

Mass conversions from American to metric units are shown in Examples 2.15 and 2.16.

Example 2.15   A 1.50-lb package of baking soda contains how many grams?

Solution   We are solving for the number of grams equivalent to 1.50 lb. Since 1 lb = 453.6 g, the conversion factor is 453.6 g/lb:

$$(1.50 \text{ lb})\left(\frac{453.6 \text{ g}}{1 \text{ lb}}\right) = 680. \text{ g}$$

Example 2.16   Suppose four ostrich feathers weigh 1.00 lb. Assuming that each feather is equal in mass, how many milligrams does a single feather weigh?

Solution   The unit conversion in this problem is from 1 lb/4 feathers to milligrams per feather. Since the unit *feathers* occurs in the denominator of both the starting unit and the desired unit, the unit conversions are

lb → g → mg

$$\left(\frac{1.00 \text{ lb}}{4 \text{ feathers}}\right)\left(\frac{453.6 \text{ g}}{1 \text{ lb}}\right)\left(\frac{1000 \text{ mg}}{1 \text{ g}}\right) = \frac{113,400 \text{ mg}}{\text{feather}} \quad (1.13 \times 10^5 \text{ mg/feather})$$

Practice 2.8

You are traveling in Europe and wake up one morning to find your mass is 75.0 kg. Determine the American equivalent to see whether you need to go on a diet before you return home.

Practice 2.9

A tennis ball has a mass of 65 g. Determine the American equivalent in pounds.

2.10   Measurement of Volume

volume   **Volume,** as used here, is the amount of space occupied by matter. The SI unit of volume is the *cubic meter* (m^3). However, the liter (pronounced *leeter* and abbreviated L) and the milliliter (abbreviated mL) are the standard units of volume used in most chemical laboratories. A **liter** is usually defined as 1 cubic decimeter (1 kg) of water at 4°C.

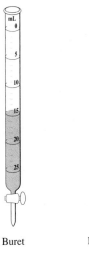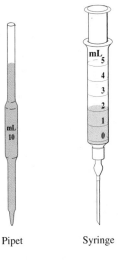

◄ **FIGURE 2.5**
Calibrated glassware for measuring the volume of liquids.

| Graduated cylinder | Volumetric flask | Buret | Pipet | Syringe |

The most common instruments or equipment for measuring liquids are the graduated cylinder, volumetric flask, buret, pipet, and syringe, which are illustrated in Figure 2.5. These pieces are usually made of glass and are available in various sizes.

The volume of a cubic or rectangular container can be determined by multiplying its length × width × height. Thus a box 10 cm on each side has a volume of $(10 \text{ cm})(10 \text{ cm})(10 \text{ cm}) = 1000 \text{ cm}^3$. Let's try some examples.

Common volume relationships:
$1 \text{ L} = 1000 \text{ mL} = 1000 \text{ cm}^3$
$1 \text{ mL} = 1 \text{ cm}^3$
$1 \text{ L} = 1.057 \text{ qt}$
$946.1 \text{ mL} = 1 \text{ qt}$

Example 2.17

Solution

How many milliliters are contained in 3.5 liters?

The conversion factor to change liters to milliliters is 1000 mL/L:

$$(3.5 \text{ \cancel{L}})\left(\frac{1000 \text{ mL}}{\cancel{L}}\right) = 3500 \text{ mL} \qquad (3.5 \times 10^3 \text{ mL})$$

Liters may be changed to milliliters by moving the decimal point three places to the right and changing the units to milliliters:

$$1.500 \text{ L} = 1500. \text{ mL}$$

Example 2.18

Solution

How many cubic centimeters are in a cube that is 11.1 inches on a side?

First we change inches to centimeters; our conversion factor is 2.54 cm/in.:

$$(11.1 \text{ \cancel{in.}})\left(\frac{2.54 \text{ cm}}{1 \text{ \cancel{in.}}}\right) = 28.2 \text{ cm on a side}$$

Then determine volume (length × width × height):

$$(28.2 \text{ cm})(28.2 \text{ cm})(28.2 \text{ cm}) = 22,426 \text{ cm}^3 \qquad (2.24 \times 10^4 \text{ cm}^3)$$

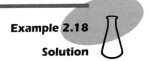

When doing problems with multiple steps, you should round only at the end of the problem. We are rounding at the end of each step in example problems to illustrate the proper significant figures.

Practice 2.10

A bottle of excellent chianti holds 750. mL. What is its volume in quarts?

Practice 2.11

Milk is often purchased by the half gallon. Determine the number of liters equal to this amount.

2.11   Measurement of Temperature

heat

Heat is a form of energy associated with the motion of small particles of matter. The term *heat* refers to the quantity of energy within a system or to a quantity of energy added to or taken away from a system. *System* as used here simply refers to the entity that is being heated or cooled. Depending on the amount of heat energy present, a given system is said to be hot or cold. **Temperature** is a measure of the intensity of heat, or how hot a system is, regardless of its size. Heat always flows from a region of higher temperature to one of lower temperature. The SI unit of temperature is the kelvin. The common laboratory instrument for measuring temperature is a thermometer (see Figure 2.6).

temperature

The temperature of a system can be expressed by several different scales. Three commonly used temperature scales are Celsius (pronounced *sell-see-us*), Kelvin (absolute), and Fahrenheit. The unit of temperature on the Celsius and Fahrenheit scales is called a *degree,* but the size of the Celsius and the Fahrenheit degree is not the same. The symbol for the Celsius and Fahrenheit degrees is °, and it is placed as a superscript after the number and before the symbol for the scale. Thus 100°C means 100 *degrees Celsius.* The degree sign is not used with Kelvin temperatures.

degrees Celsius = °C
Kelvin (absolute) = K
degrees Fahrenheit = °F

On the Celsius scale the interval between the freezing and boiling temperatures of water is divided into 100 equal parts, or degrees. The freezing point of water is assigned a temperature of 0°C and the boiling point of water a temperature of 100°C. The Kelvin temperature scale is known as the absolute temperature scale, because 0 K is the lowest temperature theoretically attainable. The Kelvin zero is 273.15 kelvins below the Celsius zero. (A kelvin is equal in size to a Celsius degree.) The freezing point of water on the Kelvin scale is 273.15 K. The Fahrenheit scale has 180 degrees between the freezing and boiling temperatures of water. On this scale the freezing point of water is 32°F and the boiling point is 212°F.

$$0°C \cong 273 \text{ K} \cong 32°F$$

The three scales are compared in Figure 2.6. Although absolute zero (0 K) is the lower limit of temperature on these scales, temperature has no upper limit. (Temperatures of several million degrees are known to exist in the Sun and in other stars.)

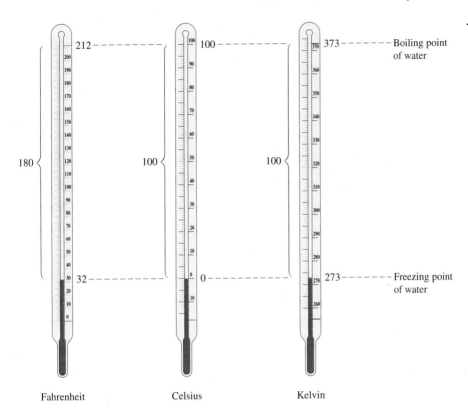

By examining Figure 2.6 we can see that there are 100 Celsius degrees and 100 kelvins between the freezing and boiling points of water, but there are 180 Fahrenheit degrees between these two temperatures. Hence, the size of the Celsius degree and the kelvin are the same, but 1 Celsius degree is equal to 1.8 Fahrenheit degrees.

$$\frac{180}{100} = 1.8$$

From these data, mathematical formulas have been derived to convert a temperature on one scale to the corresponding temperature on another scale:

$$K = °C + 273.15$$

$$°F = (1.8 × °C) + 32$$

$$°C = \frac{°F - 32}{1.8}$$

The temperature at which table salt (sodium chloride) melts is 800.°C. What is this temperature on the Kelvin and Fahrenheit scales?

Example 2.19

To calculate K from °C, we use the formula

Solution

$$K = °C + 273.15$$

$$K = 800.°C + 273.15 = 1073 K$$

To calculate °F from °C, we use the formula

$$°F = (1.8 \times °C) + 32$$

$$°F = (1.8)(800.°C) + 32$$

$$°F = 1440 + 32 = 1472°F$$

Summarizing our calculations we see

$$800.°C = 1073 \text{ K} = 1472°F$$

Remember, the original measurement of 800.°C was to the units place, so the converted temperature is also to the units place.

Example 2.20   The temperature for December 1 in Honolulu, Hawaii, was 110.°F, a new record. Convert this temperature to °C.

Solution   We use the formula

$$°C = \frac{°F - 32}{1.8}$$

$$°C = \frac{110. - 32}{1.8} = \frac{78}{1.8} = 43°C$$

Example 2.21   What temperature on the Fahrenheit scale corresponds to $-8.0°C$? (Notice the negative sign in this problem.)

Solution

$$°F = (1.8 \times °C) + 32$$

$$°F = (1.8)(-8.0) + 32 = -14 + 32$$

$$°F = 18°F$$

Temperatures used throughout this book are in degrees Celsius (°C) unless specified otherwise. The temperature after conversion should be expressed to the same precision as the original measurement.

Practice 2.12

Helium boils at 4 K. Convert this temperature to °C and then to °F.

Practice 2.13

"Normal" human body temperature is 98.6°F. Convert this to °C and K.

Taking the Temperature of Old Faithful

If you have ever struggled to take the temperature of a sick child imagine the difficulty in taking the temperature of a geyser. Such are the tasks scientists set for themselves! In 1984, James A. Westphal and Susan W. Keiffer, geologists from the California Institute of Technology, measured the temperature and pressure inside Old Faithful during several eruptions in order to learn more about how a geyser functions. The measurements, taken at eight depths along the upper part of the geyser, were so varied and complicated the researchers returned to Yellowstone in 1992 to further investigate Old Faithful's structure and functioning. To see what happened between eruptions Westphal and Keiffer lowered an insulated 2-inch video camera into the geyser. Keiffer had assumed the vent (opening in the ground) was a uniform vertical tube, but this is not the case. Instead, the geyser appears to be an east–west crack in the earth that extends downward at least 14 m. In some places it is over 1.8 m wide, and in other places it narrows to less than 15 cm. The walls of the vent contain many cracks allowing water to enter at several depths. The complicated nature of the temperature data is explained by these cracks. Cool water enters the vent at depths of 5.5 m and 7.5 m. Superheated water and steam blast into the vent 14 m underground. According to Westphal temperature increases of up to 130°C at the beginning of an eruption suggest that water and steam also surge into the vent from deeper geothermal sources.

During the first 20–30 seconds of an eruption, steam and boiling water shoot through the narrowest part of the vent at near the speed of sound. The narrow tube limits the rate at which the water can shoot from the geyser. When the pressure falls below a critical value, the process slows and Old Faithful begins to quiet down again.

The frequency of Old Faithful's eruptions are not on a precise schedule but vary from 45 to 105 minutes—the average is about 79 minutes. The variations in time between eruptions depend on the amount of boiling water left in the fissure. Westphal says, "There's no real pattern except that a short eruption is always followed by a long one." Measurements of temperatures inside Old Faithful have given scientists a better understanding of what causes a geyser to erupt.

Old Faithful eruptions shoot into the air an average of 130 feet.
▼

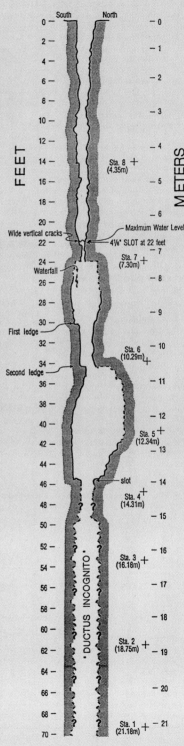

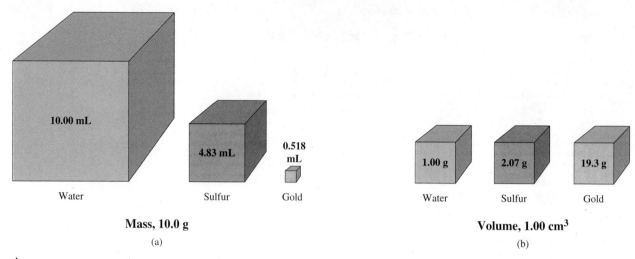

Mass, 10.0 g

(a)

Volume, 1.00 cm³

(b)

▲
FIGURE 2.7
(a) Comparison of the volumes of equal masses (10.0 g) of water, sulfur, and gold. (b) Comparison of the masses of equal volumes (1.00 cm³) of water, sulfur, and gold. (Water is at 4°C; the two solids, at 20°C.)

2.12   Density

density

Density (d) is the ratio of the mass of a substance to the volume occupied by that mass; it is the mass per unit of volume and is given by the equation

$$d = \frac{\text{mass}}{\text{volume}}$$

Density is a physical characteristic of a substance and may be used as an aid to its identification. When the density of a solid or a liquid is given, the mass is usually expressed in grams and the volume in milliliters or cubic centimeters.

$$d = \frac{\text{mass}}{\text{volume}} = \frac{\text{g}}{\text{mL}} \qquad \text{or} \qquad d = \frac{\text{g}}{\text{cm}^3}$$

Since the volume of a substance (especially liquids and gases) varies with temperature, it is important to state the temperature along with the density. For example, the volume of 1.0000 g of water at 4°C is 1.0000 mL; at 20°C, it is 1.0018 mL; and at 80°C, it is 1.0290 mL. Density therefore also varies with temperature.

The density of water at 4°C is 1.0000 g/mL, but at 80°C the density of water is 0.9718 g/mL.

$$d^{4°C} = \frac{1.0000 \text{ g}}{1.0000 \text{ mL}} = 1.0000 \text{ g/mL}$$

$$d^{80°C} = \frac{1.0000 \text{ g}}{1.0290 \text{ mL}} = 0.97182 \text{ g/mL}$$

The density of iron at 20°C is 7.86 g/mL.

$$d^{20°C} = \frac{7.86 \text{ g}}{1.00 \text{ mL}} = 7.86 \text{ g/mL}$$

The densities of a variety of materials are compared in Figure 2.7.

Densities for liquids and solids are usually represented in terms of grams per milliliter (g/mL) or grams per cubic centimeter (g/cm^3). The density of gases, however, is expressed in terms of grams per liter (g/L). Unless otherwise stated, gas densities are given for 0°C and 1 atmosphere pressure (discussed further in Chapter 13). Table 2.5 lists the densities of some common materials.

Suppose that water, Karo syrup, and vegetable oil are successively poured into a graduated cylinder. The result is a layered three-liquid system. Can we predict the order of the liquid layers? Yes, by looking up the densities in Table 2.5. Karo syrup has the greatest density (1.37 g/mL), and vegetable oil has the lowest density (0.91 g/mL). Karo syrup will be the bottom layer and vegetable oil will be the top layer. Water, with a density between the other two liquids, will form the middle layer. This information can also be determined by experiment. Vegetable oil, being less dense than water, will float when added to the graduated cylinder. Figure 2.8 shows a more complex density column made in the same way.

The density of air at 0°C is approximately 1.293 g/L. Gases with densities less than this value are said to be "lighter than air." A helium-filled balloon will rise rapidly in air because the density of helium is only 0.178 g/L.

When an insoluble solid object is dropped into water, it will sink or float, depending on its density. If the object is less dense than water, it will float, displacing a *mass* of water equal to the mass of the object. If the object is more dense than water, it will sink, displacing a *volume* of water equal to the volume of the object. This information can be used to determine the volume (and density) of irregularly shaped objects.

▲
FIGURE 2.8
Relative density of liquids. When liquids are carefully poured into a cylinder the liquid with highest density will form the bottom layer (maple syrup). The remaining liquids decreasing in density are antifreeze, dish detergent, shampoo, water, and corn oil.

TABLE 2.5	Densities of Some Selected Materials		

Liquids and solids		Gases	
Substance	Density (g/mL at 20°C)	Substance	Density (g/L at 0°C)
Wood (Douglas fir)	0.512	Hydrogen	0.090
Ethyl alcohol	0.789	Helium	0.178
Vegetable oil	0.91	Methane	0.714
Water (4°C)	**1.000***	Ammonia	0.771
Sugar	1.59	Neon	0.90
Glycerin	1.26	Carbon monoxide	1.25
Karo syrup	1.37	Nitrogen	1.251
Magnesium	1.74	**Air**	**1.293***
Sulfuric acid	1.84	Oxygen	1.429
Sulfur	2.07	Hydrogen chloride	1.63
Salt	2.16	Argon	1.78
Aluminum	2.70	Carbon dioxide	1.963
Silver	10.5	Chlorine	3.17
Lead	11.34		
Mercury	13.55		
Gold	19.3		

*For comparing densities the density of water is the reference for solids and liquids; air is the reference for gases.

▲
These ducks float since their density is less than water.

specific gravity

The **specific gravity** (sp gr) of a substance is the ratio of the density of that substance to the density of another substance, usually water at 4°C. Specific gravity has no units because the density units cancel. The specific gravity tells us how many times as heavy a liquid, a solid, or a gas is as compared to the reference material. Since the density of water at 4°C is 1.00 g/mL, the specific gravity of a solid or liquid is the same as its density in g/mL without the units.

$$\text{sp gr} = \frac{\text{density of a liquid or solid}}{\text{density of water}}$$

Sample calculations of density problems follow.

Example 2.22   What is the density of a mineral if 427 g of the mineral occupy a volume of 35.0 mL?

Solution   We need to solve for density, so we start by writing the formula for calculating density:

$$d = \frac{\text{mass}}{\text{volume}}$$

Then we substitute the data given in the problem into the equation and solve:

$$\text{mass} = 427 \text{ g} \qquad \text{volume} = 35.0 \text{ mL}$$

$$d = \frac{\text{mass}}{\text{volume}} = \frac{427 \text{ g}}{35.0 \text{ mL}} = 12.2 \text{ g/mL}$$

Example 2.23   The density of gold is 19.3 g/mL. What is the mass of 25.0 mL of gold?

Solution   There are two ways to solve this problem: (1) Solve the density equation for mass, then substitute the density and volume data into the new equation and calculate. (2) Solve by dimensional analysis.

When alternative methods of solution are available, more than one is shown in the example. Choose the method you are most comfortable with.

Method 1.   (a)   Solve the density equation for mass:

$$d = \frac{\text{mass}}{\text{volume}} \qquad d \times \text{volume} = \text{mass}$$

(b)   Substitute the data and calculate.

$$\text{mass} = \left(\frac{19.3 \text{ g}}{\text{mL}}\right)(25.0 \text{ mL}) = 483 \text{ g}$$

Method 2.   Dimensional analysis: Use density as a conversion factor, converting

$$\text{mL} \rightarrow \text{g}$$

The conversion of units is

$$\text{mL} \times \frac{\text{g}}{\text{mL}} = \text{g}$$

$$(25.0 \text{ mL})\left(\frac{19.3 \text{ g}}{\text{mL}}\right) = 483 \text{ g}$$

Calculate the volume (in mL) of 100. g of ethyl alcohol.

Example 2.24

From Table 2.5 we see that the density of ethyl alcohol is 0.789 g/mL. This density also means that 1 mL of the alcohol has a mass of 0.789 g (1 mL/0.789 g).

Method 1. Solve the density equation for volume and then substitute the data in the new equation.

$$d = \frac{mass}{volume}$$

$$volume = \frac{mass}{d}$$

$$volume = \frac{100. \cancel{g}}{0.789 \ \cancel{g}/mL} = 127 \ mL$$

Method 2. Dimensional analysis. For a conversion factor, we can use either

$$\frac{g}{mL} \quad or \quad \frac{mL}{g}$$

In this case the conversion is from g → mL, so we use mL/g. Substituting the data,

$$100. \ \cancel{g} \times \frac{1 \ mL}{0.789 \ \cancel{g}} = 127 \ mL \ of \ ethyl \ alcohol$$

The water level in a graduated cylinder stands at 20.0 mL before and at 26.2 mL after a 16.74-g metal bolt is submerged in the water. (a) What is the volume of the bolt? (b) What is the density of the bolt?

Example 2.25

(a) The bolt will displace a volume of water equal to the volume of the bolt. Thus the increase in volume is the volume of the bolt.

$$\begin{array}{rl} 26.2 \ mL = & volume \ of \ water \ plus \ bolt \\ -20.0 \ mL = & volume \ of \ water \\ \hline 6.2 \ mL = & volume \ of \ bolt \end{array}$$

(b) $d = \dfrac{mass \ of \ bolt}{volume \ of \ bolt} = \dfrac{16.74 \ g}{6.2 \ mL} = 2.7 \ g/mL$

Practice 2.14

Pure silver has a density of 10.5 g/mL. A ring sold as pure silver has a mass of 25.0 g. When placed in a graduated cylinder, the water level rises 2.0 mL. Determine whether the ring is actually pure silver or if the customer should see the Better Business Bureau.

Practice 2.15

The water level in a metric measuring cup is 0.75 L before the addition of 150. g of shortening. The water level after submerging the shortening is 0.92 L. Determine the density of the shortening.

Concepts in Review

Major concepts of the chapter are listed in this section to help you review the chapter.

1. Differentiate between mass and weight. Indicate the instruments used to measure each.
2. Know the metric units of mass, length, and volume.
3. Know the numerical equivalent for the metric prefixes *deci, centi, milli, micro, nano, kilo,* and *mega.*
4. Express any number in scientific notation.
5. Express answers to calculations to the proper number of significant figures.

6. Set up and solve problems using dimensional analysis.
7. Convert measurements of mass, length, and volume from American units to metric units, and vice versa.
8. Make temperature conversions among Fahrenheit, Celsius, and Kelvin scales.
9. Differentiate between heat and temperature.
10. Calculate the density, mass, and volume of an object.

Key Terms

The terms listed here are defined in the chapter. Section nuumbers are given in parentheses. More detailed definitions are given in the Glossary.

density (2.12)	mass (2.1)	scientific notation (2.4)	temperature (2.11)
heat (2.11)	meter (2.7)	SI (2.6)	volume (2.10)
kilogram (2.9)	metric system (SI) (2.6)	significant figures (2.2)	weight (2.1)
liter (2.10)	rounding off numbers (2.3)	specific gravity (2.12)	

Questions

Questions refer to tables, figures, and key words and concepts defined within the chapter. A particularly challenging question or exercise is indicated with an asterisk. Answers to even-numbered questions are given in Appendix VI.

1. How many centimeters make up 1 km? (Table 2.3)
2. What is the metric equivalent of 3 in.? (Figure 2.3)
3. Why is the neck of a 100-mL volumetric flask narrower than the top of a 100-mL graduated cylinder? (Figure 2.5)
4. Describe the order of the following substances (top to bottom) if these three substances were placed in a 100-mL graduated cylinder: 25 mL glycerin, 25 mL mercury, and a cube of magnesium 2.0 cm on an edge. (Table 2.5)

5. Arrange these materials in order of increasing density: salt, vegetable oil, lead, and ethyl alcohol. (Table 2.5)
6. Ice floats in vegetable oil and sinks in ethyl alcohol. The density of ice must lie between what numerical values? (Table 2.5)
7. Distinguish between heat and temperature.
8. Distinguish between density and specific gravity.
9. State the rules used in this text for rounding off numbers.
10. Compare the number of degrees between the freezing point of water and its boiling point on the Fahrenheit, Kelvin, and Celsius temperature scales. (Figure 2.6)

Paired Exercises

Answers to the even-numbered exercises are given in Appendix VI.

Metric Abbreviations

11. State the abbreviation for each of the following units:
  (a) gram
  (b) microgram
  (c) centimeter
  (d) micrometer
  (e) milliliter
  (f) deciliter

12. State the abbreviation for each of the following units:
  (a) milligram
  (b) kilogram
  (c) meter
  (d) nanometer
  (e) angstrom
  (f) microliter

Significant Figures, Rounding, Exponential Notation

13. For the following numbers, tell whether the zeros are significant:
  (a) 503
  (b) 0.007
  (c) 4200
  (d) 3.0030
  (e) 100.00
  (f) 8.00×10^2

14. Are the zeros significant in these numbers?
  (a) 63,000
  (b) 6.004
  (c) 0.00543
  (d) 8.3090
  (e) 60.
  (f) 5.0×10^{-4}

15. How many significant figures are in each of the following numbers?
  (a) 0.025
  (b) 22.4
  (c) 0.0404
  (d) 5.50×10^3

16. State the number of significant figures in each of the following numbers:
  (a) 40.0
  (b) 0.081
  (c) 129,042
  (d) 4.090×10^{-3}

17. Round each of the following numbers to three significant figures:
  (a) 93.246
  (b) 0.02857
  (c) 4.644
  (d) 34.250

18. Round each of the following numbers to three significant figures:
  (a) 8.8726
  (b) 21.25
  (c) 129.509
  (d) 1.995×10^6

19. Express each of the following numbers in exponential notation:
  (a) 2,900,000
  (b) 0.587
  (c) 0.00840
  (d) 0.0000055

20. Write each of the following numbers in exponential notation:
  (a) 0.0456
  (b) 4082.2
  (c) 40.30
  (d) 12,000,000

21. Solve the following problems, stating answers to the proper number of significant figures:
    (a) $12.62 + 1.5 + 0.25 = ?$
    (b) $(2.25 \times 10^3)(4.80 \times 10^4) = ?$
    (c) $\dfrac{(452)(6.2)}{14.3} = ?$
    (d) $(0.0394)(12.8) = ?$
    (e) $\dfrac{0.4278}{59.6} = ?$
    (f) $10.4 + 3.75(1.5 \times 10^4) = ?$

22. Evaluate each of the following expressions. State the answer to the proper number of significant figures:
    (a) $15.2 - 2.75 + 15.67$
    (b) $(4.68)(12.5)$
    (c) $\dfrac{182.6}{4.6}$
    (d) $1986 + 23.84 + 0.012$
    (e) $\dfrac{29.3}{(284)(415)}$
    (f) $(2.92 \times 10^{-3})(6.14 \times 10^5)$

23. Change these fractions into decimals. Express each answer to three significant figures:
    (a) $\dfrac{5}{6}$
    (b) $\dfrac{3}{7}$
    (c) $\dfrac{12}{16}$
    (d) $\dfrac{9}{18}$

24. Change each of the following decimals to fractions in lowest terms:
    (a) 0.25
    (b) 0.625
    (c) 1.67
    (d) 0.8888

25. Solve each of these equations for x:
    (a) $3.42x = 6.5$
    (b) $\dfrac{x}{12.3} = 7.05$
    (c) $\dfrac{0.525}{x} = 0.25$

26. Solve each equation for the variable:
    (a) $x = \dfrac{212 - 32}{1.8}$
    (b) $8.9 \dfrac{\text{g}}{\text{mL}} = \dfrac{40.90 \text{ g}}{x}$
    (c) $72°F = 1.8x + 32$

Unit Conversions

27. Complete the following metric conversions using the correct number of significant figures:
    (a) 28.0 cm to m
    (b) 1000. m to km
    (c) 9.28 cm to mm
    (d) 10.68 g to mg
    (e) 6.8×10^4 mg to kg
    (f) 8.54 g to kg
    (g) 25.0 mL to L
    (h) 22.4 L to μL

28. Complete the following metric conversions using the correct number of significant figures:
    (a) 4.5 cm to Å
    (b) 12 nm to cm
    (c) 8.0 km to mm
    (d) 164 mg to g
    (e) 0.65 kg to mg
    (f) 5.5 kg to g
    (g) 0.468 L to mL
    (h) 9.0 μL to mL

29. Complete the following American/metric conversions using the correct number of significant figures:
    (a) 42.2 in. to cm
    (b) 0.64 mi to in.
    (c) 2.00 in.^2 to cm^2
    (d) 42.8 kg to lb
    (e) 3.5 qt to mL
    (f) 20.0 gal to L

30. Make the following conversions using the correct number of significant figures:
    (a) 35.6 m to ft
    (b) 16.5 km to mi
    (c) 4.5 in.^3 to mm^3
    (d) 95 lb to g
    (e) 20.0 gal to L
    (f) $4.5 \times 10^4 \text{ ft}^3$ to m^3

31. An automobile traveling at 55 mi/hr is moving at what speed in kilometers per hour?

32. A cyclist is traveling downhill at 55 km/hr. How fast is she moving in feet per second?

33. Carl Lewis, a sprinter in the 1988 Olympic Games, ran the 100.-m dash in 9.92 s. What was his speed in feet per second?

34. Al Unser, Jr., qualified for the pole position at the 1994 Indianapolis 500 at a speed of 229 mph. What was his speed in kilometers per second?

35. When the space probe *Galileo* reached Jupiter in 1995, it was traveling at an average speed of 27,000 mi/hr. What was its speed in kilometers per second?

36. The Sun is approximately 93 million miles from the Earth. How many seconds will it take for light from the Sun to travel to the Earth if the velocity of light is 3.00×10^8 m/s?

37. How many kilograms does a 176-lb man weigh?

38. The average mass of the heart of a human baby is about 1 oz. What is its mass in milligrams?

39. A regular aspirin tablet contains 5.0 grains of aspirin. How many grams of aspirin are in one tablet? (1 grain = 1/7000. lb)

40. An adult ruby-throated hummingbird has an average mass of 3.2 g, while an adult California condor may attain a mass of 21 lb. How many hummingbirds would it take to equal the mass of one condor?

41. A bag of pretzels has a mass of 283.5 g and costs $1.49. If a bag contains 18 pretzels, what is the cost of a pound of pretzels?

42. The price of gold varies greatly and has been as high as $875 per ounce. What is the value of 250 g of gold at $350 per ounce? Gold is priced by troy ounces (14.58 troy ounces = 1 lb).

43. At 35¢/L how much will it cost to fill a 15.8-gal tank with gasoline?

44. How many liters of gasoline will be used to drive 525 miles in a car that averages 35 mi/gal?

*45. Assuming that there are 20. drops in 1.0 mL, how many drops are in 1.0 gallon?

46. How many liters of oil are in a 42-gal barrel of oil?

*47. Calculate the number of milliliters of water in a cubic foot of water.

*48. Oil spreads in a thin layer on water called an "oil slick." How much area in m^2 will 200 cm^3 of oil cover if it forms a layer 0.5 nm thick?

49. A textbook is 27 cm long, 21 cm wide, and 4.4 cm thick. What is the volume in:
    (a) cubic centimeters?
    (b) liters?
    (c) cubic inches?

50. An aquarium measures 16 in. $\times$ 8 in. $\times$ 10 in. How many liters of water does it hold? How many gallons?

Temperature Conversions

51. Normal body temperature for humans is 98.6°F. What is this temperature on the Celsius scale?

52. Driving to the grocery store you notice the temperature is 45°C. Determine what this temperature is on the Fahrenheit scale and what season of the year it might be.

53. Make the following conversions and include an equation for each one:
    (a) 162°F to °C
    (b) 0.0°F to K
    (c) −18°C to °F
    (d) 212 K to °C

54. Make the following conversions and include an equation for each one:
    (a) 32°C to °F
    (b) −8.6°F to °C
    (c) 273°C to K
    (d) 100 K to °F

*55. At what temperature are the Fahrenheit and Celsius temperatures exactly equal?

*56. At what temperature are Fahrenheit and Celsius temperatures the same in value but opposite in sign?

Density

57. Calculate the density of a liquid if 50.00 mL of the liquid has a mass of 78.26 g.

58. A 12.8-mL sample of bromine has a mass of 39.9 g. What is the density of bromine?

59. When a 32.7-g piece of chromium metal was placed into a graduated cylinder containing 25.0 mL of water, the water level rose to 29.6 mL. Calculate the density of the chromium.

60. An empty graduated cylinder has a mass of 42.817 g. When filled with 50.0 mL of an unknown liquid it has a mass of 106.773 g. What is the density of the liquid?

61. Concentrated hydrochloric acid has a density of 1.19 g/mL. Calculate the mass of 250.0 mL of this acid.

62. What mass of mercury (density 13.6 g/mL) will occupy a volume of 25.0 mL?

Additional Exercises

These exercises are not paired or labeled by topic and provide additional practice on the concepts covered in this chapter.

63. One liter of homogenized whole milk has a mass of 1032 g. What is the density of the milk in grams per milliliter? in kilograms per liter?

64. The volume of blood plasma in adults is 3.1 L. Its density is 1.03 g/cm^3. Approximately how many pounds of blood plasma are there in your body?

*65. The dashed lane markers on an interstate highway are 2.5 ft long and 4.0 in. wide. One (1.0) qt of paint covers 43 ft^2. How many dashed lane markers can be painted with 15 gal of paint?

66. Will a hollow cube with sides of length 0.50 m hold 8.5 L of solution? Depending on your answer, how much additional solution would be required to fill the container or how many times would the container need to be filled to measure the 8.5 L?

*67. The accepted toxic dose of mercury is 300 μg/day. Dental offices sometimes contain as much as 180 μg of mercury per cubic meter of air. If a nurse working in the office ingests 2×10^4 L of air per day, is he or she at risk for mercury poisoning?

68. Which is the higher temperature, 4.5°F or −15°C?

*69. A flask containing 100. mL of alcohol ($d = 0.789$ g/mL) is placed on one pan of a two-pan balance. A larger container, with a mass of 11.0 g more than the empty flask, is placed on the other pan of the balance. What volume of turpentine ($d = 0.87$ g/mL) must be added to this container to bring the two pans into balance?

70. Suppose you have samples of two metals, A and B. Use the data below to determine which sample occupies the larger volume.

	A	B
Mass	25 g	65 g
Density	10 g/mL	4 g/mL

*71. As a solid substance is heated its volume increases but its mass remains the same. Sketch a graph of density versus temperature showing the trend you expect. Briefly explain.

72. A 35.0-mL sample of ethyl alcohol (density 0.789 g/mL) is added to a graduated cylinder that has a mass of 49.28 g. What will be the mass of the cylinder plus the alcohol?

73. You are given three cubes, A, B, and C; one is magnesium, one is aluminum, and the third is silver. All three cubes have the same mass, but cube A has a volume of 25.9 mL, cube B has a volume of 16.7 mL, and cube C has a volume of 4.29 mL. Identify cubes A, B, and C.

*74. A cube of aluminum has a mass of 500. g. What will be the mass of a cube of gold of the same dimensions?

75. A 25.0-mL sample of water at 90°C has a mass of 24.12 g. Calculate the density of water at this temperature.

76. The mass of an empty container is 88.25 g. The mass of the container when filled with a liquid ($d = 1.25$ g/mL) is 150.50 g. What is the volume of the container?

77. Which liquid will occupy the greater volume, 50 g of water or 50 g of ethyl alcohol? Explain.

78. A gold bullion dealer advertised a bar of pure gold for sale. The gold bar had a mass of 3300 g and measured 2.00 cm × 15.0 cm × 6.00 cm. Was the bar pure gold? Show evidence for your answer.

79. The largest nugget of gold on record was found in 1872 in New South Wales, Australia, and had a mass of 93.3 kg. Assuming the nugget is pure gold, what is its volume in cubic centimeters? What is it worth by today's standards if gold is $345/oz? (14.58 troy oz = 1 lb)

*80. Forgetful Freddie placed 25.0 mL of a liquid in a graduated cylinder with a mass of 89.450 g when empty. When Freddie placed a metal slug with a mass of 15.454 g into the cylinder, the volume rose to 30.7 mL. Freddie was asked to calculate the density of the liquid and of the metal slug from his data, but he forgot to obtain the mass of the liquid. He was told that if he found the mass of the cylinder containing the liquid and the slug, he would have enough data for the calculations. He did so and found its mass to be 125.934 g. Calculate the density of the liquid and of the metal slug.

Answers to Practice Exercises

2.1 (a) 2; (b) 4; (c) 4; (d) 1; (e) 3;
(f) 4; (g) 1; (h) 3

2.2 (a) 42.25 (Rule 2); (b) 88.0 (Rule 1);
(c) 0.0897 (Rule 2); (d) 0.090
(Rule 2); (e) 225 (Rule 1); (f) 14.2
(Rule 2)

2.3 (a) $1200 = 1.200 \times 10^3$
(left means positive exponent)
(b) $6,600,000 = 6.6 \times 10^6$
(left means positive exponent)
(c) $0.0468 = 4.68 \times 10^{-2}$
(right means negative exponent)
(d) $0.00003 = 3 \times 10^{-5}$
(right means negative exponent)

2.4 (a) $3350 \text{ in.}^2 = 3.4 \times 10^3 \text{ in.}^2$;
(b) 50.7 mi/hr; (c) 0.79

2.5 (a) 2; (b) 2; (c) 1; (d) 2; (e) 4;
(f) 2; (g) 2; (h) 2

2.6 1.69×10^4 m

2.7 30 m^3 or 3×10^1 m^3

2.8 165 lb

2.9 0.14 lb

2.10 0.793 qt

2.11 1.89 L (the number of significant
figures is arbitrary)

2.12 $-269°C$, $-452°F$

2.13 37.0°C, 310.2 K

2.14 The density is 13 g/mL; therefore the
ring is *not* pure silver

2.15 0.88 g/mL

CHAPTER

3

Classification of Matter

▲
Flower varieties are often classified by color in the nursery.

Throughout our lives we seek to bring order into the chaos that surrounds us. To do this we classify things according to their similarities. In the library we find books grouped according to the subject, and then by author. Department stores organize their merchandise by the size and style of clothing, as well as by the type of customer. Ballparks and theaters classify their seats by price and location. Chemists also classify things according to properties they can observe. For example, chemists classify substances as liquid (like water), solid (ice), or gas (steam). Chemists are also interested in identifying the tiniest part of a substance that still has the properties of the substance.

3.1   Matter Defined

The entire universe consists of matter and energy. Every day we come into contact with countless kinds of matter. Air, food, water, rocks, soil, glass, and this book are all different types of matter. Broadly defined, **matter** is *anything* that has mass and occupies space.

Matter may be quite invisible. For example, if an apparently empty test tube is submerged mouth downward in a beaker of water, the water rises only slightly into the tube. The water cannot rise further because the tube is filled with invisible matter: air (see Figure 3.1).

To the eye, matter appears to be continuous and unbroken. However, it is actually discontinuous and is composed of discrete, tiny particles called *atoms*. The particulate nature of matter will become evident when we study atomic structure and the properties of gases.

▲
FIGURE 3.1
An apparently empty test tube is submerged, mouth downward, in water. Only a small volume of water rises into the tube, which is actually filled with invisible matter—air.

matter

3.2   Physical States of Matter

Matter exists in three physical states: solid, liquid, and gas. A **solid** has a definite shape and volume, with particles that cohere rigidly to one another. The shape of a solid can be independent of its container. For example, a crystal of sulfur has the same shape and volume whether it is placed in a beaker or simply laid on a glass plate.

Most commonly occurring solids, such as salt, sugar, quartz, and metals, are *crystalline*. The particles that form crystalline materials exist in regular, repeating, three-dimensional, geometric patterns. Because their particles do not have any regular, internal geometric pattern, such solids as plastics, glass, and gels are called **amorphous** solids. (*Amorphous* means without shape or form.)

A **liquid** has a definite volume but not a definite shape, with particles that cohere firmly but not rigidly. Although the particles are held together by strong attractive forces and are in close contact with one another, they are able to move freely. Particle mobility gives a liquid fluidity and causes it to take the shape of the container in which it is stored.

A **gas** has indefinite volume and no fixed shape, with particles that move independently of one another. Particles in the gaseous state have gained enough energy to

solid

amorphous
liquid

gas

overcome the attractive forces that held them together as liquids or solids. A gas presses continuously in all directions on the walls of any container. Because of this quality a gas completely fills a container. The particles of a gas are relatively far apart compared with those of solids and liquids. The actual volume of the gas particles is very small compared to the volume of the space occupied by the gas. A gas therefore may be compressed into a very small volume or expanded almost indefinitely. Liquids cannot be compressed to any great extent, and solids are even less compressible than liquids.

If a bottle of ammonia solution is opened in one corner of the laboratory, we can soon smell its familiar odor in all parts of the room. The ammonia gas escaping from the solution demonstrates that gaseous particles move freely and rapidly and tend to permeate the entire area into which they are released.

Although matter is discontinuous, attractive forces exist that hold the particles together and give matter its appearance of continuity. These attractive forces are strongest in solids, giving them rigidity; they are weaker in liquids but still strong enough to hold liquids to definite volumes. In gases the attractive forces are so weak that the particles of a gas are practically independent of one another. Table 3.1 lists common materials that exist as solids, liquids, and gases. Table 3.2 compares the properties of solids, liquids, and gases.

TABLE 3.1   Common Materials in the Solid, Liquid, and Gaseous States of Matter

Solids	Liquids	Gases
Aluminum	Alcohol	Acetylene
Copper	Blood	Air
Gold	Gasoline	Butane
Polyethylene	Honey	Carbon dioxide
Salt	Mercury	Chlorine
Sand	Oil	Helium
Steel	Vinegar	Methane
Sulfur	Water	Oxygen

TABLE 3.2   Physical Properties of Solids, Liquids, and Gases

State	Shape	Volume	Particles	Compressibility
Solid	Definite	Definite	Rigidly cohering; tightly packed	Very slight
Liquid	Indefinite	Definite	Mobile; cohering	Slight
Gas	Indefinite	Indefinite	Independent of each other and relatively far apart	High

◄ **Water can exist as a solid (snow), a liquid (water), and a gas (steam) as shown here at Yellowstone National Park.**

3.3 Substances and Mixtures

The term *matter* refers to all materials that make up the universe. Many thousands of different, and distinct, kinds of matter exist. A **substance** is a particular kind of matter with a definite, fixed composition. Sometimes known as *pure substances*, substances are either elements or compounds. Familiar examples of elements are copper, gold, and oxygen. Familiar compounds are salt, sugar, and water.

substance

We classify a sample of matter as either *homogeneous* or *heterogeneous* by examining it. **Homogeneous** matter is uniform in appearance and has the same properties throughout. Matter consisting of two or more physically distinct phases is **heterogeneous.** A **phase** is a homogeneous part of a system separated from other parts by physical boundaries. A **system** is simply the body of matter under consideration. Whenever we have a system in which visible boundaries exist between the parts or components, that system has more than one phase and is heterogeneous. It does not matter whether these components are in the solid, liquid, or gaseous states.

homogeneous

heterogeneous
phase
system

A pure substance may exist as different phases in a heterogeneous system. Ice floating in water, for example, is a two-phase system made up of solid water and liquid water. The water in each phase is homogeneous in composition, but because two phases are present, the system is heterogeneous.

A **mixture** is a material containing two or more substances and can be either heterogeneous or homogeneous. Mixtures are variable in composition. If we add a spoonful of sugar to a glass of water, a heterogeneous mixture is formed immediately. The two phases are a solid (sugar) and a liquid (water). But upon stirring the sugar dissolves to form a homogeneous mixture or solution. Both substances are still

mixture

Figure 3.2 ▶
Classification of matter. A pure substance is always homogeneous in composition, whereas a mixture always contains two or more substances and may be either homogeneous or heterogeneous.

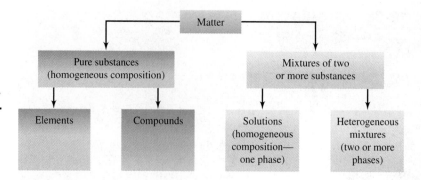

present: All parts of the solution are sweet and wet. The proportions of sugar and water can be varied simply by adding more sugar and stirring to dissolve.

Many substances do not form homogeneous mixtures. If we mix sugar and fine white sand, a heterogeneous mixture is formed. Careful examination may be needed to decide that the mixture is heterogeneous because the two phases (sugar and sand) are both white solids. Ordinary matter exists mostly as mixtures. If we examine soil, granite, iron ore, or other naturally occurring mineral deposits, we find them to be heterogeneous mixtures. Air is a homogeneous mixture (solution) of several gases. Figure 3.2 illustrates the relationships of substances and mixtures.

Flowcharts can help you to visualize the connections between concepts.

3.4   Elements

All words in English are formed from an alphabet consisting of only 26 letters. All known substances on Earth—and most probably in the universe, too—are formed from a sort of "chemical alphabet" consisting of 111 presently known elements. An
element **element** is a fundamental or elementary substance that cannot be broken down by chemical means to simpler substances. Elements are the building blocks of all substances. The elements are numbered in order of increasing complexity beginning with hydrogen, number 1. Of the first 92 elements, 88 are known to occur in nature. The other four—technetium (43), promethium (61), astatine (85), and francium (87)—either do not occur in nature or have only transitory existences during radioactive decay. With the exception of number 94, plutonium, elements above number 92 are not known to occur naturally but have been synthesized, usually in very small quantities, in laboratories. The discovery of trace amounts of element 94 (plutonium) in nature has been reported recently. The syntheses of elements 110 and 111 were reported in 1994. No elements other than those on the Earth have been detected on other bodies in the universe.

Most substances can be decomposed into two or more simpler substances. Water can be decomposed into hydrogen and oxygen. Sugar can be decomposed into carbon, hydrogen, and oxygen. Table salt is easily decomposed into sodium and chlorine. An element, however, cannot be decomposed into simpler substances by ordinary chemical changes.

If we could take a small piece of an element, say copper, and divide it and subdivide it into smaller and smaller particles, we would finally come to a single unit of copper that we could no longer divide and still have copper. This smallest particle of
atom an element that can exist is called an **atom,** which is also the smallest unit of an element that can enter into a chemical reaction. Atoms are made up of still smaller

subatomic particles. However, these subatomic particles (described in Chapter 5) do not have the properties of elements.

3.5 Distribution of Elements

Elements are distributed unequally in nature, as shown in Figure 3.3. At normal room temperature two of the elements, bromine and mercury, are liquids. Eleven elements, hydrogen, nitrogen, oxygen, fluorine, chlorine, helium, neon, argon, krypton, xenon, and radon, are gases. All the other elements are solids.

Ten elements make up about 99% of the mass of the Earth's crust, seawater, and atmosphere. Oxygen, the most abundant of these, constitutes about 50% of this mass. The distribution of the elements shown in Figure 3.3 includes the Earth's crust to a depth of about 10 miles, the oceans, fresh water, and the atmosphere but does not include the mantle and core of the Earth, which are believed to consist of metallic iron and nickel. Because the atmosphere contains relatively little matter, its inclusion has almost no effect on the distribution. But the inclusion of fresh and salt water does have an appreciable effect since water contains about 11.2% hydrogen. Nearly all of the 0.87% hydrogen shown in the table is from water.

The average distribution of the elements in the human body is shown in Figure 3.3. Note again the high percentage of oxygen.

▲
Mercury (left) and bromine are liquid elements at room temperature.

3.6 Names of the Elements

The names of the elements come to us from various sources. Many are derived from early Greek, Latin, or German words that describe some property of the element. For example, iodine is taken from the Greek word *iodes,* meaning violetlike, and iodine is certainly violet in the vapor state. The name of the metal bismuth originates from the German words *weisse masse,* which means white mass. Miners called it *wismat;* it was later changed to *bismat,* and finally to bismuth. Some elements are named for the location of their discovery—for example, germanium, discovered in 1886 by a German chemist. Others are named in commemoration of famous scientists, such as einsteinium and curium, named for Albert Einstein and Marie Curie, respectively.

3.7 Symbols of the Elements

We all recognize Mr., N.Y., and Ave. as abbreviations for mister, New York, and avenue. In a like manner each element also has an abbreviation; these are called **symbols** of the elements. Fourteen elements have a single letter as their symbol, and the rest have two letters. A symbol stands for the element itself, for one atom of the element, and (as we shall see later) for a particular quantity of the element.

symbol

Rules governing symbols of elements are as follows:

1. Symbols have either one or two letters.
2. If one letter is used, it is capitalized.
3. If two letters are used, only the first is capitalized.

Examples: Iodine I   Barium Ba

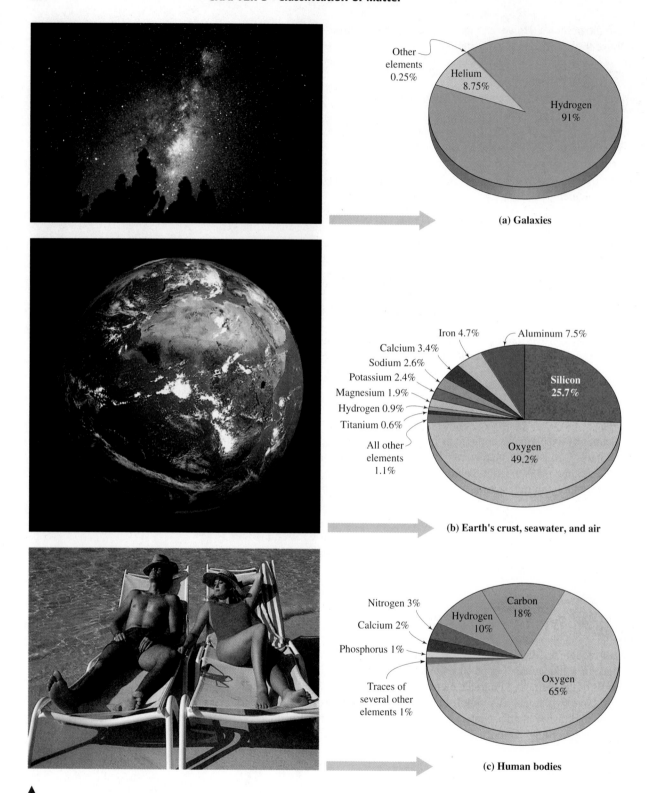

(a) Galaxies

(b) Earth's crust, seawater, and air

(c) Human bodies

▲
Figure 3.3 Distribution of the common elements in nature.

The symbols and names of all the elements are given in the table on the inside front cover of this book. Table 3.3 lists the more commonly used symbols. Examine this table carefully and you will note that most of the symbols start with the same letter as the name of the element that is represented. A number of symbols, however, appear to have no connection with the names of the elements they represent (see Table 3.4). These symbols have been carried over from earlier names (usually in Latin) of the elements and are so firmly implanted in the literature that their use is continued today.

TABLE 3.3	Symbols of the Most Common Elements				
Element	**Symbol**	**Element**	**Symbol**	**Element**	**Symbol**
Aluminum	Al	Fluorine	F	Phosphorus	P
Antimony	Sb	Gold	Au	Platinum	Pt
Argon	Ar	Helium	He	Potassium	K
Arsenic	As	Hydrogen	H	Radium	Ra
Barium	Ba	Iodine	I	Silicon	Si
Bismuth	Bi	Iron	Fe	Silver	Ag
Boron	B	Lead	Pb	Sodium	Na
Bromine	Br	Lithium	Li	Strontium	Sr
Cadmium	Cd	Magnesium	Mg	Sulfur	S
Calcium	Ca	Manganese	Mn	Tin	Sn
Carbon	C	Mercury	Hg	Titanium	Ti
Chlorine	Cl	Neon	Ne	Tungsten	W
Chromium	Cr	Nickel	Ni	Uranium	U
Cobalt	Co	Nitrogen	N	Zinc	Zn
Copper	Cu	Oxygen	O		

▲ **Iodine in its elemental form is a dark purple crystal. The symbol for iodine is I.**

TABLE 3.4	Symbols of the Elements Derived from Early Names*	
Present name	**Symbol**	**Former name**
Antimony	Sb	Stibium
Copper	Cu	Cuprum
Gold	Au	Aurum
Iron	Fe	Ferrum
Lead	Pb	Plumbum
Mercury	Hg	Hydrargyrum
Potassium	K	Kalium
Silver	Ag	Argentum
Sodium	Na	Natrium
Tin	Sn	Stannum
Tungsten	W	Wolfram

*These symbols are in use today even though they do not correspond to the current name of the element.

▲ **The metal sodium is soft enough to cut with a knife. The symbol for sodium is Na.**

▲
The colors of these varieties of quartz are the result of the presence of different metallic elements in the sample.

Special care must be taken in writing symbols. Capitalize only the first letter and use a lowercase second letter if needed. This is important. For example, consider Co, the symbol for the element cobalt. If you write CO (capital C and capital O), you will have written the two elements carbon and oxygen (the *formula* for carbon monoxide), *not* the single element cobalt. Also make sure you write the letters distinctly; otherwise Co (for cobalt) may be misread as Ca (for calcium).

Knowledge of symbols is essential for writing chemical formulas and equations, and will be needed in the remainder of this book and in any future chemistry courses you may take. One way to learn the symbols is to practice a few minutes a day by making flash cards of names and symbols and then practicing daily. Initially it is a good plan to learn the symbols of the most common elements shown in Table 3.3.

3.8 Metals, Nonmetals, and Metalloids

The elements are classified as metals, nonmetals, and metalloids. Most of the elements are metals. We are familiar with them because of their widespread use in tools, construction materials, automobiles, and so on. But nonmetals are equally useful in our everyday life as major components of clothing, food, fuel, glass, plastics, and wood. Metalloids are often used in the electronics industry.

metal

The **metals** are solids at room temperature (mercury is an exception). They have high luster, are good conductors of heat and electricity, are *malleable* (can be rolled or hammered into sheets), and are *ductile* (can be drawn into wires). Most metals

Samples of various metals, ▶ including aluminum, copper, mercury, titanium, beryllium, cadmium, calcium, and nickel.

have a high melting point and high density. Familiar metals are aluminum, chromium, copper, gold, iron, lead, magnesium, mercury, nickel, platinum, silver, tin, and zinc. Less familiar but still important metals are calcium, cobalt, potassium, sodium, uranium, and titanium.

Metals have little tendency to combine with each other to form compounds. But many metals readily combine with nonmetals such as chlorine, oxygen, and sulfur to form ionic compounds such as metallic chlorides, oxides, and sulfides. In nature minerals are composed of the more reactive metals combined with other elements. A few of the less reactive metals such as copper, gold, and silver are sometimes found in a native, or free, state.

Nonmetals, unlike metals, are not lustrous, have relatively low melting points and densities, and are generally poor conductors of heat and electricity. Carbon, phosphorus, sulfur, selenium, and iodine are solids; bromine is a liquid; the rest of the nonmetals are gases. Common nonmetals found uncombined in nature are carbon (graphite and diamond), nitrogen, oxygen, sulfur, and the noble gases (helium, neon, argon, krypton, xenon, and radon).

Nonmetals combine with one another to form molecular compounds such as carbon dioxide (CO_2), methane (CH_4), butane (C_4H_{10}), and sulfur dioxide (SO_2). Fluorine, the most reactive nonmetal, combines readily with almost all other elements.

Several elements (boron, silicon, germanium, arsenic, antimony, tellurium, and polonium) are classified as **metalloids** and have properties that are intermediate between those of metals and those of nonmetals. The intermediate position of these elements is shown in Table 3.5. Certain metalloids, such as boron, silicon, and germanium, are the raw materials for the semiconductor devices that make our modern electronics industry possible.

▲ **This tiny and powerful computer chip is made of silicon, a metalloid.**

nonmetal
metalloid

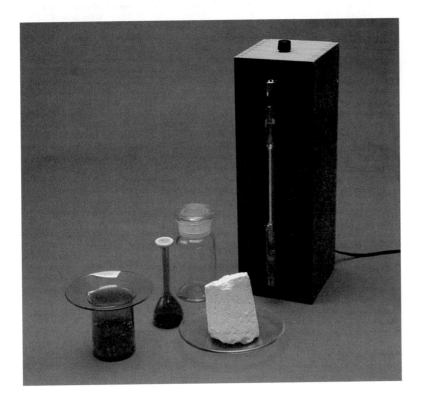

◀ **Samples of various nonmetals, including iodine (far left), bromine, oxygen, neon (in box), and sulfur.**

TABLE 3.5   Classification of the Elements into Metals, Metalloids, and Nonmetals

1 H																	2 He
3 Li	4 Be											5 B	6 C	7 N	8 O	9 F	10 Ne
11 Na	12 Mg											13 Al	14 Si	15 P	16 S	17 Cl	18 Ar
19 K	20 Ca	21 Sc	22 Ti	23 V	24 Cr	25 Mn	26 Fe	27 Co	28 Ni	29 Cu	30 Zn	31 Ga	32 Ge	33 As	34 Se	35 Br	36 Kr
37 Rb	38 Sr	39 Y	40 Zr	41 Nb	42 Mo	43 Tc	44 Ru	45 Rh	46 Pd	47 Ag	48 Cd	49 In	50 Sn	51 Sb	52 Te	53 I	54 Xe
55 Cs	56 Ba	57 La*	72 Hf	73 Ta	74 W	75 Re	76 Os	77 Ir	78 Pt	79 Au	80 Hg	81 Tl	82 Pb	83 Bi	84 Po	85 At	86 Rn
87 Fr	88 Ra	89 Ac†	104 Rf	105 Db	106 Sg	107 Bh	108 Hs	109 Mt	110 —	111 —	112 —						

Legend: □ Metals ▨ Metalloids □ Nonmetals

*	58 Ce	59 Pr	60 Nd	61 Pm	62 Sm	63 Eu	64 Gd	65 Tb	66 Dy	67 Ho	68 Er	69 Tm	70 Yb	71 Lu
†	90 Th	91 Pa	92 U	93 Np	94 Pu	95 Am	96 Cm	97 Bk	98 Cf	99 Es	100 Fm	101 Md	102 No	103 Lr

3.9   Compounds

compound   A **compound** is a distinct substance that contains two or more elements chemically combined in definite proportions by mass. Compounds, unlike elements, can be decomposed chemically into simpler substances—that is, into simpler compounds and/or elements. Atoms of the elements in a compound are combined in whole-number ratios, never as fractional parts. Compounds fall into two general types, *molecular* and *ionic*.

molecule   A **molecule** is the smallest uncharged individual unit of a compound formed by the union of two or more atoms. Water is a typical molecular compound. If we divide a drop of water into smaller and smaller particles, we finally obtain a single molecule of water consisting of two hydrogen atoms bonded to one oxygen atom. This molecule is the ultimate particle of water; it cannot be further subdivided without destroying the water molecule and forming hydrogen and oxygen.

ion   An **ion** is a positively or negatively charged atom or group of atoms. An ionic compound is held together by attractive forces that exist between positively and negatively charged ions. A positively charged ion is called a **cation** (pronounced *cat-eye-on*); a negatively charged ion is called an **anion** (pronounced *an-eye-on*). Figure 3.4 illustrates the classification of compounds.

cation
anion

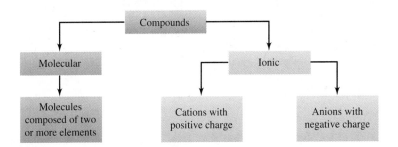

◀ **Figure 3.4**
Compounds can be classified as molecular or ionic. Ionic compounds are held together by attractive forces between their positive and negative charges. Molecular compounds are held together by covalent bonds.

Sodium chloride is a typical ionic compound. The ultimate particles of sodium chloride are positively charged sodium ions and negatively charged chloride ions. Sodium chloride is held together in a crystalline structure by the attractive forces existing between these oppositely charged ions. Although ionic compounds consist of large aggregates of cations and anions, their formulas are normally represented by the simplest possible ratio of the atoms in the compound. For example, in sodium chloride the ratio is one sodium ion to one chlorine ion, so the formula is NaCl. The two types of compounds, molecular and ionic, are illustrated in Figure 3.5.

There are more than 9 million known registered compounds, with no end in sight as to the number that will be prepared in the future. Each compound is unique and has characteristic properties. Let's consider two compounds, water and sodium chloride, in some detail. Water is a colorless, odorless, tasteless liquid that can be changed to a solid (ice) at 0°C and to a gas (steam) at 100°C. Composed of two atoms of hydrogen and one atom of oxygen per molecule, water is 11.2% hydrogen and 88.8% oxygen by mass. Water reacts chemically with sodium to produce hydrogen gas and sodium hydroxide, with lime to produce calcium hydroxide, and with sulfur trioxide to produce sulfuric acid. No other compound has all these exact physical and chemical properties; they are characteristic of water alone.

Sodium chloride is a colorless crystalline substance with a ratio of one atom of sodium to one atom of chlorine. Its composition by mass is 39.3% sodium and 60.7% chlorine. It does not conduct electricity in its solid state; it dissolves in water to produce a solution that conducts electricity. When a current is passed through molten sodium chloride, solid sodium and gaseous chlorine are produced. These specific properties belong to sodium chloride and to no other substance. Thus, a compound may be identified and distinguished from all other compounds by its characteristic properties.

(a) H_2O

Na^+ Cl^-

(b) NaCl

▲
FIGURE 3.5
Representation of molecular and ionic (nonmolecular) compounds. (a) Two hydrogen atoms combined with an oxygen atom to form a molecule of water. (b) A positively charged sodium ion and a negatively charged chloride ion form the compound sodium chloride.

3.10   Elements That Exist as Diatomic Molecules

Seven elements (all nonmetals) occur as **diatomic molecules.** These elements and their symbols, formulas, and brief descriptions are listed in Table 3.6. Whether found free in nature or prepared in the laboratory, the molecules of these elements always contain two atoms. The formulas of the free elements are therefore always written to show this molecular composition: H_2, N_2, O_2, F_2, Cl_2, Br_2, and I_2.

It is important to see that symbols can designate either an atom or a molecule of an element. Consider hydrogen and oxygen. Hydrogen gas is present in volcanic gases and can be prepared by many chemical reactions. Regardless of their source,

diatomic molecules

Hydrogen: Fuel of the Future

Hydrogen, the lightest element on the periodic table, could provide society with a fuel that is nearly inexhaustible, environmentally harmless, and available everywhere. The U.S. space program has used hydrogen for nearly 30 years to power rockets, provide electrical power, and produce drinking water for astronauts. How can we harness hydrogen's energy and use it in our daily lives?

Mercedes Benz has developed a prototype minivan that is powered by hydrogen fuel cells. These cells combine compressed hydrogen with oxygen to form water and power the car. Because of the difficulties in storing enough compressed hydrogen these vehicles have limited range. Hydrogen gas has a low density (requiring high pressure to avoid large, bulky tanks) and liquid hydrogen (boiling point, $-253°$ C) requires constant refrigeration and insulated containers. To improve range and eliminate storage problems engineers are currently considering ways to generate hydrogen as needed for fuel.

The U.S. Department of Energy and Arthur D. Little Co. of Cambridge, Massachusetts, have developed an alternative system that may solve the storage problem. Their gasoline-fueled system contains a processor that partially converts the gasoline to hydrogen gas, which is then sent to a fuel cell and used to generate power. Preliminary tests have produced hydrogen fast enough to generate 50 kW of electric power, plenty to run a mid-size car. The processor uses gasoline, ethanol, methanol, or natural gas to supply hydrogen to the fuel cell. A car using this system gets about twice the mileage of a traditional car.

This alternative fuel system doesn't require large batteries (like those needed in electric cars) or storage tanks for the hydrogen, so it may be a significant step toward a hydrogen-powered economy. This hybrid system also takes advantage of the gasoline distribution system already in place. You could soon drive a hydrogen car and fill up at your local gas station!

TABLE 3.6	Elements That Exist as Diatomic Molecules		
Element	**Symbol**	**Molecular formula**	**Normal state**
Hydrogen	H	H_2	Colorless gas
Nitrogen	N	N_2	Colorless gas
Oxygen	O	O_2	Colorless gas
Fluorine	F	F_2	Pale yellow gas
Chlorine	Cl	Cl_2	Yellow-green gas
Bromine	Br	Br_2	Reddish-brown liquid
Iodine	I	I_2	Bluish-black solid

all samples of free hydrogen gas consist of diatomic molecules. Free hydrogen is designated by the formula H_2, which also expresses its composition. Oxygen makes up about 21% by volume of the air that we breathe. This free oxygen is constantly being replenished by photosynthesis; it can also be prepared in the laboratory by several reactions. The majority of free oxygen is diatomic and is designated by the formula O_2. Now consider water, a compound designated by the formula H_2O (sometimes HOH). Water contains neither free hydrogen (H_2) nor free oxygen (O_2). The H_2 part of the formula H_2O simply indicates that two atoms of hydrogen are combined with one atom of oxygen to form water.

> **Symbols are used to designate elements, show the composition of molecules of elements, and give the elemental composition of compounds.**

3.11 Chemical Formulas

Chemical formulas are used as abbreviations for compounds. A **chemical formula** shows the symbols and the ratio of the atoms of the elements in a compound. Sodium chloride contains one atom of sodium per atom of chlorine; its formula is NaCl. The formula for water is H_2O; it shows that a molecule of water contains two atoms of hydrogen and one atom of oxygen.

chemical formula

The formula of a compound tells us which elements it is composed of and how many atoms of each element are present in a formula unit. For example, a unit of sulfuric acid is composed of two atoms of hydrogen, one atom of sulfur, and four atoms of oxygen. We could express this compound as HHSOOOO, but this is cumbersome, so we write H_2SO_4 instead. The formula may be expressed verbally as "H-two-S-O-four." Numbers that appear partially below the line and to the right of a symbol of an element are called **subscripts.** Thus the 2 and the 4 in H_2SO_4 are subscripts (see Figure 3.6). Characteristics of chemical formulas are

subscript

1. The formula of a compound contains the symbols of all the elements in the compound.
2. When the formula contains one atom of an element, the symbol of that element represents that one atom. The number one (1) is not used as a subscript to indicate one atom of an element.
3. When the formula contains more than one atom of an element, the number of atoms is indicated by a subscript written to the right of the symbol of that atom. For example, the two (2) in H_2O indicates two atoms of H in the formula.
4. When the formula contains more than one of a group of atoms that occurs as a unit, parentheses are placed around the group, and the number of units of the group is indicated by a subscript placed to the right of the parentheses. Consider the nitrate group, NO_3. The formula for sodium nitrate, $NaNO_3$, has only one nitrate group, so no parentheses are needed. Calcium nitrate, $Ca(NO_3)_2$, has two nitrate groups, as indicated by the use of parentheses and the subscript 2. $Ca(NO_3)_2$ has a total of nine atoms: one Ca, two N, and six O atoms. The formula $Ca(NO_3)_2$ is read as "C-A [pause] N-O-three taken twice."

Figure 3.6 ▶
**Explanation of the formulas
NaCl, H₂SO₄, and Ca(NO₃)₂.**

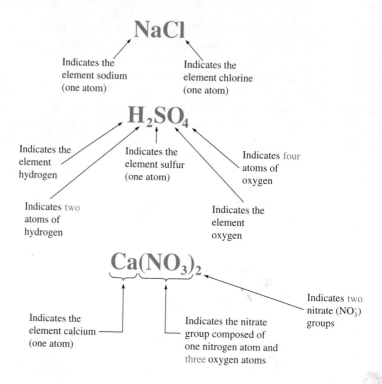

5. Formulas written as H_2O, H_2SO_4, $Ca(NO_3)_2$, and $C_{12}H_{22}O_{11}$ show only the number and kind of each atom contained in the compound; they do not show the arrangement of the atoms in the compound or how they are chemically bonded to one another.

Example 3.1   Write formulas for the following compounds; the atomic composition is given. (a) Hydrogen chloride: 1 atom hydrogen + 1 atom chlorine; (b) methane: 1 atom carbon + 4 atoms hydrogen; (c) glucose: 6 atoms carbon + 12 atoms hydrogen + 6 atoms oxygen.

Solution   (a) First write the symbols of the atoms in the formula: H Cl. Since the ratio of atoms is one to one, we bring the symbols together to give the formula for hydrogen chloride as HCl.

(b) Write the symbols of the atoms: C H. Now bring the symbols together and place a subscript 4 after the hydrogen atom. The formula is CH_4.

(c) Write the symbols of the atoms: C H O. Now write the formula, bringing together the symbols followed by the correct subscripts according to the data given (six C, twelve H, six O). The formula is $C_6H_{12}O_6$.

3.12 Mixtures

Single substances—elements or compounds—seldom occur naturally in a pure state. Air is a mixture of gases; seawater is a mixture of a variety of dissolved minerals; ordinary soil is a complex mixture of minerals and various organic materials.

(a)   (b)   (c)

▲
(a) When iron and sulfur exist as pure substances, only the iron is attracted to a magnet. (b) A mixture of iron and sulfur can be separated by using the difference in magnetic attraction. (c) The compound iron(II) sulfide cannot be separated into its elements with a magnet.

How is a mixture distinguished from a pure substance? A mixture always contains two or more substances that can be present in varying concentrations. Let's consider two examples.

Homogeneous Mixture Homogeneous mixtures (solutions) containing either 5% or 10% salt in water can be prepared simply by mixing the correct amounts of salt and water. These mixtures can be separated by boiling away the water, leaving the salt as a residue.

Heterogeneous Mixture The composition of a heterogeneous mixture of sulfur crystals and iron filings can be varied by merely blending in either more sulfur or more iron filings. This mixture can be separated physically by using a magnet to attract the iron.

Iron(II) sulfide (FeS) contains 63.5% Fe and 36.5% S by mass. If we mix iron and sulfur in this proportion, do we have iron(II) sulfide? No, it is still a mixture; the iron is still attracted by a magnet. But if this mixture is heated strongly, a chemical change (reaction) occurs in which the reacting substances, iron and sulfur, form a new substance, iron(II) sulfide. Iron(II) sulfide, FeS, is a compound of iron and sulfur and has properties that are different from those of either iron or sulfur: It is not attracted by a magnet as shown in the photo. The general characteristics of mixtures and compounds are compared in Table 3.7. The differences between the iron and sulfur *mixture* and the iron(II) sulfide *compound* are as follows:

Iron(II) sulfide is the correct name for the compound formed from iron and sulfur. We will discuss the reason for the (II) in Chapter 6 when we learn to name compounds.

	Mixture of iron and sulfur	Compound of iron and sulfur
Formula	Has no definite formula; consists of Fe and S.	FeS
Composition	Contains Fe and S in any proportion by mass.	63.5% Fe and 36.5% S by mass.
Separation	Fe and S can be separated by physical means.	Fe and S can be separated only by chemical change.

Carbon—The Chameleon

Just as a chameleon changes its color to reflect its environment, the element carbon is found in many different forms. Carbon is found in the mineral *graphite,* as *diamond,* and as *buckminsterfullerene.* The physical properties of each form of carbon are quite distinct.

Diamonds are transparent crystals that are colorless when pure—but they can range from pale blue to jet black when impurities are present. Diamond is the hardest natural substance, an excellent heat conductor, and when certain impurities are added, it becomes an electrical semiconductor. Diamonds are used in cutting and drilling tools and displayed as beautiful gems in jewelry.

Graphite consists of layered sheets of carbon atoms. Because the sheets easily slip over each other, graphite is slippery making it an excellent choice for lubricants. Mixtures of clay and graphite are molded into the "lead" used in pencils. The higher the clay content, the harder the "lead" in the pencil (and the more difficult to use in writing). An excellent conductor of electricity, graphite is mined as massive crystals or is obtained from heating coal and pitch in very high temperature furnaces. Graphite is formed into electrodes for batteries and used as a lubricant in locks.

Charcoal, consisting of tiny crystals of graphite, can adsorb large quantities of substances onto its surfaces. For this reason it is very useful in water purification systems, in the manufacture of gas masks, and in removing color from solutions (as in refining sugar). Carbon is added to iron to form steel; carbon black (formed when natural gas is burned with an insufficient quantity of oxygen) is used in ink, shoe polish, and as an additive to rubber to make tires black.

Buckminsterfullerene is composed of clusters of carbon atoms arranged in the shape of a soccer ball. This cagelike structure permits capture of other atoms, leading to some interesting applications. More information on buckminsterfullerene can be found in the Chemistry in Action in Chapter 11.

In addition, carbon combines with other elements to form millions of useful compounds. Hydrocarbons (molecules containing carbon and hydrogen) are found in petroleum and natural gas. So many hydrocarbons and their derivatives exist that an entire branch of chemistry, organic chemistry, is dedicated to studying them. Finally, carbon is an essential constituent of biomolecules. Found in all living organisms as part of carbohydrates, proteins, and fats, carbon is also a critical atom in both DNA and RNA, the molecules that determine the genetic composition of each organism.

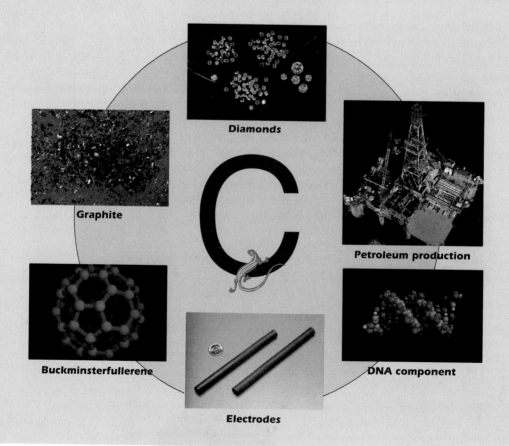

Diamonds

Graphite

Petroleum production

Buckminsterfullerene

DNA component

Electrodes

TABLE 3.7 Comparison of Mixtures and Compounds		
	Mixture	**Compound**
Composition	May be composed of elements, compounds, or both in variable composition.	Composed of two or more elements in a definite, fixed proportion by mass.
Separation of components	By physical or mechanical means.	Only by chemical changes in the elements.
Identification of components	Components do not lose their identity.	A compound does not resemble the elements from which it is formed.

Concepts in Review

1. Identify the three physical states of matter.
2. Distinguish between substances and mixtures.
3. Classify common materials as elements, compounds, or mixtures.
4. Write the symbols when given the names, or write the names when given the symbols, of the common elements listed in Table 3.3.
5. Understand how symbols, including subscripts and parentheses, are used to write chemical formulas.
6. Differentiate among atoms, molecules, and ions.
7. List the characteristics of metals, nonmetals, and metalloids.
8. List the elements that occur as diatomic molecules.

Key Terms

amorphous (3.2)
anion (3.9)
atom (3.4)
cation (3.9)
chemical formula (3.11)

compound (3.9)
diatomic molecules (3.10)
element (3.4)
gas (3.2)
heterogeneous (3.3)

homogeneous (3.3)
ion (3.9)
liquid (3.2)
matter (3.1)
metal (3.8)

metalloid (3.8)
mixture (3.3)
molecule (3.9)
nonmetal (3.8)
phase (3.3)

solid (3.2)
subscripts (3.11)
substance (3.3)
symbol (3.7)
system (3.3)

Questions

1. List four different substances in each of the three states of matter.
2. In terms of the properties of the ultimate particles of a substance, explain
   (a) why a solid has a definite shape but a liquid does not
   (b) why a liquid has a definite volume but a gas does not
   (c) why a gas can be compressed rather easily but a solid cannot be compressed appreciably
3. What evidence can you find in Figure 3.1 that gases occupy space?
4. Which liquids listed in Table 3.1 are not mixtures?
5. Which of the gases listed in Table 3.1 are not pure substances?
6. When the stopper is removed from a partly filled bottle containing solid and liquid acetic acid at 16.7°C, a strong vinegarlike odor is noticeable immediately. How many acetic acid phases must be present in the bottle? Explain.
7. Is the system enclosed in the bottle in Question 6 homogeneous or heterogeneous? Explain.
8. Is a system that contains only one substance necessarily homogeneous? Explain.

9. Is a system that contains two or more substances necessarily heterogeneous? Explain.

10. Are there more atoms of silicon or hydrogen in the Earth's crust, seawater, and atmosphere? Use Figure 3.3 and the fact that the mass of a silicon atom is about 28 times that of a hydrogen atom.

11. What does the symbol of an element stand for?

12. Write down what you believe to be the symbols for the elements phosphorus, aluminum, hydrogen, potassium, magnesium, sodium, nitrogen, nickel, and silver. Check yourself by looking up the correct symbols in Table 3.3.

13. Interpret the difference in meanings for each of these pairs:
   (a) Si and SI   (b) Pb and PB   (c) 4 P and P_4

14. List six elements and their symbols in which the first letter of the symbol is different from that of the name. (Table 3.4)

15. Write the names and symbols for the 14 elements that have only one letter as their symbol. (See tables on inside front cover.)

16. Distinguish between an element and a compound.

17. How many metals are there? nonmetals? metalloids? (Table 3.5)

18. Of the ten most abundant elements in the Earth's crust, seawater, and atmosphere, how many are metals? nonmetals? metalloids? (Figure 3.3)

19. Of the six most abundant elements in the human body, how many are metals? nonmetals? metalloids? (Figure 3.3)

20. Why is the symbol for gold Au rather than G or Go?

21. Give the names of (a) the solid diatomic nonmetal and (b) the liquid diatomic nonmetal. (Table 3.6)

22. Distinguish between a compound and a mixture.

23. What are the two general types of compounds? How do they differ from each other?

24. What is the basis for distinguishing one compound from another?

25. How many atoms are contained in (a) one molecule of hydrogen, (b) one molecule of water, and (c) one molecule of sulfuric acid?

26. What is the major difference between a cation and an anion?

27. Write the names and formulas of the elements that exist as diatomic molecules. (Table 3.6)

28. Distinguish between homogeneous and heterogeneous mixtures.

29. Tabulate the properties that characterize metals and nonmetals.

30. Which of the following are diatomic molecules?
   (a) H_2 (c) HCl   (e) NO   (g) $MgCl_2$
   (b) SO_2 (d) H_2O (f) NO_2

Paired Exercises

31. What elements are present in each compound?
   (a) potassium iodide KI
   (b) sodium carbonate Na_2CO_3
   (c) aluminum oxide Al_2O_3
   (d) calcium bromide $CaBr_2$
   (e) acetic acid $HC_2H_3O_2$

32. What elements are present in each compound?
   (a) magnesium bromide $MgBr_2$
   (b) carbon tetrachloride CCl_4
   (c) nitric acid HNO_3
   (d) barium sulfate $BaSO_4$
   (e) aluminum phosphate $AlPO_4$

33. Write the formula for each compound (the composition is given after each name):
   (a) zinc oxide   1 atom Zn, 1 atom O
   (b) potassium chlorate 1 atom K, 1 atom Cl, 3 atoms O
   (c) sodium hydroxide 1 atom Na, 1 atom O, 1 atom H
   (d) ethyl alcohol   2 atoms C, 6 atoms H, 1 atom O

34. Write the formula for each compound (the composition is given after each name):
   (a) aluminum bromide 1 atom Al, 3 atoms Br
   (b) calcium fluoride 1 atom Ca, 2 atoms F
   (c) lead(II) chromate 1 atom Pb, 1 atom Cr, 4 atoms O
   (d) benzene 6 atoms C, 6 atoms H

35. Explain the meaning of each symbol and number in these formulas:
   (a) H_2O
   (b) Na_2SO_4
   (c) $HC_2H_3O_2$

36. Explain the meaning of each symbol and number in these formulas:
   (a) $AlBr_3$
   (b) $Ni(NO_3)_2$
   (c) $C_{12}H_{22}O_{11}$ (sucrose)

37. How many atoms are represented in each formula?
   (a) KF   (d) $NaC_2H_3O_2$
   (b) $CaCO_3$ (e) $(NH_4)_2C_2O_4$
   (c) $K_2Cr_2O_7$

38. How many atoms are represented in each formula?
   (a) NaCl   (d) CCl_2F_2 (Freon)
   (b) N_2 (e) $Al_2(SO_4)_3$
   (c) $Ba(ClO_3)_2$

39. How many atoms of oxygen are represented in each formula?
   (a) H_2O **(d)** $Fe(OH)_3$
   (b) $CuSO_4$ **(e)** $Al(ClO_3)_3$
   (c) H_2O_2

40. How many atoms of hydrogen are represented in each formula?
   (a) H_2 **(d)** $HC_2H_3O_2$
   (b) $Ba(C_2H_3O_2)_2$ **(e)** $(NH_4)_2Cr_2O_7$
   (c) $C_6H_{12}O_6$

41. Classify each material as an element, compound, or mixture:
   (a) air   **(c)** sodium chloride
   (b) oxygen   **(d)** wine

42. Classify each material as an element, compound, or mixture:
   (a) platinum   **(c)** iodine
   (b) sulfuric acid   **(d)** crude oil

43. Classify each material as an element, compound, or mixture:
   (a) paint   **(c)** copper
   (b) salt   **(d)** beer

44. Classify each material as an element, compound, or mixture:
   (a) hydrochloric acid   **(c)** milk
   (b) silver   **(d)** sodium hydroxide

45. Reduce the following chemical formulas to the smallest whole-number relationship among atoms. (In chemistry this is called the empirical formula. You will learn more about it later.)
   (a) $C_6H_{12}O_6$ glucose
   (b) C_8H_{18} octane
   (c) $C_{25}H_{52}$ paraffin wax

46. Reduce the following chemical formulas to the smallest whole-number relationship among atoms. (In chemistry this is called the empirical formula. You will learn more about it later.)
   (a) H_2O_2 hydrogen peroxide
   (b) C_2H_6O ethyl alcohol
   (c) $Na_2Cr_2O_7$ sodium dichromate

47. Is there a pattern to the location of the gaseous elements on the periodic table? If so, describe it.

48. Is there a pattern to the location of the liquid elements on the periodic table? If so, describe it.

49. What percent of the first 36 elements on the periodic table are metals?

50. What percent of the first 36 elements on the periodic table are solids at room temperature?

Additional Exercises

51. The formula for vitamin B_{12} is $C_{63}H_{88}CoN_{14}O_{14}P$.
   (a) How many atoms make up one molecule of vitamin B_{12}?
   (b) What percentage of the total atoms are carbon?
   (c) What fraction of the total atoms are metallic?

52. How many total atoms are there in seven dozen molecules of nitric acid, HNO_3?

53. These formulas look similar but represent different things.

   8 S S_8

Compare and contrast them. How are they alike? How are they different?

54. Calcium dihydrogen phosphate is an important fertilizer. How many atoms of hydrogen are there in ten formula units of $Ca(H_2PO_4)_2$?

55. How many total atoms are there in one molecule of $C_{145}H_{293}O_{168}$?

56. Name the following:
   (a) three elements, all metals, beginning with the letter M
   (b) four elements, all solid nonmetals
   (c) five elements, all solids in the first five rows of the periodic table, whose symbols start with letters different than the element name

57. It has been estimated that there is 4×10^{-4} mg of gold per liter of seawater. At a price of \$19.40/g, what would be the value of the gold in 1 km^3 (1×10^{15} cm^3) of the ocean?

58. Make a graph using the data below. Plot the density of air in grams per liter along the x-axis and temperature along the y-axis.

Temperature (°C)	Density (g/L)
0	1.29
10	1.25
20	1.20
40	1.14
80	1.07

   (a) What is the relationship between density and temperature according to your graph?
   (b) From your plot find the density of air at these temperatures:

     5°C   25°C   70°C

CHAPTER 4

Properties of Matter

▲ Iron can be melted at very high temperatures and then cast into a variety of shapes.

CHAPTER 4 / OUTLINE

The world we live in is a kaleidoscope of sights, sounds, smells, and tastes. Our senses help us to describe these objects in our lives. For example, the smell of freshly baked cinnamon rolls creates a mouthwatering desire to gobble down a sample. Just as sights, sounds, smells, and tastes form the properties of the objects around us, each substance in chemistry has its own unique properties that allow us to identify it and predict its interactions.

These interactions produce both physical and chemical changes. When you eat an apple, the ultimate metabolic result is carbon dioxide and water. These same products are achieved by burning logs. Not only does a chemical change occur in these cases, but an energy change as well. Some reactions release energy (as does the apple or the log) whereas others require energy, such as the production of steel or the melting of ice. Over 90% of our current energy comes from chemical reactions.

4.1   Properties of Substances

How do we recognize substances? Each substance has a set of **properties** that are characteristic of that substance and give it a unique identity. Properties—the "personality traits" of substances—are classified as either physical or chemical. **Physical properties** are the inherent characteristics of a substance that can be determined without altering its composition; they are associated with its physical existence. Common physical properties include color, taste, odor, state of matter (solid, liquid, or gas), density, melting point, and boiling point. **Chemical properties** describe the ability of a substance to form new substances, either by reaction with other substances or by decomposition.

Let's consider a few of the physical and chemical properties of chlorine. Physically, chlorine is a gas about 2.4 times heavier than air. It is yellowish-green in color and has a disagreeable odor. Chemically, chlorine will not burn but will support the combustion of certain other substances. It can be used as a bleaching agent, as a disinfectant for water, and in many chlorinated substances such as refrigerants and insecticides. When chlorine combines with the metal sodium, it forms a salt called sodium chloride. These properties, among others, help us characterize and identify chlorine.

Substances, then, are recognized and differentiated by their properties. Table 4.1 lists four substances and several of their common physical properties. Information about physical properties, such as that given in Table 4.1, is available in handbooks of chemistry and physics. Scientists don't pretend to know all the answers or to remember voluminous amounts of data, but it is important for them to know where to look for data in the literature and on the Internet.

properties

physical properties

chemical properties

Many chemists have reference books such as Handbook of Chemistry and Physics to use as a resource.

No two substances have identical physical and chemical properties.

TABLE 4.1	Physical Properties of Chlorine, Water, Sugar, and Acetic Acid					
Substance	Color	Odor	Taste	Physical state	Melting point (°C)	Boiling point (°C)
Chlorine	Yellowish-green	Sharp, suffocating	Sharp, sour	Gas	−101.6	−34.6
Water	Colorless	Odorless	Tasteless	Liquid	0.0	100.0
Sugar	White	Odorless	Sweet	Solid	—	Decomposes 170–186
Acetic acid	Colorless	Like vinegar	Sour	Liquid	16.7	118.0

4.2   Physical Changes

physical change   Matter can undergo two types of changes, physical and chemical. **Physical changes** are changes in physical properties (such as size, shape, and density) or changes in the state of matter without an accompanying change in composition. The changing of ice into water and water into steam are physical changes from one state of matter into another (Figure 4.1). No new substances are formed in these physical changes.

When a clean platinum wire is heated in a burner flame, the appearance of the platinum changes from silvery metallic to glowing red. This change is physical because the platinum can be restored to its original metallic appearance by cooling and, more importantly, because the composition of the platinum is not changed by heating and cooling.

FIGURE 4.1 ▶
Ice melting into water or water turning into steam are physical changes from one state of matter to another.

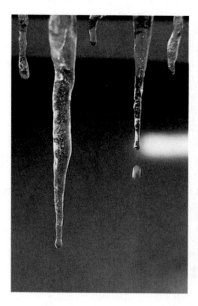

4.3   Chemical Changes

In a **chemical change,** new substances are formed that have different properties and composition from the original material. The new substances need not resemble the initial material in any way.

chemical change

When a clean copper wire is heated in a burner flame, the appearance of the copper changes from coppery metallic to glowing red. Unlike the platinum wire, the copper wire is not restored to its original appearance by cooling but becomes instead a black material. This black material is a new substance called copper(II) oxide. It was formed by chemical change when copper combined with oxygen in the air during the heating process. The unheated wire was essentially 100% copper, but the copper(II) oxide is 79.9% copper and 20.1% oxygen. One gram of copper will yield 1.251 g of copper(II) oxide (see Figure 4.2). The platinum is changed only physically when heated, but the copper is changed both physically and chemically when heated.

When 1.00 g of copper reacts with oxygen to yield 1.251 g of copper(II) oxide, the copper must have combined with 0.251 g of oxygen. The percentage of copper and oxygen can be calculated from these data—the copper and oxygen each being a percent of the total mass of copper(II) oxide.

$$1.00 \text{ g copper} + 0.251 \text{ g oxygen} \rightarrow 1.251 \text{ g copper(II) oxide}$$

$$\frac{1.00 \text{ g copper}}{1.251 \text{ g copper(II) oxide}} \times 100 = 79.9\% \text{ copper}$$

$$\frac{0.251 \text{ g oxygen}}{1.251 \text{ g copper(II) oxide}} \times 100 = 20.1\% \text{ oxygen}$$

Water can be decomposed chemically into hydrogen and oxygen. This is usually accomplished by passing electricity through the water in a process called *electrolysis.* Hydrogen collects at one electrode while oxygen collects at the other (see Figure 4.3). The composition and the physical appearance of the hydrogen and the oxygen are quite different from that of water. They are both colorless gases, but each behaves differently when a burning splint is placed into the sample: The hydrogen

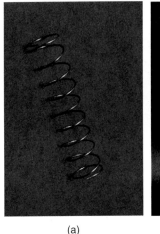
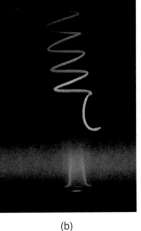

(a)   (b)   (c)

◀ FIGURE 4.2
Chemical change: Forming of copper(II) oxide from copper and oxygen
(a) **Before heating the wire is 100% copper (1.00g)**
(b) **Copper and oxygen from the air combine chemically when the wire is heated.**
(c) **After heating the wire is black copper(II) oxide (79.9% copper, 20.1% oxygen) (1.251g)**

explodes with a pop while the flame brightens considerably in the oxygen (oxygen supports and intensifies the combustion of the wood). From these observations we conclude that a chemical change has taken place.

chemical equations

Chemists have devised a shorthand method for expressing chemical changes in the form of **chemical equations.** The two previous examples of chemical changes can be represented by the following word equations:

$$\text{water} \xrightarrow[\text{energy}]{\text{electrical}} \text{hydrogen} + \text{oxygen} \tag{1}$$

$$\text{copper} + \text{oxygen} \xrightarrow{\Delta} \text{copper(II) oxide} \tag{2}$$

Equation (1) states that water decomposes into hydrogen and oxygen when electrolyzed. Equation (2) states that copper plus oxygen when heated produce copper(II) oxide. The arrow means "produces," and it points to the products. The Greek letter delta (Δ) represents heat. The starting substances (water, copper, and

reactants
products

oxygen) are called the **reactants** and the substances produced (hydrogen, oxygen, and copper(II) oxide) are called the **products.** These chemical equations can be presented in still more abbreviated form by using symbols to represent the substances:

$$2\,H_2O \xrightarrow[\text{energy}]{\text{electrical}} 2\,H_2 + O_2$$

$$2\,Cu + O_2 \xrightarrow{\Delta} 2\,CuO$$

We will learn more about writing chemical equations in later chapters.

Physical change usually accompanies a chemical change. Table 4.2 lists some common physical and chemical changes; note that wherever a chemical change occurs, a physical change occurs also. However, wherever a physical change is listed, only a physical change occurs.

TABLE 4.2 Physical or Chemical Changes of Some Common Processes		
Process taking place	**Type of change**	**Accompanying observations**
Rusting of iron	Chemical	Shiny, bright metal changes to reddish-brown rust.
Boiling of water	Physical	Liquid changes to vapor.
Burning of sulfur in air	Chemical	Yellow, solid sulfur changes to gaseous, choking sulfur dioxide.
Boiling an egg	Chemical	Liquid white and yolk change to solids.
Combustion of gasoline	Chemical	Liquid gasoline burns to gaseous carbon monoxide, carbon dioxide, and water.
Digesting food	Chemical	Food changes to liquid nutrients and partially solid wastes.
Sawing of wood	Physical	Smaller pieces of wood and sawdust are made from a larger piece of wood.
Burning of wood	Chemical	Wood burns to ashes, gaseous carbon dioxide, and water.
Heating of glass	Physical	Solid becomes pliable during heating, and the glass may change its shape.

4.4 Conservation of Mass

The **law of conservation of mass** states that no change is observed in the total mass of the substances involved in a chemical change. This law, tested by extensive laboratory experimentation, is the basis for the quantitative mass relationships among reactants and products.

law of conservation of mass

The decomposition of water into hydrogen and oxygen illustrates this law. Thus 100.0 g of water decomposes into 11.2 g of hydrogen and 88.8 g of oxygen:

water $\longrightarrow$ hydrogen + oxygen
100.0 g   11.2 g   88.8 g

$$\boxed{\text{100.0 g reactant}} \longrightarrow \boxed{\text{100.0 g products}}$$

mass of reactants = mass of products

4.5   Energy

From the early discovery that fire can warm us and cook our food to our recent discovery that nuclear reactors can be used to produce vast amounts of controlled energy, our technical progress has been directed by our ability to produce, harness, and utilize energy. **Energy** is the capacity of matter to do work. Energy exists in many forms; some of the more familiar forms are mechanical, chemical, electrical, heat, nuclear, and radiant or light energy. Matter can have both potential and kinetic energy.

energy

Potential energy (PE) is stored energy, or energy that an object possesses due to its relative position. For example, a ball located 20 ft above the ground has more potential energy than when located 10 ft above the ground and will bounce higher when allowed to fall. Water backed up behind a dam represents potential energy that can be converted into useful work in the form of electrical or mechanical energy. Gasoline is a source of chemical potential energy. When gasoline burns (combines with oxygen), the heat released is associated with a decrease in potential energy. The new substances formed by burning have less chemical potential energy than the gasoline and oxygen.

potential energy

Kinetic energy (KE) is energy that matter possesses due to its motion. When the water behind the dam is released and allowed to flow, its potential energy is changed into kinetic energy, which can be used to drive generators and produce electricity. Moving bodies possess kinetic energy. We all know the results when two moving vehicles collide: Their kinetic energy is expended in the crash that occurs. The pressure exerted by a confined gas is due to the kinetic energy of rapidly moving gas particles.

kinetic energy

Energy can be converted from one form to another form. Some kinds of energy can be converted to other forms easily and efficiently. For example, mechanical energy can be converted to electrical energy with an electric generator at better than 90% efficiency. On the other hand, solar energy has thus far been directly converted to electrical energy at an efficiency of only about 15%. In chemistry, energy is most frequently expressed as heat.

▲
The mechanical energy of falling water is converted to electrical energy at the hydroelectric plant at Niagra Falls.

4.6   Heat: Quantitative Measurement

The SI-derived unit for energy is the joule (pronounced *jool* and abbreviated J). Another unit for heat energy, which has been used for many years, is the calorie (abbreviated cal). The relationship between joules and calories is

$$4.184 \text{ J} = 1 \text{ cal} \quad \text{(exactly)}$$

To give you some idea of the magnitude of these heat units, 4.184 **joules** or 1 **calorie** is the quantity of heat energy required to change the temperature of 1 gram of water by 1°C, usually measured from 14.5°C to 15.5°C.

joule

calorie

Since joules and calories are rather small units, kilojoules (kJ) and kilocalories (kcal) are used to express heat energy in many chemical processes. The kilocalorie is

1 kJ = 1000 J

1 kcal = 1000 cal = 1 Cal

TABLE 4.3	Specific Heat of Selected Substances	
Substance	Specific heat J/g°C	Specific heat cal/g°C
Water	4.184	1.00
Ethyl alcohol	2.138	0.511
Ice	2.059	0.492
Aluminum	0.900	0.215
Iron	0.473	0.113
Copper	0.385	0.0921
Gold	0.131	0.0312
Lead	0.128	0.0305

also known as the nutritional or large Calorie (spelled with a capital C and abbreviated Cal). In this book heat energy will be expressed in joules.

The difference in the meanings of the terms *heat* and *temperature* can be seen by this example: Visualize two beakers, A and B. Beaker A contains 100 g of water at 20°C, and beaker B contains 200 g of water also at 20°C. The beakers are heated until the temperature of the water in each reaches 30°C. The temperature of the water in the beakers was raised by exactly the same amount, 10°C. But twice as much heat (8368 J) was required to raise the temperature of the water in beaker B as was required in beaker A (4184 J).

In the middle of the 18th century, Joseph Black, a Scottish chemist, was experimenting with the heating of elements. He heated and cooled equal masses of iron and lead through the same temperature range. Black noted that much more heat was needed for the iron than for the lead. He had discovered a fundamental property of matter—namely that every substance has a characteristic heat capacity. Heat capacities may be compared in terms of specific heats. The **specific heat** of a substance is the quantity of heat (lost or gained) required to change the temperature of 1 g of that substance by 1°C. It follows then that the specific heat of liquid water is 4.184 J/g°C (1.000 cal/g°C). The specific heat of water is high compared with that of most substances. Aluminum and copper, for example, have specific heats of 0.900/g°C and 0.385 J/g°C, respectively (see Table 4.3). The relation of mass, specific heat, temperature change (Δt), and quantity of heat lost or gained by a system is expressed by this general equation:

specific heat

$$\left(\begin{matrix} \text{mass of} \\ \text{substance} \end{matrix} \right) \left(\begin{matrix} \text{specific heat} \\ \text{of substance} \end{matrix} \right) \Delta t = \text{heat}$$

Thus the amount of heat needed to raise the temperature of 200. g of water by 10.0°C can be calculated as follows:

$$(200.\ g) \left(\frac{4.184\ J}{g°C} \right) (10.0°C) = 8.37 \times 10^3\ J$$

Fast Food or Fast Fat?

Fat in our diets today has become a major problem. High-fat diets have been connected with heart disease and to a variety of cancers. In the average American diet, fat supplies about 40% of the Calories (kcal). Nutritionists suggest that fat should be not more than 30% of our total daily Calories. For the average person eating about 2000 Calories per day, this is about 67 grams, 600 Calories, or 15 teaspoons of fat (e.g., a double hamburger with cheese). A small amount of fat is necessary in the diet to provide a natural source of vitamins A, D, E, and K.

Fats also supply us with energy. One gram of fat supplies 9 Calories of energy; the same amount of protein or carbohydrate supplies only 4 Calories. We use energy in a variety of ways but one of the most important is to maintain body temperature. Our bodies tend to lose heat to the surroundings since heat flows from an area of higher temperature to an area of lower temperature. Additional heat energy is used to evaporate moisture and cool our bodies as we perspire. Still more energy is demanded by our daily physical activities. The source for all this energy is the chemical oxidation of the food we eat.

The energy content of food is determined by burning it in a calorimeter and measuring the heat released. Since the initial substances and the final products of the combustion in the calorimeter are the same as those accomplished in our bodies, a calorie content can be assigned to each food. Caloric values are now found on food packages along with nutri-

tional information regarding the contents of the food we eat.

How does fast food stack up in the nutrition department? The sample meals are shown here from fast food restaurants that get less than 30% of their calories from fat. In general, chicken and turkey sandwiches that are not fried have less fat than hamburgers or roast beef. Salads with the lowest fat have little or no cheese. The best way to reduce fat in salads is to eliminate the dressing or use a low-fat dressing.

- Broiled Chicken Salad
  (200 Calories; 7 grams of fat)
- Newman's Own Light Italian Dressing
  (30 Calories; 1 gram of fat)
- Vanilla shake
  (310 Calories; 7 grams of fat)

Total—540 Calories; 15 grams of fat 25% Calories from fat

- Grilled Chicken Sandwich
  (290 Calories; 7 grams of fat)
- Black coffee
  (0 Calories; 0 grams fat)
- Chocolate Frosty Dairy Dessert (medium)
  (460 Calories; 13 grams of fat)

Total—750 Calories; 20 grams of fat 24% Calories from fa

Believe it or not, most of the milkshakes and frozen desserts served in fast-food restaurants have less than 30% Calories from fat per serving. This is primarily because they are made with skim milk. Fast food does not necessarily mean fast fat if you make careful selections. Balancing a full day's diet is more important than worrying about each and every food eaten, although choosing foods that have less than 30% Calories from fat makes maintaining the balance easier.

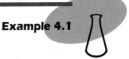

Example 4.1

Calculate the specific heat of a solid in J/g°C and cal/g°C if 1638 J raise the temperature of 125 g of the solid from 25.0°C to 52.6°C.

Solution

Substitute the data into the equation (mass)(specific heat)(Δt) = heat and solve for specific heat.

$$\text{specific heat} = \frac{\text{heat}}{\text{g} \times \Delta t}$$

heat = 1638 J   mass = 125 g Δt = 52.6°C − 25.0°C = 27.6°C

$$\text{specific heat} = \frac{1638 \text{ J}}{(125 \text{ g})(27.6°\text{C})} = 0.475 \text{ J/g}°\text{C}$$

Now convert joules to calories using 1.000 cal/4.184 J:

$$\text{specific heat} = \left(\frac{0.475 \text{ J}}{\text{g}°\text{C}}\right)\left(\frac{1.000 \text{ cal}}{4.184 \text{ J}}\right) = 0.114 \text{ cal/g}°\text{C}$$

Example 4.2

A sample of a metal with a mass of 212 g is heated to 125.0°C and then dropped into 375 g water at 24.0°C. If the final temperature of the water is 34.2°C, what is the specific heat of the metal? (Assume no heat losses to the surroundings.)

Solution

When the metal enters the water, it begins to cool, losing heat to the water. At the same time the temperature of the water rises. This process continues until the temperature of the metal and the temperature of the water are equal, at which point (34.2°C) no net flow of heat occurs.

The heat lost or gained by a system is given by (mass)(specific heat)(Δt) = energy change. First calculate the heat gained by the water and then calculate the specific heat of the metal.

temperature rise of the water   (Δt) = 34.2°C − 24.0°C = 10.2°C

$$\text{heat gained by the water} = (375 \text{ g})\left(\frac{4.184 \text{ J}}{\text{g}°\text{C}}\right)(10.2°\text{C}) = 1.60 \times 10^4 \text{ J}$$

The metal dropped into the water must have a final temperature the same as the water (34.2°C):

temperature drop by the metal   (Δt) = 125.0°C − 34.2°C = 90.8°C

heat lost by the metal = heat gained by the water = 1.60×10^4 J

To determine the specific heat of the metal we rearrange the equation (mass)(specific heat)(Δt) = heat to obtain

$$\text{specific heat} = \frac{\text{heat}}{(\text{mass})(\Delta t)}$$

$$\text{specific heat of the metal} = \frac{1.60 \times 10^4 \text{ J}}{(212 \text{ g})(90.8°\text{C})} = 0.831 \text{ J/g}°\text{C}$$

Practice 4.1

Calculate the quantity of energy needed to heat 8.0 g of water from 42.0°C to 45.0°C.

Practice 4.2

A 110.0-g sample of iron at 55.5°C raises the temperature of 150.0 mL of water from 23.0°C to 25.5°C. Determine the specific heat of the iron in J/g°C.

4.7   Energy in Chemical Changes

In all chemical changes, matter either absorbs or releases energy. Chemical changes can produce different forms of energy. For example, electrical energy to start automobiles is produced by chemical changes in the lead storage battery. Light energy is produced by the chemical change that occurs in a light stick. Heat and light are released from the combustion of fuels. All the energy needed for our life processes—breathing, muscle contraction, blood circulation, and so on—is produced by chemical changes occurring within the cells of our bodies.

Conversely, energy is used to cause chemical changes. For example, a chemical change occurs in the electroplating of metals when electrical energy is passed through a salt solution in which the metal is submerged. A chemical change occurs when radiant energy from the sun is used by green plants in the process of photosynthesis. And as we saw, a chemical change occurs when electricity is used to decompose water into hydrogen and oxygen. Chemical changes are often used primarily to produce energy rather than to produce new substances. The heat or thrust generated by the combustion of fuels is more important than the new substances formed.

▲
Energy from the sun is used to produce the chemical changes that occur during photosynthesis in the rain forests.

4.8   Conservation of Energy

An energy transformation occurs whenever a chemical change occurs (see Figure 4.4). If energy is absorbed during the change, the products will have more chemical potential energy than the reactants. Conversely, if energy is given off in a chemical change, the products will have less chemical potential energy than the reactants. Water can be decomposed in an electrolytic cell by absorbing electrical energy. The products, hydrogen and oxygen, have a greater chemical potential energy level than that of water (see Figure 4.4a). This potential energy is released in the form of heat and light when the hydrogen and oxygen are burned to form water again (see Figure 4.4b). Thus energy can be changed from one form to another or from one substance to another, and therefore is not lost.

The energy changes occurring in many systems have been thoroughly studied. No system has ever been found to acquire energy except at the expense of energy possessed by another system. This is the **law of conservation of energy**: Energy can be neither created nor destroyed, though it can be transformed from one form to another.

law of conservation of energy

Instant Relief!

Have you ever used a heat pack when you have tired, aching muscles? Remember the relief you felt when you grabbed a cold pack to reduce the inflammation and swelling of your sprained ankle? How do these hot and cold packs remain ready to use and provide the relief just when we need it? Instant hot and cold packs use the properties of matter to release or absorb energy.

In a cold pack, a small sealed package of ammonium nitrate is placed in a sealed pouch containing water. As long as the substances remain separated, nothing happens. When ammonium nitrate dissolves in water energy is absorbed. If the small package is broken (as the pack is activated) the substances mix and, as the solid dissolves, the temperature of the solution falls. The cold pack absorbs heat from its surroundings and relief from pain begins.

Instant hot packs work in one of two ways. The first type depends on a spontaneous chemical reaction that releases heat energy. In one product, an inner paper bag perforated with tiny holes is contained in a plastic envelope. The inner bag contains a mixture of powdered iron, salt, activated charcoal, and dampened sawdust. The heat pack is activated by removing the inner bag, shaking to mix the chemicals, and replacing it in the outer envelope. The heat is the result of a

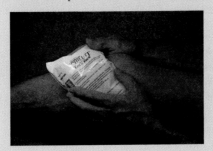

chemical change—that of iron rusting (oxidizing) very rapidly. These packs work slowly and cannot be reused. In the second type of heat pack a physical property of matter is responsible for producing the heat. This hot pack consists of a tough sealed plastic envelope containing a solution of sodium acetate or sodium thiosulfate. A small crystal of the chemical is formed in the solution by squeezing the corner of the pack or bending a small metal activator. Crystals then form throughout the solution and release heat to the surroundings. This type of hot pack does not overheat and is reusable. To reuse it simply pop it into your microwave and heat it until the crystals are all dissolved, then cool it carefully, and store it away. Instant relief for the weekend athlete or accident victim results directly from the properties of matter!

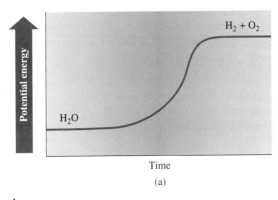

(a)

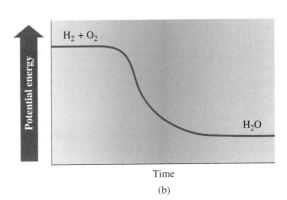

(b)

▲
FIGURE 4.4
(a) In electrolysis of water, energy is absorbed by the system so the products H₂ and O₂ have a higher potential energy. (b) When hydrogen is burned (in O₂), energy is released and the product (H₂O) has lower potential energy.

Concepts in Review

1. List the physical properties used to characterize a substance.
2. Distinguish between the physical and chemical properties of matter.
3. Classify changes undergone by matter as either physical or chemical.
4. Distinguish between kinetic and potential energy.
5. State the law of conservation of mass.
6. State the law of conservation of energy.
7. Differentiate clearly between heat and temperature.
8. Make calculations using the equation:

$$\text{heat} = (\text{mass})(\text{specific heat})(\Delta t)$$

Key Terms

calorie (4.6)
chemical change (4.3)
chemical equations (4.3)
chemical properties (4.1)
energy (4.5)

joule (4.6)
kinetic energy (4.5)
law of conservation of energy (4.8)
law of conservation of mass (4.4)

physical change (4.2)
physical properties (4.1)
potential energy (4.5)
products (4.3)

properties (4.1)
reactants (4.3)
specific heat (4.6)

Questions

1. In what physical state does acetic acid exist at 10°C? (Table 4.1)
2. In what physical state does chlorine exist at 102 K? (Table 4.1)
3. What evidence of chemical change is visible when electricity is run through water? (Figure 4.2)
4. What physical changes occur during the electrolysis of water?
5. Distinguish between chemical and physical properties.
6. What is the fundamental difference between a chemical change and a physical change?
7. Distinguish between potential and kinetic energy.
8. Calculate the boiling point of acetic acid in
   (a) Kelvins
   (b) degrees Fahrenheit (Table 4.1)

Paired Exercises

9. Classify the following as being primarily a physical or primarily a chemical change:
   (a) formation of a snowflake
   (b) freezing ice cream
   (c) boiling water
   (d) churning cream to make butter
   (e) boiling an egg
   (f) souring milk

10. Classify the following as being primarily a physical or primarily a chemical change:
   (a) lighting a candle
   (b) stirring cake batter
   (c) dissolving sugar in water
   (d) decomposition of limestone by heat
   (e) a leaf turning yellow
   (f) formation of bubbles in a pot of water long before the water boils

11. Cite the evidence that indicates that only physical changes occur when a platinum wire is heated in a Bunsen burner flame.

12. Cite the evidence that indicates that both physical and chemical changes occur when a copper wire is heated in a Bunsen burner flame.

13. Identify the reactants and products for heating a copper wire in a Bunsen burner flame.

14. Identify the reactants and products for the electrolysis of water.

15. What happens to the kinetic energy of a speeding car when the car is braked to a stop?

16. What energy transformation is responsible for the fiery reentry of the space shuttle?

17. Indicate with a plus sign (+) any of these processes that require energy, and a negative sign (−) any that release energy:
    (a) melting ice
    (b) relaxing a taut rubber band
    (c) a rocket launching
    (d) striking a match
    (e) a Slinky toy (spring) "walking" down stairs

18. Indicate with a plus sign (+) any of these processes that require energy and a negative sign (−) any that release energy:
    (a) boiling water
    (b) releasing a balloon full of air with the neck open
    (c) a race car crashing into the wall
    (d) cooking a potato in a microwave oven
    (e) ice cream freezing in an ice cream maker

19. How many joules of energy are required to raise the temperature of 75 g of water from 20.0°C to 70.0°C?

20. How many joules of energy are required to raise the temperature of 65 g of iron from 25°C to 95°C?

21. A 250.0-g metal bar requires 5.866 kJ to change its temperature from 22°C to 100.0°C. What is the specific heat of the metal?

22. A 1.00-kg sample of antimony absorbed 30.7 kJ, thus raising the temperature of the antimony from 20.0°C to its melting point (630.0°C). Calculate the specific heat of antimony.

*23. A 325-g piece of gold at 427°C is dropped into 200.0 mL of water at 22.0°C. The specific heat of gold is 0.131 J/g°C. Calculate the final temperature of the mixture. (Assume no heat loss to the surroundings.)

*24. A 500.0-g iron bar at 212°C is placed in 2.0 L of water at 24.0°C. What will be the change in temperature of the water? (Assume no heat is lost to the surroundings.)

Additional Exercises

*25. The specific heat of zinc is 0.096 cal/g°C. Determine the energy required to raise the temperature of 250.0 g of zinc from room temperature (24°C) to 150.0°C.

*26. If 40.0 kJ of energy is absorbed by 500.0 g of water at 10.0°C what is the final temperature of the water?

*27. The heat of combustion of a sample of coal is 5500 cal/g. What quantity of this coal must be burned to heat 500.0 g of water from 20.0°C to 90.0°C?

28. One gram of anthracite coal gives off 7000. cal when burned. How many joules is this? How many grams of anthracite are required to raise the temperature of 4.0 L of water from 20.0°C to 100.0°C?

29. A 100.0-g sample of copper is heated from 10.0°C to 100.0°C.
    (a) Determine the number of calories needed. (Specific heat of copper is 0.0921 cal/g°C.)
    (b) The same amount of heat is added to 100.0 g of Al at 10.0°C. (Specific heat of Al is 0.215 cal/g°C.) Which metal gets hotter, the copper or the aluminum?

30. A 500.0-g piece of iron is heated in a flame and dropped into 400.0 g of water at 10.0°C. The temperature of the water rises to 90.0°C. How hot was the iron when it was first removed from the flame? (Specific heat of iron is 0.473 J/g°C.)

*31. A 20.0-g piece of metal at 203°C is dropped into 100.0 g of water at 25.0°C. The water temperature rises to 29.0°C. Calculate the specific heat of the metal (J/g°C). Assume that all of the heat lost by the metal is transferred to the water and no heat is lost to the surroundings.

*32. Assuming no heat loss by the system, what will be the final temperature when 50.0 g of water at 10.0°C is mixed with 10.0 g of water at 50.0°C?

33. Three 500.0-g pans of iron, aluminum, and copper are each used to fry an egg. Which pan fries the egg (105°C) the quickest? Explain.

*34. At 6:00 P.M., you put a 300.0-g copper pan containing 800.0 mL of water (all at room temperature, which is 25°C) on the stove. The stove supplies 628 J/s. When will the water reach the boiling point? (Assume no heat loss.)

35. Why does blowing gently across the surface of a hot cup of coffee help to cool it? Why does inserting a spoon into the coffee do the same thing?

36. If you are boiling some potatoes in a pot of water, will they cook faster if the water is boiling vigorously than if the water is only gently boiling? Explain your reasoning.

37. Homogenized whole milk contains 4% butterfat by volume. How many milliliters of fat are there in a glass (250 mL) of milk? How many grams of butterfat ($d = 0.8$ g/mL) are in this glass of milk?

38. A 100.0-mL volume of mercury (density = 13.6 g/mL) is put into a container along with 100.0 g of sulfur. The two substances react when heated and result in 1460 g of a dark, solid matter. Is that material an element or a compound? Explain. How many grams of mercury were in the container? How does this support the law of conservation of matter?

Answers to Practice Exercises

4.1 1.0×10^2 J = 24 cal.
4.2 0.477 J/g°C.

PUTTING IT TOGETHER
Review for Chapters 1–4

Multiple Choice: *Choose the correct answer to each of the following.*

1. 1.00 cm is equal to how many meters?
   (a) 2.54 (b) 100. (c) 10.0 (d) 0.0100

2. 1.00 cm is equal to how many inches?
   (a) 0.394 (b) 0.10 (c) 12 (d) 2.54

3. 4.50 ft is how many centimeters?
   (a) 11.4 (b) 21.3 (c) 454 (d) 137

4. The number 0.0048 contains how many significant figures?
   (a) 1 (b) 2 (c) 3 (d) 4

5. Express 0.00382 in scientific notation.
   (a) 3.82×10^3 (c) 3.82×10^{-2}
   (b) 3.8×10^{-3} (d) 3.82×10^{-3}

6. 42.0°C is equivalent to
   (a) 273 K (b) 5.55°F (c) 108°F (d) 53.3°F

7. 267°F is equivalent to
   (a) 404 K (b) 116°C (c) 540 K (d) 389 K

8. An object has a mass of 62 g and a volume of 4.6 mL. Its density is
   (a) 0.074 mL/g (c) 7.4 g/mL
   (b) 285 g/mL (d) 13 g/mL

9. The mass of a block is 9.43 g and its density is 2.35 g/mL. The volume of the block is
   (a) 4.01 mL (c) 22.2 mL
   (b) 0.249 mL (d) 2.49 mL

10. The density of copper is 8.92 g/mL. The mass of a piece of copper that has a volume of 9.5 mL is
    (a) 2.6 g (b) 85 g (c) 0.94 g (d) 1.1 g

11. An empty graduated cylinder has a mass of 54.772 g. When filled with 50.0 mL of an unknown liquid, it has a mass of 101.074 g. The density of the liquid is
    (a) 0.926 g/mL (c) 2.02 g/mL
    (b) 1.08 g/mL (d) 1.85 g/mL

12. The conversion factor to change grams to milligrams is
    (a) $\dfrac{100 \text{ mg}}{1 \text{ g}}$ (c) $\dfrac{1 \text{ g}}{1000 \text{ mg}}$
    (b) $\dfrac{1 \text{ g}}{100 \text{ mg}}$ (d) $\dfrac{1000 \text{ mg}}{1 \text{ g}}$

13. What Fahrenheit temperature is twice the Celsius temperature?
    (a) 64°F (b) 320°F (c) 200°F (d) 546°F

14. A gold alloy has a density of 12.41 g/mL and contains 75.0% gold by mass. The volume of this alloy that can be made from 255 g of pure gold is
    (a) 4.22×10^3 mL (c) 27.4 mL
    (b) 2.37×10^3 mL (d) 20.5 mL

15. A lead cylinder ($V = \pi r^2 h$) has radius 12.0 cm and length 44.0 cm and a density of 11.4 g/mL. The mass of the cylinder is
    (a) 2.27×10^5 g (c) 1.78×10^3 g
    (b) 1.89×10^5 g (d) 3.50×10^5 g

16. The following units can all be used for density *except*
    (a) g/cm^3 (b) kg/m^3 (c) g/L (d) kg/m^2

17. 37.4 cm × 2.2 cm equals
    (a) 82.28 cm^2 (c) 82 cm^2
    (b) 82.3 cm^2 (d) 82.2 cm^2

18. The following elements are among the five most abundant by mass in the Earth's crust, seawater, and atmosphere *except*
    (a) oxygen (c) silicon
    (b) hydrogen (d) aluminum

19. Which of the following is a compound?
    (a) lead (c) potassium
    (b) wood (d) water

20. Which of the following is a mixture?
    (a) water (c) wood
    (b) chromium (d) sulfur

21. How many atoms are represented in the formula Na_2CrO_4?
    (a) 3 (b) (5) (c) (7) (d) 8

22. Which of the following is a characteristic of metals?
    (a) ductile (c) extremely strong
    (b) easily shattered (d) dull

23. Which of the following is a characteristic of nonmetals?
    (a) always a gas
    (b) poor conductor of electricity
    (c) shiny
    (d) combines only with metals

24. When a pure substance was analyzed, it was found to contain carbon and chlorine. This substance must be classified as
    (a) an element
    (b) a mixture
    (c) a compound
    (d) both a mixture and a compound

25. Chromium, fluorine, and magnesium have the symbols
    (a) Ch, F, Ma (c) Cr, F, Mg
    (b) Cr, Fl, Mg (d) Cr, F, Ma

26. Sodium, carbon, and sulfur have the symbols
    (a) Na, C, S (c) Na, Ca, Su
    (b) So, C, Su (d) So, Ca, Su

27. Coffee is an example of
    (a) an element (c) a homogeneous mixture
    (b) a compound (d) a heterogeneous mixture

28. The number of oxygen atoms in $Al(C_2H_3O_2)_3$ is
(a) 2 (b) 3 (c) 5 (d) 6

29. Which of the following is a mixture?
(a) water   (c) sugar solution
(b) iron(II) oxide (d) iodine

30. Which is the most compact state of matter?
(a) solid   (c) gas
(b) liquid   (d) amorphous

31. Which is not characteristic of a solution?
(a) a homogeneous mixture
(b) a heterogeneous mixture
(c) contains two or more substances
(d) has a variable composition

32. A chemical formula is a combination of
(a) symbols   (c) elements
(b) atoms   (d) compounds

33. The number of nonmetal atoms in $Al_2(SO_3)_3$ is
(a) 5 (b) 7 (c) 12 (d) 14

34. Which of the following is not a physical property?
(a) boiling point   (c) bleaching action
(b) physical state (d) color

35. Which of the following is a physical change?
(a) A piece of sulfur is burned.
(b) A firecracker explodes.
(c) A rubber band is stretched.
(d) A nail rusts.

36. Which of the following is a chemical change?
(a) Water evaporates.
(b) Ice melts.
(c) Rocks are ground to sand.
(d) A penny tarnishes.

37. When 9.44 g of calcium are heated in air, 13.22 g of calcium oxide are formed. The percent by mass of oxygen in the compound is
(a) 28.6% (b) 40.0% (c) 71.4% (d) 13.2%

38. Barium iodide, BaI_2, contains 35.1% barium by mass. An 8.50-g sample of barium iodide contains what mass of iodine?
(a) 5.52 g (b) 2.98 g (c) 3.51 g (d) 6.49 g

39. Mercury(II) sulfide, HgS, contains 86.2% mercury by mass. The mass of HgS that can be made from 30.0 g of mercury is
(a) 2586 g (b) 2.87 g (c) 25.9 g (d) 34.8 g

40. The changing of liquid water to ice is known as a
(a) chemical change
(b) heterogeneous change
(c) homogeneous change
(d) physical change

41. Which of the following does not represent a chemical change?
(a) heating of copper in air
(b) combustion of gasoline
(c) cooling of red-hot iron
(d) digestion of food

42. Heating 30·g of water from 20·°C to 50·°C requires
(a) 30. cal   (c) 3.8×10^3 J
(b) 50. cal   (d) 6.3×10^3 J

43. The specific heat of aluminum is 0.900 J/g°C. How many joules of energy are required to raise the temperature of 20.0 g of Al from 10.0°C to 15.0°C?
(a) 79 J (b) 90. J (c) 100. J (d) 112 J

44. A 100.-g iron ball (specific heat = 0.473 J/g°C) is heated to 125°C and is placed in a calorimeter holding 200. g of water at 25.0°C. What will be the highest temperature reached by the water?
(a) 43.7°C (c) 65.3°C
(b) 30.4°C (d) 35.4°C

45. Which has the highest specific heat?
(a) ice (b) lead (c) water (d) aluminum

46. When 20.0 g of mercury is heated from 10.0°C to 20.0°C, 27.6 J of energy are absorbed. What is the specific heat of mercury?
(a) 0.726 J/g°C (c) 2.76 J/g°C
(b) 0.138 J/g°C (d) no correct answer given

47. Changing hydrogen and oxygen into water is a
(a) physical change
(b) chemical change
(c) conservation reaction
(d) no correct answer given

Free Response Questions: *Answer each of the following. Be sure to include your work and explanations in a clear, logical form.*

1. You decide to go sailing in the tropics with some friends. Once there you listen to the marine forecast that predicts in-shore wave heights of 1.5 meters, offshore wave heights of 4 meters, and temperature of 27°C. Your friend the captain is unfamiliar with the metric system and he needs to know whether it is safe for your small boat and if it will be warm. He asks you to convert the measurements to feet and Fahrenheit respectively.

2. Jane is melting butter in a copper pot on the stove. If she knows how much heat her stove releases per minute, what other measurements does she need to take to determine how much heat the butter absorbed? She assumes that the stove does not lose any heat to the surroundings.

3. Julius decided to heat 75 g $CaCO_3$ to determine how much carbon dioxide was produced. (Note: When $CaCO_3$ is heated it produces CaO and carbon dioxide.) He collected the carbon dioxide in a balloon. Julius found the mass of the

CaO remaining was 42 g. If 44 g of carbon dioxide takes up 24 dm^3 of space, how many *liters* of gas were trapped in the balloon?

Use these pictures to answer the following questions.

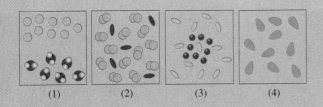

(1)   (2)   (3)   (4)

4. (a) Which picture best describes a homogeneous mixture?
   (b) How would you classify the contents of the other containers?
   (c) Which picture contains a compound? Explain how you made your choice.

Use these pictures to answer the following questions.

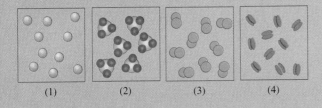

(1)   (2)   (3)   (4)

5. (a) Which picture best represents fluorine gas? Why?
   (b) Which other elements could that picture also represent?
   (c) Which of the pictures could represent SO$_3$ gas?

6. Sue and Tim each left a one-quart bowl outside one night. Sue's bowl was full of water and covered while Tim's was empty and open. The next day there was a huge snowstorm that filled Tim's bowl with snow. The temperature that night went down to 12°F.
   (a) Which bowl would require less energy to bring its contents to room temperature (25°C)? Why?
   (b) What temperature change (°C) is required to warm the bowls to 25°C?
   (c) How much heat (in kJ) is required to raise the temperature of the contents of Sue's bowl to 0°C (without converting the ice to water)?
   (d) Did the water in Sue's bowl undergo chemical or physical changes or both?

7. One cup of Raisin Bran provides 60.% of the U.S. recommended daily allowance (RDA) of iron.
   (a) If the cereal provides 11 mg of iron, what is the U.S. RDA for Fe?
   (b) When the iron in the cereal is extracted, it is found to be the pure element. What is the volume of iron in a cup of the cereal?

8. 16% of the U.S. RDA for Ca is 162 mg.
   (a) What mass of calcium phosphate, Ca$_3$(PO$_4$)$_2$, provides 162 mg of calcium?
   (b) Is Ca$_3$(PO$_4$)$_2$ an element, mixture, or compound?
   (c) Milk is a good source of calcium in the diet. If 120 mL of skim milk provides 13% of the U.S. RDA for Ca, how many cups of milk should you drink per day if that is your only source of calcium?

9. Absentminded Alfred put down a bottle containing silver on the table. When he went to retrieve it, he realized he had forgotten to label the bottle. Unfortunately there were two full bottles of the same size side-by-side. Alfred realized he had placed a bottle of mercury on the same table last week. State two ways Alfred can determine which bottle contains silver without opening the bottles.

10. Suppose 25 g of solid sulfur and 35 g of oxygen gas are placed in a sealed container.
    (a) Does the container hold a mixture or a compound?
    (b) After heating the contents, the container was weighed. From a comparison of the total mass before heating to the total mass after heating, can you tell whether a reaction took place? Explain.
    (c) After the container is heated, all the contents are gaseous. Has the density of the container including its contents changed?

CHAPTER

5

Early Atomic Theory and Structure

▲
Lightning occurs when electrons move to neutralize a charge difference between the clouds and the Earth.

CHAPTER 5 / OUTLINE

Pure substances are classified as elements or compounds, but just what makes a substance possess its unique properties? How small a piece of salt will still taste salty? Carbon dioxide puts out fires, is used by plants to produce oxygen, and forms dry ice when solidified. But how small a mass of this material still behaves like carbon dioxide? Substances are in their simplest identifiable form at the atomic, ionic, or molecular level. Further division produces a loss of characteristic properties.

What particles lie within an atom or ion? How are these tiny particles alike? How do they differ? How far can we continue to divide them? Alchemists began the quest, early chemists laid the foundation, and modern chemists continue to build and expand on models of the atom.

5.1 Early Thoughts

The structure of matter has long intrigued and engaged us. The earliest models of the atom were developed by the ancient Greek philosophers. About 440 B.C. Empedocles stated that all matter was composed of four "elements"—earth, air, water, and fire. Democritus (about 470–370 B.C.) thought that all forms of matter were divisible into tiny invisible particles, which he called atoms, derived from the Greek word *atomos,* meaning indivisible. He held that atoms were in constant motion and that they combined with one another in various ways. This purely speculative hypothesis was not based on scientific observations. Shortly thereafter, Aristotle (384–322 B.C.) opposed the theory of Democritus and instead endorsed and advanced the Empedoclean theory. So strong was the influence of Aristotle that his theory dominated the thinking of scientists and philosophers until the beginning of the 17th century.

5.2 Dalton's Model of the Atom

More than 2000 years after Democritus, the English schoolmaster John Dalton (1766–1844) revived the concept of atoms and proposed an atomic model based on facts and experimental evidence (Figure 5.1). His theory, described in a series of papers published from 1803 to 1810, rested on the idea of a different kind of atom for each element. The essence of **Dalton's atomic model** may be summed up as follows:

Dalton's atomic model

1. Elements are composed of minute, indivisible particles called atoms.
2. Atoms of the same element are alike in mass and size.
3. Atoms of different elements have different masses and sizes.
4. Chemical compounds are formed by the union of two or more atoms of different elements.
5. Atoms combine to form compounds in simple numerical ratios, such as one to one, two to two, two to three, and so on.
6. Atoms of two elements may combine in different ratios to form more than one compound.

FIGURE 5.1 ▶
(a) Dalton's atoms were individual particles; the atoms of
each element being alike in
mass and size but different in
mass and size from other elements. (b) and (c) Dalton's
atoms combine in specific
ratios to form compounds.

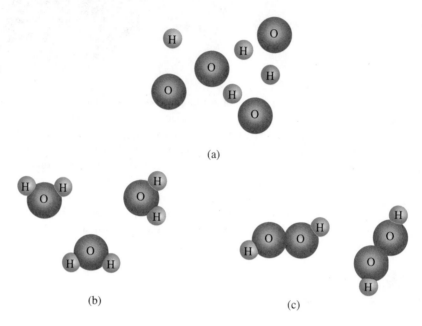

(a)

(b)   (c)

Dalton's atomic model stands as a landmark in the development of chemistry. The
major premises of his model are still valid, but some of his statements must be modified or qualified because later investigations have shown that (1) atoms are composed of subatomic particles; (2) not all the atoms of a specific element have the
same mass; and (3) atoms, under special circumstances, can be decomposed.

5.3   Composition of Compounds

A large number of experiments extending over a long period of time have established
the fact that a particular compound always contains the same elements in the same
proportions by mass. For example, water will always contain 11.2% hydrogen and
88.8% oxygen by mass. The fact that water contains hydrogen and oxygen in this
particular ratio does not mean that hydrogen and oxygen cannot combine in some
other ratio, but that a compound with a different ratio would not be water. In fact, hydrogen peroxide is made up of two atoms of hydrogen and two atoms of oxygen per
molecule and contains 5.9% hydrogen and 94.1% oxygen by mass; its properties are
markedly different from those of water.

	Water	Hydrogen peroxide
Percent H	11.2	5.9
Percent O	88.8	94.1
Atomic composition	2 H + 1 O	2 H + 2 O

The **law of definite composition** states

law of definite composition

> **A compound always contains two or more elements combined in a definite proportion by mass.**

Let's consider two elements, oxygen and hydrogen, that form more than one compound. In water, 8.0 g of oxygen are present for each gram of hydrogen. In hydrogen peroxide, 16.0 g of oxygen are present for each gram of hydrogen. The masses of oxygen are in the ratio of small whole numbers, 16:8 or 2:1. Hydrogen peroxide has twice as much oxygen (by mass) as does water. Using Dalton's atomic model, we deduce that hydrogen peroxide has twice as many oxygen atoms per hydrogen atom as water. In fact, we now write the formulas for water as H_2O and for hydrogen peroxide as H_2O_2. See Figure 5.1b and c.

The **law of multiple proportions** states

law of multiple proportions

> **Atoms of two or more elements may combine in different ratios to produce more than one compound.**

The reliability of this law and the law of definite composition is the cornerstone of the science of chemistry. In essence these laws state that (1) the composition of a particular substance will always be the same no matter what its origin or how it is formed, and (2) the composition of different compounds formed from the same elements will always be unique.

5.4 The Nature of Electric Charge

You've probably received a shock after walking across a carpeted area on a dry day. You've also experienced the static associated with combing your hair, and have had your clothing cling to you. These phenomena result from an accumulation of *electric charge*. This charge may be transferred from one object to another. The properties of electric charge are

1. Charge may be of two types, positive and negative.
2. Unlike charges attract (positive attracts negative), and like charges repel (negative repels negative and positive repels positive).
3. Charge may be transferred from one object to another, by contact or induction.
4. The less the distance between two charges, the greater the force of attraction between unlike charges (or repulsion between identical charges).

5.5 Discovery of Ions

The great English scientist Michael Faraday (1791–1867) made the discovery that certain substances when dissolved in water conduct an electric current. He also

▲
When ions are present in a solution of salt water and an electric current is passed through the solution, the light bulb glows.

noticed that certain compounds decompose into their elements by passing an electric current through the compound. Atoms of some elements are attracted to the positive electrode, while atoms of other elements are attracted to the negative electrode. Faraday concluded that these atoms are electrically charged. He called them *ions* after the Greek word meaning "wanderer."

Any moving charge is an electric current. The electrical charge must travel through a substance known as a conducting medium. The most familiar conducting media are metals formed into wires.

The Swedish scientist Svante Arrhenius (1859–1927) extended Faraday's work. Arrhenius reasoned that an ion is an atom carrying a positive or negative charge. When a compound such as sodium chloride (NaCl) is melted, it conducts electricity. Water is unnecessary. Arrhenius's explanation of this conductivity was that upon melting, the sodium chloride dissociates, or breaks up, into charged ions, Na^+ and Cl^-. The Na^+ ions move toward the negative electrode (cathode), whereas the Cl^- migrate toward the positive electrode (anode). Thus positive ions are *cations*, and negative ions are *anions*.

From Faraday's and Arrhenius's work with ions, Irish physicist G. J. Stoney (1826–1911) realized there must be some fundamental unit of electricity associated with atoms. He named this unit the *electron* in 1891. Unfortunately he had no means of supporting his idea with experimental proof. Evidence remained elusive until 1897 when English physicist Joseph Thomson (1856–1940) was able to show experimentally the existence of the electron.

5.6   Subatomic Parts of the Atom

The concept of the atom—a particle so small that until recently it could not be seen even with the most powerful microscope—and the subsequent determination of its structure stand among the greatest creative intellectual human achievements.

Any visible quantity of an element contains a vast number of identical atoms. But when we refer to an atom of an element, we isolate a single atom from the multitude in order to present the element in its simplest form. Figure 5.2 shows individual atoms as we can see them today.

FIGURE 5.2 ▶
A scanning tunneling microscope shows an array of copper atoms.

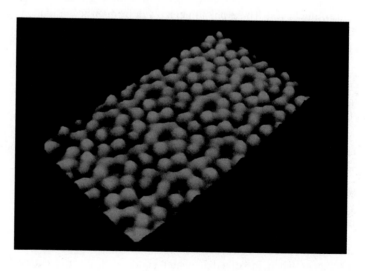

CHEMISTRY IN ACTION Triboluminescence

Some substances give off light when they are rubbed, crushed, or broken in a phenomenon called triboluminescence. Examples of substances exhibiting this property include crystals of quartz, sugar cubes, adhesive tape torn off certain surfaces, and Wintergreen Lifesavers.

Linda M. Sweeting, a chemist at Towson State University in Maryland, investigated the sparks created by a Wintergreen Lifesaver when crushed inside one's mouth in a dark room. When a sugar crystal is fractured, separate areas of positive and negative charge form on opposite sides of the crack. Electrons tend to leap the gap and neutralize the charge. When the electrons collide with nitrogen molecules in the air, small amounts of light are emitted. (Lightning is a somewhat similar phenomenon but on a much grander scale.) The addition of wintergreen molecules changes the outcome though because they absorb some of the light energy from the electron leap and reemit it as bright blue-green flashes. These are the sparks we see as we crush a Lifesaver in the dark.

▲ **Triboluminescence occurs when Wintergreen Lifesavers are crushed in the mouth.**

What is this tiny particle we call the atom? The diameter of a single atom ranges from 0.1 to 0.5 nanometers (1 nm $= 1 \times 10^{-9}$ m). Hydrogen, the smallest atom, has a diameter of about 0.1 nm. To arrive at some idea of how small an atom is, consider this dot (•), which has a diameter of about 1 mm, or 1×10^6 nm. It would take 10 million hydrogen atoms to form a line of atoms across this dot. As inconceivably small as atoms are, they contain smaller particles, the **subatomic particles,** such as electrons, protons, neutrons.

subatomic particles

The development of atomic theory was helped in large part by the invention of new instruments. For example, the Crookes tube, developed by Sir William Crookes in 1875, opened the door to the subatomic structure of the atom (Figure 5.3). The emissions generated in a Crookes tube are called *cathode rays.* Joseph Thomson demonstrated in 1897 that cathode rays (1) travel in straight lines, (2)

▲
**FIGURE 5.3
Crookes tube. Emissions generated in Crookes tube (a) travel in straight lines and are negative in charge, (b) are deflected by a magnetic field, and (c) provide a sharp shadow of the cross in the center of the tube.**

are negative in charge, (3) are deflected by electric and magnetic fields, (4) produce sharp shadows, and (5) are capable of moving a small paddle wheel. This was the experimental discovery of the fundamental unit of charge—the electron.

electron   The **electron** (e⁻) is a particle with a negative electrical charge and a mass of 9.110×10^{-28} g. This mass is 1/1837 the mass of a hydrogen atom and corresponds to 0.0005486 atomic mass unit (amu) (defined in Section 5.11). One atomic mass unit has a mass of 1.6606×10^{-24} g. Although the actual charge of an electron is known, its value is too cumbersome for practical use and has therefore been assigned a relative electrical charge of -1. The size of an electron has not been determined exactly, but its diameter is believed to be less than 10^{-12} cm.

Protons were first observed by German physicist Eugen Goldstein (1850–1930) in 1886. However, it was Thomson who discovered the nature of the proton. He showed that the proton is a particle, and he calculated its mass to be about 1837 times that of an electron. The **proton** (p) is a particle with a relative mass of 1 amu and an actual mass of 1.673×10^{-24} g. Its relative charge ($+1$) is equal in magnitude, but of opposite sign, to the charge on the electron. The mass of a proton is only very slightly less than that of a hydrogen atom.

proton

Thomson had shown that atoms contained both negatively and positively charged particles. Clearly, the Dalton model of the atom was no longer acceptable. Atoms are not indivisible but are instead composed of smaller parts. Thomson proposed a new model of the atom.

Thomson model of the atom   In the **Thomson model of the atom,** the electrons are negatively charged particles embedded in the atomic sphere. Since atoms are electrically neutral, the sphere also contains an equal number of protons, or positive charges. A neutral atom could become an ion by gaining or losing electrons.

Positive ions were explained by assuming that the neutral atom loses electrons. An atom with a net charge of $+1$ (for example, Na⁺ or Li⁺) has lost one electron. An atom with a net charge of $+3$ (for example, Al³⁺) has lost three electrons (Figure 5.4a).

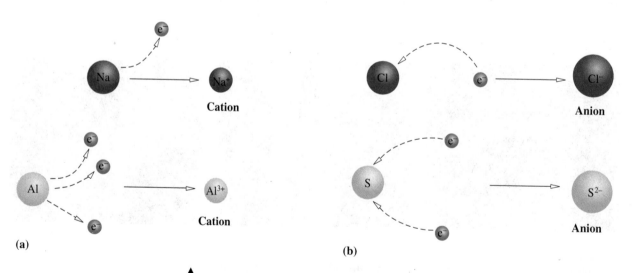

FIGURE 5.4
(a) When an electron is lost from an atom, a cation is formed. (b) When electrons are added to a neutral atom, an anion is formed.

TABLE 5.1 Electrical Charge and Relative Mass of Electrons, Protons, and Neutrons				
Particle	Symbol	Relative electrical charge	Relative mass (amu)	Actual mass (g)
Electron	e^-	-1	$\dfrac{1}{1837}$	9.110×10^{-28}
Proton	p	$+1$	1	1.673×10^{-24}
Neutron	n	0	1	1.675×10^{-24}

Negative ions were explained by assuming that extra electrons can be added to atoms. A net charge of -1 (for example, Cl^- or F^-) is produced by the addition of one electron. A net charge of -2 (for example, O^{2-} or S^{2-}) requires the addition of two electrons (Figure 5.4b).

The third major subatomic particle was discovered in 1932 by James Chadwick (1891–1974). This particle, the **neutron** (n), has neither a positive nor a negative charge and has a relative mass of about 1 amu. Its actual mass (1.675×10^{-24} g) is only very slightly greater than that of a proton. The properties of these three subatomic particles are summarized in Table 5.1.

neutron

Nearly all the ordinary chemical properties of matter can be explained in terms of atoms consisting of electrons, protons, and neutrons. The discussion of atomic structure that follows is based on the assumption that atoms contain only these principal subatomic particles. Many other subatomic particles, such as mesons, positrons, neutrinos, and antiprotons, have been discovered, but it is not yet clear whether all these particles are actually present in the atom or whether they are produced by reactions occurring within the nucleus. The fields of atomic and particle or high-energy physics have produced a long list of subatomic particles. Descriptions of the properties of many of these particles are to be found in recent physics textbooks.

The mass of a helium atom is 6.65×10^{-24} g. How many atoms are in a 4.0-g sample of helium?

Example 5.1

Solution

$$(4.0 \text{ g})\left(\frac{1 \text{ atom He}}{6.65 \times 10^{-24} \text{ g}}\right) = 6.0 \times 10^{23} \text{ atoms He}$$

Practice 5.1

The mass of an atom of hydrogen is 1.673×10^{-24} g. How many atoms are in a 10.0-g sample of hydrogen?

5.7   The Nuclear Atom

The discovery that positively charged particles are present in atoms came soon after the discovery of radioactivity by Henri Becquerel in 1896. Radioactive elements spontaneously emit alpha particles, beta particles, and gamma rays from their nuclei (see Chapter 18).

By 1907 Ernest Rutherford had established that the positively charged alpha particles emitted by certain radioactive elements are ions of the element helium. Rutherford used these alpha particles to establish the nuclear nature of atoms. In experiments performed in 1911, he directed a stream of positively charged helium ions (alpha particles) at a very thin sheet of gold foil (about 1000 atoms thick). See Figure 5.5a. He observed that most of the alpha particles passed through the foil with little or no deflection; but a few of the particles were deflected at large angles, and occasionally one even bounced back from the foil (Figure 5.5b). It was known that like charges repel each other and that an electron with a mass of 1/1837 amu could not possibly have an appreciable effect on the path of a 4-amu alpha particle, which is about 7350 times more massive than an electron. Rutherford therefore reasoned that each gold atom must contain a positively charged mass occupying a relatively tiny volume and that, when an alpha particle approaches close enough to this positive mass, it is deflected. Rutherford spoke of this positively charged mass as the *nucleus* of the atom. Because alpha particles have relatively high masses, the extent of the deflections (some actually bounced back) indicated to Rutherford that the nucleus is very heavy and dense. (The density of the nucleus of a hydrogen atom is about 10^{12} g/cm^3—about 1 trillion times the density of water.) Because most of the alpha particles passed through the thousand or so gold atoms without any apparent deflection, he further concluded that most of an atom consists of empty space.

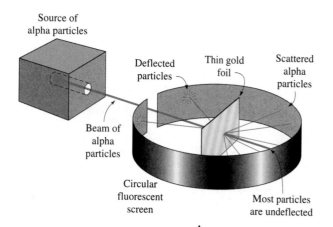

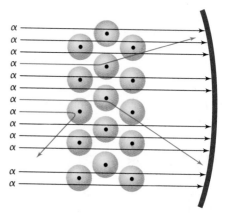

▲
FIGURE 5.5
Left: Rutherford's experiment on alpha-particle scattering, where positive alpha particles (a), emanating from a radioactive source, were directed at a thin gold foil. Right: Deflection (red) and scattering (blue) of the positive alpha particles by the positive nuclei of the gold atoms.

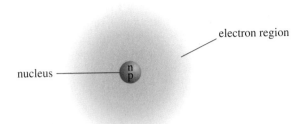

◄ FIGURE 5.6
In the nuclear model of the atom, protons (p) and neutrons (n) are located in the nucleus. The electrons are found in the remainder of the atom (which is mostly empty space because electrons are very tiny).

When we speak of the mass of an atom, we are referring primarily to the mass of the nucleus. The nucleus contains all the protons and neutrons, which represent more than 99.9% of the total mass of any atom (see Table 5.1). By way of illustration, the largest number of electrons known to exist in an atom is 112. The mass of even 112 electrons is only about 1/17 of the mass of a single proton or neutron. The mass of an atom therefore is primarily determined by the combined masses of its protons and neutrons.

5.8 General Arrangement of Subatomic Particles

The alpha-particle scattering experiments of Rutherford established that the atom contains a dense, positively charged nucleus. The later work of Chadwick demonstrated that the atom contains neutrons, which are particles with mass but no charge. Rutherford also noted that light, negatively charged electrons are present and offset the positive charges in the nucleus. Based on this experimental evidence, a model of the atom and the location of its subatomic particles was devised in which each atom consists of a **nucleus** surrounded by electrons (see Figure 5.6). The nucleus contains protons and neutrons but not electrons. In a neutral atom the positive charge of the nucleus (due to protons) is exactly offset by the negative electrons. Because the charge of an electron is equal but of opposite sign to the charge of a proton, a neutral atom must contain exactly the same number of electrons as protons. However, this model of atomic structure provides no information on the arrangement of electrons within the atom.

nucleus

A neutral atom contains the same number of protons and electrons.

5.9 Atomic Numbers of the Elements

The **atomic number** of an element is the number of protons in the nucleus of an atom of that element. The atomic number determines the identity of an atom. For

atomic number

▲
Carbon, shown here as a beautiful diamond, has 6 protons and 6 electrons in each atom.

example, every atom with an atomic number of 1 is a hydrogen atom; it contains one proton in its nucleus. Every atom with an atomic number of 6 is a carbon atom; it contains 6 protons in its nucleus. Every atom with an atomic number of 92 is a uranium atom; it contains 92 protons in its nucleus. The atomic number tells us not only the number of positive charges in the nucleus, but also the number of electrons in the neutral atom.

atomic number = number of protons in the nucleus

You don't need to memorize the atomic numbers of the elements because a periodic table is commonly provided in texts, laboratories, and on examinations. The atomic numbers of all elements are shown in the periodic table on the inside front cover of this book and are also listed in the table of atomic masses on the inside front endpapers.

5.10   Isotopes of the Elements

Shortly after Rutherford's conception of the nuclear atom, experiments were performed to determine the masses of individual atoms. These experiments showed that the masses of nearly all atoms were greater than could be accounted for by simply adding up the masses of all the protons and electrons that were known to be present in an atom. This fact led to the concept of the neutron, a particle with no charge but with a mass about the same as that of a proton. Because this particle has no charge, it was very difficult to detect, and the existence of the neutron was not proven experimentally until 1932. All atomic nuclei, except that of the simplest hydrogen atom, are now believed to contain neutrons.

All atoms of a given element have the same number of protons. Experimental evidence has shown that, in most cases, all atoms of a given element do not have identical masses. This is because atoms of the same element may have different numbers of neutrons in their nuclei.

isotopes   Atoms of an element having the same atomic number but different atomic masses are called **isotopes** of that element. Atoms of the various isotopes of an element therefore have the same number of protons and electrons but different numbers of neutrons.

Three isotopes of hydrogen (atomic number 1) are known. Each has one proton in the nucleus and one electron. The first isotope (protium), without a neutron, has a mass number of 1; the second isotope (deuterium), with one neutron in the nucleus, has a mass number of 2; the third isotope (tritium), with two neutrons, has a mass number of 3 (see Figure 5.7).

The three isotopes of hydrogen may be represented by the symbols ^1_1H, ^2_1H, ^3_1H, indicating an atomic number of 1 and mass numbers of 1, 2, and 3, respectively. This method of representing atoms is called *isotopic notation.* The subscript (Z) is the

mass number   atomic number; the superscript (A) is the **mass number,** which is the sum of the number of protons and the number of neutrons in the nucleus. The hydrogen isotopes may also be referred to as hydrogen-1, hydrogen-2, and hydrogen-3.

▲
A sample of uranium contains 92 protons in each nucleus.

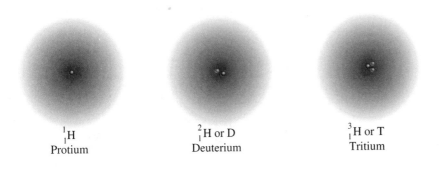

◀ FIGURE 5.7
FIGURE 5.7
The isotopes of hydrogen.
The number of protons
(purple) and neutrons
(blue) are shown within the
nucleus. The electron (e⁻)
exists outside the nucleus.

$^{1}_{1}$H
Protium

$^{2}_{1}$H or D
Deuterium

$^{3}_{1}$H or T
Tritium

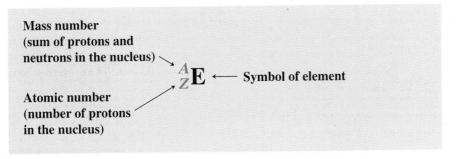

Mass number
(sum of protons and
neutrons in the nucleus)

$^{A}_{Z}$E ⟵ Symbol of element

Atomic number
(number of protons
in the nucleus)

The mass number of an
element is the sum of the
protons and neutrons in the
nucleus.

Most of the elements occur in nature as mixtures of isotopes. However, not all isotopes are stable; some are radioactive and are continuously decomposing to form other elements. For example, of the seven known isotopes of carbon, only two, carbon-12 and carbon-13, are stable. Of the seven known isotopes of oxygen, only three, $^{16}_{8}$O, $^{17}_{8}$O, and $^{18}_{8}$O, are stable. Of the fifteen known isotopes of arsenic, $^{75}_{33}$As is the only one that is stable.

5.11   Atomic Mass

The mass of a single atom is far too small to measure on a balance. But fairly precise determinations of the masses of individual atoms can be made with an instrument called a *mass spectrometer* (see Figure 5.8). The mass of a single hydrogen atom is 1.673×10^{-24} g. However, it is neither convenient nor practical to compare the actual masses of atoms expressed in grams; therefore a table of relative atomic masses using *atomic mass units* was devised. (The term *atomic weight* is sometimes used instead of atomic mass.) The carbon isotope having six protons and six neutrons and designated carbon-12, or $^{12}_{6}$C, was chosen as the standard for atomic masses. This reference isotope was assigned a value of exactly 12 atomic mass units (amu). Thus 1 **atomic mass unit** is defined as equal to exactly 1/12 of the mass of a carbon-12 atom. The actual mass of a carbon-12 atom is 1.9927×10^{-23} g, and that of one atomic mass unit is 1.6606×10^{-24} g. In the table of atomic masses, all elements then have values that are relative to the mass assigned to the reference isotope carbon-12.

atomic mass unit

1 amu = 1.6606×10^{-24} g

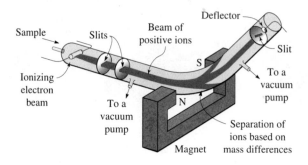

FIGURE 5.8 ▶
A modern mass spectrometer. A beam of positive ions is produced from the sample as it enters the chamber. The positive ions are then accelerated as they pass through slits in an electric field. When the ions enter the magnetic field, they are deflected differently, depending on mass and charge. The ions are then detected at the end of the tube. From intensity and position of the lines on the mass spectrogram, the different isotopes of the elements and their relative amounts can be determined.

▲
These copper plates and bracelets made in Mexico City all contain a mixture of the isotopes of copper.

A table of atomic masses is given on the inside front cover of this book. Hydrogen atoms, with a mass of about 1/12 that of a carbon atom, have an average atomic mass of 1.00797 amu on this relative scale. Magnesium atoms, which are about twice as heavy as carbon, have an average mass of 24.305 amu. The average atomic mass of oxygen is 15.9994 amu.

Since most elements occur as mixtures of isotopes with different masses, the atomic mass determined for an element represents the average relative mass of all the naturally occurring isotopes of that element. The atomic masses of the individual isotopes are approximately whole numbers, because the relative masses of the protons and neutrons are approximately 1.0 amu each. Yet we find that the atomic masses given for many of the elements deviate considerably from whole numbers.

For example, the atomic mass of rubidium is 85.4678 amu, that of copper is 63.546 amu, and that of magnesium is 24.305 amu. The deviation of an atomic mass from a whole number is due mainly to the unequal occurrence of the various isotopes of an element. The two principal isotopes of copper are $^{63}_{29}Cu$ and $^{65}_{29}Cu$. It is apparent that copper-63 atoms are the more abundant isotope, since the atomic mass of copper, 63.546 amu, is closer to 63 than to 65 amu (see Figure 5.9). The actual values of the copper isotopes observed by mass spectra determination are shown here:

Isotope	Isotopic mass (amu)	Abundance (%)	Average atomic mass (amu)
$^{63}_{29}Cu$	62.9298	69.09	63.55
$^{65}_{29}Cu$	64.9278	30.91	

The average atomic mass is a weighted average of the masses of all the isotopes present in the sample.

The average atomic mass can be calculated by multiplying the atomic mass of each isotope by the fraction of each isotope present and adding the results. The calculation for copper is

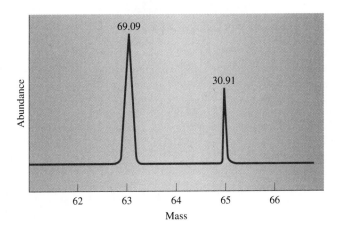

◀ FIGURE 5.9
A typical reading from a
mass spectrometer. The two
principal isotopes of copper
are shown with the abun-
dance (%) given.

$(62.9298 \text{ amu})(0.6909) = 43.48 \text{ amu}$

$(64.9278 \text{ amu})(0.3091) = \underline{20.07 \text{ amu}}$

63.55 amu

The **atomic mass** of an element is the average relative mass of the isotopes of   *atomic mass*
that element compared to the atomic mass of carbon-12 (exactly 12.0000 . . . amu).

The relationship between mass number and atomic number is such that, if we
subtract the atomic number from the mass number of a given isotope, we obtain the
number of neutrons in the nucleus of an atom of that isotope. Table 5.2 shows this
method of determining the number of neutrons. For example, the fluorine atom
($^{19}_{9}F$), atomic number 9, having a mass of 19 amu, contains 10 neutrons:

mass number − atomic number = number of neutrons

19   −   9   =   10

The atomic masses given in the table on the front endpapers of this book are
values accepted by international agreement. You need not memorize atomic masses.
In the calculations in this book, the use of atomic masses to four significant figures
will give results of sufficient accuracy.

TABLE 5.2	Determination of the Number of Neutrons in an Atom by Subtracting Atomic Number from Mass Number				
	Hydrogen ($^{1}_{1}H$)	Oxygen ($^{16}_{8}O$)	Sulfur ($^{32}_{16}S$)	Fluorine ($^{19}_{9}F$)	Iron ($^{56}_{26}Fe$)
Mass number	1	16	32	19	56
Atomic number	(−)1	(−)8	(−)16	(−)9	(−)26
Number of neutrons	0	8	16	10	30

CHEMISTRY IN ACTION Real or Imitation?

What's the difference between vanilla extract and imitation vanilla extract? Both substances taste the same because they contain the same molecules, and taste is related to molecular structure.

Vanilla extract is an alcohol and water solution of materials extracted from vanilla beans. Imitation vanilla extract is also an alcohol and water solution, but it is made from lignin, a waste product in the wood pulp industry, which is converted to vanillin—the same molecule found in vanilla extract. Why would anyone want to use wood pulp in place of vanilla beans? The main reason is cost—natural vanilla costs twice as much to produce as synthetic vanilla.

One role of a government chemist is to prevent mislabeling of products. Because the compounds in both products are chemically the same, detecting the source of the molecule is done by "inspecting" the carbon atoms in the vanilla. Two isotopes in organic compounds are carbon-12 and carbon-13. The ratio of these isotopes is slightly different for natural vanillin and for imitation vanillin. It is from those slight differences that government scientists distinguish whether your vanilla comes from the bean or from wood pulp. Chemists also work as consumer advocates—making sure you get what you pay for in your vanilla extract!

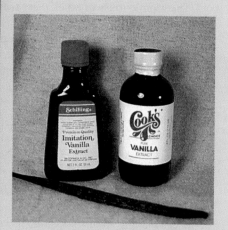

▲ The source of "real" vanilla is the vanilla bean; imitation vanilla comes from wood pulp.

Example 5.2   How many protons, neutrons, and electrons are found in an atom of $^{14}_{6}C$?

Solution   The element is carbon, atomic number 6. The number of protons or electrons equals the atomic number and is 6. The number of neutrons is determined by subtracting the atomic number from the mass number: $14 - 6 = 8$.

Practice 5.2

How many protons, neutrons, and electrons are in each element?
(a) $^{16}_{8}O$, (b) $^{80}_{35}Br$, (c) $^{235}_{92}U$, (d) $^{64}_{29}Cu$

Chlorine is found in nature as two isotopes, $^{37}_{17}Cl$ (24.47%) and $^{35}_{17}Cl$ (75.53%). The atomic masses are 36.96590 and 34.96885 amu, respectively. Determine the average atomic mass of chlorine.

Example 5.3

Multiply each mass by its percentage and add the results to find the average.

Solution

$$(0.2447)(36.96590 \text{ amu}) + (0.7553)(34.96885 \text{ amu})$$

$$= 35.4575 \text{ amu}$$

$$= 35.46 \text{ amu (4 significant figures)}$$

Practice 5.3

Silver is found in two isotopes with atomic masses 106.9041 and 108.9047 amu, respectively. The first isotope represents 51.82% and the second 48.18%. Determine the average atomic mass of silver.

Concepts in Review

1. State the major provisions of Dalton's atomic model.
2. State the law of definite composition and indicate its significance.
3. State the law of multiple proportions and indicate its significance.
4. Give the names, symbols, and relative masses of the three principal subatomic particles.
5. Describe the Thomson model of the atom.
6. Describe the atom as conceived by Ernest Rutherford after his alpha-scattering experiment.

7. Determine the atomic number, mass number, or number of neutrons of an isotope when given the values of any two of these three items.
8. Name and distinguish among the three isotopes of hydrogen.
9. Calculate the average atomic mass of an element, given the isotopic masses and the abundance of its isotopes.
10. Determine the number of protons, neutrons, and electrons from the atomic number and atomic mass of an atom.

Key Terms

atomic mass (5.11)
atomic mass unit (5.11)
atomic number (5.9)
Dalton's atomic model (5.2)

electron (5.6)
isotopes (5.10)
law of definite composition (5.3)
law of multiple proportions (5.3)

mass number (5.10)
neutron (5.6)
nucleus (5.8)

proton (5.6)
subatomic particles (5.6)
Thomson model of the atom (5.6)

Questions

1. What are the atomic numbers of (a) copper, (b) nitrogen, (c) phosphorus, (d) radium, and (e) zinc?

2. A neutron is approximately how many times heavier than an electron?

3. From the chemist's point of view, what are the essential differences among a proton, a neutron, and an electron?

4. Distinguish between an atom and an ion.

5. What letters are used to designate atomic number and mass number in isotopic notation of atoms?

6. In what ways are isotopes alike? In what ways are they different?

Paired Exercises

7. Explain why, in Rutherford's experiments, some alpha particles were scattered at large angles by the gold foil or even bounced back.

8. What experimental evidence led Rutherford to conclude the following?
   (a) The nucleus of the atom contains most of the atomic mass.
   (b) The nucleus of the atom is positively charged.
   (c) The atom consists of mostly empty space.

9. Describe the general arrangement of subatomic particles in the atom.

10. What part of the atom contains practically all its mass?

11. What contribution did these scientists make to atomic models of the atom?
    (a) Dalton
    (b) Thomson
    (c) Rutherford

12. Consider the following models of the atom: (a) Dalton, (b) Thomson, (c) Rutherford. How does the location of the electrons in an atom vary? How does the location of the atom's positive matter compare?

13. Explain why the atomic masses of elements are not whole numbers.

14. Is the isotopic mass of a given isotope ever an exact whole number? Is it always? (Consider the masses of $^{12}_{6}C$ and $^{63}_{29}Cu$.)

15. What special names are given to the isotopes of hydrogen?

16. List the similarities and differences in the three isotopes of hydrogen.

17. What is the symbol and name of the element that has an atomic number of 24 and a mass number of 52?

18. An atom of an element has a mass number of 201 and has 121 neutrons in its nucleus.
    (a) What is the electrical charge of the nucleus?
    (b) What is the symbol and name of the element?

19. What is the nuclear composition of the six naturally occurring isotopes of calcium having mass numbers of 40, 42, 43, 44, 46, and 48?

20. Which of the isotopes of calcium in the previous exercise is the most abundant isotope? Can you be sure? Explain your choice.

21. Write isotopic notation symbols for the following:
    (a) $Z = 26$, $A = 55$
    (b) $Z = 12$, $A = 26$
    (c) $Z = 3$, $A = 6$
    (d) $Z = 79$, $A = 188$

22. Give the nuclear composition and isotopic notation for
    (a) an atom containing 27 protons, 32 neutrons, and 27 electrons
    (b) an atom containing 15 protons, 16 neutrons, and 15 electrons
    (c) an atom containing 110 neutrons, 74 electrons, and 74 protons
    (d) an atom containing 92 electrons, 143 neutrons, and 92 protons

23. Naturally occurring lead exists as four stable isotopes: ^{204}Pb with a mass of 203.973 amu (1.480%); ^{206}Pb, 205.974 amu (23.60%); ^{207}Pb, 206.9759 amu (22.60%); and ^{208}Pb, 207.9766 amu (52.30%). Calculate the average atomic mass of lead.

24. Naturally occurring magnesium consists of three stable isotopes: ^{24}Mg, 23.985 amu (78.99%); ^{25}Mg, 24.986 amu (10.00%); and ^{26}Mg, 25.983 amu (11.01%). Calculate the average atomic mass of magnesium.

25. 68.9257 amu is the mass of 60.4% of the atoms of an element with only two naturally occurring isotopes. The atomic mass of the other isotope is 70.9249 amu. Determine the average atomic mass of the element. Identify the element.

26. A sample of enriched lithium contains 30.00% ^6Li (6.015 amu) and 70.00% ^7Li (7.016 amu). What is the average atomic mass of the sample?

*27. An average dimension for the radius of an atom is 1.0×10^{-8} cm, and the average radius of the nucleus is 1.0×10^{-13} cm. Determine the ratio of atomic volume to nuclear volume. Assume that the atom is spherical ($V = (4/3)\pi r^3$ for a sphere).

*28. An aluminum atom has an average diameter of about 3.0×10^{-8} cm. The nucleus has a diameter of about 2.0×10^{-13} cm. Calculate the ratio of the atom's diameter to its nucleus.

Additional Exercises

These exercises are not paired and provide additional practice on the concepts covered in this chapter.

29. What experimental evidence supports these statements?
   (a) The nucleus of an atom is small.
   (b) The atom consists of both positive and negative charges.
   (c) The nucleus of the atom is positive.

30. What is the relationship between the following two atoms:
   (a) one atom with 10 protons, 10 neutrons, and 10 electrons; and another atom with 10 protons, 11 neutrons, and 10 electrons
   (b) one atom with 10 protons, 11 neutrons, and 10 electrons; and another atom with 11 protons, 10 neutrons, and 11 electrons

31. How will an atom's nucleus differ if it loses an alpha particle?

32. The radius of a carbon atom in many compounds is 0.77×10^{-8} cm. If the radius of a Styrofoam ball used to represent the carbon atom in a molecular model is 1.5 cm, how much of an enlargement is this?

33. How is it possible for there to be more than one kind of atom of the same element?

34. Which element contains the largest number of neutrons per atom: ^{210}Bi, ^{210}Po, ^{210}At, or ^{211}At?

*35. An unknown element Q has two known isotopes: ^{60}Q and ^{63}Q. If the average atomic mass is 61.5 amu, what are the relative percentages of the isotopes?

*36. The actual mass of one atom of an unknown isotope is 2.18×10^{-22} g. Calculate the atomic mass of this isotope.

37. The mass of an atom of argon is 6.63×10^{-24} g. How many atoms are in a 40.0-g sample of argon?

38. Using the periodic table inside the front cover of the book, determine which of the first 20 elements have isotopes that you would expect to have the same number of protons, neutrons, and electrons.

39. Complete this table with the appropriate data for each isotope given:

Atomic number	Mass number	Symbol of element	Number of protons	Number of neutrons
(a) 8	16			
(b)		Ni		30
(c)	199		80	

40. Complete this table (all are neutral atoms):

Element	Symbol	Atomic number	Number of protons	Number of neutrons	Number of electrons
(a) Platinum					
(b)	^{30}P				
(c)		53			
(d)			36		
(e)				45	34
(f)	^{40}Ca				

Answers to Practice Exercises

5.1 5.98×10^{24} atoms

5.2

	protons	neutrons	electrons
(a)	8	8	8
(b)	35	45	35
(c)	92	143	92
(d)	29	35	29

5.3 107.9 amu

CHAPTER 6

Nomenclature of Inorganic Compounds

CHAPTER 6 / OUTLINE

▲
This exquisite chambered nautilus shell is formed from calcium carbonate, commonly called limestone.

CHAPTER

6

As children, we begin to communicate with other people in our lives by learning the names of objects around us. As we continue to develop, we learn to speak and use language to complete a wide variety of tasks. As we enter school, we begin to learn of other languages—the languages of mathematics, of other cultures, of computers. In each case, we begin by learning the names of the building blocks, and then proceed to more abstract concepts. Chemistry has a language all its own—a whole new way of describing the objects so familiar to us in our daily lives. Only after learning the language are we able to understand the complexities of the modern model of the atom and its applications.

6.1 Common and Systematic Names

Chemical nomenclature is the system of names that chemists use to identify compounds. When a new substance is formulated, it must be named in order to distinguish it from all other substances (see Figure 6.1). In this chapter, we will restrict our discussion to the nomenclature of inorganic compounds—compounds that do not generally contain carbon.

Common names are arbitrary names that are not based on the chemical composition of compounds. Before chemistry was systematized, a substance was given a name that generally associated it with one of its outstanding physical or chemical properties. For example, *quicksilver* is a common name for mercury, and nitrous oxide (N_2O), used as an anesthetic in dentistry, has been called *laughing gas* because it induces laughter when inhaled. Water and ammonia are also common names because neither provides any information about the chemical composition of the compounds. If every substance were assigned a common name, the amount of memorization required to learn over 9 million names would be astronomical.

Common names have distinct limitations, but they remain in frequent use. Common names continue to be used in industry because the systematic name is too long or too technical for everyday use. For example, calcium oxide (CaO) is called *lime* by plasterers; photographers refer to *hypo* rather than sodium thiosulfate

Water (H_2O) and ammonia (NH_3) are almost always referred to by their common names.

FIGURE 6.1 ▶
Where to find rules for naming inorganic substances in this book.

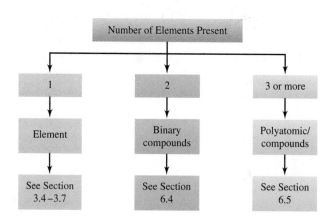

TABLE 6.1 **Common Names, Formulas, and Chemical Names of Familiar Substances**

Common name	Formula	Chemical name
Acetylene	C_2H_2	ethyne
Lime	CaO	calcium oxide
Slaked lime	$Ca(OH)_2$	calcium hydroxide
Water	H_2O	water
Galena	PbS	lead(II) sulfide
Alumina	Al_2O_3	aluminum oxide
Baking soda	$NaHCO_3$	sodium hydrogen carbonate
Cane or beet sugar	$C_{12}H_{22}O_{11}$	sucrose
Borax	$Na_2B_4O_7 \cdot 10\,H_2O$	sodium tetraborate decahydrate
Brimstone	S	sulfur
Calcite, marble, limestone	$CaCO_3$	calcium carbonate
Cream of tartar	$KHC_4H_4O_6$	potassium hydrogen tartrate
Epsom salts	$MgSO_4 \cdot 7\,H_2O$	magnesium sulfate heptahydrate
Gypsum	$CaSO_4 \cdot 2\,H_2O$	calcium sulfate dihydrate
Grain alcohol	C_2H_5OH	ethanol, ethyl alcohol
Hypo	$Na_2S_2O_3$	sodium thiosulfate
Laughing gas	N_2O	dinitrogen monoxide
Lye, caustic soda	NaOH	sodium hydroxide
Milk of magnesia	$Mg(OH)_2$	magnesium hydroxide
Muriatic acid	HCl	hydrochloric acid
Plaster of Paris	$CaSO_4 \cdot \frac{1}{2}\,H_2O$	calcium sulfate hemihydrate
Potash	K_2CO_3	potassium carbonate
Pyrite (fool's gold)	FeS_2	iron disulfide
Quicksilver	Hg	mercury
Saltpeter (chile)	$NaNO_3$	sodium nitrate
Table salt	NaCl	sodium chloride
Vinegar	$HC_2H_3O_2$	acetic acid
Washing soda	$Na_2CO_3 \cdot 10\,H_2O$	sodium carbonate decahydrate
Wood alcohol	CH_3OH	methanol, methyl alcohol

($Na_2S_2O_3$); and nutritionists use the name *vitamin D_3*, instead of 9,10-secocholesta-5,7,10(19)-trien-3-β-ol ($C_{27}H_{44}O$). Table 6.1 lists the common names, formulas, and systematic names of some familiar substances.

Chemists prefer systematic names that precisely identify the chemical composition of chemical compounds. The system for inorganic nomenclature was devised by the International Union of Pure and Applied Chemistry (IUPAC), which was founded in 1921. The IUPAC meets regularly and constantly reviews and updates the system.

6.2 Elements and Ions

In Chapter 3, we studied the names and symbols for the elements as well as their general location on the periodic table. In Chapter 5, we investigated the composition of the atom and learned that all atoms are composed of protons, electrons, and neutrons; that a particular element is defined by the number of protons it contains; and that atoms are uncharged because they contain equal numbers of protons and electrons.

What's in a Name?

When a scientist discovered a new element in the early days of chemistry, he or she had the honor of naming it. Now researchers must submit their choices for a name to an international committee called the International Union of Pure and Applied Chemistry before they can be placed on the periodic table. In September 1997, the IUPAC decided on names for the elements from 104 through109. These six elements are now called rutherfordium (Rf), dubnium (Db), seaborgium (Sg), bohrium (Bh), hassium (Hs), and meitnerium (Mt).

The new names are a compromise among choices presented by different research teams. The Russians gained recognition for work done at a laboratory in Dubna. Americans gained recognition for Glenn Seaborg, the first living scientist to have an element named after him. The British recognized Ernest Rutherford, who discovered the atomic nucleus. The Germans won recognition both for Lise Meitner, the woman who co-discovered atomic fission, and for one of their labs in the German state of Hesse. Both the Germans and the Russians won recognition for Niels Bohr, whose model of the atom led the way toward modern ideas about atomic structure. As researchers continue to discover elements and expand the periodic table, the job of deciding on a name and symbol becomes an increasingly complex task.

Russian lab in Dubna

Niels Bohr

Ernest Rutherford

104	105	106	107	108	109
Rf	Db	Sg	Bh	Hs	Mt

Lise Meitner

Glenn Seaborg

German lab in Hesse

The formula for most elements is simply the symbol of the element. In chemical reactions or mixtures the element behaves as though it were a collection of individual particles. A small number of elements have formulas that are not single atoms at normal temperatures. Seven of the elements are *diatomic* molecules—that is, two atoms bonded together to form a molecule. These diatomic elements are hydrogen, H_2; oxygen, O_2; nitrogen, N_2; fluorine, F_2; chlorine, Cl_2; bromine, Br_2; and iodine, I_2. Two other elements that are commonly polyatomic are S_8, sulfur; and P_4, phosphorus.

Elements Occurring as Molecules		
Hydrogen H_2	Chlorine Cl_2	Sulfur S_8
Oxygen O_2	Fluorine F_2	Phosphorus P_4
Nitrogen N_2	Bromine Br_2	
	Iodine I_2	

A charged particle, known as an *ion,* can be produced by adding or removing one or more electrons from a neutral atom. For example, potassium atoms contain 19 protons and 19 electrons. To make a potassium ion, we remove one electron leaving 19 protons and only 18 electrons. This gives an ion with a positive one (+1) charge:

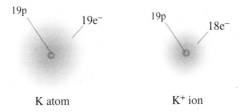

K atom K^+ ion

Written in the form of an equation, $K \rightarrow K^+ + e^-$. A positive ion is called a *cation.* Any neutral atom that *loses* an electron will form a cation. Sometimes an atom may lose one electron, as in the potassium example. Other atoms may lose more than one electron:

$$Mg \rightarrow Mg^{2+} + 2e^-$$

or

$$Al \rightarrow Al^{3+} + 3e^-$$

Cations are named the same as their parent atoms, as shown here:

Atom		Ion	
K	potassium	K^+	potassium ion
Mg	magnesium	Mg^{2+}	magnesium ion
Al	aluminum	Al^{3+}	aluminum ion

Ions can also be formed by adding electrons to a neutral atom. For example, the chlorine atom contains 17 protons and 17 electrons. The equal number of positive charges and negative charges results in a net charge of zero for the atom. If one electron is added to the chlorine atom, it now contains 17 protons and 18 electrons resulting in a net charge of negative one (-1) on the ion:

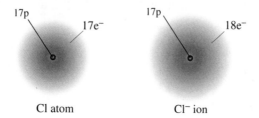

Cl atom Cl^- ion

In a chemical equation this process is summarized as $Cl + e^- \rightarrow Cl^-$. A negative ion is called an *anion*. Any neutral atom that *gains* an electron will form an anion. Atoms may gain more than one electron to form anions with different charges:

$$O + 2e^- \rightarrow O^{2-}$$

$$N + 3e^- \rightarrow N^{3-}$$

Anions are named differently than cations. To name an anion consisting of only one element, use the stem of the parent element name and change the ending to *-ide*. For example, the Cl^- ion is named by using the stem *chlor-* from chlorine and adding *-ide* to form chloride ion. Here are some examples:

Symbol	Name of atom	Ion	Name of ion
F	fluorine	F^-	fluoride ion
Br	bromine	Br^-	bromide ion
Cl	chlorine	Cl^-	chloride ion
I	iodine	I^-	iodide ion
O	oxygen	O^{2-}	oxide ion
N	nitrogen	N^{3-}	nitride ion

Ions are always formed by adding or removing electrons from an atom. Atoms do not form ions on their own. Most often ions are formed when metals combine with non-metals.

The charge on an ion can often be predicted from the position of the element on the periodic table. Figure 6.2 shows the charges of selected ions from several groups on the periodic table. Notice that all the metals in the far left column (Group IA) are ($1+$), all those in the next column (Group IIA) are ($2+$), and the metals in the next tall column (Group IIIA) form ($3+$) ions. The elements in the lower center part of the table are called *transition metals*. These elements tend to form cations with various positive charges. There is no easy way to predict the charges on these cations. All metals lose electrons to form positive ions.

IA													IIIA	IVA	VA	VIA	VIIA	
H^+	IIA																	
Li^+	Be^{2+}														N^{3-}	O^{2-}	F^-	
Na^+	Mg^{2+}												Al^{3+}		P^{3-}	S^{2-}	Cl^-	
K^+	Ca^{2+}																Br^-	
Rb^+	Sr^{2+}			Transition metals													I^-	
Cs^+	Ba^{2+}																	

▲ **FIGURE 6.2**
Charges of selected ions in the periodic table.

In contrast, the nonmetals form anions by gaining electrons. On the right side of the periodic table in Figure 6.2, you can see that the atoms in Group VIIA form $(1-)$ ions. The nonmetals in Group VIA form $(2-)$ ions. It's important to learn the charges on the ions shown in Figure 6.2 and their relationship to the group number at the top of the column. We will learn more about why these ions carry their particular charges later in the course.

6.3 Writing Formulas from Names of Compounds

In Chapters 3 and 5, we learned that compounds can be composed of ions. These substances are called *ionic compounds* and will conduct electricity when dissolved in water. An excellent example of an ionic compound is ordinary table salt. It's composed of crystals of sodium chloride. When dissolved in water, sodium chloride conducts electricity very well, as shown in Figure 6.3.

A chemical compound must have a net charge of zero. If it contains ions, the charges on the ions must add up to zero in the formula for the compound. This is relatively easy in the case of sodium chloride. The sodium ion $(1+)$ and the chloride ion $(1-)$ add to zero, resulting in the formula NaCl. Now consider an ionic compound containing calcium (Ca^{2+}) and fluoride (F^-) ions. How can we write a formula with a net charge of zero? To do this we need one Ca^{2+} and two F^- ions. The correct formula is CaF_2. The subscript 2 indicates two fluoride ions are needed for each calcium ion.

FIGURE 6.3
(a) Distilled water does not conduct electricity. (b) Salt is added to the distilled water. (c) The solution of salt water contains ions and conducts electricity.
▼

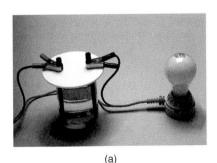

(a)

(b)

(c)

Aluminum oxide is a bit more complicated because it consists of Al^{3+} and O^{2-} ions. Since 6 is the least common multiple of 3 and 2, we have $2(3+) + 3(2-) = 0$ or a formula containing $2\ Al^{3+}$ ions and $3\ O^{2-}$ ions for Al_2O_3. Here are a few more examples of formula writing for ionic compounds:

Name of compound	Ions	Lowest common multiple	Sum of charges on ions	Formula
Sodium bromide	Na^+, Br^-	1	$(+1) + (-1) = 0$	$NaBr$
Potassium sulfide	K^+, S^{2-}	2	$2(+1) + (-2) = 0$	K_2S
Zinc sulfate	Zn^{2+}, SO_4^{2-}	2	$(+2) + (-2) = 0$	$ZnSO_4$
Ammonium phosphate	NH_4^+, PO_4^{3-}	3	$3(+1) + (-3) = 0$	$(NH_4)_3PO_4$
Aluminum chromate	Al^{3+}, CrO_4^{2-}	6	$2(+3) + 3(-2) = 0$	$Al_2(CrO_4)_3$

Example 6.1   Write formulas for (a) calcium chloride, (b) magnesium oxide, and (c) barium phosphide.

Solution   (a) Use the following steps for calcium chloride.

> **Step 1.** From the name we know that calcium chloride is composed of calcium and chloride ions. First write down the formulas of these ions:
>
> Ca^{2+} Cl^-

> **Step 2.** To write the formula of the compound, combine the smallest numbers of Ca^{2+} and Cl^- ions to give the charge sum equal to zero. In this case the lowest common multiple of the charges is 2:
>
> $(Ca^{2+}) + 2(Cl^-) = 0$
>
> $(2+)\ \ + 2(1-) = 0$
>
> Therefore the formula is $CaCl_2$.

(b) Use the same procedure for magnesium oxide.

> **Step 1.** From the name we know that magnesium oxide is composed of magnesium and oxide ions. First write down the formulas of these ions:
>
> Mg^{2+} O^{2-}

> **Step 2.** To write the formula of the compound, combine the smallest numbers of Mg^{2+} and O^{2-} ions to give the charge sum equal to zero:
>
> $(Mg^{2+}) + (O^{2-}) = 0$
>
> $(2+)\ \ + (2-) = 0$
>
> The formula is MgO.

◄ **Compounds of transition elements are typically very colorful and are useful as paint pigments.**

1. Cobalt(II) chloride, $CoCl_2$
2. Sodium chloride, $NaCl$
3. Lead(II) sulfide, PbS
4. Sulfur, S
5. Zinc, Zn
6. Marble chips, $CaCO_3$
7. Logwood chips
8. Charcoal, C
9. Mercury(II) iodide, HgI_2
10. Pyrite, FeS
11. Chromium(III) oxide, Cr_2O_3
12. Iron(II) sulfate, $FeSO_4$
13. Sodium sulfite, Na_2SO_3
14. Rosin
15. Sodium thiosulfate, $Na_2S_2O_3$
16. Iron, Fe
17. Aluminum, Al
18. Potassium hexacyanoferrate, $K_3Fe(CN)_6$
19. Potassium chromium(III) sulfate, $KCr(SO_4)_2$
20. Menthol, $C_{10}H_{19}OH$
21. Potassium permanganate, $KMnO_4$
22. Ammonium nickel(II) sulfate, $(NH_4)_2Ni(SO_4)_2$
23. Copper(II) sulfate pentahydrate, $CuSO_4 \cdot 5H_2O$
24. Sodium chromate, Na_2CrO_4
25. Trilead tetraoxide, Pb_3O_4
26. Hydroquinone, $C_6H_4(OH)_2$
27. Copper, Cu

(c) Use the same procedure for barium phosphide.

Step 1. From the name we know that barium phosphide is composed of barium and phosphide ions. First write down the formulas of these ions:

$$Ba^{2+} \qquad P^{3-}$$

Step 2. To write the formula of the compound, combine the smallest numbers of Ba^{2+} and P^{3-} ions to give the charge sum equal to zero. In this case the lowest common multiple of the charges is 6:

$$3(Ba^{2+}) + 2(P^{3-}) = 0$$

$$3(2+) \quad + 2(3-) = 0$$

The formula is Ba_3P_2.

Practice 6.1

Write formulas for compounds containing the following ions:

(a) K^+ and F^- (d) Na^+ and S^{2-}

(b) Ca^{2+} and Br^- (e) Ba^{2+} and O^{2-}

(c) Mg^{2+} and N^{3-}

6.4   Binary Compounds

Binary compounds contain only two different elements. Many binary compounds are formed when a metal combines with a nonmetal to form a *binary ionic compound.* The metal loses one or more electrons to become a cation while the nonmetal gains one or more electrons to become an anion. The cation is written first in the formula, followed by the anion.

A. Binary Ionic Compounds Containing a Metal Forming Only One Type of Cation

The chemical name is composed of the name of the metal followed by the name of the nonmetal, which has been modified to an identifying stem plus the suffix *-ide.*

For example, sodium chloride, NaCl, is composed of one atom each of sodium and chlorine. The name of the metal, sodium, is written first and is not modified. The second part of the name is derived from the nonmetal, chlorine, by using the stem *chlor-* and adding the ending *-ide*; it is named *chloride.* The compound name is *sodium chloride.*

To name these compounds:
1. **Write the name of the cation.**
2. **Write the stem for the anion with the suffix -ide.**

NaCl
Elements: Sodium (metal)
                       Chlorine (nonmetal)
                            name modified to the stem *chlor-* + *-ide*
Name of compound: Sodium chloride

Stems of the more common negative-ion-forming elements are shown in Table 6.2. Table 6.3 lists some compounds with names ending in *-ide.*

Compounds may contain more than one atom of the same element, but as long as they contain only two different elements and only one compound of these two elements exists, the name follows the rules for binary compounds:

Examples:

$CaBr_2$ Mg_3N_2 Li_2O

calcium bromide   magnesium nitride   lithium oxide

Example 6.2 Name the compound CaF_2.

Solution

Step 1. From the formula it is a two-element compound and follows the rules for binary compounds.

Step 2. The compound is composed of Ca, a metal, and F, a nonmetal. Elements in the IIA column form only one type of cation. Thus we name the positive part of the compound *calcium.*

Step 3. Modify the name of the second element to the stem *fluor-* and add the binary ending *-ide* to form the name of the negative part, *fluoride*.

Step 4. The name of the compound is therefore *calcium fluoride*.

▲
The mineral fluorite contains CaF₂.

Practice 6.2

Write formulas for these compounds:
(a) strontium chloride (b) potassium iodide (c) aluminum nitride
(d) calcium sulfide (e) sodium oxide

B. Binary Ionic Compounds Containing a Metal That Can Form Two or More Types of Cations

The metals in the center of the periodic table (including the transition metals) often form more than one type of cation. For example, iron can form Fe^{2+} and Fe^{3+} ions, and copper can form Cu^{+} and Cu^{2+} ions. This can be confusing when you are naming compounds. For example, copper chloride could be $CuCl_2$ or $CuCl$. To

TABLE 6.2   Examples of Elements Forming Anions

Symbol	Element	Stem	Anion name
Br	bromine	brom	bromide
Cl	chlorine	chlor	chloride
F	fluorine	fluor	fluoride
H	hydrogen	hydr	hydride
I	iodine	iod	iodide
N	nitrogen	nitr	nitride
O	oxygen	ox	oxide
P	phosphorus	phosph	phosphide
S	sulfur	sulf	sulfide

TABLE 6.3   Examples of Compounds with Names Ending in -ide

Formula	Name	Formula	Name
$AlCl_3$	aluminum chloride	BaS	barium sulfide
Al_2O_3	aluminum oxide	LiI	lithium iodide
CaC_2	calcium carbide	$MgBr_2$	magnesium bromide
HCl	hydrogen chloride	NaH	sodium hydride
HI	hydrogen iodide	Na_2O	sodium oxide

Stock System

resolve this difficulty the IUPAC devised a system, known as the **Stock System,** to name these compounds. This system is currently recognized as the official system to name these compounds, although another older system is sometimes used. In the Stock System, when a compound contains a metal that can form more than one type of cation, the charge on the cation of the metal is designated by a Roman numeral placed in parentheses immediately following the name of the metal. The negative element is treated in the usual manner for binary compounds.

To name these compounds:
1. Write the name of the cation.
2. Write the charge on the cation as a Roman numeral in parentheses.
3. Write the stem of the anion with the suffix -ide.

Cation charge	+1	+2	+3	+4	+5
Roman numeral	(I)	(II)	(III)	(IV)	(V)

Examples: $FeCl_2$ iron(II) chloride Fe^{2+}

           $FeCl_3$ iron(III) chloride Fe^{3+}

           $CuCl$ copper(I) chloride Cu^+

           $CuCl_2$ copper(II) chloride Cu^{2+}

The fact that $FeCl_2$ has two chloride ions, each with a -1 charge, establishes that the charge of Fe is $+2$. To distinguish between the two iron chlorides, $FeCl_2$ is named iron(II) chloride and $FeCl_3$ is named iron(III) chloride.

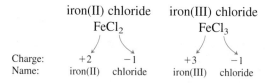

When a metal forms only one possible cation, we need not distinguish one cation from another, so Roman numerals are not needed. Thus we do not say calcium(II) chloride for $CaCl_2$, but rather calcium chloride, since the charge of calcium is understood to be $+2$.

In classical nomenclature, when the metallic ion has only two cation types, the name of the metal (usually the Latin name) is modified with the suffixes *-ous* and *-ic* to distinguish between the two. The lower charge cation is given the *-ous* ending, and the higher one, the *-ic* ending.

Examples: $FeCl_2$ ferrous chloride Fe^{2+} (lower charge cation)
              $FeCl_3$ ferric chloride Fe^{3+} (higher charge cation)
              $CuCl$ cuprous chloride Cu^+ (lower charge cation)
              $CuCl_2$ cupric chloride Cu^{2+} (higher charge cation)

Table 6.4 lists some common metals that have more than one type of cation.

In this book we will use only the Stock System.

Notice that the *ous–ic* naming system does not give the charge of the cation of an element but merely indicates that at least two types of cations exist. The Stock System avoids any possible uncertainty by clearly stating the charge on the cation.

Example 6.3   Name the compound FeS.

Solution

     Step 1. This compound follows the rules for a binary compound.

     Step 2. It is a compound of Fe, a metal, and S, a nonmetal, and Fe is a transition metal that has more than one type of cation. In sulfides, the charge on

TABLE 6.4 Names and Charges of Some Common Metal Ions That Have More Than One Type of Cation		
Formula	Stock System name	Classical name
Cu^{1+}	copper(I)	cuprous
Cu^{2+}	copper(II)	cupric
Hg^{1+} $(Hg_2)^{2+}$	mercury(I)	mercurous
Hg^{2+}	mercury(II)	mercuric
Fe^{2+}	iron(II)	ferrous
Fe^{3+}	iron(III)	ferric
Sn^{2+}	tin(II)	stannous
Sn^{4+}	tin(IV)	stannic
Pb^{2+}	lead(II)	plumbous
Pb^{4+}	lead(IV)	plumbic
As^{3+}	arsenic(III)	arsenous
As^{5+}	arsenic(V)	arsenic
Ti^{3+}	titanium(III)	titanous
Ti^{4+}	titanium(IV)	titanic

the S is -2. Therefore, the charge on Fe must be $+2$, and the name of the positive part of the compound is *iron(II)*.

Step 3. We have already determined that the name of the negative part of the compound will be *sulfide*.

Step 4. The name of FeS is *iron(II) sulfide*.

Practice 6.3

Write the name for each of the following compounds using the Stock System:
(a) PbI_2 (b) SnF_4 (c) Fe_2O_3 (d) CuO

C. Binary Compounds Containing Two Nonmetals

Compounds between nonmetals are molecular, not ionic. Therefore a different system for naming them is used. In a compound formed between two nonmetals, the element that occurs first in this series is written and named first:

Si, B, P, H, C, S, I, Br, N, Cl, O, F

The name of the second element retains the *-ide* ending as though it were an anion. A Latin or Greek prefix (*mono-, di-, tri-*, and so on) is attached to the name of each element to indicate the number of atoms of that element in the molecule. The prefix *mono-* is never used for naming the first element. Some common prefixes and their numerical equivalences follow.

mono = 1	*tetra* = 4	*hepta* = 7	*nona* = 9
di = 2	*penta* = 5	*octa* = 8	*deca* = 10
tri = 3	*hexa* = 6		

Here are some examples of compounds that illustrate this system:

$$N_2O_3$$

(di)nitrogen ← → (tri)oxide

Indicates two   Indicates three
nitrogen atoms   oxygen atoms

To name these compounds:

1. **Write the name for the first element using a prefix if there is more than one atom of this element.**
2. **Write the stem for the second element with the suffix -ide. Use a prefix to indicate the number of atoms for the second element.**

CO	carbon monoxide	N_2O	dinitrogen monoxide
CO_2	carbon dioxide	N_2O_4	dinitrogen tetroxide
PCl_3	phosphorus trichloride	NO	nitrogen monoxide
SO_2	sulfur dioxide	N_2O_3	dinitrogen trioxide
P_2O_5	diphosphorus pentoxide	S_2Cl_2	disulfur dichloride
CCl_4	carbon tetrachloride	S_2F_{10}	disulfur decafluoride

These examples illustrates that we sometimes drop the final *o* (mono) or *a* (penta) of the prefix when the second element is oxygen. This avoids creating a name that is awkward to pronounce. For example, CO is carbon monoxide instead of carbon mon*oo*xide.

Example 6.4   Name the compound PCl_5.

Solution

Step 1. Phosphorus and chlorine are nonmetals, so the rules for naming binary compounds containing two nonmetals apply. Phosphorus is named first. Therefore the compound is a chloride.

Step 2. No prefix is needed for phosphorus because each molecule has only one atom of phosphorus. The prefix *penta-* is used with chloride to indicate the five chlorine atoms. (PCl_3 is also a known compound.)

Step 3. The name for PCl_5 is *phosphorus pentachloride*.

Practice 6.4

Name these compounds:
(a) Cl_2O (b) SO_2 (c) CBr_4 (d) N_2O_5 (e) NH_3

D. Acids Derived from Binary Compounds

Certain binary hydrogen compounds, when dissolved in water, form solutions that have *acid* properties. Because of this property, these compounds are given acid names in addition to their regular *-ide* names. For example, HCl is a gas and is called *hydrogen chloride,* but its water solution is known as *hydrochloric acid.* Binary acids are composed of hydrogen and one other nonmetallic element. However, not all binary hydrogen compounds are acids. To express the formula of a binary acid, it's customary to write the symbol of hydrogen first, followed by the symbol of the second element (e.g., HCl, HBr, H_2S). When we see formulas such as CH_4 or NH_3, we understand that these compounds are not normally considered to be acids.

TABLE 6.5	Names and Formulas of Selected Binary Acids		
Formula	Acid name	Formula	Acid name
HF	Hydrofluoric acid	HI	Hydroiodic acid
HCl	Hydrochloric acid	H_2S	Hydrosulfuric acid
HBr	Hydrobromic acid	H_2Se	Hydroselenic acid

To name a binary acid, place the prefix *hydro-* in front of, and the suffix *-ic* after, the stem of the nonmetal name. Then add the word *acid*.

HCl H_2S

Examples: *Hydro-chlor-ic acid*   *Hydro-sulfur-ic acid*
(hydrochloric acid)   (hydrosulfuric acid)

To name these compounds:
1. **Write the prefix hydro- with the stem of the second element and add the suffix -ic.**
2. **Write the word acid.**

Acids are hydrogen-containing substances that liberate hydrogen ions when dissolved in water. The same formula is often used to express binary hydrogen compounds, such as HCl, regardless of whether or not they are dissolved in water. Table 6.5 shows several examples of binary acids.

Naming binary compounds is summarized in Figure 6.4.

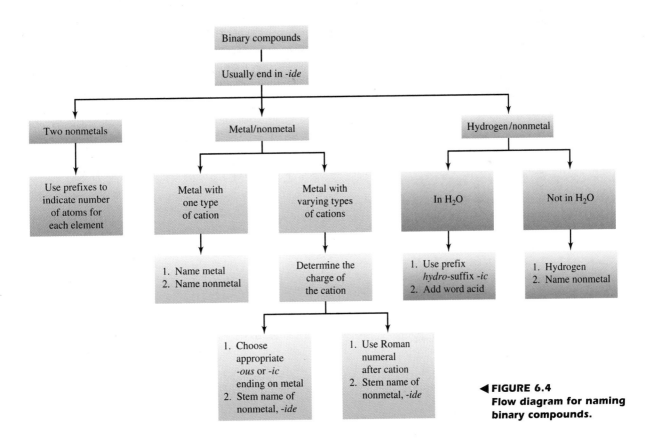

◄ FIGURE 6.4
Flow diagram for naming binary compounds.

Practice 6.5

Name these binary compounds:
(a) KBr (b) Ca_3N_2 (c) SO_3 (d) SnF_2 (e) $CuCl_2$ (f) N_2O_4

6.5 Naming Compounds Containing Polyatomic Ions

polyatomic ion

A **polyatomic ion** is an ion that contains two or more elements. Compounds containing polyatomic ions are composed of three or more elements and usually consist of one or more cations combined with a negative polyatomic ion. In general, naming compounds containing polyatomic ions is similar to naming binary compounds. The cation is named first, followed by the name for the negative polyatomic ion.

To name these compounds you must learn to recognize the common polyatomic ions (Table 6.6) and know their charges. Consider the formula $KMnO_4$. You must be able to recognize that it consists of two parts $KMnO_4$. These parts are composed of a K^+ ion and a MnO_4^- ion. The correct name for this compound is potassium permanganate. Many polyatomic ions that contain oxygen are called *oxy-anions*, and generally have the suffix *-ate* or *-ite*. Unfortunately, the suffix doesn't indicate the number of oxygen atoms present. The *-ate* form contains more oxygen atoms than the *-ite* form. Examples include sulfate (SO_4^{2-}), sulfite (SO_3^{2-}), nitrate (NO_3^-), and nitrite (NO_2^-).

Some elements form more than two different polyatomic ions containing oxygen. To name these ions, prefixes are used in addition to the suffix. To indicate more oxygen than in the *-ate* form, we add the prefix *per-*, which is a short form of *hyper-*, meaning more. The prefix *hypo-*, meaning less (oxygen in this case), is used for the ion containing less oxygen than the *-ite* form. An example of this system is shown for the polyatomic ions containing chlorine and oxygen in Table 6.7. The prefixes are also used with other similar ions, such as iodate (IO_3^-), bromate (BrO_3^-), and phosphate (PO_4^{3-}).

▲
Potassium permanganate crystals are dark purple.

To name these compounds:
1. **Write the name of the cation.**
2. **Write the name of the anion.**

TABLE 6.6	Names, Formulas, and Charges of Some Common Polyatomic Ions				
Name	**Formula**	**Charge**	**Name**	**Formula**	**Charge**
Acetate	$C_2H_3O_2^-$	-1	Cyanide	CN^-	-1
Ammonium	NH_4^+	$+1$	Dichromate	$Cr_2O_7^{2-}$	-2
Arsenate	AsO_4^{3-}	-3	Hydroxide	OH^-	-1
Hydrogen carbonate	HCO_3^-	-1	Nitrate	NO_3^-	-1
Hydrogen sulfate	HSO_4^-	-1	Nitrite	NO_2^-	-1
Bromate	BrO_3^-	-1	Permanganate	MnO_4^-	-1
Carbonate	CO_3^{2-}	-2	Phosphate	PO_4^{3-}	-3
Chlorate	ClO_3^-	-1	Sulfate	SO_4^{2-}	-2
Chromate	CrO_4^{2-}	-2	Sulfite	SO_3^{2-}	-2

Only three of the common negatively charged polyatomic ions do not use the *ate/ite* system. These exceptions are hydroxide (OH^-), hydrogen sulfide (HS^-), and cyanide (CN^-). Care must be taken with these, as their endings can easily be confused with the *-ide* ending for binary compounds (Section 6.4).

There are two common positively charged polyatomic ions as well—the ammonium and the hydronium ions. The ammonium ion (NH_4^+) is frequently found in polyatomic compounds (Section 6.5), whereas the hydronium ion (H_3O^+) is usually seen in aqueous solutions of acids (Chapter 15).

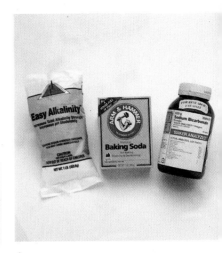

▲
All three of these samples (sodium hydrogen carbonate, baking soda, and sodium bicarbonate) are the same chemical, named three different ways.

Practice 6.6

Name these compounds:
(a) $NaNO_3$ (b) $Ca_3(PO_4)_2$ (c) KOH (d) Li_2CO_3 (e) $NaClO_3$

Inorganic compounds are also formed from more than three elements (see Table 6.8). In these cases one or more of the ions is often a polyatomic ion. Once you have

TABLE 6.7 **Oxy-Anions and Oxy-Acids of Chlorine**

Anion formula	Anion name	Acid formula	Acid name
ClO^-	*hypo*chlor*ite*	$HClO$	*hypo*chlor*ous* acid
ClO_2^-	chlor*ite*	$HClO_2$	chlor*ous* acid
ClO_3^-	chlor*ate*	$HClO_3$	chlor*ic* acid
ClO_4^-	*per*chlor*ate*	$HClO_4$	*per*chloric acid

TABLE 6.8 **Names of Selected Compounds That Contain More Than One Kind of Positive Ion**

Formula	Name of compound
$KHSO_4$	potassium hydrogen sulfate
$Ca(HSO_3)_2$	calcium hydrogen sulfite
NH_4HS	ammonium hydrogen sulfide
$MgNH_4PO_4$	magnesium ammonium phosphate
NaH_2PO_4	sodium dihydrogen phosphate
Na_2HPO_4	disodium hydrogen phosphate
KHC_2O_4	potassium hydrogen oxalate
$KAl(SO_4)_2$	potassium aluminum sulfate
$Al(HCO_3)_3$	aluminum hydrogen carbonate

learned to recognize the polyatomic ions, naming these compounds follows the patterns we have already learned. First identify the ions. Name the cations in the order given, and follow them with the names of the anions. Study the following examples:

Compound	Ions	Name
$NaHCO_3$	Na^+; HCO_3^-	sodium hydrogen carbonate
$NaHS$	Na^+; HS^-	sodium hydrogen sulfide
$MgNH_4PO_4$	Mg^{2+}; NH_4^+; PO_4^{3-}	magnesium ammonium phosphate
$NaKSO_4$	Na^+; K^+; SO_4^{2-}	sodium potassium sulfate

6.6 Acids

While we will learn much more about acids later (see Chapter 15), it is helpful to be able to recognize and name common acids both in the laboratory and in class. The simplest way to recognize many acids is to know that acid formulas often begin with hydrogen. Naming binary acids was covered in Section 6.4D. Inorganic compounds containing hydrogen, oxygen, and one other element are called *oxy-acids*. The element other than hydrogen or oxygen in these acids is often a nonmetal, but it can also be a metal.

The first step in naming these acids is to determine that the compound in question is really an oxy-acid. The keys to identification are (1) hydrogen is the first element in the compound's formula and (2) the second part of the formula consists of a polyatomic ion containing oxygen.

Hydrogen in an oxy-acid is not specifically designated in the acid name. The presence of hydrogen in the compound is indicated by the use of the word *acid* in the name of the substance. To determine the particular type of acid, the polyatomic ion following hydrogen must be examined. The name of the polyatomic ion is modified in the following manner: (1) *-ate* changes to an *-ic* ending; (2) *-ite* changes to an *-ous* ending. (See Table 6.9.) The compound with the *-ic* ending contains more oxygen than the one with the *-ous* ending. Consider these examples:

H_2SO_4 sulf*ate* ⟶ sulfur*ic* acid (contains 4 oxygens)

H_2SO_3 sulf*ite* ⟶ sulfur*ous* acid (3 oxygens)

HNO_3 nitr*ate* ⟶ nitr*ic* acid (3 oxygens)

HNO_2 nitr*ite* ⟶ nitr*ous* acid (2 oxygens)

The complete system for naming oxy-acids is shown in Table 6.7 for the various acids containing chlorine. Examples of other oxy-acids and their names are shown in Table 6.10.

Naming polyatomic compounds is summarized in Figure 6.5 on page 120. We have now looked at ways of naming a variety of inorganic compounds—binary compounds consisting of a metal and a nonmetal and of two nonmetals, binary acids and polyatomic compounds. These compounds are just a small part of the classified chemical compounds. Most of the remaining classes are in the broad field of organic chemistry under such categories as hydrocarbons, alcohols, ethers, aldehydes, ketones, phenols, and carboxylic acids.

Practice 6.7

Name these compounds:
(a) Cu_2CO_3 (b) $Fe(ClO)_3$ (c) $Sn(C_2H_3O_2)_2$ (d) $HBrO_3$

Practice 6.8

Write formulas for
(a) lead(II) nitrate (b) potassium phosphate (c) mercury(II) cyanide
(d) ammonium chromate

TABLE 6.9   Comparison of Acid and Anion Names for Selected Oxy-Acids

Acid	Anion	Acid	Anion
H_2SO_4 Sulfuric acid	SO_4^{2-} Sulfate ion	H_3PO_4 Phosphoric acid	PO_4^{3-} Phosphate ion
H_2SO_3 Sulfurous acid	SO_3^{2-} Sulfite ion	H_3PO_3 Phosphorous acid	PO_3^{3-} Phosphite ion
HNO_3 Nitric acid	NO_3^- Nitrate ion	HIO_3 Iodic acid	IO_3^- Iodate ion
HNO_2 Nitrous acid	NO_2^- Nitrite ion	$HC_2H_3O_2$ Acetic acid	$C_2H_3O_2^-$ Acetate ion
H_2CO_3 Carbonic acid	CO_3^{2-} Carbonate ion	$H_2C_2O_4$ Oxalic acid	$C_2O_4^{2-}$ Oxalate ion

TABLE 6.10   Formulas and Names of Selected Oxy-Acids

Formula	Acid name	Formula	Acid name
H_2SO_3	sulfurous acid	$HC_2H_3O_2$	acetic acid
H_2SO_4	sulfuric acid	$H_2C_2O_4$	oxalic acid
HNO_2	nitrous acid	H_2CO_3	carbonic acid
HNO_3	nitric acid	$HBrO_3$	bromic acid
H_3PO_3	phosphorous acid	HIO_3	iodic acid
H_3PO_4	phosphoric acid	H_3BO_3	boric acid

Charges in Your Life

Ions are used in living organisms to perform many important functions. For example, electrical neutrality must be maintained both inside and outside the body cells. Within the cell, neutrality is maintained by potassium ions (K^+) and hydrogen phosphate ions (HPO_4^{2-}). Outside the cell in the intercellular fluid the ions responsible for neutrality are sodium (Na^+) and chloride (Cl^-).

Another important ion within organisms is magnesium (Mg^{2+}), found in chlorophyll and used during nerve and muscle activity, as well as in conjunction with certain enzymes. Iron ions (Fe^{2+}) are incorporated within the hemoglobin molecule and are an integral part of the oxygen transport system within the body. Calcium ions (Ca^{2+}) are part of the matrix of both bones and teeth and play a significant role in the clotting of blood.

Ions also have a significant role in the detergent industry. Water is said to be "hard" when it contains relatively high concentrations of Ca^{2+} and Mg^{2+} ions. In solution, soaps combine with these ions to form an insoluble scum. This material forms the common bathtub ring and, in laundry, settles on the clothes

leaving them gray and dingy. Although soaps and detergents are similar in their cleansing actions, detergents are less likely to form this scum. For this reason, many people who live in areas with "hard" water use detergents instead of soaps.

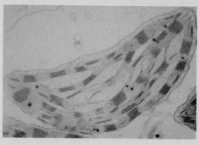

▲ **This chloroplast from a corn plant contains Mg^{2+} ions.**

▲ **Calcium ions (Ca^{2+}) are important in the formation of bones.**

FIGURE 6.5 ▶
Flow diagram for naming polyatomic compounds.

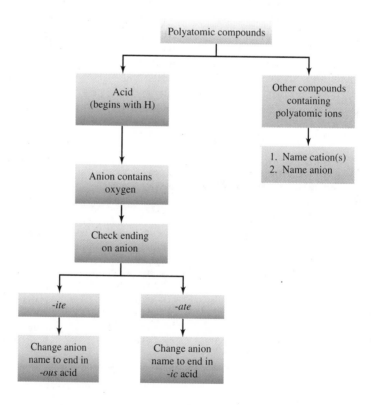

Concepts in Review

1. Write the formulas of compounds formed by combining the ions from Figure 6.2 (or from the inside front cover of this book) in the correct ratios.

2. Write the names or formulas for inorganic binary compounds in which the metal has only one type of cation.

3. Write the names or formulas for inorganic binary compounds that contain metals with multiple types of cations, using the Stock System.

4. Write the names or formulas for inorganic binary compounds that contain two nonmetals.

5. Write the names or formulas for binary acids.

6. Write the names or formulas for oxy-acids.

7. Write the names or formulas for compounds that contain polyatomic ions.

Key Terms

polyatomic ion (6.5)
Stock System (6.4)

Questions

1. Use the common ion table on the inside front cover of your text to determine the formulas for compounds composed of the following ions:
   (a) sodium and chlorate
   (b) hydrogen and sulfate
   (c) tin(II) and acetate
   (d) copper(I) and oxide
   (e) zinc and hydrogen carbonate
   (f) iron(III) and carbonate

2. Does the fact that two elements combine in a one-to-one atomic ratio mean that the charges on their ions are both 1? Explain.

3. Write the names and formulas for the four oxy-acids containing (a) bromine, (b) iodine. (Table 6.7)

4. Explain why N_2O_5 is called dinitrogen pentoxide.

5. Write the formulas for the compounds formed when a chromium(III) ion is combined with the following anions: (a) hydroxide, (b) nitrate, (c) nitrite, (d) hydrogen carbonate, (e) carbonate, (f) dichromate, (g) phosphate, (h) oxalate, (i) oxide, (j) fluoride. (See Table of Names, Formulas, and Charges of Common Ions on the inside front cover.)

6. Explain why the name for $MgCl_2$ is magnesium chloride but the name for $CuCl_2$ is copper(II) chloride.

Paired Exercises

7. Write the formula of the compound that would be formed between these elements:
   (a) Na and I   (d) K and S
   (b) Ba and F   (e) Cs and Cl
   (c) Al and O   (f) Sr and Br

8. Write the formula of the compound that would be formed between these elements:
   (a) Ba and O   (d) Be and Br
   (b) H and S   (e) Li and Se
   (c) Al and Cl   (f) Mg and P

9. Write formulas for the following cations (don't forget to include the charges): sodium, magnesium, aluminum, copper(II), iron(II), iron(III), lead(II), silver, cobalt(II), barium, hydrogen, mercury(II), tin(II), chromium(III), tin(IV), manganese(II), bismuth(III).

10. Write formulas for the following anions (don't forget to include the charges): chloride, bromide, fluoride, iodide, cyanide, oxide, hydroxide, sulfide, sulfate, hydrogen sulfate, hydrogen sulfite, chromate, carbonate, hydrogen carbonate, acetate, chlorate, permanganate, oxalate.

11. Complete the table, filling in each box with the proper formula.

Anions

Cations	Br^-	O^{2-}	NO_3^-	PO_4^{3-}	CO_3^{2-}
K^+	KBr				
Mg^{2+}					
Al^{3+}					
Zn^{2+}				$Zn_3(PO_4)_2$	
H^+					

12. Complete the table, filling in each box with the proper formula.

Anions

Cations	SO_4^{2-}	Cl^-	AsO_4^{3-}	$C_2H_3O_2^-$	CrO_4^{2-}
NH_4^+			$(NH_4)_3AsO_4$		
Ca^{2+}					
Fe^{3+}	$Fe_2(SO_4)_3$				
Ag^+					
Cu^{2+}					

13. Write formulas for these binary compounds, all of which are composed of nonmetals:
- **(a)** carbon monoxide
- **(b)** sulfur trioxide
- **(c)** carbon tetrabromide
- **(d)** phosphorus trichloride
- **(e)** nitrogen dioxide
- **(f)** dinitrogen pentoxide
- **(g)** iodine monobromide
- **(h)** silicon tetrachloride
- **(i)** phosphorus pentaiodide
- **(j)** diboron trioxide

14. Name these binary compounds, all of which are composed of nonmetals:
- **(a)** CO_2
- **(b)** N_2O
- **(c)** PCl_5
- **(d)** CCl_4
- **(e)** SO_2
- **(f)** N_2O_4
- **(g)** P_2O_5
- **(h)** OF_2
- **(i)** NF_3
- **(j)** CS_2

15. Write formulas for these compounds:
- **(a)** sodium nitrate
- **(b)** magnesium fluoride
- **(c)** barium hydroxide
- **(d)** ammonium sulfate
- **(e)** silver carbonate
- **(f)** calcium phosphate
- **(g)** potassium nitrite
- **(h)** strontium oxide

16. Name these compounds:
- **(a)** K_2O
- **(b)** NH_4Br
- **(c)** CaI_2
- **(d)** $BaCO_3$
- **(e)** Na_3PO_4
- **(f)** Al_2O_3
- **(g)** $Zn(NO_3)_2$
- **(h)** Ag_2SO_4

17. Name these compounds by the Stock (IUPAC) System:
- **(a)** $CuCl_2$
- **(b)** $CuBr$
- **(c)** $Fe(NO_3)_2$
- **(d)** $FeCl_3$
- **(e)** SnF_2
- **(f)** $HgCO_3$

18. Write formulas for these compounds:
- **(a)** tin(IV) bromide
- **(b)** copper(I) sulfate
- **(c)** iron(III) carbonate
- **(d)** mercury(II) nitrite
- **(e)** titanium(IV) sulfide
- **(f)** iron(II) acetate

19. Write formulas for these acids:
- **(a)** hydrochloric acid
- **(b)** chloric acid
- **(c)** nitric acid
- **(d)** carbonic acid
- **(e)** sulfurous acid
- **(f)** phosphoric acid

20. Write formulas for these acids:
- **(a)** acetic acid
- **(b)** hydrofluoric acid
- **(c)** hypochlorous acid
- **(d)** boric acid
- **(e)** nitrous acid
- **(f)** hydrosulfuric acid

21. Name these acids:
- **(a)** HNO_2
- **(b)** H_2SO_4
- **(c)** $H_2C_2O_4$
- **(d)** HBr
- **(e)** H_3PO_3
- **(f)** $HC_2H_3O_2$
- **(g)** HF
- **(h)** $HBrO_3$

22. Name these acids:
- **(a)** H_3PO_4
- **(b)** H_2CO_3
- **(c)** HIO_3
- **(d)** HCl
- **(e)** $HClO$
- **(f)** HNO_3
- **(g)** HI
- **(h)** $HClO_4$

23. Write formulas for these compounds:
- **(a)** silver sulfite
- **(b)** cobalt(II) bromide
- **(c)** tin(II) hydroxide
- **(d)** aluminum sulfate
- **(e)** manganese(II) fluoride
- **(f)** ammonium carbonate
- **(g)** chromium(III) oxide
- **(h)** copper(II) chloride
- **(i)** potassium permanganate
- **(j)** barium nitrite
- **(k)** sodium peroxide
- **(l)** iron(II) sulfate
- **(m)** potassium dichromate
- **(n)** bismuth(III) chromate

24. Write formulas for these compounds:
- **(a)** sodium chromate
- **(b)** magnesium hydride
- **(c)** nickel(II) acetate
- **(d)** calcium chlorate
- **(e)** lead(II) nitrate
- **(f)** potassium dihydrogen phosphate
- **(g)** manganese(II) hydroxide
- **(h)** cobalt(II) hydrogen carbonate
- **(i)** sodium hypochlorite
- **(j)** arsenic(V) carbonate
- **(k)** chromium(III) sulfite
- **(l)** antimony(III) sulfate
- **(m)** sodium oxalate
- **(n)** potassium thiocyanate

25. Write the name of each compound.
  (a) $ZnSO_4$
  (b) $HgCl_2$
  (c) $CuCO_3$
  (d) $Cd(NO_3)_2$
  (e) $Al(C_2H_3O_2)_3$
  (f) CoF_2
  (g) $Cr(ClO_3)_3$
  (h) Ag_3PO_4
  (i) NiS
  (j) $BaCrO_4$

26. Write the name of each compound.
  (a) $Ca(HSO_4)_2$
  (b) $As_2(SO_3)_3$
  (c) $Sn(NO_2)_2$
  (d) $FeBr_3$
  (e) $KHCO_3$
  (f) $BiAsO_4$
  (g) $Fe(BrO_3)_2$
  (h) $(NH_4)_2HPO_4$
  (i) $NaClO$
  (j) $KMnO_4$

27. Write the chemical formula for these substances:
  (a) baking soda
  (b) lime
  (c) epsom salts
  (d) muriatic acid
  (e) vinegar
  (f) potash
  (g) lye

28. Write the chemical formula for these substances:
  (a) fool's gold
  (b) saltpeter
  (c) limestone
  (d) cane sugar
  (e) milk of magnesia
  (f) washing soda
  (g) grain alcohol

Additional Exercises

29. Name these compounds:
  (a) $Ba(NO_3)_2$
  (b) $NaC_2H_3O_2$
  (c) PbI_2
  (d) $MgSO_4$
  (e) $CdCrO_4$
  (f) $BiCl_3$
  (g) NiS
  (h) $Sn(NO_3)_4$
  (i) $Ca(OH)_2$

30. State how each of the following is used in naming inorganic compounds: *ide, ous, ic, hypo, per, ite, ate,* Roman numerals.

31. Translate the following sentences into unbalanced formula chemical equations:
  (a) Silver nitrate and sodium chloride react to form silver chloride and sodium nitrate.
  (b) Iron(III) sulfate and calcium hydroxide react to form iron(III) hydroxide and calcium sulfate.
  (c) Potassium hydroxide and sulfuric acid react to form potassium sulfate and water.

32. How many of each type of subatomic particle (protons and electrons) is in
  (a) an atom of tin?
  (b) an Sn^{2+} ion?
  (c) an Sn^{4+} ion?

33. The compound X_2Y_3 is a stable solid. What ionic charge do you expect for X and Y? Explain.

*34. The ferricyanide ion has the formula $Fe(CN)_6^{3-}$. Write the formula for the compounds that ferricyanide would form with the cations of elements 3, 13, and 30.

35. Compare and contrast the formulas of
  (a) nitride with nitrite
  (b) nitrite with nitrate
  (c) nitrous acid with nitric acid

36. In the beaker below there is a pool of ions. Write all possible formulas for ionic compounds that could form using these ions. Write the name of each compound next to its formula.

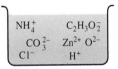

Answers to Practice Exercises

6.1 (a) KF; (b) $CaBr_2$; (c) Mg_3N_2; (d) Na_2S; (e) BaO.
6.2 (a) $SrCl_2$; (b) KI; (c) AlN; (d) CaS; (e) Na_2O.
6.3 (a) lead(II) iodide; (b) tin(IV) fluoride; (c) iron(III) oxide; (d) copper(II) oxide.
6.4 (a) dichlorine monoxide; (b) sulfur dioxide; (c) carbon tetrabromide; (d) dinitrogen pentoxide; (e) ammonia.
6.5 (a) potassium bromide; (b) calcium nitride; (c) sulfur trioxide; (d) tin(II) fluoride; (e) copper(II) chloride; (f) dinitrogen tetroxide.

6.6 (a) sodium nitrate; (b) calcium phosphate; (c) potassium hydroxide; (d) lithium carbonate; (e) sodium chlorate.
6.7 (a) copper(I) carbonate; (b) iron(III) hypochlorite; (c) tin(II) acetate; (d) bromic acid.
6.8 (a) $Pb(NO_3)_2$; (b) K_3PO_4; (c) $Hg(CN)_2$; (d) $(NH_4)_2CrO_4$.

PUTTING IT TOGETHER
Review for Chapters 5–6

Multiple Choice: *Choose the correct answer to each of the following.*

1. The concept of positive charge and a small, "heavy" nucleus surrounded by electrons was the contribution of
   (a) Dalton   (c) Thomson
   (b) Rutherford   (d) Chadwick

2. The neutron was discovered in 1932 by
   (a) Dalton   (c) Thomson
   (b) Rutherford   (d) Chadwick

3. An atom of atomic number 53 and mass number 127 contains how many neutrons?
   (a) 53   (c) 127
   (b) 74   (d) 180

4. How many electrons are in an atom of $^{40}_{18}Ar$?
   (a) 20   (c) 40
   (b) 22   (d) no correct answer given

5. The number of neutrons in an atom of $^{139}_{56}Ba$ is
   (a) 56   (c) 139
   (b) 83   (d) no correct answer given

6. The name of the isotope containing one proton and two neutrons is
   (a) protium   (c) deuterium
   (b) tritium   (d) helium

7. Each atom of a specific element has the same
   (a) number of protons
   (b) atomic mass
   (c) number of neutrons
   (d) no correct answer given

8. Which pair of symbols represents isotopes?
   (a) $^{23}_{11}Na$ and $^{23}_{12}Na$ (c) $^{63}_{29}Cu$ and $^{29}_{64}Cu$
   (b) $^{7}_{3}Li$ and $^{6}_{3}Li$ (d) $^{12}_{24}Mg$ and $^{12}_{26}Mg$

9. Two naturally occurring isotopes of an element have masses and abundance as follows: 54.00 amu (20.00%) and 56.00 amu (80.00%). What is the relative atomic mass of the element?
   (a) 54.20   (c) 54.80
   (b) 54.40   (d) 55.60

10. Substance X has 13 protons, 14 neutrons, and 10 electrons. Determine its identity.
    (a) ^{27}Mg (c) $^{27}Al^{3+}$
    (b) ^{27}Ne (d) ^{27}Al

11. The mass of a chlorine atom is 5.90×10^{-23} g. How many atoms are in a 42.0-g sample of chlorine?
    (a) 2.48×10^{-21} (c) 1.40×10^{-24}
    (b) 7.12×10^{23} (d) no correct answer given

12. The number of neutrons in an atom of $^{108}_{47}Ag$ is
    (a) 47   (c) 155
    (b) 108   (d) no correct answer given

13. The number of electrons in an atom of $^{27}_{13}Al$ is
    (a) 13   (c) 27
    (b) 14   (d) 40

14. The number of protons in an atom of $^{65}_{30}Zn$ is
    (a) 65   (c) 30
    (b) 35   (d) 95

15. The number of electrons in the nucleus of an atom of $^{24}_{12}Mg$ is
    (a) 12   (c) 36
    (b) 24   (d) no correct answer given

Names and Formulas: *In which of the following is the formula correct for the name given?*

1. copper(II) sulfate, $CuSO_4$
2. ammonium hydroxide, NH_4OH
3. mercury(I) carbonate, $HgCO_3$
4. phosphorus triiodide, PI_3
5. calcium acetate, $Ca(C_2H_3O_2)_2$
6. hypochlorous acid, $HClO$
7. dichlorine heptoxide, Cl_2O_7
8. magnesium iodide, MgI
9. sulfurous acid, H_2SO_3
10. potassium manganate, $KMnO_4$
11. lead(II) chromate, $PbCrO_4$
12. ammonium hydrogen carbonate, NH_4HCO_3
13. iron(II) phosphate, $FePO_4$
14. calcium hydrogen sulfate, $CaHSO_4$
15. mercury(II) sulfate, $HgSO_4$
16. dinitrogen pentoxide, N_2O_5
17. sodium hypochlorite, $NaClO$
18. sodium dichromate, $Na_2Cr_2O_7$
19. cadmium cyanide, $Cd(CN)_2$
20. bismuth(III) oxide, Bi_3O_2
21. carbonic acid, H_2CO_3
22. silver oxide, Ag_2O
23. ferric iodide, FeI_2

24. tin(II) fluoride, TiF_2
25. carbon monoxide, CO
26. phosphoric acid, H_3PO_3
27. sodium bromate, Na_2BrO_3
28. hydrosulfuric acid, H_2S
29. potassium hydroxide, POH
30. sodium carbonate, Na_2CO_3
31. zinc sulfate, $ZnSO_3$
32. sulfur trioxide, SO_3
33. tin(IV) nitrate, $Sn(NO_3)_4$
34. ferrous sulfate, $FeSO_4$
35. chloric acid, HCl
36. aluminum sulfide, Al_2S_3
37. cobalt(II) chloride, $CoCl_2$
38. acetic acid, $HC_2H_3O_2$
39. zinc oxide, ZnO_2
40. stannous fluoride, SnF_2

Free Response Questions: *Answer each of the following. Be sure to include your work and explanations in a clear, logical form.*

1. (a) What is an ion?
   (b) The average mass of a calcium atom is 40.08 amu. Why do we also use 40.08 amu as the average mass of a calcium ion (Ca^{2+})?

2. Congratulations! You discover a new element you name wyzzlebium (Wz). The average atomic mass of Wz was found to be 303.001 amu and its atomic number is 120.
   (a) If the masses of the two isotopes of wyzzlebium are 300.9326 amu and 303.9303 amu, what is the relative abundance of each isotope?
   (b) What are the isotopic notations of the two isotopes? (e.g., $_Z^A E$)
   (c) How many neutrons are in one atom of the more abundant isotope?

3. How many protons are in one molecule of dichlorine heptoxide? Is it possible to determine precisely how many electrons and neutrons are in a molecule of dichlorine heptoxide? Why or why not?

4. An unidentified metal forms an ionic compound with phosphate. The metal forms a 2+ cation. If the minimum ratio of protons in the metal to the phosphorus is 6:5, what metal is it? (*Hint:* First write the formula for the ionic compound formed with phosphate anion?)

5. For each of the following compounds, indicate what is wrong with the name and why. If possible, fix the name.
   (a) iron hydroxide
   (b) dipotassium dichromium heptoxide
   (c) sulfur oxide

6. Sulfur dioxide is a gas formed as a by-product of burning coal. Sulfur trioxide is a significant contributor to acid rain. Does the existence of these two substances violate the law of multiple proportions? Explain.

7. (a) Which subatomic particles are not in the nucleus?
   (b) What happens to the size of an atom when it becomes an anion?
   (c) What do an ion of Ca and an atom of Ar have in common?

8. An unidentified atom is found to have an atomic mass 7.18 times that of the carbon-12 isotope.
   (a) What is the mass of the unidentified atom?
   (b) What are the possible identities of this atom?
   (c) Why are you unable to positively identify the element based on the atomic mass and the periodic table?
   (d) If the element formed a compound M_2O, where M is the unidentified atom, identify M by writing the isotopic notation for the atom.

9. Scientists such as Dalton, Thomson, and Rutherford proposed important models, which were ultimately challenged by later technology. What do we know to be false in Dalton's atomic model? What was missing in Thomson's model of the atom? What was Rutherford's experiment that led to the current model of the atom?

CHAPTER 7 / OUTLINE

CHAPTER 7

Cereals, cleaning products, and pain remedies all list their ingredients on the package label. The ingredients are listed in order from most to least but the amounts are rarely given. However, it is precisely these amounts that give products their desired properties and distinguish them from the competition. Understandably, manufacturers carefully regulate the amounts of ingredients to maintain quality and hopefully their customers' loyalty. In the medicines we purchase these quantities are especially important for safety reasons—for example, they determine whether a medicine is given to children or is safe only for adults.

The composition of compounds is an important concept in chemistry. Determining numerical relationships among the elements in compounds and measuring exact quantities of particles are fundamental tasks that chemists routinely perform in their daily work.

7.1   The Mole

The atom is an incredibly tiny object. Its mass is far too small to measure on an ordinary balance. In Chapter 5 (Section 5.11), we learned to compare atoms using a table of atomic mass units. These units are valuable when we compare the masses of individual atoms (mentally), but they have no practical use in the laboratory. The mass in grams for an "average" carbon atom (atomic mass 12.00 amu) would be 2.00×10^{-23} g, which is much too tiny for the best laboratory balance.

So how can we confidently measure masses for these very tiny atoms? We increase the number of atoms in a sample until we have an amount large enough to measure on a laboratory balance. The problem then is how to count our sample of atoms.

Consider for a moment the produce in a supermarket. Frequently apples and oranges are sorted by size and then sold by weight, not by the piece of fruit. The grocer is counting by weighing. To do this he needs to know the mass of an "average" apple (235 g) and the mass of an "average" orange (186 g). Now suppose he has an order from the local college for 275 apples and 350 oranges. It would take a long time to count and package this order. The grocer can quickly count fruit by weighing—that is,

$$(275 \text{ apples})\left(\frac{235 \text{ g}}{\text{apple}}\right) = 6.46 \times 10^4 \text{ g} = 64.6 \text{ kg}$$

$$(350 \text{ oranges})\left(\frac{186 \text{ g}}{\text{orange}}\right) = 6.51 \times 10^4 \text{ g} = 65.1 \text{ kg}$$

He can now weigh 64.6 kg of apples and 65.1 kg of oranges and pack them without actually counting them. Manufacturers and suppliers often count by weighing. Other examples of counting by weighing include nuts, bolts, and candy.

Chemists also count atoms by weighing. We know the "average" masses of atoms so we can count atoms by defining a unit to represent a larger number of

▲
Oranges can be "counted" by weighing them in the store.

Units of measurement need ▶
to be appropriate for the
object being measured. (a)
Eggs are measured by the
dozen, (b) paper is measured
by the ream (500 sheets), and
(c) pencils are measured by
the gross (144).

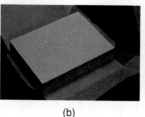
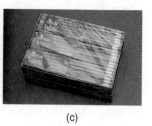

(a)   (b)   (c)

atoms. Chemists have chosen the mole (mol) as the unit for counting atoms. The mole is a unit for counting just as a dozen or a ream or a gross is used to count:

1 dozen = 12 objects
1 ream = 500 objects
1 gross = 144 objects
1 mole = 6.022×10^{23} objects

Note that we use a unit only when it is appropriate. A dozen eggs is practical in our kitchen, a gross might be practical for a restaurant, but a ream of eggs would not be very practical. Chemists can't use dozens, grosses, or reams because atoms are so tiny that a dozen, gross, or ream of atoms still couldn't be measured in the laboratory.

Avogadro's number

The number represented by 1 mol, 6.022×10^{23}, is called **Avogadro's number** in honor of Amadeo Avogadro (1776–1856), who investigated several quantitative aspects in chemistry. It's difficult to imagine how large Avogadro's number really is, but this example may help: If 10,000 people started to count Avogadro's number and each counted at the rate of 100 numbers per minute each minute of the day, it would take them over 1 trillion (10^{12}) years to count the total number. Even the tiniest amount of matter contains extremely large numbers of atoms.

Remember that ^{12}C is the
reference isotope for atomic
masses.

Avogadro's number has been experimentally determined by several methods. How does it relate to atomic mass units? Remember that the atomic mass for an element is the average relative mass of all the isotopes for the elements. The atomic mass (expressed in grams) of 1 mole of any element contains the same number of particles (Avogadro's number) as there are in exactly 12 g of ^{12}C. Thus 1 **mole** of anything is the amount of the substance that contains the same number of items as there are atoms in exactly 12 g of ^{12}C.

mole

$$1 \text{ mole} = 6.022 \times 10^{23} \text{ items}$$

From the definition of mole, we can see that the atomic mass in grams of any element contains 1 mol of atoms. The term *mole* is so commonplace in chemistry that chemists use it as freely as the words *atom* or *molecule*. A mole of atoms, molecules, ions, or electrons represent Avogadro's number of these particles. If we can speak of a mole of atoms, we can also speak of a mole of molecules, a mole of electrons, or a mole of ions, understanding that in each case we mean 6.022×10^{23} particles.

1 mol of atoms = 6.022×10^{23} atoms
1 mol of molecules = 6.022×10^{23} molecules
1 mol of ions = 6.022×10^{23} ions

The atomic mass of an element in grams contains Avogadro's number of atoms and is defined as the **molar mass** of that element. To determine molar mass, we change the units of the atomic mass (found in the periodic table) from atomic mass units to grams. For example, sulfur has an atomic mass of 32.07 amu, so 1 mol of sulfur has a molar mass of 32.07 g and contains 6.022×10^{23} atoms of sulfur. Here are some other examples:

molar mass

Element	Atomic mass	Molar mass	Number of atoms
H	1.008 amu	1.008 g	6.022×10^{23}
Mg	24.31 amu	24.31 g	6.022×10^{23}
Na	22.99 amu	22.99 g	6.022×10^{23}

To summarize:

1. The atomic mass expressed in grams is the *molar mass* of an element. It is different for each element. In this text, molar masses are expressed to four significant figures.

   1 molar mass = atomic mass of element in grams

2. One mole of any element contains Avogadro's number of atoms.

   1 mol of atoms = 6.022×10^{23} atoms

We can use these relationships to make conversions between number of atoms, mass, and moles, as shown in the following examples.

How many moles of iron does 25.0 g of iron, Fe, represent?

Example 7.1

Solution

We need to change grams of Fe to moles of Fe. The atomic mass of Fe (from either the periodic table or the table of atomic masses) is 55.85. Use the proper conversion factor to obtain moles:

grams Fe $\longrightarrow$ moles Fe $(\text{grams Fe})\left(\dfrac{1 \text{ mol Fe}}{55.85 \text{ g Fe}}\right)$

$(25.0 \text{ g Fe})\left(\dfrac{1 \text{ mol Fe}}{55.85 \text{ g Fe}}\right) = 0.448 \text{ mol Fe}$

How many magnesium atoms are contained in 5.00 g of Mg?

Example 7.2

Solution

We need to change grams of Mg to atoms of Mg.

grams Mg $\longrightarrow$ atoms Mg

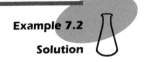

We find the atomic mass of magnesium to be 24.31 and set up the calculation using a conversion factor between atoms and grams:

$$(\text{grams Mg})\left(\frac{6.022 \times 10^{23} \text{ atoms Mg}}{24.31 \text{ g Mg}}\right)$$

$$(5.00 \text{ g Mg})\left(\frac{6.022 \times 10^{23} \text{ atoms Mg}}{24.31 \text{ g Mg}}\right) = 1.24 \times 10^{23} \text{ atoms Mg}$$

Alternatively we could first convert grams of Mg to moles of Mg, which are then changed to atoms of Mg.

$$\text{grams Mg} \longrightarrow \text{moles Mg} \longrightarrow \text{atoms Mg}$$

Use conversion factors for each step. The calculation setup is

$$(5.00 \text{ g Mg})\left(\frac{1 \text{ mol Mg}}{24.31 \text{ g Mg}}\right)\left(\frac{6.022 \times 10^{23} \text{ atoms Mg}}{1 \text{ mol Mg}}\right) = 1.24 \times 10^{23} \text{ atoms Mg}$$

Thus 1.24×10^{23} atoms of Mg are contained in 5.00 g of Mg.

Example 7.3

Solution

What is the mass, in grams, of one atom of carbon, C?

The molar mass of C is 12.01 g. First we create a conversion factor between molar mass and atoms:

$$\text{atoms C} \longrightarrow \text{grams C}$$

$$(\text{atoms C})\left(\frac{12.01 \text{ g C}}{6.022 \times 10^{23} \text{ atoms C}}\right)$$

$$(1 \text{ atom C})\left(\frac{12.01 \text{ g C}}{6.022 \times 10^{23} \text{ atoms C}}\right) = 1.994 \times 10^{-23} \text{ g C}$$

Example 7.4

Solution

What is the mass of 3.01×10^{23} atoms of sodium, Na?

We need the molar mass of Na (22.99 g) and a conversion factor between molar mass and atoms.

$$\text{atoms Na} \longrightarrow \text{grams Na}$$

$$(\text{atoms Na})\left(\frac{22.99 \text{ g Na}}{6.022 \times 10^{23} \text{ atoms Na}}\right)$$

$$(3.01 \times 10^{23} \text{ atoms Na})\left(\frac{22.99 \text{ g Na}}{6.022 \times 10^{23} \text{ atoms Na}}\right) = 11.5 \text{ g Na}$$

Example 7.5

Solution

What is the mass of 0.252 mol of copper (Cu)?

We need the molar mass of Cu (63.55 g) and a conversion factor between molar mass and moles.

$$\text{moles Cu} \longrightarrow \text{grams Cu}$$

$$(\text{moles Cu})\left(\frac{1 \text{ molar mass Cu}}{1 \text{ mol Cu}}\right)$$

$$(0.252 \text{ mol Cu})\left(\frac{63.55 \text{ g Cu}}{1 \text{ mol Cu}}\right) = 16.0 \text{ g Cu}$$

How many oxygen atoms are present in 1.00 mol of oxygen molecules?

Example 7.6

Oxygen is a diatomic molecule with the formula O_2. Therefore a molecule of oxygen contains two oxygen atoms.

Solution

$$\frac{2 \text{ atoms O}}{1 \text{ molecule O}_2}$$

The sequence of conversions is

$$\text{moles O}_2 \longrightarrow \text{molecules O}_2 \longrightarrow \text{atoms O}$$

The two conversion factors needed are

$$\frac{6.022 \times 10^{23} \text{ molecules O}_2}{1 \text{ mol O}_2} \quad \text{and} \quad \frac{2 \text{ atoms O}}{1 \text{ molecule O}_2}$$

The calculation is

$$(1.00 \text{ mol O}_2)\left(\frac{6.022 \times 10^{23} \text{ molecules O}_2}{1 \text{ mol O}_2}\right)\left(\frac{2 \text{ atoms O}}{1 \text{ molecules O}_2}\right)$$

$$= 1.20 \times 10^{24} \text{ atoms O}$$

Practice 7.1

What is the mass of 2.50 mol of helium, He?

Practice 7.2

How many atoms are present in 0.025 mol of iron?

7.2 Molar Mass of Compounds

One mole of a compound contains Avogadro's number of *formula units* of that compound. The terms *molecular weight, molecular mass, formula weight,* and *formula mass* have been used in the past to refer to the mass of 1 mol of a compound. However, the term *molar mass* is more inclusive, because it can be used for all types of compounds.

A formula unit is indicated by the formula, e.g., Mg, MgS, H_2O, NaCl.

If the formula of a compound is known, its molar mass can be determined by adding the molar masses of all the atoms in the formula. If more than one atom of any element is present, its mass must be added as many times as it is used.

A mole of table salt (in ▶ front of a salt shaker) and a mole of water (in the film container) have different sizes but both contain Avogadro's number of formula units.

Example 7.7   The formula for water is H_2O. What is its molar mass?

Solution   First we look up the molar masses of H (1.008 g) and O (16.00 g), then we add the masses of all the atoms in the formula unit. Water contains two H atoms and one O atom. Thus

$$2\ H = 2(1.008\ g) = 2.016\ g$$

$$1\ O = 1(16.00\ \ g) = \underline{16.00\ \ g}$$

$$18.02\ \ g = \text{molar mass of } H_2O$$

Example 7.8   Calculate the molar mass of calcium hydroxide, $Ca(OH)_2$.

Solution   The formula of this substance contains one atom of Ca and two atoms each of O and H. We proceed as in Example 7.7:

$$1\ Ca = 1(40.08\ g) = 40.08\ \ g$$

$$2\ O = 2(16.00\ g) = 32.00\ \ g$$

$$2\ H = 2(1.008\ g) = \underline{2.016\ g}$$

$$74.10\ \ g = \text{molar mass of } Ca(OH)_2$$

In this text we round all molar masses to four significant figures, although you may need to use a different number of significant figures for other work (in the lab).

Practice 7.3

Calculate the molar mass of KNO_3.

The mass of 1 mol of a compound contains Avogadro's number of formula units. Consider the compound hydrogen chloride, HCl. One atom of H combines with one atom of Cl to form HCl. When 1 mol of H (1.008 g of H or 6.022×10^{23} H atoms) combines with 1 mol of Cl (35.45 g of Cl or 6.022×10^{23} Cl atoms), 1 mol of HCl (36.46 g of HCl or 6.022×10^{23} HCl molecules) is produced. These relationships are summarized in the following table:

H	Cl	HCl
6.022×10^{23} H *atoms*	6.022×10^{23} Cl *atoms*	6.022×10^{23} HCl *molecules*
1 mol H *atoms*	1 mol Cl *atoms*	1 mol HCl *molecules*
1.008 g H	35.45 g Cl	36.46 g HCl
1 molar mass H *atoms*	1 molar mass Cl *atoms*	1 molar mass HCl

In dealing with diatomic elements (H_2, O_2, N_2, F_2, Cl_2, Br_2, and I_2), we must take special care to distinguish between a mole of atoms and a mole of molecules. For example consider 1 mol of oxygen molecules, which has a mass of 32.00 g. This quantity is equal to 2 mol of oxygen atoms. Remember that 1 mol represents Avogadro's number of the particular chemical entity that is under consideration.

$$1 \text{ mol } H_2O = 18.02 \text{ g } H_2O = 6.022 \times 10^{23} \text{ molecules}$$

$$1 \text{ mol NaCl} = 58.44 \text{ g NaCl} = 6.022 \times 10^{23} \text{ formula units}$$

$$1 \text{ mol } H_2 = 2.016 \text{ g } H_2 = 6.022 \times 10^{23} \text{ molecules}$$

$$1 \text{ mol } HNO_3 = 63.02 \text{ g } HNO_3 = 6.022 \times 10^{23} \text{ molecules}$$

$$1 \text{ mol } K_2SO_4 = 174.3 \text{ g } K_2SO_4 = 6.022 \times 10^{23} \text{ formula units}$$

> **Formula units are often used in place of molecules for substances that contain ions.**

> **$1 \text{ mol} = 6.022 \times 10^{23}$ formula units or molecules**
> **$= 1$ molar mass of a compound**

> **Create the appropriate conversion factor by placing the unit desired in the numerator and the unit to be eliminated in the denominator.**

What is the mass of 1 mol of sulfuric acid, H_2SO_4?

We look up the masses of hydrogen, sulfur, and oxygen and solve in a manner similar to Examples 7.7 and 7.8.

Example 7.9

Solution

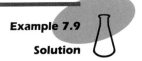

$$2 \text{ H} = 2(\ 1.008 \text{ g}) = \ 2.016 \text{ g}$$

$$1 \text{ S} = 1(32.07 \ \text{ g}) = 32.07 \ \text{ g}$$

$$4 \text{ O} = 4(16.00 \ \text{ g}) = \underline{64.00 \ \text{ g}}$$

$$98.09 \ \text{ g} = \text{mass of 1 mol of } H_2SO_4$$

Example 7.10 How many moles of sodium hydroxide, NaOH, are there in 1.00 kg of sodium hydroxide?

Solution   First we know that

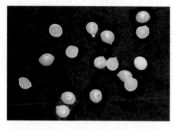

$$\overset{Na}{} \quad \overset{O}{} \quad \overset{H}{}$$
$$molar\ mass = (22.99\ g + 16.00\ g + 1.008\ g)\ or\ 40.00\ g\ NaOH$$

To convert grams to moles we use the conversion factor $\dfrac{1\ mol\ NaOH}{40.00\ g\ NaOH}$ and this conversion sequence:

$$kg\ NaOH \longrightarrow g\ NaOH \longrightarrow mol\ NaOH$$

The calculation is

▲
Sodium hydroxide pellets gain moisture from the atmosphere, making them difficult to weigh.

$$(1.00\ kg\ \cancel{NaOH})\left(\frac{1000\ g\ \cancel{NaOH}}{kg\ \cancel{NaOH}}\right)\left(\frac{1\ mol\ NaOH}{40.00\ g\ \cancel{NaOH}}\right) = 25.0\ mol\ NaOH$$

$$1.00\ kg\ NaOH = 25.0\ mol\ NaOH$$

Example 7.11 What is the mass of 5.00 mol of water?

Solution   First we know that

$$1\ mol\ H_2O = 18.02\ g\ (Example\ 7.7)$$

The conversion is

$$mol\ H_2O \longrightarrow g\ H_2O$$

To convert moles to grams, we use the conversion factor $\dfrac{18.02\ g\ H_2O}{1\ mol\ H_2O}$

The calculation is

$$(5.00\ mol\ \cancel{H_2O})\left(\frac{18.02\ g\ H_2O}{1\ mol\ \cancel{H_2O}}\right) = 90.1\ g\ H_2O$$

Example 7.12 How many molecules of hydrogen chloride, HCl, are there in 25.0 g of hydrogen chloride?

Solution   From the formula, we find that the molar mass of HCl is 36.46 g ($\overset{H}{1.008}\,g + \overset{Cl}{35.45}\,g$). The sequence of conversions is

$$g\ HCl \longrightarrow mol\ HCl \longrightarrow molecules\ HCl$$

Using the conversion factors

$$\frac{1\ mol\ HCl}{36.46\ g\ HCl} \quad and \quad \frac{6.022 \times 10^{23}\ molecules\ HCl}{1\ mol\ HCl}$$

$$(25.0 \text{ g HCl})\left(\frac{1 \text{ mol HCl}}{36.46 \text{ g HCl}}\right)\left(\frac{6.022 \times 10^{23} \text{ molecules HCl}}{1 \text{ mol HCl}}\right)$$

$$= 4.13 \times 10^{23} \text{ molecules HCl}$$

Practice 7.4

What is the mass of 0.150 mol of Na_2SO_4?

Practice 7.5

How many moles are there in 500.0 g of $AlCl_3$?

7.3   Percent Composition of Compounds

Percent means parts per 100 parts. Just as each piece of pie is a percent of the whole pie, each element in a compound is a percent of the whole compound. The **percent composition of a compound** is the *mass percent* of each element in the compound. The molar mass represents the total mass, or 100%, of the compound. Thus the percent composition of water, H_2O, is 11.19% H and 88.79% O by mass. According to the Law of Definite Composition, the percent composition must be the same no matter what size sample is taken.

The percent composition of a compound can be determined (1) from knowing its formula or (2) from experimental data.

percent composition of a compound

Percent Composition from Formula

If the formula is known, a two-step process is needed to determine the percent composition.

Step 1. Calculate the molar mass (Section 7.2).

Step 2. Divide the total mass of each element in the formula by the molar mass and multiply by 100. This gives the percent composition.

$$\frac{\text{total mass of the element}}{\text{molar mass}} \times 100 = \text{percent of the element}$$

Calculate the percent composition of sodium chloride, NaCl.

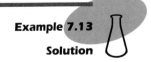

Example 7.13

Solution

Step 1. Calculate the molar mass of NaCl:

$$1 \text{ Na} = 1(22.99 \text{ g}) = 22.99 \text{ g}$$

$$1 \text{ Cl} = 1(35.45 \text{ g}) = \underline{35.45 \text{ g}}$$

$$58.44 \text{ g (molar mass)}$$

Step 2. Now calculate the percent composition. We know there are 22.99 g Na and 35.45 g Cl in 58.44 g NaCl.

$$\text{Na:} \left(\frac{22.99 \text{ g Na}}{58.44 \text{ g}} \right)(100) = 39.34\% \text{ Na}$$

$$\text{Cl:} \left(\frac{35.45 \text{ g Cl}}{58.44 \text{ g}} \right)(100) = \underline{60.66\% \text{ Cl}}$$

$$100.00\% \text{ total}$$

In any two-component system, if the percent of one component is known, the other is automatically defined by the difference; that is, if Na is 39.34%, then Cl is $100\% - 39.34\% = 60.66\%$. However, the calculation of the percent of each component should be carried out, since this provides a check against possible error. The percent composition data should add up to $100 \pm 0.2\%$.

Example 7.14   Calculate the percent composition of potassium chloride, KCl.

Solution

Step 1. Calculate the molar mass of KCl:

$$1 \text{ K } = 1(39.10 \text{ g}) = 39.10 \text{ g}$$
$$1 \text{ Cl} = 1(35.45 \text{ g}) = \underline{35.45 \text{ g}}$$
$$74.55 \text{ g (molar mass)}$$

Step 2. Now calculate the percent composition. We know there are 39.10 g K and 35.45 g Cl in 74.55 g KCl.

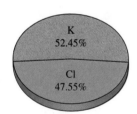

$$\text{K:} \left(\frac{39.10 \text{ g K}}{74.55 \text{ g}} \right)(100) = \quad 52.45\% \text{ K}$$

$$\text{Cl:} \left(\frac{35.45 \text{ g Cl}}{74.55 \text{ g}} \right)(100) = \quad \underline{47.55\% \text{ Cl}}$$

$$100.00\% \text{ total}$$

Comparing the results calculated for NaCl and for KCl, we see that NaCl contains a higher percentage of Cl by mass, although each compound has a one-to-one atom ratio of Cl to Na and Cl to K. The reason for this mass percent difference is that Na and K do not have the same atomic masses.

It is important to realize that, when we compare 1 mol of NaCl with 1 mol of KCl, each quantity contains the same number of Cl atoms—namely, 1 mol of Cl atoms. However, if we compare equal masses of NaCl and KCl, there will be more Cl atoms in the mass of NaCl since NaCl has a higher mass percent of Cl.

1 mol NaCl contains	100.00 g NaCl contains
1 mol Na	39.34 g Na
1 mol Cl	60.66 g Cl
	60.66%Cl

1 mol KCl contains	100.00 g KCl contains
1 mol K	52.45 g K
1 mol Cl	47.55 g Cl
	47.55%Cl

Calculate the percent composition of potassium sulfate, K_2SO_4.

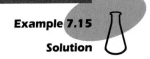

Example 7.15

Solution

Step 1. Calculate the molar mass of K_2SO_4:

$$2\ K = 2(39.10\ g) = \quad 78.20\ g$$

$$1\ S = 1(32.07\ g) = \quad 32.07\ g$$

$$4\ O = 4(16.00\ g) = \quad \underline{64.00\ g}$$

$$\qquad\qquad\qquad\qquad 174.3\quad g\ (molar\ mass)$$

Step 2. Now calculate the percent composition. We know there are 78.20 g of K, 32.07 g of S, and 64.00 g of O in 174.3 g of K_2SO_4.

$$K:\left(\frac{78.20\ g\ K}{174.3\ g}\right)(100) = 44.87\%\ K$$

$$S:\left(\frac{32.07\ g\ S}{174.3\ g}\right)(100) = 18.40\%\ S$$

$$O:\left(\frac{64.00\ g\ O}{174.3\ g}\right)(100) = \underline{36.72\%\ O}$$

$$\qquad\qquad\qquad\qquad\qquad 99.99\%\ total$$

Practice 7.6

Calculate the percent composition of $Ca(NO_3)_2$.

Practice 7.7

Calculate the percent composition of K_2CrO_4.

Percent Composition from Experimental Data

The percent composition can be determined from experimental data without knowing the formula of a compound.

Step 1. Calculate the mass of the compound formed.

Step 2. Divide the mass of each element by the total mass of the compound and multiply by 100.

The Taste of Chemistry

Flavorings, seasonings, and preservatives have been added to foods since ancient times. Spices originally served as preservatives when refrigeration was unavailable. These spices contained mild antiseptics and antioxidants and were effective in prolonging the time during which food could be eaten. Now, a wide variety of food additives are used—preservatives, colorings, flavorings, antioxidants, sweeteners, and so on. Many of these substances are regarded as virtual necessities for processing foods. However, there is genuine concern that some additives, particularly those not naturally present in foods, may be detrimental to our health. This has led to the passage of federal and state laws and regulations.

The U.S. Food and Drug Administration (FDA) is the principal agency charged with enforcing federal laws concerning food and must approve the use of all food additives. The FDA divides food additives into several cate-gories. These include a classification known as GRAS (Generally Regarded As Safe). The GRAS designation includes substances that were in use in 1958 and that met certain specifications for safety. All substances introduced after 1958 have been approved on an individual basis.

Before commercial use of a new food additive, a company must provide the FDA with satisfactory evidence that the chemical is safe for the proposed usage. Providing this evidence of safety is complex and expensive. This task requires the services of many people trained in a variety of disciplines—biochemistry, microbiology, medicine, physiology, and so on. Research and testing may have to be done in several laboratories before final FDA approval (or rejection) is obtained.

The amount of an additive considered to be safe in foods is determined by the maximum tolerable daily intake (MTDI). This is the amount of the food additive that can be eaten daily for a lifetime without adverse effects. It is calculated on the basis of body mass (mg/kg/day). If there is any doubt regarding the safety of an additive, a time limit is determined, and a conditional MTDI is issued and reviewed, with further testing at the end of the initial time period. To establish the MTDI, experiments are run on animals, increasing the quantities of the additive in successive experiments until acute and chronic toxicity occurs. A minimum of two species of animals must be tested, the most sensitive species forming the basis for the appropriate level for the additive. This quantity is then divided by 100 (for additional safety) to set the MTDI.

Once the MTDI is established, all foods in which the use of the additive is proposed must be considered. An estimate is made of the maximum amount of the additive that could be ingested. On the basis of this information, the use may then be restricted to only certain foods (as in the exclusion of sulfites from most meats) or broadened to include additional foods.

Unfortunately, it is extremely difficult to determine safe levels of additives because the majority of toxicological data is from animal testing. Results in animals are often not the same as those in humans—chemicals toxic to one species may not be so for humans.

Children add a further complication. They cannot be considered simply as small adults. A child has a much greater energy demand per kilogram than an adult, and children's chemical defense mechanisms are often much different than those of adults.

◀ **These are a few foods that contain additives.**

When heated in the air, 1.63 g of zinc, Zn, combines with 0.40 g of oxygen, O_2, to form zinc oxide. Calculate the percent composition of the compound formed.

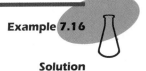

Example 7.16

Step 1. First calculate the total mass of the compound formed.

Solution

$$\begin{array}{l} 1.63 \text{ g Zn} \\ \underline{0.40 \text{ g } O_2} \\ 2.03 \text{ g} \quad = \text{total mass of product} \end{array}$$

Step 2. Divide the mass of each element by the total mass (2.03 g) and multiply by 100.

$$\left(\frac{1.63 \text{ g Zn}}{2.03 \text{ g}}\right)(100) = 80.3\% \text{ Zn} \qquad \left(\frac{0.40 \text{ g O}}{2.03 \text{ g}}\right)(100) = 19.7\% \text{ O}$$

The compound formed contains 80.3% Zn and 19.7% O.

Practice 7.8

Aluminum chloride is formed by reacting 13.43 g aluminum with 53.18 g chlorine. What is the percent composition of the compound?

7.4   Empirical Formula versus Molecular Formula

The **empirical formula,** or *simplest formula,* gives the smallest whole-number ratio of atoms present in a compound. This formula gives the relative number of atoms of each element in the compound.

empirical formula

The **molecular formula** is the true formula, representing the total number of atoms of each element present in one molecule of a compound. It is entirely possible that two or more substances will have the same percent composition yet be distinctly different compounds. For example, acetylene, C_2H_2, is a common gas used in welding; benzene, C_6H_6, is an important solvent obtained from coal tar and is used in the synthesis of styrene and nylon. Both acetylene and benzene contain 92.3% C and 7.7% H. The smallest ratio of C and H corresponding to these percentages is CH (1:1). Therefore the *empirical* formula for both acetylene and benzene is CH, even though the *molecular* formulas are C_2H_2 and C_6H_6, respectively. Often the molecular formula is the same as the empirical formula. If the molecular formula is not the same, it will be an integral (whole number) multiple of the empirical formula. For example,

molecular formula

CH = empirical formula

$(CH)_2 = C_2H_2$ = acetylene   (molecular formula)

$(CH)_6 = C_6H_6$ = benzene   (molecular formula)

Table 7.1 compares the formulas of these substances. Table 7.2 shows empirical and molecular formula relationships of other compounds.

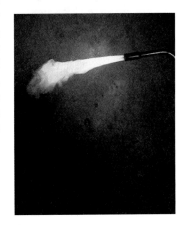

▲
Acetylene is used as a fuel in welding torches.

TABLE 7.1	Molecular Formulas of Two Compounds Having an Empirical Formula with a 1:1 Ratio of Carbon and Hydrogen Atoms		
	Composition		
Formula	**% C**	**% H**	**Molar mass**
CH (empirical)	92.3	7.7	13.02 (empirical)
C_2H_2 (acetylene)	92.3	7.7	26.04 (2×13.02)
C_6H_6 (benzene)	92.3	7.7	78.12 (6×13.02)

TABLE 7.2	Some Empirical and Molecular Formulas				
Compound	**Empirical formula**	**Molecular formula**	**Compound**	**Empirical formula**	**Molecular formula**
Acetylene	CH	C_2H_2	Diborane	BH_3	B_2H_6
Benzene	CH	C_6H_6	Hydrazine	NH_2	N_2H_4
Ethylene	CH_2	C_2H_4	Hydrogen	H	H_2
Formaldehyde	CH_2O	CH_2O	Chlorine	Cl	Cl_2
Acetic acid	CH_2O	$C_2H_4O_2$	Bromine	Br	Br_2
Glucose	CH_2O	$C_6H_{12}O_6$	Oxygen	O	O_2
Hydrogen chloride	HCl	HCl	Nitrogen	N	N_2
Carbon dioxide	CO_2	CO_2			

7.5   Calculating Empirical Formulas

We can establish empirical formulas because (1) individual atoms in a compound are combined in whole-number ratios, and (2) each element has a specific atomic mass.

To calculate an empirical formula, we need to know (1) the elements that are combined, (2) their atomic masses, and (3) the ratio by mass or percentage in which they are combined. If elements A and B form a compound, we may represent the empirical formula as A_xB_y, where x and y are small whole numbers that represent the atoms of A and B. To write the empirical formula we must determine x and y:

Step 1. Assume a definite starting quantity (usually 100.0 g) of the compound, if not given, and express the mass of each element in grams.

Step 2. Convert the grams of each element into moles using each element's molar mass. This conversion gives the number of moles of atoms of each element in the quantity assumed in Step 1. At this point these numbers will usually not be whole numbers.

Step 3. Divide each value obtained in Step 2 by the smallest of these values. If the numbers obtained are whole numbers, use them as subscripts and write the empirical formula. If the numbers obtained are not whole numbers, go on to Step 4.

Step 4. Multiply the values obtained in Step 3 by the smallest number that will convert them to whole numbers. Use these whole numbers as the subscripts in the empirical formula. For example, if the ratio of A to B is 1.0:1.5, multiply both numbers by 2 to obtain a ratio of 2:3. The empirical formula then is A_2B_3.

In many of these calculations results will vary somewhat from an exact whole number; this can be due to experimental errors in obtaining the data or from rounding off numbers. Calculations that vary by no more than ±0.1 from a whole number usually are rounded off to the nearest whole number. Deviations greater than about 0.1 unit usually mean that the calculated ratios need to be multiplied by a factor to make them all whole numbers. For example, an atom ratio of 1:1.33 should be multiplied by 3 to make the ratio 3:4.

Some common fractions and their decimal equivalents are
¼ = 0.25
⅓ = 0.333 . . .
⅔ = 0.666 . . .
½ = 0.5
¾ = 0.75
Multiply the decimal equivalent by the number in the denominator of the fraction to get a whole number: 4(0.75) = 3.

Calculate the empirical formula of a compound containing 11.19% hydrogen, H, and 88.79% oxygen, O.

Example 7.17

Solution

Step 1. Express each element in grams. Assuming 100.00 g of material, the percent of each element equals the grams of each element, and the percent sign can be omitted:

H = 11.19 g

O = 88.79 g

Step 2. Convert the grams of each element to moles:

H: $(11.19 \text{ g H})\left(\dfrac{1 \text{ mol H atoms}}{1.008 \text{ g H}}\right) = 11.10 \text{ mol H atoms}$

O: $(88.79 \text{ g O})\left(\dfrac{1 \text{ mol O atoms}}{16.00 \text{ g O}}\right) = 5.549 \text{ mol O atoms}$

The formula could be expressed as $H_{11.10}O_{5.549}$. However, it's customary to use the smallest whole-number ratio of atoms. This ratio is calculated in Step 3.

Step 3. Change these numbers to whole numbers by dividing them by the smaller number.

$H = \dfrac{11.10 \text{ mol}}{5.549 \text{ mol}} = 2.000 \qquad O = \dfrac{5.549 \text{ mol}}{5.549 \text{ mol}} = 1.000$

In this step the ratio of atoms has not changed, because we divided the number of moles of each element by the same number.

The simplest ratio of H to O is 2:1.
Empirical formula = H_2O

Example 7.18   The analysis of a salt shows that it contains 56.58% potassium, K; 8.68% carbon, C; and 34.73% oxygen, O. Calculate the empirical formula for this substance.

Solution

Steps 1 and 2. After changing the percentage of each element to grams, we find the relative number of moles of each element by multiplying by the proper mole/molar mass factor:

$$K: (56.58 \text{ g K})\left(\frac{1 \text{ mol K atoms}}{39.10 \text{ g K}}\right) = 1.447 \text{ mol K atoms}$$

$$C: (8.68 \text{ g C})\left(\frac{1 \text{ mol C atoms}}{12.01 \text{ g C}}\right) = 0.723 \text{ mol C atoms}$$

$$O: (34.73 \text{ g O})\left(\frac{1 \text{ mol O atoms}}{16.00 \text{ g O}}\right) = 2.171 \text{ mol O atoms}$$

Step 3. Divide each number of moles by the smallest value:

$$K = \frac{1.447 \text{ mol}}{0.723 \text{ mol}} = 2.00$$

$$C = \frac{0.723 \text{ mol}}{0.723 \text{ mol}} = 1.00$$

$$O = \frac{2.171 \text{ mol}}{0.723 \text{ mol}} = 3.00$$

The simplest ratio of K:C:O is 2:1:3.
Empirical formula = K_2CO_3

Example 7.19   A sulfide of iron was formed by combining 2.233 g of iron, Fe, with 1.926 g of sulfur, S. What is the empirical formula of the compound?

Solution

Steps 1 and 2. The grams of each element are given, so we use them directly in our calculations. Calculate the relative number of moles of each element by multiplying the grams of each element by the proper mole/molar mass factor:

$$Fe: (2.233 \text{ g Fe})\left(\frac{1 \text{ mol Fe atoms}}{55.85 \text{ g Fe}}\right) = 0.03998 \text{ mol Fe atoms}$$

$$S: (1.926 \text{ g S})\left(\frac{1 \text{ mol S atoms}}{32.07 \text{ g S}}\right) = 0.06006 \text{ mol S atoms}$$

Step 3. Divide each number of moles by the smaller of the two numbers:

$$Fe = \frac{0.03998 \text{ mol}}{0.03998 \text{ mol}} = 1.000 \qquad S = \frac{0.06006 \text{ mol}}{0.03998 \text{ mol}} = 1.502$$

Step 4. We still have not reached a ratio that gives whole numbers in the formula so we double each value to obtain a ratio of 2.000 Fe atoms to

3.000 S atoms. Doubling both values does not change the ratio of Fe and S atoms.

Fe: (1.000)2 = 2.000

S: (1.502)2 = 3.004

Empirical formula = Fe_2S_3

Practice 7.9

Calculate the empirical formula of a compound containing 52.14% C, 13.12% H, and 34.73% O.

Practice 7.10

Calculate the empirical formula of a compound that contains 43.7% phosphorus and 56.3% O by mass.

7.6 Calculating the Molecular Formula from the Empirical Formula

The molecular formula can be calculated from the empirical formula if the molar mass is known. The molecular formula, as stated in Section 7.4, will be equal either to the empirical formula or some multiple of it. For example, if the empirical formula of a compound of hydrogen and fluorine is HF, the molecular formula can be expressed as $(HF)_n$, where $n = 1, 2, 3, 4, \ldots$. This n means that the molecular formula could be HF, H_2F_2, H_3F_3, H_4F_4, and so on. To determine the molecular formula, we must evaluate n.

$$n = \frac{\text{molar mass}}{\text{mass of empirical formula}} = \text{number of empirical formula units}$$

What we actually calculate is the number of units of the empirical formula contained in the molecular formula.

Example 7.20

A compound of nitrogen and oxygen with a molar mass of 92.00 g was found to have an empirical formula of NO_2. What is its molecular formula?

Solution

Let n be the number of NO_2 units in a molecule; then the molecular formula is $(NO_2)_n$. Each NO_2 unit has a mass of [14.01 g + 2(16.00 g)] or 46.01 g. The molar mass of $(NO_2)_n$ is 92.00 g and the number of 46.01 units in 92.00 is 2:

$$n = \frac{92.00 \text{ g}}{46.01 \text{ g}} = 2 \quad \text{(empirical formula units)}$$

The molecular formula is $(NO_2)_2$ or N_2O_4.

Example 7.21   The hydrocarbon propylene has a molar mass of 42.00 g and contains 14.3% H and 85.7% C. What is its molecular formula?

Solution   First find the empirical formula:

$$C: \quad (85.7 \; \cancel{g \, C})\left(\frac{1 \; \text{mol C atoms}}{12.01 \; \cancel{g \, C}}\right) = 7.14 \; \text{mol C atoms}$$

$$H: \quad (14.3 \; \cancel{g \, H})\left(\frac{1 \; \text{mol H atoms}}{1.008 \; \cancel{g \, H}}\right) = 14.2 \; \text{mol H atoms}$$

Divide each value by the smaller number of moles:

$$C = \frac{7.14 \; \text{mol}}{7.14 \; \text{mol}} = 1.00$$

▲
When propylene is polymerized into polypropylene it can be used for rope and other recreational products.

$$H = \frac{14.2 \; \text{mol}}{7.14 \; \text{mol}} = 1.99$$

Empirical formula = CH_2

We determine the molecular formula from the empirical formula and the molar mass:

Molecular formula = $(CH_2)_n$

Molar mass = 42.00 g

Each CH_2 unit has a mass of (12.01 g + 2.016 g) or 14.03 g. The number of CH_2 units in 42.00 g is 3:

$$n = \frac{42.00 \; g}{14.03 \; g} = 3 \quad (\text{empirical formula units})$$

The molecular formula is $(CH_2)_3$, or C_3H_6.

Practice 7.11

Calculate the empirical and molecular formulas of a compound that contains 80.0% C, 20.0% H, and has a molar mass of 30.00 g.

Concepts in Review

1. Explain the meaning of the mole.

2. Discuss the relationship between a mole and Avogadro's number.

3. Convert grams, atoms, molecules, and molar masses to moles, and vice versa.

4. Determine the molar mass of a compound from its formula.

5. Calculate the percent composition of a compound from its formula.

6. Calculate the percent composition of a compound from experimental data.

7. Explain the relationship between an empirical formula and a molecular formula.

8. Determine the empirical formula for a compound from its percent composition.

9. Calculate the molecular formula of a compound from its percent composition and molar mass.

Key Terms

Avogadro's number (7.1)
empirical formula (7.4)
molar mass (7.1)
mole (7.1)
molecular formula (7.4)
percent composition of a compound (7.3)

Questions

1. What is a mole?

2. Which would have a higher mass: a mole of K atoms or a mole of Au atoms?

3. Which would contain more atoms: a mole of K atoms or a mole of Au atoms?

4. Which would contain more electrons: a mole of K atoms or a mole of Au atoms?

*5. If the atomic mass scale had been defined differently, with an atom of $^{12}_{6}C$ defined as a mass of 50 amu, would this have any effect on the value of Avogadro's number? Explain.

6. What is the numerical value of Avogadro's number?

7. What is the relationship between Avogadro's number and the mole?

8. Complete these statements, supplying the proper quantity.
   (a) A mole of O atoms contains _____ atoms.
   (b) A mole of O_2 molecules contains _____ molecules.
   (c) A mole of O_2 molecules contains _____ atoms.
   (d) A mole of O atoms has a mass of _____ grams.
   (e) A mole of O_2 molecules has a mass of _____ grams.

9. How many molecules are present in 1 molar mass of sulfuric acid (H_2SO_4)? How many atoms are present?

10. In calculating the empirical formula of a compound from its percent composition, why do we choose to start with 100.0 g of the compound?

Paired Exercises

Molar Masses

11. Determine the molar masses of these compounds:
   (a) KBr
   (b) Na_2SO_4
   (c) $Pb(NO_3)_2$
   (d) C_2H_5OH
   (e) $HC_2H_3O_2$
   (f) Fe_3O_4
   (g) $C_{12}H_{22}O_{11}$
   (h) $Al_2(SO_4)_3$
   (i) $(NH_4)_2HPO_4$

12. Determine the molar masses of these compounds:
   (a) NaOH
   (b) Ag_2CO_3
   (c) Cr_2O_3
   (d) $(NH_4)_2CO_3$
   (e) $Mg(HCO_3)_2$
   (f) C_6H_5COOH
   (g) $C_6H_{12}O_6$
   (h) $K_4Fe(CN)_6$
   (i) $BaCl_2 \cdot 2\, H_2O$

Moles and Avogadro's Number

13. How many moles of atoms are contained in the following?
   (a) 22.5 g Zn
   (b) 0.688 g Mg
   (c) 4.5×10^{22} atoms Cu
   (d) 382 g Co
   (e) 0.055 g Sn
   (f) 8.5×10^{24} molecules N_2

14. How many moles are contained in the following?
   (a) 25.0 g NaOH
   (b) 44.0 g Br_2
   (c) 0.684 g $MgCl_2$
   (d) 14.8 g CH_3OH
   (e) 2.88 g Na_2SO_4
   (f) 4.20 lb ZnI_2

15. Calculate the number of grams in each of the following:
   (a) 0.550 mol Au
   (b) 15.8 mol H_2O
   (c) 12.5 mol Cl_2
   (d) 3.15 mol NH_4NO_3

16. Calculate the number of grams in each of the following:
   (a) 4.25×10^{-4} mol H_2SO_4
   (b) 4.5×10^{22} molecules CCl_4
   (c) 0.00255 mol Ti
   (d) 1.5×10^{16} atoms S

17. How many molecules are contained in each of the following:
    (a) 1.26 mol O_2
    (b) 0.56 mol C_6H_6
    (c) 16.0 g CH_4
    (d) 1000. g HCl

18. How many molecules are contained in each of the following:
    (a) 1.75 mol Cl_2
    (b) 0.27 mol C_2H_6O
    (c) 12.0 g CO_2
    (d) 100. g CH_4

19. Calculate the mass in grams of each of the following:
    (a) 1 atom Pb
    (b) 1 atom Ag
    (c) 1 molecule H_2O
    (d) 1 molecule $C_3H_5(NO_3)_3$

20. Calculate the mass in grams of each of the following:
    (a) 1 atom Au
    (b) 1 atom U
    (c) 1 molecule NH_3
    (d) 1 molecule $C_6H_4(NH_2)_2$

21. Make the following conversions:
    (a) 8.66 mol Cu to grams Cu
    (b) 125 mol Au to kilograms Au
    (c) 10 atoms C to moles C
    (d) 5000 molecules CO_2 to moles CO_2

22. Make the following conversions:
    (a) 28.4 g S to moles S
    (b) 2.50 kg NaCl to moles NaCl
    (c) 42.4 g Mg to atoms Mg
    (d) 485 mL Br_2 ($d = 3.12$ g/mL) to moles Br_2

23. Exactly 1 mol of carbon disulfide contains
    (a) how many carbon disulfide molecules?
    (b) how many carbon atoms?
    (c) how many sulfur atoms?
    (d) how many total atoms of all kinds?

24. One mole of ammonia contains
    (a) how many ammonia molecules?
    (b) how many nitrogen atoms?
    (c) how many hydrogen atoms?
    (d) how many total atoms of all kinds?

25. How many atoms of oxygen are contained in each of the following?
    (a) 16.0 g O_2
    (b) 0.622 mol MgO
    (c) 6.00×10^{22} molecules $C_6H_{12}O_6$

26. How many atoms of oxygen are contained in each of the following?
    (a) 5.0 mol MnO_2
    (b) 255 g $MgCO_3$
    (c) 5.0×10^{18} molecules H_2O

27. Calculate the number of
    (a) grams of silver in 25.0 g AgBr
    (b) grams of nitrogen in 6.34 mol $(NH_4)_3PO_4$
    (c) grams of oxygen in 8.45×10^{22} molecules SO_3

28. Calculate the number of
    (a) grams of chlorine in 5.0 g $PbCl_2$
    (b) grams of hydrogen in 4.50 mol H_2SO_4
    (c) grams of hydrogen in 5.45×10^{22} molecules NH_3

Percent Composition

29. Calculate the percent composition by mass of these compounds:
    (a) NaBr   (d) $SiCl_4$
    (b) $KHCO_3$ (e) $Al_2(SO_4)_3$
    (c) $FeCl_3$ (f) $AgNO_3$

30. Calculate the percent composition by mass of these compounds:
    (a) $ZnCl_2$ (d) $(NH_4)_2SO_4$
    (b) $NH_4C_2H_3O_2$ (e) $Fe(NO_3)_3$
    (c) MgP_2O_7 (f) ICl_3

31. Calculate the percent of iron in the following compounds:
    (a) FeO
    (b) Fe_2O_3
    (c) Fe_3O_4
    (d) $K_4Fe(CN)_6$

32. Which of the following chlorides has the highest and which has the lowest percentage of chlorine, by mass, in its formula?
    (a) KCl
    (b) $BaCl_2$
    (c) $SiCl_4$
    (d) LiCl

33. A 6.20-g sample of phosphorus was reacted with oxygen to form an oxide with a mass of 14.20 g. Calculate the percent composition of the compound.

34. A sample of ethylene chloride was analyzed to contain 6.00 g of C, 1.00 g of H, and 17.75 g of Cl. Calculate the percent composition of ethylene chloride.

35. Examine the following formulas. Which compound has the
    (a) higher percent by mass of hydrogen, H_2O or H_2O_2?
    (b) lower percent by mass of nitrogen, NO or N_2O_3?
    (c) higher percent by mass of oxygen, NO_2 or N_2O_4?
    Check your answers by calculation if you wish.

Empirical and Molecular Formulas

37. Calculate the empirical formula of each compound from
    the percent compositions given:
    (a) 63.6% N, 36.4% O
    (b) 46.7% N, 53.3% O
    (c) 25.9% N, 74.1% O
    (d) 43.4% Na, 11.3% C, 45.3% O
    (e) 18.8% Na, 29.0% Cl, 52.3% O
    (f) 72.02% Mn, 27.98% O

39. A sample of tin having a mass of 3.996 g was oxidized and
    found to have combined with 1.077 g of oxygen. Calculate
    the empirical formula of this oxide of tin.

41. Hydroquinone is an organic compound commonly used as
    a photographic developer. It has a molar mass of
    110.1 g/mol and a composition of 65.45% C, 5.45% H,
    and 29.09% O. Calculate the molecular formula of
    hydroquinone.

36. Examine the following formulas. Which compound has the
    (a) lower percent by mass of chlorine, $NaClO_3$ or $KClO_3$?
    (b) higher percent by mass of sulfur, $KHSO_4$ or K_2SO_4?
    (c) lower percent by mass of chromium, Na_2CrO_4 or
    $Na_2Cr_2O_7$?
    Check your answers by calculation if you wish.

38. Calculate the empirical formula of each compound from
    the percent compositions given:
    (a) 64.1% Cu, 35.9% Cl
    (b) 47.2% Cu, 52.8% Cl
    (c) 51.9% Cr, 48.1% S
    (d) 55.3% K, 14.6% P, 30.1% O
    (e) 38.9% Ba, 29.4% Cr, 31.7% O
    (f) 3.99% P, 82.3% Br, 13.7% Cl

40. A 3.054-g sample of vanadium (V) combined with oxygen
    to form 5.454 g of product. Calculate the empirical formula
    for this compound.

42. Fructose is a very sweet natural sugar that is present in
    honey, fruits, and fruit juices. It has a molar mass of
    180.1 g/mol and a composition of 40.0% C, 6.7% H, and
    53.3% O. Calculate the molecular formula of fructose.

Additional Exercises

43. White phosphorus is one of several forms of phosphorus
    and exists as a waxy solid consisting of P_4 molecules.
    How many atoms are present in 0.350 mol of P_4?

44. How many grams of sodium contain the same number of
    atoms as 10.0 g of potassium?

45. One atom of an unknown element is found to have a
    mass of 1.79×10^{-23} g. What is the molar mass of this
    element?

*46. If a stack of 500 sheets of paper is 4.60 cm high, what will
    be the height, in meters, of a stack of Avogadro's number
    of sheets of paper?

47. There are about 5.0 billion (5.0×10^9) people on earth. If
    exactly 1 mol of dollars were distributed equally among
    these people, how many dollars would each person
    receive?

*48. If 20. drops of water equal 1.0 mL (1.0 cm^3),
    (a) how many drops of water are there in a cubic mile of
    water?
    (b) what would be the volume in cubic miles of a mole of
    drops of water?

*49. Silver has a density of 10.5 g/cm^3. If 1.00 mol of silver
    were shaped into a cube,
    (a) what would be the volume of the cube?
    (b) what would be the length of one side of the cube?

*50. A sulfuric acid solution contains 65.0% H_2SO_4 by mass
    and has a density of 1.55 g/mL. How many moles of the
    acid are present in 1.00 L of the solution?

*51. A nitric acid solution containing 72.0% HNO_3 by mass
    has a density of 1.42 g/mL. How many moles of HNO_3
    are present in 100. mL of the solution?

52. Given 1.00-g samples of each of the following com-
    pounds, CO_2, O_2, H_2O, and CH_3OH,
    (a) which sample will contain the largest number of mol-
    ecules?
    (b) which sample will contain the largest number of
    atoms?
    Show proof for your answers.

53. How many grams of Fe_2S_3 will contain a total number of
    atoms equal to Avogadro's number?

54. How many grams of lithium will combine with 20.0 g of sulfur to form the compound Li_2S?

55. Calculate the percentage of
(a) mercury in $HgCO_3$
(b) oxygen in $Ca(ClO_3)_2$
(c) nitrogen in $C_{10}H_{14}N_2$ (nicotine)
(d) Mg in $C_{55}H_{72}MgN_4O_5$ (chlorophyll)

*56. Zinc and sulfur react to form zinc sulfide, ZnS. If we mix 19.5 g of zinc and 9.40 g of sulfur, have we added sufficient sulfur to fully react all the zinc? Show evidence for your answer.

57. Aspirin is well known as a pain reliever (analgesic) and as a fever reducer (antipyretic). It has a molar mass of 180.2 g/mol and a composition of 60.0% C, 4.48% H, and 35.5% O. Calculate the molecular formula of aspirin.

58. How many grams of oxygen are contained in 8.50 g of $Al_2(SO_4)_3$?

59. Gallium arsenide is one of the newer materials used to make semiconductor chips for use in supercomputers. Its composition is 48.2% Ga and 51.8% As. What is the empirical formula?

60. The compositions of four different compounds of carbon and chlorine follow. Determine the empirical formula and the molecular formula for each compound.

Percent C	Percent Cl	Molar mass (g)
(a) 7.79	92.21	153.8
(b) 10.13	89.87	236.7
(c) 25.26	74.74	284.8
(d) 11.25	88.75	319.6

61. How many years is a mole of seconds?

62. A normal penny has a mass of about 2.5 g. If we assume the penny to be pure copper (which means the penny is very old since newer pennies are a mixture of copper and zinc), how many atoms of copper does it contain?

63. What would be the mass (in grams) of one thousand trillion molecules of glycerin, $C_3H_8O_3$?

64. If we assume there are 5.0 billion people on the Earth, how many moles of people is this?

65. An experimental catalyst used to make polymers has the following composition: Co, 23.3%; Mo, 25.3%; and Cl, 51.4%. What is the empirical formula for this compound?

66. If a student weighs 18 g of aluminum and needs twice as many atoms of magnesium as she has of aluminum, how many grams of Mg does she need?

*67. If 10.0 g of an unknown compound composed of carbon, hydrogen, and nitrogen contains 17.7% N and 3.8×10^{23} atoms of hydrogen, what is its empirical formula?

68. A substance whose formula is A_2O (A is a mystery element) is 60.0% A and 40.0% O. Identify the element A.

69. For the following compounds whose molecular formulas are given, indicate the empirical formula:
(a) $C_6H_{12}O_6$ glucose
(b) C_8H_{18} octane
(c) $C_3H_6O_3$ lactic acid
(d) $C_{25}H_{52}$ paraffin
(e) $C_{12}H_4Cl_4O_2$ dioxin (a powerful poison)

Answers to Practice Exercises

7.1 10.0 g helium

7.2 1.5×10^{22} atoms

7.3 101.1 g KNO_3

7.4 21.3 g Na_2SO_4

7.5 3.751 mol $AlCl_3$

7.6 24.42% Ca; 17.07% N; 58.50% O

7.7 40.27% K; 26.78% Cr; 32.96% O

7.8 20.16% Al; 79.84% Cl

7.9 C_2H_6O

7.10 P_2O_5

7.11 The empirical formula is CH_3; the molecular formula is C_2H_6.

CHAPTER

8

Chemical Equations

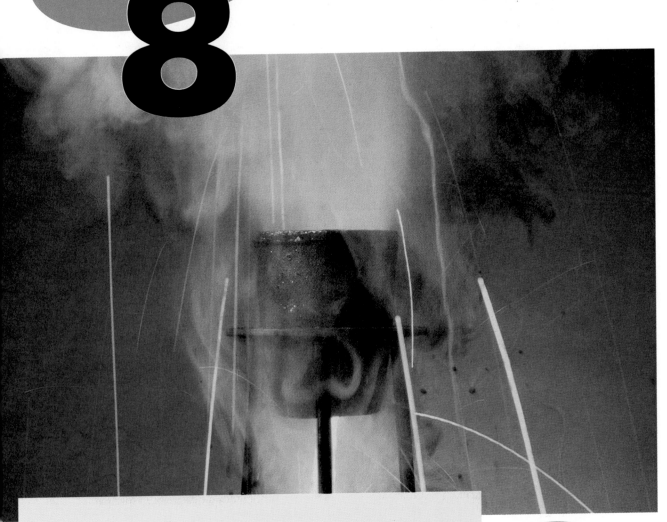

CHAPTER 8 / OUTLINE

▲
The thermite reaction is a reaction between elemental aluminum and iron oxide. This reaction produces so much energy the iron becomes molten. The thermite reaction is used to weld railroad rails.

In the world today, we continually strive to express information in a concise, useful manner. From early childhood, we are taught to translate our ideas and desires into sentences. In mathematics, we learn to describe numerical relationships and situations through mathematical expressions and equations. Historians describe thousands of years of history in 500-page textbooks. Filmmakers translate entire events, such as the Olympics, into a few hours of entertainment.

Chemists use chemical equations to describe reactions they observe in the laboratory or in nature. Chemical equations provide us with the means to (1) summarize the reaction, (2) display the substances that are reacting, (3) show the products, and (4) indicate the amounts of all component substances in a reaction.

8.1   The Chemical Equation

Chemical reactions always involve change. Atoms, molecules, or ions rearrange to form new substances, sometimes in a spectacular manner. For example, the thermite reaction (shown in the chapter opening photo) is a reaction between aluminum metal and iron(III) oxide, which produces molten iron and aluminum oxide. The substances entering the reaction are called the *reactants* and the substances formed are called the *products*. In our example,

reactants   aluminum
iron(III) oxide

products   iron
aluminum oxide

During reactions chemical bonds are broken and new bonds are formed. The reactants and products may be present as solids, liquids, or gases, or in solution.

In a chemical reaction atoms are neither created nor destroyed.
All atoms present in the reactants must also be present in the products.

chemical equation   A **chemical equation** is a shorthand expression for a chemical change or reaction. A chemical equation uses the chemical symbols and formulas of the reactants and products and other symbolic terms to represent a chemical reaction. The equations are written according to this general format:

1. Reactants are separated from products by an arrow ($\longrightarrow$) that indicates the direction of the reaction. The reactants are placed to the left and the products to the right of the arrow. A plus sign ($+$) is placed between reactants and between products when needed.

$$Al + Fe_2O_3 \longrightarrow Fe + Al_2O_3$$
reactants   products

2. Coefficients (whole numbers) are placed in front of substances to balance the equation and to indicate the number of units (atoms, molecules, moles, ions)

TABLE 8.1	Symbols Commonly Used in Chemical Equations	

Symbol	Meaning
+	Plus or added to (placed between substances)
$\longrightarrow$	Yields; produces (points to products)
(s)	Solid state (written after a substance)
(l)	Liquid state (written after a substance)
(g)	Gaseous state (written after a substance)
(aq)	Aqueous solution (substance dissolved in water)
Δ	Heat (written above arrow)

of each substance reacting or being produced. When no number is shown, it is understood that one unit of the substance is indicated.

$$2\ Al + Fe_2O_3 \longrightarrow 2\ Fe + Al_2O_3$$

3. **Conditions required to carry out the reaction** may, if desired, be placed above or below the arrow or equality sign. For example, a delta sign placed over the arrow ($\xrightarrow{\Delta}$) indicates that heat is supplied to the reaction.

$$2\ Al + Fe_2O_3 \xrightarrow{\Delta} 2\ Fe + Al_2O_3$$

4. **The physical state of a substance** is indicated by the following symbols: (s) for solid state; (l) for liquid state; (g) for gaseous state; and (aq) for substances in aqueous solution. States are not always given in chemical equations.

$$2\ Al(s) + Fe_2O_3(s) \xrightarrow{\Delta} Fe(l) + Al_2O_3(s)$$

Commonly used symbols are given in Table 8.1.

8.2 Writing and Balancing Equations

To represent the quantitative relationships of a reaction, the chemical equation must be balanced. A **balanced equation** contains the same number of each kind of atom on each side of the equation. The balanced equation therefore obeys the law of conservation of mass.

balanced equation

Every chemistry student must learn to *balance* equations. Many equations are balanced by trial and error, but care and attention to detail are still required. The way to balance an equation is to adjust the number of atoms of each element so that they are the same on each side of the equation, but a correct formula is never changed in order to balance an equation. The general procedure for balancing equations is:

Correct formulas are not changed to balance an equation.

Step 1. Identify the reaction. Write a description or word equation for the reaction. For example, let's consider mercury(II) oxide decomposing into mercury and oxygen.

Study this procedure carefully and refer to it when you work examples.

mercury (II) oxide $\longrightarrow$ mercury + oxygen.

Step 2. Write the unbalanced (skeleton) equation. Make sure that the formula for each substance is correct and that reactants are written to the left and products to the right of the arrow. For our example

$$HgO \longrightarrow Hg + O_2$$

The correct formulas must be known or determined from the periodic table, lists of ions, or experimental data.

Step 3. Balance the equation. Use the following process as necessary:
(a) Count and compare the number of atoms of each element on each side of the equation and determine those that must be balanced.

Hg is balanced   (1 on each side)

O needs to be balanced   (1 on reactant side, 2 on product side)

(b) Balance each element, one at a time, by placing whole numbers (coefficients) in front of the formulas containing the unbalanced element. It is usually best to balance metals first, then nonmetals, then hydrogen and oxygen. Select the smallest coefficients that will give the same number of atoms of the element on each side. A coefficient placed before a formula multiplies every atom in the formula by that number (e.g., 2 H_2SO_4 means two molecules of sulfuric acid and also means four H atoms, two S atoms, and eight O atoms). Place a 2 in front of HgO to balance O.

$$2\ HgO \longrightarrow Hg + O_2$$

(c) Check all other elements after each individual element is balanced to see whether, in balancing one element, other elements have become unbalanced. Make adjustments as needed. Now Hg is not balanced. To adjust this we write a 2 in front of Hg.

$$2\ HgO \longrightarrow 2\ Hg + O_2 \quad \text{(balanced)}$$

(d) Do a final check, making sure that each element and/or polyatomic ion is balanced and that the smallest possible set of whole-number coefficients has been used.

$$2\ HgO \longrightarrow 2\ Hg + O_2 \quad \text{(correct form)}$$

$$4\ HgO \longrightarrow 4\ Hg + 2\ O_2 \quad \text{(incorrect form)}$$

Not all chemical equations can be balanced by the simple method of inspection just described. The following examples show *stepwise* sequences leading to balanced equations. Study each one carefully.

> Leave elements that are in two or more formulas (on the same side of the equation) unbalanced until just before balancing hydrogen and oxygen.

Example 8.1   Write the balanced equation for the reaction that takes place when magnesium metal is burned in air to produce magnesium oxide.

Solution

Step 1. *Word equation:*

$$\text{magnesium} + \text{oxygen} \longrightarrow \text{magnesium oxide}$$

reactants (R)   products (P)

Step 2. *Skeleton equation:*

Mg + O$_2$ $\longrightarrow$ MgO   (unbalanced)

| | R = reactant | R | 1 Mg | 2 O |
| | P = product | P | 1 Mg | 1 O |

Step 3. *Balance:*

(a) Mg is balanced.
Oxygen is not balanced. Two O atoms appear on the left side and one on the right side.

(b) Place the coefficient 2 in front of MgO:

Mg + O$_2$ $\longrightarrow$ 2 MgO   (unbalanced)

| R | 1 Mg | 2 O |
| P | 2 Mg | 2 O |

(c) Now Mg is not balanced. One Mg atom appears on the left side and two on the right side. Place a 2 in front of Mg:

2 Mg + O$_2$ $\longrightarrow$ 2 MgO   (balanced)

(d) *Check:* Each side has two Mg and two O atoms.

| R | 2 Mg | 2 O |
| P | 2 Mg | 2 O |

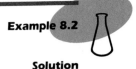

Example 8.2

When methane, CH$_4$, undergoes complete combustion, it reacts with oxygen to produce carbon dioxide and water. Write the balanced equation for this reaction.

Solution

Step 1. *Word equation:*

methane + oxygen $\longrightarrow$ carbon dioxide + water

Step 2. *Skeleton equation:*

CH$_4$ + O$_2$ $\longrightarrow$ CO$_2$ + H$_2$O   (unbalanced)

Step 3. *Balance:*

(a) Carbon is balanced.
Hydrogen and oxygen are not balanced.

| R | 1 C | 4 H | 2 O |
| P | 1 C | 2 H | 3 O |

(b) Balance H atoms by placing a 2 in front of H$_2$O:

CH$_4$ + O$_2$ $\longrightarrow$ CO$_2$ + 2 H$_2$O   (unbalanced)

| R | 1 C | 4 H | 2 O |
| P | 1 C | 4 H | 4 O |

Each side of the equation has four H atoms; oxygen is still not balanced. Place a 2 in front of O$_2$ to balance the oxygen atoms:

CH$_4$ + 2 O$_2$ $\longrightarrow$ CO$_2$ + 2 H$_2$O   (balanced)

| R | 1 C | 4 H | 4 O |
| P | 1 C | 4 H | 4 O |

(c) The other atoms remain balanced.

(d) *Check:* The equation is correctly balanced; it has one C, four O, and four H atoms on each side.

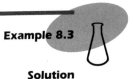

Example 8.3

Oxygen and potassium chloride are formed by heating potassium chlorate. Write a balanced equation for this reaction.

Solution

Step 1. *Word equation:*

potassium chlorate $\xrightarrow{\Delta}$ potassium chloride + oxygen

Step 2. *Skeleton equation:*

$$KClO_3 \xrightarrow{\Delta} KCl + O_2 \quad \text{(unbalanced)}$$

Step 3. *Balance:*

(a) Potassium and chlorine are balanced.
Oxygen is unbalanced (three O atoms on the left and two on the right side).

(b) How many oxygen atoms are needed? The subscripts of oxygen (3 and 2) in $KClO_3$ and O_2 have a least common multiple of 6. Therefore coefficients for $KClO_3$ and O_2 are needed to get six O atoms on each side. Place a 2 in front of $KClO_3$ and a 3 in front of O_2:

R	1 K	1 Cl	3 O
P	1 K	1 Cl	2 O

R	2 K	2 Cl	6 O
P	1 K	1 Cl	6 O

$$2\,KClO_3 \xrightarrow{\Delta} KCl + 3\,O_2 \quad \text{(unbalanced)}$$

(c) Now K and Cl are not balanced. Place a 2 in front of KCl, which balances both K and Cl at the same time:

$$2\,KClO_3 \xrightarrow{\Delta} 2\,KCl + 3\,O_2 \quad \text{(balanced)}$$

R	2 K	2 Cl	6 O
P	2 K	2 Cl	6 O

(d) *Check:* Each side now contains two K, two Cl, and six O atoms.

Example 8.4   Silver nitrate reacts with hydrogen sulfide to produce silver sulfide and nitric acid. Write a balanced equation for this reaction.

Solution

Step 1. *Word equation:*

silver nitrate + hydrogen sulfide $\longrightarrow$ silver sulfide + nitric acid

Step 2. *Skeleton equation:*

$$AgNO_3 + H_2S \longrightarrow Ag_2S + HNO_3 \quad \text{(unbalanced)}$$

R	1 Ag	2 H	1 S	1 NO_3
P	2 Ag	1 H	1 S	1 NO_3

Step 3. *Balance:*

(a) Ag and H are unbalanced.

(b) Place a 2 in front of $AgNO_3$ to balance Ag:

$$2\,AgNO_3 + H_2S \longrightarrow Ag_2S + HNO_3 \quad \text{(unbalanced)}$$

R	2 Ag	2 H	1 S	2 NO_3
P	2 Ag	1 H	1 S	1 NO_3

(c) H and NO_3^- are still unbalanced. Balance by placing a 2 in front of HNO_3:

$$2\,AgNO_3 + H_2S \longrightarrow Ag_2S + 2\,HNO_3 \quad \text{(balanced)}$$

In this example N and O atoms are balanced by balancing the NO_3^- ion as a unit.

R	2 Ag	2 H	1 S	2 NO_3
P	2 Ag	2 H	1 S	2 NO_3

(d) The other atoms remain balanced.

(e) *Check:* Each side has two Ag, two H, and one S atom. Also, each side has two NO_3^- ions.

When aluminum hydroxide is mixed with sulfuric acid the products are aluminum sulfate and water. Write a balanced equation for this reaction.

Example 8.5

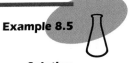

Step 1. *Word equation:*

aluminum hydroxide + sulfuric acid $\longrightarrow$ aluminum sulfate + water

Solution

Step 2. *Skeleton equation:*

$Al(OH)_3 + H_2SO_4 \longrightarrow Al_2(SO_4)_3 + H_2O$ (unbalanced)

Step 3. *Balance:*

(a) All elements are unbalanced.

(b) Balance Al by placing a 2 in front of $Al(OH)_3$. Treat the unbalanced SO_4^{2-} ion as a unit and balance by placing a 3 in front of H_2SO_4:

R	1 Al	1 SO$_4$	3 O	5 H
P	2 Al	3 SO$_4$	1 O	2 H

$2\,Al(OH)_3 + 3\,H_2SO_4 \longrightarrow Al_2(SO_4)_3 + H_2O$ (unbalanced)

Balance the unbalanced H and O by placing a 6 in front of H_2O:

R	2 Al	3 SO$_4$	6 O	12 H
P	2 Al	3 SO$_4$	1 O	2 H

$2\,Al(OH)_3 + 3\,H_2SO_4 \longrightarrow Al_2(SO_4)_3 + 6\,H_2O$ (balanced)

(c) The other atoms remain balanced.

R	2 Al	3 SO$_4$	6 O	12 H
P	2 Al	3 SO$_4$	6 O	12 H

(d) *Check:* Each side has two Al, twelve H, three S, and eighteen O atoms.

When the fuel in a butane gas stove undergoes complete combustion, it reacts with oxygen to form carbon dioxide and water. Write the balanced equation for this reaction.

Example 8.6

Step 1. *Word equation:*

butane + oxygen $\longrightarrow$ carbon dioxide + water

Solution

Step 2. *Skeleton equation:*

$C_4H_{10} + O_2 \longrightarrow CO_2 + H_2O$ (unbalanced)

Step 3. *Balance:*

(a) All elements are unbalanced.

(b) Balance C by placing a 4 in front of CO_2:

R	4 C	10 H	2 O
P	1 C	2 H	3 O

$C_4H_{10} + O_2 \longrightarrow 4\,CO_2 + H_2O$ (unbalanced)

Balance H by placing a 5 in front of H_2O:

R	4 C	10 H	2 O
P	4 C	10 H	13 O

$C_4H_{10} + O_2 \longrightarrow 4\,CO_2 + 5\,H_2O$ (unbalanced)

Oxygen remains unbalanced. The oxygen atoms on the right side are fixed, because 4 CO_2 and 5 H_2O are derived from the single C_4H_{10} molecule on the left. When we try to balance the O atoms, we find there is no whole number that can be placed in front of O_2 to bring about a balance. So we double the coefficients of each substance, and then balance the oxygen:

R	8 C	20 H	26 O
P	8 C	20 H	26 O

$2\,C_4H_{10} + 13\,O_2 \longrightarrow 8\,CO_2 + 10\,H_2O$ (balanced)

(c) The other atoms remain balanced.
(d) *Check:* Each side now has eight C, twenty H, and twenty-six O atoms.

Practice 8.1

Write a balanced formula equation:

aluminum + oxygen $\longrightarrow$ aluminum oxide

Practice 8.2

Write a balanced formula equation:

magnesium hydroxide + phosphoric acid $\longrightarrow$

magnesium phosphate + water

8.3   What Information Does an Equation Tell Us?

Depending on the particular context in which it is used, a formula can have different meanings. A formula can refer to an individual chemical entity (atom, ion, molecule, or formula unit) or to a mole of that chemical entity. For example, the formula H_2O can mean any of the following:

1. 2 H atoms and 1 O atom
2. 1 molecule of water
3. 1 mol of water
4. 6.022×10^{23} molecules of water
5. 18.02 g of water

Formulas used in equations can represent units of individual chemical entities or moles, the latter being more commonly used. For example, in the reaction of hydrogen and oxygen to form water,

$$2\,H_2 \quad + \quad O_2 \quad \longrightarrow \quad 2\,H_2O$$

| 2 molecules hydrogen | 1 molecule oxygen | 2 molecules water |
| 2 mol hydrogen | 1 mol oxygen | 2 mol water |

We generally use moles in equations because molecules are so small.

As indicated earlier, a chemical equation is a shorthand description of a chemical reaction. Interpretation of a balanced equation gives us the following information:

1. What the reactants are and what the products are
2. The formulas of the reactants and products
3. The number of molecules or formula units of reactants and products in the reaction
4. The number of atoms of each element involved in the reaction
5. The number of moles of each substance

CHEMISTRY IN ACTION Colorful Trees

The wondrous array of color we see in the fall is the result of chemical reactions. These reactions contribute to the metabolic cycle in plants while they give us our annual autumn show of foliage. Three chemical substances play key roles in this process: chlorophylls, carotenoids, and anthocyanins.

The green pigments in leaves come from chlorophylls, which are necessary for the tree to produce food in the process called *photosynthesis*. Photosynthesis involves hundreds of chemical reactions but can be summarized by the following chemical equation:

$$6 \text{ CO}_2 + 6 \text{ H}_2\text{O} \xrightarrow[\text{chlorophyll}]{\text{sunlight}}$$

carbon   water   chlorophyll
dioxide

$$\text{C}_6\text{H}_{12}\text{O}_6 + 6 \text{ O}_2$$

glucose   oxygen

There are several types of chlorophyll. All photosynthetic plants contain chlorophyll *a*, and some contain chlorophyll *b* and chlorophyll *c*, known as *accessory pigments*. The accessory pigments extend the range of colored light trees can use for photosynthesis. Chlorophyll *b* and *c* absorb light of slightly different colors and pass the energy on to chlorophyll *a*. This allows the trees to produce more food during the summer months.

The carotenoids are responsible for the yellow, orange, and some of the reds in the autumn leaves. These substances protect the plant from the destructive potential of chlorophyll. When chlorophyll absorbs energy (from light or the other chlorophylls) it produces a high energy electron that is normally passed on to other compounds in the chemical reactions of photosynthesis. Sometimes there is a shortage of these secondary compounds and the high energy electron can then combine with oxygen (in other chemical reactions) and destroy the cell. The carotenoid pigments act to save the cell from destruction by combining with the high energy oxygen and releasing it later when it can be used successfully to produce sugars. Carotenoids and anthocyanins (red and blue pigment molecules) also help the tree to attract insects for pollination and make fruits noticeable to birds and animals, which then spread seeds for reproduction of the tree. The carotenoids and anthocyanins are masked by the green of the chlorophylls during the summer and can be seen only as the chlorophyll disintegrates in the fall.

What conditions and chemical reactions are necessary to make the autumn show spectacular? The colors of autumn leaves are produced by the carotenoids and anthocyanins. Trees need lots of each molecule to produce a fall show. Trees produce the carotenoids and anthocyanins during photosynthesis. A long and moist summer provides the proper reactants and time to make plenty of sugar. When the tree signals its leaves to stop photosynthesis the chlorophylls are released from the leaf proteins. The chlorophylls break down and the protein nutrients are carried to the roots for winter storage. Now the carotenoids that have been hard at work protecting the leaf from too much chlorophyll activity can show their color! The more productive the leaf has been (more photosynthesis) the greater the concentration of carotenoids and the more beautiful the orange and yellow colors.

But what about the anthocyanins? What is their role in leaf color? As fall approaches the tree stops producing new wood from the sugar made in photosynthesis. Sugar manufactured in the leaves is now sent to the roots for storage. If the weather is chilly the sugary fluid becomes slow flowing (viscous). Some of the sugar made during the day cannot get out of the leaf to the roots during the night. The leftover sugar is changed into anthocyanins. The greater the amount of these pigment molecules the brighter the leaf color.

In the hills of Pennsylvania a parade of chemical reactions sets the stage for the brilliant colors of fall. The chlorophyll molecules begin to disintegrate, the critical biomolecules migrate to the roots, and the carotenoids begin by showing their yellow and orange colors. The sugar that remains in the leaves produces red anthocyanins and the leaves of the maples change from yellow and orange to flaming red. At last all of the pigments disintegrate in decomposition reactions and the leaves become brown, the color of the woody fibers of their skeletons, and eventually fall to the ground. If we are fortunate, we are treated to a spectacular show of color before the leaves die!

▲ **Vivid colors are reflected in Tuscarora Lake in central Pennsylvania.**

▲
Propane is burned in many outdoor grills.

Consider the reaction that occurs when propane gas (C_3H_8) is burned in air; the products are carbon dioxide, CO_2, and water, H_2O. The balanced equation and its interpretation are as follows:

Propane		Oxygen		Carbon dioxide		Water
$C_3H_8(g)$	+	$5\,O_2(g)$	→	$3\,CO_2(g)$	+	$4\,H_2O(g)$
1 molecule		5 molecules		3 molecules		4 molecules
3 atoms C 8 atoms H		10 atoms O		3 atoms C 6 atoms O		8 atoms H 4 atoms O
1 mol 44.09 g		5 mol 5 (32.00 g) (160.0 g)		3 mol 3 (44.01 g) (132.0 g)		4 mol 4 (18.02 g) (72.08 g)

Practice 8.3

Consider the reaction that occurs when hydrogen gas reacts with chlorine gas to produce gaseous hydrogen chloride.
(a) Write a word equation for this reaction.
(b) Write a balanced formula equation including state for each substance.
(c) Label each reactant and product to show relative amounts of each substance.
    1) number of molecules
    2) number of atoms
    3) number of moles
    4) mass
(d) What mass of HCl would be produced if you reacted 2 mol hydrogen gas with 2 mol chlorine gas?

The quantities involved in chemical reactions are important when working in industry or the laboratory. We will study the relationship among quantities of reactants and products in the next chapter.

8.4   Types of Chemical Equations

Chemical equations represent chemical changes or reactions. Reactions are classified into types to assist in writing equations and in predicting other reactions. Many chemical reactions fit one or another of the four principal reaction types that we discuss in the following paragraphs. Reactions are also classified as oxidation–reduction. Special methods are used to balance complex oxidation–reduction equations. (See Chapter 17.)

Combination Reaction

In a **combination reaction,** two reactants combine to give one product. The general form of the equation is

$$A + B \longrightarrow AB$$

in which A and B are either elements or compounds and AB is a compound. The formula of the compound in many cases can be determined from a knowledge of the ionic charges of the reactants in their combined states. Some reactions that fall into this category are given here.

combination reaction

1. metal + oxygen $\longrightarrow$ metal oxide

   $$2 \, Mg(s) + O_2(g) \xrightarrow{\Delta} 2 \, MgO(s)$$
   $$4 \, Al(s) + 3 \, O_2(g) \xrightarrow{\Delta} 2 \, Al_2O_3(s)$$

2. nonmetal + oxygen $\longrightarrow$ nonmetal oxide

   $$S(s) + O_2(g) \xrightarrow{\Delta} SO_2(g)$$
   $$N_2(g) + O_2(g) \xrightarrow{\Delta} 2 \, NO(g)$$

3. metal + nonmetal $\longrightarrow$ salt

   $$2 \, Na(s) + Cl_2(g) \longrightarrow 2 \, NaCl(s)$$
   $$2 \, Al(s) + 3 \, Br_2(l) \longrightarrow 2 \, AlBr_3(s)$$

4. metal oxide + water $\longrightarrow$ metal hydroxide

   $$Na_2O(s) + H_2O(l) \longrightarrow 2 \, NaOH(aq)$$
   $$CaO(s) + H_2O(l) \longrightarrow Ca(OH)_2(aq)$$

5. nonmetal oxide + water $\longrightarrow$ oxy-acid

   $$SO_3(g) + H_2O(l) \longrightarrow H_2SO_4(aq)$$
   $$N_2O_5(s) + H_2O(l) \longrightarrow 2 \, HNO_3(aq)$$

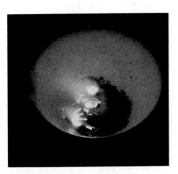

▲
Flames and sparks result when aluminum foil is dropped into liquid bromine.

Decomposition Reaction

In a **decomposition reaction,** a single substance is decomposed or broken down to give two or more different substances. This reaction may be considered the reverse of combination. The starting material must be a compound, and the products may be elements or compounds. The general form of the equation is

$$AB \longrightarrow A + B$$

decomposition reaction

Predicting the products of a decomposition reaction can be difficult and requires an understanding of each individual reaction. Heating oxygen-containing compounds often results in decomposition. Some reactions that fall into this category are

1. Metal oxides. Some metal oxides decompose to yield the free metal plus oxygen; others give another oxide, and some are very stable, resisting decomposition by heating:

   $$2 \, HgO(s) \xrightarrow{\Delta} 2 \, Hg(l) + O_2(g)$$
   $$2 \, PbO_2(s) \xrightarrow{\Delta} 2 \, PbO(s) + O_2(g)$$

When a strip of zinc is placed in HCl, hydrogen gas bubbles form immediately.

2. Carbonates and hydrogen carbonates decompose to yield CO_2 when heated:

$$CaCO_3(s) \xrightarrow{\Delta} CaO(s) + CO_2(g)$$

$$2\ NaHCO_3(s) \xrightarrow{\Delta} Na_2CO_3(s) + H_2O(g) + CO_2(g)$$

3. Miscellaneous reactions in this category:

$$2\ KClO_3(s) \xrightarrow{\Delta} 2\ KCl(s) + 3\ O_2(g)$$

$$2\ NaNO_3(s) \xrightarrow{\Delta} 2\ NaNO_2(g) + O_2(g)$$

$$2\ H_2O_2(l) \xrightarrow{\Delta} 2\ H_2O(l) + O_2(g)$$

Single-Displacement Reaction

In a **single-displacement reaction,** one element reacts with a compound to replace one of the elements of that compound, yielding a different element and a different compound. The general form of the equation is

$$A + BC \longrightarrow B + AC \qquad \text{or} \qquad A + BC \longrightarrow C + BA$$

If A is a metal, A will replace B to form AC, provided A is a more reactive metal than B. If A is a halogen, it will replace C to form BA, provided A is a more reactive halogen than C.

single-displacement reaction

A brief activity series of selected metals (and hydrogen) and halogens is shown in Table 8.2. This series is listed in descending order of chemical activity, with the most active metals and halogens at the top. Many chemical reactions can be predicted from an activity series because the atoms of any element in the series will replace the atoms of those elements below it. For example, zinc metal will replace hydrogen from a hydrochloric acid solution. But copper metal, which is below hydrogen on the list and thus less reactive than hydrogen, will not replace hydro-gen from a hydrochloric acid-solution. Here are some reactions that fall into this category:

1. metal + acid $\longrightarrow$ hydrogen + salt

$$Zn(s) + 2\ HCl(aq) \longrightarrow H_2(g) + ZnCl_2(aq)$$

$$2\ Al(s) + 3\ H_2SO_4(aq) \longrightarrow 3\ H_2(g) + Al_2(SO_4)_3(aq)$$

2. metal + water $\longrightarrow$ hydrogen + metal hydroxide or metal oxide

$$2\ Na(s) + 2\ H_2O \longrightarrow H_2(g) + 2\ NaOH(aq)$$

$$Ca(s) + 2\ H_2O \longrightarrow H_2(g) + Ca(OH)_2(aq)$$

$$3\ Fe(s) + 4\ H_2O(g) \longrightarrow 4\ H_2(g) + Fe_3O_4(s)$$
steam

3. metal + salt $\longrightarrow$ metal + salt

$$Fe(s) + CuSO_4(aq) \longrightarrow Cu(s) + FeSO_4(aq)$$

$$Cu(s) + 2\ AgNO_3(aq) \longrightarrow 2\ Ag(s) + Cu(NO_3)_2(aq)$$

4. halogen + halide salt $\longrightarrow$ halogen + halide salt

$$Cl_2(g) + 2\ NaBr(aq) \longrightarrow Br_2(l) + 2\ NaCl(aq)$$

$$Cl_2(g) + 2\ KI(aq) \longrightarrow I_2(s) + 2\ KCl(aq)$$

TABLE 8.2	Activity Series	
Metals	**Halogens**	
K	F_2	
Ca	Cl_2	
Na	Br_2	
Mg	I_2	
Al		
Zn		
Fe		
Ni		
Sn		
Pb		
H		
Cu		
Ag		
Hg		
Au		

increasing activity

A common chemical reaction is the displacement of hydrogen from water or acids. This reaction is a good illustration of the relative reactivity of metals and the use of the activity series. For example,

- K, Ca, and Na displace hydrogen from cold water, steam, and acids.
- Mg, Al, Zn, and Fe displace hydrogen from steam and acids.
- Ni, Sn, and Pb displace hydrogen only from acids.
- Cu, Ag, Hg, and Au do not displace hydrogen.

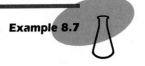

Example 8.7

Will a reaction occur between (a) nickel metal and hydrochloric acid and (b) tin metal and a solution of aluminum chloride? Write balanced equations for the reactions.

Solution

(a) Nickel is more reactive than hydrogen, so it will displace hydrogen from hydrochloric acid. The products are hydrogen gas and a salt of Ni^{2+} and Cl^- ions:

$$Ni(s) + 2\ HCl(aq) \longrightarrow H_2(g) + NiCl_2(aq)$$

(b) According to the activity series, tin is less reactive than aluminum, so no reaction will occur:

$$Sn(s) + AlCl_3(aq) \longrightarrow \text{no reaction}$$

Practice 8.4

Write balanced equations for these reactions:
(a) iron metal and a solution of magnesium chloride
(b) zinc metal and a solution of lead(II) nitrate

Double-Displacement Reaction

In a **double-displacement reaction,** two compounds exchange partners with each other to produce two different compounds. The general form of the equation is

double-displacement reaction

$$AB + CD \longrightarrow AD + CB$$

This reaction can be thought of as an exchange of positive and negative groups, in which A combines with D and C combines with B. In writing formulas for the products, we must account for the charges of the combining groups.

It's also possible to write an equation in the form of a double-displacement reaction when a reaction has not occurred. For example, when solutions of sodium chloride and potassium nitrate are mixed, the following equation can be written:

$$NaCl(aq) + KNO_3(aq) \longrightarrow NaNO_3(aq) + KCl(aq)$$

When the procedure is carried out, no physical changes are observed, indicating that no chemical reaction has taken place.

A double-displacement reaction is accompanied by evidence of such reactions as the evolution of heat, the formation of an insoluble precipitate, or the production of gas bubbles. Let's look at some of these reactions more closely.

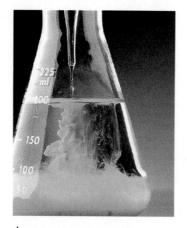

A double-displacement
reaction results from pouring
a clear, colorless solution of
$Pb(NO_3)_2$ into a clear, colorless solution of KI forming a
yellow precipitate of PbI_2.

Neutralization of an acid and a base   The production of a molecule of water from an H^+ and an OH^- ion is accompanied by a release of heat, which can be detected by touching the reaction container. For neutralization reactions, $H^+ + OH^- \longrightarrow H_2O$.

$$\text{acid} + \text{base} \longrightarrow \text{salt} + \text{water}$$

$$HCl(aq) + NaOH(aq) \longrightarrow NaCl(aq) + H_2O(l)$$

$$H_2SO_4(aq) + Ba(OH)_2(aq) \longrightarrow BaSO_4(s) + 2\ H_2O(l)$$

Formation of an insoluble precipitate   The solubilities of the products can be determined by consulting the Solubility Table in Appendix V to see whether one or both of the products are insoluble in water. An insoluble product (precipitate) is indicated by placing an (*s*) after its formula in the equation.

$$BaCl_2(aq) + 2\ AgNO_3(aq) \longrightarrow 2\ AgCl(s) + Ba(NO_3)_2(aq)$$

$$Pb(NO_3)_2(aq) + 2\ KI(aq) \longrightarrow PbI_2(s) + 2\ KNO_3(aq)$$

Metal oxide + acid   Heat is released by the production of a molecule of water.

$$\text{metal oxide} + \text{acid} \longrightarrow \text{salt} + \text{water}$$

$$CuO(s) + 2\ HNO_3(aq) \longrightarrow Cu(NO_3)_2(aq) + H_2O(l)$$

$$CaO(s) + 2\ HCl(aq) \longrightarrow CaCl_2(aq) + H_2O(l)$$

Formation of a gas   A gas such as HCl or H_2S may be produced directly, as in these two examples:

$$H_2SO_4(l) + NaCl(s) \longrightarrow NaHSO_4(s) + HCl(g)$$

$$2\ HCl(aq) + ZnS(s) \longrightarrow ZnCl_2(aq) + H_2S(g)$$

A gas can also be produced indirectly. Some unstable compounds formed in a double-displacement reaction, such as H_2CO_3, H_2SO_3, and NH_4OH, will decompose to form water and a gas:

$$2\ HCl(aq) + Na_2CO_3(aq) \longrightarrow 2\ NaCl(aq) + H_2CO_3(aq) \longrightarrow 2\ NaCl(aq) + H_2O(l) + CO_2(g)$$

$$2\ HNO_3(aq) + K_2SO_3(aq) \longrightarrow 2\ KNO_3(aq) + H_2SO_3(aq) \longrightarrow 2\ KNO_3(aq) + H_2O(l) + SO_2(g)$$

$$NH_4Cl(aq) + NaOH(aq) \longrightarrow NaCl(aq) + NH_4OH(aq) \longrightarrow NaCl(aq) + H_2O(l) + NH_3(g)$$

Example 8.8

Write the equation for the reaction between aqueous solutions of hydrobromic acid and potassium hydroxide.

Solution

First we write the formulas for the reactants. (They are HBr and KOH.) Then we classify the type of reaction that would occur between them. Because the reactants are compounds, one an acid and the other a base, the reaction will be of the neutralization type:

$$\text{acid} + \text{base} \longrightarrow \text{salt} + \text{water}$$

Now rewrite the equation using the formulas for the known substances:

$$HBr(aq) + KOH(aq) \longrightarrow \text{salt} + H_2O$$

In this reaction, which is a double-displacement type, the H^+ from the acid combines with the OH^- from the base to form water. The ionic compound must be composed of the other two ions, K^+ and Br^-. We determine the formula of the ionic compound to be KBr from the fact that K is a $+1$ cation and Br is a -1 anion. The final balanced equation is

$$HBr(aq) + KOH(aq) \longrightarrow KBr(aq) + H_2O(l)$$

Example 8.9

Complete and balance the equation for the reaction between aqueous solutions of barium chloride and sodium sulfate.

Solution

First determine the formula for the reactants. (They are $BaCl_2$ and Na_2SO_4.) Then classify these substances as acids, bases, or ionic compounds. (Both substances are salts.) Since both substances are compounds, the reaction will be of the double-displacement type. Start writing the equation with the reactants:

$$BaCl_2(aq) + Na_2SO_4(aq) \longrightarrow$$

If the reaction is double-displacement, Ba^{2+} will be written combined with SO_4^{2-}, and Na^+ with Cl^- as the products. The balanced equation is

$$BaCl_2(aq) + Na_2SO_4(aq) \longrightarrow BaSO_4 + 2\ NaCl$$

The final step is to determine the nature of the products, which controls whether or not the reaction will take place. If both products are soluble, we have a mixture of all the ions in solution. But if an insoluble precipitate is formed, the reaction will definitely occur. We know from experience that NaCl is fairly soluble in water, but what about $BaSO_4$? Consulting the Solubility Table in Appendix V we see that $BaSO_4$ is insoluble in water, so it will be a precipitate in the reaction. Thus the reaction will occur, forming a precipitate. The equation is

$$BaCl_2(aq) + Na_2SO_4(aq) \longrightarrow BaSO_4(s) + 2\ NaCl(aq)$$

▲
When barium chloride is poured into a solution of sodium sulfate, a white precipitate of barium sulfate forms.

Practice 8.5

Complete and balance the equations for these reactions:
(a) potassium phosphate + barium chloride
(b) hydrochloric acid + nickel carbonate
(c) ammonium chloride + sodium nitrate

Some of the reactions you attempt may fail because the substances are not reactive or because the proper conditions for reaction are not present. For example, mercury(II) oxide does not decompose until it is heated; magnesium does not burn in air or oxygen until the temperature reaches a certain point. When silver is placed in a solution of copper(II) sulfate, no reaction occurs. When a strip of copper is placed

A copper wire is placed in ▶
a solution of silver nitrate
(left). After 24 hours crystals
of silver are seen hanging on
the copper wire and the solution has turned blue indicating copper ions are present
there.

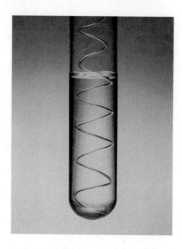

A copper wire is placed in ▶
a solution of silver nitrate
(left). After 24 hours crystals
of silver are seen hanging on
the copper wire and the solution has turned blue indicating copper ions are present
there.

in a solution of silver nitrate, a single-displacement reaction takes place, because copper is a more reactive metal than silver.

The successful prediction of the products of a reaction is not always easy. The ability to predict products correctly comes with knowledge and experience. Although you may not be able to predict many reactions at this point, as you continue to experiment you will find that reactions can be categorized, and that prediction of the products thereby becomes easier, if not always certain.

8.5   Heat in Chemical Reactions

Energy changes always accompany chemical reactions. One reason why reactions occur is that the products attain a lower, more stable energy state than the reactants. When the reaction leads to a more stable state, energy is released to the surroundings as heat (and/or work). When a solution of a base is neutralized by the addition of an acid, the liberation of heat energy is signaled by an immediate rise in the temperature of the solution. When an automobile engine burns gasoline, heat is certainly liberated; at the same time, part of the liberated energy does the work of moving the automobile.

exothermic reaction   Reactions are either exothermic or endothermic. **Exothermic reactions** liberate
endothermic reaction   heat; **endothermic reactions** absorb heat. In an exothermic reaction, heat is a product and may be written on the right side of the equation for the reaction. In an endothermic reaction, heat can be regarded as a reactant and is written on the left side of the equation. Here are two examples:

$$H_2(g) + Cl_2(g) \longrightarrow 2\ HCl(g) + 185\ kJ \quad \textit{(exothermic)}$$

$$N_2(g) + O_2(g) + 181\ kJ \longrightarrow 2\ NO(g) \quad \textit{(endothermic)}$$

heat of reaction   The quantity of heat produced by a reaction is known as the **heat of reaction.** The units used can be kilojoules or kilocalories. Consider the reaction represented by this equation:

$$C(s) + O_2(g) \longrightarrow CO_2(g) + 393\ kJ$$

When the heat liberated is expressed as part of the equation, the substances are expressed in units of moles. Thus when 1 mol (12.01 g) of C combines with 1 mol (32.00 g) of O_2, 1 mol (44.01 g) of CO_2 is formed and 393 kJ of heat is liberated. In

this reaction, as in many others, the heat energy is more useful than the chemical products.

Aside from relatively small amounts of energy from nuclear processes, the sun is the major provider of energy for life on Earth. The Sun maintains the temperature necessary for life and also supplies light energy for the endothermic photosynthetic reactions of green plants. In photosynthesis, carbon dioxide and water are converted to free oxygen and glucose:

$$6\ CO_2 + 6\ H_2O + 2519\ kJ \longrightarrow C_6H_{12}O_6 + 6\ O_2$$
$$\text{glucose}$$

Nearly all of the chemical energy used by living organisms is obtained from glucose or compounds derived from glucose.

The major source of energy for modern technology is fossil fuel—coal, petroleum, and natural gas. The energy is obtained from the combustion (burning) of these fuels, which are converted to carbon dioxide and water. Fossil fuels are mixtures of **hydrocarbons,** compounds containing only hydrogen and carbon.

hydrocarbon

Natural gas is primarily methane, CH_4. Petroleum is a mixture of hydrocarbons (compounds of carbon and hydrogen). Liquefied petroleum gas (LPG) is a mixture of propane (C_3H_8) and butane (C_4H_{10}).

The combustion of these fuels releases a tremendous amount of energy, but reactions won't occur to a significant extent at ordinary temperatures. A spark or a flame must be present before methane will ignite. The amount of energy that must be supplied to start a chemical reaction is called the **activation energy.** In an exothermic reaction once this activation energy is provided, enough energy is then generated to keep the reaction going.

activation energy

Here are some examples:

$$CH_4(g) + 2\ O_2(g) \longrightarrow CO_2(g) + 2\ H_2O(g) + 890\ kJ$$

$$C_3H_8(g) + 5\ O_2(g) \longrightarrow 3\ CO_2(g) + 4\ H_2O(g) + 2200\ kJ$$

$$2\ C_8H_{18}(l) + 25\ O_2(g) \longrightarrow 16\ CO_2(g) + 18\ H_2O(g) + 10,900\ kJ$$

Be careful not to confuse an exothermic reaction that requires heat (activation energy) to get it *started* with an endothermic process that requires energy to keep it going. The combustion of magnesium, for example, is highly exothermic, yet magnesium must be heated to a fairly high temperature in air before combustion begins. Once started, however, the combustion reaction goes very vigorously until either the magnesium or the available supply of oxygen is exhausted. The electrolytic decomposition of water to hydrogen and oxygen is highly endothermic. If the electric current is shut off when this process is going on, the reaction stops instantly. The relative energy levels of reactants and products in exothermic and in endothermic processes are presented graphically in Figures 8.1 and 8.2.

Examples of endothermic and exothermic processes can be easily demonstrated in the laboratory. In Figure 8.1, the products are at a higher potential energy than the reactants. Energy has therefore been absorbed, and the reaction is endothermic. Here, solid $Ba(OH)_2$ and solid NH_4SCN are mixed in a beaker, which is standing in a puddle of water. The solids liquefy and absorb heat from the surroundings causing the beaker to freeze to the board. In Figure 8.2, the products are at a lower potential energy than the reactants. Energy (heat) is given off, producing an exothermic reaction. Here, potassium chlorate ($KClO_3$) and sugar are mixed and placed into a pile. A drop of concentrated sulfuric acid is added, creating a spectacular exothermic reaction.

Outside the laboratory you can experience an endothermic process when you apply a cold pack to an injury. In this case ammonium chloride (NH_4Cl) dissolves in water. For example, temperature changes from 24.5°C to 18.1°C result when 10 g of NH_4Cl are added to 100 mL of water. Energy, in the form of heat, is taken from the immediate surroundings (water) causing the salt solution to become cooler. (See Chemistry in Action, p. 75.)

FIGURE 8.1
Endothermic reaction between Ba(OH)₂ and NH₄SCN. Equal amounts of the two solids are placed in separate beakers. After placing a beaker on a wet board the two solids are mixed in the beaker, causing the board to freeze to the beaker.
▼

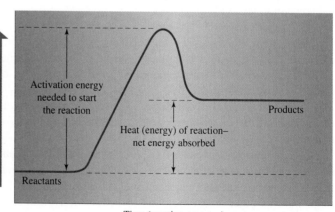

Endothermic reaction

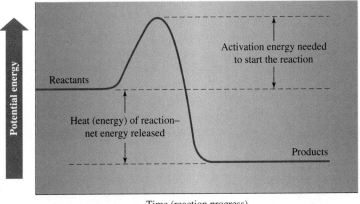

Exothermic reaction

▲
**FIGURE 8.2
Exothermic Reaction between
KClO₃ and sugar. A sample of
KClO₃ and sugar are well
mixed and placed on a
fireproof pad. Several drops
of concentrated H₂SO₄ are
used to ignite the mixture.**

<table>
<tr><td></td></tr>
</table>

8.6   Global Warming: The Greenhouse Effect

Fossil fuels, derived from coal and petroleum, provide the energy we use to power our industries, heat and light our homes and workplaces, and run our cars. As these fuels are burned they produce carbon dioxide and water, releasing over 50 billion tons of carbon dioxide into our atmosphere each year.

The concentration of carbon dioxide has been monitored by scientists since 1958. Analysis of the air trapped in a core sample of snow from Antartica provides data on carbon dioxide levels for the past 160,000 years. The results of this study show that as the carbon dioxide increased, the global temperature increased as well. The levels of carbon dioxide remained reasonably constant from the last ice age, 100,000 years ago, until the industrial revolution. Since then the concentration of carbon dioxide in our atmosphere has risen 15% to an all-time high.

Carbon dioxide is a minor component in our atmosphere and is not usually considered to be a pollutant. The concern expressed by scientists arises from the dramatic increase occurring in the Earth's atmosphere. Without the influence of humans in the environment, the exchange of carbon dioxide between plants and animals would be relatively balanced. Our continued use of fossil fuels has led to an increase of 7.4% in carbon dioxide between 1900 and 1970 and an additional 3.5% increase during the 1980s. Continued increases are being seen in the 1990s.

Besides our growing consumption of fossil fuels, other factors contribute to increased carbon dioxide levels in the atmosphere: Rain forests are being destroyed by cutting and burning to make room for increased population and agricultural needs. Carbon dioxide is added to the atmosphere during the burning, and the loss of trees diminishes the uptake of carbon dioxide by plants.

About half of all the carbon dioxide released into our atmosphere each year remains there, thus increasing its concentration. The other half is absorbed by plants during photosynthesis or is dissolved in the ocean to form hydrogen carbonates and carbonates.

Carbon dioxide and other greenhouse gases, such as methane and water, act to warm our atmosphere by trapping heat near the surface of the Earth. Solar radiation strikes the Earth and warms the surface. The warmed surface then reradiates this

Burning rainforests and ▶ deforestation throughout the world are adding to the greenhouse problem.

FIGURE 8.3 ▶
Elements of a global temperature warming are caused by the greenhouse effect.

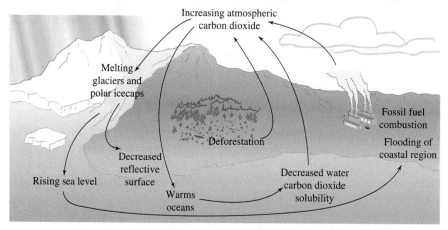

energy as heat. The greenhouse gases absorb some of this heat energy from the surface, which then warms our atmosphere. A similar principle is illustrated in a greenhouse where sunlight comes through the glass yet heat cannot escape. The air in the greenhouse warms, producing a climate considerably different than in nature. In the atmosphere these greenhouse gases may produce dramatic changes in our climate. See Figure 8.3.

Long-term effects of global warming are still a matter of speculation and debate. One consideration is whether the polar ice caps would melt; this would cause a rise in sea level and lead to major flooding on the coasts of our continents. Further effects could include shifts in rainfall patterns, producing droughts and extreme seasonal changes in such major agricultural regions as California. To reverse these trends will require major efforts in the following areas:

- The development of new energy sources to cut our dependence on fossil fuels
- An end to deforestation worldwide
- Intense efforts to improve conservation

On an individual basis each of us can play a significant role. For example, the simple conversion of a 100-watt incandescent light bulb to a compact fluorescent bulb can reduce the electrical consumption for that light by 20%, and the bulb can last ten times longer. Recycling, switching to more fuel-efficient cars, and energy-efficient appliances, heaters, and air conditioners all would result in decreased energy consumption and less carbon dioxide being released into our atmosphere.

Concepts in Review

1. Know the format used in setting up chemical equations.
2. Recognize the various symbols commonly used in writing chemical equations.
3. Be able to balance simple chemical equations.
4. Interpret a balanced equation in terms of the relative numbers or amounts of molecules, atoms, grams, or moles of each substance represented.
5. Classify equations as combination, decomposition, single-displacement, or double-displacement reactions.

6. Use the activity series to predict whether a single-displacement reaction will occur.
7. Complete and balance equations for simple combination, decomposition, single-displacement, and double-displacement reactions when given the reactants.
8. Distinguish between exothermic and endothermic reactions, and relate the quantity of heat to the amounts of substances involved in the reaction.
9. Identify the major sources of chemical energy and their uses.

Key Terms

activation energy (8.5)
balanced equation (8.2)
chemical equation (8.1)
combination reaction (8.4)

decomposition reaction (8.4)
double-displacement reaction (8.4)
endothermic reaction (8.5)
exothermic reaction (8.5)

heat of reaction (8.5)
hydrocarbon (8.5)
single-displacement reaction (8.4)

Questions

1. What is the purpose of balancing equations?
2. What is represented by the numbers (coefficients) that are placed in front of the formulas in a balanced equation?

3. In a balanced chemical equation:
   (a) are atoms conserved?
   (b) are molecules conserved?
   (c) are moles conserved?
   Explain yours answers briefly.
4. Explain how endothermic reactions differ from exothermic reactions.

Paired Exercises

5. Balance these equations:
   (a) $H_2 + O_2 \longrightarrow H_2O$
   (b) $C + Fe_2O_3 \longrightarrow Fe + CO$
   (c) $H_2SO_4 + NaOH \longrightarrow H_2O + Na_2SO_4$
   (d) $Al_2(CO_3)_3 \xrightarrow{\Delta} Al_2O_3 + CO_2$
   (e) $NH_4I + Cl_2 \longrightarrow NH_4Cl + I_2$

7. Classify the reactions in Exercise 5 as combination, decomposition, single-displacement, or double-displacement.

6. Balance these equations:
   (a) $H_2 + Br_2 \longrightarrow HBr$
   (b) $Al + C \xrightarrow{\Delta} Al_4C_3$
   (c) $Ba(ClO_3)_2 \xrightarrow{\Delta} BaCl_2 + O_2$
   (d) $CrCl_3 + AgNO_3 \longrightarrow Cr(NO_3)_3 + AgCl$
   (e) $H_2O_2 \longrightarrow H_2O + O_2$

8. Classify the reactions in Exercise 6 as combination, decomposition, single-displacement, or double-displacement.

9. Balance the following equations:
   (a) $MnO_2 + CO \longrightarrow Mn_2O_3 + CO_2$
   (b) $Mg_3N_2 + H_2O \longrightarrow Mg(OH)_2 + NH_3$
   (c) $C_3H_5(NO_3)_3 \longrightarrow CO_2 + H_2O + N_2 + O_2$
   (d) $FeS + O_2 \longrightarrow Fe_2O_3 + SO_2$
   (e) $Cu(NO_3)_2 \longrightarrow CuO + NO_2 + O_2$
   (f) $NO_2 + H_2O \longrightarrow HNO_3 + NO$
   (g) $Al + H_2SO_4 \longrightarrow Al_2(SO_4)_3 + H_2$
   (h) $HCN + O_2 \longrightarrow N_2 + CO_2 + H_2O$
   (i) $B_5H_9 + O_2 \longrightarrow B_2O_3 + H_2O$

10. Balance the following equations:
   (a) $SO_2 + O_2 \longrightarrow SO_3$
   (b) $Al + MnO_2 \xrightarrow{\Delta} Mn + Al_2O_3$
   (c) $Na + H_2O \longrightarrow NaOH + H_2$
   (d) $AgNO_3 + Ni \longrightarrow Ni(NO_3)_2 + Ag$
   (e) $Bi_2S_3 + HCl \longrightarrow BiCl_3 + H_2S$
   (f) $PbO_2 \xrightarrow{\Delta} PbO + O_2$
   (g) $LiAlH_4 \xrightarrow{\Delta} LiH + Al + H_2$
   (h) $KI + Br_2 \longrightarrow KBr + I_2$
   (i) $K_3PO_4 + BaCl_2 \longrightarrow KCl + Ba_3(PO_4)_2$

11. Change these word equations into formula equations and balance them:
   (a) water $\longrightarrow$ hydrogen + oxygen
   (b) acetic acid + potassium hydroxide $\longrightarrow$
     potassium acetate + water
   (c) phosphorus + iodine $\longrightarrow$ phosphorus triiodide
   (d) aluminum + copper(II) sulfate $\longrightarrow$
     copper + aluminum sulfate
   (e) ammonium sulfate + barium chloride $\longrightarrow$
     ammonium chloride + barium sulfate
   (f) sulfur tetrafluoride + water $\longrightarrow$
     sulfur dioxide + hydrogen fluoride
   (g) chromium(III) carbonate $\xrightarrow{\Delta}$
     chromium(III) oxide + carbon dioxide

12. Change these word equations into formula equations and balance them:
   (a) copper + sulfur $\xrightarrow{\Delta}$ copper(I) sulfide
   (b) phosphoric acid + calcium hydroxide $\longrightarrow$
     calcium phosphate + water
   (c) silver oxide $\xrightarrow{\Delta}$ silver + oxygen
   (d) iron(III) chloride + sodium hydroxide $\longrightarrow$
     iron(III) hydroxide + sodium chloride
   (e) nickel(II) phosphate + sulfuric acid $\longrightarrow$
     nickel(II) sulfate + phosphoric acid
   (f) zinc carbonate + hydrochloric acid $\longrightarrow$
     zinc chloride + water + carbon dioxide
   (g) silver nitrate + aluminum chloride $\longrightarrow$
     silver chloride + aluminum nitrate

13. Use the activity series to predict which of the following reactions will occur. Complete and balance the equations. Where no reaction will occur, write "no reaction" as the product.
   (a) $Ag(s) + H_2SO_4(aq) \longrightarrow$
   (b) $Cl_2(g) + NaBr(aq) \longrightarrow$
   (c) $Mg(s) + ZnCl_2(aq) \longrightarrow$
   (d) $Pb(s) + AgNO_3(aq) \longrightarrow$

14. Use the activity series to predict which of the following reactions will occur. Complete and balance the equations. Where no reaction will occur, write "no reaction" as the product.
   (a) $Cu(s) + FeCl_3(aq) \longrightarrow$
   (b) $H_2(g) + Al_2O_3(aq) \longrightarrow$
   (c) $Al(s) + HBr(aq) \longrightarrow$
   (d) $I_2(s) + HCl(aq) \longrightarrow$

15. Complete and balance the equations for these reactions. All reactions yield products.
   (a) $H_2 + I_2 \longrightarrow$
   (b) $CaCO_3 \xrightarrow{\Delta}$
   (c) $Mg + H_2SO_4 \longrightarrow$
   (d) $FeCl_2 + NaOH \longrightarrow$

16. Complete and balance the equations for these reactions. All reactions yield products.
   (a) $SO_2 + H_2O \longrightarrow$
   (b) $SO_3 + H_2O \longrightarrow$
   (c) $Ca + H_2O \longrightarrow$
   (d) $Bi(NO_3)_3 + H_2S \longrightarrow$

17. Complete and balance the equations for these reactions. All reactions yield products.
   (a) $Ba + O_2 \longrightarrow$
   (b) $NaHCO_3 \xrightarrow{\Delta} Na_2CO_3 +$
   (c) $Ni + CuSO_4 \longrightarrow$
   (d) $MgO + HCl \longrightarrow$
   (e) $H_3PO_4 + KOH \longrightarrow$

18. Complete and balance the equations for these reactions. All reactions yield products.
   (a) $C + O_2 \longrightarrow$
   (b) $Al(ClO_3)_3 \xrightarrow{\Delta} O_2 +$
   (c) $CuBr_2 + Cl_2 \longrightarrow$
   (d) $SbCl_3 + (NH_4)_2S \longrightarrow$
   (e) $NaNO_3 \xrightarrow{\Delta} NaNO_2 +$

19. Interpret these chemical reactions in terms of the number of moles of each reactant and product:
   (a) $MgBr_2 + 2\,AgNO_3 \longrightarrow Mg(NO_3)_2 + 2\,AgBr$
   (b) $N_2 + 3\,H_2 \longrightarrow 2\,NH_3$
   (c) $2\,C_3H_7OH + 9\,O_2 \longrightarrow 6\,CO_2 + 8\,H_2O$

20. Interpret these equations in terms of the relative number of moles of each substance involved and indicate whether the reaction is exothermic or endothermic:
   (a) $2\,Na + Cl_2 \longrightarrow 2\,NaCl + 822\ kJ$
   (b) $PCl_5 + 92.9\ kJ \longrightarrow PCl_3 + Cl_2$

21. Write balanced equations for each of these reactions, including the heat term:
    (a) Lime (CaO) is converted to slaked lime $Ca(OH)_2$ by reaction with water. The reaction liberates 65.3 kJ of heat for each mole of lime reacted.
    (b) The industrial production of aluminum metal from aluminum oxide is an endothermic electrolytic process requiring 1630 kJ per mole of Al_2O_3. Oxygen is also a product.

22. Write a balanced equation for these reactions. Include a heat term on the appropriate side of the equation.
    (a) Powdered aluminum will react with crystals of iodine when moistened with dishwashing detergent. The reaction produces violet sparks and flaming aluminum. The major product is aluminum iodide (AlI_3). The detergent is not a reactant.
    (b) Copper(II) oxide (CuO), a black powder, can be decomposed to produce pure copper by heating the powder in the presence of methane gas (CH_4). The products are copper, carbon dioxide, and water vapor.

Additional Exercises

23. Name one piece of evidence that a chemical reaction is actually taking place in each of these situations:
    (a) making a piece of toast
    (b) frying an egg
    (c) striking a match

24. Balance this equation, using the smallest possible whole numbers. Then determine how many atoms of oxygen appear on each side of the equation:

    $$P_4O_{10} + HClO_4 \longrightarrow Cl_2O_7 + H_3PO_4$$

25. Suppose that in a balanced equation the term 7 $Al_2(SO_4)_3$ appears.
    (a) How many atoms of aluminum are represented?
    (b) How many atoms of sulfur are represented?
    (c) How many atoms of oxygen are represented?
    (d) How many atoms of all kinds are represented?

26. Name two pieces of information that can be obtained from a balanced chemical equation. Name two pieces of information that the reaction does not provide.

27. Make a drawing to show six molecules of ammonia gas decomposing to form hydrogen and nitrogen gases.

28. Explain briefly why this single-displacement reaction will not take place:

    $$Zn + Mg(NO_3)_2 \longrightarrow \text{no reaction}$$

29. A student does an experiment to determine where titanium metal should be placed on the activity series chart. He places newly cleaned pieces of titanium into solutions of nickel(II) nitrate, lead(II) nitrate, and magnesium nitrate. He finds that the titanium reacts with the nickel(II) nitrate and lead(II) nitrate solutions, but not with the magnesium nitrate solution. From this information place titanium in the activity series in a position relative to these ions.

30. Complete and balance the equations for these combination reactions:
    (a) $K + O_2 \longrightarrow$
    (b) $Al + Cl_2 \longrightarrow$
    (c) $CO_2 + H_2O \longrightarrow$
    (d) $CaO + H_2O \longrightarrow$

31. Complete and balance the equations for these decomposition reactions:
    (a) $HgO \overset{\Delta}{\longrightarrow}$
    (b) $NaClO_3 \overset{\Delta}{\longrightarrow}$
    (c) $MgCO_3 \overset{\Delta}{\longrightarrow}$
    (d) $PbO_2 \overset{\Delta}{\longrightarrow} PbO +$

32. Complete and balance the equations for these single-displacement reactions:
    (a) $Zn + H_2SO_4 \longrightarrow$
    (b) $AlI_3 + Cl_2 \longrightarrow$
    (c) $Mg + AgNO_3 \longrightarrow$
    (d) $Al + CoSO_4 \longrightarrow$

33. Complete and balance the equations for these double-displacement reactions:
    (a) $ZnCl_2 + KOH \rightarrow$
    (b) $CuSO_4 + H_2S \rightarrow$
    (c) $Ca(OH)_2 + H_3PO_4 \rightarrow$
    (d) $(NH_4)_3PO_4 + Ni(NO_3)_2 \rightarrow$
    (e) $Ba(OH)_2 + HNO_3 \rightarrow$
    (f) $(NH_4)_2S + HCl \rightarrow$

34. Predict which of the following double-displacement reactions will occur. Complete and balance the equations. Where no reaction will occur, write "no reaction" as the product.
    (a) $AgNO_3(aq) + KCl(aq) \longrightarrow$
    (b) $Ba(NO_3)_2(aq) + MgSO_4(aq) \longrightarrow$
    (c) $H_2SO_4(aq) + Mg(OH)_2(aq) \longrightarrow$
    (d) $MgO(s) + H_2SO_4(aq) \longrightarrow$
    (e) $Na_2CO_3(aq) + NH_4Cl(aq) \longrightarrow$

35. Write balanced equations for the combustion of these hydrocarbons:
    (a) ethane, C_2H_6
    (b) benzene, C_6H_6
    (c) heptane, C_7H_{16}

36. List the factors that contribute to an increase in carbon dioxide in our atmosphere.

37. List three gases considered to be greenhouse gases. Explain why they are given this name.

38. How can the effects of global warming be reduced?

39. What happens to carbon dioxide released into our atmosphere?

Answers to Practice Exercises

8.1 $4\,Al + 3\,O_2 \longrightarrow 2\,Al_2O_3$

8.2 $3\,Mg(OH)_2 + 2\,H_3PO_4 \longrightarrow Mg_3(PO_4)_2 + 6\,H_2O$

8.3 (a) hydrogen gas + chlorine gas $\longrightarrow$ hydrogen chloride

     (b) $H_2(g) + Cl_2(g) \longrightarrow 2\,HCl(g)$

     (c)

$H_2(g)$	+	$Cl_2(g)$	$\longrightarrow$	$2\,HCl(g)$
1 molecule		1 molecule		2 molecules
2 atoms H		2 atoms Cl		2 atoms H
				2 atoms Cl
1 mol		1 mol		2 mol
2.016 g		70.90 g		2 (36.46 g)
				(72.92 g)

     (d) $2\,mol\ H_2 + 2\,mol\ Cl_2 \longrightarrow 4\,mol\ HCl$ (145.8 g)

8.4 (a) $Fe + MgCl_2 \longrightarrow$ no reaction

     (b) $Zn(s) + Pb(NO_3)_2(aq) \longrightarrow Pb(s) + Zn(NO_3)_2(aq)$

8.5 (a) $2\,K_3PO_4(aq) + 3\,BaCl_2(aq) \longrightarrow Ba_3(PO_4)_2(s) + 6\,KCl(aq)$

     (b) $2\,HCl(aq) + NiCO_3(aq) \longrightarrow NiCl_2(aq) + H_2O(l) + CO_2(g)$

     (c) $NH_4Cl(aq) + NaNO_3(aq) \longrightarrow$ no reaction

CHAPTER

9

Calculations from Chemical Equations

▲ A chemist must measure exact quantities of each reactant to produce new chemical compounds.

The old adage "waste not, want not" is appropriate in our daily life and in the laboratory. Determining correct amounts comes into play in most all professions. A seamstress determines the amount of material, lining, and trim necessary to produce a gown for her client by relying on a pattern or her own experience to guide the selection. A carpet layer determines the correct amount of carpet and padding necessary to recarpet a customer's house by calculating the floor area. The IRS determines the correct deduction for federal income taxes from your paycheck based on your expected annual income.

The chemist also finds it necessary to calculate amounts of products or reactants by using a balanced chemical equation. With these calculations, the chemist can control the amount of product by scaling the reaction up or down to fit the needs of the laboratory, and can thereby minimize waste or excess materials formed during the reaction.

9.1   A Short Review

Molar mass   Molar mass is the sum of the atomic masses of all the atoms in a molecule. Molar mass also applies to the mass of a mole of any formula unit—atoms, molecules, or ions; it is the atomic mass of an atom, or the sum of the atomic masses in a molecule or an ion (in grams).

Relationship between molecule and mole   A molecule is the smallest unit of a molecular substance (e.g., Br_2), and a mole is Avogadro's number (6.022×10^{23}) of molecules of that substance. A mole of bromine (Br_2) has the same number of molecules as a mole of carbon dioxide, a mole of water, or a mole of any other molecular substance. When we relate molecules to molar mass, 1 molar mass is equivalent to 1 mol, or 6.022×10^{23} molecules.

The term *mole* also refers to any chemical species. It represents a quantity (6.022×10^{23} particles) and may be applied to atoms, ions, electrons, and formula units of nonmolecular substances. In other words,

$$1 \text{ mole} = \begin{cases} 6.022 \times 10^{23} \text{ molecules} \\ 6.022 \times 10^{23} \text{ formula units} \\ 6.022 \times 10^{23} \text{ atoms} \\ 6.022 \times 10^{23} \text{ ions} \end{cases}$$

Other useful mole relationships are

$$\text{molar mass} = \frac{\text{grams of a substance}}{\text{number of moles of the substance}}$$

$$\text{molar mass} = \frac{\text{grams of a monatomic element}}{\text{number of moles of the element}}$$

$$\text{number of moles} = \frac{\text{number of molecules}}{6.022 \times 10^{23} \text{ molecules/mole}}$$

▲
A mole of water, salt and any gas all have the same number of particles (6.022×10^{23}).

Balanced equations When using chemical equations for calculations of mole–mass–volume relationships between reactants and products, the equations must be balanced. *Remember*: The number in front of a formula in a balanced chemical equation represents the number of moles of that substance in the chemical reaction.

9.2   Introduction to Stoichiometry: The Mole-Ratio Method

We often need to calculate the amount of a substance that is either produced from, or needed to react with, a given quantity of another substance. The area of chemistry that deals with quantitative relationships among reactants and products is known as **stoichiometry** (*stoy-key-ah-meh-tree*). The *mole* or *mole-ratio* method is generally best for solving problems in stoichiometry.

stoichiometry

A **mole ratio** is a ratio between the number of moles of any two species involved in a chemical reaction. For example, in the reaction

mole ratio

$$2\ H_2 + O_2 \longrightarrow 2\ H_2O$$

2 mol   1 mol   2 mol

six mole ratios can be written:

$$\frac{2\ \text{mol}\ H_2}{1\ \text{mol}\ O_2} \qquad \frac{2\ \text{mol}\ H_2}{2\ \text{mol}\ H_2O} \qquad \frac{1\ \text{mol}\ O_2}{2\ \text{mol}\ H_2}$$

$$\frac{1\ \text{mol}\ O_2}{2\ \text{mol}\ H_2O} \qquad \frac{2\ \text{mol}\ H_2O}{2\ \text{mol}\ H_2} \qquad \frac{2\ \text{mol}\ H_2O}{1\ \text{mol}\ O_2}$$

We use the mole ratio to convert the number of moles of one substance to the corresponding number of moles of another substance in a chemical reaction. For example, if we want to calculate the number of moles of H_2O that can be obtained from 4.0 mol of O_2, we use the mole ratio 2 mol H_2O/1 mol O_2:

$$(4.0\ \cancel{\text{mol}\ O_2})\left(\frac{2\ \text{mol}\ H_2O}{1\ \cancel{\text{mol}\ O_2}}\right) = 8.0\ \text{mol}\ H_2O$$

Since you will encounter stoichiometric problems frequently, master the mole-ratio method now. There are three basic steps:

1. Convert the quantity of starting substance to moles (if it is not given in moles).
2. Convert the moles of starting substance to moles of desired substance.
3. Convert the moles of desired substance to the units specified in the problem.

Like balancing chemical equations, making stoichiometric calculations requires practice. A detailed description of the mole-ratio method and several worked examples follow. Study this material and practice on the problems at the end of this chapter.

Write and balance the equation before you begin the problem.

Use a balanced equation.

Step 1.   Determine the number of moles of starting substance.
Identify the starting substance from the data given in the problem state-

ment. Convert the quantity of the starting substance to moles, if it is not already done.

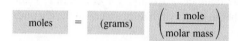

$$\text{moles} = (\text{grams})\left(\frac{1 \text{ mole}}{\text{molar mass}}\right)$$

As in all problems with units, the desired quantity is in the numerator, and the quantity to be eliminated is in the denominator.

Step 2. Determine the mole ratio of the desired substance to the starting substance.

The number of moles of each substance in the balanced equation is indicated by the coefficient in front of each substance. Use these coefficients to set up the mole ratio:

$$\text{mole ratio} = \frac{\text{moles of } \boxed{\text{desired substance}} \text{ in the equation}}{\text{moles of } \boxed{\text{starting substance}} \text{ in the equation}}$$

Multiply the number of moles of starting substance (from Step 1) by the mole ratio to obtain the number of moles of desired substance:

Units of moles of starting substance cancel in the numerator and denominator.

$$\begin{matrix} & & \text{From Step 1} & \\ \text{moles of desired substance} & = & \left(\begin{matrix}\text{moles of starting}\\ \text{substance}\end{matrix}\right) & \left(\dfrac{\text{moles of desired substance in the equation}}{\text{moles of starting substance in the equation}}\right) \end{matrix}$$

Step 3. Calculate the desired substance in the units specified in the problem.

If the answer is to be in moles, the calculation is complete. If units other than moles are wanted, multiply the moles of the desired substance (from Step 2) by the appropriate factor to convert moles to the units required. For example, if grams of the desired substance are wanted,

$$\begin{matrix} & & \text{From Step 2} & \\ \text{grams} & = & (\text{moles}) & \left(\dfrac{\text{molar mass}}{1 \text{ mol}}\right) \end{matrix}$$

$$\text{If moles} \longrightarrow \text{atoms, use } \frac{6.022 \times 10^{23} \text{ atoms}}{1 \text{ mol}}$$

$$\text{If moles} \longrightarrow \text{molecules, use } \frac{6.022 \times 10^{23} \text{ molecules}}{1 \text{ mol}}$$

The steps for converting the mass of a starting substance A to either the mass, atoms, or molecules of desired substance B are summarized in Figure 9.1.

FIGURE 9.1 ▶
Steps for converting starting substance A to mass, atoms, or molecules of desired substance B.

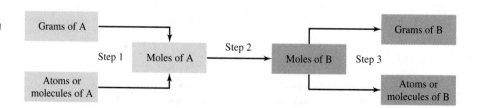

9.3 Mole–Mole Calculations

Let's use the mole-ratio method to solve stoichiometric problems for mole–mole calculations. The quantity of starting substance is given in moles and the quantity of desired substance is requested in moles.

How many moles of carbon dioxide will be produced by the complete reaction of 2.0 mol of glucose ($C_6H_{12}O_6$), according to the following reaction?

Example 9.1

$$C_6H_{12}O_6 + 6\ O_2 \longrightarrow 6\ CO_2 + 6\ H_2O$$

1 mol   6 mol   6 mol   6 mol

The balanced equation states that 6 mol of CO_2 will be produced from 1 mol of $C_6H_{12}O_6$. Even though we can readily see that 12 mol of CO_2 will be formed from 2.0 mol of $C_6H_{12}O_6$, let's use the mole-ratio method to solve the problem.

Solution

Step 1. The number of moles of starting substance is 2.0 mol $C_6H_{12}O_6$.

Step 2. The conversion needed is

moles $C_6H_{12}O_6 \longrightarrow$ moles CO_2

Multiply 2.0 mol of glucose (given in the problem) by this mole ratio:

$$(2.0\ \text{mol}\ C_6H_{12}O_6) \left(\frac{6\ \text{mol}\ CO_2}{1\ \text{mol}\ C_6H_{12}O_6} \right) = 12\ \text{mol}\ CO_2$$

The mole ratio (shown in color) is exact and does not affect the number of significant figures in the answer.

Note how the units work; the moles of $C_6H_{12}O_6$ cancel, leaving the answer in units of moles of CO_2.

How many moles of ammonia can be produced from 8.00 mol of hydrogen reacting with nitrogen? The balanced equation is

Example 9.2

$$3\ H_2 + N_2 \longrightarrow 2\ NH_3$$

Step 1. The starting substance is 8.00 mol of H_2.

Solution

Step 2. The conversion needed is

moles $H_2 \longrightarrow$ moles NH_3

The balanced equation states that we get 2 mol of NH_3 for every 3 mol of H_2 that react. Set up the mole ratio of desired substance (NH_3) to starting substance (H_2):

$$\text{mole ratio} = \frac{2\ \text{mol}\ NH_3}{3\ \text{mol}\ H_2} \quad \text{(from equation)}$$

Multiply the 8.00 mol H_2 by the mole ratio:

$$(8.00\ \text{mol}\ H_2) \left(\frac{2\ \text{mol}\ NH_3}{3\ \text{mol}\ H_2} \right) = 5.33\ \text{mol}\ NH_3$$

Example 9.3   Given the balanced equation

$$K_2Cr_2O_7 + 6\ KI + 7\ H_2SO_4 \longrightarrow Cr_2(SO_4)_3 + 4\ K_2SO_4 + 3\ I_2 + 7\ H_2O$$

1 mol   6 mol   3 mol

calculate (a) the number of moles of potassium dichromate ($K_2Cr_2O_7$) that will react with 2.0 mol of potassium iodide (KI) and (b) the number of moles of iodine (I_2) that will be produced from 2.0 mol of potassium iodide.

Solution   Since the equation is balanced, we are concerned only with $K_2Cr_2O_7$, KI, and I_2, and we can ignore all the other substances. The equation states that 1 mol of $K_2Cr_2O_7$ will react with 6 mol of KI to produce 3 mol of I_2.

(a) Calculate the number of moles of $K_2Cr_2O_7$.

Step 1. The starting substance is 2.0 mol of KI.

Step 2. The conversion needed is

moles KI $\longrightarrow$ moles $K_2Cr_2O_7$

Set up the mole ratio of desired substance to starting substance:

$$\text{mole ratio} = \frac{1\ \text{mol}\ K_2Cr_2O_7}{6\ \text{mol}\ KI} \quad \text{(from equation)}$$

Multiply the moles of starting material by this ratio:

$$(2.0\ \text{mol KI}) \left(\frac{1\ \text{mol}\ K_2Cr_2O_7}{6\ \text{mol KI}} \right) = 0.33\ \text{mol}\ K_2Cr_2O_7$$

(b) Calculate the number of moles of I_2.

Step 1. The moles of starting substance are 2.0 mol KI as in part (a).

Step 2. The conversion needed is

moles KI $\longrightarrow$ moles I_2

Set up the mole ratio of desired substance to starting substance:

$$\text{mole ratio} = \frac{3\ \text{mol}\ I_2}{6\ \text{mol}\ KI} \quad \text{(from equation)}$$

Multiply the moles of starting material by this ratio:

$$(2.0\ \text{mol KI}) \left(\frac{3\ \text{mol}\ I_2}{6\ \text{mol KI}} \right) = 1.0\ \text{mol}\ I_2$$

Example 9.4   How many molecules of water can be produced by reacting 0.010 mol of oxygen with hydrogen?

Solution   The balanced equation is $2\ H_2 + O_2 \longrightarrow 2\ H_2O$.

The sequence of conversions needed in the calculation is

$$\text{moles } O_2 \longrightarrow \text{moles } H_2O \longrightarrow \text{molecules } H_2O$$

Step 1. The starting substance is 0.010 mol O_2.

Step 2. The conversion needed is moles $O_2 \longrightarrow$ moles H_2O. Set up the mole ratio of desired substance to starting substance:

$$\text{mole ratio} = \frac{2 \text{ mol } H_2O}{1 \text{ mol } O_2} \quad \text{(from equation)}$$

Multiply 0.010 mol O_2 by the mole ratio:

$$(0.010 \text{ mol } O_2) \left(\frac{2 \text{ mol } H_2O}{1 \text{ mol } O_2} \right) = 0.020 \text{ mol } H_2O$$

Step 3. Since the problem asks for molecules instead of moles of H_2O, we must convert moles to molecules. Use the conversion factor $(6.022 \times 10^{23}$ molecules)/mole:

$$(0.020 \text{ mol } H_2O) \left(\frac{6.022 \times 10^{23} \text{ molecules}}{1 \text{ mol}} \right) = 1.2 \times 10^{22} \text{ molecules } H_2O$$

Note that 0.020 mol is still quite a large number of water molecules.

▲
The space shuttle is powered by H_2 and O_2, which react to produce H_2O.

Practice 9.1

How many moles of aluminum oxide will be produced from 0.50 mol of oxygen?

$$4 \text{ Al} + 3 \text{ O}_2 \longrightarrow 2 \text{ Al}_2\text{O}_3$$

Practice 9.2

How many moles of aluminum hydroxide are required to produce 22.0 mol of water?

$$2 \text{ Al(OH)}_3 + 3 \text{ H}_2\text{SO}_4 \longrightarrow \text{Al}_2(\text{SO}_4)_3 + 6 \text{ H}_2\text{O}$$

9.4 Mole–Mass Calculations

The object of this type of problem is to calculate the mass of one substance that reacts with or is produced from a given number of moles of another substance in a chemical reaction. If the mass of the starting substance is given, we need to convert it to moles. We use the mole ratio to convert moles of starting substance to moles of desired substance. If required, we can then change moles of desired substance to mass. Each example is solved in two ways.

Method 1: Step-by-step.

Method 2: Continuous calculation, where the individual steps are combined in a single line.

Select the method that is easiest for you and use it to practice solving problems.

Example 9.5   What mass of hydrogen can be produced by reacting 6.0 mol of aluminum with hydrochloric acid?

Solution   The balanced equation is $2\ Al(s) + 6\ HCl(aq) \longrightarrow 2\ AlCl_3(aq) + 3\ H_2(g)$.

Method 1. Step-by-step

First calculate the moles of hydrogen produced, using the mole-ratio method; then calculate the mass of hydrogen by multiplying the moles of hydrogen by its grams per mole.

Step 1. The starting substance is 6.0 mol of aluminum.

Step 2. Calculate moles of H_2 by the mole-ratio method:

$$\text{moles Al} \longrightarrow \text{moles } H_2$$

$$(6.0\ \text{mol Al}) \left(\frac{3\ \text{mol } H_2}{2\ \text{mol Al}} \right) = 9.0\ \text{mol } H_2$$

Step 3. Convert moles of H_2 to grams:

$$\text{moles } H_2 \longrightarrow \text{grams } H_2$$

$$(9.0\ \text{mol } H_2) \left(\frac{2.016\ \text{g } H_2}{1\ \text{mol } H_2} \right) = 18\ \text{g } H_2$$

We see that 18 g of H_2 can be produced by reacting 6.0 mol of Al with HCl.

Method 2. Continuous calculation

$$\text{moles Al} \longrightarrow \text{moles } H_2 \longrightarrow \text{grams } H_2$$

$$(6.0\ \text{mol Al}) \left(\frac{3\ \text{mol } H_2}{2\ \text{mol Al}} \right) \left(\frac{2.016\ \text{g } H_2}{1\ \text{mol } H_2} \right) = 18\ \text{g } H_2$$

▲
Al wire in HCl.

Example 9.6   How many moles of water can be produced by burning 325 g of octane (C_8H_{18})? The balanced equation is $2\ C_8H_{18}(g) + 25\ O_2(g) \longrightarrow 16\ CO_2(g) + 18\ H_2O(g)$.

Solution   **Method 1. Step-by-step**

Step 1. The starting substance is 325 g C_8H_{18}. Convert 325 g of C_8H_{18} to moles:

$$\text{grams } C_8H_{18} \longrightarrow \text{moles } C_8H_{18}$$

$$(325\ \text{g } C_8H_{18}) \left(\frac{1\ \text{mol } C_8H_{18}}{114.2\ \text{g } C_8H_{18}} \right) = 2.85\ \text{mol } C_8H_{18}$$

Step 2. Calculate the moles of water by the mole-ratio method:

moles C_8H_{18} ⟶ moles H_2O

$$(2.85 \text{ mol } C_8H_{18}) \left(\frac{18 \text{ mol } H_2O}{2 \text{ mol } C_8H_{18}} \right) = 25.7 \text{ mol } H_2O$$

Method 2. Continuous calculation

grams C_8H_{18} ⟶ moles C_8H_{18} ⟶ moles H_2O

$$(325 \text{ g } C_8H_{18}) \left(\frac{1 \text{ mol } C_8H_{18}}{114.2 \text{ g } C_8H_{18}} \right) \left(\frac{18 \text{ mol } H_2O}{2 \text{ mol } C_8H_{18}} \right) \doteq 25.6 \text{ mol } H_2O$$

The answers for the different methods vary in the last digit. This results from rounding off at different times in the calculation. Check with your instructor to find the appropriate rules for your course.

Practice 9.3

How many moles of potassium chloride can be produced from 100.0 g of potassium chlorate?

$$2 \text{ KClO}_3 \longrightarrow 2 \text{ KCl} + 3 \text{ O}_2$$

Practice 9.4

How many grams of silver nitrate are required to produce 0.25 mol of silver sulfide?

$$2 \text{ AgNO}_3 + H_2S \longrightarrow Ag_2S + 2 \text{ HNO}_3$$

9.5   Mass–Mass Calculations

Solving mass–mass stoichiometry problems requires all the steps of the mole-ratio method. The mass of starting substance is converted to moles. The mole ratio is then used to determine moles of desired substance, which in turn is converted to mass.

Use either the step-by-step method or the continuous calculation method to solve these problems.

What mass of carbon dioxide is produced by the complete combustion of 100. g of the hydrocarbon pentane, C_5H_{12}? The balanced equation is

Example 9.7

$$C_5H_{12} + 8 \text{ O}_2 \qquad 5 \text{ CO}_2 + 6 \text{ H}_2O$$

Method 1. Step-by-step

Solution

Step 1. The starting substance is 100. g of C_5H_{12}. Convert 100. g of C_5H_{12} to moles:

grams ⟶ moles

$$(100. \text{ g } C_5H_{12}) \left(\frac{1 \text{ mol } C_5H_{12}}{72.15 \text{ g } C_5H_{12}} \right) = 1.39 \text{ mol } C_5H_{12}$$

Step 2. Calculate the moles of CO_2 by the mole-ratio method:

moles C_5H_{12} ⟶ moles CO_2

$$(1.39 \text{ mol } C_5H_{12}) \left(\frac{5 \text{ mol } CO_2}{1 \text{ mol } C_5H_{12}} \right) = 6.95 \text{ mol } CO_2$$

Step 3. Convert moles of CO_2 to grams:

moles CO_2 ⟶ grams CO_2

$$(6.95 \text{ mol } CO_2) \left(\frac{44.01 \text{ g } CO_2}{1 \text{ mol } CO_2} \right) = 306 \text{ g } CO_2$$

Method 2. Continuous calculation

Remember: Round off as appropriate for your particular course.

grams C_5H_{12} ⟶ moles C_5H_{12} ⟶ moles CO_2 ⟶ grams CO_2

$$(100. \text{ g } C_5H_{12}) \left(\frac{1 \text{ mol } C_5H_{12}}{72.15 \text{ g } C_5H_{12}} \right) \left(\frac{5 \text{ mol } CO_2}{1 \text{ mol } C_5H_{12}} \right) \left(\frac{44.01 \text{ g } CO_2}{1 \text{ mol } CO_2} \right) = 305 \text{ g } CO_2$$

Example 9.8   How many grams of nitric acid, HNO_3, are required to produce 8.75 g of dinitrogen monoxide (N_2O) according to the following equation?

$$4 \text{ Zn}(s) + 10 \text{ HNO}_3(aq) \longrightarrow 4 \text{ Zn(NO}_3)_2(aq) + N_2O(g) + 5 H_2O(l)$$
$$\phantom{4 \text{ Zn}(s) + \,}10 \text{ mol} \phantom{\longrightarrow 4 \text{ Zn(NO}_3)_2(aq) + \,}1 \text{ mol}$$

Solution

Method 1. Step-by-step

Step 1. The starting substance is 8.75 g of N_2O.

grams N_2O ⟶ moles N_2O

$$(8.75 \text{ g } N_2O) \left(\frac{1 \text{ mol } N_2O}{44.02 \text{ g } N_2O} \right) = 0.199 \text{ mol } N_2O$$

Step 2. Calculate the moles of HNO_3 by the mole-ratio method:

moles N_2O ⟶ moles HNO_3

$$(0.199 \text{ mol } N_2O) \left(\frac{10 \text{ mol } HNO_3}{1 \text{ mol } N_2O} \right) = 1.99 \text{ mol } HNO_3$$

Step 3. Convert moles of HNO_3 to grams:

moles HNO_3 ⟶ grams HNO_3

$$(1.99 \text{ mol } HNO_3) \left(\frac{63.02 \text{ g } HNO_3}{1 \text{ mol } HNO_3} \right) = 125 \text{ g } HNO_3$$

Method 2. Continuous calculation

grams N_2O ⟶ moles N_2O ⟶ moles HNO_3 ⟶ grams HNO_3

$$(8.75 \text{ g } N_2O)\left(\frac{1 \text{ mol } N_2O}{44.02 \text{ g } N_2O}\right)\left(\frac{10 \text{ mol } HNO_3}{1 \text{ mol } N_2O}\right)\left(\frac{63.02 \text{ g } HNO_3}{1 \text{ mol } HNO_3}\right) = 125 \text{ g } HNO_3$$

Practice 9.5

How many grams of chromium(III) chloride are required to produce 75.0 g of silver chloride?

$$CrCl_3 + 3 AgNO_3 \longrightarrow Cr(NO_3)_3 + 3 AgCl$$

Practice 9.6

What mass of water is produced by the complete combustion of 225.0 g of butane (C_4H_{10})?

$$2 C_4H_{10} + 13 O_2 \longrightarrow 8 CO_2 + 10 H_2O$$

9.6   Limiting-Reactant and Yield Calculations

In many chemical processes the quantities of the reactants used are such that one re-actant is in excess. The amount of the product(s) formed in such a case depends on the reactant that is not in excess. This is the **limiting reactant**—it limits the amount of product that can be formed.

limiting reactant

Consider the case illustrated in Figure 9.2. How many bicycles can be assem-bled from the parts shown? The limiting part in this case is the number of pedal as-sembles; only three bicycles can be built because there are only three pedal assem-blies. The wheels and frames are parts in excess.

Let's consider a chemical example at the molecular level in which seven mole-cules of H_2 are combined with four molecules of Cl_2 (Figure 9.3a). How many mol-ecules of HCl can be produced according to this reaction?

$$H_2 + Cl_2 \longrightarrow 2 \, HCl$$

If the molecules of H_2 and Cl_2 are taken apart and recombined as HCl (Figure 9.3b), we see that eight molecules of HCl can be formed before we run out of Cl_2. Therefore the Cl_2 is the limiting reactant and H_2 is in excess—three molecules of H_2 remain unreacted.

FIGURE 9.2 ▶
The number of bicycles that can be built from these parts is determined by the "limiting reactant" (the pedal assemblies).

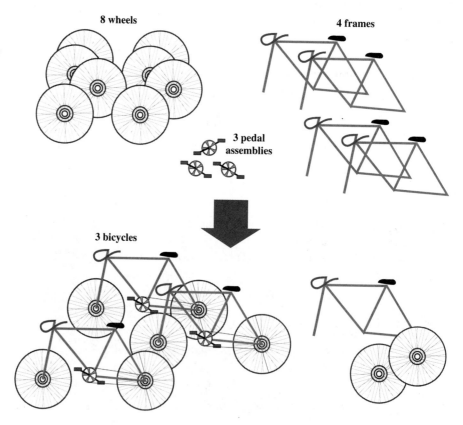

8 wheels

4 frames

3 pedal assemblies

3 bicycles

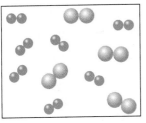

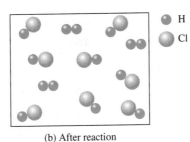

H
Cl

(a) Before reaction   (b) After reaction

When problem statements give the amounts of two reactants, one of them is usually a limiting reactant. We can identify the limiting reactant using the following method:

1. Calculate the amount of product (moles or grams, as needed) formed from each reactant.
2. Determine which reactant is limiting. (The reactant that gives the least amount of product is the limiting reactant; the other reactant is in excess.)

Sometimes we must find the amount of excess reactant needed in the reaction, so we add a third step:

3. Calculate the amount of excess reactant required to react with the limiting reactant, then subtract this amount from the starting quantity of the excess reactant. This gives the amount of that substance that remains unreacted.

How many moles of Fe_3O_4 can be obtained by reacting 16.8 g Fe with 10.0 g H_2O? Which substance is the limiting reactant? Which substance is in excess?

Example 9.9

$$3\ Fe(s) + 4\ H_2O(g) \xrightarrow{\Delta} Fe_3O_4(s) + 4\ H_2(g)$$

Solution

Step 1. Calculate the moles of Fe_3O_4 that can be formed from each reactant:

g reactant ⟶ mol reactant ⟶ mol Fe_3O_4

The continuous calculation method is shown here. You can also use the step-by-step method to determine the requested substance.

$$(16.8\ g\ \text{Fe})\left(\frac{1\ \text{mol Fe}}{55.85\ g\ \text{Fe}}\right)\left(\frac{1\ \text{mol } Fe_3O_4}{3\ \text{mol Fe}}\right) = 0.100\ \text{mol } Fe_3O_4$$

$$(10.0\ g\ H_2O)\left(\frac{1\ \text{mol } H_2O}{18.02\ g\ H_2O}\right)\left(\frac{1\ \text{mol } Fe_3O_4}{4\ \text{mol } H_2O}\right) = 0.139\ \text{mol } Fe_3O_4$$

Step 2. Determine the limiting reactant. The limiting reactant is Fe because it produces less Fe_3O_4; the H_2O is in excess. The yield of product is 0.100 mol of Fe_3O_4.

How many grams of silver bromide (AgBr) can be formed when solutions containing 50.0 g of $MgBr_2$ and 100.0 g of $AgNO_3$ are mixed together? How many grams of the excess reactant remain unreacted?

Example 9.10

Solution

$$MgBr_2(aq) + 2\, AgNO_3(aq) \longrightarrow 2\, AgBr(s) + Mg(NO_3)_2(aq)$$

Step 1. Calculate the grams of AgBr that can be formed from each reactant.

$$\text{g reactant} \longrightarrow \text{mol reactant} \longrightarrow \text{mol AgBr} \longrightarrow \text{g AgBr}$$

$$(50.0\text{ g MgBr}_2)\left(\frac{1\text{ mol MgBr}_2}{184.1\text{ g MgBr}_2}\right)\left(\frac{2\text{ mol AgBr}}{1\text{ mol MgBr}_2}\right)\left(\frac{187.8\text{ g AgBr}}{1\text{ mol AgBr}}\right) = 102\text{ g AgBr}$$

$$(100.0\text{ g AgNO}_3)\left(\frac{1\text{ mol AgNO}_3}{169.9\text{ g AgNO}_3}\right)\left(\frac{2\text{ mol AgBr}}{2\text{ mol AgNO}_3}\right)\left(\frac{187.8\text{ g AgBr}}{1\text{ mol AgBr}}\right) = 110.5\text{ g AgBr}$$

Step 2. Determine the limiting reactant. The limiting reactant is $MgBr_2$ because it gives less AgBr; $AgNO_3$ is in excess. The yield is 102 g AgBr.

Step 3. Calculate the grams of unreacted $AgNO_3$. Calculate the grams of $AgNO_3$ that will react with 50.0 g of $MgBr_2$:

$$\text{g MgBr}_2 \longrightarrow \text{mol MgBr}_2 \longrightarrow \text{mol AgNO}_3 \longrightarrow \text{g AgNO}_3$$

$$(50.0\text{ g MgBr}_2)\left(\frac{1\text{ mol MgBr}_2}{184.1\text{ g MgBr}_2}\right)\left(\frac{2\text{ mol AgNO}_3}{1\text{ mol MgBr}_2}\right)\left(\frac{169.9\text{ g AgNO}_3}{1\text{ mol AgNO}_3}\right) = 92.3\text{ g AgNO}_3$$

Thus 92.3 g of $AgNO_3$ reacts with 50.0 g of $MgBr_2$. The amount of $AgNO_3$ that remains unreacted is

$$100.0\text{ g AgNO}_3 - 92.3\text{ g AgNO}_3 = 7.7\text{ g AgNO}_3 \text{ unreacted}$$

The final mixture will contain 102 g AgBr(s), 7.7 g $AgNO_3$, and an undetermined amount of $Mg(NO_3)_2$ in solution.

Practice 9.7

How many grams of hydrogen chloride can be produced from 0.490 g of hydrogen and 50.0 g of chlorine?

$$H_2(g) + Cl_2(g) \longrightarrow 2\, HCl(g)$$

Practice 9.8

How many grams of barium sulfate will be formed from 200.0 g of barium nitrate and 100.0 grams of sodium sulfate?

$$Ba(NO_3)_2(aq) + Na_2SO_4(aq) \longrightarrow BaSO_4(s) + 2\, NaNO_3(aq)$$

The quantities of the products we have been calculating from equations represent the maximum yield (100%) of product according to the reaction represented by the equation. Many reactions, especially those involving organic substances, fail to give a 100% yield of product. The main reasons for this failure are the side reactions that give products other than the main product and the fact that many reactions are reversible. In addition, some product may be lost in handling and transferring from one vessel to another. The **theoretical yield** of a reaction is the calculated amount of product that can be obtained from a given amount or reactant, according to the

theoretical yield

chemical equation. The **actual yield** is the amount of product that we finally obtain.

actual yield

The **percent yield** is the ratio of the actual yield to the theoretical yield multiplied by 100. Both the theoretical and the actual yields must have the same units to obtain a percent:

percent yield

$$\frac{\text{actual yield}}{\text{theoretical yield}} \times 100 = \text{percent yield}$$

For example, if the theoretical yield calculated for a reaction is 14.8 g and the amount of product obtained is 9.25 g, the percent yield is

Round off as appropriate for your particular course.

$$\text{percent yield} = \left(\frac{9.25 \text{ g}}{14.8 \text{ g}}\right)(100) = 62.5\%$$

Carbon tetrachloride (CCl_4) was prepared by reacting 100. g of carbon disulfide and 100. g of chlorine. Calculate the percent yield if 65.0 g of CCl_4 was obtained from the reaction:

Example 9.11

$$CS_2 + 3 Cl_2 \longrightarrow CCl_4 + S_2Cl_2$$

We need to determine the limiting reactant to calculate the quantity of CCl_4 (theoretical yield) that can be formed. Then we can compare that amount with the 65.0 g CCl_4 (actual yield) to calculate the percent yield.

Solution

Step 1. Determine the theoretical yield. Calculate the grams of CCl_4 that can be formed from each reactant:

$$\text{g reactant} \longrightarrow \text{mol reactant} \longrightarrow \text{mol } CCl_4 \longrightarrow \text{g } CCl_4$$

$$(100. \text{ g } CS_2)\left(\frac{1 \text{ mol } CS_2}{76.15 \text{ g } CS_2}\right)\left(\frac{1 \text{ mol } CCl_4}{1 \text{ mol } CS_2}\right)\left(\frac{153.8 \text{ g } CCl_4}{1 \text{ mol } CCl_4}\right) = 202 \text{ g } CCl_4$$

$$(100. \text{ g } Cl_2)\left(\frac{1 \text{ mol } Cl_2}{70.90 \text{ g } Cl_2}\right)\left(\frac{1 \text{ mol } CCl_4}{3 \text{ mol } Cl_2}\right)\left(\frac{153.8 \text{ g } CCl_4}{1 \text{ mol } CCl_4}\right) = 72.3 \text{ g } CCl_4$$

Step 2. Determine the limiting reactant. The limiting reactant is Cl_2 because it gives less CCl_4. The CS_2 is in excess. The theoretical yield is 72.3 g CCl_4.

Step 3. Calculate the percent yield. According to the equation, 72.3 g of CCl_4 is the maximum amount or theoretical yield of CCl_4 possible from 100. g of Cl_2. Actual yield is 65.0 g of CCl_4:

$$\text{percent yield} = \left(\frac{65.0 \text{ g}}{72.3 \text{ g}}\right)(100) = 89.9\%$$

Silver bromide was prepared by reacting 200.0 g of magnesium bromide and an adequate amount of silver nitrate. Calculate the percent yield if 375.0 g of silver bromide was obtained from the reaction:

Example 9.12

$$MgBr_2 + 2 AgNO_3 \longrightarrow Mg(NO_3)_3 + 2 AgBr$$

Solution

Step 1. Determine the theoretical yield. Calculate the grams of AgBr that can be formed:

$$g\ MgBr_2 \longrightarrow mol\ MgBr_2 \longrightarrow mol\ AgBr \longrightarrow g\ AgBr$$

$$(200.0\ g\ MgBr_2)\left(\frac{1\ mol\ MgBr_2}{184.1\ g\ MgBr_2}\right)\left(\frac{2\ mol\ AgBr}{1\ mol\ MgBr_2}\right)\left(\frac{187.8\ g\ AgBr}{1\ mol\ AgBr}\right) = 408.0\ g\ AgBr$$

The theoretical yield is 408.0 g AgBr.

Step 2. Calculate the percent yield. According to the equation, 408.0 g AgBr is the maximum amount of AgBr possible from 200.0 g $MgBr_2$. Actual yield is 375.0 g AgBr:

$$percent\ yield = \frac{375.0\ g\ AgBr}{408.0\ g\ AgBr} \times 100 = 91.91\%$$

Practice 9.9

Aluminum oxide was prepared by heating 225 g of chromium(II) oxide with 125 g of aluminum. Calculate the percent yield if 100.0 g of aluminum oxide was obtained:

$$2\ Al + 3\ CrO \longrightarrow Al_2O_3 + 3\ Cr$$

When solving problems, you will achieve better results if you work in an organized manner.

1. Write data and numbers in a logical, orderly manner.
2. Make certain that the equations are balanced and that the computations are accurate and expressed to the correct number of significant figures.
3. Check the units. Units are very important; a number without units has little or no meaning.

Concepts in Review

1. Write mole ratios for any two substances involved in a chemical reaction.
2. Outline the mole-ratio method for making stoichiometric calculations.
3. Calculate the number of moles of a desired substance obtainable from a given number of moles of a starting substance in a chemical reaction (mole–mole calculations).
4. Calculate the mass of a desired substance obtainable from a given number of moles of a starting substance in a chemical reaction, and vice versa (mole–mass and mass–mole calculations).

5. Calculate the mass of a desired substance involved in a chemical reaction from a given mass of a starting substance (mass–mass calculation).
6. Deduce the limiting reactant when given the amounts of starting substances, and then calculate the moles or mass of desired substance obtainable from a given chemical reaction (limiting reactant calculation).
7. Apply theoretical yield or actual yield to any of the foregoing types of problems, or calculate theoretical and actual yields of a chemical reaction.

Key Terms

actual yield (9.6)   percent yield (9.6)
limiting reactant (9.6)   stoichiometry (9.2)
mole ratio (9.2)   theoretical yield (9.6)

Questions

1. Phosphine (PH_3) can be prepared by the hydrolysis of calcium phosphide, Ca_3P_2:

$$Ca_3P_2 + 6 H_2O \longrightarrow 3 Ca(OH)_2 + 2 PH_3$$

Based on this equation, which of the following statements are correct? Show evidence to support your answer.
(a) One mole of Ca_3P_2 produces 2 mol of PH_3.
(b) One gram of Ca_3P_2 produces 2 g of PH_3.
(c) Three moles of $Ca(OH)_2$ are produced for each 2 mol of PH_3 produced.
(d) The mole ratio between phosphine and calcium phosphide is
$$\frac{2 \text{ mol } PH_3}{1 \text{ mol } Ca_3P_2}.$$
(e) When 2.0 mol of Ca_3P_2 and 3.0 mol of H_2O react, 4.0 mol of PH_3 can be formed.
(f) When 2.0 mol of Ca_3P_2 and 15.0 mol of H_2O react, 6.0 mol of $Ca(OH)_2$ can be formed.
(g) When 200. g of Ca_3P_2 and 100. g of H_2O react, Ca_3P_2 is the limiting reactant.
(h) When 200. g of Ca_3P_2 and 100. g of H_2O react, the theoretical yield of PH_3 is 57.4 g.

2. The equation representing the reaction used for the commercial preparation of hydrogen cyanide is

$$2 CH_4 + 3 O_2 + 2 NH_3 \longrightarrow 2 HCN + 6 H_2O$$

Based on this equation, which of the following statements are correct? Rewrite incorrect statements to make them correct.
(a) Three moles of O_2 are required for 2 mol of NH_3.
(b) Twelve moles of HCN are produced for every 16 mol of O_2 that react.
(c) The mole ratio between H_2O and CH_4 is
$$\frac{6 \text{ mol } H_2O}{2 \text{ mol } CH_4}.$$
(d) When 12 mol of HCN are produced, 4 mol of H_2O will be formed.
(e) When 10 mol of CH_4, 10 mol of O_2, and 10 mol of NH_3 are mixed and reacted, O_2 is the limiting reactant.
(f) When 3 mol each of CH_4, O_2, and NH_3 are mixed and reacted, 3 mol of HCN will be produced.

Paired Exercises

Mole Review Exercises

3. Calculate the number of moles in these quantities:
(a) 25.0 g KNO_3
(b) 56 millimol NaOH
(c) 5.4×10^2 g $(NH_4)_2C_2O_4$
(d) 16.8 mL H_2SO_4 solution ($d = 1.727$ g/mL, 80.0% H_2SO_4 by mass)

5. Calculate the number of grams in these quantities:
(a) 2.55 mol $Fe(OH)_3$
(b) 125 kg $CaCO_3$
(c) 10.5 mol NH_3
(d) 72 millimol HCl
(e) 500.0 mL of liquid Br_2 ($d = 3.119$ g/mL)

4. Calculate the number of moles in these quantities:
(a) 2.10 kg $NaHCO_3$
(b) 525 mg $ZnCl_2$
(c) 9.8×10^{24} molecules CO_2
(d) 250 mL ethyl alcohol, C_2H_5OH ($d = 0.789$ g/mL)

6. Calculate the number of grams in these quantities:
(a) 0.00844 mol $NiSO_4$
(b) 0.0600 mol $HC_2H_3O_2$
(c) 0.725 mol Bi_2S_3
(d) 4.50×10^{21} molecules glucose, $C_6H_{12}O_6$
(e) 75 mL K_2CrO_4 solution ($d = 1.175$ g/mL, 20.0% K_2CrO_4 by mass)

7. Which contains the larger number of molecules, 10.0 g H_2O or 10.0 g H_2O_2? Show evidence for your answer.

8. Which contains the larger numbers of molecules, 25.0 g HCl or 85.0 g $C_6H_{12}O_6$? Show evidence for your answer.

Mole-Ratio Exercises

9. Given the equation for the combustion of isopropyl alcohol

$$2 C_3H_7OH + 9 O_2 \longrightarrow 6 CO_2 + 8 H_2O$$

set up the mole ratio of
(a) CO_2 to C_3H_7OH (d) H_2O to C_3H_7OH
(b) C_3H_7OH to O_2 (e) CO_2 to H_2O
(c) O_2 to CO_2 (f) H_2O to O_2

10. For the reaction

$$3 CaCl_2 + 2 H_3PO_4 \longrightarrow Ca_3(PO_4)_2 + 6 HCl$$

set up the mole ratio of
(a) $CaCl_2$ to $Ca_3(PO_4)_2$ (d) $Ca_3(PO_4)_2$ to H_3PO_4
(b) HCl to H_3PO_4 (e) HCl to $Ca_3(PO_4)_2$
(c) $CaCl_2$ to H_3PO_4 (f) H_3PO_4 to HCL

11. How many moles of CO_2 can be produced from 7.75 mol C_2H_5OH? (Balance the equation first.)

$$C_2H_5OH + O_2 \longrightarrow CO_2 + H_2O$$

12. How many moles of Cl_2 can be produced from 5.60 mol HCl?

$$4 HCl + O_2 \longrightarrow 2 Cl_2 + 2 H_2O$$

13. An early method of producing chlorine was by reacting pyrolusite (MnO_2) with hydrochloric acid. How many moles of HCl will react with 1.05 mol of MnO_2? (Balance the equation first.)

$$MnO_2 (s) + HCl (aq) \longrightarrow Cl_2 (g) + $$
$$MnCl_2 (aq) + H_2O (l)$$

14. Given the equation

$$Al_4C_3 + 12 H_2O \longrightarrow 4 Al(OH)_3 + 3 CH_4$$

(a) How many moles of water are needed to react with 100. g of Al_4C_3?
(b) How many moles of $Al(OH)_3$ will be produced when 0.600 mol of CH_4 is formed?

15. How many grams of sodium hydroxide can be produced from 500 g of calcium hydroxide according to this equation?

$$Ca(OH)_2 + Na_2CO_3 \longrightarrow 2 NaOH + CaCO_3$$

16. How many grams of zinc phosphate, $Zn_3(PO_4)_2$, are formed when 10.0 g of Zn are reacted with phosphoric acid?

$$3 Zn + 2 H_3PO_4 \longrightarrow Zn_3(PO_4)_2 + 3 H_2$$

17. In a blast furnace, iron(III) oxide reacts with coke (carbon) to produce molten iron and carbon monoxide:

$$Fe_2O_3 + 3 C \xrightarrow{\Delta} 2 Fe + 3 CO$$

How many kilograms of iron would be formed from 125 kg of Fe_2O_3?

18. How many grams of steam and iron must react to produce 375 g of magnetic iron oxide, Fe_3O_4?

$$3 Fe(s) + 4 H_2O(g) \longrightarrow Fe_3O_4(s) + 4 H_2(g)$$

19. Ethane gas, C_2H_6, burns in air (i.e., reacts with the oxygen in air) to form carbon dioxide and water:

$$2 C_2H_6 + 7 O_2 \longrightarrow 4 CO_2 + 6 H_2O$$

(a) How many moles of O_2 are needed for the complete combustion of 15.0 mol of ethane?
(b) How many grams of CO_2 are produced for each 8.00 g of H_2O produced?
(c) How many grams of CO_2 will be produced by the combustion of 75.0 g of C_2H_6?

20. Given the equation

$$4 FeS_2 + 11 O_2 \longrightarrow 2 Fe_2O_3 + 8 SO_2$$

(a) How many moles of Fe_2O_3 can be made from 1.00 mol of FeS_2?
(b) How many moles of O_2 are required to react with 4.50 mol of FeS_2?
(c) If the reaction produces 1.55 mol of Fe_2O_3, how many moles of SO_2 are produced?
(d) How many grams of SO_2 can be formed from 0.512 mol of FeS_2?
(e) If the reaction produces 40.6 g of SO_2, how many moles of O_2 were reacted?
(f) How many grams of FeS_2 are needed to produce 221 g of Fe_2O_3?

Limiting-Reactant and Percent-Yield Exercises

21. In the following equations, determine which reactant is the limiting reactant and which reactant is in excess. The amounts used are given below each reactant. Show evidence for your answers.
   (a) $KOH + HNO_3 \longrightarrow KNO_3 + H_2O$
       16.0 g   12.0 g
   (b) $2 NaOH + H_2SO_4 \longrightarrow Na_2SO_4 + 2 H_2O$
       10.0 g   10.0 g

23. The reaction for the combustion of propane is

    $$C_3H_8 + 5 O_2 \longrightarrow 3 CO_2 + 4 H_2O$$

   (a) If 20.0 g of C_3H_8 and 20.0 g of O_2 are reacted, how many moles of CO_2 can be produced?
   (b) If 20.0 g of C_3H_8 and 80.0 g of O_2 are reacted, how many moles of CO_2 are produced?
   (c) If 2.0 mol of C_3H_8 and 14.0 mol of O_2 are placed in a closed container and they react to completion (until one reactant is completely used up), what compounds will be present in the container after the reaction, and how many moles of each compound are present?

*25. When a certain nonmetal whose formula is X_8 burns in air, XO_3 forms. Write a balanced equation for this reaction. If 120.0 g of oxygen gas is consumed completely, along with 80.0 g of X_8, identify element X.

27. Aluminum reacts with bromine to form aluminum bromide:

    $$2 Al + 3 Br_2 \longrightarrow 2 AlBr_3$$

   If 25.0 g of Al and 100. g of Br_2 are reacted, and 64.2 g of $AlBr_3$ product is recovered, what is the percent yield for the reaction?

*29. Carbon disulfide, CS_2, can be made from coke, C, and sulfur dioxide, SO_2:

    $$3 C + 2 SO_2 \longrightarrow CS_2 + 2 CO_2$$

   If the actual yield of CS_2 is 86.0% of the theoretical yield, what mass of coke is needed to produce 950 g of CS_2?

22. In the following equations, determine which reactant is the limiting reactant and which reactant is in excess. The amounts used are given below each reactant. Show evidence for your answers.
   (a) $2 Bi(NO_3)_3 + 3 H_2S \longrightarrow Bi_2S_3 + 6 HNO_3$
       50.0 g   6.00 g
   (b) $3 Fe + 4 H_2O \longrightarrow Fe_3O_4 + 4 H_2$
       40.0 g   16.0 g

24. The reaction for the combustion of propane is

    $$C_3H_8 + 5 O_2 \longrightarrow 3 CO_2 + 4 H_2O$$

   (a) If 5.0 mol of C_3H_8 and 5.0 mol of O_2 are reacted, how many moles of CO_2 can be produced?
   (b) If 3.0 mol of C_3H_8 and 20.0 mol of O_2 are reacted, how many moles of CO_2 are produced?
   (c) If 20.0 mol of C_3H_8 and 3.0 mol of O_2 are reacted, how many moles of CO_2 can be produced?

*26. When a particular metal X reacts with HCl, the resulting products are XCl_2 and H_2. Write and balance the equation. When 78.5 g of the metal reacts completely, 2.42 g of hydrogen gas results. Identify the element X.

28. Iron was reacted with a solution containing 400. g of copper(II) sulfate. The reaction was stopped after 1 hour, and 151 g of copper was obtained. Calculate the percent yield of copper obtained.

    $$Fe(s) + CuSO_4(aq) \longrightarrow Cu(s) + FeSO_4(aq)$$

*30. Acetylene (C_2H_2) can be manufactured by the reaction of water and calcium carbide, CaC_2:

    $$CaC_2(s) + 2 H_2O(l) \longrightarrow C_2H_2(g) + Ca(OH)_2(s)$$

   When 44.5 g of commercial grade (impure) calcium carbide is reacted, 0.540 mol of C_2H_2 is produced. Assuming that all of the CaC_2 was reacted to C_2H_2, what is the percent of CaC_2 in the commercial grade material?

Additional Exercises

31. A tool set contains 6 wrenches, 4 screwdrivers, and 2 pliers. The manufacturer has 1000 pliers, 2000 screwdrivers, and 3000 wrenches in stock. Can an order for 600 tool sets be filled? Explain briefly.

32. What is the difference between using a number as a subscript and using a number as a coefficient in a chemical equation?

33. Oxygen masks for producing O_2 in emergency situations contain potassium superoxide (KO_2). It reacts according to this equation:

$$4 KO_2 + 2 H_2O + 4 CO_2 \longrightarrow 4 KHCO_3 + 3 O_2$$

(a) If a person wearing such a mask exhales 0.85 g of CO_2 every minute, how many moles of KO_2 are consumed in 10.0 minutes?

(b) How many grams of oxygen are produced in 1.0 hour?

34. Ethyl alcohol is the result of fermentation of sugar, $C_6H_{12}O_6$:

$$C_6H_{12}O_6 \longrightarrow 2 C_2H_5OH + 2 CO_2$$

(a) How many grams of ethyl alcohol and how many grams of carbon dioxide can be produced from 750 g of sugar?

(b) How many milliliters of alcohol ($d = 0.79$ g/mL) can be produced from 750 g of sugar?

35. Phosphoric acid, H_3PO_4, can be synthesized from phosphorus, oxygen, and water according to these two reactions:

$$4 P + 5 O_2 \longrightarrow P_4O_{10}$$

$$P_4O_{10} + 6 H_2O \longrightarrow 4 H_3PO_4$$

Starting with 20.0 g P, 30.0 g O_2, and 15.0 g H_2O, what is the mass of phosphoric acid that can be formed?

36. The methyl alcohol (CH_3OH) used in alcohol burners combines with oxygen gas to form carbon dioxide and water. How many grams of oxygen are required to burn 60.0 mL of methyl alcohol ($d = 0.72$ g/mL)?

37. Hydrazine (N_2H_4) and hydrogen peroxide (H_2O_2) have been used as rocket propellants. They react according to the following equation:

$$7 H_2O_2 + N_2H_4 \longrightarrow 2 HNO_3 + 8 H_2O$$

(a) How many moles of HNO_3 are formed from 0.33 mol of hydrazine?

(b) How many moles of hydrogen peroxide are required if 2.75 mol of water are to be produced?

(c) How many moles of water are produced if 8.72 mol of HNO_3 are also produced?

(d) How many grams of hydrogen peroxide are needed to react completely with 120 g of hydrazine?

38. Silver tarnishes in the presence of hydrogen sulfide (which smells like rotten eggs) and oxygen because of the reaction

$$4 Ag + 2 H_2S + O_2 \longrightarrow 2 Ag_2S + 2 H_2O$$

How many grams of silver sulfide can be formed from a mixture of 1.1 g Ag, 0.14 g H_2S, and 0.080 g O_2?

39. After 180.0 g of zinc was dropped into a beaker of hydrochloric acid and the reaction ceased, 35 g of unreacted zinc remained in the beaker:

$$Zn + HCl \longrightarrow ZnCl_2 + H_2$$

Balance the equation first.

(a) How many moles of hydrogen gas were produced?

(b) How many grams of HCl were reacted?

40. Use this equation to answer (a) and (b):

$$Fe(s) + CuSO_4(aq) \longrightarrow Cu(s) + FeSO_4(aq)$$

(a) When 2.0 mol of Fe and 3.0 mol of $CuSO_4$ are reacted, what substances will be present when the reaction is over? How many moles of each substance are present?

(b) When 20.0 g of Fe and 40.0 g of $CuSO_4$ are reacted, what substances will be present when the reaction is over? How many grams of each substance are present?

41. Methyl alcohol (CH_3OH) is made by reacting carbon monoxide and hydrogen in the presence of certain metal oxide catalysts. How much alcohol can be obtained by reacting 40.0 g of CO and 10.0 g of H_2? How many grams of excess reactant remain unreacted?

$$CO(g) + 2 H_2(g) \longrightarrow CH_3OH(l)$$

*42. Ethyl alcohol (C_2H_5OH), also called grain alcohol, can be made by the fermentation of sugar:

$$\underset{\text{glucose}}{C_6H_{12}O_6} \longrightarrow \underset{\text{ethyl alcohol}}{2 C_2H_5OH} + 2 CO_2$$

If an 84.6% yield of ethyl alcohol is obtained,

(a) what mass of ethyl alcohol will be produced from 750 g of glucose?

(b) what mass of glucose should be used to produce 475 g of C_2H_5OH?

*43. Both $CaCl_2$ and $MgCl_2$ react with $AgNO_3$ to precipitate AgCl. When solutions containing equal masses of $CaCl_2$ and $MgCl_2$ are reacted with $AgNO_3$, which salt solution will produce the larger amount of AgCl? Show proof.

*44. An astronaut excretes about 2500 g of water a day. If lithium oxide (Li_2O) is used in the spaceship to absorb this water, how many kilograms of Li_2O must be carried for a 30-day space trip for three astronauts?

$$Li_2O + H_2O \longrightarrow 2 LiOH$$

*45. Much commercial hydrochloric acid is prepared by the reaction of concentrated sulfuric acid with sodium chloride:

$$H_2SO_4 + 2 NaCl \longrightarrow Na_2SO_4 + 2 HCl$$

How many kilograms of concentrated H_2SO_4, 96% H_2SO_4 by mass, are required to produce 20.0 L of concentrated hydrochloric acid ($d = 1.20$ g/mL, 42.0% HCl by mass)?

*46. Gastric juice contains about 3.0 g HCl per liter. If a person produces about 2.5 L of gastric juice per day, how many antacid tablets, each containing 400. mg of $Al(OH)_3$, are needed to neutralize all the HCl produced in 1 day?

$$Al(OH)_3(s) + 3\,HCl(aq) \longrightarrow AlCl_3(aq) + 3\,H_2O(l)$$

*47. When 12.82 g of a mixture of $KClO_3$ and NaCl is heated strongly, the $KClO_3$ reacts according to this equation:

$$2\,KClO_3(s) \longrightarrow 2\,KCl(s) + 3\,O_2(g)$$

The NaCl does not undergo any reaction. After the heating, the mass of the residue (KCl and NaCl) is 9.45 g. Assuming that all the loss of mass represents loss of oxygen gas, calculate the percent of $KClO_3$ in the original mixture.

Answers to Practice Exercises

9.1 0.33 mol Al_2O_3

9.2 7.33 mol $Al(OH)_3$

9.3 0.8157 mol KCl

9.4 85 g $AgNO_3$

9.5 27.6 g $CrCl_3$

9.6 348.8 g H_2O

9.7 17.7 g HCl

9.8 164.3 g $BaSO_4$

9.9 88.5% yield

PUTTING IT TOGETHER
Review for Chapters 7–9

Multiple Choice: *Choose the correct answer to each of the following.*

1. 4.0 g of oxygen contains
   (a) 1.5×10^{23} atoms of oxygen
   (b) 4.0 molar masses of oxygen
   (c) 0.50 mol of oxygen
   (d) 6.022×10^{23} atoms of oxygen

2. One mole of hydrogen atoms contains
   (a) 2.0 g of hydrogen
   (b) 6.022×10^{23} atoms of hydrogen
   (c) 1 atom of hydrogen
   (d) 12 g of carbon-12

3. The mass of one atom of magnesium is
   (a) 24.31 g (c) 12.00 g
   (b) 54.94 g (d) 4.037×10^{-23} g

4. Avogardro's number of magnesium atoms
   (a) has a mass of 1.0 g
   (b) has the same mass as Avogadro's number of sulfur atoms
   (c) has a mass of 12.0 g
   (d) is 1 mol of magnesium atoms

5. Which of the following contains the largest number of moles?
   (a) 1.0 g Li (c) 1.0 g Al
   (b) 1.0 g Na (d) 1.0 g Ag

6. The number of moles in 112 g of acetylsalicylic acid (aspirin), $C_9H_8O_4$, is
   (a) 1.61 (b) 0.619 (c) 112 (d) 0.161

7. How many moles of aluminum hydroxide are in one antacid tablet containing 400 mg of $Al(OH)_3$?
   (a) 5.13×10^{-3} (c) 5.13
   (b) 0.400 (d) 9.09×10^{-3}

8. How many grams of Au_2S can be obtained from 1.17 mol of Au?
   (a) 182 g (b) 249 g (c) 364 g (d) 499 g

9. The molar mass of $Ba(NO_3)_2$ is
   (a) 199.3 (b) 261.3 (c) 247.3 (d) 167.3

10. A 16-g sample of O_2
    (a) is 1 mol of O_2
    (b) contains 6.022×10^{23} molecules of O_2
    (c) is 0.50 molecule of O_2
    (d) is 0.50 molar mass of O_2

11. What is the percent composition for a compound formed from 8.15 g of zinc and 2.00 g of oxygen?
    (a) 80.3% Zn, 19.7% O (c) 70.3% Zn, 29.7% O
    (b) 80.3% O, 19.7% Zn (d) 65.3% Zn, 34.7% O

12. Which of these compounds contains the largest percentage of oxygen?
    (a) SO_2 (c) N_2O_3
    (b) SO_3 (d) N_2O_5

13. 2.00 mol of CO_2
    (a) has a mass of 56.0 g
    (b) contains 1.20×10^{24} molecules
    (c) has a mass of 44.0 g
    (d) contains 6.00 molar masses of CO_2

14. In Ag_2CO_3, the percent by mass of
    (a) carbon is 43.5% (c) oxygen is 17.4%
    (b) silver is 64.2% (d) oxygen is 21.9%

15. The empirical formula of the compound containing 31.0% Ti and 69.0% Cl is
    (a) TiCl (c) $TiCl_3$
    (b) $TiCl_2$ (d) $TiCl_4$

16. A compound contains 54.3% C, 5.6% H, and 40.1% Cl. The empirical formula is
    (a) CH_3Cl (c) $C_2H_4Cl_2$
    (b) C_2H_5Cl (d) C_4H_5Cl

17. A compound contains 40.0% C, 6.7% H, and 53.3% O. The molar mass is 60.0 g/mol. The molecular formula is
    (a) $C_2H_3O_2$ (c) C_2HO
    (b) C_3H_8O (d) $C_2H_4O_2$

18. How many chlorine atoms are in 4.0 mol of PCl_3?
    (a) 3 (c) 12
    (b) 7.2×10^{24} (d) 2.4×10^{24}

19. What is the mass of 4.53 mol of Na_2SO_4?
    (a) 142.1 g (c) 31.4 g
    (b) 644 g (d) 3.19×10^{-2} g

20. The percent composition of Mg_3N_2 is
    (a) 72.2% Mg, 27.8% N
    (b) 63.4% Mg, 36.6% N
    (c) 83.9% Mg, 16.1% N
    (d) no correct answer given

21. How many grams of oxygen are contained in 0.500 mol of Na_2SO_4?
    (a) 16.0 g (c) 64.0 g
    (b) 32.0 g (d) no correct answer given

22. The empirical formula of a compound is CH. If the molar mass of this compound is 78.11, then the molecular formula is
    (a) C_2H_2 (c) C_6H_6
    (b) C_5H_{18} (d) no correct answer given

23. The reaction

    $BaCl_2 + (NH_4)_2CO_3 \longrightarrow BaCO_3 + 2 NH_4Cl$

    is an example of
    (a) combination
    (b) decomposition
    (c) single displacement
    (d) double displacement

24. When the equation

    $Al + O_2 \longrightarrow Al_2O_3$

    is properly balanced, which of the following terms appears?
    (a) 2 Al (b) $2 Al_2O_3$ (c) 3 Al (d) $2 O_2$

25. Which equation is *incorrectly* balanced?
    (a) $2 KNO_3 \xrightarrow{\Delta} 2 KNO_2 + O_2$
    (b) $H_2O_2 \longrightarrow H_2O + O_2$
    (c) $2 Na_2O_2 + 2 H_2O \longrightarrow 4 NaOH + O_2$
    (d) $2 H_2O \xrightarrow[H_2SO_4]{\text{electrical energy}} 2 H_2 + O_2$

26. The reaction

    $2 Al + 3 Br_2 \longrightarrow 2 AlBr_3$

    is an example of
    (a) combination
    (b) single displacement
    (c) decomposition
    (d) double displacement

27. When the equation

    $PbO_2 \xrightarrow{\Delta} PbO + O_2$

    is balanced, one term in the balanced equation is
    (a) PbO_2 (b) $3 O_2$ (c) 3 PbO (d) O_2

28. When the equation

    $Cr_2S_3 + HCl \longrightarrow CrCl_3 + H_2S$

    is balanced, one term in the balanced equation is
    (a) 3 HCl (b) $CrCl_3$ (c) $3 H_2S$ (d) $2 Cr_2S_3$

29. When the equation

    $F_2 + H_2O \longrightarrow HF + O_2$

    is balanced, one term in the balanced equation is
    (a) 2 HF (b) $3 O_2$ (c) 4 HF (d) $4 H_2O$

30. When the equation

    $NH_4OH + H_2SO_4 \longrightarrow$

    is completed and balanced, one term in the balanced equation is
    (a) NH_4SO_4 (c) H_2OH
    (b) $2 H_2O$ (d) $2 (NH_4)_2SO_4$

31. When the equation

    $H_2 + V_2O_5 \longrightarrow V +$

    is completed and balanced, one term in the balanced equation is
    (a) $2 V_2O_5$ (b) $3 H_2O$ (c) 2 V (d) $8 H_2$

32. When the equation

    $Al(OH)_3 + H_2SO_4 \longrightarrow Al_2(SO_4)_3 + H_2O$

    is balanced, the sum of the coefficients will be
    (a) 9 (c) 13
    (b) 11 (d) 15

33. When the equation

    $H_3PO_4 + Ca(OH)_2 \longrightarrow H_2O + Ca_3(PO_4)_2$

    is balanced, the proper sequence of coefficients is
    (a) 3, 2, 1, 6 (c) 2, 3, 1, 6
    (b) 2, 3, 6, 1 (d) 2, 3, 3, 1

34. When the equation

    $Fe_2(SO_4)_3 + Ba(OH)_2 \longrightarrow$

    is completed and balanced, one term in the balanced equation is
    (a) $Ba_2(SO_4)_3$ (c) $2 Fe_2(SO_4)_3$
    (b) $2 Fe(OH)_2$ (d) $2 Fe(OH)_3$

35. For the reaction

    $2 H_2 + O_2 \longrightarrow 2 H_2O + 572.4 \text{ kJ}$

    which of the following is not true?
    (a) The reaction is exothermic.
    (b) 572.4 kJ of heat are liberated for each mole of water formed.
    (c) 2 mol of hydrogen react with 1 mol of oxygen.
    (d) 572.4 kJ of heat are liberated for each 2 mol of hydrogen reacted.

36. How many moles is 20.0 g of Na_2CO_3?
    (a) 1.89 mol (c) 212 mol
    (b) 2.12×10^3 mol (d) 0.189 mol

37. What is the mass in grams of 0.30 mol of $BaSO_4$?
    (a) 7.0×10^3 g (c) 70. g
    (b) 0.13 g (d) 700.20 g

38. How many molecules are in 5.8 g of acetone, C_3H_6O?
    (a) 0.10 molecules
    (b) 6.0×10^{22} molecules
    (c) 3.5×10^{24} molecules
    (d) 6.0×10^{23} molecules

Problems 39–45 refer to the reaction

$$2\,C_2H_4 + 6\,O_2 \longrightarrow 4\,CO_2 + 4\,H_2O$$

39. If 6.0 mol of CO_2 are produced, how many moles of O_2 were reacted?
   (a) 4.0 mol   (c) 9.0 mol
   (b) 7.5 mol   (d) 15.0 mol

40. How many moles of O_2 are required for the complete reaction of 45 g of C_2H_4?
   (a) 1.3×10^2 mol   (c) 112.5 mol
   (b) 0.64 mol   (d) 4.8 mol

41. If 18.0 g of CO_2 are produced, how many grams of H_2O are produced?
   (a) 7.37 g   (b) 3.68 g   (c) 9.00 g   (d) 14.7 g

42. How many moles of CO_2 can be produced by the reaction of 5.0 mol of C_2H_4 and 12.0 mol of O_2?
   (a) 4.0 mol   (c) 8.0 mol
   (b) 5.0 mol   (d) 10. mol

43. How many moles of CO_2 can be produced by the reaction of 0.480 mol of C_2H_4 and 1.08 mol of O_2?
   (a) 0.240 mol   (c) 0.720 mol
   (b) 0.960 mol   (d) 0.864 mol

44. How many grams of CO_2 can be produced from 2.0 g of C_2H_4 and 5.0 g of O_2?
   (a) 5.5 g   (b) 4.6 g   (c) 7.6 g   (d) 6.3 g

45. If 14.0 g of C_2H_4 is reacted and the actual yield of H_2O is 7.84 g, the percent yield in the reaction is
   (a) 0.56%   (b) 43.6%   (c) 87.1%   (d) 56.0%

Problems 46–48 refer to the equation

$$H_3PO_4 + MgCO_3 \longrightarrow Mg_3(PO_4)_2 + CO_2 + H_2O$$

46. The sequence of coefficients for the balanced equation is
   (a) 2, 3, 1, 3, 3   (c) 2, 2, 1, 2, 3
   (b) 3, 1, 3, 2, 3   (d) 2, 3, 1, 3, 2

47. If 20.0 g of carbon dioxide is produced, the number of moles of magnesium carbonate used is
   (a) 0.228 mol   (c) 0.910 mol
   (b) 1.37 mol   (d) 0.454 mol

48. If 50.0 g of magnesium carbonate reacts completely with H_3PO_4, the number of grams of carbon dioxide produced is
   (a) 52.2 g   (c) 13.1 g
   (b) 26.1 g   (d) 50.0 g

49. When 10.0 g of $MgCl_2$ and 10.0 g of Na_2CO_3 are reacted in

$$MgCl_2 + Na_2CO_3 \longrightarrow MgCO_3 + 2\,NaCl$$

the limiting reactant is
   (a) $MgCl_2$ (c) $MgCO_3$
   (b) Na_2CO_3 (d) NaCl

50. When 50.0 g of copper is reacted with silver nitrate solution in

$$Cu + 2\,AgNO_3 \longrightarrow Cu(NO_3)_2 + 2\,Ag$$

148 g of silver is obtained. What is the percent yield of silver obtained?
   (a) 87.1%   (c) 55.2%
   (b) 84.9%   (d) no correct answer given

Free Response Questions

1. Compound X requires 104 g of O_2 to produce 2 moles of CO_2 and 2.5 moles of H_2O.
   (a) What is the empirical formula for X?
   (b) What additional information would you need to determine the molecular formula for X?

2. Consider the reaction of sulfur dioxide with oxygen to form sulfur trioxide taking place in a closed container.

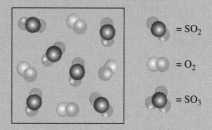

   (a) Draw what you would expect to see in the box at the completion of the reaction.
   (b) If you begin with 25 g of SO_2 and 5 g of oxygen gas, which is the limiting reagent?
   (c) Is the following statement true or false? "When SO_2 is converted into SO_3, the percent composition of S in the compounds changes from 33% to 25%." Explain.

3. The percent composition of compound Z is 63.16% C and 8.77% H. When compound Z burns in air, the only products are carbon dioxide and water. The molar mass for Z is 114.
   (a) What is the molecular formula for compound Z?
   (b) What is the balanced reaction for Z burning in air?

4. Compound A decomposes at room temperature in an exothermic reaction while compound B requires heating before it will decompose in an endothermic reaction.
   (a) Draw reaction profiles (potential energy vs. reaction progress) for both reactions.
   (b) (i) The decomposition of 0.500 moles of $NaHCO_3$ to form sodium carbonate, water, and carbon dioxide requires 85.5 kJ of heat. How many grams of water could be collected and how many kJ of heat were absorbed when 24.0 g of CO_2 is produced?
       (ii) Could $NaHCO_3$ be compound A or compound B? Explain your reasoning.

5. Aqueous ammonium hydroxide reacts with aqueous cobalt (II) sulfate to produce aqueous ammonium sulfate and solid cobalt (II) hydroxide. When 38.0 g of one of the reactants was fully reacted with enough of the other reactant, 8.09 g of ammonium sulfate was obtained, which corresponded to a 25.0% yield.
   (a) What type of reaction took place?
   (b) Write the balanced chemical equation.
   (c) What is the theoretical yield of ammonium sulfate?
   (d) Which reactant was the limiting reagent?

6. $C_6H_{12}O_6 \longrightarrow 2\ C_2H_5OH + 2\ CO_2(g)$
   (a) If 25.0 g of $C_6H_{12}O_6$ were used, and only 11.2 g of C_2H_5OH were produced, how much reactant was left over and what volume of gas was produced? (*Note:* One mole of gas occupies 24.0 L of space at room temperature.) What assumption are you making?
   (b) What was the yield of the reaction?
   (c) What type of reaction took place?

7. When solutions containing 25 g each of lead(II) nitrate and potassium iodide are mixed, a yellow precipitate results.
   (a) What type of reaction occurred?
   (b) What is the name and formula for the solid product?
   (c) If after filtration and drying, the solid product weighed 7.66 g, what was the percent yield for the reaction?

8. Consider the following unbalanced reaction:

   $$XNO_3 + CaCl_2 \longrightarrow XCl + Ca(NO_3)_2$$

   (a) If 30.8 g of $CaCl_2$ produced 79.6 g of XCl, what is X?
   (b) Would X be able to displace hydrogen from an acid?

9. Consider the following reaction: $H_2O_2 \longrightarrow H_2O + O_2$
   (a) If at the end of the reaction there are eight water molecules and eight oxygen molecules, what was in the flask at the start of the reaction?
   (b) Does the following reaction profile indicate the reaction is exothermic or endothermic?

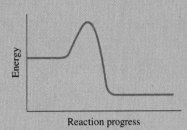

Reaction progress

   (c) What type of reaction is given above?
   (d) What is the empirical formula for hydrogen peroxide?

CHAPTER

10

Modern Atomic Theory and the Periodic Table

▲
The brilliance of these neon signs in Las Vegas is the result of electrons being transferred between energy levels.

CHAPTER 10 / OUTLINE

How do we go about studying an object that is too small to see? Think back to that birthday present you could look at but not yet open. Judging from the wrapping and size of the box was not very useful, but shaking, turning, and lifting the package all gave indirect clues to its contents. After all your experiments were done you could make a fairly good guess about the contents. But was your guess correct? The only way to know for sure would be to open the package.

Chemists have the same dilemma when they study the atom. Atoms are so very small that it isn't possible to use the normal senses to describe them. We are essentially working in the dark with this package we call the atom. However, our improvements in instruments (X-ray machines and scanning-tunneling microscopes) and measuring devices (spectrophotometers and magnetic resonance imaging, MRI) as well as in our mathematical skills are bringing us closer to revealing the secrets of the atom.

10.1 A Brief History

In the last 200 years, vast amounts of data have been accumulated to support atomic theory. When atoms were originally suggested by the early Greeks, no physical evidence existed to support their ideas. Early chemists did a variety of experiments, which culminated in Dalton's model of the atom. Because of the limitations of Dalton's model, modifications were proposed first by Thomson and then by Rutherford, which eventually led to our modern concept of the nuclear atom. These early models of the atom work reasonably well—in fact, we continue to use them to visualize a variety of chemical concepts. There remain questions that these models cannot answer, including an explanation of how atomic structure relates to the periodic table. In this chapter we will present our modern model of the atom; we will see how it varies from and improves upon the earlier atomic models.

10.2 Electromagnetic Radiation

▲
Surfers judge the wavelength, frequency, and speed of waves to get the best ride.

Scientists have studied energy and light for centuries, and several models have been proposed to explain how energy is transferred from place to place. One way energy travels through space is by *electromagnetic radiation*. Examples of electromagnetic radiation include light from the sun, X-rays in your dentist's office, microwaves from your microwave oven, radio and television waves, and radiant heat from your fireplace. While these examples seem quite different, they are all similar in some important ways. Each shows wavelike behavior, and all travel at the same speed in a vacuum (3.00×10^8 m/s).

The study of wave behavior is a topic for another course, but we need some basic terminology to understand atoms. Waves have three basic characteristics: wavelength, frequency, and speed. **Wavelength** (lambda, λ) is the distance between consecutive peaks (or troughs) in a wave, as shown in Figure 10.1. **Frequency** (nu, μ)

wavelength

frequency

FIGURE 10.1 ▶
The wavelength of this wave is shown by λ. It can be measured from peak to peak or trough to trough.

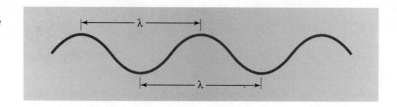

FIGURE 10.2 ▶
The electromagnetic spectrum.

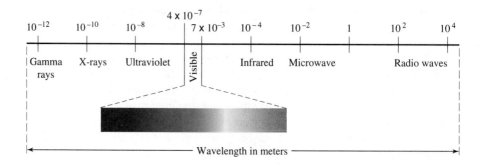

speed tells how many waves pass a particular point per second. **Speed** (v) tells how fast a wave moves through space.

Light is one form of electromagnetic radiation and is usually classified by its wavelength, as shown in Figure 10.2. Visible light, as you can see, is only a tiny part of the electromagnetic spectrum. Some examples of electromagnetic radiation involved in energy transfer outside the visible region are hot coals in your backyard grill, which transfer infrared radiation to cook your food; and microwaves, which transfer energy to water molecules in the food, causing them to move more quickly and thus raise the temperature of your food.

photons We have evidence for the wavelike nature of light. We also know that a beam of light behaves like a stream of tiny packets of energy called **photons.** So what is light exactly? Is it a particle? Is it a wave? Scientists have agreed to explain the properties of electromagnetic radiation by using both wave and particle properties. Neither explanation is ideal, but currently these are our best models.

10.3   The Bohr Atom

As scientists struggled to understand the properties of electromagnetic radiation, evidence began to accumulate that atoms could radiate light. At high temperatures, or when subjected to high voltages, elements in the gaseous state give off colored light. Brightly colored neon signs illustrate this property of matter very well. When the light emitted by a gas is passed through a prism or diffraction grating, a set of
line spectrum brightly colored lines called a **line spectrum** results (Figure 10.3). These colored lines indicate that the light is being emitted only at certain wavelengths, or frequencies, that correspond to specific colors. Each element possesses a unique set of these spectral lines that is different from the sets of all the other elements.

In 1912–1913, while studying the line spectrum of hydrogen, Niels Bohr (1885–1962), a Danish physicist, made a significant contribution to the rapidly growing knowledge of atomic structure. His research led him to believe that elec-

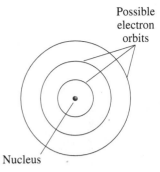

▲
FIGURE 10.3
Line spectrum of hydrogen. Each line corresponds to the wavelength of the energy emitted when the electron of a hydrogen atom, which has absorbed energy, falls back to a lower principal energy level.

▲
FIGURE 10.4
The Bohr model of the hydrogen atom described the electron revolving in certain allowed circular orbits around the nucleus.

trons exist in specific regions at various distances from the nucleus. He also visualized the electrons as revolving in orbits around the nucleus, like planets rotating around the Sun as shown in Figure 10.4.

Bohr's first paper in this field dealt with the hydrogen atom, which he described as a single electron revolving in an orbit about a relatively heavy nucleus. He applied the concept of energy quanta, proposed in 1900 by the German physicist Max Planck (1858–1947), to the observed line spectrum of hydrogen. Planck stated that energy is never emitted in a continuous stream but only in small discrete packets called **quanta** (Latin, *quantus,* how much). From this Bohr theorized that electrons have several possible energies corresponding to several possible orbits at different distances from the nucleus. Therefore an electron has to be in one specific energy level; it cannot exist between energy levels. In other words, the energy of the electron is said to be quantized. Bohr also stated that when a hydrogen atom absorbed one or more quanta of energy, its electron would "jump" to a higher energy level.

Bohr was able to account for spectral lines of hydrogen this way. A number of energy levels are available, the lowest of which is called the **ground state.** When an electron falls from a high-energy level to a lower one (say, from the fourth to the second), a quantum of energy is emitted as light at a specific frequency, or wavelength (Figure 10.5). This light corresponds to one of the lines visible in the hydrogen spectrum (Figure 10.3). Several lines are visible in this spectrum, each one corresponding to a specific electron energy-level shift within the hydrogen atom.

The chemical properties of an element and its position in the periodic table depend on electron behavior within the atoms. In turn much of our knowledge of the behavior of electrons within atoms is based on spectroscopy. Niels Bohr contributed a great deal to our knowledge of atomic structure by (1) suggesting quantized energy levels for electrons, and (2) showing that spectral lines result from the radiation of small increments of energy (Planck's quanta) when electrons shift from one energy level to another. Bohr's calculations succeeded very well in correlating the experimentally observed spectral lines with electron energy levels for the hydrogen atom. However, Bohr's methods of calculation did not succeed for heavier atoms. More theoretical work on atomic structure was needed.

In 1924, the French physicist Louis de Broglie suggested a surprising hypothesis: All objects have wave properties. De Broglie used sophisticated mathematics to show that the wave properties for an object of ordinary size, such as a baseball, are too small to be observed. But for small objects, such as an electron, the wave prop-

quanta

ground state

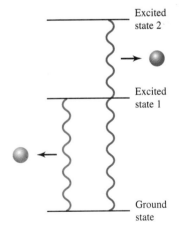

▲
FIGURE 10.5
When an excited electron returns to the ground state, energy is emitted as a photon is released. The color (wavelength) of the light is determined by the difference in energy between the two states (excited and ground).

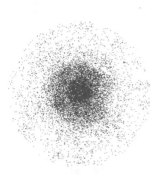

▲
FIGURE 10.6
An orbital for a hydrogen atom. The intensity of the dots shows that the electron spends more time closer to the nucleus.

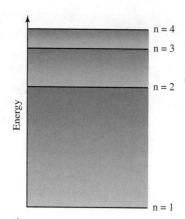

▲
FIGURE 10.7
The first four principal energy levels in the hydrogen atom. Each level is assigned a principal quantum number n.

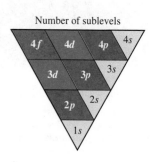

▲
FIGURE 10.8
The types of orbitals on each of the first four principal energy levels.

erties become significant. Other scientists confirmed de Broglie's hypothesis, showing that electrons do exhibit wave properties. In 1926, Erwin Schrödinger, an Austrian physicist, created a mathematical model that described electrons as waves. Using Schrödinger's wave mechanics, we can determine the *probability* of finding an electron in a certain region around the atom.

This treatment of the atom led to a new branch of physics called *wave mechanics* or *quantum mechanics,* which forms the basis for our modern understanding of atomic structure. Although the wave-mechanical description of the atom is mathematical, it can be translated, at least in part, into a visual model. It's important to recognize that we cannot locate an electron precisely within an atom; however, it is clear that electrons are not revolving around the nucleus in orbits as Bohr postulated. The electrons are instead found in *orbitals.* An **orbital** is pictured in Figure 10.6 as a region in space around the nucleus where there is a high probability of finding a given electron. We will have more to say about the meaning of orbitals in the next section.

orbital

10.4   Energy Levels of Electrons

One of the ideas Bohr contributed to the modern concept of the atom was that the energy of the electron is quantized—that is, the electron is restricted to only certain allowed energies. The wave-mechanical model of the atom also predicts discrete **principal energy levels** within the atom. These energy levels are designated by the letter *n,* where *n* is a positive integer (Figure 10.7). The lowest principal energy level corresponds to $n = 1$, the next to $n = 2$, and so on. As *n* increases, the energy of the electron increases, and the electron is found on average farther from the nucleus.

principal energy levels

Each principal energy level is divided into **sublevels,** which are illustrated in Figure 10.8. The first principal energy level has one sublevel. The second principal energy level has two sublevels, the third energy level has three sublevels, and so on. Each of these sublevels contains spaces for electrons called orbitals.

sublevel

In each sublevel the electrons are found within specified orbitals. Let's consider each principal energy level in turn. The first principal energy level ($n = 1$) has one sublevel or type of orbital. It is spherical in shape and is designated as $1s$. It is important to understand what the spherical shape of the $1s$ orbital means. The electron does *not* move around on the surface of the sphere, but rather the surface encloses a space where there is a 90% probability that the electron may be found. It might help to consider these orbital shapes in the same way we consider the atmosphere. There is no distinct dividing line between the atmosphere and "space." The boundary is quite fuzzy. The same is true for atomic orbitals. Each has a region of highest density roughly corresponding to its shape. The probability of finding the electron outside this region drops rapidly but never quite reaches zero. Scientists often speak of orbitals as electron "clouds" to emphasize the fuzzy nature of their boundaries.

How many electrons can fit into a $1s$ orbital? To answer this question we need to consider one more property of electrons. This property is called **spin.** Each electron appears to be spinning on an axis, like a globe. It can only spin in two directions. We represent this spin with an arrow, ↑ or ↓. In order to occupy the same orbital, electrons must have *opposite* spins. That is, two electrons with the same spin cannot occupy the same orbital. This gives us the answer to our question: An atomic orbital can hold a maximum of two electrons, which must have opposite spins. This rule is called the **Pauli exclusion principle.** The first principal energy level contains one type of orbital ($1s$) that holds a maximum of two electrons.

What happens with the second principal energy level ($n = 2$)? Here we find two sublevels, $2s$ and $2p$. Like $1s$ in the first principal energy level the $2s$ orbital is spherical in shape, but is larger in size and higher in energy. It also holds a maximum of two electrons. The second type of orbital is designated by $2p$. The $2p$ sublevel consists of three orbitals $2p_x$, $2p_y$, and $2p_z$. The shape of p orbitals is quite different from the s orbitals as shown in Figure 10.9.

Each p orbital has two "lobes." Remember, the space enclosed by these surfaces represents the regions of probability for finding the electrons 90% of the time. There are three separate p orbitals, each oriented in a different direction, and each p orbital can hold a maximum of two electrons. Thus the total number of electrons that can reside in all three p orbitals is six. To summarize our model, the first principal energy level of an atom has a $1s$ orbital. The second principal energy level has a $2s$ and three $2p$ orbitals labeled $2p_x$, $2p_y$, and $2p_z$ as shown in Figure 10.10.

The third principal energy level has three sublevels labeled $3s$, $3p$, and $3d$. The $3s$ orbital is spherical and larger than the $1s$ and $2s$ orbitals. The $3p_x$, $3p_y$, $3p_z$ orbitals are shaped like those of the second level, only larger. The five $3d$ orbitals have the

The light in this neon sign is the result of electrons falling from one principal energy level to another.

spin

Pauli exclusion principle

Notice that there is a correspondence between the energy level and number of sublevels.

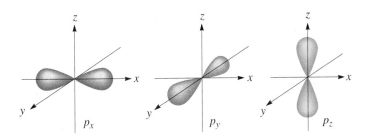

◀**FIGURE 10.9**
Perspective representation of the p_x, p_y, and p_z atomic orbitals.

FIGURE 10.10 ▶
Orbitals on the second principal energy level are one 2s and three 2p orbitals.

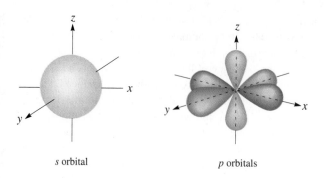

s orbital   p orbitals

FIGURE 10.11
The five d orbitals are found in the third principal energy level along with one 3s orbital and three 3p orbitals.
▼

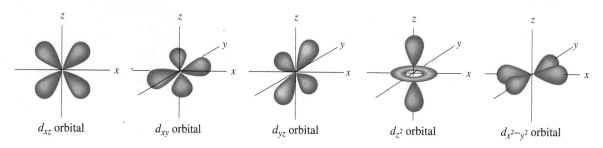

d_{xz} orbital d_{xy} orbital d_{yz} orbital d_{z^2} orbital $d_{x^2-y^2}$ orbital

shapes shown in Figure 10.11. You don't need to memorize these shapes, but notice that they look different from the s or p orbitals.

Each time a new principal energy level is added, we also add a new sublevel. This makes sense because each energy level corresponds to a larger average distance from the nucleus, which provides more room on each level for new sublevels containing more orbitals.

The pattern continues with the fourth principal energy level. It has 4s, 4p, 4d, and 4f orbitals. There are one 4s, three 4p, five 4d, and seven 4f orbitals. The shapes of the s, p, and d orbitals are the same as those for lower levels, only larger. We will not consider the shapes of the f orbitals. We summarize each principal energy level:

$n = 1$	$1s$			
$n = 2$	$2s$	$2p\ 2p\ 2p$		
$n = 3$	$3s$	$3p\ 3p\ 3p$	$3d\ 3d\ 3d\ 3d\ 3d$	
$n = 4$	$4s$	$4p\ 4p\ 4p$	$4d\ 4d\ 4d\ 4d\ 4d$	$4f\ 4f\ 4f\ 4f\ 4f\ 4f\ 4f$

The hydrogen atom consists of a nucleus (containing one proton) and an electron moving around the outside of the nucleus. In its ground state the electron occupies a 1s orbital, but by absorbing energy the electron can become *excited* and move to a higher energy level.

The hydrogen atom can be represented as shown in Figure 10.12. The diameter of the nucleus is about 10^{-13} cm, and the diameter of the electron orbital is about 10^{-8} cm. The diameter of the electron cloud of a hydrogen atom is about 100,000 times greater than the diameter of the nucleus.

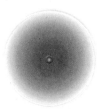

▲
FIGURE 10.12
The modern concept of a hydrogen atom consists of a proton and an electron in an s orbital. The shaded area represents a region where the electron may be found with 90% probability.

CHEMISTRY IN ACTION Atomic Clocks

Imagine a clock that keeps time to within 1 second over a million years. The National Institute of Standards and Technology in Boulder, Colorado, has an atomic clock that does just that—a little better than your average alarm, grandfather, or cuckoo clock! This atomic clock serves as the international standard for time and frequency. How does it work?

Within the glistening case are several layers of magnetic shielding. In the heart of the clock is a small oven that heats cesium metal to release cesium atoms, which are collected into a narrow beam (1 mm wide). The beam of atoms passes down a long evacuated tube while being excited by a laser until all the cesium atoms are in the same electron state.

The atoms then enter another chamber filled with reflecting microwaves. The frequency of the microwaves (9,192,631,770 cycles per second) is exactly the same frequency required to excite a cesium atom from its ground state to the next higher energy level. These excited cesium atoms then release electromagnetic radiation in a process known as fluorescence. Electronic circuits maintain the microwave frequency at precisely the right level to keep the cesium atoms moving from one level to the next. One second is equal to 9,192,631,770 of these

vibrations. The clock is set to this frequency and can keep accurate time for over a million years.

This clock automatically updates ▶ itself by comparing time with an atomic clock by radio signal.

10.5 Atomic Structures of the First 18 Elements

We have seen that hydrogen has one electron that can occupy a variety of orbitals in different principal energy levels. Now let's consider the structure of atoms with more than one electron. Because all atoms contain orbitals similar to those found in hydrogen, we can describe the structures of atoms beyond hydrogen by systematically placing electrons in these hydrogenlike orbitals. We use the following guidelines:

1. No more than two electrons can occupy one orbital.
2. Electrons occupy the lowest energy orbitals available. They enter a higher energy orbital only when the lower orbitals are filled. For the atoms beyond hydrogen, orbital energies vary as $s < p < d < f$ for a given value of n.
3. Each orbital on a sublevel is occupied by a single electron before a second electron enters. For example, all three p orbitals must contain one electron before a second electron enters a p orbital.

We can use several methods to represent the atomic structures of atoms, depending on what we are trying to illustrate. When we want to show both the nuclear makeup and the electron structure of each principal energy level (without orbital detail), we can use a diagram such as Figure 10.13.

FIGURE 10.13
Atomic structure diagrams of fluorine, sodium, and magnesium atoms. The number of protons and neutrons is shown in the nucleus. The number of electrons is shown in each principal energy level ◀ outside the nucleus.

$\begin{array}{c} \text{9p} \\ \text{10n} \end{array}$ $\begin{array}{ccc} n=1 & 2 \\ 2e^- & 7e^- \end{array}$

$\begin{array}{c} \text{11p} \\ \text{12n} \end{array}$ $\begin{array}{ccc} n=1 & 2 & 3 \\ 2e^- & 8e^- & 1e^- \end{array}$

$\begin{array}{c} \text{12p} \\ \text{12n} \end{array}$ $\begin{array}{ccc} n=1 & 2 & 3 \\ 2e^- & 8e^- & 2e^- \end{array}$

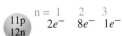
Fluorine atom

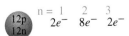
Sodium atom

Magnesium atom

CHEMISTRY IN ACTION **Ripples on the Surface**

Modern physicists have astounded us with the idea that elementary particles have wave properties. Protons, electrons, and all other elementary particles sometimes behave like waves and sometimes like particles. Now we have pictures to actually show this wave behavior of particles. Donald Eigler at the IBM Almaden Research Center in San Jose, California, used a scanning-tunneling microscope at 4 kelvins to produce the picture shown here. The surface of this copper crystal surprised even the physicists. The waves in the photo are produced by electrons moving around on the surface of the crystal and bouncing off impurities (the two pits in the photo) in the copper. Since each electron behaves like a wave, it interferes with itself after

reflecting off an impurity in the copper. The interference pattern is called a stand-

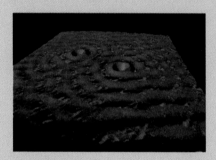

▲
This scanning-tunneling microscope photo of the surface of a copper crystal beautifully illustrates the wave nature of matter. Magnification is 215,000,000×.

ing wave, that is, a wave that vibrates up and down without visible transverse movement, similar to a plucked violin string. The crests of the waves represent regions where the electron is most probably in its particle form.

Metal atoms easily lose one or more electrons, which move about freely within the metal crystal, forming what is often called an "electron sea." At the surface of the metal these loose electrons are confined to a single layer, which is free to move in only two dimensions. Within these constraints the particles behave like waves. The electron layer responsible for this beautiful pattern is only 0.02 Å thick. Similar images of standing electron waves have also been produced with gold at room temperature.

electron configuration

Often we are interested in showing the arrangement of the electrons in an atom in their orbitals. There are two ways to do this. The first method is called the **electron configuration.** In this method we list each type of orbital, showing the number of electrons in it as an exponent. An electron configuration is read like this:

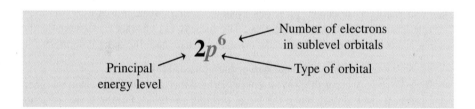

orbital diagram

We can also represent this configuration with an **orbital diagram** in which boxes represent the orbitals (containing small arrows indicating the electrons). When the orbital contains one electron, an arrow, pointing upward ($\uparrow$), is placed in the box. A second arrow, pointing downward ($\downarrow$), indicates the second electron in that orbital.

Let's consider each of the first 18 elements on the periodic table in turn. The order of filling for the orbitals in these elements is $1s$, $2s$, $2p$, $3s$, $3p$, and $4s$. Hydrogen, the first element, has only one electron. The electron will be in the $1s$ orbital because this is the most favorable position (where it will have the greatest attraction for the nucleus). Both representations are shown here:

H $\boxed{\uparrow}$ $1s^1$

Orbital   Electron
diagram   configuration

Helium, with two electrons, can be shown as

He   [↑↓] $1s^2$
     Orbital   Electron
     diagram   configuration

The first energy level is now full. An atom with three electrons will have its third electron in the second energy level. Thus in lithium (atomic number 3), the first two electrons are in the $1s$ orbital, and the third electron is in the $2s$ orbital of the second energy level. Lithium has the following structure:

Li   [↑↓] [↑] $1s^2 2s^1$

All four electrons of beryllium are s electrons:

Be   [↑↓] [↑↓] $1s^2 2s^2$

The next six elements illustrate the filling of the p orbitals. Boron has the first p electron. Because p orbitals all have the same energy, it doesn't matter which of these orbitals fills first:

B   [↑↓] [↑↓] [↑][ ][ ] $1s^2 2s^2 2p^1$

Carbon is the sixth element. It has two electrons in the $1s$ orbital, two electrons in the $2s$ orbital, and two electrons to place in the $2p$ orbitals. Because it is more difficult for the p electrons to pair up than to occupy a second p orbital, the second p electron is located in a different p orbital. We could show this by writing $2p_x^1 2p_y^1$, but we usually write it as $2p^2$; it is *understood* that the electrons are in different p orbitals. The spins on these electrons are alike for reasons we will not explain here.

C   [↑↓] [↑↓] [↑][↑][ ] $1s^2 2s^2 2p^2$

Nitrogen has seven electrons. They occupy the $1s$, $2s$, and $2p$ orbitals. The third p electron in nitrogen is still unpaired and is found in the $2p_z$ orbital:

N   [↑↓] [↑↓] [↑][↑][↑] $1s^2 2s^2 2p^3$

Oxygen is the eighth element. It has two electrons in both the $1s$ and $2s$ orbitals, and four electrons in the $2p$ orbitals. One of the $2p$ orbitals is now occupied by a second electron, which has a spin opposite the electron already in that orbital:

O   [↑↓] [↑↓] [↑↓][↑][↑] $1s^2 2s^2 2p^4$

The next two elements are fluorine with nine electrons and neon with ten electrons:

F   [↑↓] [↑↓] [↑↓][↑↓][↑] $1s^2 2s^2 2p^5$

Ne   [↑↓] [↑↓] [↑↓][↑↓][↑↓] $1s^2 2s^2 2p^6$

With neon, the first and second energy levels are filled as shown in Table 10.1.
   Sodium, element 11, has two electrons in the first energy level and eight electrons in the second energy level, with the remaining electron occupying the $3s$ orbital in the third energy level:

Na   [↑↓] [↑↓] [↑↓][↑↓][↑↓] [↑] $1s^2 2s^2 2p^6 3s^1$
        $1s$ $2s$ $2p$ $3s$

Yes, We Can See Atoms!

▲
Chromosomes are visible in an onion cell under an optical microscope.

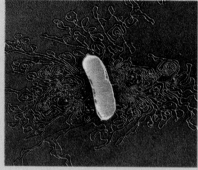

▲
Electron microscopes reveal the DNA of an E. coli bacterium (yellow strands).

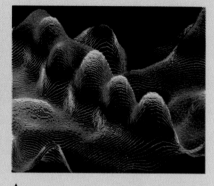

▲
DNA molecule, scanning-tunneling microscope.

For centuries, scientists have argued and theorized over the nature and existence of atoms. Today, physicists and chemists can produce pictures of atoms and even move them individually from place to place. This new-found ability to see atoms, molecules, and even watch chemical reactions occur is the direct result of the evolution of the microscope.

An optical microscope is capable of viewing objects as small as the size of a cell. To see smaller objects, an electron microscope is necessary. Since the eye cannot respond to a beam of electrons, the image is produced on a fluorescent screen or on film. These microscopes have been used for some time to photograph large molecules. To see tiny objects, however, the objects must be placed in a vacuum and the electrons must be in a high energy state. If the sample is fragile, as are most molecules, it can be destroyed before the image is formed.

In 1981, Gerd Binnig and Heinrich Rohrer, two scientists from IBM,

invented the first scanning-probe microscope. These instruments are fundamentally different from previous microscopes. In a scanning-probe microscope a probe is placed near the surface of a sample and a parameter of some sort (voltage, magnetic field, etc.) is measured. As the probe is moved across the surface, an image is produced—in the same manner a child would determine the identity of an object sealed in an opaque bag. The first of the scanning-probe instruments was called a scanning-tunneling microscope. It produced the first clear pictures of silicon atoms in January 1983. The greatest limitation for the scanning-tunneling microscope is that, in order to be viewed, organic molecules must be given a thin coating of metal so that electrons are free to jump from the surface to the probe.

In 1985, a team of physicists from Stanford University and IBM found a solution to this problem. In a new instrument known as an atomic-force

microscope, the probe measures tiny electric forces between electrons instead of the actual movement of electrons from surface to probe. The great advantage of this approach is that the probe is so gentle that even very fragile molecules remain intact. The probe is a tiny shard of a diamond attached to a tiny piece of silicon, and it works like a phonograph needle. At the University of California, Santa Barbara, a group of scientists has even succeeded in making a movie of the formation of a blood clot on the molecular level.

In industry, another type of scanning-probe instrument has been developed to check the quality of microelectronic equipment. Researchers at IBM have developed a laser-force microscope in which a tiny wire probe measures small attractive forces (surface tension of water on the sample) to show imperfections as small as 25 atoms across.

Magnesium (12), aluminum (13), silicon (14), phosphorus (15), sulfur (16), chlorine (17), and argon (18) follow in order. Table 10.2 summarizes the filling of the orbitals for elements 11–18.

The electrons in the outermost (highest) energy level of an atom are called the **valence electrons.** For example, oxygen, which has the electron configuration of $1s^2 2s^2 2p^4$, has electrons in the first and second energy levels. Therefore the second level is the valence level for oxygen. The $2s$ and $2p$ electrons are the valence electrons. In the case of magnesium ($1s^2 2s^2 2p^6 3s^2$), the valence electrons are in the

valence electrons

TABLE 10.1   Orbital Filling for the First Ten Elements*

Number	Element	Orbitals	Electron configuration
		$1s$ $2s$ $2p$	
1	H	↑	$1s^1$
2	He	↑↓	$1s^2$
3	Li	↑↓ ↑	$1s^2 2s^1$
4	Be	↑↓ ↑↓	$1s^2 2s^2$
5	B	↑↓ ↑↓ ↑	$1s^2 2s^2 2p^1$
6	C	↑↓ ↑↓ ↑ ↑	$1s^2 2s^2 2p^2$
7	N	↑↓ ↑↓ ↑ ↑ ↑	$1s^2 2s^2 2p^3$
8	O	↑↓ ↑↓ ↑↓ ↑ ↑	$1s^2 2s^2 2p^4$
9	F	↑↓ ↑↓ ↑↓ ↑↓ ↑	$1s^2 2s^2 2p^5$
10	Ne	↑↓ ↑↓ ↑↓ ↑↓ ↑↓	$1s^2 2s^2 2p^6$

*Boxes represent the orbitals grouped by sublevel. Electrons are shown by arrows.

TABLE 10.2   Orbital Diagrams and Electron Configurations for Elements 11–18

Number	Element	Orbital	Electron configuration
		$1s$ $2s$ $2p$ $3s$ $3p$	
11	Na	↑↓ ↑↓ ↑↓↑↓↑↓ ↑	$1s^2 2s^2 2p^6 3s^1$
12	Mg	↑↓ ↑↓ ↑↓↑↓↑↓ ↑↓	$1s^2 2s^2 2p^6 3s^2$
13	Al	↑↓ ↑↓ ↑↓↑↓↑↓ ↑↓ ↑	$1s^2 2s^2 2p^6 3s^2 3p^1$
14	Si	↑↓ ↑↓ ↑↓↑↓↑↓ ↑↓ ↑ ↑	$1s^2 2s^2 2p^6 3s^2 3p^2$
15	P	↑↓ ↑↓ ↑↓↑↓↑↓ ↑↓ ↑ ↑ ↑	$1s^2 2s^2 2p^6 3s^2 3p^3$
16	S	↑↓ ↑↓ ↑↓↑↓↑↓ ↑↓ ↑↓ ↑ ↑	$1s^2 2s^2 2p^6 3s^2 3p^4$
17	Cl	↑↓ ↑↓ ↑↓↑↓↑↓ ↑↓ ↑↓ ↑↓ ↑	$1s^2 2s^2 2p^6 3s^2 3p^5$
18	Ar	↑↓ ↑↓ ↑↓↑↓↑↓ ↑↓ ↑↓ ↑↓ ↑↓	$1s^2 2s^2 2p^6 3s^2 3p^6$

3s orbital since these are outermost electrons. Valence electrons are involved in bonding atoms to form compounds and are of particular interest to chemists, as we will see in Chapter 11.

Practice 10.1

State the valence electrons in these elements.
(a) B
(b) N
(c) Na
(d) Cl

10.6   Electron Structures and the Periodic Table

We have seen how the electrons are assigned for the atoms of elements 1–18. How do the electron structures of these atoms relate to their position on the periodic table? To answer this question, we need to look at the periodic table more closely.

The periodic table represents the efforts of chemists to organize the elements logically. Chemists of the early 19th century had sufficient knowledge of the properties of elements to recognize similarities among groups of elements. In 1869, Dimitri Mendeleev (1834–1907) of Russia and Lothar Meyer (1830–1895) of Germany independently published periodic arrangements of the elements based on increasing atomic masses. Mendeleev's arrangement is the precursor to the modern periodic table and his name is associated with it. The modern periodic table is shown in Figure 10.14.

FIGURE 10.14
The periodic table of the elements.
▼

Group number																	Noble gases
IA																	

Atomic number — 9
Symbol — F

Period	IA	IIA	IIIB	IVB	VB	VIB	VIIB	VIII			IB	IIB	IIIA	IVA	VA	VIA	VIIA	Noble gases
1	1 H																	2 He
2	3 Li	4 Be											5 B	6 C	7 N	8 O	9 F	10 Ne
3	11 Na	12 Mg											13 Al	14 Si	15 P	16 S	17 Cl	18 Ar
4	19 K	20 Ca	21 Sc	22 Ti	23 V	24 Cr	25 Mn	26 Fe	27 Co	28 Ni	29 Cu	30 Zn	31 Ga	32 Ge	33 As	34 Se	35 Br	36 Kr
5	37 Rb	38 Sr	39 Y	40 Zr	41 Nb	42 Mo	43 Tc	44 Ru	45 Rh	46 Pd	47 Ag	48 Cd	49 In	50 Sn	51 Sb	52 Te	53 I	54 Xe
6	55 Cs	56 Ba	57–71 La–Lu	72 Hf	73 Ta	74 W	75 Re	76 Os	77 Ir	78 Pt	79 Au	80 Hg	81 Tl	82 Pb	83 Bi	84 Po	85 At	86 Rn
7	87 Fr	88 Ra	89–103 Ac–Lr	104 Rf	105 Db	106 Sg	107 Bh	108 Hs	109 Mt	110 Uun	111 Uuu	112 Uub						

Each horizontal row in the periodic table is called a **period,** as shown in Figure 10.14. The number of each period corresponds to the outermost energy level that contains electrons for elements in that period. Those in Period 1 contain electrons only in energy level 1 while those in Period 2 contain electrons in levels 1 and 2. In Period 3, electrons are found in levels 1, 2, and 3, and so on.

Elements that behave in a similar manner are found in **groups** or **families.** These form the vertical columns on the periodic table. Several systems exist for numbering the groups. In one system the columns are numbered from left to right using the numbers 1–18. However, we use a system that numbers the columns with Roman numerals and the letters A and B, as shown in Figure 10.14. The A groups are known as the **representative elements.** The B groups and Group VIII are called the **transition elements.** In this book we will focus on the representative elements. The groups (columns) of the periodic table often have family names. For example, the group on the far right side of the periodic table (He, Ne, Ar, Kr, Xe, and Rn) is called the *noble gases.* Group IA is called the *alkali metals,* Group IIA the *alkaline earth metals,* and Group VIIA the *halogens.*

How is the structure of the periodic table related to the atomic structures of the elements? We've just seen that the periods of the periodic table are associated with the energy level of the outermost electrons of the atoms in that period. Look at the valence electron configurations of the elements we have just examined (Figure 10.15). Do you see a pattern? The valence configuration for the elements in columns is the same. The chemical behavior and properties of elements in a particular family must therefore be associated with the electron configuration of the elements. The number for the energy level is different. This is expected since each new period is associated with a different energy level for the valence electrons.

The electron configurations for elements beyond these first 18 become long and tedious to write. We often abbreviate the electron configuration using the following notation:

Na $[Ne]3s^1$

Look carefully at Figure 10.15 and you will see that the *p* orbitals are full at the noble gases. By placing the symbol for the noble gas in square brackets we can abbreviate the complete electron configuration and focus our attention on the valence electrons (the electrons we will be interested in when we discuss bonding in Chapter 11). To write the abbreviated electron configuration for any element, go back to the previous

(margin notes)
period

groups, families

representative elements
transition elements

IA							Noble gases
1 **H** $1s^1$	IIA	IIIA	IVA	VA	VIA	VIIA	2 **He** $1s^2$
3 **Li** $2s^1$	4 **Be** $2s^2$	5 **B** $2s^22p^1$	6 **C** $2s^22p^2$	7 **N** $2s^22p^3$	8 **O** $2s^22p^4$	9 **F** $2s^22p^5$	10 **Ne** $2s^22p^6$
11 **Na** $3s^1$	12 **Mg** $3s^2$	13 **Al** $3s^23p^1$	14 **Si** $3s^23p^2$	15 **P** $3s^23p^3$	16 **S** $3s^23p^4$	17 **Cl** $3s^23p^5$	18 **Ar** $3s^23p^6$

◀ **FIGURE 10.15**
Valence electron configurations for the first 18 elements.

noble gas and place its symbol in square brackets. Then list the valence electrons. Here are some examples:

B	$1s^2\,2s^2\,2p^1$	$[He]2s^22p^1$
Cl	$1s^22s^22p^63s^23p^5$	$[Ne]3s^23p^5$
Na	$1s^22s^22p^63s^1$	$[Ne]3s^1$

The sequence for filling the orbitals is exactly as we would expect up through the $3p$ orbitals. The third energy level might be expected to fill with $3d$ electrons before electrons enter the $4s$ orbital, but this is not the case. The behavior and properties of the next two elements, potassium (19) and calcium (20), are very similar to the elements in Groups IA and IIA. They clearly belong in these groups. The other elements in Group IA and Group IIA have electron configurations that indicate valence electrons in the s orbitals. For example, since the electron configuration is connected to the element's properties, we should place the last electrons for potassium and calcium in the $4s$ orbital. Their electron configurations are

K	$1s^22s^22p^63s^23p^64s^1$	or	$[Ar]4s^1$
Ca	$1s^22s^22p^63s^23p^64s^2$	or	$[Ar]4s^2$

Elements 21–30 belong to the elements known as *transition elements*. Electrons are placed in the $3d$ orbitals for each of these elements. When the $3d$ orbitals are full, the electrons fill the $4p$ orbitals to complete the fourth period. Let's consider the overall relationship between orbital filling and the periodic table. Figure 10.16 illustrates the type of orbital filling and its location on the periodic table. The tall columns on the table (labeled IA–VIIA, and noble gases) are often called the *representative ele-*

FIGURE 10.16
Arrangement of elements according to the sublevel being filled.
▼

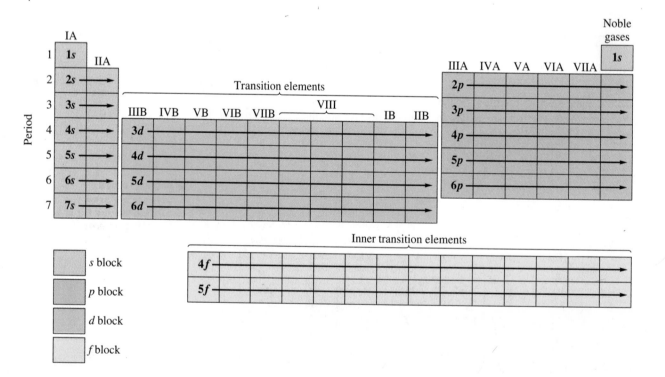

ments. Valence electrons in these elements occupy *s* and *p* orbitals. The period number corresponds to the energy level of the valence electrons. The elements in the center of the periodic table (shown in ▓) are the transition elements where the *d* orbitals are being filled. Notice that the number for the *d* orbitals is one behind the period number. The two rows shown at the bottom of the table in Figure 10.16 are called the *inner transition elements*. The last electrons in these elements are placed in the *f* orbitals. The number for the *f* orbitals is always two behind that of the *s* and *p* orbitals. A periodic table is almost always available to you, so if you understand the relationship between the orbitals and the periodic table, you can write the electron configuration for any element. There are several minor variations to these rules, but we won't concern ourselves with them in this course.

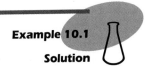

Example 10.1

Solution

Use the periodic table to write the electron configuration for phosphorus and tin.

Phosphorus is element 15 and is located in Period 3, Group VA. The electron configuration must have a full first and second energy level:

P $1s^2 2s^2 2p^6 3s^2 3p^3$ or $[Ne]3s^2 3p^3$

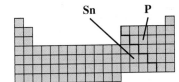

You can determine the electron configuration by looking across the period and counting the element blocks.

Tin is element 50 in Period 5, Group IVA, two places after the transition metals. It must have two electrons in the 5*p* series. Its electron configuration is

Sn $1s^2 2s^2 2p^6 3s^2 3p^6 4s^2 3d^{10} 4p^6 5s^2 4d^{10} 5p^2$ or $[Kr]5s^2 4d^{10} 5p^2$

Notice that the *d* series of electrons is always one energy level behind the period number.

Practice 10.2

Use the periodic table to write the electron configuration for (a) O, (b) Ca, and (c) Ti.

The early chemists classified the elements based only on their observed properties, but modern atomic theory gives us an explanation for why the properties of elements vary periodically. For example, as we "build" atoms by filling orbitals with electrons, the same orbitals occur on each energy level. This means that the same electron configuration reappears regularly for each level. Groups of elements show similar chemical properties because of the similarity of these outermost electron configurations.

In Figure 10.17, only the electron configuration of the outermost electrons is given. This periodic table illustrates these important points:

1. The number of the period corresponds with the highest energy level occupied by electrons.
2. The group numbers for the representative elements are equal to the total number of outermost electrons in the atoms of the group. For example, ele-

Group number

IA	IIA	IIIB	IVB	VB	VIB	VIIB		VIII		IB	IIB	IIIA	IVA	VA	VIA	VIIA	Noble gases
1 **H** $1s^1$																	2 **He** $1s^2$
3 **Li** $2s^1$	4 **Be** $2s^2$											5 **B** $2s^2 2p^1$	6 **C** $2s^2 2p^2$	7 **N** $2s^2 2p^3$	8 **O** $2s^2 2p^4$	9 **F** $2s^2 2p^5$	10 **Ne** $2s^2 2p^6$
11 **Na** $3s^1$	12 **Mg** $3s^2$											13 **Al** $3s^2 3p^1$	14 **Si** $3s^2 3p^2$	15 **P** $3s^2 3p^3$	16 **S** $3s^2 3p^4$	17 **Cl** $3s^2 3p^5$	18 **Ar** $3s^2 3p^6$
19 **K** $4s^1$	20 **Ca** $4s^2$	21 **Sc** $4s^2 3d^1$	22 **Ti** $4s^2 3d^2$	23 **V** $4s^2 3d^3$	24 **Cr** $4s^1 3d^5$	25 **Mn** $4s^2 3d^5$	26 **Fe** $4s^2 3d^6$	27 **Co** $4s^2 3d^7$	28 **Ni** $4s^2 3d^8$	29 **Cu** $4s^1 3d^{10}$	30 **Zn** $4s^2 3d^{10}$	31 **Ga** $4s^2 4p^1$	32 **Ge** $4s^2 4p^2$	33 **As** $4s^2 4p^3$	34 **Se** $4s^2 4p^4$	35 **Br** $4s^2 4p^5$	36 **Kr** $4s^2 4p^6$
37 **Rb** $5s^1$	38 **Sr** $5s^2$	39 **Y** $5s^2 4d^1$	40 **Zr** $5s^2 4d^2$	41 **Nb** $5s^1 4d^4$	42 **Mo** $5s^1 4d^5$	43 **Tc** $5s^1 4d^6$	44 **Ru** $5s^1 4d^7$	45 **Rh** $5s^1 4d^8$	46 **Pd** $5s^0 4d^{10}$	47 **Ag** $5s^1 4d^{10}$	48 **Cd** $5s^2 4d^{10}$	49 **In** $5s^2 5p^1$	50 **Sn** $5s^2 5p^2$	51 **Sb** $5s^2 5p^3$	52 **Te** $5s^2 5p^4$	53 **I** $5s^2 5p^5$	54 **Xe** $5s^2 5p^6$
55 **Cs** $6s^1$	56 **Ba** $6s^2$	57 **La** $6s^2 5d^1$	72 **Hf** $6s^2 5d^2$	73 **Ta** $6s^2 5d^3$	74 **W** $6s^2 5d^4$	75 **Re** $6s^2 5d^5$	76 **Os** $6s^2 5d^6$	77 **Ir** $6s^2 5d^7$	78 **Pt** $6s^1 5d^9$	79 **Au** $6s^1 5d^{10}$	80 **Hg** $6s^2 5d^{10}$	81 **Tl** $6s^2 6p^1$	82 **Pb** $6s^2 6p^2$	83 **Bi** $6s^2 6p^3$	84 **Po** $6s^2 6p^4$	85 **At** $6s^2 6p^5$	86 **Rn** $6s^2 6p^6$
87 **Fr** $7s^1$	88 **Ra** $7s^2$	89 **Ac** $7s^2 6d^1$	104 **Rf** $7s^2 6d^2$	105 **Db** $7s^2 6d^3$	106 **Sg** $7s^2 6d^4$	107 **Bh** $7s^2 6d^5$	108 **Hs** $7s^2 6d^6$	109 **Mt** $7s^2 6d^7$	110 **Uun** $7s^1 6d^9$	111 **Uuu** $7s^1 6d^{10}$	112 **Uub** $7s^2 6d^{10}$						

Period (1–7, rows above)

▲ FIGURE 10.17
Outermost electron configurations.

ments in Group VIIA always have the electron configuration $ns^2\, np^5$. The d and f electrons are always in a lower energy level than the highest energy level and so are not considered as outermost electrons.

3. The elements of a family have the same outermost electron configuration except that the electrons are in different energy levels.

4. The elements within each of the s, p, d, f blocks are filling the s, p, d, f orbitals, as shown in Figure 10.16.

5. Within the transition elements some discrepancies in the order of filling occur. (Explanation of these discrepancies and similar ones in the inner transition elements are beyond the scope of this book.)

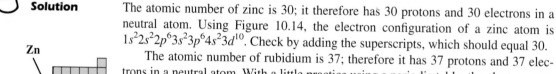

Example 10.2   Write the electron configuration of a zinc atom and a rubidium atom.

Solution   The atomic number of zinc is 30; it therefore has 30 protons and 30 electrons in a neutral atom. Using Figure 10.14, the electron configuration of a zinc atom is $1s^2 2s^2 2p^6 3s^2 3p^6 4s^2 3d^{10}$. Check by adding the superscripts, which should equal 30.

The atomic number of rubidium is 37; therefore it has 37 protons and 37 electrons in a neutral atom. With a little practice using a periodic table, the electron configuration may be written directly. The electron configuration of a rubidium atom is $1s^2 2s^2 2p^6 3s^2 3p^6 4s^2 3d^{10} 4p^6 5s^1$. Check by adding the superscripts, which should equal 37.

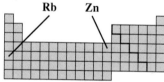

Rb   Zn

Concepts in Review

1. Describe the atom as conceived by Niels Bohr.

2. Discuss the contributions to atomic theory made by Dalton, Thomson, Rutherford, Bohr, and Schrödinger. (See Chapter 5 as well.)

3. Explain what is meant by an electron orbital.

4. Explain how an electron configuration can be determined from the periodic table.

5. Write the electron configuration for any of the first 56 elements.

6. Indicate the locations of the metals, nonmetals, metalloids, and noble gases in the periodic table. (See Chapter 3 for help.)

7. Indicate the areas in the periodic table where the s, p, d, and f orbitals are being filled.

8. Determine the number of valence electrons in any atom in the Group A elements.

9. Distinguish between representative elements and transition elements.

10. Identify groups of elements by their special family names.

11. Describe the changes in valence electron structure (a) when moving from left to right in a period and (b) when going from top to bottom in a group.

12. Explain the relationship between group number and the number of valence electrons for the representative elements.

Key Terms

electron configuration (10.5)
frequency (10.2)
ground state (10.3)
groups, families (10.6)
line spectrum (10.3)

orbital (10.3, 10.4)
orbital diagram (10.5)
Pauli exclusion principle (10.4)
period (10.6)
photons (10.2)

principal energy levels (10.4)
quanta (10.3)
representative elements (10.6)
speed (10.2)

spin (10.4)
transition elements (10.6)
valence electrons (10.5)
wavelength (10.2)

Questions

1. What is an orbital?

2. Under what conditions can a second electron enter an orbital already containing one electron?

3. What is meant when we say the electron structure of an atom is in its ground state?

4. How do $1s$ and $2s$ orbitals differ? How are they alike?

5. What letters are used to designate the types of orbitals?

6. List the following orbitals in order of increasing energy: $2s$, $2p$, $4s$, $1s$, $3d$, $3p$, $4p$, $3s$.

7. How many s electrons, p electrons, and d electrons are possible in any energy level?

8. What is the major difference between an orbital and a Bohr orbit?

9. Explain how and why Bohr's model of the atom was modified to include the cloud model of the atom.

10. Sketch the s, p_x, p_y, and p_z orbitals.

11. In the designation $3d^7$, give the significance of 3, d, and 7.

12. Describe the difference between transition and representative elements.

13. From the standpoint of electron structure, what do the elements in the s block have in common?

14. Write symbols for elements with atomic numbers 8, 16, 34, 52, and 84. What do these elements have in common?

15. Write the symbols of the family of elements that have seven electrons in their outermost energy level.

16. What is the greatest number of elements to be found in any period? Which periods have this number?

17. From the standpoint of energy level, how does the placement of the last electron in the Group A elements differ from that of the Group B elements?

18. Find the places in the periodic table where elements are not in proper sequence according to atomic mass. (See inside of front cover for periodic table.)

Paired Exercises

19. How many protons are in the nucleus of an atom of these elements?
    (a) H
    (b) B
    (c) Sc
    (d) U

20. How many protons are in the nucleus of an atom of these elements?
    (a) F
    (b) Ag
    (c) Br
    (d) Sb

21. Give the electron configuration for
    (a) B
    (b) Ti
    (c) Zn
    (d) Br
    (e) Sr

22. Give the electron configuration for
    (a) chlorine
    (b) silver
    (c) lithium
    (d) iron
    (e) iodine

23. Explain how the spectral lines of hydrogen occur.

24. Explain how Bohr used the data from the hydrogen spectrum to support his model of the atom.

25. How many orbitals exist in the third principal energy level? What are they?

26. How many electrons can be present in the fourth principal energy level?

27. Write orbital diagrams for
    (a) N
    (b) Cl
    (c) Zn
    (d) Zr
    (e) I

28. Write orbital diagrams for
    (a) Si
    (b) S
    (c) Ar
    (d) V
    (e) P

29. Which elements have these electron configurations?
    (a) $1s^2 2s^2 2p^6 3s^2$
    (b) $1s^2 2s^2 2p^6 3s^2 3p^1$
    (c) $1s^2 2s^2 2p^6 3s^2 3p^6 4s^2 3d^8$
    (d) $1s^2 2s^2 2p^6 3s^2 3p^6 4s^2 3d^5$

30. Which elements have these electron configurations?
    (a) $[Ar]4s^2 3d^1$
    (b) $[Ar]4s^2 3d^{10}$
    (c) $[Kr]5s^2 4d^{10} 5p^2$
    (d) $[Xe]6s^1$

31. Show the electron configurations for elements with these atomic numbers:
    (a) 8
    (b) 11
    (c) 17
    (d) 23
    (e) 28
    (f) 34

32. Show the electron configurations for elements with these atomic numbers:
    (a) 9
    (b) 26
    (c) 31
    (d) 39
    (e) 52
    (f) 10

33. Identify these elements from their atomic structure diagrams:
    (a) $\left(\begin{array}{c}16p\\16n\end{array}\right)$ $2e^-$ $8e^-$ $6e^-$
    (b) $\left(\begin{array}{c}28p\\32n\end{array}\right)$ $2e^-$ $8e^-$ $16e^-$ $2e^-$

34. Diagram the atomic structures (as you see in Exercise 33) for these elements:
    (a) $^{27}_{13}Al$
    (b) $^{48}_{22}Ti$

35. Why is the 11th electron of the sodium atom located in the third energy level rather than in the second energy level?

36. Why is the last electron in potassium located in the fourth energy level rather than in the third energy level?

37. What electron structure do the noble gases have in common?

38. What is unique about the noble gases, from an electron point of view?

39. How are elements in a period related to one another?

40. How are elements in a group related to one another?

41. What do the electron structures of the alkali metals have in common?

43. Pick the electron structures that represent elements in the same chemical family:
(a) $1s^2 2s^1$
(b) $1s^2 2s^2 2p^4$
(c) $1s^2 2s^2 2p^2$
(d) $1s^2 2s^2 2p^6 3s^2 3p^4$
(e) $1s^2 2s^2 2p^6 3s^2 3p^6$
(f) $1s^2 2s^2 2p^6 3s^2 3p^6 4s^2$
(g) $1s^2 2s^2 2p^6 3s^2 3p^6 4s^1$
(h) $1s^2 2s^2 2p^6 3s^2 3p^6 4s^2 3d^1$

45. In the periodic table, calcium, element 20, is surrounded by elements 12, 19, 21, and 38. Which of these have physical and chemical properties most resembling calcium?

47. Classify the following elements as metals, nonmetals, or metalloids (review Chapter 3 if you need help):
(a) potassium
(b) plutonium
(c) sulfur
(d) antimony

49. In which period and group does an electron first appear in an f orbital?

51. How many electrons occur in the valence level of Group VIIA and VIIB elements? Why are they different?

42. Why would you expect the elements zinc, cadmium, and mercury to be in the same chemical family?

44. Pick the electron structures that represent elements in the same chemical family:
(a) $[\text{He}]2s^2 2p^6$
(b) $[\text{Ne}]3s^1$
(c) $[\text{Ne}]3s^2$
(d) $[\text{Ne}]3s^2 3p^3$
(e) $[\text{Ar}]4s^2 3d^{10}$
(f) $[\text{Ar}]4s^2 3d^{10} 4p^6$
(g) $[\text{Ar}]4s^2 3d^5$
(h) $[\text{Kr}]5s^2 4d^{10}$

46. In the periodic table, phosphorus, element 15, is surrounded by elements 14, 7, 16, and 33. Which of these have physical and chemical properties most resembling phosphorus?

48. Classify the following elements as metals, nonmetals, or metalloids (review Chapter 3 if you need help):
(a) iodine
(b) tungsten
(c) molybdenum
(d) germanium

50. In which period and group does an electron first appear in a d orbital?

52. How many electrons occur in the valence level of Group IIIA and IIIB elements? Why are they different?

Additional Exercises

53. If the orbitals in an atom could each hold three electrons rather than two, what would be the atomic numbers of the first three noble gases?

54. Why does the emission spectrum for nitrogen reveal many more spectral lines than that for hydrogen?

55. Among the first 100 elements on the periodic table, how many have at least
(a) one s electron?
(b) one p electron?
(c) one d electron?
(d) one f electron?

56. For the following elements, what percent of their electrons occupy s orbitals?
(a) He   (d) Se
(b) Be   (e) Cs
(c) Xe

57. How many pairs of valence electrons do these elements have?
(a) O   (d) Xe
(b) P   (e) Rb
(c) I

58. Suppose we use a foam ball to represent a typical atom. If the radius of the ball is 1.5 cm and the radius of a typical atom is 1.0×10^{-8} cm, how much of an enlargement is this? Use a ratio to express your answer.

59. List the first element on the periodic table that satisfies each of these conditions:
(a) a completed set of p orbitals
(b) two $4p$ electrons
(c) seven valence electrons
(d) three unpaired electrons

60. Oxygen is a gas. Sulfur is a solid. What is it about their electron structures that causes them to be grouped in the same chemical family?

61. In which groups are the transition elements located?

62. How do the electron structures of the transition elements differ from the representative elements?

63. The atomic numbers of the noble gases are 2, 10, 18, 36, 54, and 86. What are the atomic numbers for the elements with six electrons in their outermost electron configuration?

64. Element number 87 is in Group IA, Period 7. Describe its outermost energy level. How many energy levels of electrons does it have?

65. If element 36 is a noble gas, in which groups would you expect elements 35 and 37 to occur?

66. Write a paragraph describing the general features of the periodic table.

67. Some scientists have proposed the existence of element 117. If it were to exist,
   (a) what would its electron configuration be?
   (b) how many valence electrons would it have?
   (c) what element would it likely resemble?
   (d) to what family and period would it belong?

68. What is the relationship between two elements if
   (a) one of them has 10 electrons, 10 protons, and 10 neutrons and the other has 10 electrons, 10 protons, and 12 neutrons?
   (b) one of them has 23 electrons, 23 protons, and 27 neutrons and the other has 24 electrons, 24 protons, and 26 neutrons?

69. Is there any pattern for the location of gases on the periodic table? for the location of liquids? for the location of solids?

Answers to Practice Exercises

10.1 (a) $2s^2 2p^1$
(b) $2s^2 2p^3$
(c) $3s^1$
(d) $3s^2 3p^5$

10.2 (a) O $1s^2 2s^2 2p^4$
(b) Ca $1s^2 2s^2 2p^6 3s^2 3p^6 4s^2$
(c) Ti $1s^2 2s^2 2p^6 3s^2 3p^6 4s^2 3d^2$

CHAPTER

11

Chemical Bonds: The Formation of Compounds from Atoms

▲
This colorfully lighted limestone cave reveals dazzling stalactites and stalagmites, formed from calcium carbonate.

For centuries we've been aware that certain metals cling to a magnet. We've seen balloons sticking to walls. Why? Superconductors floating in air are seen in television commercials. High-speed levitation trains are heralded to be the wave of the future. How do they function? In each case, forces of attraction and repulsion are at work.

Human interactions also suggest that "opposites attract" and "likes repel." Attractions draw us into friendships and significant relationships, whereas repulsive forces may produce debate and antagonism. We form and break apart interpersonal bonds throughout our lives.

In chemistry, we also see this phenomenon. Substances form chemical bonds as a result of electrical attractions. These bonds provide the tremendous diversity of compounds found in nature.

11.1   Periodic Trends in Atomic Properties

Although atomic theory and electron configuration help us understand the arrangement and behavior of the elements, it's important to remember that the design of the periodic table is based on observing properties of the elements. Before we use the concept of atomic structure to explain how and why atoms combine to form compounds, we need to understand the characteristic properties of the elements and the trends that occur in these properties on the periodic table. These trends allow us to use the periodic table to accurately predict properties and reactions of a wide variety of substances.

FIGURE 11.1
The elements are classified as metals, nonmetals, and metalloids.
▼

1 H																	2 He
3 Li	4 Be											5 B	6 C	7 N	8 O	9 F	10 Ne
11 Na	12 Mg											13 Al	14 Si	15 P	16 S	17 Cl	18 Ar
19 K	20 Ca	21 Sc	22 Ti	23 V	24 Cr	25 Mn	26 Fe	27 Co	28 Ni	29 Cu	30 Zn	31 Ga	32 Ge	33 As	34 Se	35 Br	36 Kr
37 Rb	38 Sr	39 Y	40 Zr	41 Nb	42 Mo	43 Tc	44 Ru	45 Rh	46 Pd	47 Ag	48 Cd	49 In	50 Sn	51 Sb	52 Te	53 I	54 Xe
55 Cs	56 Ba	57 La*	72 Hf	73 Ta	74 W	75 Re	76 Os	77 Ir	78 Pt	79 Au	80 Hg	81 Tl	82 Pb	83 Bi	84 Po	85 At	86 Rn
87 Fr	88 Ra	89 Ac†	104 Rf	105 Db	106 Sg	107 Bh	108 Hs	109 Mt	110 Uun	111 Uuu	112 Uub						

Legend:
- Metals
- Metalloids
- Nonmetals

*	58 Ce	59 Pr	60 Nd	61 Pm	62 Sm	63 Eu	64 Gd	65 Tb	66 Dy	67 Ho	68 Er	69 Tm	70 Yb	71 Lu
†	90 Th	91 Pa	92 U	93 Np	94 Pu	95 Am	96 Cm	97 Bk	98 Cf	99 Es	100 Fm	101 Md	102 No	103 Lr

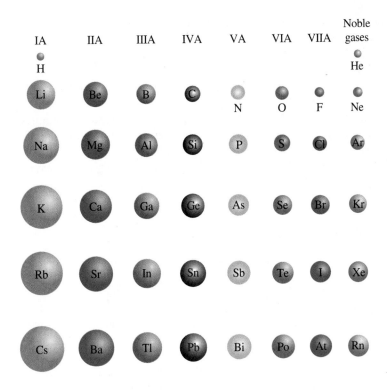

◀ FIGURE 11.2
Relative atomic radii for the representative elements. Atomic radius decreases across a period and increases down a group in the periodic table.

Metals and Nonmetals

In Section 3.8, we classified elements as metals, nonmetals, or metalloids. The heavy stair-step line beginning at boron and running diagonally down the periodic table separates the elements into metals and nonmetals. Metals are usually lustrous, malleable, and good conductors of heat and electricity. Nonmetals are just the opposite—nonlustrous, brittle, and poor conductors. Metalloids are found bordering the heavy diagonal line and may have properties of both metals and nonmetals.

Most elements are classified as metals (see Figure 11.1). Metals are found on the left side of the stair-step line, while the nonmetals are located toward the upper right of the table. Note that hydrogen does not fit into the division of metals and nonmetals. It displays nonmetallic properties under normal conditions, even though it has only one outermost electron like the alkali metals. Hydrogen is considered to be a unique element.

It is the chemical properties of metals and nonmetals that interest us most. Metals tend to lose electrons and form positive ions, while nonmetals tend to gain electrons and form negative ions. When a metal reacts with a nonmetal, electrons are often transferred from the metal to the nonmetal.

Atomic Radius

The relative radii of the representative elements are shown in Figure 11.2. Notice that the radii of the atoms tend to increase down each group and that they tend to decrease from left to right across a period.

The increase in radius down a group can be understood if we consider the electron structure of the atoms. For each step down a group, an additional energy level is added to the atom. The average distance from the nucleus to the outside edge of the atom must increase as each new energy level is added. The atoms get bigger as electrons are placed in these new higher energy levels.

Understanding the decrease in atomic radius across a period requires more thought, however. As we move from left to right across a period, electrons within the same block are being added to the same energy level. Within a given energy level we expect the orbitals to have about the same size. We would then expect the atoms to be about the same size across the period. But each time an electron is added, a proton is added to the nucleus as well. The increase in positive charge (in the nucleus) pulls the electrons closer to the nucleus, which results in a gradual decrease in atomic radius across a period.

Ionization Energy

ionization energy

The **ionization energy** of an atom is the energy required to remove an electron from the atom. For example,

$$\text{Na} + \text{ionization energy} \longrightarrow \text{Na}^+ + \text{e}^-$$

The first ionization energy is the amount of energy required to remove the first electron from an atom, the second is the amount required to remove the second electron from that atom, and so on.

Table 11.1 gives the ionization energies for the removal of one to five electrons from several elements. The table shows that even higher amounts of energy are needed to remove the second, third, fourth, and fifth electrons. This makes sense because removing electrons leaves fewer electrons attracted to the positive charge in the nucleus. The data in Table 11.1 also show that an extra large ionization energy (red) is needed when an electron is removed from a noble gas structure, clearly showing the stability of this configuration.

TABLE 11.1   Ionization Energies for Selected Elements*

Element	Required amounts of energy (kJ/mol)				
	1st e^-	2nd e^-	3rd e^-	4th e^-	5th e^-
H	1,314				
He	2,372	5,247			
Li	520	7,297	11,810		
Be	900	1,757	14,845	21,000	
B	800	2,430	3,659	25,020	32,810
C	1,088	2,352	4,619	6,222	37,800
Ne	2,080	3,962	6,276	9,376	12,190
Na	496	4,565	6,912	9,540	13,355

*Values are expressed in kilojoules per mole, showing energies required to remove 1 to 5 electrons per atom. Red type indicates the energy needed to remove an electron from a noble gas electron structure.

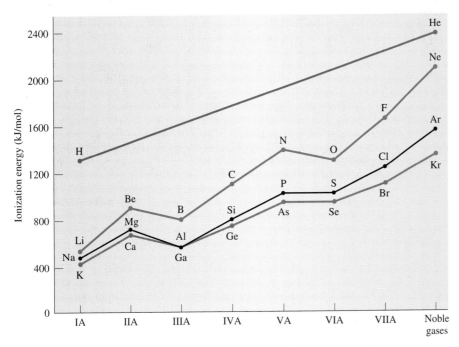

◄ FIGURE 11.3
Periodic relationship of the first ionization energy for representative elements in the first four periods.

First ionization energies have been experimentally determined for most elements. Figure 11.3 plots these energies for representative elements in the first four periods. Note these important points:

1. Ionization energy in Group A elements decreases from top to bottom in a group. For example, in Group IA the ionization energy changes from 520 kJ/mol for Li to 419 kJ/mol for K.
2. Ionization energy gradually increases from left to right across a period. Noble gases have a relatively high value, confirming the nonreactive nature of these elements.

Metals don't behave in exactly the same manner. Some metals give up electrons much more easily than others. In the alkali metal family, cesium gives up its $6s$ electron much more easily than the metal lithium gives up its $2s$ electron. This makes sense when we consider that the size of the atoms increases down the group. The distance between the nucleus and the outer electrons increases and the ionization energy decreases. The most chemically active metals are located at the lower left of the periodic table.

Nonmetals have relatively large ionization energies compared to metals. Nonmetals tend to gain electrons and form anions. Since the nonmetals are located at the right side of the periodic table, it is not surprising that ionization energies tend to increase from left to right across a period. The most active nonmetals are found in the *upper* right corner of the periodic table.

11.2 Lewis Structures of Atoms

Metals tend to form cations (positively charged ions) and nonmetals form anions (negatively charged ions) in order to attain a stable valence electron structure. For many elements this stable valence level contains eight electrons (two s and six p),

FIGURE 11.4 ▶
Lewis structures of the first 20 elements. Dots represent electrons in the outermost energy level only.

IA	IIA	IIIA	IVA	VA	VIA	VIIA	Noble Gases
H·							He:
Li·	Be:	:Ḃ	:Ċ·	:N̈·	·Ö:	:F̈:	:N̈e:
Na·	Mg:	:Äl	:S̈i·	:P̈·	·S̈:	:C̈l·	:Är:
K·	Ca:						

identical to the valence electron configuration of the noble gases. Atoms undergo rearrangements of electron structure to lower their chemical potential energy (or to become more stable). These rearrangements are accomplished by losing, gaining, or sharing electrons with other atoms. For example, a hydrogen atom could accept a second electron and attain an electron structure the same as the noble gas helium. A fluorine atom could gain an electron and attain an electron structure like neon. A sodium atom could lose one electron to attain an electron structure like neon.

Lewis structure

The valence electrons in the outermost energy level of an atom are responsible for the electron activity that occurs to form chemical bonds. The **Lewis structure** of an atom is a representation that shows the valence electrons for that atom. American chemist Gilbert N. Lewis (1875–1946) proposed using the symbol for the element and dots for electrons. The number of dots placed around the symbol equals the number of s and p electrons in the outermost energy level of the atom. Paired dots represent paired electrons; unpaired dots represent unpaired electrons. For example, **H·** is the Lewis symbol for a hydrogen atom, $1s^1$; **:Ḃ** is the Lewis symbol for a boron atom, with valence electrons $2s^2 2p^1$. In the case of boron, the symbol B represents the boron nucleus and the $1s^2$ electrons; the dots represent only the $2s^2 2p^1$ electrons.

▲
A sample of boron.

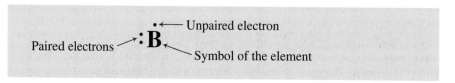

Paired electrons ⟶ :**B** ⟵ Unpaired electron
Symbol of the element

The Lewis method is used not only because of its simplicity of expression but also because much of the chemistry of the atom is directly associated with the electrons in the outermost energy level. Figure 11.4 shows Lewis structures for the elements hydrogen through calcium.

Example 11.1   Write the Lewis structure for a phosphorus atom.

Solution   First establish the electron structure for a phosphorus atom, which is $1s^2 2s^2 2p^6 3s^2 3p^3$. Note that there are five electrons in the outermost energy level; they are $3s^2 3p^3$. Write the symbol for phosphorus and place the five electrons as dots around it.

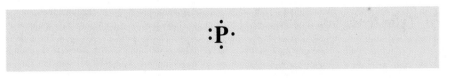

$$:\overset{\displaystyle\cdot}{\underset{\displaystyle\cdot}{P}}\cdot$$

A quick way to determine the correct number of dots (electrons) for a Lewis structure is to use the group number. For the A groups on the periodic table, the Roman numeral is the same as the number of electrons in the Lewis structure.

The $3s^2$ electrons are paired and are represented by the paired dots. The $3p^3$ electrons, which are unpaired, are represented by the single dots.

Practice 11.1

Write the Lewis structure for the following elements:
(a) N (b) Al (c) Sr (d) Br

11.3   The Ionic Bond: Transfer of Electrons from One Atom to Another

The chemistry of many elements, especially the representative ones, is to attain an outer electron structure like that of the chemically stable noble gases. With the exception of helium, this stable structure consists of eight electrons in the outermost energy level (see Table 11.2).

Let's look at the electron structures of sodium and chlorine to see how each element can attain a structure of eight electrons in its outermost energy level. A sodium atom has 11 electrons: 2 in the first energy level, 8 in the second energy level, and 1 in the third energy level. A chlorine atom has 17 electrons: 2 in the first energy level, 8 in the second energy level, and 7 in the third energy level. If a sodium atom transfers or loses its $3s$ electron, its third energy level becomes vacant, and it becomes a sodium ion with an electron configuration identical to that of the noble gas neon. This process requires energy:

11^+ $11e^-$ $\longrightarrow$ 11^+ $10e^-$

                                                       $+$ $1e^-$

Na atom Na^+ ion
$(1s^2 2s^2 2p^6 3s^1)$ $(1s^2 2s^2 2p^6)$

An atom that has lost or gained electrons will have a positive or negative charge, depending on which particles (protons or electrons) are in excess. Remember that a charged particle or group of particles is called an *ion*.

By losing a negatively charged electron, the sodium atom becomes a positively charged particle known as a sodium ion. The charge, $+1$, results because the nucleus still contains 11 positively charged protons, and the electron orbitals contain only 10

TABLE 11.2	Arrangement of Electrons in the Noble Gases*						
		Electron structure					
Noble gas	Symbol	$n = 1$	2	3	4	5	6
Helium	He	$1s^2$					
Neon	Ne	$1s^2$	$2s^22p^6$				
Argon	Ar	$1s^2$	$2s^22p^6$	$3s^23p^6$			
Krypton	Kr	$1s^2$	$2s^22p^6$	$3s^23p^63d^{10}$	$4s^24p^6$		
Xenon	Xe	$1s^2$	$2s^22p^6$	$3s^23p^63d^{10}$	$4s^24p^64d^{10}$	$5s^25p^6$	
Radon	Rn	$1s^2$	$2s^22p^6$	$3s^23p^63d^{10}$	$4s^24p^64d^{10}4f^{14}$	$5s^25p^65d^{10}$	$6s^26p^6$

*Each gas except helium has eight electrons in its outermost energy level.

negatively charged electrons. The charge is indicated by a plus sign (+) and is written as a superscript after the symbol of the element (Na^+).

A chlorine atom with seven electrons in the third energy level needs one electron to pair up with its one unpaired $3p$ electron to attain the stable outer electron structure of argon. By gaining one electron, the chlorine atom becomes a chloride ion (Cl^-), a negatively charged particle containing 17 protons and 18 electrons. This process releases energy:

17+   17e⁻   + 1e⁻ ⟶   17+   18e⁻

Cl atom Cl^- ion
($1s^22s^22p^63s^23p^5$)   ($1s^22s^22p^63s^23p^6$)

Consider sodium and chlorine atoms reacting with each other. The $3s$ electron from the sodium atom transfers to the half-filled $3p$ orbital in the chlorine atom to form a positive sodium ion and a negative chloride ion. The compound sodium chloride results because the Na^+ and Cl^- ions are strongly attracted to each other by their opposite electrostatic charges. The force holding the oppositely charged ions together is an ionic bond:

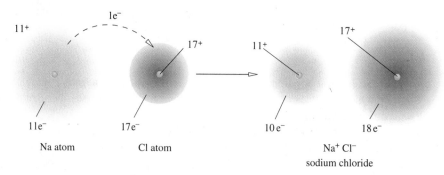

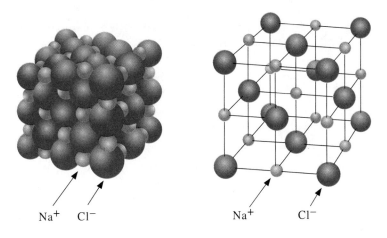

◀ **FIGURE 11.5**
**Sodium chloride crystal.
Diagram represents a small
fragment of sodium
chloride, which forms cubic
crystals. Each sodium ion is
surrounded by six chloride
ions, and each chloride ion
is surrounded by six sodium
ions. The tiny NaCl crystals
show the cubic crystal
structure of salt.**

The Lewis representation of sodium chloride formation is:

$$Na\cdot + \cdot \ddot{\underset{..}{Cl}}: \longrightarrow [Na]^+ \left[:\ddot{\underset{..}{Cl}}:\right]^-$$

The chemical reaction between sodium and chlorine is a very vigorous one, producing considerable heat in addition to the salt formed. When energy is released in a chemical reaction, the products are more stable than the reactants. Note that in NaCl both atoms attain a noble gas electron structure.

Sodium chloride is made up of cubic crystals in which each sodium ion is surrounded by six chloride ions and each chloride ion by six sodium ions, except at the crystal surface. A visible crystal is a regularly arranged aggregate of millions of these ions, but the ratio of sodium to chloride ions is 1:1, hence the formula NaCl. The cubic crystalline lattice arrangement of sodium chloride is shown in Figure 11.5.

Figure 11.6 contrasts the relative sizes of sodium and chlorine atoms with those of their ions. The sodium ion is smaller than the atom due primarily to two factors: (1) The sodium atom has lost its outermost electron, thereby reducing its size; and (2) the 10 remaining electrons are now attracted by 11 protons and are thus drawn closer to the nucleus. Conversely, the chloride ion is larger than the atom because (1) it has 18 electrons but only 17 protons; and (2) the nuclear attraction on each electron is thereby decreased, allowing the chlorine atom to expand as it forms an ion.

We've seen that when sodium reacts with chlorine, each atom becomes an ion. Sodium chloride, like all ionic substances, is held together by the attraction existing between positive and negative charges. An **ionic bond** is the attraction between oppositely charged ions.

▲
**These tiny NaCl crystals (on a
penny) show the cubic structure illustrated in Figure 11.5.**

Remember: A cation is always
smaller than its parent atom
whereas an anion is always
larger than its parent atom.

ionic bond

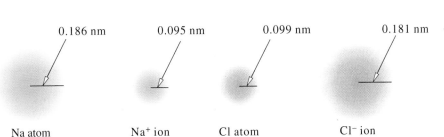

0.186 nm   0.095 nm   0.099 nm   0.181 nm

Na atom   Na⁺ ion   Cl atom   Cl⁻ ion

◀ **FIGURE 11.6**
**Relative radii of sodium
and chlorine atoms and
their ions.**

The reaction between sodium metal and chlorine gas releases both heat and light in the production of salt.

TABLE 11.3	Change in Atomic Radii (nm) of Selected Metals and Nonmetals*						
Atomic radius		**Ionic radius**		**Atomic radius**		**Ionic radius**	
Li	0.152	Li$^+$	0.060	F	0.071	F$^-$	0.136
Na	0.186	Na$^+$	0.095	Cl	0.099	Cl$^-$	0.181
K	0.227	K$^+$	0.133	Br	0.114	Br$^-$	0.195
Mg	0.160	Mg^{2+}	0.065	O	0.074	O^{2-}	0.140
Al	0.143	Al^{3+}	0.050	S	0.103	S^{2-}	0.184

*The metals shown lose electrons to become positive ions. The nonmetals gain electrons to become negative ions.

Ionic bonds are formed whenever one or more electrons are transferred from one atom to another. Metals, which have relatively little attraction for their valence electrons, tend to form ionic bonds when they combine with nonmetals.

It's important to recognize that substances with ionic bonds do not exist as molecules. In sodium chloride, for example, the bond does not exist solely between a single sodium ion and a single chloride ion. Each sodium ion in the crystal attracts six near-neighbor negative chloride ions; in turn, each negative chloride ion attracts six near-neighbor positive sodium ions (see Figure 11.5).

A metal will usually have one, two, or three electrons in its outer energy level. In reacting, metal atoms characteristically lose these electrons, attain the electron structure of a noble gas, and become positive ions. A nonmetal, on the other hand, is only a few electrons short of having a noble gas electron structure in its outer energy level and thus has a tendency to gain electrons. In reacting with metals, nonmetal atoms characteristically gain one, two, or three electrons; attain the electron structure of a noble gas; and become negative ions. The ions formed by loss of electrons are much smaller than the corresponding metal atoms; the ions formed by gaining electrons are larger than the corresponding nonmetal atoms. The dimensions of the atomic and ionic radii of several metals and nonmetals are given in Table 11.3.

Study the following examples. Note the loss and gain of electrons between atoms; also note that the ions in each compound have a noble gas electron structure.

Example 11.2   Explain how magnesium and chlorine combine to form magnesium chloride, $MgCl_2$.

Solution   A magnesium atom of electron structure $1s^2 2s^2 2p^6 3s^2$ must lose two electrons or gain six to reach a stable electron structure. If magnesium reacts with chlorine and each chlorine atom can accept only one electron, two chlorine atoms will be needed for the two electrons from each magnesium atom. The compound formed will contain one magnesium ion and two chloride ions. The magnesium atom, having lost two electrons, becomes a magnesium ion with a +2 charge. Each chloride ion will have a -1 charge. The transfer of electrons from a magnesium atom to two chlorine atoms is shown in the following illustration:

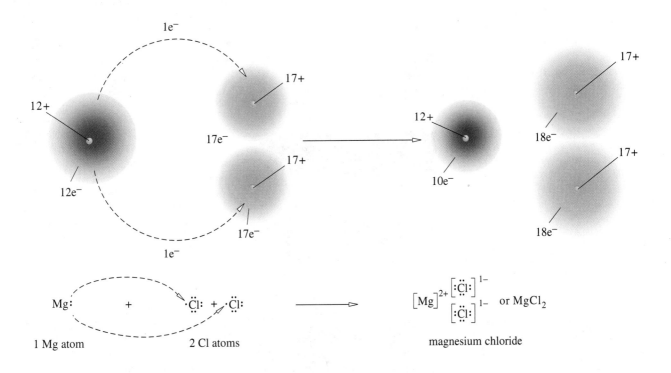

$$\left[Mg\right]^{2+} \begin{bmatrix} :\ddot{\underset{..}{C}l}: \end{bmatrix}^{1-} \quad or \ MgCl_2$$

1 Mg atom   2 Cl atoms   magnesium chloride

Explain the formation of sodium fluoride, NaF, from its elements.   **Example 11.3**

$$\left[Na\right]^{+} \begin{bmatrix} :\ddot{\underset{..}{F}}: \end{bmatrix}^{-}$$

Na atom   F atom   Sodium fluoride

The fluorine atom, with seven electrons in its outer energy level, behaves similarly to the chlorine atom.   **Solution**

CHEMISTRY IN ACTION

Superconductors—A New Frontier

When electric current flows through a wire, resistance slows the current and heats the wire. To keep the current flowing, this electrical "friction" must be overcome by adding energy to the system. In fact, the existence of electrical resistance limits the efficiency of all electrical devices.

In 1911, a Dutch scientist, Heike Kamerlingh Onnes, discovered that at very cold temperatures (near 0 K), electrical resistance disappears. Onnes named this phenomenon *superconductivity*. Scientists have been fascinated by it ever since. Unfortunately, because such low temperatures were required, liquid helium was necessary to cool the wires. With helium costing $3.50 per liter, commercial applications were far too expensive to be considered.

For many years scientists were convinced that superconductivity wasn't possible at higher temperatures (not even as high as 77 K, the boiling point of liquid nitrogen, a bargain at $0.39 per liter). The first high-temperature superconductor, developed in 1986, was superconducting at 30 K. The material was a complex metal oxide that has a sandwichlike crystal structure with copper and oxygen atoms on the inside and barium and lanthanum atoms on the outside.

Scientists immediately tried to develop materials that would be super-conducting at even higher temperatures. To do this they relied on their knowledge of the periodic table and the properties of chemical families. Paul Chu at the University of Houston, Texas, found that critical temperature could be raised by compressing the superconducting oxide. The pressure was too intense to be useful

▲ **When chilled in liquid nitrogen, the superconductor acts as a perfect mirror to the magnet, causing it to levitate as it "sees" its reflection in the superconductor.**

commercially, so Chu looked for another way to bring the atoms closer together. He accomplished this by replacing the barium with strontium, an element in the same family with similar chemical properties and a smaller ionic radius. The idea was successful—the critical temperature changed from 30 K to 40 K.

Chu then tried replacing the strontium with calcium (same family, smaller still) but to no avail. The new material had a lower critical temperature! But Chu persisted, and in 1987, by substituting yttrium for lanthanum (same family, smaller radius) he produced a new superconductor having a critical temperature of 95 K, well above the 77 K boiling point of liquid nitrogen. This material has the formula $YBa_2Cu_3O_7$ and is a good candidate for commercial applications.

Several barriers must be crossed before superconductors are in wide use. The materials developed so far are brittle and easily broken, nonmalleable, and do not carry as high a current per unit cross section as conventional conductors. Many researchers are currently working to surmount these problems and develop potential uses for superconductors, including high-speed levitation trains, tiny efficient electric motors, and smaller, faster computers.

Example 11.4   Explain the formation of aluminum fluoride, AlF_3, from its elements.

Solution

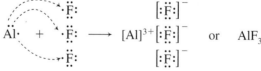

Each fluorine atom can accept only one electron. Therefore three fluorine atoms are needed to combine with the three outer electrons of one aluminum atom. The aluminum atom has lost three electrons to become an aluminum ion, Al^{3+}, with a +3 charge.

Explain the formation of magnesium oxide, MgO, from its elements.

Example 11.5
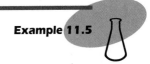

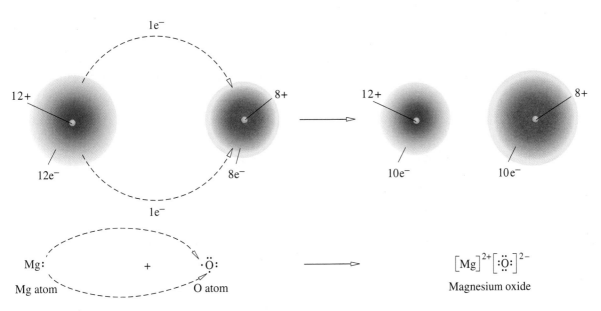

$$\text{Mg:} \qquad + \qquad \cdot\ddot{\text{O}}: \qquad \longrightarrow \qquad \left[\text{Mg}\right]^{2+}\left[:\ddot{\text{O}}:\right]^{2-}$$

Mg atom   O atom   Magnesium oxide

Solution

The magnesium atom, with two electrons in the outer energy level, exactly fills the need of one oxygen atom for two electrons. The resulting compound has a ratio of one magnesium atom to one oxygen atom. The oxygen (oxide) ion has a -2 charge, having gained two electrons. In combining with oxygen, magnesium behaves the same way as when it combines with chlorine—it loses two electrons.

Explain the formation of sodium sulfide, Na_2S, from its elements.

Example 11.6

Solution
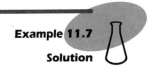

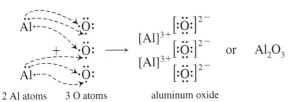

$$\begin{array}{l}\text{Na}\cdot \\ \quad + \quad \cdot\ddot{\text{S}}: \\ \text{Na}:\end{array} \longrightarrow \begin{array}{l}[\text{Na}]^+ \\ \quad \left[:\ddot{\text{S}}:\right]^{2-} \quad \text{or} \quad Na_2S \\ [\text{Na}]^+\end{array}$$

2 Na atoms   1 S atom   sodium sulfide

Two sodium atoms supply the electrons that one sulfur atom needs to make eight in its outer energy level.

Explain the formation of aluminum oxide, Al_2O_3, from its elements.

Example 11.7

Solution

$$\begin{array}{l}\text{Al}\cdot \quad \ddot{\text{O}}: \\ \quad \quad \ddot{\text{O}}: \\ \quad + \quad \ddot{\text{O}}: \longrightarrow \\ \text{Al}\cdot \quad \ddot{\text{O}}:\end{array} \begin{array}{l}[\text{Al}]^{3+} \quad \left[:\ddot{\text{O}}:\right]^{2-} \\ \quad \quad \left[:\ddot{\text{O}}:\right]^{2-} \quad \text{or} \quad Al_2O_3 \\ [\text{Al}]^{3+} \quad \left[:\ddot{\text{O}}:\right]^{2-}\end{array}$$

2 Al atoms   3 O atoms   aluminum oxide

One oxygen atom, needing two electrons, cannot accommodate the three electrons from one aluminum atom. One aluminum atom falls one electron short of the four electrons needed by two oxygen atoms. A ratio of two atoms of aluminum to three atoms of oxygen, involving the transfer of six electrons (two to each oxygen atom), gives each atom a stable electron configuration.

Note that in each of these examples, outer energy levels containing eight electrons were formed in all the negative ions. This formation resulted from the pairing of all the s and p electrons in these outer levels.

11.4   Predicting Formulas of Ionic Compounds

In the previous examples we've seen that when a metal and a nonmetal react to form an ionic compound, the metal loses one or more electrons to the nonmetal. In Chapter 6, where we learned to name compounds and write formulas, we saw that Group IA metals always formed $+1$ cations, whereas Group IIA formed $+2$ cations. Group VIIA elements formed -1 anions and Group VIA formed -2 anions.

It stands to reason then that this pattern is directly related to the stability of the noble gas configuration. Metals lose electrons to attain the electron configuration of a noble gas (the previous one on the periodic table). A nonmetal forms an ion by gaining enough electrons to achieve the electron configuration of the noble gas following it on the periodic table. These observations lead us to an important chemical principle:

> In almost all stable chemical compounds of representative elements, each atom attains a noble gas electron configuration. This concept forms the basis for our understanding of chemical bonding.

We can apply this principle in predicting the formulas of ionic compounds. To predict the formula of an ionic compound, we must recognize that chemical compounds are always electrically neutral. In addition, the metal will lose electrons to achieve noble gas configuration and the nonmetal will gain electrons to achieve noble gas configuration. Consider the compound formed between barium and sulfur. Barium has two valence electrons, whereas sulfur has six valence electrons:

Ba $[Xe]6s^2$ S $[Ne]3s^23p^4$

If barium loses two electrons, it will achieve the configuration of xenon. By gaining two electrons, sulfur achieves the configuration of argon. Consequently a pair of electrons is transferred between atoms. Now we have Ba^{2+} and S^{2-}. Since compounds are electrically neutral, there must be a ratio of one Ba to one S, giving the formula BaS.

The same principle works for many other cases. Since the key lies in the electron configuration, the periodic table can be used to extend the prediction even further. Because of similar electron structures, the elements in a family generally form compounds with the same atomic ratios. In general if we know the atomic ratio

TABLE 11.4	Formulas of Compounds Formed by Alkali Metals			
Lewis structure	Oxides	Chlorides	Bromides	Sulfates
Li·	Li_2O	LiCl	LiBr	Li_2SO_4
Na·	Na_2O	NaCl	NaBr	Na_2SO_4
K·	K_2O	KCl	KBr	K_2SO_4
Rb·	Rb_2O	RbCl	RbBr	Rb_2SO_4
Cs·	Cs_2O	CsCl	CsBr	Cs_2SO_4

of a particular compound, say NaCl, we can predict the atomic ratios and formulas of the other alkali metal chlorides. These formulas are LiCl, KCl, RbCl, CsCl, and FrCl (see Table 11.4).

Similarly if we know that the formula of the oxide of hydrogen is H_2O, we can predict that the formula of the sulfide will be H_2S, because sulfur has the same valence electron structure as oxygen. Recognize, however, that these are only predictions; it doesn't necessarily follow that every element in a group will behave like the others or even that a predicted compound will actually exist. For example, knowing the formulas for potassium chlorate, bromate, and iodate to be $KClO_3$, $KBrO_3$, and KIO_3, we can correctly predict the corresponding sodium compounds to have the formulas $NaClO_3$, $NaBrO_3$, and $NaIO_3$. Fluorine belongs to the same family of elements (Group VIIA) as chlorine, bromine, and iodine, so it would appear that fluorine should combine with potassium and sodium to give fluorates with the formulas KFO_3 and $NaFO_3$. However, potassium and sodium fluorates are not known to exist.

In the discussion in this section we refer only to representative metals (Groups IA, IIA, IIIA). The transition metals (Group B) show more complicated behavior (they form multiple ions), and their formulas are not as easily predicted.

The formula for calcium sulfide is CaS and that for lithium phosphide is Li_3P. Predict formulas for (a) magnesium sulfide, (b) potassium phosphide, and (c) magnesium selenide.

Example 11.8

(a) Look up calcium and magnesium in the periodic table; they are both in Group IIA. The formula for calcium sulfide is CaS, so it's reasonable to predict that the formula for magnesium sulfide is MgS.
(b) Find lithium and potassium in the periodic table; they are in Group IA. Since the formula for lithium phosphide is Li_3P, it's reasonable to predict that K_3P is the formula for potassium phosphide.
(c) Find selenium in the periodic table; it is in Group VIA just below sulfur. Therefore it's reasonable to assume that selenium forms selenide in the same way that sulfur forms sulfide. Since MgS was the predicted formula for magnesium sulfide in part (a), we can reasonably assume that the formula for magnesium selenide is MgSe.

Solution

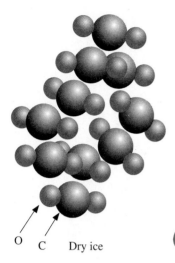

O   C   Dry ice

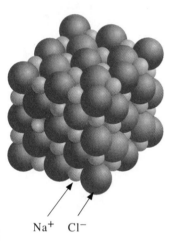

Na$^+$ Cl$^-$

Sodium chloride

▲
FIGURE 11.7
Solid carbon dioxide (dry ice) is composed of individual co-valently bonded molecules of CO_2 closely packed together. Table salt is a large aggregate of Na$^+$ and Cl$^-$ ions instead of molecules.

Practice 11.2

The formula for sodium oxide is Na_2O. Predict the formula for
(a) sodium sulfide
(b) rubidium oxide

Practice 11.3

The formula for barium phosphide is Ba_3P_2. Predict the formula for
(a) magnesium nitride
(b) barium arsenide

11.5   The Covalent Bond: Sharing Electrons

Some atoms do not transfer electrons from one atom to another to form ions. Instead they form a chemical bond by sharing pairs of electrons between them.

A **covalent bond** consists of a pair of electrons shared between two atoms. This bonding concept was introduced in 1916 by G. N. Lewis. In the millions of known compounds, the covalent bond is the predominant chemical bond.

True molecules exist in substances in which the atoms are covalently bonded. It is proper to refer to molecules of such substances as hydrogen, chlorine, hydrogen chloride, carbon dioxide, water, or sugar (Figure 11.7). These substances contain only covalent bonds and exist as aggregates of molecules. We don't use the term *molecule* when talking about ionically bonded compounds such as sodium chloride, because such substances exist as large aggregates of positive and negative ions, not as molecules (Figure 11.7).

A study of the hydrogen molecule gives us an insight into the nature of the co-valent bond and its formation. The formation of a hydrogen molecule, H_2, involves the overlapping and pairing of $1s$ electron orbitals from two hydrogen atoms, shown in Figure 11.8. Each atom contributes one electron of the pair that is shared jointly by two hydrogen nuclei. The orbital of the electrons now includes both hydrogen nuclei, but probability factors show that the most likely place to find the electrons (the point of highest electron density) is between the two nuclei. The two nuclei are shielded from each other by the pair of electrons, allowing the two nuclei to be drawn very close to each other.

The formula for chlorine gas is Cl_2. When the two atoms of chlorine combine to form this molecule, the electrons must interact in a manner similar to that shown in the hydrogen example. Each chlorine atom would be more stable with eight elec-trons in its outer energy level. But chlorine atoms are identical, and neither is able to

covalent bond

FIGURE 11.8 ▶
The formation of a hydrogen molecule from two hydrogen atoms. The two 1s orbitals overlap, forming the H_2 mole-cule. In this molecule the two electrons are shared between the atoms, forming a covalent bond.

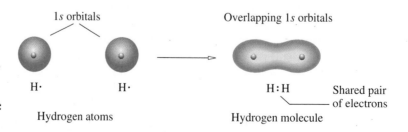

$1s$ orbitals

H·   H·

Hydrogen atoms

Overlapping $1s$ orbitals

H:H   Shared pair
              of electrons

Hydrogen molecule

p orbitals   Overlap of p orbitals   Paired p orbital

Chlorine atoms   Chlorine molecule

:C̈l·   +   ·C̈l: :C̈l ⦂ C̈l:

Unshared p orbitals   Shared pair of p electrons

◀ **FIGURE 11.9**
Pairing of p electrons in the formation of a chlorine molecule.

pull an electron away from the other. What happens is this: The unpaired 3p electron orbital of one chlorine atom overlaps the unpaired 3p electron orbital of the other atom, resulting in a pair of electrons that are mutually shared between the two atoms. Each atom furnishes one of the pair of shared electrons. Thus each atom attains a stable structure of eight electrons by sharing an electron pair with the other atom. The pairing of the p electrons and the formation of a chlorine molecule are illustrated in Figure 11.9. Neither chlorine atom has a positive or negative charge, because both contain the same number of protons and have equal attraction for the pair of electrons being shared. Other examples of molecules in which electrons are equally shared between two atoms are hydrogen, H_2; oxygen, O_2; nitrogen, N_2; fluorine, F_2; bromine, Br_2; and iodine, I_2. Note that more than one pair of electrons may be shared between atoms:

H:H   :F̈:F̈:   :B̈r:B̈r: :Ï:Ï: :Ö::Ö: :N⋮⋮⋮N:

hydrogen   fluorine   bromine   iodine   oxygen   nitrogen

The Lewis structure given for oxygen does not adequately account for all the properties of the oxygen molecule. Other theories explaining the bonding in oxygen molecules have been advanced, but they are complex and beyond the scope of this book.

In writing structures we commonly replace the pair of dots used to represent a shared pair of electrons with a dash (—). One dash represents a single bond; two dashes, a double bond; and three dashes, a triple bond. The six structures just shown may be written thus:

H—H   :F̈—F̈:   :B̈r—B̈r: :Ï—Ï: :Ö=Ö: :N≡N:

The ionic bond and the covalent bond represent two extremes. In ionic bonding the atoms are so different that electrons are transferred between them, forming a charged pair of ions. In covalent bonding two identical atoms share electrons equally. The bond is the mutual attraction of the two nuclei for the shared electrons. Between these extremes lie many cases in which the atoms are not different enough for a transfer of electrons, but are different enough that the electron pair cannot be shared equally. This unequal sharing of electrons results in the formation of a **polar covalent bond.**

Remember: A dash represents a shared pair of electrons.

▲
Molecular models for F_2 (green, single bond), O_2, (black, double bond), and N_2 (blue, triple bond).

polar covalent bond

11.6   Electronegativity

When two *different* kinds of atoms share a pair of electrons, a bond forms in which electrons are shared unequally. One atom assumes a partial positive charge and the other a partial negative charge with respect to each other. This difference in charge occurs because the two atoms exert unequal attraction for the pair of shared electrons. The attractive force that an atom of an element has for shared electrons in a

electronegativity   molecule or polyatomic ion is known as its **electronegativity.** Elements differ in their electronegativities. For example, both hydrogen and chlorine need one electron to form stable electron configurations. They share a pair of electrons in hydrogen chloride, HCl. Chlorine is more electronegative and therefore has a greater attraction for the shared electrons than does hydrogen. As a result the pair of electrons is displaced toward the chlorine atom, giving it a partial negative charge and leaving the hydrogen atom with a partial positive charge. Note that the electron is not transferred entirely to the chlorine atom (as in the case of sodium chloride) and that no ions are formed. The entire molecule, HCl, is electrically neutral. A partial charge is usually indicated by the Greek letter delta, δ. Thus a partial positive charge is represented by $\delta+$ and a partial negative charge by $\delta-$.

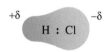

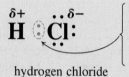

The pair of shared electrons in HCl is closer to the more electronegative chlorine atom than to the hydrogen atom, giving chlorine a partial negative charge with respect to the hydrogen atom.

hydrogen chloride

A scale of relative electronegativities, in which the most electronegative element, fluorine, is assigned a value of 4.0, was developed by the Nobel laureate (1954 and 1962) Linus Pauling (1901–1994). Table 11.5 shows that the relative electronegativity of the nonmetals is high and that of the metals is low. These electronegativities indicate that atoms of metals have a greater tendency to lose electrons than do atoms of nonmetals and that nonmetals have a greater tendency to gain electrons than do metals. The higher the electronegativity value, the greater the attraction for electrons. Note that electronegativity generally increases from left to right across a period and decreases down a group for the representative elements. The highest electronegativity is 4.0 for fluorine, and the lowest is 0.7 for francium and cesium. It's important to remember that the higher the electronegativity, the stronger an atom attracts electrons.

The polarity of a bond is determined by the difference in electronegativity values of the atoms forming the bond (see Figure 11.10). If the electronegativities

nonpolar covalent bond   are the same, the bond is **nonpolar covalent** and the electrons are shared equally. If the atoms have greatly different electronegativities, the bond is very *polar.* At the extreme, one or more electrons are actually transferred and an ionic bond results.

dipole   A **dipole** is a molecule that is electrically asymmetrical, causing it to be oppositely charged at two points. A dipole is often written as ⊕—⊝ A hydrogen chloride molecule is polar and behaves as a small dipole. The HCl dipole may be written as

TABLE 11.5	Three-Dimensional Representation of Electronegativity

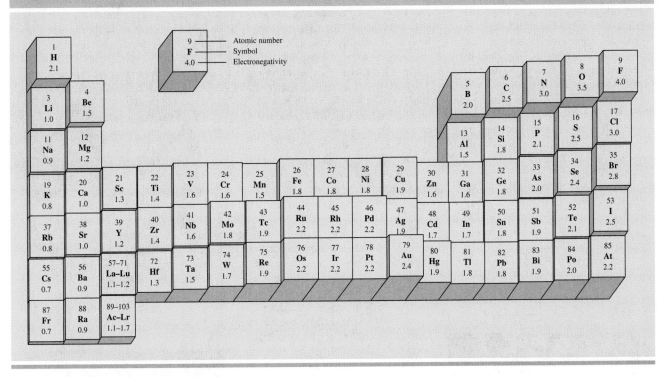

H$+\!\!\longrightarrow$Cl. The arrow points toward the negative end of the dipole. Molecules of H_2O, HBr, and ICl are polar:

$$H+\!\!\longrightarrow Cl \qquad H+\!\!\longrightarrow Br \qquad I+\!\!\longrightarrow Cl \qquad H\overset{\displaystyle O}{\underset{}{\nearrow\!\!\!\nwarrow}} H$$

How do we know whether a bond between two atoms is ionic or covalent? The difference in electronegativity between the two atoms determines the character of the bond formed between them. As the difference in electronegativity increases, the polarity of the bond (or percent ionic character) increases.

> If the electronegativity difference between two bonded atoms is greater than 1.7–1.9, the bond will be more ionic than covalent.

H_2 Cl_2	HCl	NaCl
Nonpolar molecules	Polar covalent molecule	Ionic compound

◄ FIGURE 11.10
Nonpolar, polar covalent, and ionic compounds.

CHEMISTRY IN ACTION Goal! A Spherical Molecule

One of the most diverse elements in the periodic table is carbon. Graphite and diamond, two well-known forms of elemental carbon, both contain extended arrays of carbon atoms. In graphite the carbon atoms are arranged in sheets, and the bonding between the sheets is very weak. This property makes graphite useful as a lubricant and as a writing material. Diamond consists of transparent octahedral crystals in which each carbon atom is bonded to four other carbon atoms. This three-dimensional network of bonds gives diamond the property of hardness for which it is noted. In the 1980s, a new form of carbon was discovered in which the atoms are arranged in relatively small clusters.

When Harold Kroto of the University of Sussex, England, and Richard Smalley of Rice University, Texas, dis-

covered a strange carbon molecule of formula C_{60}, they deduced that the most stable arrangement for the atoms would be in the shape of a soccer ball. In thinking about possible arrangements, the scientists considered the geodesic domes designed by R. Buckminster Fuller in the 1960s. This cluster form of carbon was thus named *buckminsterfullerene* and is commonly called buckyballs.

Buckyballs have captured the imagination of a variety of chemists. Research on buckyballs has led to a host of possible applications for these molecules. If metals are bound to the carbon atoms, the fullerenes become superconducting; that is, they conduct electrical current without resistance, at very low temperature. Scientists are now able to make buckyball compounds that superconduct at temperatures of 45 K. Other fullerenes are being used in lubricants and optical materials.

Chemists at Yale University have managed to trap helium and neon inside buckyballs. This is the first time chemists have ever observed helium or neon in a compound of any kind. They found that at temperatures from 1000°F to 1500°F one of the covalent bonds linking neighboring carbon atoms in the buckyball breaks. This opens a window in the fullerene molecule through which a helium or neon atom can enter the buckyball. When the fullerene is allowed to cool, the broken bond between carbon atoms re-forms, shutting the window and trapping the helium or neon atom inside the buckyball. Since the trapped helium or neon cannot react or share electrons

with its host, the resulting compound has forced scientists to invent a new kind of chemical formula to describe the compound. The relationship between the "prisoner" helium or neon and the host buckyball is shown with an @ sign. A helium fullerene containing 60 carbon atoms would thus be He@C_{60}.

Buckyballs can be tailored to fit a particular size requirement. Raymond Schinazi of the Emory University School of Medicine, Georgia, made a buckyball to fit the active site of a key HIV enzyme that paralyzes the virus, making it non-infectious in human cells. The key to making this compound was preparing a water-soluble buckyball that would fit in the active site of the enzyme. Eventually, scientists created a water-soluble fullerene molecule that has two charged arms to grasp the binding site of the enzyme. It is toxic to the virus but doesn't appear to harm the host cells.

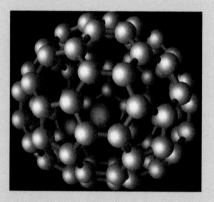

▲
Scandium atoms trapped in a buckyball.

▲
"Raspberries" of fullerene lubricant.

If the electronegativity difference is greater than 2.0, the bond is strongly ionic. If the electronegativity difference is less than 1.5, the bond is strongly covalent.

Care must be taken to distinguish between polar bonds and polar molecules. A covalent bond between different kinds of atoms is always polar. But a molecule containing different kinds of atoms may or may not be polar, depending on its shape or geometry. Molecules of HF, HCl, HBr, HI, and ICl are all polar because each contains a single polar bond. However, CO_2, CH_4, and CCl_4 are nonpolar molecules despite the fact that all three contain polar bonds. The carbon dioxide molecule

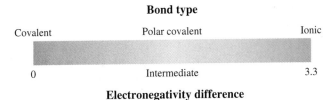

◀ **FIGURE 11.11**
**Relating bond type to
electronegativity difference
between atoms.**

$O{=}C{=}O$ is nonpolar because the carbon–oxygen dipoles cancel each other by acting in opposite directions.

$$\overset{\longleftarrow + \;\; + \longrightarrow}{O{=}C{=}O}$$

dipoles in opposite directions

Carbon tetrachloride (CCl_4) is nonpolar because the four C—Cl polar bonds are identical, and since these bonds emanate from the center to the corners of a tetrahedron in the molecule, their polarities cancel one another. Methane has the same molecular structure and is also nonpolar. We will discuss the shapes of molecules later in this chapter.

We have said that water is a polar molecule. If the atoms in water were linear like those in carbon dioxide, the two O—H dipoles would cancel each other, and the molecule would be nonpolar. However, water is definitely polar and has a nonlinear (bent) structure with an angle of 105° between the two O—H bonds.

The relationships among types of bonds are summarized in Figure 11.11. It is important to realize that bonding is a continuum; that is, the difference between ionic and covalent is a gradual change.

11.7 Lewis Structures of Compounds

As we have seen, Lewis structures are a convenient way of showing the covalent bonds in many molecules or ions of the representative elements. In writing Lewis structures, the most important consideration for forming a stable compound is that the atoms attain a noble gas configuration.

The most difficult part of writing Lewis structures is determining the arrangement of the atoms in a molecule or an ion. In simple molecules with more than two atoms, one atom will be the central atom surrounded by the other atoms. Thus Cl_2O has two possible arrangements, C—Cl—O or Cl—O—Cl. Usually, but not always, the single atom in the formula (except H) will be the central atom.

Although Lewis structures for many molecules and ions can be written by inspection, the following procedure is helpful for learning to write them:

Step 1. Obtain the total number of valence electrons to be used in the structure by adding the number of valence electrons in all the atoms in the molecule or ion. If you are writing the structure of an ion, add one electron for each negative charge or subtract one electron for each positive charge on the ion.

Step 2. Write the skeletal arrangement of the atoms and connect them with a single covalent bond (two dots or one dash). Hydrogen, which contains only one bonding electron, can form only one covalent bond. Oxygen atoms are not normally bonded to each other, except in compounds

Remember: The number of valence electrons of Group A elements is the same as their group number in the periodic table.

known to be peroxides. Oxygen atoms normally have a maximum of two covalent bonds, (two single bonds, or one double bond).

Step 3. Subtract two electrons for each single bond you used in Step 2 from the total number of electrons calculated in Step 1. This gives you the net number of electrons available for completing the structure.

Step 4. Distribute pairs of electrons (pairs of dots) around each atom (except hydrogen) to give each atom a noble gas structure.

Step 5. If there are not enough electrons to give these atoms eight electrons, change single bonds between atoms to double or triple bonds by shifting unbonded pairs of electrons as needed. Check to see that each atom has a noble gas electron structure (two electrons for hydrogen and eight for the others). A double bond counts as four electrons for each atom to which it is bonded.

Example 11.9   How many valence electrons are in each of these atoms: Cl, H, C, O, N, S, P, I?

Solution   You can look at the periodic table to determine the electron structure, or, if the element is in Group A of the periodic table, the number of valence electrons is equal to the group number:

Atom	Group	Valence electrons
Cl	VIIA	7
H	IA	1
C	IVA	4
O	VIA	6
N	VA	5
S	VIA	6
P	VA	5
I	VIIA	7

Example 11.10 Write the Lewis structure for water, H_2O.

Solution   **Step 1.** The total number of valence electrons is eight, two from the two hydrogen atoms and six from the oxygen atom.

Step 2. The two hydrogen atoms are connected to the oxygen atom. Write the skeletal structure:

H O   or   H O H
　H

Place two dots between the hydrogen and oxygen atoms to form the covalent bonds:

H:O   or   H:O:H
‥
Ḧ

Step 3. Subtract the four electrons used in Step 2 from eight to obtain four electrons yet to be used.

Step 4. Distribute the four electrons around the oxygen atom. Hydrogen atoms cannot accommodate any more electrons:

$$H{-}\ddot{O}{:} \qquad or \qquad H{-}\ddot{\underset{..}{O}}{-}H$$
$$\underset{H}{\vert}$$

These arrangements are Lewis structures because each atom has a noble gas electron structure. Note that the shape of the molecule is not shown by the Lewis structure.

Write Lewis structures for a molecule of methane, CH₄.   **Example 11.11**

Solution

Step 1. The total number of valence electrons is eight, one from each hydrogen atom and four from the carbon atom.

Step 2. The skeletal structure contains four H atoms around a central C atom. Place two electrons between the C and each H.

$$\underset{H}{\overset{H}{H\ C\ H}} \qquad \underset{\ddot{H}}{\overset{H}{H{:}\ddot{C}{:}H}}$$

Step 3. Subtract the eight electrons used in Step 2 from eight to obtain zero electrons yet to be placed. Therefore the Lewis structure must be as written in Step 2:

$$\underset{\ddot{H}}{\overset{H}{H{:}\ddot{C}{:}H}} \qquad or \qquad H{-}\underset{\vert}{\overset{\vert}{\underset{H}{\overset{H}{C}}}}{-}H$$

Write the Lewis structure for a molecule of carbon tetrachloride, CCl₄.   **Example 11.12**

Solution

Step 1. The total number of valence electrons to be used is 32, four from the carbon atom and seven from each of the four chlorine atoms.

Step 2. The skeletal structure contains the four Cl atoms around a central C atom. Place two electrons between the C and each Cl:

$$\underset{Cl}{\overset{Cl}{Cl\ C\ Cl}} \qquad \underset{\ddot{Cl}}{\overset{\ddot{Cl}}{Cl{:}\ddot{C}{:}Cl}}$$

Step 3. Subtract the eight electrons used in Step 2 from 32 to obtain 24 electrons yet to be placed.

Step 4. Distribute the 24 electrons (12 pairs) around the Cl atoms so that each Cl atom has eight electrons around it:

$$\ddot{\underset{\ddot{..}}{Cl}}\!:\;\; :\!\ddot{Cl}\!:\!\ddot{C}\!:\!\ddot{Cl}\!:\;\;\;\; \text{or} \;\;\;\; :\!\ddot{Cl}\!-\!C\!-\!\ddot{Cl}\!:$$

This arrangement is the Lewis structure; CCl_4 contains four covalent bonds.

Example 11.13 Write a Lewis structure for CO_2.

Solution

Step 1. The total number of valence electrons is 16, four from the C atom and six from each O atom.

Step 2. The two O atoms are bonded to a central C atom. Write the skeletal structure and place two electrons between the C and each O atom.

O:C:O

Step 3. Subtract the four electrons used in Step 2 from 16 to obtain 12 electrons yet to be placed.

Step 4. Distribute the 12 electrons around the C and O atoms. Several possibilities exist:

$$:\ddot{O}:C:\ddot{O}:\;\;\;\;\; :\ddot{O}:C:\ddot{O}:\;\;\;\;\; :\ddot{O}:\ddot{C}:O:$$
$$\;\;\;I\;\;\;\;\;\;\;\;\;\;\;\;\;\;II\;\;\;\;\;\;\;\;\;\;\;\;\;\;III$$

Step 5. Not all the atoms have eight electrons around them (noble gas structure). Remove one pair of unbonded electrons from each O atom in structure I and place one pair between each O and the C atom forming two double bonds:

$$:\ddot{O}::C::\ddot{O}:\;\;\;\; \text{or} \;\;\;\; :\ddot{O}=C=\ddot{O}:$$

Each atom now has eight electrons around it. The carbon is sharing four pairs of electrons, and each oxygen is sharing two pairs. These bonds are known as double bonds because each involves sharing two pairs of electrons.

Practice 11.4

Write the Lewis structures for the following:
(a) PBr_3 (b) $CHCl_3$ (c) HF (d) H_2CO

Although many compounds attain a noble gas structure in covalent bonding, there are numerous exceptions. Sometimes it's impossible to write a structure in which each atom has eight electrons around it. For example, in BF_3 the boron atom has only six electrons around it, and in SF_6 the sulfur atom has 12 electrons around it.

CHEMISTRY IN ACTION
Cleaner Showers through Chemistry

Keeping the shower area sparkling clean and free of mildew is a job none of us enjoy. Now thanks to chemist Bob Black of Jacksonville, Florida, there is a new product that cleans the shower without any scrubbing! Black struggled with his home shower until he finally decided he really needed a new product that would solve the mildew and scrubbing problem.

His search was based on the following needs:

(1) a molecule to lift deposits off the walls of the shower
(2) a way to prevent hard water deposits from forming
(3) a wetting agent to wet the walls and rinse off deposits.

In all cases he limited his search to substances nontoxic and environmentally safe.

Black used a molecule called a glycol ether to lift deposits off the shower wall. This molecule is a long chain with a polar end and a nonpolar end. Substances that are nonpolar (such as grease, oils, and organic material) are attracted to the nonpolar end of the molecules. The molecules cluster together to form micelles (with the polar end pointed out). The polar sphere dissolves in the polar water from the shower washing off the organic deposits.

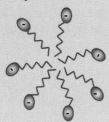

A micelle

Preventing the hard water deposits from forming on the shower walls required use of a molecule called EDTA. This molecule bonds to ions (like Ca^{2+}, Mg^{2+}, or Fe^{3+}) and prevents the formation of soap scum or hard water deposits on the walls.

Lastly Black added isopropyl alcohol to wet the shower wall and also to disturb the mildew fungus. He mixed all the ingredients in just the right pro-portions and found he no longer needed to work so hard to clean the shower.

Fortunately for all of us Black shared his solution with friends who also liked it. Black patented his new product and began mass production. You can now find Clean Shower in your local grocery store!

▲
Simply spraying your shower regularly with Clean Shower will keep it free of deposits and mildew.

Although there are exceptions, many molecules can be described using Lewis structures where each atom has a noble gas electron configuration. This is a useful model for understanding chemistry.

11.8 Complex Lewis Structures

Most Lewis structures give bonding pictures that are consistent with experimental information on bond strength and length. There are some molecules and polyatomic ions for which no single Lewis structure consistent with all characteristics and bonding information can be written. For example, consider the nitrate ion, NO_3^-. To write a Lewis structure for this polyatomic ion, we use the following steps.

Step 1. The total number of valence electrons is 24, five from the nitrogen atom, and six from each oxygen atom, plus one electron from the -1 charge.

Step 2. The three O atoms are bonded to a central N atom. Write the skeletal structure and place two electrons between each pair of atoms:

$$\left[\begin{array}{c} O \\ O:\ddot{N}:O \end{array} \right]^-$$

Step 3. Subtract the six electrons used in Step 2 from 24 to obtain 18 electrons yet to be placed.

Step 4. Distribute the 18 electrons around the N and O atoms:

$$:\ddot{\text{O}} \longleftarrow \text{electron deficient}$$
$$:\ddot{\text{O}}:\ddot{\text{N}}:\ddot{\text{O}}:$$

Step 5. One pair of electrons is still needed to give all the N and O atoms a noble gas structure. Move the unbonded pair of electrons from the N atom and place it between the N and the electron-deficient O atom, making a double bond.

$$\left[\;\ddot{\text{O}}\!\!\!\overset{:\ddot{\text{O}}}{\underset{\|}{\text{N}}}\!\!\!-\ddot{\text{O}}:\;\right]^{-} \quad \text{or} \quad \left[\;\ddot{\text{O}}-\overset{:\ddot{\text{O}}:}{\underset{|}{\text{N}}}=\ddot{\text{O}}:\;\right]^{-} \quad \text{or} \quad \left[\;\ddot{\text{O}}=\overset{:\ddot{\text{O}}:}{\underset{|}{\text{N}}}-\ddot{\text{O}}:\;\right]^{-}$$

Are these all valid Lewis structures? Yes, so there really are three possible Lewis structures for NO_3^-.

A molecule or ion that has multiple correct Lewis structures shows *resonance*. Each of these Lewis structures is called a **resonance structure.** In this book, however, we will not be concerned with how to choose the correct resonance structure for a molecule or ion. Therefore any of the possible resonance structures may be used to represent the ion or molecule.

resonance structure

Example 11.14 Write the Lewis structure for a carbonate ion, CO_3^{2-}.

Solution

Step 1. These four atoms have 22 valence electrons plus two electrons from the -2 charge, which makes 24 electrons to be placed.

Step 2. In the carbonate ion the carbon is the central atom surrounded by the three oxygen atoms. Write the skeletal structure and place two electrons between each pair of atoms:

$$
\begin{array}{c}
\text{O} \\
| \\
\text{C}-\text{O} \\
| \\
\text{O}
\end{array}
$$

Step 3. Subtract the six electrons used in Step 2 from 24 to give 18 electrons yet to be placed.

Step 4. Distribute the 18 electrons around the three oxygen atoms and indicate that the carbonate ion has a -2 charge:

$$\left[\;\overset{:\ddot{\text{O}}:}{\underset{\underset{:\ddot{\text{O}}.\quad.\ddot{\text{O}}:}{\diagup\;\;\diagdown}}{|}}{\text{C}}\;\right]^{2-}$$

The difficulty with this structure is that the carbon atom has only six electrons around it instead of a noble gas octet.

Step 5. Move one of the nonbonding pairs of electrons from one of the oxygens and place them between the carbon and the oxygen. Three Lewis structures are possible:

$$\left[\begin{array}{c} :\ddot{O}: \\ | \\ C \\ \end{array}\right]^{2-} \quad \text{or} \quad \left[\begin{array}{c} :\ddot{O}: \\ | \\ C \\ \end{array}\right]^{2-} \quad \text{or} \quad \left[\begin{array}{c} :O: \\ \| \\ C \\ \end{array}\right]^{2-}$$

Practice 11.5

Write the Lewis structure for each of the following: (a) NH_3 (b) H_3O^+ (c) NH_4^+ (d) HCO_3^-.

11.9 Compounds Containing Polyatomic Ions

A polyatomic ion is a stable group of atoms that has either a positive or a negative charge and behaves as a single unit in many chemical reactions. Sodium carbonate, Na_2CO_3, contains two sodium ions and a carbonate ion. The carbonate ion, CO_3^{2-}, is a polyatomic ion composed of one carbon atom and three oxygen atoms and has a charge of -2. One carbon and three oxygen atoms have a total of 22 electrons in their outer energy levels. The carbonate ion contains 24 outer electrons and therefore has a charge of -2. In this case the two additional electrons come from the two sodium atoms, which are now sodium ions:

$$[Na]^+ \quad \left[\begin{array}{c} :\ddot{O}: \\ | \\ C \\ \end{array}\right]^{2-} \quad [Na]^+ \qquad\qquad \left[\begin{array}{c} :\ddot{O}: \\ | \\ C \\ \end{array}\right]^{2-}$$

$\qquad\qquad$ sodium carbonate $\qquad\qquad\qquad\qquad$ carbonate ion

Sodium carbonate has both ionic and covalent bonds. Ionic bonds exist between each of the sodium ions and the carbonate ion. Covalent bonds are present between the carbon and oxygen atoms within the carbonate ion. One important difference between the ionic and covalent bonds in this compound can be demonstrated by dissolving sodium carbonate in water. It dissolves in water forming three charged particles, two sodium ions and one carbonate ion, per formula unit of Na_2CO_3:

$$Na_2CO_3 \xrightarrow{\text{water}} 2\,Na^+ + CO_3^{2-}$$

$\quad$ sodium carbonate $\qquad\quad$ sodium ions $\quad$ carbonate ion

The CO_3^{2-} ion remains as a unit, held together by covalent bonds; but where the bonds are ionic, dissociation of the ions takes place. Do not think, however, that polyatomic ions are so stable that they cannot be altered. Chemical reactions by which polyatomic ions can be changed to other substances do exist.

FIGURE 11.12▶
Geometric shapes of common molecules.

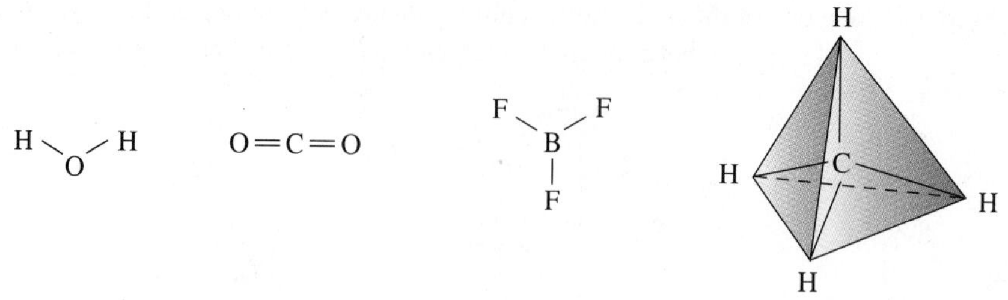

Water	Carbon dioxide	Boron trifluoride	Methane
H_2O	CO_2	BF_3	CH_4
(V-shaped)	(linear shape)	(trigonal planar)	(tetrahedral)

11.10 Molecular Shape

So far in our discussion of bonding, we have used Lewis structures to represent valence electrons in molecules and ions, but they don't indicate anything regarding the molecular or geometric shape of a molecule. The three-dimensional arrangement of the atoms within a molecule is a significant feature in understanding molecular interactions. Let's consider several examples illustrated in Figure 11.12.

Water is known to have the geometric shape known as "bent" or "V-shaped." Carbon dioxide exhibits a linear shape. BF_3 forms a third molecular shape called *trigonal planar* since all the atoms lie in one plane in a triangular arrangement. One of the more common molecular shapes is the tetrahedron illustrated by the molecule methane, CH_4.

How do we predict the geometric shape of a molecule? We will now study a model developed to assist in making predictions from the Lewis structure.

11.11 The Valence Shell Electron Pair Repulsion (VSEPR) Model

The chemical properties of a substance are closely related to the structure of its molecules. A change in a single site on a large biomolecule can make a difference in whether or not a particular reaction occurs.

Instrumental analysis can be used to determine exact spatial arrangements of atoms. Quite often, though, we only need to be able to predict the approximate structure of a molecule. A relatively simple model has been developed to allow us to make predictions of shape from Lewis structures.

Nonbonding pairs of electrons are not shown here so you can focus your attention on the shapes, not electron arrangement.

The VSEPR model is based on the idea that electron pairs will repel each other electrically and will seek to minimize this repulsion. To accomplish this minimization, the electron pairs will be arranged around a central atom as far apart as possible. Consider $BeCl_2$, a molecule with only two pairs of electrons surrounding the central atom. These electrons are arranged 180° apart for maximum separation:

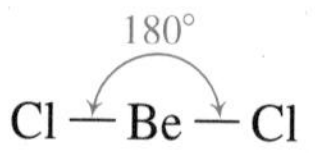

$$Cl \overset{180°}{\frown} Be \frown Cl$$

linear structure   This molecular structure can now be labeled as a **linear structure.** When only two pairs of electrons surround a central atom, they should be placed 180° apart to give a linear structure.

What occurs when there are three pairs of electrons on the central atom? Consider the BF_3 molecule. The greatest separation of electron pairs occurs when the angles between atoms are 120°:

F   120°   F
    B
120°   120°
    F

This arrangement of atoms is flat (planar) and, as noted earlier, is called **trigonal planar.** When three pairs of electrons surround an atom, they should be placed 120° apart to show the trigonal planar structure.

trigonal planar

Now consider the most common situation, CH_4, with four pairs of electrons on the central carbon atom. In this case the central atom exhibits a noble gas electron structure. What arrangement best minimizes the electron pair repulsions? At first it seems that an obvious choice is a 90° angle with all the atoms in a single plane:

H
90°   90°
H—C—H
90°   90°
H

However, we must consider that molecules are three-dimensional. This concept results in a structure in which the electron pairs are actually 109.5° apart:

H   109.5°   H
      C
H   H

In this diagram the wedged line seems to protrude from the page whereas the dashed line recedes. Two representations of this arrangement, known as **tetrahedral structure,** are illustrated in Figure 11.13. When four pairs of electrons surround an atom, they should be placed 109.5° apart to give them a tetrahedral structure.

tetrahedral structure

The VSEPR model is based on the premise that we are counting electron pairs. It's quite possible that one or more of these electron pairs may be nonbonding or lone pairs. What happens to the molecular structure in these cases? Consider the ammonia

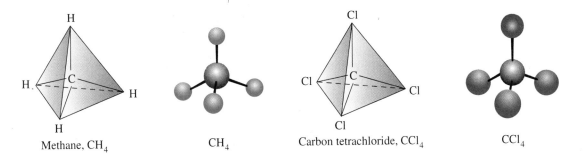

Methane, CH_4 CH_4 Carbon tetrachloride, CCl_4 CCl_4

▲
FIGURE 11.13
Ball-and-stick models of methane and carbon tetrachloride. Methane and carbon tetrachloride are nonpolar molecules because their polar bonds cancel each other in the tetrahedral arrangement of their atoms. The carbon atoms are located in the centers of the tetrahedrons.

FIGURE 11.14 ▶
(a) The tetrahedral arrangement of electron pairs around the N atom in the NH₃ molecule. (b) Three pairs are shared and one is unshared. (c) The NH₃ molecule is pyramidal.

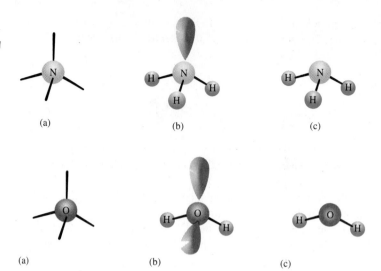

(a)   (b)   (c)

FIGURE 11.15 ▶
(a) The tetrahedral arrangement of the four electron pairs around oxygen in the H₂O molecule. (b) Two of the pairs are shared and two are unshared. (c) The H₂O molecule is bent.

(a)   (b)   (c)

molecule. First we draw the Lewis structure to determine the number of electron pairs around the central atom:

$$H:\overset{\cdot\cdot}{\underset{\overset{|}{H}}{N}}:H$$

Since there are four pairs of electrons, the arrangement of electrons around the central atom will be tetrahedral (Figure 11.14a). However, only three of the pairs are bonded to another atom, so the molecule itself is pyramidal. It is important to understand that the placement of the electron pairs determines the shape but the name for the molecule is determined by the position of the atoms themselves. Therefore, ammonia is pyramidal. See Figure 11.14c.

Now consider the effect of two unbonded pairs of electrons in the water molecule. The Lewis structure for water is

$$H\!-\!\overset{\cdot\cdot}{\underset{\overset{|}{H}}{O}}:$$

The arrangement of electron pairs around an atom determines its shape but we name the shape of molecules by the position of the atoms.

The four electron pairs indicate a tetrahedral arrangement is necessary (see Figure 11.15a). The molecule is not called tetrahedral because two of the electron pairs are unbonded pairs. The water molecule is "bent" as shown in Figure 11.15c. Using the VSEPR model helps to explain some of the unique properties of the water molecule. Because it is bent and not linear we can see that the molecule is polar.

The properties of water that cause it to be involved in so many interesting and important roles are largely a function of its shape and polarity. We will consider water in greater detail in Chapter 13.

Example 11.15 Predict the molecular structure for these molecules: H_2S, CCl_4, AlF_3.

Solution

1. Draw the Lewis structure.
2. Count the electron pairs and determine the arrangement that will minimize repulsions.
3. Determine the positions of the atoms and name the structure.

Strong Enough to Stop a Bullet?

What do color-changing pens, bullet-resistant vests, and calculators have in common? The chemicals that make each of them work are liquid crystals. These chemicals find numerous applications; you are probably most familiar with liquid crystal displays (LCDs) and color-changing products, but these chemicals are also used to make superstrong synthetic fibers.

Molecules in a normal crystal remain in an orderly arrangement, but in a liquid crystal the molecules can flow *and* maintain an orderly arrangement at the same time. Liquid crystal molecules are linear and polar. Since the atoms tend to lie in a relatively straight line, the molecules are generally much longer than they are wide. These polar molecules are attracted to each other and are able to line up in an orderly fashion, without solidifying.

Liquid crystals with twisted arrangements of molecules give us novelty color-changing products. In these liquid crystals the molecules lie side by side in a nearly flat layer. The next layer is similar but at an angle to the one below. The closely packed flat layers have a special effect on light. As the light strikes the surface, some of it is reflected from the top layer and some from lower layers. When the

same wavelength is reflected from many layers, we see a color. (This is similar to the rainbow of colors formed by oil in a puddle on the street or the film of a soap bubble.) As the temperature is increased, the molecules move faster, causing a change in the angle and the space between the layers. This results in a color change in the reflected light. Different compounds change color within different

▲ **Kevlar is used to make protective vests for police.**

temperature ranges, allowing a variety of practical and amusing applications.

Liquid crystal (nematic) molecules that lie parallel to one another are used to manufacture very strong synthetic fibers. Perhaps the best example of these liquid crystals is Kevlar, a synthetic fiber used in bullet-resistant vests, canoes, and parts of the space shuttle. Kevlar is a synthetic polymer, like nylon or polyester, that gains strength by passing through a liquid crystal state during its manufacture.

In a typical polymer the long molecular chains are jumbled together, somewhat like spaghetti. The strength of the material is limited by the disorderly arrangement. The trick is to get the molecules to line up parallel to each other. Once the giant molecules have been synthesized, they are dissolved in sulfuric acid. At the proper concentration the molecules align, and the solution is forced through tiny holes in a nozzle and further aligned. The sulfuric acid is removed in a water bath, thereby forming solid fibers in near perfect alignment. One strand of Kevlar is stronger than an equal-sized strand of steel. It has a much lower density as well, making it a material of choice in bullet-resistant vests.

Molecule	Lewis structure	Number of electron pairs	Electron pair arrangement	Molecular shape
H_2S	H:S̈:H	4	tetrahedral	bent
CCl_4	:C̈l:C:C̈l: with :C̈l: above and :C̈l: below	4	tetrahedral	tetrahedral
AlF_3	:F̈:Al:F̈: with :F̈: above	3	trigonal planar	trigonal planar

Practice 11.6

Predict the shape for CF_4, NF_3, and BeF_2.

Concepts in Review

1. Describe how atomic radii vary (a) from left to right in a period and (b) from top to bottom in a group.

2. Describe how the ionization energies of elements vary with respect to (a) the position in the periodic table and (b) the removal of successive electrons.

3. Write Lewis structures for the representative elements from their position in the periodic table.

4. Describe (a) the formation of ions by electron transfer and (b) the nature of the chemical bond formed by electron transfer.

5. Use Lewis structures to show how ionic compounds are formed from atoms.

6. Describe a crystal of sodium chloride.

7. Predict the relative sizes of an atom and a monatomic ion for a given element.

8. Describe the covalent bond and predict whether a given covalent bond will be polar or nonpolar.

9. Draw Lewis structures for the diatomic elements.

10. Identify single, double, and triple covalent bonds.

11. Describe the changes in electronegativity in (a) moving across a period and (b) moving down a group in the periodic table.

12. Predict formulas of simple compounds formed between the representative (Group A) elements using the periodic table.

13. Describe the effect of electronegativity on the type of chemical bonds in a compound.

14. Draw Lewis structures for (a) the molecules of covalent compounds and (b) polyatomic ions.

15. Describe the difference between polar and nonpolar bonds.

16. Distinguish clearly between ionic and molecular substances.

17. Predict whether the bonding in a compound will be primarily ionic or covalent.

18. Describe the VSEPR model for molecular shape.

19. Use the VSEPR model to determine molecular structure from Lewis structure for given compounds.

Key Terms

covalent bond (11.5)
dipole (11.6)
electronegativity (11.6)
ionic bond (11.3)

ionization energy (11.1)
Lewis structure (11.2)
linear structure (11.11)
nonpolar covalent bond (11.6)

polar covalent bond (11.5)
resonance structure (11.8)
tetrahedral structure (11.11)
trigonal planar (11.11)

Questions

1. Rank these elements according to the radii of their atoms, from smallest to largest: Na, Mg, Cl, K, and Rb. (Figure 11.2)

2. Explain why much more ionization energy is required to remove the first electron from neon than from sodium. (Table 11.1)

3. Explain the large increase in ionization energy needed to remove the third electron from beryllium compared with that needed for the second electron. (Table 11.1)

4. Does the first ionization energy increase or decrease from top to bottom in the periodic table for the alkali metal family? Explain. (Figure 11.3)

5. Does the first ionization energy increase or decrease from top to bottom in the periodic table for the noble gas family? Explain. (Figure 11.3)

6. Why does barium (Ba) have a lower ionization energy than beryllium (Be)? (Figure 11.3)

7. Why is there such a large increase in the ionization energy required to remove the second electron from a sodium atom as opposed to the first? (Table 11.1)

8. Which element in the pair has the larger atomic radius? (Figure 11.2)
   (a) Na or K   (c) O or F   (e) Ti or Zr
   (b) Na or Mg   (d) Br or I

9. In Groups IA–VIIA, which element in each group has the smallest atomic radius? (Figure 11.2)

10. Why does the atomic size increase in going down any family of the periodic table?

11. All the atoms within each Group A family of elements can be represented by the same Lewis structure. Complete the following table, expressing the Lewis structure for each group. (Use E to represent the elements.) (Figure 11.4)

Group	IA	IIA	IIIA	IVA	VA	VIA	VIIA
E·							

12. Draw the Lewis structure for Cs, Ba, Tl, Pb, Po, At, and Rn. How do these structures correlate with the group in which each element occurs?

13. In which general areas of the periodic table are found the elements with (a) the highest and (b) the lowest electronegativities?

14. What are valence electrons?

15. Explain why potassium usually forms a K^+ ion but not a K^{2+} ion.

16. Why does an aluminum ion have a +3 charge?

Paired Exercises

17. Which is larger, a magnesium atom or a magnesium ion? Explain.

18. Which is smaller, a bromine atom or a bromide ion? Explain.

19. Using the table of electronegativity values (Table 11.5), indicate which element is more positive and which is more negative in these compounds:
    (a) H_2O (c) NH_3 (e) NO
    (b) NaF (d) PbS (f) CH_4

20. Using the table of electronegativity values (Table 11.5), indicate which element is more positive and which is more negative in these compounds:
    (a) HCl (c) CCl_4 (e) MgH_2
    (b) LiH (d) IBr (f) OF_2

21. Classify the bond between these pairs of elements as principally ionic or principally covalent (use Table 11.5):
    (a) sodium and chlorine
    (b) carbon and hydrogen
    (c) chlorine and carbon
    (d) calcium and oxygen

22. Classify the bond between these pairs of elements as principally ionic or principally covalent (use Table 11.5):
    (a) hydrogen and sulfur
    (b) barium and oxygen
    (c) fluorine and fluorine
    (d) potassium and fluorine

23. Explain what happens to the electron structures of Mg and Cl atoms when they react to form $MgCl_2$.

24. Write an equation representing
    (a) the change of a fluorine atom to a fluoride ion and
    (b) the change of a calcium atom to a calcium ion

25. Use Lewis structures to show the electron transfer that enables these ionic compounds to be formed:
    (a) MgF_2 (b) K_2O

26. Use Lewis structures to show the electron transfer that enables these ionic compounds to be formed:
    (a) CaO (b) NaBr

27. How many valence electrons are in each of these atoms? H, K, Mg, He, Al

28. How many valence electrons are in each of these atoms? Si, N, P, O, Cl

29. How many electrons must be gained or lost for the following to achieve a noble gas electron structure?
    (a) a calcium atom
    (b) a sulfur atom
    (c) a helium atom

30. How many electrons must be gained or lost for the following to achieve a noble gas electron structure?
    (a) a chloride ion
    (b) a nitrogen atom
    (c) a potassium atom

31. Which is larger? Explain.
    (a) a magnesium ion or an aluminum ion
    (b) Fe^{2+} or Fe^{3+}

32. Which is larger? Explain.
    (a) a potassium atom or a potassium ion
    (b) a bromine atom or a bromide ion

33. Let E be any representative element. Following the pattern in the table, write formulas for the hydrogen and oxygen compounds of
(a) Na   (c) Al
(b) Ca   (d) Sn

Group IA	IIA	IIIA	IVA	VA	VIA	VIIA
EH	EH_2	EH_3	EH_4	EH_3	H_2E	HE
E_2O	EO	E_2O_3	EO_2	E_2O_5	EO_3	E_2O_7

35. The formula for sodium sulfate is Na_2SO_4. Write the names and formulas for the other alkali metal sulfates.

37. Write Lewis structures for
(a) Na (b) Br^- (c) O^{2-}

39. Classify the bonding in each compound as ionic or covalent:
(a) H_2O (c) MgO
(b) NaCl   (d) Br_2

41. Predict the type of bond that would be formed between the following pairs of atoms:
(a) Na and N
(b) N and S
(c) Br and I

43. Draw Lewis structures for
(a) H_2 (b) N_2 (c) Cl_2

45. Draw Lewis structures for
(a) NCl_3 (c) C_2H_6
(b) H_2CO_3 (d) $NaNO_3$

47. Draw Lewis structures for
(a) Ba^{2+} (d) CN^-
(b) Al^{3+} (e) HCO_3^-
(c) SO_3^{2-}

49. Classify these molecules as polar or nonpolar:
(a) H_2O
(b) HBr
(c) CF_4

51. Give the number and arrangement of the electron pairs around the central atom:
(a) C in CCl_4
(b) S in H_2S
(c) Al in AlH_3

53. Use VSEPR theory to predict the structure of these polyatomic ions:
(a) sulfate ion
(b) chlorate ion
(c) periodate ion

34. Let E be any representative element. Following the pattern in the table, write formulas for the hydrogen and oxygen compounds of
(a) Sb   (c) Cl
(b) Se   (d) C

Group IA	IIA	IIIA	IVA	VA	VIA	VIIA
EH	EH_2	EH_3	EH_4	EH_3	H_2E	HE
E_2O	EO	E_2O_3	EO_2	E_2O_5	EO_3	E_2O_7

36. The formula for calcium bromide is $CaBr_2$. Write the names and formulas for the other alkaline earth metal bromides.

38. Write Lewis structures for
(a) Ga (b) Ga^{3+} (c) Ca^{2+}

40. Classify the bonding in each compound as ionic or covalent:
(a) HCl   (c) NH_3
(b) $BaCl_2$ (d) SO_2

42. Predict the type of bond that would be formed between the following pairs of atoms:
(a) H and Si
(b) O and F
(c) Ca and I

44. Draw Lewis structures for
(a) O_2 (b) Br_2 (c) I_2

46. Draw Lewis structures for
(a) H_2S (c) NH_3
(b) CS_2 (d) NH_4Cl

48. Draw Lewis structures for
(a) I^- (d) ClO_3^-
(b) S^{2-} (e) NO_3^-
(c) CO_3^{2-}

50. Classify these molecules as polar or nonpolar:
(a) F_2
(b) CO_2
(c) NH_3

52. Give the number and arrangement of the electron pairs around the central atom:
(a) Be in BeF_2
(b) N in NF_3
(c) Cl in HCl

54. Use VSEPR theory to predict the structure of these polyatomic ions:
(a) ammonium ion
(b) sulfite ion
(c) phosphate ion

55. Use VSEPR theory to predict the shape of these molecules:
    (a) SiH_4
    (b) PH_3
    (c) SeF_2

56. Use VSEPR theory to predict the shape of these molecules:
    (a) SiF_4
    (b) OF_2
    (c) Cl_2O

57. Identify this element: Element X reacts with sodium to form the compound Na_2X and is in the second period on the periodic table.

58. Identify this element: Element Y reacts with oxygen to form the compound Y_2O and has the lowest ionization energy of any fourth-period element on the periodic table.

Additional Exercises

59. Identify the element on the periodic table that satisfies each description:
    (a) transition metal with the largest atomic radius
    (b) alkaline earth metal with the greatest ionization energy
    (c) least dense member of the nitrogen family
    (d) alkali metal with the greatest ratio of neutrons to protons
    (e) most electronegative transition metal

60. Choose the element that fits each description:
    (a) the lower electronegativity   As or Zn
    (b) the lower chemical reactivity   Ba or Be
    (c) the fewer valence electrons   N or Ne

61. Identify two reasons why fluorine has a much higher electronegativity than neon.

62. When one electron is removed from an atom of Li, it has two left. Helium atoms also have two electrons. Why is more energy required to remove the second electron from Li than to remove the first from He?

63. Group IB elements (see the periodic table on the inside cover of your book) have one electron in their outer energy level, as do Group IA elements. Would you expect them to form compounds such as CuCl, AgCl, and AuCl? Explain.

64. The formula for lead(II) bromide is $PbBr_2$: predict formulas for tin(II) and germanium(II) bromides.

65. Why is it not proper to speak of sodium chloride molecules?

66. What is a covalent bond? How does it differ from an ionic bond?

67. Briefly comment on the structure $Na:\ddot{O}:Na$ for the compound Na_2O.

68. What are the four most electronegative elements?

69. Rank these elements from highest electronegativity to lowest: Mg, S, F, H, O, Cs.

70. Is it possible for a molecule to be nonpolar even though it contains polar covalent bonds? Explain.

71. Why is CO_2 a nonpolar molecule, whereas CO is a polar molecule?

72. Estimate the bond angle between atoms in these molecules.
    (a) H_2S
    (b) NH_3
    (c) NH_4^+
    (d) $SiCl_4$

73. Consider the two molecules BF_3 and NF_3. Compare and contrast them in terms of
    (a) valence level orbitals on the central atom that are used for bonding
    (b) shape of the molecule
    (c) number of lone electron pairs on the central atom
    (d) type and number of bonds found in the molecule

74. With respect to electronegativity, why is fluorine such an important atom? What combination of atoms on the periodic table results in the most ionic bond?

75. Why does the Lewis structure of each element in a given group of representative elements on the periodic table have the same number of dots?

76. A sample of an air pollutant composed of sulfur and oxygen was found to contain 1.40 g sulfur and 2.10 g oxygen. What is the empirical formula for this compound? Draw a Lewis structure to represent it.

77. A dry-cleaning fluid composed of carbon and chlorine was found to have the composition 14.5% carbon and 85.5% chlorine. Its known molar mass is 166 g/mol. Draw a Lewis structure to represent it.

Answers to Practice Exercises

11.1 (a) $:\overset{.}{\underset{.}{N}}\cdot$

(b) $:\overset{.}{Al}$

(c) $Sr:$

(d) $:\overset{..}{Br}\cdot$

11.2 (a) Na_2S

(b) Rb_2O

11.3 (a) Mg_3N_2

(b) Ba_3As_2

11.4 (a) 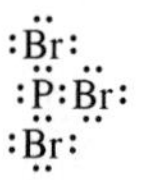 $\begin{matrix} :\overset{..}{Br}: \\ :\overset{..}{P}:\overset{..}{Br}: \\ :\overset{..}{Br}: \end{matrix}$ (b) 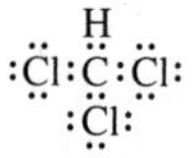 $\begin{matrix} H \\ :\overset{..}{Cl}:\overset{..}{C}:\overset{..}{Cl}: \\ :\overset{..}{Cl}: \end{matrix}$

(c) $H:\overset{..}{\underset{..}{F}}:$ (d) 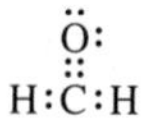 $\begin{matrix} \overset{..}{\underset{..}{O}}: \\ H:C:H \end{matrix}$

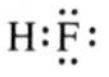

11.5 (a) 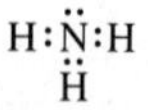 $H:\overset{..}{\underset{H}{N}}:H$ (b) 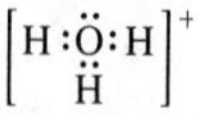 $\left[H:\overset{..}{\underset{H}{O}}:H \right]^+$

(c) 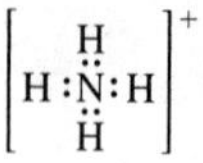 $\left[\begin{matrix} H \\ H:\overset{}{\underset{H}{N}}:H \end{matrix} \right]^+$ (d) 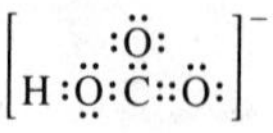 $\left[\begin{matrix} :\overset{..}{\underset{..}{O}}: \\ H:\overset{..}{\underset{..}{O}}:C::\overset{..}{O}: \end{matrix} \right]^-$

11.6 CF_4, tetrahedral; NF_3, pyramidal; BeF_2, linear

PUTTING IT TOGETHER
Review for Chapters 10–11

Multiple Choice: *Choose the correct answer to each of the following.*

1. The concept of electrons existing in specific orbits around the nucleus was the contribution of
   (a) Thomson   (c) Bohr
   (b) Rutherford   (d) Schrödinger

2. The correct electron structure for a fluorine atom, F, is
   (a) $1s^2 2s^2 2p^5$ (c) $1s^2 2s^2 2p^4 3s^1$
   (b) $1s^2 2s^2 2p^2 3s^2 3p^1$ (d) $1s^2 2s^2 2p^3$

3. The correct electron structure for $_{48}$Cd is
   (a) $1s^2 2s^2 2p^6 3s^2 3p^6 4s^2 3d^{10}$
   (b) $1s^2 2s^2 2p^6 3s^2 3p^6 4s^2 3d^{10} 4p^6 5s^2 4d^{10}$
   (c) $1s^2 2s^2 2p^6 3s^2 3p^6 4s^2 3d^{10} 4p^6 4d^4$
   (d) $1s^2 2s^2 2p^6 3s^2 3p^6 4s^2 4p^6 4d^{10} 5s^2 5d^{10}$

4. The correct electron structure of $_{23}$V is
   (a) $[Ar]4s^2 3d^3$ (c) $[Ar]4s^2 4d^3$
   (b) $[Ar]4s^2 4p^3$ (d) $[Kr]4s^2 3d^3$

5. Which of the following is the correct atomic structure for $_{22}^{48}$Ti?

   (a)
   22+ / 26n
   22e−

   (b)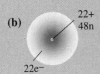
   22+ / 48n
   22e−

   (c)
   26+ / 22n
   22e−

   (d)
   22+ / 26n
   48e−

6. The number of orbitals in a d sublevel is
   (a) 3   (c) 7
   (b) 5   (d) no correct answer given

7. The number of electrons in the third principal energy level in an atom having the electron structure $1s^2 2s^2 2p^6 3s^2 3p^2$ is
   (a) 2   (b) 4   (c) 6   (d) 8

8. The total number of orbitals that contain at least one electron in an atom having the structure $1s^2 2s^2 2p^6 3s^2 3p^2$ is
   (a) 5   (c) 14
   (b) 8   (d) no correct answer given

9. Which of these elements has two s and six p electrons in its outer energy level?
   (a) He   (c) Ar
   (b) O   (d) no correct answer given

10. Which element is not a noble gas?
    (a) Ra   (b) Xe   (c) He   (d) Ar

11. Which element has the largest number of unpaired electrons?
    (a) F   (b) S   (c) Cu   (d) N

12. How many unpaired electrons are in the electron structure of $_{24}$Cr, $[Ar]4s^1 3d^5$?
    (a) 2   (b) 4   (c) 5   (d) 6

13. Groups IIIA–VIIA plus the noble gases form the area of the periodic table where the electron sublevels being filled are
    (a) p sublevels   (c) d sublevels
    (b) s and p sublevels   (d) f sublevels

14. In moving down an A group on the periodic table, the number of electrons in the outermost energy level
    (a) increases regularly
    (b) remains constant
    (c) decreases regularly
    (d) changes in an unpredictable manner

15. Which of the following is an incorrect formula?
    (a) NaCl   (b) K_2O (c) AlO   (d) BaO

16. Elements of the noble gas family
    (a) form no compounds at all
    (b) have no valence electrons
    (c) have an outer electron structure of $ns^2 np^6$ (helium excepted), where n is the period number
    (d) no correct answer given

17. The lanthanide and actinide series of elements are
    (a) representative elements
    (b) transition elements
    (c) filling in d level electrons
    (d) no correct answer given

18. The element having the structure $1s^2 2s^2 2p^6 3s^2 3p^2$ is in Group
    (a) IIA   (b) IIB   (c) IVA   (d) IVB

19. In Group VA, the element having the smallest atomic radius is
    (a) Bi   (b) P   (c) As   (d) N

20. In Group IVA, the most metallic element is
    (a) C   (b) Si   (c) Ge   (d) Sn

21. Which group in the periodic table contains the least reactive elements?
    (a) IA (b) IIA (c) IIIA (d) noble gases

22. Which group in the periodic table contains the alkali metals?
    (a) IA (b) IIA (c) IIIA (d) IVA

23. An atom of fluorine is smaller than an atom of oxygen. One possible explanation is that, compared to oxygen, fluorine has
    (a) a larger mass number
    (b) a smaller atomic number
    (c) a greater nuclear charge
    (d) more unpaired electrons

24. If the size of the fluorine atom is compared to the size of the fluoride ion,
    (a) they would both be the same size
    (b) the atom is larger than the ion
    (c) the ion is larger than the atom
    (d) the size difference depends on the reaction

25. Sodium is a very active metal because
    (a) it has a low ionization energy
    (b) it has only one outermost electron
    (c) it has a relatively small atomic mass
    (d) all of the above

26. Which of the following formulas is not correct?
    (a) Na^+ (b) S^- (c) Al^{3+} (d) F^-

27. Which of the following molecules does not have a polar covalent bond?
    (a) CH_4 (b) H_2O (c) CH_3OH (d) Cl_2

28. Which of the following molecules is a dipole?
    (a) HBr (b) CH_4 (c) H_2 (d) CO_2

29. Which of the following has bonding that is ionic?
    (a) H_2 (b) MgF_2 (c) H_2O (d) CH_4

30. Which of the following is a correct Lewis structure?

    (a) $:\overset{..}{\underset{..}{O}}:C:\overset{..}{\underset{..}{O}}:$ (b) $:\overset{..}{\underset{..}{Cl}}:\overset{:\overset{..}{Cl}:}{\underset{:\overset{..}{Cl}:}{C}}:\overset{..}{\underset{..}{Cl}}:$

    (c) $\overset{..}{\underset{..}{Cl}}::\overset{..}{\underset{..}{Cl}}$ (d) $:\overset{..}{N}:\overset{..}{N}:$

31. Which of the following is an incorrect Lewis structure?

    (a) $H:\overset{..}{N}:H$ (b) $:\overset{..}{\underset{..}{O}}:H$
                                    $\overset{|}{H}$

    (c) $H:\overset{\overset{H}{..}}{\underset{..}{C}}:H$ (d) $:N:::N:$
        $\overset{|}{H}$

32. The correct Lewis structure for SO_2 is
    (a) $:\overset{..}{\underset{..}{O}}:\overset{..}{\underset{..}{S}}:\overset{..}{\underset{..}{O}}:$ (b) $:\overset{..}{\underset{..}{O}}:\overset{..}{S}::\overset{..}{\underset{..}{O}}:$
    (c) $:\overset{..}{\underset{..}{O}}::S::\overset{..}{\underset{..}{O}}:$ (d) $:\overset{..}{\underset{..}{O}}:\overset{..}{\underset{..}{S}}:\overset{..}{\underset{..}{O}}:$

33. Carbon dioxide, CO_2, is a nonpolar molecule because
    (a) oxygen is more electronegative than carbon
    (b) the two oxygen atoms are bonded to the carbon atom
    (c) the molecule has a linear structure with the carbon atom in the middle
    (d) the carbon–oxygen bonds are polar covalent

34. When a magnesium atom participates in a chemical reaction, it is most likely to
    (a) lose 1 electron (c) lose 2 electrons
    (b) gain 1 electron (d) gain 2 electrons

35. If X represents an element of Group IIIA, what is the general formula for its oxide?
    (a) X_3O_4 (b) X_3O_2 (c) XO (d) X_2O_3

36. Which of the following has the same electron structure as an argon atom?
    (a) Ca^{2+} (b) Cl^0 (c) Na^+ (d) K^0

37. As the difference in electronegativity between two elements decreases, the tendency for the elements to form a covalent bond
    (a) increases
    (b) decreases
    (c) remains the same
    (d) sometimes increases and sometimes decreases

38. Which compound forms a tetrahedral molecule?
    (a) NaCl (b) CO_2 (c) CH_4 (d) $MgCl_2$

39. Which compound has a bent (V-shaped) molecular structure?
    (a) NaCl (b) CO_2 (c) CH_4 (d) H_2O

40. Which compound has double bonds within its molecular structure?
    (a) NaCl (b) CO_2 (c) CH_4 (d) H_2O

41. The total number of valence electrons in a nitrate ion, NO_3^-, is
    (a) 12 (b) 18 (c) 23 (d) 24

42. The number of electrons in a triple bond is
    (a) 3 (b) 4 (c) 6 (d) 8

43. The number of unbonded pairs of electrons in H_2O is
    (a) 0 (b) 1 (c) 2 (d) 4

44. Which of the following does not have a noble gas electron structure?
    (a) Na (b) Sc^{3+} (c) Ar (d) O^{2-}

Free Response Questions

1. An alkaline earth metal, M, combines with a halide, X. Will the resulting compound be ionic or covalent? Why? What is the Lewis structure for the compound?

2. Is the following statement true or false? "All electrons in atoms with even atomic numbers are paired." Explain your answer using an example.

3. Discuss whether the following statement is true or false. "All nonmetals have two valence electrons in an s sublevel with the exception of the noble gases, which have at least one unpaired electron in a p sublevel."

4. The first ionization energy, IE, of potassium is lower than the first IE for calcium but the second IE of calcium is lower than the second IE of potassium. Use an electron configuration or size argument to explain this trend in IEs.

5. Chlorine has a very large first ionization energy yet it forms a chloride ion relatively easily. Explain.

6. Three particles have the same electron configuration. One is a cation of an alkali metal, one is an anion of the halide in the third period, and the third particle is an atom of a noble gas. What are the identities of the three particles (including

charges)? Which particle should have the smallest atomic/ionic radius, which should have the largest, and why?

7. Why is the Lewis structure of $AlCl_3$ not written as

$$\begin{array}{c} \ddot{:}\ddot{Cl}: \\ / \\ :\ddot{Cl}-Al \\ \backslash \\ :\ddot{Cl}: \end{array}$$

What is the correct Lewis structure and which electrons are shown in a Lewis structure?

8. Why does carbon have a maximum of four covalent bonds?

9. Both NCl_3 and BF_3 have a central atom bonded to three other atoms yet one is pyramidal and the other is trigonal planar. Explain.

10. Draw the Lewis structure of the atom whose electron configuration is $1s^2 2s^2 2p^6 3s^2 3p^6 4s^2 3d^{10} 4p^5$. Would you expect this atom to form an ionic, nonpolar covalent, or polar covalent bond with sulfur?

CHAPTER 12

The Gaseous State of Matter

▲
Properties of gases are transformed to art and sport at the hot air balloon festival in Albuquerque, New Mexico.

Our atmosphere is composed of a mixture of gases, including nitrogen, oxygen, carbon dioxide, ozone, and trace amounts of others. These gases are essential to life yet they can also create hazards to us. For example, carbon dioxide is valuable when it is taken in by plants and converted to carbohydrates, but it also is associated with the potentially hazardous greenhouse effect. Ozone surrounds the earth at high altitudes and protects us from harmful ultraviolet rays, but it also destroys rubber and plastics. We require air to live, yet scuba divers must be concerned about oxygen poisoning, and the "bends."

In chemistry the study of the behavior of gases allows us to understand our atmosphere and the effects that gases have on our lives.

12.1 General Properties

In Chapter 3, solids, liquids, and gases were described briefly. In this chapter we consider the behavior of gases in greater detail.

Gases are the least dense and most mobile of the three states of matter. A solid has a rigid structure, and its particles remain in essentially fixed positions. When a solid absorbs sufficient heat, it melts and changes into a liquid. Melting occurs because the molecules (or ions) have absorbed enough energy to break out of the rigid crystal lattice structure of the solid. The molecules or ions in the liquid are more energetic than they were in the solid, as indicated by their increased mobility. Molecules in the liquid state cling to one another. When the liquid absorbs additional heat, the more energetic molecules break away from the liquid surface and go into the gaseous state—the most mobile state of matter. Gas molecules move at very high velocities and have high kinetic energy. The average velocity of hydrogen molecules at 0°C is over 1600 meters (1 mile) per second. Mixtures of gases are uniformly distributed within the container in which they are confined.

The same quantity of a substance occupies a much greater volume as a gas than it does as a liquid or a solid. For example, 1 mol of water (18.02 g) has a volume of 18 mL at 4°C. This same amount of water would occupy about 22,400 mL in the gaseous state—more than a 1200-fold increase in volume. We may assume from this difference in volume that (1) gas molecules are relatively far apart, (2) gases can be greatly compressed, and (3) the volume occupied by a gas is mostly empty space.

▲
A mole of water occupies 18 mL as a liquid but would fill this box (22.4 L) as a gas at the same temperature (25°C).

12.2 The Kinetic-Molecular Theory

Careful scientific studies of the behavior and properties of gases were begun in the 17th century by Robert Boyle (1627–1691). His work was carried forward by many investigators, and the accumulated data were used in the second half of the 19th century to formulate a general theory to explain the behavior and properties of gases. This theory is called the **kinetic-molecular theory (KMT).** The KMT has since

kinetic-molecular theory (KMT)

been extended to cover, in part, the behavior of liquids and solids. It ranks today with the atomic theory as one of the greatest generalizations of modern science.

The KMT is based on the motion of particles, particularly gas molecules. A gas *ideal gas* that behaves exactly as outlined by the theory is known as an **ideal gas.** No ideal gases exist, but under certain conditions of temperature and pressure, real gases approach ideal behavior, or at least show only small deviations from it. Under extreme conditions, such as very high pressure and low temperature, real gases deviate greatly from ideal behavior. For example, at low temperature and high pressure many gases become liquids.

The principal assumptions of the kinetic-molecular theory are:

1. Gases consist of tiny (submicroscopic) particles.
2. The distance between particles is large compared with the size of the particles themselves. The volume occupied by a gas consists mostly of empty space.
3. Gas particles have no attraction for one another.
4. Gas particles move in straight lines in all directions, colliding frequently with one another and with the walls of the container.
5. No energy is lost by the collision of a gas particle with another gas particle or with the walls of the container. All collisions are perfectly elastic.
6. The average kinetic energy for particles is the same for all gases at the same temperature, and its value is directly proportional to the Kelvin temperature.

The kinetic energy (KE) of a particle is one-half its mass times its velocity squared. It is expressed by the equation

$$KE = \frac{1}{2}mv^2$$

where m is the mass and v is the velocity of the particle.

All gases have the same kinetic energy at the same temperature. Therefore, from the kinetic energy equation we can see that, if we compare the velocities of the molecules of two gases, the lighter molecules will have a greater velocity than the heavier ones. For example, calculations show that the velocity of a hydrogen molecule is four times the velocity of an oxygen molecule.

diffusion   Due to their molecular motion, gases have the property of **diffusion,** the ability of two or more gases to mix spontaneously until they form a uniform mixture. The diffusion of gases may be illustrated by the use of the apparatus shown in Figure 12.1. Two large flasks, one containing reddish brown bromine vapors and the other dry air, are connected by a side tube. When the stopcock between the flasks is

FIGURE 12.1 ▶
Diffusion of gases. When the stopcock between the two flasks is opened, colored bromine molecules can be seen diffusing into the flask containing air.

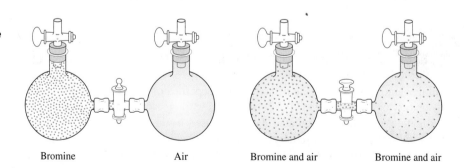

Bromine   Air   Bromine and air   Bromine and air

opened, the bromine and air will diffuse into each other. After standing awhile, both flasks will contain bromine and air.

If we put a pinhole in a balloon, the gas inside will effuse or flow out of the balloon. **Effusion** is a process by which gas molecules pass through a very small orifice (opening) from a container at higher pressure to one at lower pressure.

effusion

Thomas Graham (1805–1869), a Scottish chemist, observed that the rate of effusion was dependent on the density of a gas. This observation led to **Graham's law of effusion.**

Graham's law of effusion

> The rates of effusion of two gases at the same temperature and pressure are inversely proportional to the square roots of their densities or molar masses:
>
> $$\frac{\text{rate of effusion of gas } A}{\text{rate of effusion of gas } B} = \sqrt{\frac{d\text{B}}{d\text{A}}} = \sqrt{\frac{\text{molar mass } B}{\text{molar mass } A}}$$

A major application of Graham's law occurred during World War II with the separation of the isotopes of uranium-235 (U-235) and uranium-238 (U-238). Naturally occurring uranium consists of 0.7% U-235, 99.3% U-238, and a trace of U-234. However, only U-235 is useful as fuel for nuclear reactors and atomic bombs, so the concentration of U-235 in the mixture of isotopes had to be increased.

Uranium was first changed to uranium hexafluoride, UF_6, a white solid that readily goes into the gaseous state. The gaseous mixture of $^{235}UF_6$ and $^{238}UF_6$ was then allowed to effuse through porous walls. Although the effusion rate of the lighter gas is only slightly faster than that of the heavier one,

$$\frac{\text{effusion rate } ^{235}UF_6}{\text{effusion rate } ^{238}UF_6} = \sqrt{\frac{\text{molar mass } ^{238}UF_6}{\text{molar mass } ^{235}UF_6}} = \sqrt{\frac{352}{349}} = 1.0043$$

the separation and enrichment of U-235 was accomplished by subjecting the gaseous mixture to several thousand stages of effusion.

12.3 Measurement of Pressure of Gases

pressure

Pressure is defined as force per unit area. When a rubber balloon is inflated with air, it stretches and maintains its larger size because the pressure on the inside is greater than that on the outside. Pressure results from the collisions of gas molecules with the walls of the balloon (see Figure 12.2). When the gas is released, the force or pressure of the air escaping from the small neck propels the balloon in a rapid, irregular flight. If the balloon is inflated until it bursts, the gas escaping all at once causes an explosive noise.

The effects of pressure are also observed in the mixture of gases surrounding Earth—our atmosphere, which is composed of about 78% nitrogen, 21% oxygen, 1% argon, and other minor constituents by volume (see Table 12.1). The outer boundary of the atmosphere is not known precisely, but more than 99% of the atmosphere is below an altitude of 20 miles (32 km). Thus, the concentration of gas molecules in the atmosphere decreases with altitude, and at about 4 miles the amount of oxygen is insufficient to sustain human life. The gases in the atmosphere exert a

▲ **FIGURE 12.2**
The pressure resulting from the collisions of gas molecules with the walls of the balloon keeps the balloon inflated.

TABLE 12.1	Average Composition of Dry Air		
Gas	Percent by volume	Gas	Percent by volume
N_2	78.08	He	0.0005
O_2	20.95	CH_4	0.0002
Ar	0.93	Kr	0.0001
CO_2	0.033	Xe, H_2, and N_2O	Trace
Ne	0.0018		

atmospheric pressure

pressure known as **atmospheric pressure.** The pressure exerted by a gas depends on the number of molecules of gas present, the temperature, and the volume in which the gas is confined. Gravitational forces hold the atmosphere relatively close to Earth and prevent air molecules from flying off into outer space. Thus the atmospheric pressure at any point is due to the mass of the atmosphere pressing downward at that point.

barometer

The pressure of the gases in the atmosphere can be measured with a **barometer.** A mercury barometer may be prepared by completely filling a long tube with pure, dry mercury and inverting the open end into an open dish of mercury. If the tube is longer than 760 mm, the mercury level will drop to a point at which the column of mercury in the tube is just supported by the pressure of the atmosphere. If the tube is properly prepared, a vacuum will exist above the mercury column. The weight of mercury, per unit area, is equal to the pressure of the atmosphere. The column of mercury is supported by the pressure of the atmosphere, and the height of the column is a measure of this pressure (see Figure 12.3). The mercury barometer was invented in 1643 by the Italian physicist E. Torricelli (1608–1647), for whom the unit of pressure *torr* was named.

1 atmosphere

Air pressure is measured and expressed in many units. The standard atmospheric pressure, or simply **1 atmosphere** (atm), is the pressure exerted by a column

FIGURE 12.3 ▶
Preparation of a mercury barometer. The full tube of mercury at the left is inverted and placed in a dish of mercury.

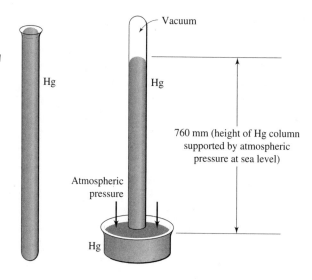

of mercury 760 mm high at a temperature of 0°C. The normal pressure of the atmosphere at sea level is 1 atm or 760 torr or 760 mm Hg. The SI unit for pressure is the pascal (Pa), where 1 atm = 101,325 Pa or 101.3 kPa. Other units for expressing pressure are inches of mercury, centimeters of mercury, the millibar (mbar), and pounds per square inch (lb/in.^{2} or psi). The values of these units equivalent to 1 atm are summarized in Table 12.2.

Atmospheric pressure varies with altitude. The average pressure at Denver, Colorado, 1.61 km (1 mile) above sea level, is 630 torr (0.83 atm). Atmospheric pressure is 0.5 atm at about 5.5 km (3.4 miles) altitude.

Pressure is often measured by reading the heights of mercury columns in millimeters on a barometer. Thus pressure may be recorded as mm Hg. But in many applications the torr is superceding mm Hg as a unit of pressure. In problems dealing with gases it is necessary to make interconversions among the various pressure units. Since atm, torr, and mm Hg are common pressure units, we give examples involving all three of these units.

1 atm = 760 torr = 760 mm Hg

TABLE 12.2	Pressure Units Equivalent to 1 Atmosphere
1 atm	
	760 torr
	760 mm Hg
	76 cm Hg
	101.325 kPa
	1013 mbar
	29.9 in. Hg
	14.7 lb/in.^{2}

The average atmospheric pressure at Walnut, California, is 740. mm Hg. Calculate this pressure in (a) torr and (b) atmospheres.   **Example 12.1**

Let's use conversion factors that relate one unit of pressure to another.   **Solution**

(a) To convert mm Hg to torr, use the conversion factor 760 torr/760 mm Hg (1 torr/1 mm Hg):

$$(740. \text{ mm Hg})\left(\frac{1 \text{ torr}}{1 \text{ mm Hg}}\right) = 740. \text{ torr}$$

(b) To convert mm Hg to atm, use the conversion factor 1 atm/760. mm Hg:

$$(740. \text{ mm Hg})\left(\frac{1 \text{ atm}}{760. \text{ mm Hg}}\right) = 0.974 \text{ atm}$$

Practice 12.1

A barometer reads 1.12 atm. Calculate the corresponding pressure in (a) torr and (b) mm Hg.

12.4   Dependence of Pressure on Number of Molecules and Temperature

Pressure is produced by gas molecules colliding with the walls of a container. At a specific temperature and volume the number of collisions depends on the number of gas molecules present. The number of collisions can be increased by increasing the number of gas molecules present. If we double the number of molecules, the frequency of collisions and the pressure should double. We find, for an ideal gas, that

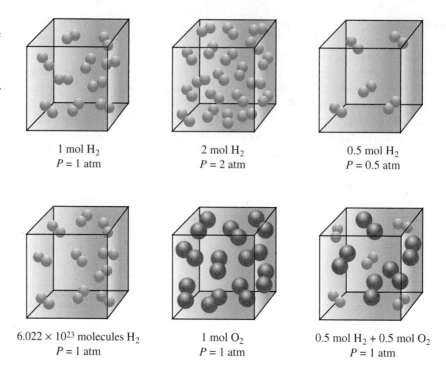

1 mol H₂   2 mol H₂   0.5 mol H₂
$P = 1$ atm $P = 2$ atm $P = 0.5$ atm

6.022×10^{23} molecules H_2 1 mol O_2 0.5 mol H_2 + 0.5 mol O_2
$P = 1$ atm $P = 1$ atm $P = 1$ atm

this doubling is actually what happens. When the temperature and mass are kept constant, the pressure is directly proportional to the number of moles or molecules of
gas present. Figure 12.4 illustrates this concept.

A good example of this molecule–pressure relationship may be observed in an
ordinary cylinder of compressed gas equipped with a pressure gauge. When the
valve is opened, gas escapes from the cylinder. The volume of the cylinder is con-

 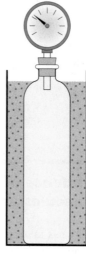

0°C   100°C
Volume = 1 liter   Volume = 1 liter
0.1 mole gas   0.1 mole gas
$P = 2.24$ atm $P = 3.06$ atm

stant, and the decrease in quantity (moles) of gas is registered by a drop in pressure indicated on the gauge.

The pressure of a gas in a fixed volume also varies with temperature. When the temperature is increased, the kinetic energy of the molecules increases, causing more frequent and more energetic collisions of the molecules with the walls of the container. This increase in collision frequency and energy results in a pressure increase (see Figure 12.5).

12.5 Boyle's Law

Through a series of experiments, Robert Boyle (1627–1691) determined the relationship between the pressure (P) and volume (V) of a particular quantity of a gas. This relationship of P and V is known as **Boyle's law.**

Boyle's law

> **At constant temperature (T), the volume (V) of a fixed mass of a gas is inversely proportional to the pressure (P), which may be expressed as**
>
> $$V \propto \frac{1}{P} \qquad \text{or} \qquad P_1V_1 = P_2V_2$$

This equation says that the volume varies ($\propto$) inversely with the pressure, at constant mass and temperature. When the pressure on a gas is increased, its volume will decrease, and vice versa. The inverse relationship of pressure and volume is graphed in Figure 12.6.

When Boyle doubled the pressure on a specific quantity of a gas, keeping the temperature constant, the volume was reduced to one-half the original volume; when he tripled the pressure on the system, the new volume was one-third the original volume; and so on. His work showed that the product of volume and pressure is constant if the temperature is not changed:

$$PV = \text{constant} \qquad \text{or} \qquad PV = k \qquad \text{(mass and temperature are constant)}$$

Let's demonstrate this law using a cylinder with a movable piston so that the volume of gas inside the cylinder may be varied by changing the external pressure (see Figure 12.7). Assume that the temperature and the number of gas molecules do not change. We start with a volume of 1000 mL and a pressure of 1 atm. When we change the pressure to 2 atm, the gas molecules are crowded closer together, and the volume is reduced to 500 mL. When we increase the pressure to 4 atm, the volume becomes 250 mL.

Note that the product of the pressure times the volume is the same number in each case, substantiating Boyle's law. We may then say that

$$P_1V_1 = P_2V_2$$

where P_1V_1 is the pressure–volume product at one set of conditions, and P_2V_2 is the product at another set of conditions. In each case the new volume may be calculated by multiplying the starting volume by a ratio of the two pressures involved. Of course, the ratio of pressures used must reflect the direction in which the volume

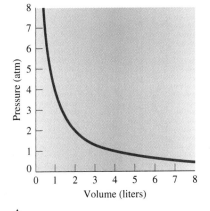

▲
FIGURE 12.6
Graph of pressure versus volume showing the inverse PV relationship of an ideal gas.

FIGURE 12.7 ▶
The effect of pressure on the
volume of a gas.

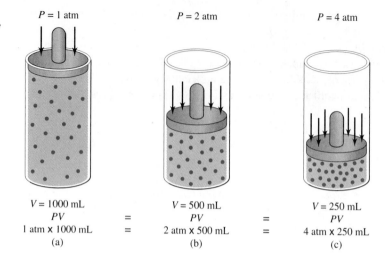

$P = 1$ atm $P = 2$ atm $P = 4$ atm

$V = 1000$ mL $V = 500$ mL $V = 250$ mL

$$\frac{PV}{1\ \text{atm} \times 1000\ \text{mL}} \quad = \quad = \quad \frac{PV}{2\ \text{atm} \times 500\ \text{mL}} \quad = \quad = \quad \frac{PV}{4\ \text{atm} \times 250\ \text{mL}}$$

(a)   (b)   (c)

should change. When the pressure is changed from 1 atm to 2 atm, the ratio to be used is 1 atm/2 atm. Now we can verify the results given in Figure 12.7:

(a) Starting volume, 1000 mL; pressure change, 1 atm ⟶ 2 atm

$$(1000\ \text{mL})\left(\frac{1\ \text{atm}}{2\ \text{atm}}\right) = 500\ \text{mL}$$

(b) Starting volume, 1000 mL; pressure change, 1 atm ⟶ 4 atm

$$(1000\ \text{mL})\left(\frac{1\ \text{atm}}{4\ \text{atm}}\right) = 250\ \text{mL}$$

(c) Starting volume, 500 mL; pressure change, 2 atm ⟶ 4 atm

$$(500\ \text{mL})\left(\frac{2\ \text{atm}}{4\ \text{atm}}\right) = 250\ \text{mL}$$

In summary, a change in the volume of a gas due to a change in pressure can be calculated by multiplying the original volume by a ratio of the two pressures. If the pressure is increased, the ratio should have the smaller pressure in the numerator and the larger pressure in the denominator. If the pressure is decreased, the larger pressure should be in the numerator and the smaller pressure in the denominator.

new volume = original volume × ratio of pressures

We use Boyle's law in the following examples. If no mention is made of temperature, assume that it remains constant.

Example 12.2   What volume will 2.50 L of a gas occupy if the pressure is changed from 760. mm Hg to 630. mm Hg?

Solution   **Method A. Conversion Factors**

Step 1. Determine whether pressure is being increased or decreased:

pressure decreases ⟶ volume increases

Step 2. Multiply the original volume by a ratio of pressures that will result in an increase in volume:

$$V = (2.50 \text{ L})\left(\frac{760. \text{ mm Hg}}{630. \text{ mm Hg}}\right) = 3.02 \text{ L (new volume)}$$

Method B. Algebraic Equation

Decide which method is the best for you and stick with it.

Step 1. Organize the given information:

$P_1 = 760. \text{ mm Hg}$ $V_1 = 2.50 \text{ L}$

$P_2 = 630. \text{ mm Hg}$ $V_2 = ?$

Step 2. Write and solve this equation for the unknown:

$$P_1V_1 = P_2V_2 \qquad V_2 = \frac{P_1V_1}{P_2}$$

Step 3. Put the given information into this equation and calculate:

$$V_2 = \frac{(760. \text{ mm Hg})(2.50 \text{ L})}{630. \text{ mm Hg}} = 3.02 \text{ L}$$

A given mass of hydrogen occupies 40.0 L at 700. torr. What volume will it occupy at 5.00 atm pressure?

Example 12.3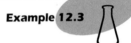

Method A. Conversion Factors

Solution

Step 1. Determine whether the pressure is being increased or decreased. Note that in order to compare the values the units must be the same. We'll convert 700. torr to atm:

$$(700. \text{ torr})\left(\frac{1 \text{ atm}}{760 \text{ torr}}\right) = 0.921 \text{ atm}$$

The pressure is going from 0.921 atm to 5.00 atm.

pressure increases ⟶ volume decreases

Step 2. Multiply the original volume by a ratio of pressures that will result in a decrease in volume:

$$V = (40.0 \text{ L})\left(\frac{0.921 \text{ atm}}{5.00 \text{ atm}}\right) = 7.37 \text{ L}$$

Method B. Algebraic Equation

Step 1. Organize the given information. Remember to make the pressure units the same.

$P_1 = 700. \text{ torr} = 0.921 \text{ atm}$ $V_1 = 40.0 \text{ L}$

$P_2 = 5.00 \text{ atm}$ $V_2 = ?$

Step 2. Write and solve this equation for the unknown:

$$P_1V_1 = P_2V_2 \qquad V_2 = \frac{P_1V_1}{P_2}$$

Step 3. Put the given information into this equation and calculate:

$$V_2 = \frac{(0.921 \text{ atm})(40.0 \text{ L})}{5.00 \text{ atm}} = 7.37 \text{ L}$$

Example 12.4   A gas occupies a volume of 200. mL at 400. torr pressure. To what pressure must the gas be subjected in order to change the volume to 75.0 mL?

Solution

Method A. Conversion Factors

Step 1. Determine whether volume is being increased or decreased:

volume decreases ⟶ pressure increases

Step 2. Multiply the original pressure by a ratio of volumes that will result in an increase in pressure:

new pressure = original pressure × ratio of volumes

$$P = (400. \text{ torr})\left(\frac{200. \text{ mL}}{75.0 \text{ mL}}\right) = 1067 \text{ torr} \quad \text{or} \quad 1.07 \times 10^3 \text{ torr (new pressure)}$$

Method B. Algebraic Equation

Step 1. Organize the given information. Remember to make units the same.

$$P_1 = 400. \text{ torr} \qquad V_1 = 200. \text{ mL}$$
$$P_2 = ? \qquad V_2 = 75.0 \text{ mL}$$

Step 2. Write and solve this equation for the unknown:

$$P_1V_1 = P_2V_2 \qquad P_2 = \frac{P_1V_1}{V_2}$$

Step 3. Put the given information into the equation and calculate:

$$P_2 = \frac{P_1V_1}{V_2} = \frac{(400. \text{ torr})(200. \text{ mL})}{75.0 \text{ mL}} = 1.07 \times 10^3 \text{ torr}$$

Practice 12.2

A gas occupies a volume of 3.86 L at 0.750 atm. At what pressure will the volume be 4.86 L?

12.6   Charles' Law

The effect of temperature on the volume of a gas was observed in about 1787 by the French physicist J. A. C. Charles (1746–1823). Charles found that various gases expanded by the same fractional amount when they underwent the same change in temperature. Later it was found that if a given volume of any gas initially at 0°C was cooled by 1°C, the volume decreased by $\frac{1}{273}$; if cooled by 2°C, it decreased by $\frac{2}{273}$; if cooled by 20°C, by $\frac{20}{273}$; and so on. Since each degree of cooling reduced the volume by $\frac{1}{273}$, it was apparent that any quantity of any gas would have zero volume if it could be cooled to −273°C. Of course, no real gas can be cooled to −273°C for the simple reason that it would liquefy before that temperature is reached. However, −273°C (more precisely −273.15°C) is referred to as **absolute zero;** this temperature is the zero point on the Kelvin (absolute) temperature scale—the temperature at which the volume of an ideal, or perfect, gas would become zero.

absolute zero

The volume–temperature relationship for methane is shown graphically in Figure 12.8. Experimental data show the graph to be a straight line that, when extrapolated, crosses the temperature axis at −273.15°C, or absolute zero. This is characteristic for all gases.

In modern form, **Charles' law** is as follows:

Charles' law

At *constant pressure* the volume of a fixed mass of any gas is directly proportional to the absolute temperature, which may be expressed as:

$$V \propto T \qquad \text{or} \qquad \frac{V_1}{T_1} = \frac{V_2}{T_2}$$

Mathematically this states that the volume of a gas varies directly with the absolute temperature when the pressure remains constant. In equation form Charles' law may be written as

A capital T is usually used for absolute temperature, K, and a small t for °C.

$$V = kT \qquad \text{or} \qquad \frac{V}{T} = k \quad \text{(at constant pressure)}$$

where k is a constant for a fixed mass of the gas. If the absolute temperature of a gas is doubled, the volume will double.

To illustrate, let's return to the gas cylinder with the movable or free-floating piston (see Figure 12.9). Assume that the cylinder labeled (a) contains a quantity of gas and the pressure on it is 1 atm. When the gas is heated, the molecules move faster, and their kinetic energy increases. This action should increase the number of

FIGURE 12.8 Volume–temperature relationship of methane (CH_4). Extrapolated portion of the graph is shown by the broken line.

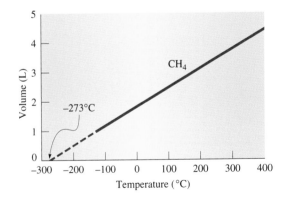

FIGURE 12.9 ▶
The effect of temperature on the volume of a gas. The gas in cylinder (a) is heated from T₁ to T₂. With the external pressure constant at 1 atm, the free-floating piston rises, resulting in an increased volume, shown in cylinder (b).

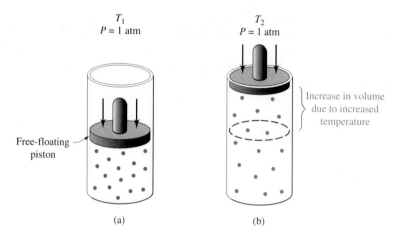

(a)   (b)

collisions per unit of time and therefore increase the pressure. However, the increased internal pressure will cause the piston to rise to a level at which the internal and external pressures again equal 1 atm, as we see in cylinder (b). The net result is an increase in volume due to an increase in temperature.

Another equation relating the volume of a gas at two different temperatures is

$$\frac{V_1}{T_1} = \frac{V_2}{T_2} \text{ (constant } P)$$

where V_1 and T_1 are one set of conditions and V_2 and T_2 are another set of conditions.

A simple experiment showing the variation of the volume of a gas with temperature is illustrated in Figure 12.10. A balloon is placed in a beaker and liquid N_2 is poured over it. The volume is reduced, as shown by the collapse of the balloon; when the balloon is removed from the liquid N_2, the gas expands and the balloon increases in size.

Example 12.5   Three liters of hydrogen at $-20.°C$ are allowed to warm to a room temperature of $27°C$. What is the volume at room temperature if the pressure remains constant?

Solution   **Method A. Conversion Factors**

Step 1. Determine whether temperature is being increased or decreased.

Remember temperature must be changed to Kelvin in gas law problems. Note that we use 273 to convert instead of 273.15 since our original measurements are to the nearest degree.

$$-20.°C + 273 = 253 \text{ K}$$
$$27°C + 273 = 300. \text{ K}$$

temperature increases ⟶ volume increases

Step 2. Multiply the original volume by a ratio of temperatures that will result in an increase in volume.

$$V = (3.00 \text{ L})\left(\frac{300. \text{ K}}{253 \text{ K}}\right) = 3.56 \text{ L} \quad \text{(new volume)}$$

(a)

(b)

(c)

FIGURE 12.10
The air-filled balloons in (a) are placed in liquid nitrogen (b). The volume of the air decreases tremendously at this temperature. In (c) the balloons are removed from the beaker and are beginning to return to their original volume as they warm back to room temperature.

Method B. Algebraic Equation

Step 1. Organize the given information. Remember to make units the same.

$$V_1 = 3.00 \text{ L} \qquad T_1 = 20.°C = 253 \text{ K}$$

$$V_2 = ? \qquad T_2 = 27°C = 300. \text{ K}$$

Step 2. Write and solve the equation for the unknown:

$$\frac{V_1}{T_1} = \frac{V_2}{T_2} \qquad V_2 = \frac{V_1 T_2}{T_1}$$

Step 3. Put the given information into the equation and calculate:

$$V_2 = \frac{V_1 T_2}{T_1} = \frac{(3.00 \text{ L})(300. \text{ K})}{253 \text{ K}} = 3.56 \text{ L}$$

If 20.0 L of oxygen are cooled from 100.°C to 0.°C, what is the new volume?

Since no mention is made of pressure, assume that pressure does not change.

Example 12.6

Solution

Method A. Conversion Factors

Step 1. Change °C to K:

$$100.°C + 273 = 373. \text{ K}$$

$$0.°C + 273 = 273. \text{ K}$$

Step 2. The ratio of temperature to be used is 273 K/373 K, because the final volume should be smaller than the original volume. The calculation is

$$V = (20.0 \text{ L})\left(\frac{273 \text{ K}}{373 \text{ K}}\right) = 14.6 \text{ L} \quad \text{(new volume)}$$

Method B. Algebraic Equation

Step 1. Organize the given information. Remember to make units coincide.

$$V_1 = 20.0 \text{ L} \qquad T_1 = 100.°C = 373. \text{ K}$$

$$V_2 = ? \qquad T_2 = 0.°C = 273. \text{ K}$$

Step 2. Write and solve the equation for the unknown:

$$\frac{V_1}{T_1} = \frac{V_2}{T_2} \qquad V_2 = \frac{V_1 T_2}{T_1}$$

Step 3. Put the given information into the equation and calculate:

$$V_2 = \frac{V_1 T_2}{T_1} = \frac{(20.0 \text{ L})(273 \text{ K})}{373 \text{ K}} = 14.6 \text{ L}$$

Practice 12.3

A 4.50-L container of nitrogen gas at 28.0°C is heated to 56.0°C. Assuming the volume of the container can vary, what is the new volume of the gas?

12.7 Gay-Lussac's Law

J. L. Gay-Lussac (1778–1850) was a French chemist involved in the study of volume relationships of gases. The three variables (pressure, P; volume, V; and temperature, T) are needed to describe a fixed amount of a gas. Boyle's law, $PV = k$, relates pressure and volume at constant temperature; Charles' law, $V = kT$, relates volume and temperature at constant pressure. A third relationship involving pressure and temperature at constant volume is a modification of Charles' law and is some-times called **Gay-Lussac's law:**

Gay-Lussac's law

The pressure of a fixed mass of a gas, at constant volume, is directly proportional to the Kelvin temperature:

$$P = kT \qquad \text{or} \qquad \frac{P_1}{T_1} = \frac{P_2}{T_2}$$

Example 12.7   The pressure of a container of helium is 650. torr at 25°C. If the sealed container is cooled to 0°C, what will the pressure be?

Solution   **Method A. Conversion Factors**

Step 1. Determine whether temperature is being increased or decreased.

temperature decreases ⟶ pressure decreases

Step 2. Multiply the original pressure by a ratio of Kelvin temperatures that will result in a decrease in pressure:

$$(650. \text{ torr})\left(\frac{273 \text{ K}}{298 \text{ K}}\right) = 595 \text{ torr}$$

Method B. Algebraic Equation

Step 1. Organize the given information. Remember to make units the same.

$P_1 = 650.$ torr $T_1 = 25°C = 298$ K

$P_2 = ?$ $T_2 = 0.°C = 273$ K

Step 2. Write and solve equation for the unknown:

$$\frac{P_1}{T_1} = \frac{P_2}{T_2} \qquad P_2 = \frac{P_1 T_2}{T_1}$$

Step 3. Put given information into equation and calculate:

$$P_2 = \frac{(650. \text{ torr})(273 \text{ K})}{298 \text{ K}} = 595 \text{ torr}$$

Practice 12.4

A gas cylinder contains 40.0 L of gas at 45.0°C and has a pressure of 650. torr. What will the pressure be if the temperature is changed to 100.°C?

We may summarize the effects of changes in pressure, temperature, and quantity of a gas as follows:

1. In the case of a constant volume,
   (a) when the temperature is increased, the pressure increases.
   (b) when the quantity of a gas is increased, the pressure increases (T remaining constant).
2. In the case of a variable volume,
   (a) when the external pressure is increased, the volume decreases (T remaining constant).
   (b) when the temperature of a gas is increased, the volume increases (P remaining constant).
   (c) when the quantity of a gas is increased, the volume increases (P and T remaining constant).

12.8 Standard Temperature and Pressure

In order to compare volumes of gases, common reference points of temperature and pressure were selected and called **standard conditions** or **standard temperature and pressure** (abbreviated **STP**). Standard temperature is 273.15 K (0°C), and standard pressure is 1 atm or 760 torr or 760 mm Hg or 101.325 kPa. For purposes of comparison, volumes of gases are usually changed to STP conditions.

standard conditions

standard temperature

and pressure (STP)

In this text we'll use 273 K for temperature conversions and calculations. Check with your instructor for rules in your class.

---

standard temperature = 273.15 K or 0.00°C
standard pressure = 1 atm or 760 torr or 760 mm Hg or 101.325 kPa

---

12.9   Combined Gas Laws

When temperature and pressure change at the same time, the new volume may be calculated by multiplying the initial volume by the correct ratios of both pressure and temperature, as follows:

$$\text{final volume} = (\text{initial volume})\left(\begin{array}{c}\text{ratio of}\\\text{pressures}\end{array}\right)\left(\begin{array}{c}\text{ratio of}\\\text{temperatures}\end{array}\right)$$

This equation combines Boyle's and Charles' laws, and the same considerations for the pressure and temperature ratios should be used in the calculation. The four possible variations are as follows:

1. Both T and P cause an increase in volume.
2. Both T and P cause a decrease in volume.
3. T causes an increase and P causes a decrease in volume.
4. T causes a decrease and P causes an increase in volume.

The P, V, and T relationships for a given mass of any gas, in fact, may be expressed as a single equation, $PV/T = k$. For problem solving, this equation is usually written

Note, in the examples below, the use of 273 K does not change the number of significant figures in the temperature. The converted temperature is expressed to the same precision as the original measurement.

$$\frac{P_1 V_1}{T_1} = \frac{P_2 V_2}{T_2}$$

where P_1, V_1, and T_1 are the initial conditions and P_2, V_2, and T_2 are the final conditions.

This equation can be solved for any one of the six variables and is useful in dealing with the pressure–volume–temperature relationships of gases. Note that when T is constant ($T_1 = T_2$), Boyle's law is represented; when P is constant ($P_1 = P_2$), Charles' law is represented; and when V is constant ($V_1 = V_2$), Gay-Lussac's law is represented.

Example 12.8   Given 20.0 L of ammonia gas at 5°C and 730. torr, calculate the volume at 50.°C and 800. torr.

Solution   **Step 1.** Organize the given information, putting temperatures in Kelvin:

$$P_1 = 730.\text{ torr} \qquad\qquad P_2 = 800.\text{ torr}$$

$$V_1 = 20.0\text{ L} \qquad\qquad V_2 = ?$$

$$T_1 = 5°C = 278\text{ K} \qquad T_2 = 50.°C = 323\text{ K}$$

Method A. Conversion Factors

Step 2. Set up ratios of T and P:

$$T\text{ ratio} = \frac{323\text{ K}}{278\text{ K}} \quad (\text{increase in } T \text{ should increase } V)$$

$$P\text{ ratio} = \frac{730.\text{ torr}}{800.\text{ torr}} \quad (\text{increase in } P \text{ should decrease } V)$$

Step 3. Multiply the original pressure by the ratios:

$$V_2 = (20.0 \text{ L})\left(\frac{730. \text{ torr}}{800. \text{ torr}}\right)\left(\frac{323 \text{ K}}{278 \text{ K}}\right) = 21.2 \text{ L}$$

Method B. Algebraic Equation

Step 2. Write and solve the equation for the unknown. Solve

$$\frac{P_1V_1}{T_1} = \frac{P_2V_2}{T_2}$$

for V_2 by multiplying both sides of the equation by T_2/P_2 and rearranging to obtain

$$V_2 = \frac{V_1P_1T_2}{P_2T_1}$$

Step 3. Put the given information into the equation and calculate:

$$V_2 = \frac{(20.0 \text{ L})(730. \text{ torr})(323 \text{ K})}{(800. \text{ torr}) (278 \text{ K})} = 21.2 \text{ L}$$

To what temperature (°C) must 10.0 L of nitrogen at 25°C and 700. torr be heated in order to have a volume of 15.0 L and a pressure of 760. torr?   **Example 12.9**

Solution

Step 1. Organize the given information, putting temperatures in Kelvin:

$P_1 = 700.$ torr	$P_2 = 760.$ torr
$V_1 = 10.0$ L	$V_2 = 15.0$ L
$T_1 = 25°C = 298$ K	$T_2 = ?$

Method A. Conversion Factors

Step 2. Set up ratios of V and P.

$$P \text{ ratio} = \frac{760. \text{ torr}}{700. \text{ torr}} \quad (\text{increase in } P \text{ should increase } T)$$

$$V \text{ ratio} = \frac{15.0 \text{ L}}{10.0 \text{ L}} \quad (\text{increase in } V \text{ should increase } T)$$

Step 3. Multiply the original temperature by the ratios:

$$T_2 = (298 \text{ K})\left(\frac{760. \text{ torr}}{700. \text{ torr}}\right)\left(\frac{15.0 \text{ L}}{10.0 \text{ L}}\right) = 485 \text{ K}$$

Method B. Algebraic Equation

Step 2. Write and solve the equation for the unknown:

$$\frac{P_1V_1}{T_1} = \frac{P_2V_2}{T_2} \qquad T_2 = \frac{T_1P_2V_2}{P_1V_1}$$

Step 3. Put the given information into the equation and calculate:

$$T_2 = \frac{(298 \text{ K})(760. \text{ torr})(15.0 \text{ L})}{(700. \text{ torr})(10.0 \text{ L})} = 485 \text{ K}$$

In either method, since the problem asks for °C, we must subtract 273 from the Kelvin answer:

$$485 \text{ K} - 273 = 212°\text{C}$$

Example 12.10 The volume of a gas-filled balloon is 50.0 L at 20.°C and 742 torr. What volume will it occupy at standard temperature and pressure (STP)?

Solution

Step 1. Organize the given information, putting temperatures in Kelvin.

$P_1 = 742$ torr	$P_2 = 760.$ torr (standard pressure)
$V_1 = 50.0$ L	$V_2 = ?$
$T_1 = 20.°\text{C} = 293$ K	$T_2 = 273$ K (standard temperature)

Method A. Conversion Factors

Step 2. Set up ratios of T and P:

$$T \text{ ratio} = \frac{273 \text{ K}}{293 \text{ K}} \text{ (decrease in } T \text{ should decrease } V)$$

$$P \text{ ratio} = \frac{742 \text{ torr}}{760. \text{ torr}} \text{ (increase in } P \text{ should decrease } V)$$

Step 3. Multiply the original volume by the ratios:

$$V_2 = (50.0 \text{ L})\left(\frac{273 \text{ K}}{293 \text{ K}}\right)\left(\frac{742 \text{ torr}}{760. \text{ torr}}\right) = 45.5 \text{ L}$$

Method B. Algebraic Equation

Step 2. Write and solve the equation for the unknown:

$$\frac{P_1 V_1}{T_1} = \frac{P_2 V_2}{T_2} \qquad V_2 = \frac{P_1 V_1 T_2}{P_2 T_1}$$

Step 3. Put the given information into the equation and calculate:

$$V_2 = \frac{(742 \text{ torr})(50.0 \text{ L})(273 \text{ K})}{(760. \text{ torr})(293 \text{ K})} = 45.5 \text{ L}$$

Practice 12.5

15.00 L of gas at 45.0°C and 800. torr is heated to 400.°C, and the pressure changed to 300. torr. What is the new volume?

Practice 12.6

To what temperature must 5.00 L of oxygen at 50.°C and 600. torr be heated in order to have a volume of 10.0 L and a pressure of 800. torr?

CHEMISTRY IN ACTION Messenger Molecules

Traditional "messenger" molecules are amino acids (also known as the building blocks for proteins). You've probably heard of endorphins, which are the messenger molecules associated with "runner's high." Until recently these molecules (called *neurotransmitters*) were thought to be specific; that is, each neurotransmitter fits into the target cell like a key in a lock. It was also thought that neurotransmitters were stored in tiny pouches where they are manufactured and then released when needed. But then two gases, nitrogen monoxide (NO) and carbon monoxide (CO), were found to act as neurotransmitters. These gases break all the "rules" for neurotransmission because as gases they are nonspecific. Gases freely diffuse into nearby cells, so NO and CO neurotransmitters must be made "on demand" since they cannot be stored. Gas neurotransmitters cannot use the lock-and-key model to act on cells, so they must use their chemical properties.

Nitrogen monoxide has been used for nearly a century to dilate blood vessels and increase blood flow, lowering blood pressure. In the late 1980s, scientists discovered that biologically produced NO is an important signaling

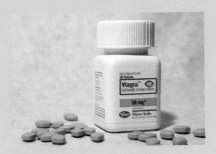

▲ **Viagra, the newest medication for male impotence, is the latest application of our understanding of the role of NO in the body.**

molecule for nerve cells. Nitrogen monoxide mediates certain neurons that do not respond to traditional neurotransmitters. These NO-sensitive neurons are found in the cardiovascular, respiratory, digestive, and urogenital systems. Nitrogen monoxide also appears to play a role in regulating blood pressure, blood clotting, and neurotransmission. The 1998 Nobel prize in medicine was awarded to three Americans for their work on nitrogen monoxide's role in the body.

Nitrogen monoxide is a key molecule in producing erections in men.

When a man is sexually stimulated, NO is released into the penis where it activates the release of an enzyme that increases the level of a molecule called cGMP and ultimately produces relaxation of smooth muscles, allowing blood to flow into the penis. Viagra, the new impotence pill, enhances the effect of NO, inhibiting the enzyme that breaks down the cGMP molecule. This results in more cGMP molecules, and smooth muscle relaxation begins producing an erection.

After NO transmitters were discovered, researchers at Johns Hopkins Medical School reasoned that if one gas acted as a neurotransmitter so might others. They proposed CO as a possible transmitter because the enzyme used to make it is localized in specific parts of the brain (those responsible for smell and long-term memory). The enzyme used to make messengers was found in exactly the same locations! Researchers then showed that nerve cells made the messenger molecule when stimulated with CO. An inhibitor for CO blocked the messenger molecule production. Researchers are exploring the possibility of still other roles for these gaseous messenger molecules.

12.10 Dalton's Law of Partial Pressures

If gases behave according to the kinetic-molecular theory, there should be no difference in the pressure–volume–temperature relationships whether the gas molecules are all the same or different. This similarity in the behavior of gases is the basis for an understanding of **Dalton's law of partial pressures:**

Dalton's law of partial pressures

> **The total pressure of a mixture of gases is the sum of the partial pressures exerted by each of the gases in the mixture.**

Each gas in the mixture exerts a pressure that is independent of the other gases present. These pressures are called **partial pressures.** Thus if we have a mixture of

partial pressure

277

FIGURE 12.11 ▶
Oxygen collected over water.

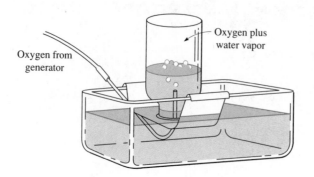

three gases, *A*, *B*, and *C*, exerting partial pressures of 50. torr, 150. torr, and 400. torr, respectively, the total pressure will be 600. torr:

$$P_{Total} = P_A + P_B + P_C$$

$$P_{Total} = 50.\ torr + 150.\ torr + 400.\ torr = 600.\ torr$$

We can see an application of Dalton's law in the collection of insoluble gases over water. When prepared in the laboratory, oxygen is commonly collected by the downward displacement of water. Thus the oxygen is not pure but is mixed with water vapor (see Figure 12.11). When the water levels are adjusted to the same height inside and outside the bottle, the pressure of the oxygen plus water vapor inside the bottle is equal to the atmospheric pressure:

$$P_{atm} = P_{O_2} + P_{H_2O}$$

To determine the amount of O_2 or any other gas collected over water, we subtract the pressure of the water vapor from the total pressure of the gas. The vapor pressure of water at various temperatures is tabulated in Appendix III.

$$P_{O_2} = P_{atm} - P_{H_2O}$$

Example 12.11 A 500.-mL sample of oxygen was collected over water at 23°C and 760. torr. What volume will the dry O_2 occupy at 23°C and 760. torr? The vapor pressure of water at 23°C is 21.2 torr.

Solution To solve this problem, we must first determine the pressure of the oxygen alone, by subtracting the pressure of the water vapor present.

Step 1. Determine the pressure of dry O_2:

$$P_{Total} = 760.\ torr = P_{O_2} + P_{H_2O}$$

$$P_{O_2} = 760.\ torr - 21.2\ torr = 739\ torr \qquad (dry\ O_2)$$

Step 2. Organize the given information:

$$P_1 = 739\ torr \qquad\qquad P_2 = 760.\ torr$$

$$V_1 = 500.\ mL \qquad\qquad V_2 = ?$$

$$T\ is\ constant$$

Step 3. Solve as a Boyle's law problem:

$$V = \frac{(500.\ \text{mL})(739\ \text{torr})}{760.\ \text{torr}} = 486\ \text{mL dry O}_2$$

Practice 12.7

Hydrogen gas was collected by downward displacement of water. A volume of 600.0 mL of gas was collected at 25.0°C and 740.0 torr. What volume will the dry hydrogen occupy at STP?

12.11 Avogadro's Law

Early in the 19th century, Gay-Lussac studied the volume relationships of reacting gases. His results, published in 1809, were summarized in a statement known as **Gay-Lussac's law of combining volumes:**

Gay-Lussac's law of combining volumes

> **When measured at the same temperature and pressure, the ratios of the volumes of reacting gases are small whole numbers.**

Thus H_2 and O_2 combine to form water vapor in a volume ratio of 2:1 (Figure 12.12); H_2 and Cl_2 react to form HCl in a volume ratio of 1:1; and H_2 and N_2 react to form NH_3 in a volume ratio of 3:1.

Two years later, in 1811, Amedeo Avogadro (1776–1856) used the law of combining volumes of gases to make a simple but significant and far-reaching generalization concerning gases. **Avogadro's law** states:

Avogadro's law

> **Equal volumes of different gases at the same temperature and pressure contain the same number of molecules.**

This law was a real breakthrough in understanding the nature of gases.

1. It offered a rational explanation of Gay-Lussac's law of combining volumes of gases and indicated the diatomic nature of such elemental gases as hydrogen, chlorine, and oxygen.

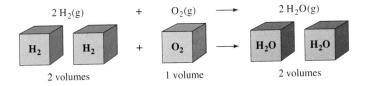

FIGURE 12.12
Gay-Lussac's law of combining volumes of gases applied to the reaction of hydrogen and oxygen. When measured at the same temperature and pressure, hydrogen and oxygen react in a volume ◀ ratio of 2:1.

2. It provided a method for determining the molar masses of gases and for comparing the densities of gases of known molar mass (see Sections 12.12 and 12.13).

3. It afforded a firm foundation for the development of the kinetic-molecular theory.

By Avogadro's law, equal volumes of hydrogen and chlorine at the same temperature and pressure contain the same number of molecules. On a volume basis, hydrogen and chlorine react thus:

hydrogen + chlorine ⟶ hydrogen chloride

| 1 volume | 1 volume | | 2 volumes |

Therefore, hydrogen molecules react with chlorine molecules in a 1:1 ratio. Since two volumes of hydrogen chloride are produced, one molecule of hydrogen and one molecule of chlorine must produce two molecules of hydrogen chloride. Therefore, each hydrogen molecule and each chlorine molecule must be made up of two atoms. The coefficients of the balanced equation for the reaction give the correct ratios for volumes, molecules, and moles of reactants and products:

$$H_2 \quad + \quad Cl_2 \quad \longrightarrow \quad 2\ HCl$$

1 volume	1 volume	2 volumes
1 molecule	1 molecule	2 molecules
1 mol	1 mol	2 mol

By like reasoning, oxygen molecules also must contain at least two atoms because one volume of oxygen reacts with two volumes of hydrogen to produce two volumes of water vapor.

The volume of a gas depends on the temperature, the pressure, and the number of gas molecules. Different gases at the same temperature have the same average kinetic energy. Hence, if two different gases are at the same temperature, occupy equal volumes, and exhibit equal pressures, each gas must contain the same number of molecules. This statement is true because systems with identical *PVT* properties can be produced only by equal numbers of molecules having the same average kinetic energy.

12.12 Mole–Mass–Volume Relationships of Gases

molar volume

As with many constants, the molar volume is known more exactly to be 22.414 L. We use 22.4 L in our calculations since the extra figures don't often affect the result, given the other measurements in the calculation.

Because a mole contains 6.022×10^{23} molecules (Avogadro's number), a mole of any gas will have the same volume as a mole of any other gas at the same temperature and pressure. It has been experimentally determined that the volume occupied by a mole of any gas is 22.4 L at STP. This volume, 22.4 L, is known as the **molar volume** of a gas. The molar volume is a cube about 28.2 cm (11.1 in.) on a side. The molar masses of several gases, each occupying 22.4 L at STP, are shown in Figure 12.13.

One mole of a gas occupies 22.4 L at STP.

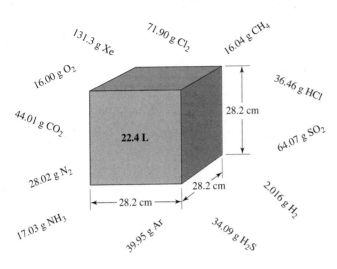

▲ **FIGURE 12.13**
One mole of a gas occupies 22.4 L at STP. The mass given for each gas is the mass of 1 mol.

The molar volume is useful for determining the molar mass of a gas or of substances that can be easily vaporized. If the mass and the volume of a gas at STP are known, we can calculate its molar mass. For example, 1 L of pure oxygen at STP has a mass of 1.429 g. The molar mass of oxygen may be calculated by multiplying the mass of 1 L by 22.4 L/mol:

$$\left(\frac{1.429 \text{ g}}{1 \text{ L}}\right)\left(\frac{22.4 \text{ L}}{1 \text{ mol}}\right) = 32.0 \text{ g/mol} \qquad \text{(molar mass)}$$

If the mass and volume are at other than standard conditions, we change the volume to STP and then calculate the molar mass.

The molar volume, 22.4 L/mol, is used as a conversion factor to convert grams per liter to grams per mole (molar mass) and also to convert liters to moles. The two conversion factors are

$$\frac{22.4 \text{ L}}{1 \text{ mol}} \qquad \text{and} \qquad \frac{1 \text{ mol}}{22.4 \text{ L}}$$

These conversions must be done at STP except under certain special circumstances. Examples follow.

Standard conditions apply only to pressure, temperature, and volume. Mass is not affected.

If 2.00 L of a gas measured at STP has a mass of 3.23 g, what is the molar mass of the gas?

The unit of molar mass is g/mol; the conversion is from

$$\frac{\text{g}}{\text{L}} \longrightarrow \frac{\text{g}}{\text{mol}}$$

The starting amount is $\dfrac{3.23 \text{ g}}{2.00 \text{ L}}$. The conversion factor is $\dfrac{22.4 \text{ L}}{1 \text{ mol}}$.

The calculation is $\left(\dfrac{3.23 \text{ g}}{2.00 \text{ L}}\right)\left(\dfrac{22.4 \text{ L}}{1 \text{ mol}}\right) = 36.2 \text{ g/mol}$ (molar mass)

Example 12.12

Solution

Example 12.13 Measured at 40°C and 630. torr, the mass of 691 mL of ethyl ether is 1.65 g. Calculate the molar mass of ethyl ether.

Solution

Step 1. Organize the given information, converting temperatures to Kelvin. Note that we must change to STP in order to determine molar mass.

$$P_1 = 630. \text{ torr} \qquad\qquad P_2 = 760. \text{ torr}$$

$$V_1 = 691 \text{ mL} \qquad\qquad V_2 = ?$$

$$T_1 = 313 \text{ K } (40.°C) \qquad T_2 = 273 \text{ K}$$

Step 2. Use either the conversion factor method or the algebraic method and the combined gas law to correct the volume (V_2) to STP:

$$V_2 = \frac{(691 \text{ mL})(273 \text{ K})(630. \text{ torr})}{(313 \text{ K})(760. \text{ torr})} = 500. \text{ mL} = 0.500 \text{ L} \quad \text{(at STP)}$$

Step 3. In the example, V_2 is the volume for 1.65 g of the gas, so we can now find the molar mass by converting g/L to g/mol:

$$\left(\frac{1.65 \text{ g}}{0.500 \text{ L}}\right)\left(\frac{22.4 \text{ L}}{\text{mol}}\right) = 73.9 \text{ g/mol}$$

Practice 12.8

A gas with a mass of 86 g occupies 5.00 L at 25°C and 3.00 atm pressure. What is the molar mass of the gas?

12.13 Density of Gases

The density, d, of a gas is its mass per unit volume, which is generally expressed in grams per liter as follows:

$$d = \frac{\text{mass}}{\text{volume}} = \frac{\text{g}}{\text{L}}$$

Because the volume of a gas depends on temperature and pressure, both should be given when stating the density of a gas. The volume of a solid or liquid is hardly affected by changes in pressure and is changed only slightly when the temperature is varied. Increasing the temperature from 0°C to 50°C will reduce the density of a gas by about 18% if the gas is allowed to expand, whereas a 50°C rise in the temperature of water (0°C ⟶ 50°C) will change its density by less than 0.2%.

The density of a gas at any temperature and pressure can be determined by calculating the mass of gas present in 1 L. At STP, in particular,

TABLE 12.3	Density of Common Gases at STP					
Gas	Molar mass (g/mol)	Density (g/L at STP)		Gas	Molar mass (g/mol)	Density (g/L at STP)
H_2	2.016	0.0900		H_2S	34.09	1.52
CH_4	16.04	0.716		HCl	36.46	1.63
NH_3	17.03	0.760		F_2	38.00	1.70
C_2H_2	26.04	1.16		CO_2	44.01	1.96
HCN	27.03	1.21		C_3H_8	44.09	1.97
CO	28.01	1.25		O_3	48.00	2.14
N_2	28.02	1.25		SO_2	64.07	2.86
air	(28.9)	(1.29)		Cl_2	70.90	3.17
O_2	32.00	1.43				

the density can be calculated by multiplying the molar mass of the gas by 1 mol/22.4 L:

$$d_{STP} = \text{molar mass} \left(\frac{1 \text{ mol}}{22.4 \text{ L}} \right)$$

$$\text{molar mass} = d_{STP} \left(\frac{22.4 \text{ L}}{1 \text{ mol}} \right)$$

Table 12.3 lists the densities of some common gases.

Example 12.14

Solution

Calculate the density of Cl_2 at STP.

First calculate the molar mass of Cl_2. It is 70.90 g/mol. Since $d = $ g/L, the conversion is

$$\frac{g}{mol} \longrightarrow \frac{g}{L}$$

The conversion factor is $\dfrac{1 \text{ mol}}{22.4 \text{ L}}$:

$$d = \left(\frac{70.90 \text{ g}}{1 \text{ mol}} \right) \left(\frac{1 \text{ mol}}{22.4 \text{ L}} \right) = 3.17 \text{ g/L}$$

Practice 12.9

The molar mass of a gas is 20. g/mol. Calculate the density of the gas at STP.

Physiological Effects of Pressure Changes

The human body has a variety of methods for coping with the changes in atmospheric pressure. As we travel to the mountains, fly in an airplane, or take a high-speed elevator to the top of a skyscraper, the pressure around us decreases. Our ears are sensitive to this because the eardrum (tympanic membrane) has air on both sides of it. The difference in pressure is relieved by yawning or moving the jaw to open the (Eustachian) tubes that connect the middle ear and throat and allow the pressure inside the eardrum to equalize with the outside.

Divers must also contend with the effects of pressure, most notably in body cavities containing air, such as the lungs, ears, and sinuses. Scuba divers don't experience a crushing effect of pressure at increased depths because the tank regulators deliver air at the same pressure as that of the surroundings. The diver must always breathe out regularly while ascending to the surface. Failure to do so can cause the lungs to expand, thus rupturing some of the alveoli, and resulting in loss of consciousness, brain damage, or heart attack. This is a clear application of Boyle's law.

Divers are also affected by consequences of **Henry's law,** which states that the amount of gas that will dissolve

▲ **If a diver returns too quickly to the surface, the pressure reduction may produce bubbles in the blood— a condition known as "the bends."**

in a liquid varies directly with the pressure above the liquid. This means that during a dive the gases entering the lungs are absorbed into the blood to a greater extent than at the water's surface. If the diver returns too rapidly to the surface, the swift pressure reduction can cause dissolved gases to produce bubbles in the blood, resulting in a condition known as *decompression sickness* or "the bends." The only successful method of treatment for this involves the use of a decompression chamber to increase the pressure once again and slowly decompress the diver back to normal pressure.

In the medical field, *hyperbaric units* are used to treat patients who have cells starved for oxygen. In these units the whole room may be placed at high pressure (2 or 3 atm), and the entire staff as well as the patient undergo gradual compression and, following treatment, decompression. These units are widely used to treat carbon monoxide poisoning. Oxygen is dissolved directly into the plasma giving the tissues temporary relief from oxygen deprivation. Hyperbaric units are also effective in treating other problems such as skin grafts, severe thermal burns, and radiation tissue damage.

12.14 Ideal Gas Equation

We've used four variables in calculations involving gases: the volume, V; the pressure, P; the absolute temperature, T; and the number of molecules or moles, (abbreviated n). Combining these variables into a single expression, we obtain

$$V \propto \frac{nT}{P} \quad \text{or} \quad V = \frac{nRT}{P}$$

where R is a proportionality constant known as the *ideal gas constant.* The equation is commonly written as

$$PV = nRT$$

and is known as the **ideal gas equation.** This equation states in a single expression   *ideal gas equation*
what we have considered in our earlier discussions: The volume of a gas varies di-
rectly with the number of gas molecules and the absolute temperature, and varies inversely with the pressure. The value and units of R depend on the units of P, V, and T.
We can calculate one value of R by taking 1 mol of a gas at STP conditions. Solve the
equation for R:

$$R = \frac{PV}{nT} = \frac{(1 \text{ atm})(22.4 \text{ L})}{(1 \text{ mol})(273 \text{ K})} = 0.0821 \frac{\text{L-atm}}{\text{mol-K}}$$

The units of R in this case are liter-atmospheres (L-atm) per mole Kelvin (mol-K).
When the value of $R = 0.0821$ L-atm/mol-K, P is in atmospheres, n is in moles, V is
in liters, and T is in Kelvin.

The ideal gas equation can be used to calculate any one of the four variables
when the other three are known.

What pressure will be exerted by 0.400 mol of a gas in a 5.00-L container   **Example 12.15**
at 17°C?

Step 1. Organize the given information, converting temperatures to Kelvin: **Solution**

$P = ?$

$V = 5.00$ L

$T = 290.$ K

$n = 0.400$ mol

Step 2. Write and solve the ideal gas equation for the unknown:

$$PV = nRT \qquad \text{or} \qquad P = \frac{nRT}{V}$$

Step 3. Substitute the given information into the equation and calculate:

$$P = \frac{(0.400 \text{ mol})(0.0821 \text{ L·atm/mol·K})(290. \text{ K})}{5.00 \text{ L}} = 1.90 \text{ atm}$$

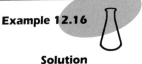

How many moles of oxygen gas are in a 50.0-L tank at 22.0°C if the pressure   **Example 12.16**
gauge reads 2000. lb/in.^{2}?

Step 1. Organize the given information, converting temperature to kelvins and   **Solution**
pressure to atmospheres:

$$P = \left(\frac{2000. \text{ lb}}{\text{in.}^2}\right)\left(\frac{1 \text{ atm}}{14.7 \text{ lb/in.}^2}\right) = 136.1 \text{ atm}$$

$V = 50.0$ L

$T = 295$ K

$n = ?$

Step 2. Write and solve the ideal gas equation for the unknown:

$$PV = nRT \qquad \text{or} \qquad n = \frac{PV}{RT}$$

Step 3. Substitute the given information into the equation and calculate:

$$n = \frac{(136.1 \text{ atm})(50.0 \text{ L})}{(0.0821 \text{ L·atm/mol·K})(295 \text{ K})} = 281 \text{ mol } O_2$$

Practice 12.10

A 23.8-L cylinder contains oxygen gas at 20.0°C and 732 torr. How many moles of oxygen are in the cylinder?

The molar mass of a gaseous substance can be determined using the ideal gas equation. Since molar mass = g/mol, then mol = g/molar mass. Using M for molar mass and g for grams, we can substitute g/M for n (moles) in the ideal gas equation to get

This form of the ideal gas equation is most useful in problems containing mass instead of moles.

$$PV = \frac{g}{M}RT \qquad \text{or} \qquad M = \frac{gRT}{PV} \quad \text{(modified ideal gas equation)}$$

which allows us to calculate the molar mass, M, for any substance in the gaseous state.

Example 12.17 Calculate the molar mass of butane gas, if 3.69 g occupies 1.53 L at 20.0°C and 1.00 atm.

Solution Change 20°C to 293 K and substitute the data into the modified ideal gas equation:

$$M = \frac{gRT}{PV} = \frac{(3.69 \text{ g})(0.0821 \text{ L·atm/mol·K})(293 \text{ K})}{(1.00 \text{ atm})(1.53 \text{ L})} = 58.0 \text{ g/mol}$$

Practice 12.11

A sample of 0.286 g of a certain gas occupies 50.0 mL at standard temperature and 76.0 cm Hg. Determine the molar mass of the gas.

12.15   Gas Stoichiometry

Mole–Volume and Mass–Volume Calculations

Stoichiometric problems involving gas volumes can be solved by the general mole-ratio method outlined in Chapter 9. The factors 1 mol/22.4 L and 22.4 L/1 mol are

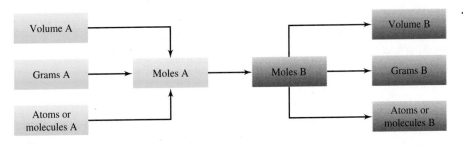

used for converting volume to moles and moles to volume, respectively. (See Figure 12.14.) These conversion factors are used under the assumption that the gases are at STP and that they behave as ideal gases. In actual practice, gases are measured at other than STP conditions, and the volumes are converted to STP for stoichiometric calculations.

In a balanced equation, the number preceding the formula of a gaseous substance represents the number of moles or molar volumes (22.4 L at STP) of that substance.

The following are examples of typical problems involving gases and chemical equations.

What volume of oxygen (at STP) can be formed from 0.500 mol of potassium **Example 12.18** chlorate?

Solution

Step 1. Write the balanced equation:

$$2 \text{ KClO}_3 \longrightarrow 2 \text{ KCl} + 3 \text{ O}_2(g)$$

Step 2. The starting amount is 0.500 mol KClO_3. The conversion is from

$$\text{moles KClO}_3 \longrightarrow \text{moles O}_2 \longrightarrow \text{liters O}_2$$

Step 3. Calculate the moles of O_2, using the mole-ratio method:

$$(0.500 \text{ mol KClO}_3)\left(\frac{3 \text{ mol O}_2}{2 \text{ mol KClO}_3}\right) = 0.750 \text{ mol O}_2$$

Step 4. Convert moles of O_2 to liters of O_2. The moles of a gas at STP are converted to liters by multiplying by the molar volume, 22.4 L/mol:

$$(0.750 \text{ mol O}_2)\left(\frac{22.4 \text{ L}}{1 \text{ mol}}\right) = 16.8 \text{ L O}_2$$

Setting up a continuous calculation, we obtain

$$(0.500 \text{ mol KClO}_3)\left(\frac{3 \text{ mol O}_2}{2 \text{ mol KClO}_3}\right)\left(\frac{22.4 \text{ L}}{1 \text{ mol}}\right) = 16.8 \text{ L O}_2$$

Example 12.19 How many grams of aluminum must react with sulfuric acid to produce 1.25 L of hydrogen gas at STP?

Solution

Step 1. The balanced equation is

$$2 \, Al(s) + 3 \, H_2SO_4(aq) \longrightarrow Al_2(SO_4)_3(aq) + 3 \, H_2(g)$$

Step 2. We first convert liters of H_2 to moles of H_2. Then the familiar stoichiometric calculation from the equation is used. The conversion is

$$L \, H_2 \longrightarrow mol \, H_2 \longrightarrow mol \, Al \longrightarrow g \, Al$$

$$1.25 \, \cancel{L \, H_2} \left(\frac{1 \, \cancel{mol}}{22.4 \, \cancel{L}} \right) \left(\frac{2 \, \cancel{mol \, Al}}{3 \, \cancel{mol \, H_2}} \right) \left(\frac{26.98 \, g \, Al}{1 \, \cancel{mol \, Al}} \right) = 1.00 \, g \, Al$$

Example 12.20 What volume of hydrogen, collected at 30.°C and 700. torr, will be formed by reacting 50.0 g of aluminum with hydrochloric acid?

$$2 \, Al(s) + 6 \, HCl(aq) \longrightarrow 2 \, AlCl_3(aq) + 3 \, H_2(g)$$

Solution

In this problem the conditions are not at STP, so we cannot use the method shown in Example 12.18. Either we need to calculate the volume at STP from the equation and then convert this volume to the conditions given in the problem, or we can use the ideal gas equation. Let's use the ideal gas equation.

First calculate the moles of H_2 obtained from 50.0 g of Al. Then, using the ideal gas equation, calculate the volume of H_2 at the conditions given in the problem.

Step 1. Moles of H_2: The conversion is

$$grams \, Al \longrightarrow moles \, Al \longrightarrow moles \, H_2$$

$$50.0 \, \cancel{g \, Al} \left(\frac{1 \, \cancel{mol \, Al}}{26.98 \, \cancel{g \, Al}} \right) \left(\frac{3 \, mol \, H_2}{2 \, \cancel{mol \, Al}} \right) = 2.78 \, mol \, H_2$$

Step 2. Liters of H_2: Solve $PV = nRT$ for V and substitute the data into the equation.

Convert °C to K: 30.°C + 273 = 303 K.
Convert torr to atm: (700. $\cancel{torr}$)(1 atm/760. $\cancel{torr}$) = 0.921 atm.

$$V = \frac{nRT}{P} = \frac{(2.78 \, \cancel{mol \, H_2})(0.0821 \, L\text{-}atm)(303 \, \cancel{K})}{(0.921 \, \cancel{atm})(\cancel{mol\text{-}K})} = 75.1 \, L \, H_2$$

Note: The volume at STP is 62.3 L H_2.

Practice 12.12

If 10.0 g of sodium peroxide, Na_2O_2, react with water to produce sodium hydroxide and oxygen, how many liters of oxygen will be produced at 20°C and 750. torr?

$$2 \, Na_2O_2(s) + 2 \, H_2O(l) \longrightarrow 4 \, NaOH(aq) + O_2(g)$$

Volume–Volume Calculations

When all substances in a reaction are in the gaseous state, simplifications in the calculation can be made. These are based on Avogadro's law, which states that gases under identical conditions of temperature and pressure contain the same number of molecules and occupy the same volume. Using this same law, we can also state that, under the standard conditions of temperature and pressure, the volumes of gases reacting are proportional to the numbers of moles of the gases in the balanced equation. Consider the reaction:

$$H_2(g) \ + \ Cl_2(g) \ \longrightarrow \ 2 \ HCl(g)$$

1 mol	1 mol	2 mol
22.4 L	22.4 L	2 × 22.4 L
1 volume	1 volume	2 volumes
Y volume	Y volume	2 Y volumes

In this reaction 22.4 L of hydrogen will react with 22.4 L of chlorine to give 2 (22.4) = 44.8 L of hydrogen chloride gas. This statement is true because these volumes are equivalent to the number of reacting moles in the equation. Therefore, Y volume of H_2 will combine with Y volume of Cl_2 to give 2 Y volumes of HCl. For example, 100 L of H_2 react with 100 L of Cl_2 to give 200 L of HCl; if the 100 L of H_2 and of Cl_2 are at 50°C, they will give 200 L of HCl at 50°C. When the temperature and pressure before and after a reaction are the same, volumes can be calculated without changing the volumes to STP.

> **For reacting gases at constant temperature and pressure:**
> **Volume–volume relationships are the same as mole–mole relationships.**

What volume of oxygen will react with 150. L of hydrogen to form water vapor? **Example 12.21** What volume of water vapor will be formed?

Assume that both reactants and products are measured at standard conditions. Calculate by using reacting volumes: **Solution**

$$2 \ H_2(g) \ + \ O_2(g) \ \longrightarrow \ 2 \ H_2O(g)$$

2 mol	1 mol	2 mol
2 × 22.4 L	22.4 L	2 × 22.4 L
2 volumes	1 volume	2 volumes
150. L	75 L	150. L

For every two volumes of H_2 that react, one volume of O_2 reacts and two volumes of $H_2O(g)$ are produced:

$$(150. \ L \ H_2) \left(\frac{1 \ volume \ O_2}{2 \ volumes \ H_2} \right) = 75 \ L \ O_2$$

$$(150. \ L \ H_2) \left(\frac{2 \ volumes \ H_2O}{2 \ volumes \ H_2} \right) = 150. \ L \ H_2O$$

Example 12.22 The equation for the preparation of ammonia is

$$3H_2(g) + N_2(g) \xrightarrow{400°C} 2 NH_3(g)$$

Assuming that the reaction goes to completion,

(a) what volume of H_2 will react with 50.0 L of N_2?
(b) what volume of NH_3 will be formed from 50.0 L of N_2?
(c) what volume of N_2 will react with 100. mL of H_2?
(d) what volume of NH_3 will be produced from 100. mL of H_2?
(e) if 600. mL of H_2 and 400. mL of N_2 are sealed in a flask and allowed to react, what amounts of H_2, N_2, and NH_3 are in the flask at the end of the reaction?

Solution   The answers to parts (a)–(d) are shown in the boxes and can be determined from the equation by inspection, using the principle of reacting volumes:

$$3 H_2(g) + N_2(g) \longrightarrow 2 NH_3(g)$$
   3 volumes   1 volume   2 volumes

(a) $\boxed{150.\ L}$ 50.0 L
(b)   50.0 L $\boxed{100.\ L}$
(c) 100. mL $\boxed{33.3\ mL}$
(d) 100. mL $\boxed{66.7\ mL}$

(e) Volume ratio from the equation $= \dfrac{3\ \text{volumes } H_2}{1\ \text{volume } N_2}$

   Volume ratio used $= \dfrac{600.\ \text{mL } H_2}{400.\ \text{mL } N_2} = \dfrac{3\ \text{volumes } H_2}{2\ \text{volumes } N_2}$

Comparing these two ratios, we see that an excess of N_2 is present in the gas mixture. Therefore, the reactant limiting the amount of NH_3 that can be formed is H_2:

$$3 H_2(g) + N_2(g) \longrightarrow 2 NH_3(g)$$
   600 mL   200 mL   400 mL

To have a $3:1$ ratio of volumes reacting, 600. mL of H_2 will react with 200 mL of N_2 to produce 400. mL of NH_3, leaving 200. mL of N_2 unreacted. At the end of the reaction the flask will contain 400. mL of NH_3 and 200. mL of N_2.

Practice 12.13

What volume of oxygen will react with 15.0 L of propane (C_3H_8) to form carbon dioxide and water? What volume of carbon dioxide will be formed? What volume of water vapor will be formed?

$$C_3H_8(g) + 5 O_2(g) \longrightarrow 3 CO_2(g) + 4 H_2O(g)$$

12.16 Real Gases

All the gas laws are based on the behavior of an ideal gas—that is, a gas with a behavior that is described exactly by the gas laws for all possible values of P, V, and T. Most real gases actually do behave very nearly as predicted by the gas laws over a fairly wide range of temperatures and pressures. However, when conditions are such that the gas molecules are crowded closely together (high pressure and/or low temperature), they show marked deviations from ideal behavior. Deviations occur because molecules have finite volumes and also have intermolecular attractions, which result in less compressibility at high pressures and greater compressibility at low temperatures than predicted by the gas laws. Many gases become liquids at high pressure and low temperature.

12.17 Air Pollution

Chemical reactions occur among the gases that are emitted into our atmosphere. In recent years, there has been growing concern over the effects these reactions have on our environment and our lives.

The outer portion (stratosphere) of the atmosphere plays a significant role in determining the conditions for life at the surface of the Earth. This stratosphere protects the surface from the intense radiation and particles bombarding our planet. Some of the high energy radiation from the sun acts upon oxygen molecules in the stratosphere, converting them into ozone, O_3. Different molecular forms of an element are called **allotropes** of that element. Thus oxygen and ozone are allotropic forms of oxygen:

allotrope

$$O_2 \xrightarrow{\text{sunlight}} O + O$$
$$\text{oxygen atoms}$$

$$O_2 + O \longrightarrow O_3$$
$$\text{ozone}$$

Ultraviolet radiation from the sun is highly damaging to living tissues of plants and animals. The ozone layer, however, shields the Earth by absorbing ultraviolet radiation and thus prevents most of this lethal radiation from reaching the Earth's surface. The reaction that occurs is the reverse of the preceding one:

$$O_3 \xrightarrow[\text{radiation}]{\text{ultraviolet}} O_2 + O + \text{heat}$$

Scientists have become concerned about a growing hazard to the ozone layer. Chlorofluorocarbon propellants, such as the Freons, CCl_3F and CCl_2F_2, which were used in aerosol spray cans and are used in refrigeration and air-conditioning units, are stable compounds and remain unchanged in the lower atmosphere. But when these chlorofluorocarbons are carried by convection currents to the stratosphere, they absorb ultraviolet radiation and produce chlorine atoms (chlorine free radicals), which in turn react with ozone. The following reaction sequence involving free radicals has been proposed to explain the partial destruction of the ozone layer by chlorofluorocarbons.

A free radical is a species containing an odd number of electrons. Free radicals are highly reactive.

$$CCl_3F \xrightarrow[\text{radiation}]{\text{ultraviolet}} \cdot CCl_2F + Cl\cdot \tag{1}$$

fluorocarbon   fluorocarbon   chlorine free
molecule   free radical   radical (atom)

$$Cl\cdot + O_3 \longrightarrow ClO\cdot + O_2 \tag{2}$$

$$ClO\cdot + O \longrightarrow O_2 + Cl\cdot \tag{3}$$

Because a chlorine atom is generated for each ozone molecule that is destroyed (reactions 2 and 3 can proceed repeatedly), a single chlorofluorocarbon molecule can be responsible for the destruction of many ozone molecules. During the past decade, scientists have discovered an annual thinning in the ozone layer over Antarctica. This is what we call the "hole" in the ozone layer. If this hole were to occur over populated regions of the world, severe effects would result, including a rise in the cancer rate, increased climatic temperatures, and vision problems. See Figure 12.15.

Ozone can be prepared by passing air or oxygen through an electrical discharge:

$$3 O_2(g) + 286 \text{ kJ} \xrightarrow[\text{discharge}]{\text{electrical}} 2 O_3(g)$$

The characteristic pungent odor of ozone is noticeable in the vicinity of electrical machines and power transmission lines. Ozone is formed in the atmosphere during electrical storms and by the photochemical action of ultraviolet radiation on a mixture of nitrogen dioxide and oxygen. Areas with high air pollution are subject to high atmospheric ozone concentrations.

Ozone is not a desirable low-altitude constituent of the atmosphere because it is known to cause extensive plant damage, cracking of rubber, and the formation of eye-irritating substances. Concentrations of ozone greater than 0.1 part per million (ppm) of air cause coughing, choking, headache, fatigue, and reduced resistance to respiratory infection. Concentrations between 10 and 20 ppm are fatal to humans.

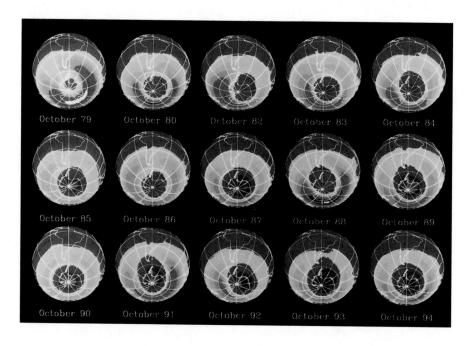

FIGURE 12.15 ▶
Satellite map showing a severe depletion or "hole" in the ozone layer over Antarctica from October 1979 to 1994. The hole is believed to be due to pollution of the atmosphere by chlorofluorocarbons used in aerosols and refrigerants.

In addition to ozone, the air in urban areas contains nitrogen oxides, which are components of smog. The term *smog* refers to air pollution in urban environments. Often the chemical reactions occur as part of a *photochemical process*. Nitrogen monoxide (NO) is oxidized in the air or in automobile engines to produce nitrogen dioxide (NO_2). In the presence of light,

$$NO_2 \xrightarrow{\text{light}} NO + O$$

In addition to nitrogen oxides, combustion of fossil fuels releases CO_2, CO, and sulfur oxides. Incomplete combustion releases unburned and partially burned hydrocarbons.

Society is continually attempting to discover, understand, and control emissions that contribute to this sort of atmospheric chemistry. It is a problem that each one of us faces as we look to the future if we want to continue to support life as we know it on our planet.

Concepts in Review

1. State the principal assumptions of the kinetic-molecular theory.

2. Estimate the relative rates of effusion of two gases of known molar mass.

3. Sketch and explain the operation of a mercury barometer.

4. List two factors that determine gas pressure in a vessel of fixed volume.

5. State Boyle's, Charles' and Gay-Lussac's laws. Use all of them in problems.

6. State the combined gas law. Indicate when it is used.

7. Use Dalton's law of partial pressures and the combined gas law to determine the dry STP volume of a gas collected over water.

8. State Avogadro's law.

9. Understand the mole–mass–volume relationship of gases.

10. Determine the density of any gas at STP.

11. Determine the molar mass of a gas from its density at a known temperature and pressure.

12. Solve problems involving the ideal gas equation.

13. Make mole–volume, mass–volume, and volume–volume stoichiometric calculations from balanced chemical equations.

14. State two reasons why real gases may deviate from the behavior predicted for an ideal gas.

Key Terms

absolute zero (12.6)	Dalton's law of partial pressures (12.10)	ideal gas equation (12.14)
allotrope (12.17)	diffusion (12.2)	kinetic-molecular theory (KMT) (12.2)
1 (one) atmosphere (12.3)	effusion (12.2)	molar volume (12.12)
atmospheric pressure (12.3)	Gay Lussac's law (12.7)	partial pressure (12.10)
Avogadro's law (12.11)	Gay Lussac's law of combining volumes (12.11)	pressure (12.3)
barometer (12.3)	Graham's law of effusion (12.2)	standard conditions (12.8)
Boyle's law (12.5)	Henry's law (CIA)	standard temperature and pressure (STP) (12.8)
Charles' law (12.6)	ideal gas (12.2)	

Questions

1. What evidence is used to show diffusion in Figure 12.1? If H_2 and O_2 were in the two flasks, how could you prove that diffusion had taken place?

2. How does the air pressure inside the balloon shown in Figure 12.2 compare with the air pressure outside the balloon? Explain.

3. According to Table 12.1, what two gases are the major constituents of dry air?

4. How does the pressure represented by 1 torr compare in magnitude to the pressure represented by 1 mm Hg? See Table 12.2.

5. In which container illustrated in Figure 12.5 are the molecules of gas moving faster? Assume both gases to be hydrogen.

6. In Figure 12.6, what gas pressure corresponds to a volume of 4 L?

7. How do the data illustrated in Figure 12.6 substantiate Boyle's law?

8. What effect would you observe in Figure 12.9 if T_2 were lower than T_1?

9. In the diagram shown in Figure 12.11, is the pressure of the oxygen plus water vapor inside the bottle equal to, greater than, or less than the atmospheric pressure outside the bottle? Explain.

10. List five gases in Table 12.3 that are more dense than air. Explain the basis for your selection.

11. What are the basic assumptions of the kinetic-molecular theory?

12. Arrange the following gases, all at standard temperature, in order of increasing relative molecular velocities: H_2, CH_4, Rn, N_2, F_2, He. What is your basis for determining the order?

13. List, in descending order, the average kinetic energies of the molecules in Question 12.

14. What are the four parameters used to describe the behavior of a gas?

15. What are the characteristics of an ideal gas?

16. Under what condition of temperature, high or low, is a gas least likely to exhibit ideal behavior? Explain.

17. Under what condition of pressure, high or low, is a gas least likely to exhibit ideal behavior? Explain.

18. Compare, at the same temperature and pressure, equal volumes of H_2 and O_2 as to
    (a) number of molecules
    (b) mass
    (c) number of moles
    (d) average kinetic energy of the molecules
    (e) rate of effusion
    (f) density

19. How does the kinetic-molecular theory account for the behavior of gases as described by
    (a) Boyle's law?
    (b) Charles' law?
    (c) Dalton's law of partial pressures?

20. Explain how the reaction

    $$N_2(g) + O_2(g) \xrightarrow{\Delta} 2\, NO(g)$$

    proves that nitrogen and oxygen are diatomic molecules.

21. What is the reason for comparing gases to STP?

22. Is the conversion of oxygen to ozone an exothermic or endothermic reaction? How do you know?

23. When constant pressure is maintained, what effect does heating a mole of N_2 gas have on
    (a) its density?
    (b) its mass?
    (c) the average kinetic energy of its molecules?
    (d) the average velocity of its molecules?
    (e) the number of N_2 molecules in the sample?

24. Write formulas for an oxygen atom, an oxygen molecule, and an ozone molecule. How many electrons are in an oxygen molecule?

Paired Exercises

Pressure Units

25. The barometer reads 715 mm Hg. Calculate the corresponding pressure in
    (a) atmospheres
    (b) inches of Hg
    (c) lb/in.^{2}

26. The barometer reads 715 mm Hg. Calculate the corresponding pressure in
    (a) torrs
    (b) millibars
    (c) kilopascals

27. Express the following pressures in atmospheres:
    (a) 28 mm Hg
    (b) 6000. cm Hg
    (c) 795 torr
    (d) 5.00 kPa

28. Express the following pressures in atmospheres:
    (a) 62 mm Hg
    (b) 4250. cm Hg
    (c) 225 torr
    (d) 0.67 kPa

Boyle's and Charles' Laws

29. A gas occupies a volume of 400. mL at 500. mm Hg pressure. What will be its volume, at constant temperature, if the pressure is changed to (a) 760 mm Hg? (b) 250 torr?

30. A gas occupies a volume of 400. mL at 500. mm Hg pressure. What will be its volume, at constant temperature, if the pressure is changed to (a) 2.00 atm? (b) 325 torr?

31. A 500.-mL sample of a gas is at a pressure of 640. mm Hg. What must be the pressure, at constant temperature, if the volume is changed to 855 mL?

32. A 500.-mL sample of a gas is at a pressure of 640. mm Hg. What must be the pressure, at constant temperature, if the volume is changed to 450. mL?

33. Given 6.00 L of N_2 gas at $-25°C$, what volume will the nitrogen occupy at (a) 0.0°C? (b) 100. K? (Assume constant pressure.)

34. Given 6.00 L of N_2 gas at $-25°C$, what volume will the nitrogen occupy at (a) 0.0°F? (b) 345. K? (Assume constant pressure.)

Combined Gas Laws

35. A gas occupies a volume of 410 mL at 27°C and 740 mm Hg pressure. Calculate the volume the gas would occupy at STP.

36. A gas occupies a volume of 410 mL at 27°C and 740 mm Hg pressure. Calculate the volume the gas would occupy at 250.°C and 680 mm Hg pressure.

37. An expandable balloon contains 1400. L of He at 0.950 atm pressure and 18°C. At an altitude of 22 miles (temperature 2.0°C and pressure 4.0 torr), what will be the volume of the balloon?

38. A gas occupies 22.4 L at 2.50 atm and 27°C. What will be its volume at 1.50 atm and $-5.00°C$?

Dalton's Law of Partial Pressures

39. What would be the partial pressure of N_2 gas collected over water at 20°C and 720. torr pressure? (Check Appendix III for the vapor pressure of water.)

40. What would be the partial pressure of N_2 gas collected over water at 25°C and 705. torr pressure? (Check Appendix III for the vapor pressure of water.)

41. A mixture contains H_2 at 600. torr pressure, N_2 at 200. torr pressure, and O_2 at 300. torr pressure. What is the total pressure of the gases in the system?

42. A mixture contains H_2 at 325. torr pressure, N_2 at 475. torr pressure, and O_2 at 650. torr pressure. What is the total pressure of the gases in the system?

43. A sample of methane gas, CH_4, was collected over water at 25.0°C and 720. torr. The volume of the wet gas is 2.50 L. What will be the volume of the dry methane at standard pressure?

44. A sample of propane gas, C_3H_8, was collected over water at 22.5°C and 745 torr. The volume of the wet gas is 1.25 L. What will be the volume of the dry propane at standard pressure?

Mole–Mass–Volume Relationships

45. What volume will 2.5 mol of Cl_2 occupy at STP?

46. What volume will 1.25 mol of N_2 occupy at STP?

47. How many grams of CO_2 are present in 2500 mL of CO_2 at STP?

48. How many grams of NH_3 are present in 1.75 L of NH_3 at STP?

49. What volume will each of the following occupy at STP?
(a) 1.0 mol of NO_2
(b) 17.05 g of NO_2
(c) 1.20×10^{24} molecules of NO_2

50. What volume will each of the following occupy at STP?
(a) 0.50 mol of H_2S
(b) 22.41 g of H_2S
(c) 8.55×10^{23} molecules of H_2S

51. How many molecules of NH_3 gas are present in a 1.00-L flask of NH_3 gas at STP?

52. How many molecules of CH_4 gas are present in a 1.00-L flask of CH_4 gas at STP?

Density of Gases

53. Calculate the density of the following gases at STP:
(a) Kr
(b) SO_3

54. Calculate the density of the following gases at STP:
(a) He
(b) C_4H_8

55. Calculate the density of
(a) F_2 gas at STP
(b) F_2 gas at 27°C and 1.00 atm pressure

56. Calculate the density of
(a) Cl_2 gas at STP
(b) Cl_2 gas at 22°C and 0.500 atm pressure

Ideal Gas Equation and Stoichiometry

57. At 27°C and 750 torr pressure, what will be the volume of 2.3 mol of Ne?

58. At 25°C and 725 torr pressure, what will be the volume of 0.75 mol of Kr?

59. What volume will a mixture of 5.00 mol of H_2 and 0.500 mol of CO_2 occupy at STP?

60. What volume will a mixture of 2.50 mol of N_2 and 0.750 mol of HCl occupy at STP?

61. Given the equation:

$$4\,NH_3(g) + 5\,O_2(g) \longrightarrow 4\,NO(g) + 6\,H_2O(g)$$

(a) How many moles of NH_3 are required to produce 5.5 mol of NO?
(b) How many liters of NO can be made from 12 L of O_2 and 10. L of NH_3 at STP?
(c) At constant temperature and pressure, what is the maximum volume, in liters, of NO that can be made from 3.0 L of NH_3 and 3.0 L of O_2?

62. Given the equation:

$$4\,NH_3(g) + 5\,O_2(g) \longrightarrow 4\,NO(g) + 6\,H_2O(g)$$

(a) How many moles of NH_3 will react with 7.0 mol of O_2?
(b) At constant temperature and pressure, how many liters of NO can be made by the reaction of 800. mL of O_2?
(c) How many grams of O_2 must react to produce 60. L of NO measured at STP?

63. Given the equation:

$$4\,FeS(s) + 7\,O_2(g) \xrightarrow{\Delta} 2\,Fe_2O_3(s) + 4\,SO_2(g)$$

how many liters of O_2, measured at STP, will react with 0.600 kg of FeS?

64. Given the equation:

$$4\,FeS(s) + 7\,O_2(g) \xrightarrow{\Delta} 2\,Fe_2O_3(s) + 4\,SO_2(g)$$

how many liters of SO_2, measured at STP, will be produced from 0.600 kg of FeS?

Additional Exercises

65. Sketch a graph to show each of the following relationships:
   (a) P vs. V at constant temperature and number of moles
   (b) T vs. V at constant pressure and number of moles
   (c) T vs. P at constant volume and number of moles
   (d) n vs. V at constant temperature and pressure

66. Why is it dangerous to incinerate an aerosol can?

67. What volume does 1 mol of an ideal gas occupy at standard conditions?

68. Which of these occupies the greatest volume?
   (a) 0.2 mol of chlorine gas at 48°C and 80 cm Hg
   (b) 4.2 g of ammonia at 0.65 atm and −112°C
   (c) 21 g of sulfur trioxide at room temperature and 110 kPa

69. Which of these has the greatest density?
   (a) SF_6 at STP
   (b) C_2H_6 at room conditions
   (c) He at −80°C and 2.15 atm

70. A chemist carried out a chemical reaction that produced a gas. It was found that the gas contained 80.0% carbon and 20.0% hydrogen. It was also noticed that 1500 mL of the gas at STP had a mass of 2.01 g.
   (a) What is the empirical formula of the compound?
   (b) What is the molecular formula of the compound?
   (c) What Lewis structure fits this compound?

*71. Three gases were added to the same 2.0-L container. The total pressure of the gases was 790 torr at room temperature (25.0°C). If the mixture contained 0.65 g of oxygen gas, 0.58 g of carbon dioxide, and an unknown amount of nitrogen gas, determine:
   (a) the total number of moles of gas in the container
   (b) the number of grams of nitrogen in the container
   (c) the partial pressure of each gas in the mixture

*72. When carbon monoxide and oxygen gas react, carbon dioxide results. If 500. mL of O_2 at 1.8 atm and 15°C are mixed with 500. mL of CO at 800 mm Hg and 60°C, how many milliliters of CO_2 at STP could possibly result?

73. One of the methods for estimating the temperature at the center of the Sun is based on the ideal gas equation. If the center is assumed to be a mixture of gases whose average molar mass is 2.0 g/mol, and if the density and pressure are 1.4 g/cm^3 and 1.3×10^9 atm, respectively, calculate the temperature.

74. A soccer ball of constant volume 2.24 L is pumped up with air to a gauge pressure of 13 lb/in.^{2} at 20.0°C. The molar mass of air is about 29 g/mol.
   (a) How many moles of air are in the ball?
   (b) What mass of air is in the ball?
   (c) During the game, the temperature rises to 30.0°C. What mass of air must be allowed to escape to bring the gauge pressure back to its original value?

75. A balloon will burst at a volume of 2.00 L. If it is partially filled at 20.0°C and 65 cm Hg to occupy 1.75 L, at what temperature will it burst if the pressure is exactly 1 atm at the time that it breaks?

76. At constant temperature, what pressure would be required to compress 2500 L of hydrogen gas at 1.0 atm pressure into a 25-L tank?

77. Given a sample of a gas at 27°C, at what temperature would the volume of the gas sample be doubled, the pressure remaining constant?

78. A gas sample at 22°C and 740 torr pressure is heated until its volume is doubled. What pressure would restore the sample to its original volume?

79. A gas occupies 250 mL at 700. torr and 22°C. When the pressure is changed to 500. torr, what temperature (°C) is needed to maintain the same volume?

80. Hydrogen stored in a metal cylinder has a pressure of 252 atm at 25°C. What will be the pressure in the cylinder when the cylinder is lowered into liquid nitrogen at −196°C?

81. The tires on an automobile were filled with air to 30. psi at 71.0°F. When driving at high speeds, the tires become hot. If the tires have a bursting pressure of 44 psi, at what temperature (°F) will the tires "blow out"?

82. What volume would 5.30 L of H_2 gas at STP occupy at 70°C and 830 torr pressure?

83. What pressure will 800. mL of a gas at STP exert when its volume is 250. mL at 30°C?

84. How many gas molecules are present in 600. mL of N_2O at 40°C and 400. torr pressure? How many atoms are present? What would be the volume of the sample at STP?

85. 5.00 L of CO_2 at 500. torr and 3.00 L of CH_4 at 400. torr are put into a 10.0-L container. What is the pressure exerted by the gases in the container?

86. A steel cylinder contains 60.0 mol of H_2 at a pressure of 1500 lb/in.^{2}.
   (a) How many moles of H_2 are in the cylinder when the pressure reads 850 lb/in.^{2}?
   (b) How many grams of H_2 were initially in the cylinder?

*87. How many moles of Cl_2 are in one cubic meter (1.00 m^3) of Cl_2 gas at STP?

88. At STP, 560. mL of a gas have a mass of 1.08 g. What is the molar mass of the gas?

*89. At what temperature (°C) will the density of methane, CH_4, be 1.0 g/L at 1.0 atm pressure?

90. A gas has a density at STP of 1.78 g/L. What is its molar mass?

91. Using the ideal gas equation, $PV = nRT$, calculate:
   (a) the volume of 0.510 mol of H_2 at 47°C and 1.6 atm pressure
   (b) the number of grams in 16.0 L of CH_4 at 27°C and 600. torr pressure
   *(c) the density of CO_2 at 4.00 atm pressure and −20.0°C
   *(d) the molar mass of a gas having a density of 2.58 g/L at 27°C and 1.00 atm pressure.

92. What is the molar mass of a gas if 1.15 g occupy 0.215 L at 0.813 atm and 30.0°C?

93. What is the Kelvin temperature of a system in which 4.50 mol of a gas occupy 0.250 L at 4.15 atm?

94. How many moles of N_2 gas occupy 5.20 L at 250 K and 0.500 atm?

*95. Acetylene, C_2H_2, and hydrogen fluoride, HF, react to give difluoroethane.

$$C_2H_2(g) + 2\ HF(g) \longrightarrow C_2H_4F_2(g)$$

When 1.0 mol of C_2H_2 and 5.0 mol of HF are reacted in a 10.0-L flask, what will be the pressure in the flask at 0°C when the reaction is complete?

96. What volume of hydrogen at STP can be produced by reacting 8.30 mol of Al with sulfuric acid? The equation is

$$2\ Al(s) + 3\ H_2SO_4(aq) \longrightarrow Al_2(SO_4)_3(aq) + 3\ H_2(g)$$

97. What are the relative rates of effusion of N_2 and He?

*98. (a) What are the relative rates of effusion of CH_4 and He?
   (b) If these two gases are simultaneously introduced into opposite ends of a 100.-cm tube and allowed to diffuse toward each other, at what distance from the helium end will molecules of the two gases meet?

*99. A gas has a percent composition by mass of 85.7% carbon and 14.3% hydrogen. At STP the density of the gas is 2.50 g/L. What is the molecular formula of the gas?

*100. Assume that the reaction

$$2\ CO(g) + O_2(g) \longrightarrow 2\ CO_2(g)$$

goes to completion. When 10. mol of CO and 8.0 mol of O_2 react in a closed 10.-L vessel,
   (a) how many moles of CO, O_2, and CO_2 are present at the end of the reaction?
   (b) what will be the total pressure in the flask at 0°C?

*101. If 250 mL of O_2, measured at STP, is obtained by the decomposition of the $KClO_3$ in a 1.20-g mixture of KCl and $KClO_3$,

$$2\ KClO_3(s) \longrightarrow 2\ KCl(s) + 3\ O_2(g)$$

what is the percent by mass of $KClO_3$ in the mixture?

*102. Examine the apparatus shown. When a small amount of water is squirted into the flask containing ammonia gas (by squeezing the bulb of the medicine dropper), water from the beaker fills the flask through the long glass tubing. Explain this phenomenon. (Remember that ammonia dissolves in water.)

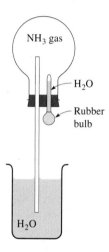

*103. Determine the pressure of the gas in each of the figures below:

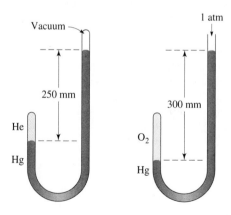

*104. Air has a density of 1.29 g/L at STP. Calculate the density of air on Pikes Peak, where the pressure is 450 torr and the temperature is 17°C.

*105. Consider the arrangement of gases shown below. If the valve between the gases is opened and the temperature is held constant:
   (a) determine the pressure of each gas
   (b) determine the total pressure in the system

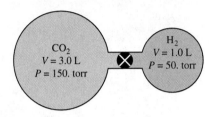

*106. A steel cylinder contained 50.0 L of oxygen gas under a pressure of 40.0 atm and a temperature of 25°C. What was the pressure in the cylinder during a storeroom fire that caused the temperature to rise 152°C? (Be careful!)

Answers to Practice Exercises

12.1 (a) 851 torr, (b) 851 mm Hg
12.2 0.596 atm
12.3 4.92 L
12.4 762 torr
12.5 84.7 L
12.6 861 K (588°C)
12.7 518 mL

12.8 1.4×10^2 g/mol
12.9 0.89 g/L
12.10 0.953 mol
12.11 128 g/mol
12.12 1.56 L O_2
12.13 75.0 L O_2, 45.0 L CO_2, 60.0 L H_2O

CHAPTER 13

Water and the Properties of Liquids

CHAPTER 13 / OUTLINE

Planet Earth, that magnificent blue sphere we enjoy viewing from space, is spectacular. Over 75% of Earth is covered with water. We are born from it, drink it, bathe in it, cook with it, enjoy its beauty in waterfalls and rainbows, and stand in awe of the majesty of icebergs. Water supports and enhances life.

In chemistry, water provides the medium for numerous reactions. The shape of the water molecule is the basis for hydrogen bonds. These bonds determine the unique properties and reactions of water. The tiny water molecule holds the answers to many of the mysteries of chemical reactions.

13.1   What Is a Liquid?

In the last chapter we found that gases contain particles that are far apart, in rapid random motion, and independent of each other. The kinetic molecular theory, along with the ideal gas equation, summarizes the behavior of most gases at relatively high temperatures and low pressures.

Solids are obviously very different from gases. Solids contain particles that are very close together; solids have a high density, compress negligibly, and maintain their shape regardless of container. These characteristics indicate large attractive forces between particles. The model for solids is very different from the one for gases.

Liquids, on the other hand, lie somewhere between the extremes of gases and solids. Liquids contain particles that are close together; liquids are essentially incompressible, and have definite volume. These properties are very similar to solids. But liquids also take the shape of their containers; this is closer to the model of a gas.

Although liquids and solids show similar properties, they differ tremendously from gases. No simple mathematical relationship, like the ideal gas equation, works well for liquids or solids. Instead these models are directly related to the forces of attraction between molecules. With these general statements in mind, let's consider some specific properties of liquids.

13.2   Evaporation

When beakers of water, ethyl ether, and ethyl alcohol are allowed to stand uncovered, their volumes gradually decrease. The process by which this change takes place is called *evaporation.*

Attractive forces exist between molecules in the liquid state. Not all of these molecules, however, have the same kinetic energy. Molecules that have greater-than-average kinetic energy can overcome the attractive forces and break away from the surface of the liquid to become a gas. **Evaporation** or **vaporization** is the escape of molecules from the liquid state to the gas or vapor state.

evaporation
vaporization

In evaporation, molecules of higher-than-average kinetic energy escape from a liquid, leaving it cooler than it was before they escaped. For this reason, evaporation

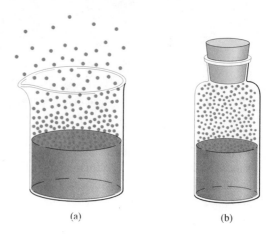

(a)   (b)

of perspiration is one way the human body cools itself and keeps its temperature constant. When volatile liquids such as ethyl chloride, C_2H_5Cl, are sprayed on the skin, they evaporate rapidly, cooling the area by removing heat. The numbing effect of the low temperature produced by evaporation of ethyl chloride allows it to be used as a local anesthetic for minor surgery.

Solids such as iodine, camphor, naphthalene (moth balls), and, to a small extent, even ice will go directly from the solid to the gaseous state, bypassing the liquid state. This change is a form of evaporation and is called **sublimation**:

sublimation

$$\text{liquid} \xrightarrow{\text{evaporation}} \text{vapor}$$

$$\text{solid} \xrightarrow{\text{sublimation}} \text{vapor}$$

13.3   Vapor Pressure

When a liquid vaporizes in a closed system like that shown in Figure 13.1b, some of the molecules in the vapor or gaseous state strike the surface and return to the liquid state by the process of **condensation.** The rate of condensation increases until it's equal to the rate of vaporization. At this point, the space above the liquid is said to be saturated with vapor, and an equilibrium, or steady state, exists between the liquid and the vapor. The equilibrium equation is

condensation

$$\text{liquid} \underset{\text{condensation}}{\overset{\text{vaporization}}{\rightleftharpoons}} \text{vapor}$$

This equilibrium is dynamic; both processes—vaporization and condensation—are taking place, even though we cannot see or measure a change. The number of molecules leaving the liquid in a given time interval is equal to the number of molecules returning to the liquid.

At equilibrium the molecules in the vapor exert a pressure like any other gas. The pressure exerted by a vapor in equilibrium with its liquid is known as the **vapor pressure** of the liquid. The vapor pressure may be thought of as a measure of the "escaping" tendency of molecules to go from the liquid to the vapor state. The vapor pressure of a liquid is independent of the amount of liquid and vapor present, but it

vapor pressure

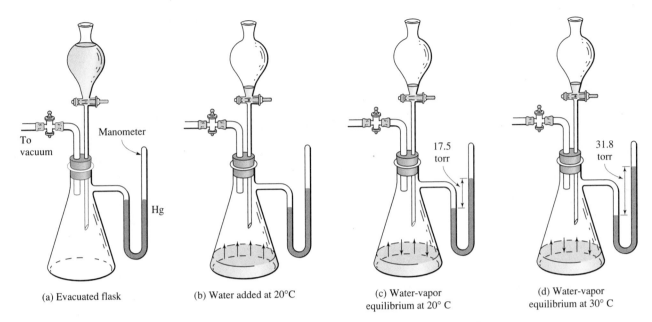

(a) Evacuated flask

(b) Water added at 20°C

(c) Water-vapor equilibrium at 20° C

(d) Water-vapor equilibrium at 30° C

▲
FIGURE 13.2
Measurement of the vapor pressure of water at 20°C and 30°C. (a) The system is evacuated. The mercury manometer attached to the flask shows equal pressure in both legs. (b) Water has been added to the flask and begins to evaporate, exerting pressure as indicated by the manometer. (c) When equilibrium is established, the pressure inside the flask remains constant at 17.5 torr. (d) The temperature is changed to 30°C, and equilibrium is reestablished with the vapor pressure at 31.8 torr.

increases as the temperature rises. Figure 13.2 illustrates a liquid–vapor equilibrium and the measurement of vapor pressure.

When equal volumes of water, ethyl ether, and ethyl alcohol are placed in separate beakers and allowed to evaporate at the same temperature, we observe that the ether evaporates faster than the alcohol, which evaporates faster than the water. This order of evaporation is consistent with the fact that ether has a higher vapor pressure at any particular temperature than ethyl alcohol or water. One reason for this higher vapor pressure is that the attraction is less between ether molecules than between alcohol or water molecules. The vapor pressures of these three compounds at various temperatures are compared in Table 13.1.

Substances that evaporate readily are said to be **volatile.** A volatile liquid has a relatively high vapor pressure at room temperature. Ethyl ether is a very volatile liquid, water is not too volatile, and mercury, which has a vapor pressure of 0.0012 torr at 20°C, is essentially a nonvolatile liquid. Most substances that are normally in a solid state are nonvolatile (solids that sublime are exceptions).

volatile

13.4 Surface Tension

Have you ever observed water and mercury in the form of small drops? These liquids form drops because liquids have *surface tension*. A droplet of liquid that is not falling or under the influence of gravity (as on the space shuttle) will form a sphere. Spheres minimize the ratio of surface area to volume. The molecules within the liquid are attracted to the surrounding liquid molecules, but at the liquid's surface, the attraction is nearly all inward. This pulls the surface into a spherical shape. The resistance of a liquid to an increase in its surface area is called the **surface tension** of the liquid. Substances with large attractive forces between molecules have high surface tensions. The effect of surface tension in water is illustrated by floating a

surface tension

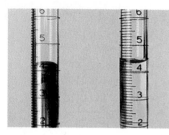

▲
FIGURE 13.3
The meniscus of mercury (left) and water (right). The meniscus is the characteristic curve of the surface of a liquid in a narrow tube.

TABLE 13.1	Vapor Pressure of Water, Ethyl Alcohol, and Ethyl Ether at Various Temperatures		
	Vapor pressure (torr)		
Temperature (°C)	Water	Ethyl alcohol	Ethyl ether*
0	4.6	12.2	185.3
10	9.2	23.6	291.7
20	17.5	43.9	442.2
30	31.8	78.8	647.3
40	55.3	135.3	921.3
50	92.5	222.2	1276.8
60	152.9	352.7	1729.0
70	233.7	542.5	2296.0
80	355.1	812.6	2993.6
90	525.8	1187.1	3841.0
100	760.0	1693.3	4859.4
110	1074.6	2361.3	6070.1

*Note that the vapor pressure of ethyl ether at temperatures of 40°C and higher exceeds standard pressure, 760 torr, which indicates that the substance has a low boiling point and should therefore be stored in a cool place in a tightly sealed container.

needle on the surface of still water. Other examples include a water strider walking across a calm pond and water beading on a freshly waxed car.

capillary action

Liquids also exhibit a phenomenon called **capillary action,** the spontaneous rising of a liquid in a narrow tube. This action results from the *cohesive forces* within the liquid, and the *adhesive forces* between the liquid and the walls of the container. If the forces between the liquid and the container are greater than those within the liquid itself, the liquid will climb the walls of the container. For example, consider the California sequoia, a tree that reaches over 200 feet in height. Although water rises only 33 feet in a glass tube (under atmospheric pressure), capillary action causes water to rise from the sequoia's roots to all its parts.

The meniscus in liquids is further evidence of cohesive and adhesive forces. When a liquid is placed in a glass cylinder, the surface of the liquid shows a curve

A water strider skims the ▶ surface of the water as a result of surface tension. At the molecular level the surface tension results from the net attraction of the water molecules toward the liquid below. In the interior of the water, the forces are balanced in all directions.

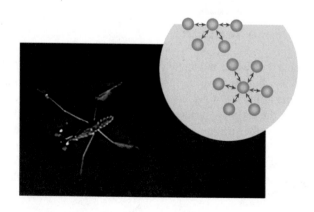

TABLE 13.2	Physical Properties of Ethyl Chloride, Ethyl Ether, Ethyl Alcohol, and Water			
Substance	Boiling point (°C)	Melting point (°C)	Heat of vaporization J/g (cal/g)	Heat of fusion J/g (cal/g)
Ethyl chloride	12.3	−139	387 (92.5)	—
Ethyl ether	34.6	−116	351 (83.9)	—
Ethyl alcohol	78.4	−112	855 (204.3)	104 (24.9)
Water	100.0	0	2259 (540)	335 (80)

called the **meniscus** (see Figure 13.3). The concave shape of water's meniscus shows that the adhesive forces between the glass and water are stronger than the cohesive forces within the water. In a nonpolar substance such as mercury, the meniscus is convex, indicating that the cohesive forces within mercury are greater than the adhesive forces between the glass wall and the mercury.

meniscus

13.5 Boiling Point

The boiling temperature of a liquid is related to its vapor pressure. We've seen that vapor pressure increases as temperature increases. When the internal or vapor pressure of a liquid becomes equal to the external pressure, the liquid boils. (By external pressure we mean the pressure of the atmosphere above the liquid.) The boiling temperature of a pure liquid remains constant as long as the external pressure does not vary.

The boiling point (bp) of water is 100°C at 1 atm pressure. Table 13.1 shows that the vapor pressure of water at 100°C is 760 torr, a figure we have seen many times before. The significant fact here is that the boiling point is the temperature at which the vapor pressure of the water or other liquid is equal to standard, or atmospheric, pressure at sea level. These relationships lead to the following definition: **Boiling point** is the temperature at which the vapor pressure of a liquid is equal to the external pressure above the liquid.

boiling point

We can readily see that a liquid has an infinite number of boiling points. When we give the boiling point of a liquid, we should also state the pressure. When we express the boiling point without stating the pressure, we mean it to be the **normal boiling point** at standard pressure (760 torr). Using Table 13.1 again, we see that the normal boiling point of ethyl ether is between 30°C and 40°C, and for ethyl alcohol it is between 70°C and 80°C because for each compound 760 torr lies within these stated temperature ranges. At the normal boiling point, 1 g of a liquid changing to a vapor (gas) absorbs an amount of energy equal to its heat of vaporization (see Table 13.2).

normal boiling point

The boiling point at various pressures can be evaluated by plotting the data of Table 13.2 on the graph in Figure 13.4, where temperature is plotted horizontally along the x-axis and vapor pressure is plotted vertically along the y-axis. The resulting curves are known as **vapor-pressure curves.** Any point on these curves represents a vapor–liquid equilibrium at a particular temperature and pressure. We can

vapor-pressure curves

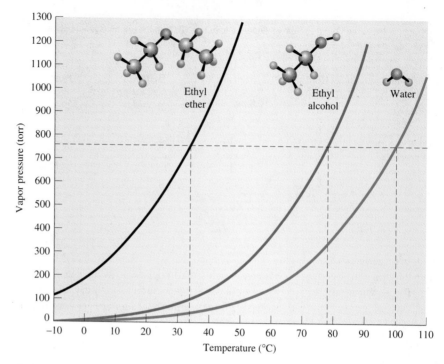

find the boiling point at any pressure by tracing a horizontal line from the designated pressure to a point on the vapor-pressure curve. From this point we draw a vertical line to obtain the boiling point on the temperature axis. Three such points are shown in Figure 13.4; they represent the normal boiling points of the three compounds at 760 torr. By reversing this process, you can ascertain at what pressure a substance will boil at a specific temperature. The boiling point is one of the most commonly used physical properties for characterizing and identifying substances.

Practice 13.1

Use the graph in Figure 13.4 to determine the boiling points of ethyl ether, ethyl alcohol, and water at 600 torr.

Practice 13.2

The average atmospheric pressure in Denver is 0.83 atm. What is the boiling point of water in Denver?

13.6   Freezing Point or Melting Point

As heat is removed from a liquid, the liquid becomes colder and colder, until a temperature is reached at which it begins to solidify. A liquid changing into a solid is said to be *freezing*, or *solidifying*. When a solid is heated continuously, a temperature is reached at which the solid begins to liquefy. A solid that is changing into a liquid is said to be *melting*. The temperature at which the solid phase of a substance is in equilibrium with its liquid phase is known as the **freezing point** or **melting point** of that substance. The equilibrium equation is

$$\text{solid} \underset{\text{freezing}}{\overset{\text{melting}}{\rightleftarrows}} \text{liquid}$$

When a solid is slowly and carefully heated so that a solid–liquid equilibrium is achieved and then maintained, the temperature will remain constant as long as both phases are present. The energy is used solely to change the solid to the liquid. The melting point is another physical property that is commonly used for characterizing substances.

The most common example of a solid–liquid equilibrium is ice and water. In a well-stirred system of ice and water, the temperature remains at 0°C as long as both phases are present. The melting point changes only slightly with pressure unless the pressure change is very large.

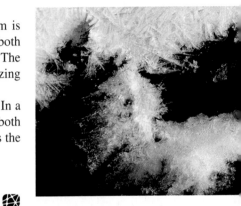

▲
Ice crystals forming at the edge of a pond.

13.7 Changes of State

The majority of solids undergo two changes of state upon heating. A solid changes to a liquid at its melting point, and a liquid changes to a gas at its boiling point. This warming process can be represented by a graph called a *heating curve* (Figure 13.5). This figure shows ice being heated at a constant rate. As energy flows into the ice, the vibrations within the crystal increase and the temperature rises ($A \longrightarrow B$). Eventually, the molecules begin to break free from the crystal and melting occurs ($B \longrightarrow C$). During the melting process all energy goes into breaking down the crystal structure; the temperature remains constant.

The energy required to change exactly one gram of a solid at its melting point into a liquid is called the **heat of fusion.** When the solid has completely melted, the temperature once again rises ($C \longrightarrow D$); the energy input is increasing the molecular

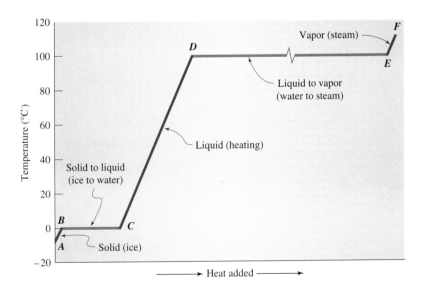

◀**FIGURE 13.5**
Heating curve for a pure substance—the absorption of heat by a substance from the solid state to the vapor state. Using water as an example, the AB interval represents the ice phase; BC interval, the melting of ice to water; CD interval, the elevation of the temperature of water from 0°C to 100°C; DE interval, the boiling of water to steam; and EF interval, the heating of steam.

motion within the water. At 100°C, the water reaches its boiling point; the temperature remains constant while the added energy is used to vaporize the water to steam ($D \longrightarrow E$). The **heat of vaporization** is the energy required to change exactly one gram of liquid to vapor at its normal boiling point. The attractive forces between the liquid molecules are overcome during vaporization. Beyond this temperature all the water exists as steam and is being heated further ($E \longrightarrow F$).

heat of vaporization

Example 13.1 How many joules of energy are needed to change 10.0 g of ice at 0.00°C to water at 20.0°C?

Solution Ice will absorb 335 J/g (heat of fusion) in going from a solid at 0°C to a liquid at 0°C. An additional 4.184 J/g°C (specific heat of water) is needed to raise the temperature of the water by 1°C.

Joules needed to melt the ice:

$$(10.0 \ g)\left(\frac{335 \ J}{1 \ g}\right) = 3.35 \times 10^3 \ J$$

Joules needed to heat the water from 0.00°C to 20.0°C:

$$(10.0 \ g)\left(\frac{4.184 \ J}{1 \ g°C}\right)(20.0°C) = 837 \ J$$

Thus 3350 J + 837 J = 4.19×10^3 J is needed.

Example 13.2 How many kilojoules of energy are needed to change 20.0 g of water at 20.°C to steam at 100.°C?

Solution

The specific heat of water is 4.184 J/g°C so the kilojoules needed to heat the water from 20.°C to 100.°C are:

$$(20.0 \text{ g})\left(\frac{4.184 \text{ J}}{1 \text{ g}°\text{C}}\right)\left(\frac{1 \text{ kJ}}{1000 \text{ J}}\right)(100. - 20.0°\text{C}) = 6.7 \text{ kJ}$$

Heat of vaporization of water is 2.26 kJ/g, so kilojoules needed to change water at 100.°C to steam at 100.°C are

$$(20.0 \text{ g})\left(\frac{2.26 \text{ kJ}}{1 \text{ g}}\right) = 45.2 \text{ kJ}$$

Thus 6.7 kJ + 45.2 kJ = 51.9 kJ is needed.

Practice 13.3

How many kilojoules of energy are required to change 50.0 g of ethyl alcohol from 60.0°C to vapor at 78.4°C? The specific heat of ethyl alcohol is 2.138 J/g°C.

13.8 Occurrence of Water

Water is our most common natural resource. It covers about 75% of Earth's surface. Not only is it found in the oceans and seas, in lakes, rivers, streams, and in glacial ice deposits, it is always present in the atmosphere and in cloud formations.

About 97% of Earth's water is in the oceans. This *saline* water contains vast amounts of dissolved minerals. More than 70 elements have been detected in the mineral content of seawater. Only four of these—chlorine, sodium, magnesium, and bromine—are now commercially obtained from the sea. The world's *fresh* water comprises the other 3%, of which about two-thirds is locked up in polar ice caps and glaciers. The remaining fresh water is found in groundwater, lakes, and the atmosphere.

Water is an essential constituent of all living matter. It is the most abundant compound in the human body, making up about 70% of total body mass. About 92% of blood plasma is water; about 80% of muscle tissue is water; and about 60% of a red blood cell is water. Water is more important than food in the sense that we can survive much longer without food than without water.

13.9 Physical Properties of Water

Water is a colorless, odorless, tasteless liquid with a melting point of 0°C and a boiling point of 100°C at 1 atm. The heat of fusion of water is 335 J/g (80 cal/g). The heat of vaporization of water is 2.26 kJ/g (540 cal/g). The values for water for both the heat of fusion and the heat of vaporization are high compared with those for other substances; this indicates strong attractive forces between the molecules.

FIGURE 13.6 ▶
Water equilibrium systems. In the beaker on the left, ice and water are in equilibrium at 0°C; in the beaker on the right, boiling water and steam are in equilibrium at 100°C.

Ice and water exist together in equilibrium at 0°C, as shown in Figure 13.6. When ice at 0°C melts, it absorbs 335 J/g in changing into a liquid; the temperature remains at 0°C. To refreeze the water, 335 J/g must be removed from the liquid at 0°C.

In Figure 13.6, both boiling water and steam are shown to have a temperature of 100°C. It takes 418 J to heat 1 g of water from 0°C to 100°C, but water at its boiling point absorbs 2.26 kJ/g in changing to steam. Although boiling water and steam are both at the same temperature, steam contains considerably more heat per gram and can cause more severe burns than hot water. Table 13.3 compares the physical properties of water with those of other hydrogen compounds of Group VIA elements.

The maximum density of water is 1.000 g/mL at 4°C. Water has the unusual property of contracting in volume as it is cooled to 4°C and then expanding when cooled from 4°C to 0°C. Therefore 1 g of water occupies a volume greater than 1 mL at all temperatures except 4°C. Although most liquids contract in volume all the way

TABLE 13.3   Physical Properties of Water and Other Hydrogen Compounds of Group VIA Elements

Formula	Color	Molar mass (g/mol)	Melting point (°C)	Boiling point, 1 atm (°C)	Heat of fusion J/g (cal/g)	Heat of vaporization J/g (cal/g)
H_2O	Colorless	18.02	0.00	100.0	335 (80.0)	2.26×10^3 (540)
H_2S	Colorless	34.09	−85.5	−60.3	69.9 (16.7)	548 (131)
H_2Se	Colorless	80.98	−65.7	−41.3	31 (7.4)	238 (57.0)
H_2Te	Colorless	129.6	−49	−2	—	179 (42.8)

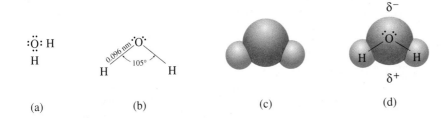

down to the point at which they solidify, a large increase (about 9%) in volume occurs when water changes from a liquid at 0°C to a solid (ice) at 0°C. The density of ice at 0°C is 0.917 g/mL, which means that ice, being less dense than water, will float in water.

13.10 Structure of the Water Molecule

A single water molecule consists of two hydrogen atoms and one oxygen atom. Each hydrogen atom is attached to the oxygen atom by a single covalent bond. This bond is formed by the overlap of the $1s$ orbital of hydrogen with an unpaired $2p$ orbital of oxygen. The average distance between the two nuclei is known as the *bond length*. The O—H bond length in water is 0.096 nm. The water molecule is nonlinear and has a bent structure with an angle of about 105 degrees between the two bonds (see Figure 13.7).

Oxygen is the second most electronegative element. As a result, the two covalent OH bonds in water are polar. If the three atoms in a water molecule were aligned in a linear structure, such as H ⟶ O ⟵ H, the two polar bonds would be acting in equal and opposite directions and the molecule would be nonpolar. However, water is a highly polar molecule. It therefore does not have a linear structure. When atoms are bonded in a nonlinear fashion, the angle formed by the bonds is called the *bond angle*. In water the HOH bond angle is 105°. The two polar covalent bonds and the bent structure result in a partial negative charge on the oxygen atom and a partial positive charge on each hydrogen atom. The polar nature of water is responsible for many of its properties, including its behavior as a solvent.

13.11 The Hydrogen Bond

Table 13.3 compares the physical properties of H_2O, H_2S, H_2Se, and H_2Te. From this comparison it is apparent that four physical properties of water—melting point, boiling point, heat of fusion, and heat of vaporization—are extremely high and do not fit the trend relative to the molar masses of the four compounds. If the properties of water followed the progression shown by the other three compounds, we would expect the melting point of water to be below $-85°C$ and the boiling point to be below $-60°C$.

Why does water exhibit these anomalies? Because liquid water molecules are held together more strongly than other molecules in the same family. The intermolecular force acting between water molecules is called a **hydrogen bond,** which acts like a very weak bond between two polar molecules. A hydrogen bond is formed between polar molecules that contain hydrogen covalently bonded to a small, highly electronegative atom such as fluorine, oxygen, or nitrogen (F—H, O—H, N—H). A

hydrogen bond

Did you think artificial sweeteners were a product of the post–World War II chemical industry? Not so—many of them have been around a long time, and several of the important ones were discovered quite by accident. In 1878, Ira Remsen was working late in his laboratory and realized he was about to miss a dinner with friends. In his haste to leave the lab, he forgot to wash his hands. Later at dinner he broke a piece of bread and tasted it only to discover that it was very sweet. The sweet taste had to be the chemical he had been working with in the lab. Back at the lab, he isolated saccharin—the first of the artificial sweeteners.

N—H or —OH group
(seeking to H-bond with
O or N on a taste bud)

O or N atom
(seeking to H-bond
with polar H on a
taste bud)

Any hydrophobic group
(e.g., CH_3, C_6H_5)

▲
Triangle of sweetness

In 1937, Michael Sveda was smoking a cigarette in his laboratory (a very dangerous practice to say the least!). He touched the cigarette to his lips and was surprised by the exceedingly sweet taste. The chemical on his hands turned out to be cyclamate, which soon became a staple of the artificial sweetener industry.

In 1965, James Schlatter was researching anti-ulcer drugs for the pharmaceutical firm G. D. Searle. In the course of his work he accidentally ingested a small amount of a preparation and found to his surprise that it had an extremely sweet taste. He had discovered aspartame, a molecule consisting of two amino acids joined together. Since only very small quantities of aspartame are necessary to produce sweetness, it proved to be an excellent low-calorie artificial sweetener. Today under the trade names of "Equal" and "Nutrasweet," aspartame is a cornerstone of the artificial sweetener industry.

More than 50 different molecules have a sweet taste, and all of them have similar molecular shapes. The triangle of sweetness theory, developed by Lamont Kier (Massachusetts College of Pharmacy) indicates that three sites on these molecules produce a structure that at-

taches to the taste bud and triggers the response that registers "sweet" in our brains.

Our taste buds are composed of proteins that can form hydrogen bonds with other molecules. The proteins contain —N—H and —OH groups (with hydrogen available to bond) as well as C=O groups (providing oxygen for hydrogen bonding). "Sweet molecules" also contain H-bonding groups including —OH, —NH_2, and O or N. These molecules not only must have the proper atoms to form hydrogen bonds, they must also contain a hydrophobic region (repels H_2O). The triangle of sweetness in the diagram shows the three necessary sites that must be located at just the proper distances.

hydrogen bond is actually the dipole–dipole attraction between polar molecules containing these three types of polar bonds.

> **Compounds that have significant hydrogen-bonding ability are those that contain H covalently bonded to F, O, or N.**

Because a hydrogen atom has only one electron, it forms only one covalent bond. When it is attached to a strong electronegative atom such as oxygen, a hydrogen atom will also be attracted to an oxygen atom of another molecule, forming a dipole–dipole attraction (H-bond) between the two molecules. Water has two types of bonds: covalent bonds that exist between hydrogen and oxygen atoms within a molecule and hydrogen bonds that exist between hydrogen and oxygen atoms in *different* water molecules.

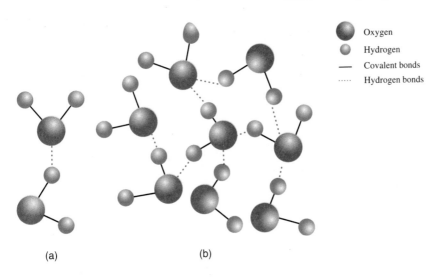

◄ **FIGURE 13.8**
Hydrogen bonding. Water in the liquid and solid states exists as aggregates in which the water molecules are linked together by hydrogen bonds.

Oxygen
Hydrogen
— Covalent bonds
..... Hydrogen bonds

(a)   (b)

Hydrogen bonds are *intermolecular* bonds; that is, they are formed between atoms in different molecules. They are somewhat ionic in character because they are formed by electrostatic attraction. Hydrogen bonds are much weaker than the ionic or covalent bonds that unite atoms to form compounds. Despite their weakness, they are of great chemical importance.

The oxygen atom in water can form two hydrogen bonds—one through each of the unbonded pairs of electrons. Figure 13.8 shows (a) two water molecules linked by a hydrogen bond and (b) six water molecules linked by hydrogen bonds. A dash (—) is used for the covalent bond and a dotted line (••••) for the hydrogen bond. In water each molecule is linked to others through hydrogen bonds to form a three-dimensional aggregate of water molecules. This intermolecular hydrogen bonding effectively gives water the properties of a much larger, heavier molecule, explaining in part its relatively high melting point, boiling point, heat of fusion, and heat of vaporization. As water is heated and energy absorbed, hydrogen bonds are continually being broken until at 100°C, with the absorption of an additional 2.26 kJ/g, water separates into individual molecules, going into the gaseous state. Sulfur, selenium, and tellurium are not sufficiently electronegative for their hydrogen compounds to behave like water. The lack of hydrogen bonding is one reason why H_2S is a gas and not a liquid at room temperature.

Fluorine, the most electronegative element, forms the strongest hydrogen bonds. This bonding is strong enough to link hydrogen fluoride molecules together as *dimers*, H_2F_2, or as larger $(HF)_n$ molecular units. The dimer structure may be represented in this way:

$$H \diagdown \ddot{\underset{..}{F}}: \diagup \overset{..}{\underset{..}{F}}:$$
H-bond

Hydrogen bonding can occur between two different atoms that are capable of forming H-bonds. Thus we may have an O••••H—N or O—H••••N linkage in which the hydrogen atom forming the H-bond is between an oxygen and a nitrogen atom. This form of H-bond exists in certain types of protein molecules and many biologically active substances.

Example 13.3   Would you expect hydrogen bonding to occur between molecules of these substances?

(a)

$$H-\overset{\displaystyle H}{\underset{\displaystyle H}{C}}-\overset{\displaystyle H}{\underset{\displaystyle H}{C}}-\ddot{O}-H$$

ethyl alcohol

(b)

$$H-\overset{\displaystyle H}{\underset{\displaystyle H}{C}}-\ddot{O}-\overset{\displaystyle H}{\underset{\displaystyle H}{C}}-H$$

dimethyl ether

Solution   (a) Hydrogen bonding should occur in ethyl alcohol because one hydrogen atom is bonded to an oxygen atom:

$$H-\overset{H}{\underset{H}{C}}-\overset{H}{\underset{H}{C}}-\ddot{O}-H\cdots\overset{H}{\underset{H}{\underset{\text{H-bond}}{O}}}-\overset{H}{\underset{H}{C}}-\overset{H}{\underset{H}{C}}-H$$

(b) There is no hydrogen bonding in dimethyl ether because all the hydrogen atoms are bonded only to carbon atoms.

Both ethyl alcohol and dimethyl ether have the same molar mass (46.07). Although both compounds have the same molecular formula, C_2H_6O, ethyl alcohol has a much higher boiling point (78.4°C) than dimethyl ether (−23.7°C) because of hydrogen bonding between the alcohol molecules.

Practice 13.4

Would you expect hydrogen bonding to occur between molecules of these substances?

(a)

$$H-\overset{H}{\underset{H}{C}}-\overset{H}{\underset{H}{C}}-\overset{H}{\underset{\cdot\cdot}{N}}-H$$

(b)

$$H-\overset{H}{\underset{H}{C}}-\overset{H}{\underset{\cdot\cdot}{N}}-\overset{H}{\underset{H}{C}}-H$$

(c)

$$H-\overset{H}{\underset{H}{C}}-\overset{H}{\underset{H}{C}}\overset{\overset{\displaystyle H}{|}}{\underset{\underset{\displaystyle H}{|}}{\overset{H-C-H}{\underset{H-C-H}{N\colon}}}}$$

13.12   Formation and Chemical Properties of Water

Water is very stable to heat; it decomposes to the extent of only about 1% at temperatures up to 2000°C. Pure water is a nonconductor of electricity. But when a small amount of sulfuric acid or sodium hydroxide is added, the solution is readily decomposed into hydrogen and oxygen by an electric current. Two volumes of hydrogen are produced for each volume of oxygen:

$$2 \ H_2O(l) \xrightarrow[\text{H}_2\text{SO}_4 \text{ or NaOH}]{\text{electrical energy}} 2 \ H_2(g) + O_2(g)$$

Formation

Water is formed when hydrogen burns in air. Pure hydrogen burns very smoothly in air, but mixtures of hydrogen and air or oxygen explode when ignited. The reaction is strongly exothermic:

$$2 \ H_2(g) + O_2(g) \longrightarrow 2 \ H_2O(g) + 484 \ \text{kJ}$$

Water is produced by a variety of other reactions, especially by (1) acid–base neutralizations, (2) combustion of hydrogen-containing materials, and (3) metabolic oxidation in living cells:

1. $HCl(aq) + NaOH(aq) \longrightarrow NaCl(aq) + H_2O(l)$

2. $2 \ C_2H_2(g) + 5 \ O_2(g) \longrightarrow 4 \ CO_2(g) + 2 \ H_2O(g) + 1212 \ \text{kJ}$
    acetylene

    $CH_4(g) + 2 \ O_2(g) \longrightarrow CO_2(g) + 2 \ H_2O(g) + 803 \ \text{kJ}$
    methane

3. $C_6H_{12}O_6(aq) + 6 \ O_2(g) \xrightarrow{\text{enzymes}} 6 \ CO_2(g) + 6 \ H_2O(l) + 2519 \ \text{kJ}$
    glucose

The combustion of acetylene shown in (2) is strongly exothermic and is capable of producing very high temperatures. It is used in oxygen–acetylene torches to cut and weld steel and other metals. Methane is known as natural gas and is commonly used as fuel for heating and cooking. The reaction of glucose with oxygen shown in (3) is the reverse of photosynthesis. It is the overall reaction by which living cells obtain needed energy by metabolizing glucose to carbon dioxide and water.

Reactions of Water with Metals and Nonmetals

The reactions of metals with water at different temperatures show that these elements vary greatly in their reactivity. Metals such as sodium, potassium, and calcium react with cold water to produce hydrogen and a metal hydroxide. A small piece of sodium added to water melts from the heat produced by the reaction, forming a silvery metal ball, which rapidly flits back and forth on the surface of the water. Caution must be used when experimenting with this reaction, because the hydrogen produced is frequently ignited by the sparking of the sodium, and it will explode, spattering sodium. Potassium reacts even more vigorously than sodium. Calcium sinks in water and liberates a gentle stream of hydrogen. The equations for these reactions are

$$2 \ Na(s) + 2 \ H_2O(l) \longrightarrow H_2(g) + 2 \ NaOH(aq)$$

$$2 \ K(s) + 2 \ H_2O(l) \longrightarrow H_2(g) + 2 \ KOH(aq)$$

$$Ca(s) + 2 \ H_2O(l) \longrightarrow H_2(g) + Ca(OH)_2(aq)$$

▲
Sodium reacts vigorously with water to produce hydrogen gas and sodium hydroxide.

Zinc, aluminum, and iron do not react with cold water but will react with steam at high temperatures, forming hydrogen and a metallic oxide. The equations are

$$Zn(s) + H_2O(g) \longrightarrow H_2(g) + ZnO(s)$$

$$2\ Al(s) + 3\ H_2O(g) \longrightarrow 3\ H_2(g) + Al_2O_3(s)$$

$$3\ Fe(s) + 4\ H_2O(g) \longrightarrow 4\ H_2(g) + Fe_3O_4(s)$$

Copper, silver, and mercury are examples of metals that do not react with cold water or steam to produce hydrogen. We conclude that sodium, potassium, and calcium are chemically more reactive than zinc, aluminum, and iron, which are more reactive than copper, silver, and mercury.

Certain nonmetals react with water under various conditions. For example, fluorine reacts violently with cold water, producing hydrogen fluoride and free oxygen. The reactions of chlorine and bromine are much milder, producing what is commonly known as "chlorine water" and "bromine water," respectively. Chlorine water contains HCl, HOCl, and dissolved Cl_2; the free chlorine gives it a yellow-green color. Bromine water contains HBr, HOBr, and dissolved Br_2; the free bromine gives it a reddish-brown color. Steam passed over hot coke (carbon) produces a mixture of carbon monoxide and hydrogen that is known as "water gas." Since water gas is combustible, it is useful as a fuel. It's also the starting material for the commercial production of several alcohols. The equations for these reactions are

$$2\ F_2(g) + 2\ H_2O(l) \longrightarrow 4\ HF(aq) + O_2(g)$$

$$Cl_2(g) + H_2O(l) \longrightarrow HCl(aq) + HOCl(aq)$$

$$Br_2(l) + H_2O(l) \longrightarrow HBr(aq) + HOBr(aq)$$

$$C(s) + H_2O(g) \xrightarrow{1000°C} CO(g) + H_2(g)$$

Reactions of Water with Metal and Nonmetal Oxides

basic anhydride

Metal oxides that react with water to form hydroxides are known as **basic anhydrides.** Examples are

$$CaO(s) + H_2O(l) \longrightarrow \underset{\text{calcium hydroxide}}{Ca(OH)_2(aq)}$$

$$Na_2O(s) + H_2O(l) \longrightarrow \underset{\text{sodium hydroxide}}{2\ NaOH(aq)}$$

Certain metal oxides, such as CuO and Al_2O_3, do not form solutions containing OH^- ions because the oxides are insoluble in water.

acid anhydride

Nonmetal oxides that react with water to form acids are known as **acid anhydrides.** Examples are

$$CO_2(g) + H_2O(l) \rightleftharpoons \underset{\text{carbonic acid}}{H_2CO_3(aq)}$$

$$SO_2(g) + H_2O(l) \rightleftharpoons \underset{\text{sulfurous acid}}{H_2SO_3(aq)}$$

$$N_2O_5(s) + H_2O(l) \longrightarrow \underset{\text{nitric acid}}{2\ HNO_3(aq)}$$

The word *anhydrous* means "without water." An anhydride is a metal oxide or a nonmetal oxide derived from a base or an oxy-acid by the removal of water. To determine the formula of an anhydride, the elements of water are removed from an acid or base formula until all the hydrogen is removed. Sometimes more than one formula unit is needed to remove all the hydrogen as water. The formula of the anhydride then consists of the remaining metal or nonmetal and the remaining oxygen atoms. In calcium hydroxide, removal of water as indicated leaves CaO as the anhydride:

$$Ca \underset{\displaystyle OH}{\overset{\displaystyle OH}{\Big<}} \quad \overset{\Delta}{\longrightarrow} \quad CaO + H_2O$$

In sodium hydroxide, H_2O cannot be removed from one formula unit, so two formula units of $NaOH$ must be used, leaving Na_2O as the formula of the anhydride:

$$\begin{matrix} NaO|H| \\ Na|OH| \end{matrix} \quad \overset{\Delta}{\longrightarrow} \quad Na_2O + H_2O$$

The removal of H_2O from H_2SO_4 gives the acid anhydride SO_3:

$$H_2SO_4 \overset{\Delta}{\longrightarrow} SO_3 + H_2O$$

The foregoing are examples of typical reactions of water but are by no means a complete list of the known reactions of water.

13.13 Hydrates

When certain solutions containing ionic compounds are allowed to evaporate, some water molecules remain as part of the crystalline compound that is left after evaporation is complete. Solids that contain water molecules as part of their crystalline structure are known as **hydrates.** Water in a hydrate is known as **water of hydration,** or **water of crystallization.**

hydrate
water of hydration
water of crystallization

Formulas for hydrates are expressed by first writing the usual anhydrous (without water) formula for the compound and then adding a dot followed by the number of water molecules present. An example is $BaCl_2 \cdot 2\ H_2O$. This formula tells us that each formula unit of this compound contains one barium ion, two chloride ions, and two water molecules. A crystal of the compound contains many of these units in its crystalline lattice.

TABLE 13.4 Selected Hydrates

Hydrate	Name	Hydrate	Name
$CaCl_2 \cdot 2\ H_2O$	calcium chloride dihydrate	$Na_2CO_3 \cdot 10\ H_2O$	sodium carbonate decahydrate
$Ba(OH)_2 \cdot 8\ H_2O$	barium hydroxide octahydrate	$(NH_4)_2C_2O_4 \cdot H_2O$	ammonium oxalate monohydrate
$MgSO_4 \cdot 7\ H_2O$	magnesium sulfate heptahydrate	$NaC_2H_3O_2 \cdot 3\ H_2O$	sodium acetate trihydrate
$SnCl_2 \cdot 2\ H_2O$	tin(II) chloride dihydrate	$Na_2B_4O_7 \cdot 10\ H_2O$	sodium tetraborate decahydrate
$CoCl_2 \cdot 6\ H_2O$	cobalt(II) chloride hexahydrate	$Na_2S_2O_3 \cdot 5\ H_2O$	sodium thiosulfate pentahydrate

Hot Ice—It's a Gas!

Ice could well become the most abundant source of energy in the 21st century. This particular type of ice has unusual properties. Even its existence has been the subject of debate since 1810 when Humphrey Day synthesized the first hydrate of chlorine. Chemists have long debated whether water could crystallize around a gas to form an unusual state of matter known as a gas hydrate. The material looks just like a chunk of ice left from a winter storm, gray and ugly. But unlike ice this gas hydrate ice pops and sizzles and a lit match causes it to burst into flames. Left alone it melts quickly into a puddle of water.

Just what is gas hydrate ice and how does it form? It turns out that gas hydrates form when water and some gases (like propane, ethane, or methane) are present at high pressure and low temperature. These conditions occur in nature at the bottom of the ocean and under permafrost in the Arctic north. Oil companies struggle with hydrate ice since it

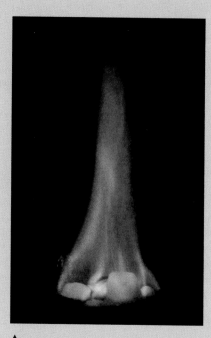

▲ **Nuggets of gas hydrate burn as they revert to water (or ice) and gas.**

plugs pipelines and damages oil rigs in oil fields in the Arctic. A gas hydrate is a gas molecule trapped inside the crystal lattice of water molecules. The most common gas is methane produced from the decomposition of organic matter. Sonar mapping has revealed large gas hydrate deposits in polar regions such as Alaska and Siberia and off our southeastern coast.

Scientists think that gas hydrates trap huge amounts of natural gas. If this gas could be brought to the surface economically it could supply 1000 years of fuel for the United States. In Russia production of natural gas from gas hydrates has already begun. As prices increase for energy sources the costs for liberating gas hydrates become more feasible. At this point most American research is still focused on prevention of gas hydrate formation. But one day in the not too distant future we could be heating our homes and running our cars by mining gas hydrates from ice!

In naming hydrates, we first name the compound exclusive of the water and then add the term *hydrate,* with the proper prefix representing the number of water molecules in the formula. For example, $BaCl_2 \cdot 2\ H_2O$ is called *barium chloride dihydrate.* Hydrates are true compounds and follow the law of definite composition. The molar mass of $BaCl_2 \cdot 2\ H_2O$ is 244.2 g/mol; it contains 56.22% barium, 29.03% chlorine, and 14.76% water.

Water molecules in hydrates are bonded by electrostatic forces between polar water molecules and the positive or negative ions of the compound. These forces are not as strong as covalent or ionic chemical bonds. As a result water of crystallization can be removed by moderate heating of the compound. A partially dehydrated or completely anhydrous compound may result. When $BaCl_2 \cdot 2\ H_2O$ is heated, it loses its water at about 100°C:

$$BaCl_2 \cdot 2\ H_2O(s) \xrightarrow{100°C} BaCl_2(s) + 2\ H_2O(g)$$

When a solution of copper(II) sulfate ($CuSO_4$) is allowed to evaporate, beautiful blue crystals containing 5 moles water per 1 mole $CuSO_4$ are formed (Figure 13.9a). The formula for this hydrate is $CuSO_4 \cdot 5\ H_2O$; it is called copper(II) sulfate pentahydrate. When $CuSO_4 \cdot 5\ H_2O$ is heated, water is lost, and a pale green-white powder, anhydrous $CuSO_4$, is formed:

$$CuSO_4 \cdot 5\ H_2O(s) \xrightarrow{250°C} CuSO_4(s) + 5\ H_2O(g)$$

(b)

(a)

FIGURE 13.9
(a) When these blue crystals of $CuSO_4 \cdot 5 H_2O$ are dissolved in water a blue solution forms. (b) The anhydrous crystals of $CuSO_4$ are pale green. When water is added they immediately change color to blue $CuSO_4 \cdot 5 H_2O$ crystals

When water is added to anhydrous copper(II) sulfate, the foregoing reaction is reversed, and the compound turns blue again (Figure 13.9b). Because of this outstanding color change, anhydrous copper(II) sulfate has been used as an indicator to detect small amounts of water. The formation of the hydrate is noticeably exothermic.

The formula for plaster of paris is $(CaSO_4)_2 \cdot H_2O$. When mixed with the proper quantity of water, plaster of paris forms a dihydrate and sets to a hard mass. It is therefore useful for making patterns for the production of art objects, molds, and surgical casts. The chemical reaction is

$$(CaSO_4)_2 \cdot H_2O(s) + 3 H_2O(l) \longrightarrow 2 CaSO_4 \cdot 2 H_2O(s)$$

Table 13.4 lists a number of common hydrates.

13.14 Hygroscopic Substances

Many anhydrous compounds and other substances readily absorb water from the atmosphere. Such substances are said to be **hygroscopic.** This property can be observed in this simple experiment: Spread a 10–20 g sample of anhydrous copper(II) sulfate on a watch glass and set it aside so that the compound is exposed to the air. Then determine the mass of the sample periodically for 24 hours, noting the increase in mass and the change in color. Over time water is absorbed from the atmosphere, forming the blue pentahydrate $CuSO_4 \cdot 5 H_2O$.

Some compounds continue to absorb water beyond the hydrate stage to form solutions. A substance that absorbs water from the air until it forms a solution is said to

hygroscopic substance

deliquescence

▲
Silica gel

be **deliquescent.** A few granules of anhydrous calcium chloride or pellets of sodium hydroxide exposed to the air will appear moist in a few minutes, and within an hour will absorb enough water to form a puddle of solution. Diphosphorus pentoxide (P_2O_5) picks up water so rapidly that its mass cannot be determined accurately except in an anhydrous atmosphere.

Compounds that absorb water are useful as drying agents (desiccants). Refrigeration systems must be kept dry with such agents or the moisture will freeze and clog the tiny orifices in the mechanism. Bags of drying agents are often enclosed in packages containing iron or steel parts to absorb moisture and prevent rusting. Anhydrous calcium chloride, magnesium sulfate, sodium sulfate, calcium sulfate, silica gel, and diphosphorus pentoxide are some of the compounds commonly used for drying liquids and gases that contain small amounts of moisture.

13.15 Natural Waters

Natural fresh waters are not pure, but contain dissolved minerals, suspended matter, and sometimes harmful bacteria. The water supplies of large cities are usually drawn from rivers or lakes. Such water is generally unsafe to drink without treatment. To make such water safe to drink, it is treated by some or all of the following processes (see Figure 13.10):

1. *Screening.* Removal of relatively large objects, such as trash, fish, and so on.
2. *Flocculation and sedimentation.* Chemicals, usually lime, CaO, and alum (aluminum sulfate), $Al_2(SO_4)_3$, are added to form a flocculent jellylike precipitate of aluminum hydroxide. This precipitate traps most of the fine suspended matter in the water and carries it to the bottom of the sedimentation basin.
3. *Sand filtration.* Water is drawn from the top of the sedimentation basin and passed downward through fine sand filters. Nearly all the remaining suspended matter and bacteria are removed by the sand filters.
4. *Aeration.* Water is drawn from the bottom of the sand filters and is aerated by spraying. The purpose of this process is to remove objectionable odors and tastes.
5. *Disinfection.* In the final stage chlorine gas is injected into the water to kill harmful bacteria before the water is distributed to the public. Ozone is also used in some countries to disinfect water. In emergencies water may be disinfected by simply boiling it for a few minutes.

If the drinking water of children contains an optimum amount of fluoride ion, their teeth will be more resistant to decay.

FIGURE 13.10▶
Typical municipal water treatment plant.

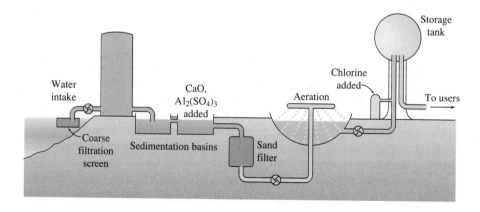

CHEMISTRY IN ACTION Moisturizers

Moisturizers have long been used to protect and rehydrate the skin. These products contain compounds called humectants and emollients that increase the water content of the skin in different ways. Emollients cover the skin with a layer of material that is immiscible with water. This prevents water from within the skin from evaporating. In contrast, humectants add water to the skin by attracting water vapor from the air.

The most common humectants are sorbitol, glycerin, and polypropylene glycol. Each molecule is polar, containing multiple —OH groups (see formulas).

The oxygen atom in each —OH group is considerably more electronegative than the hydrogen atom. This elec-

tronegativity difference results in a partial negative charge on the oxygen, whereas the hydrogen carries a partial positive charge. This polarity and the polarity of water molecules are the basis for the attraction between the humectant and water molecules.

Emollients are composed of hydrophobic (water-insoluble) molecules. These products are made of nonpolar molecules. A great diversity of compounds fall into this category, including animal oils, vegetable oils, exotic oils (such as jojoba and aloe vera), and synthetic oils. In each case the molecules form a water-insoluble layer on the skin, which traps the skin's own moisture and feels smooth to the touch.

Many consumers believe that alcohols can dry the skin and shouldn't be used in skin care products. Is this concern justified? What is the purpose of alcohols in skin care products? The problem is not so simple as we might expect. There are two different categories of alcohol with distinctly different properties. Fatty acid alcohols are large, essentially nonpolar molecules that behave like emollients. They form a water-insoluble layer on the skin that traps the moisture. A second type of alcohol, the simple alcohol, includes ethyl and isopropyl alcohols. These substances act as solvents in the skin care product and can be drying. They absorb excess oil, dissolve one ingredient into another, and make products evaporate. Some also act to keep products from spoiling and separating. The problem with alcohol in skin care products results from overuse or use by people who don't need them.

Glycerin

Propylene glycol

Sorbitol

One of the best tests for whether a skin care product will tend to dry the skin is to examine the texture. Liquids are typically the most drying formulations and are best used on oily skin. Gels containing lightweight emollients are best for skin with varying degrees of oiliness. Creams tend to contain heavier moisturizers and are best for normal to dry skin. Ointments are very heavy, creamy products that form a barrier to the skin, acting as a humectant. Ointments are for use on severely dry or damaged skin. The key to selecting the proper moisturizer lies in understanding the way in which the product functions.

Water that contains dissolved calcium and magnesium salts is called *hard water.* Ordinary soap does not lather well in hard water; the soap reacts with the calcium and magnesium ions to form an insoluble greasy scum. However, synthetic soaps, known as detergents, have excellent cleaning qualities and do not form precipitates with hard water. Hard water is also undesirable because it causes "scale" to form on the walls of water heaters, teakettles, coffee pots, and steam irons, which greatly reduces their efficiency.

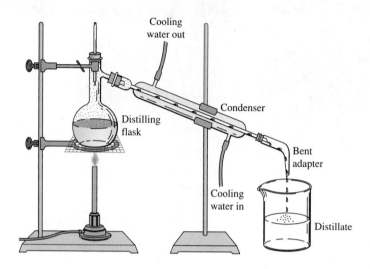

Four techniques are used to "soften" hard water:

1. **Distillation** The water is boiled, and the steam formed is condensed into a liquid again, leaving the minerals behind in the distilling vessel. Figure 13.11 illustrates a simple laboratory distillation apparatus. Commercial stills are capable of producing hundreds of liters of distilled water per hour.

2. **Calcium and magnesium precipitation** Calcium and magnesium ions are precipitated from hard water by adding sodium carbonate and lime. Insoluble calcium carbonate and magnesium hydroxide are precipitated and are removed by filtration or sedimentation.

3. **Ion exchange** Hard water is effectively softened as it is passed through a bed or tank of zeolite—a complex sodium aluminum silicate. In this process sodium ions replace objectionable calcium and magnesium ions, and the water is thereby softened:

$$Na_2(zeolite)(s) + Ca^{2+}(aq) \longrightarrow Ca(zeolite)(s) + 2\,Na^+\,(aq)$$

The zeolite is regenerated by back-flushing with concentrated sodium chloride solution, reversing the foregoing reaction.

4. **Demineralization** Both cations and anions are removed by a two-stage ion-exchange system. Special synthetic organic resins are used in the ion-exchange beds. In the first stage metal cations are replaced by hydrogen ions. In the second stage anions are replaced by hydroxide ions. The hydrogen and hydroxide ions react, and essentially pure, mineral-free water leaves the second stage.

Our oceans are an enormous source of water, but seawater contains about 3.5 lb of salts per 100 lb of water. This 35,000 ppm of dissolved salts makes seawater unfit for agricultural and domestic uses. Water that contains less than 1000 ppm of salts is considered reasonably good for drinking, and safe drinking water is already being obtained from the sea in many parts of the world. Continuous research is being done in an effort to make usable water from our oceans more abundant and economical. See Figure 13.12.

◄ FIGURE 13.12
Catalina Island, California,
gets most of its water from
a desalinization plant.

13.16 Water Pollution

Polluted water was formerly thought of as water that was unclear, had a bad odor or taste, and contained disease-causing bacteria. However, such factors as increased population, industrial requirements for water, atmospheric pollution, toxic waste dumps, and use of pesticides have greatly expanded the problem of water pollution.

Many of the newer pollutants are not removed or destroyed by the usual water-treatment processes. For example, among the 66 organic compounds found in the drinking water of a major city on the Mississippi River, 3 are labeled slightly toxic, 17 moderately toxic, 15 very toxic, 1 extremely toxic, and 1 supertoxic. Two are known carcinogens (cancer-producing agents), 11 are suspect, and 3 are metabolized to carcinogens. The U.S. Public Health Service classifies water pollutants under eight broad categories. These categories are shown in Table 13.5.

Many outbreaks of disease or poisoning, such as typhoid, dysentery, and cholera have been attributed directly to drinking water. Rivers and streams are an easy means for municipalities to dispose of their domestic and industrial waste products. Much of this water is used again by people downstream, and then discharged back into the water source. Then another community still farther downstream draws the same water and discharges its own wastes. Thus along waterways such as the Mississippi and Delaware rivers, water is withdrawn and discharged many times. If this water is not properly treated, harmful pollutants can build up, causing epidemics of various diseases.

The disposal of hazardous waste products adds to the water pollution problem. These products are unavoidable in the manufacture of many products we consider indispensable today. One common way to dispose of these wastes is to place them in toxic waste dumps. What has been found after many years of disposing of wastes in

TABLE 13.5	Classification of Water Pollutants	
Type of pollutant	**Examples**
Oxygen-demanding wastes	Decomposable organic wastes from domestic sewage and industrial wastes of plant and animal origin
Infectious agents	Bacteria, viruses, and other organisms from domestic sewage, animal wastes, and animal process wastes
Plant nutrients	Principally compounds of nitrogen and phosphorus
Organic chemicals	Large numbers of chemicals synthesized by industry, pesticides, chlorinated organic compounds
Other minerals and chemicals	Inorganic chemicals from industrial operations, mining, oil field operations, and agriculture
Radioactive substances	Waste products from mining and processing radioactive materials, airborne radioactive fallout, increased use of radioactive materials in hospitals and research
Heat from industry	Large quantities of heated water returned to water bodies from power plants and manufacturing facilities after use for cooling
Sediment from land erosion	Solid matter washed into streams and oceans by erosion, rain, and water runoff

this manner is that toxic substances have seeped into the groundwater deposits. As a result many people have become ill, and water wells have been closed until satisfactory methods of detoxifying this water are found. This problem is serious, because one-half the United States population gets its drinking water from groundwater. Cleaning up the thousands of industrial dumps and finding and implementing new and safe methods of disposing of wastes is ongoing and costly.

Many major water pollutants have been recognized and steps have been taken to eliminate them. Three that pose serious problems are lead, detergents, and chlorine-containing organic compounds. Lead poisoning, for example, has been responsible for many deaths in past years. One major toxic action of lead in the body is the inhibition of the enzyme necessary for the production of hemoglobin in the blood. The usual intake of lead into the body is through food. However, extraordinary amounts of lead can be ingested from water running through lead pipes and by using lead-containing ceramic containers for storage of food and beverages.

It has been clearly demonstrated that waterways rendered so polluted that the water is neither fit for human use nor able to sustain marine life can be successfully restored. However, keeping our lakes and rivers free from pollution is a very costly and complicated process.

Concepts in Review

1. List the common properties of liquids and solids. Explain how they are different from gases.

2. Explain the process of evaporation from the standpoint of kinetic energy.

3. Relate vapor-pressure data or vapor-pressure curves of different substances to their relative rates of evaporation and to their relative boiling points.

4. Explain the forces involved in surface tension of a liquid. Give two examples.

5. Explain why a meniscus forms on the surface of liquids in a container.

6. Explain what is occurring throughout the heating curve for water.

7. Describe a water molecule with respect to the Lewis structure, bond angle, and polarity.

8. Sketch hydrogen bonding (a) between water molecules, (b) between hydrogen fluoride molecules, and (c) between ammonia molecules.

9. Explain the effect of hydrogen bonding on the physical properties of water.

10. Determine whether a compound will or will not form hydrogen bonds.

11. Identify metal oxides as basic anhydrides and write balanced equations for their reactions with water.

12. Identify nonmetal oxides as acid anhydrides and write balanced equations for their reactions with water.

13. Deduce the formula of the acid anhydride or basic anhydride when given the formula of the corresponding acid or base.

14. Identify the product, name each reactant and product, and write equations for the complete dehydration of hydrates.

15. Outline the processes necessary to prepare safe drinking water from a contaminated river source.

16. Describe how water may be softened by distillation, chemical precipitation, ion exchange, and demineralization.

17. Complete and balance equations for (a) the reaction of water with Na, K, and Ca; (b) the reaction of steam with Zn, Al, Fe, and C; and (c) the reaction of water with halogens.

Key Terms

acid anhydride (13.12)
basic anhydride (13.12)
boiling point (13.5)
capillary action (13.4)
condensation (13.3)
deliquescence (13.14)
evaporation (13.2)
freezing or melting point (13.6)

heat of fusion (13.7)
heat of vaporization (13.7)
hydrate (13.13)
hydrogen bond (13.11)
hygroscopic substance (13.14)
meniscus (13.4)
normal boiling point (13.5)
sublimation (13.2)

surface tension (13.4)
vapor pressure (13.3)
vapor-pressure curves (13.5)
vaporization (13.2)
volatile (13.3)
water of crystallization (13.13)
water of hydration (13.13)

Questions

1. Compare the potential energy of the three states of water shown in Figure 13.6.

2. In what state (solid, liquid, or gas) would H_2S, H_2Se, and H_2Te be at $0°C$? (Table 13.3)

3. The temperature of the warer in the beaker on the hotplate (Figure 13.6) reads $100°C$. What is the pressure of the atmosphere?

4. Diagram a water molecule and point out the negative and positive ends of the dipole.

5. If the water molecule were linear, with all three atoms in a straight line rather than in the shape of a V, as shown in Figure 13.7, what effect would this have on the physical properties of water?

6. How do we specify 1, 2, 3, 4, 5, 6, 7, and 8 molecules of water in the formulas of hydrates? (Table 13.4)

7. Would the distillation setup in Figure 13.11 be satisfactory for separating salt and water? for separating ethyl alcohol and water? Explain.

8. If the liquid in the flask in Figure 13.11 is ethyl alcohol and the atmospheric pressure is 543 torr, what temperature will show on the thermometer? (Use Figure 13.4.)

9. If water were placed in both containers in Figure 13.1, would both have the same vapor pressure at the same temperature? Explain.

10. In Figure 13.1, in which case, (a) or (b), will the atmosphere above the liquid reach a point of saturation?

11. Suppose a solution of ethyl ether and ethyl alcohol is placed in the closed bottle in Figure 13.1. (Use Figure 13.4 for information on the substances.)
    (a) Are both substances present in the vapor?
    (b) If the answer to part (a) is yes, which has more molecules in the vapor?

12. In Figure 13.2, if 50% more water is added in part (b), what equilibrium vapor pressure will be observed in (c)?

13. At approximately what temperature would each of the substances shown in Figure 13.4 boil when the pressure is 30 torr?

14. Use the graph in Figure 13.4 to find the
    (a) boiling point of water at 500 torr
    (b) normal boiling point of ethyl alcohol
    (c) boiling point of ethyl ether at 0.50 atm

15. Consider Figure 13.5.
    (a) Why is line *BC* horizontal? What is happening in this interval?
    (b) What phases are present in the interval *BC*?
    (c) When heating is continued after point *C*, another horizontal line, *DE*, is reached at a higher temperature. What does this line represent?

16. List six physical properties of water.

17. What condition is necessary for water to have its maximum density? What is its maximum density?

18. Account for the fact that an ice–water mixture remains at 0°C until all the ice is melted, even though heat is applied to it.

19. Which contains less heat, ice at 0°C or water at 0°C? Explain.

20. Why does ice float in water? Would ice float in ethyl alcohol ($d = 0.789$ g/mL)? Explain.

21. If water molecules were linear instead of bent, would the heat of vaporization be higher or lower? Explain.

22. The heat of vaporization for ethyl ether is 351 J/g and that for ethyl alcohol is 855 J/g. Which of these compounds has hydrogen bonding? Explain.

23. Would there be more or less H-bonding if water molecules were linear instead of bent? Explain.

24. Which would you expect to show hydrogen bonding: ammonia, NH_3, or methane, CH_4? Explain.

25. In which condition are there fewer hydrogen bonds between molecules: water at 40°C or water at 80°C?

26. Which compound,

    $$H_2NCH_2CH_2NH_2 \quad \text{or} \quad CH_3CH_2CH_2NH_2,$$

    would you expect to have the higher boiling point? Explain. (Both compounds have similar molar masses.)

27. Explain why rubbing alcohol warmed to body temperature still feels cold when applied to your skin.

28. The vapor pressure at 20°C for the following compounds is

methyl alcohol	96 torr
acetic acid	11.7 torr
benzene	74.7 torr
bromine	173 torr
water	17.5 torr
carbon tetrachloride	91 torr
mercury	0.0012 torr
toluene	23 torr

    (a) Arrange these compounds in order of increasing rate of evaporation.
    (b) Which substance listed has the highest boiling point? the lowest?

29. Suggest a method whereby water could be made to boil at 50°C.

30. Explain why a higher temperature is obtained in a pressure cooker than in an ordinary cooking pot.

31. What is the relationship between vapor pressure and boiling point?

32. On the basis of the kinetic molecular theory, explain why vapor pressure increases with temperature.

33. Why does water have such a relatively high boiling point?

34. The boiling point of ammonia, NH_3, is −33.4°C and that of sulfur dioxide, SO_2, is −10.0°C. Which has the higher vapor pressure at −40°C?

35. Explain what is occurring physically when a substance is boiling.

36. Explain why HF (bp = 19.4°C) has a higher boiling point than HCl (bp = −85°C), whereas F_2 (bp = −188°C) has a lower boiling point than Cl_2 (bp = −34°C).

37. Why does a boiling liquid maintain a constant temperature when heat is continuously being added?

38. At what specific temperature will ethyl ether have a vapor pressure of 760 torr?

39. Why does a lake freeze from the top down?

40. What water temperature would you theoretically expect to find at the bottom of a very deep lake? Explain.

41. Is the formation of hydrogen and oxygen from water an exothermic or an endothermic reaction? How do you know?

42. (a) What is an anhydride?
    (b) What type of compound will be an acid anhydride?
    (c) What type of compound will be a basic anhydride?

Paired Exercises

43. Write the formulas for the anhydrides of these acids: $HClO_4$, H_2CO_3, H_3PO_4.

44. Write the formulas for the anhydrides of these acids: H_2SO_3, H_2SO_4, HNO_3.

45. Write the formulas for the anhydrides of these bases: $LiOH$, $NaOH$, $Mg(OH)_2$.

46. Write the formulas for the anhydrides of these bases: KOH, $Ba(OH)_2$, $Ca(OH)_2$.

47. Complete and balance these equations:
    (a) $Ba(OH)_2 \xrightarrow{\Delta}$
    (b) $CH_3OH + O_2 \longrightarrow$
        methyl alcohol
    (c) $Rb + H_2O \longrightarrow$
    (d) $SnCl_2 \cdot 2\,H_2O \xrightarrow{\Delta}$
    (e) $HNO_3 + NaOH \longrightarrow$
    (f) $CO_2 + H_2O \longrightarrow$

48. Complete and balance these equations:
    (a) $Li_2O + H_2O \longrightarrow$
    (b) $KOH \xrightarrow{\Delta}$
    (c) $Ba + H_2O \longrightarrow$
    (d) $Cl_2 + H_2O \longrightarrow$
    (e) $SO_3 + H_2O \longrightarrow$
    (f) $H_2SO_3 + KOH \longrightarrow$

49. Name these hydrates:
    (a) $BaBr_2 \cdot 2\,H_2O$
    (b) $AlCl_3 \cdot 6\,H_2O$
    (c) $FePO_4 \cdot 4\,H_2O$

50. Name these hydrates:
    (a) $MgNH_4PO_4 \cdot 6\,H_2O$
    (b) $FeSO_4 \cdot 7\,H_2O$
    (c) $SnCl_4 \cdot 5\,H_2O$

51. Distinguish between deionized water and
    (a) hard water
    (b) soft water

52. Distinguish between deionized water and
    (a) distilled water
    (b) natural water

53. How many moles of compound are in 100. g of $CoCl_2 \cdot 6\,H_2O$?

54. How many moles of compound are in 100. g of $FeI_2 \cdot 4\,H_2O$?

55. How many moles of water can be obtained from 100. g of $CoCl_2 \cdot 6\,H_2O$?

56. How many moles of water can be obtained from 100. g of $FeI_2 \cdot 4\,H_2O$?

57. When a person purchases epsom salts, $MgSO_4 \cdot 7\,H_2O$, what percent of the compound is water?

58. Calculate the mass percent of water in the hydrate $Al_2(SO_4)_3 \cdot 18\,H_2O$.

59. Sugar of lead, a hydrate of lead(II) acetate, $Pb(C_2H_3O_2)_2$, contains 14.2% H_2O. What is the empirical formula for the hydrate?

60. A 25.0-g sample of a hydrate of $FePO_4$ was heated until no more water was driven off. The mass of anhydrous sample is 16.9 g. What is the empirical formula of the hydrate?

61. How many joules are needed to change 120. g of water at 20.°C to steam at 100.°C?

62. How many joules of energy must be removed from 126 g of water at 24°C to form ice at 0°C?

*63. Suppose 100. g of ice at 0°C is added to 300. g of water at 25°C. Is this sufficient ice to lower the temperature of the system to 0°C and still have ice remaining? Show evidence for your answer.

*64. Suppose 35.0 g of steam at 100.°C is added to 300. g of water at 25°C. Is this sufficient steam to heat all the water to 100.°C and still have steam remaining? Show evidence for your answer.

*65. If 75 g of ice at 0.0°C were added to 1.5 L of water at 75°C, what would be the final temperature of the mixture?

*66. If 9560 J of energy were absorbed by 500. g of ice at 0.0°C, what would be the final temperature?

67. How many grams of water will react with the following?
    (a) 1.00 g Na
    (b) 1.00 g MgO
    (c) 1.00 g N_2O_5

68. How many grams of water will react with the following?
    (a) 1.00 mol K
    (b) 1.00 mol Ca
    (c) 1.00 mol SO_3

Additional Exercises

69. Which causes a more severe burn, liquid water at 100°C or steam at 100°C? Why?

70. You have a shallow dish of alcohol set into a tray of water. If you blow across the tray, the alcohol evaporates, while the water cools significantly and eventually freezes. Explain why.

71. Regardless of how warm the outside temperature may be, we always feels cool when stepping out of a swimming pool, the ocean, or a shower. Why is this so?

72. Sketch a heating curve for a substance X whose melting point is 40°C and whose boiling point is 65°C.
    (a) Describe what you will observe as a 60.-g sample of X is warmed from 0°C to 100°C.
    (b) If the heat of fusion of X is 80. J/g, the heat of vaporization is 190. J/g, and if 3.5 J is required to warm 1 g of X each degree, how much energy will be needed to accomplish the change in (a)?

73. Why does the vapor pressure of a liquid increase as the temperature of it is increased?

74. At the top of Mount Everest, which is just about 29,000 feet above sea level, the atmospheric pressure is about 270 torr. Use Figure 13.4 to determine the approximate boiling temperature of water on Mount Everest.

75. Explain how anhydrous copper(II) sulfate, $CuSO_4$, can act as an indicator for moisture.

76. Write formulas for magnesium sulfate heptahydrate and disodium hydrogen phosphate dodecahydrate.

77. How can soap make soft water from hard water? What objections are there to using soap for this purpose?

78. What substance is commonly used to destroy bacteria in water?

79. What chemical, other than chlorine or chlorine compounds, can be used to disinfect water for domestic use?

80. Some organic pollutants in water can be oxidized by dissolved molecular oxygen. What harmful effect can result from this depletion of oxygen in the water?

81. Why should you not drink liquids that are stored in ceramic containers, especially unglazed ones?

82. Write the chemical equation showing how magnesium ions are removed by a zeolite water softener.

83. Write an equation to show how hard water containing calcium chloride, $CaCl_2$, is softened by using sodium carbonate, Na_2CO_3.

84. How many calories are required to change 225 g of ice at 0°C to steam at 100.°C?

85. The molar heat of vaporization is the number of joules required to change 1 mol of a substance from liquid to vapor at its boiling point. What is the molar heat of vaporization of water?

*86. The specific heat of zinc is 0.096 cal/g°C. Determine the energy required to raise the temperature of 250. g of zinc from room temperature (20.0°C) to 150.°C.

*87. Suppose 150. g of ice at 0.0°C is added to 0.120 L of water at 45°C. If the mixture is stirred and allowed to cool to 0.0°C, how many grams of ice remain?

88. How many joules of energy would be liberated by condensing 50.0 mol of steam at 100.0°C and allowing the liquid to cool to 30.0°C?

89. How many kilojoules of energy are needed to convert 100. g of ice at −10.0°C to water at 20.0°C? (The specific heat of ice at −10.0°C is 2.01 J/g°C.)

90. What mass of water must be decomposed to produce 25.0 L of oxygen at STP?

*91. Suppose 1.00 mol of water evaporates in 1.00 day. How many water molecules, on the average, leave the liquid each second?

92. Compare the volume occupied by 1.00 mol of liquid water at 0°C and 1.00 mol of water vapor at STP.

*93. A quantity of sulfuric acid is added to 100. mL of water. The final volume of the solution is 122 mL and has a density of 1.26 g/mL. What mass of acid was added? Assume the density of the water is 1.00 g/mL.

94. A mixture of 80.0 mL of hydrogen and 60.0 mL of oxygen is ignited by a spark to form water.
    (a) Does any gas remain unreacted? Which one, H_2 or O_2?
    (b) What volume of which gas (if any) remains unreacted? (Assume the same conditions before and after the reaction.)

95. A student (with slow reflexes) puts his hand in a stream of steam at 100.°C until 1.5 g of water has condensed. If the water then cools to room temperature (20.0°C), how many joules have been absorbed by the student's hand?

Answers to Practice Exercises

13.1 8.5°C, 28°C, 73°C, 93°C

13.2 approximately 95°C

13.3 44.8 kJ

13.4 (a) yes, (b) yes, (c) no

CHAPTER

14

Solutions

▲
The ocean is a salt solution covering the majority of the Earth's surface, punctuated by tropical retreats such as Bay Islands in the Caribbean.

CHAPTER 14 / OUTLINE

Most substances we encounter in our daily lives are mixtures. Often they are homogeneous mixtures, which are called *solutions*. Some solutions we commonly encounter are shampoo, soft drinks, or wine. Blood plasma is a complex mixture composed of compounds and ions dissolved in water and proteins suspended in the solution. These solutions all have water as a main component, but many common items, such as air, gasoline, and steel, are also solutions that do not contain water. What are the necessary components of a solution? Why do some substances dissolve while others do not? What effect does a dissolved substance have on the properties of the solution? Answering these questions is the first step in understanding the solutions we encounter in our daily lives.

14.1 General Properties of Solutions

The term **solution** is used in chemistry to describe a system in which one or more substances are homogeneously mixed or dissolved in another substance. A simple solution has two components, a solute and a solvent. The **solute** is the component that is dissolved or is the least abundant component in the solution. The **solvent** is the dissolving agent or the most abundant component in the solution. For example, when salt is dissolved in water to form a solution, salt is the solute and water is the solvent. Complex solutions containing more than one solute and/or more than one solvent are common.

    The three states of matter—solid, liquid, and gas—give us nine different types of solutions: solid dissolved in solid, solid dissolved in liquid, solid dissolved in gas, liquid dissolved in liquid, and so on. Of these, the most common solutions are solid dissolved in liquid, liquid dissolved in liquid, gas dissolved in liquid, and gas dissolved in gas. Some common types of solutions are listed in Table 14.1.

solution

solute
solvent

TABLE 14.1	Common Types of Solutions		
Phase of solution	**Solute**	**Solvent**	**Example**
Gas	gas	gas	air
Liquid	gas	liquid	soft drinks
Liquid	liquid	liquid	antifreeze
Liquid	solid	liquid	salt water
Solid	gas	solid	H_2 in Pt
Solid	solid	solid	brass

▲
Note the beautiful purple trails of KMnO₄ as the crystals dissolve.

A true solution is one in which the particles of dissolved solute are molecular or ionic in size, generally in the range of 0.1 to 1 nm (10^{-8} to 10^{-7} cm). The properties of a true solution are as follows:

1. It is the mixture of two or more components—solute and solvent—is homogeneous, and has a variable composition; that is, the ratio of solute to solvent can be varied.
2. The dissolved solute is molecular or ionic in size.
3. It is either colored or colorless, and is usually transparent.
4. The solute remains uniformly distributed throughout the solution and will not settle out with time.
5. The solute can generally be separated from the solvent by purely physical means (e.g., by evaporation).

Let's illustrate these properties using water solutions of sugar and of potassium permanganate. We prepare two sugar solutions, the first containing 10 g of sugar added to 100 mL of water and the second containing 20 g of sugar added to 100 mL of water. Each solution is stirred until all the solute dissolves, demonstrating that we can vary the composition of a solution. Every portion of the solution has the same sweet taste because the sugar molecules are uniformly distributed throughout. If confined so that no solvent is lost, the solution will taste and appear the same a week or a month later. A solution cannot be separated into its components by filtering it. But by carefully evaporating the water, we can recover the sugar from the solution.

To observe the dissolving of potassium permanganate, $KMnO_4$, we affix a few crystals of it to paraffin wax or rubber cement at the end of a glass rod and submerge the entire rod, with the wax-permanganate end up, in a cylinder of water. Almost at once the beautiful purple color of dissolved permanganate ions, MnO_4^-, appears at the top of the rod and streams to the bottom of the cylinder as the crystals dissolve. The purple color is at first mostly at the bottom of the cylinder because $KMnO_4$ is denser than water. But after a while the purple color disperses until it's evenly distributed throughout the solution. This dispersal demonstrates that molecules and ions move about freely and spontaneously (diffuse) in a liquid or solution.

Solution permanency is explained in terms of the kinetic-molecular theory (see Section 12.2). According to the KMT both the solute and solvent particles (molecules and/or ions) are in constant random motion. This motion is energetic enough to prevent the solute particles from settling out under the influence of gravity.

14.2   Solubility

solubility

The term **solubility** describes the amount of one substance (solute) that will dissolve in a specified amount of another substance (solvent) under stated conditions. For example, 36.0 g of sodium chloride will dissolve in 100 g of water at 20°C. We say then that the solubility of NaCl in water is 36.0 g/100 g H_2O at 20°C.

Solubility is often used in a relative way. For instance, we say that a substance is very soluble, moderately soluble, slightly soluble, or insoluble. Although these terms do not accurately indicate how much solute will dissolve, they are frequently used to describe the solubility of a substance qualitatively.

Two other terms often used to describe solubility are *miscible* and *immiscible*.

miscible
immiscible

Liquids that are capable of mixing and forming a solution are **miscible**; those that do not form solutions or are generally insoluble in each other are **immiscible**.

Methyl alcohol and water are miscible in each other in all proportions. Oil and water are immiscible, forming two separate layers when they are mixed, as shown in Figure 14.1.

The general guidelines for the solubility of common ionic compounds (salts) are given in Figure 14.2. These guidelines have some exceptions, but they provide a solid foundation for the compounds considered in this course. The solubilities of over 200 compounds are given in the Solubility Table in Appendix V. Solubility data for thousands of compounds can be found by consulting standard reference sources.*

The quantitative expression of the amount of dissolved solute in a particular quantity of solvent is known as the **concentration of a solution.** Several methods of expressing concentration are described in Section 14.6.

The term salt is used interchangeably with ionic compound by many chemists.

concentration of a solution

14.3 Factors Related to Solubility

Predicting solubilities is complex and difficult. Many variables, such as size of ions, charge on ions, interaction between ions, interaction between solute and solvent, and temperature, complicate the problem. Because of the factors involved, the general rules of solubility given in Figure 14.2 have many exceptions. However, these rules are useful because they do apply to many of the more common compounds that we encounter in the study of chemistry. Keep in mind that these are rules, not laws, and are therefore subject to exceptions. Fortunately the solubility of a solute

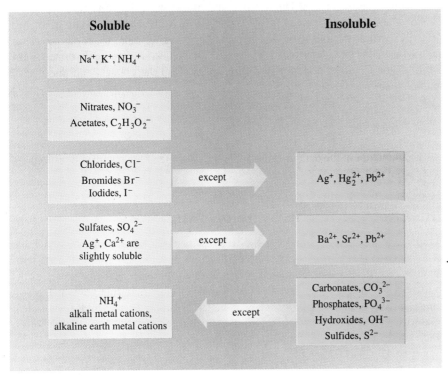

Soluble		Insoluble
Na⁺, K⁺, NH₄⁺		
Nitrates, NO_3^- Acetates, $C_2H_3O_2^-$		
Chlorides, Cl^- Bromides Br^- Iodides, I^-	except →	Ag^+, Hg_2^{2+}, Pb^{2+}
Sulfates, SO_4^{2-} Ag^+, Ca^{2+} are slightly soluble	except →	Ba^{2+}, Sr^{2+}, Pb^{2+}
NH_4^+ alkali metal cations, alkaline earth metal cations	← except	Carbonates, CO_3^{2-} Phosphates, PO_4^{3-} Hydroxides, OH^- Sulfides, S^{2-}

▲
FIGURE 14.1
An immiscible mixture of oil and water.

◀ **FIGURE 14.2**
The solubility of various common ions. Substances containing the ions on the left are generally soluble in cold water, while those substances containing the ions on the right are insoluble in cold water. The arrows point to the exceptions.

*Two commonly used handbooks are *Lange's Handbook of Chemistry,* 14th ed. (New York: McGraw-Hill, 1992), and *Handbook of Chemistry and Physics,* 79th ed. (Cleveland: Chemical Rubber Co., 1999).

is relatively easy to determine experimentally. Now let's examine the factors related to solubility.

The Nature of the Solute and Solvent

The old adage "like dissolves like" has merit, in a general way. Polar or ionic substances tend to be more miscible with other polar substances. Nonpolar substances tend to be miscible with other nonpolar substances and less miscible with polar substances. Thus ionic compounds, which are polar, tend to be much more soluble in water, which is polar, than in solvents such as ether, hexane, or benzene, which are essentially nonpolar. Sodium chloride, an ionic substance, is soluble in water, slightly soluble in ethyl alcohol (less polar than water), and insoluble in ether and benzene. Pentane, C_5H_{12}, a nonpolar substance, is only slightly soluble in water but is very soluble in benzene and ether.

At the molecular level the formation of a solution from two nonpolar substances, such as hexane and benzene, can be visualized as a process of simple mixing. The nonpolar molecules, having little tendency to either attract or repel one another, easily intermingle to form a homogeneous mixture.

Solution formation between polar substances is much more complex. See, for example, the process by which sodium chloride dissolves in water (Figure 14.3). Water molecules are very polar and are attracted to other polar molecules or ions. When salt crystals are put into water, polar water molecules become attracted to the sodium and chloride ions on the crystal surfaces and weaken the attraction between Na^+ and Cl^- ions. The positive end of the water dipole is attracted to the Cl^- ions, and the negative end of the water dipole to the Na^+ ions. The weakened attraction permits the ions to move apart, making room for more water dipoles. Thus the surface ions are surrounded by water molecules, becoming hydrated ions, $Na^+(aq)$ and $Cl^-(aq)$, and slowly diffuse away from the crystals and dissolve in solution:

$$NaCl(\text{crystal}) \xrightarrow{\text{H}_2\text{O}} Na^+(aq) + Cl^-(aq)$$

Examination of the data in Table 14.2 reveals some of the complex questions relating to solubility.

The Effect of Temperature on Solubility

Temperature affects the solubility of most substances as shown by the data in Table 14.2. Most solutes have a limited solubility in a specific solvent at a fixed temperature. For most solids dissolved in a liquid, an increase in temperature results in increased solubility (see Figure 14.4). However, no single rule governs the solubility of solids in liquids with change in temperature. Some solids increase in solubility only slightly with increasing temperature (see NaCl in Figure 14.4); other solids decrease in solubility with increasing temperature (see Li_2SO_4 in Figure 14.4).

On the other hand, the solubility of a gas in water usually decreases with increasing temperature (see HCl and SO_2 in Figure 14.4). The tiny bubbles that form when water is heated are due to the decreased solubility of air at higher temperatures. The decreased solubility of gases at higher temperatures is explained in terms

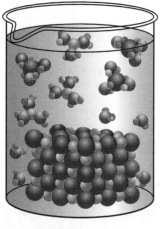

= Water

= Na^+

= Cl^-

▲
FIGURE 14.3
Dissolution of sodium chloride in water. Polar water molecules are attracted to Na^+ and Cl^- ions in the salt crystal, weakening the attraction between the ions. As the attraction between the ions weakens, the ions move apart and become surrounded by water dipoles. The hydrated ions slowly diffuse away from the crystal to become dissolved in solution.

of the KMT by assuming that, in order to dissolve, the gas molecules must form bonds of some sort with the molecules of the liquid. An increase in temperature decreases the solubility of the gas because it increases the kinetic energy (speed) of the gas molecules and thereby decreases their ability to form "bonds" with the liquid molecules.

TABLE 14.2	Solubility of Alkali Metal Halides in Water	
	Solubility (g salt/100 g H_2O)	
Salt	0°C	100°C
LiF	0.12	0.14 (at 35°C)
LiCl	67	127.5
LiBr	143	266
LiI	151	481
NaF	4	5
NaCl	35.7	39.8
NaBr	79.5	121
NaI	158.7	302
KF	92.3 (at 18°C)	Very soluble
KCl	27.6	57.6
KBr	53.5	104
KI	127.5	208

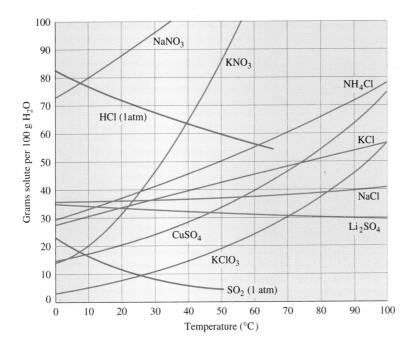

FIGURE 14.4
Solubility of various compounds in water. Solids are shown in red and gases are shown in blue.

▲
Pouring root beer into a glass illustrates the effect of pressure on solubility. The escaping CO_2 produces the foam.

The Effect of Pressure on Solubility

Small changes in pressure have little effect on the solubility of solids in liquids or liquids in liquids but have a marked effect on the solubility of gases in liquids. The solubility of a gas in a liquid is directly proportional to the pressure of that gas above the solution. Thus the amount of a gas dissolved in solution will double if the pressure of that gas over the solution is doubled. For example, carbonated beverages contain dissolved carbon dioxide under pressures greater than atmospheric pressure. When a can of carbonated soda is opened, the pressure is immediately reduced to the atmospheric pressure, and the excess dissolved carbon dioxide bubbles out of the solution.

Saturated, Unsaturated, and Supersaturated Solutions

At a specific temperature there is a limit to the amount of solute that will dissolve in a given amount of solvent. When this limit is reached, the resulting solution is said to be *saturated*. For example, when we put 40.0 g of KCl into 100 g of H_2O at 20°C, we find that 34.0 g of KCl dissolves and 6.0 g of KCl remains undissolved. The solution formed is a saturated solution of KCl.

Two processes are occurring simultaneously in a saturated solution. The solid is dissolving into solution and, at the same time, the dissolved solute is crystallizing out of solution. This may be expressed as

$$\text{solute (undissolved)} \rightleftharpoons \text{solute (dissolved)}$$

When these two opposing processes are occurring at the same rate, the amount of solute in solution is constant, and a condition of equilibrium is established between dissolved and undissolved solute. Therefore, a **saturated solution** contains dissolved solute in equilibrium with undissolved solute.

It's important to state the temperature of a saturated solution, because a solution that is saturated at one temperature may not be saturated at another. If the temperature of a saturated solution is changed, the equilibrium is disturbed, and the amount of dissolved solute will change to reestablish equilibrium.

A saturated solution may be either dilute or concentrated, depending on the solubility of the solute. A saturated solution can be conveniently prepared by dissolving a little more than the saturated amount of solute at a temperature somewhat higher than room temperature. Then the amount of solute in solution will be in excess of its solubility at room temperature, and, when the solution cools, the excess solute will crystallize, leaving the solution saturated. (In this case, the solute must be more soluble at higher temperatures and must not form a supersaturated solution.) Examples expressing the solubility of saturated solutions at two different temperatures are given in Table 14.3.

An **unsaturated solution** contains less solute per unit of volume than does its corresponding saturated solution. In other words, additional solute can be dissolved in an unsaturated solution without altering any other conditions. Consider a solution made by adding 40 g of KCl to 100 g of H_2O at 20°C (see Table 14.3). The solution formed will be saturated and will contain about 6 g of undissolved salt, because the maximum amount of KCl that can dissolve in 100 g of H_2O at 20°C is 34 g. If the solution is now heated and maintained at 50°C,

saturated solution

unsaturated solution

TABLE 14.3	Saturated Solutions at 20°C and 50°C	
	Solubility (g solute/100 g H₂O)	
Solute	20°C	50°C
NaCl	36.0	37.0
KCl	34.0	42.6
$NaNO_3$	88.0	114.0
$KClO_3$	7.4	19.3
$AgNO_3$	222.0	455.0
$C_{12}H_{22}O_{11}$	203.9	260.4

▲ The heat released in this hot pack results from the crystal-lization of a supersaturated solution of sodium acetate.

all the salt will dissolve and even more can be dissolved. The solution at 50°C is unsaturated.

In some circumstances, solutions can be prepared that contain more solute than that needed for a saturated solution at a particular temperature. These solutions are said to be **supersaturated.** However, we must qualify this definition by noting that a supersaturated solution is unstable. Disturbances, such as jarring, stirring, scratching the walls of the container, or dropping in a "seed" crystal, cause the supersaturation to return to saturation. When a supersaturated solution is disturbed, the excess solute crystallizes out rapidly, returning the solution to a saturated state.

Supersaturated solutions are not easy to prepare but may be made from certain substances by dissolving, in warm solvent, an amount of solute greater than that needed for a saturated solution at room temperature. The warm solution is then allowed to cool very slowly. With the proper solute and careful work, a supersatu-rated solution will result.

supersaturated solution

Example 14.1

Will a solution made by adding 2.5 g of $CuSO_4$ to 10 g of H_2O be saturated or un-saturated at 20°C?

Solution

We first need to know the solubility of $CuSO_4$ at 20°C. From Figure 14.4, we see that the solubility of $CuSO_4$ at 20°C is about 21 g per 100 g of H_2O. This amount is equivalent to 2.1 g of $CuSO_4$ per 10 g of H_2O.

Since 2.5 g per 10 g of H_2O is greater than 2.1 g per 10 g of H_2O, the solution will be saturated and 0.4 g of $CuSO_4$ will be undissolved.

Practice 14.1

Will a solution made by adding 9.0 g NH_4Cl to 20 g of H_2O be saturated or un-saturated at 50°C?

14.4   Rate of Dissolving Solids

The rate at which a solid dissolves is governed by (1) the size of the solute particles, (2) the temperature, (3) the concentration of the solution, and (4) agitation or stirring. Let's look at each of these conditions.

1. *Particle Size.* A solid can dissolve only at the surface that is in contact with the solvent. Because the surface-to-volume ratio increases as size decreases, smaller crystals dissolve faster than large ones. For example, if a salt crystal 1 cm on a side (6-cm^2 surface area) is divided into 1000 cubes, each 0.1 cm on a side, the total surface of the smaller cubes is 60 cm^2—a tenfold increase in surface area (see Figure 14.5).

2. *Temperature.* In most cases the rate of dissolving of a solid increases with temperature. This increase is due to kinetic effects. The solvent molecules move more rapidly at higher temperatures and strike the solid surfaces more often and harder, causing the rate of dissolving to increase.

3. *Concentration of the Solution.* When the solute and solvent are first mixed, the rate of dissolving is at its maximum. As the concentration of the solution increases and the solution becomes more nearly saturated with the solute, the rate of dissolving decreases greatly. The rate of dissolving is graphed in Figure 14.6. Note that about 17 g dissolves in the first 5-minute interval, but only about 1 g dissolves in the fourth 5-minute interval. Although different solutes show different rates, the rate of dissolving always becomes very slow as the concentration approaches the saturation point.

4. *Agitation or Stirring.* The effect of agitation or stirring is kinetic. When a solid is first put into water, it comes in contact only with solvent in its immediate vicinity. As the solid dissolves, the amount of dissolved solute around the solid becomes more and more concentrated, and the rate of dissolving slows down. If the mixture is not stirred, the dissolved solute diffuses very slowly through the solution; weeks may pass before the solid is entirely dis-

FIGURE 14.5 ▶
Surface area of crystals. A crystal 1 cm on a side has a surface area of 6 cm². Subdivided into 1000 smaller crystals, each 0.1 cm on a side, the total surface area is increased to 60 cm².

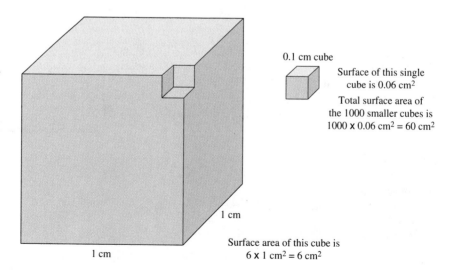

0.1 cm cube

Surface of this single cube is 0.06 cm^2

Total surface area of the 1000 smaller cubes is 1000 x 0.06 cm^2 = 60 cm^2

1 cm

1 cm

Surface area of this cube is 6 x 1 cm^2 = 6 cm^2

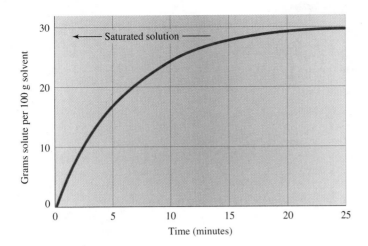

solved. Stirring distributes the dissolved solute rapidly through the solution, and more solvent is brought into contact with the solid, causing it to dissolve more rapidly.

14.5 Solutions: A Reaction Medium

Many solids must be put into solution to undergo appreciable chemical reaction. We can write the equation for the double displacement reaction between sodium chloride and silver nitrate:

$$NaCl + AgNO_3 \longrightarrow AgCl + NaNO_3$$

But suppose we mix solid NaCl and solid $AgNO_3$ and look for a chemical change. If any reaction occurs, it is slow and virtually undetectable. In fact, the crystalline structures of NaCl and $AgNO_3$ are so different that we could separate them by tediously picking out each kind of crystal from the mixture. But if we dissolve the NaCl and $AgNO_3$ separately in water and mix the two solutions, we observe the immediate formation of a white, curdlike precipitate of silver chloride.

Molecules or ions must collide with one another in order to react. In the foregoing example, the two solids did not react because the ions were securely locked within their crystal structures. But when the NaCl and $AgNO_3$ are dissolved, their crystal lattices are broken down and the ions become mobile. When the two solutions are mixed, the mobile Ag^+ and Cl^- ions come into contact and react to form insoluble AgCl, which precipitates out of solution. The soluble Na^+ and NO_3^- ions remain mobile in solution but form the crystalline salt $NaNO_3$ when the water is evaporated:

$$NaCl(aq) + AgNO_3(aq) \longrightarrow AgCl(s) + NaNO_3(aq)$$

$$Na^+(aq) + Cl^-(aq) + Ag^+(aq) + NO_3^-(aq) \longrightarrow AgCl(s) + Na^+(aq) + NO_3^-(aq)$$

| sodium chloride solution | silver nitrate solution | silver chloride | sodium nitrate in solution |

The mixture of the two solutions provides a medium or space in which the Ag^+ and Cl^- ions can react. (See Chapter 15 for further discussion of ionic reactions.)

Killer Lakes

In the tiny African nation Cameroon, two towns border on lakes that are people killers. The townspeople don't die by drowning in the lakes or by drinking contaminated water. Instead they die from carbon dioxide asphyxiation. The lakes give off clouds of carbon dioxide at irregular intervals. Thirty-seven people died near Lake Monoun in August, 1984. Just two years later 1700 people died at nearby Lake Nyos.

Scientists studying these volcanic crater lakes have found that CO_2 percolates upward from groundwater into the bottom of these lakes. The CO_2 accumulates to dangerous levels because the water is naturally stratified into layers that do not mix. A boundary called a *chemocline* separates the layers, keeping fresh water at the surface of the lake. The lower layers of the lake contain dissolved minerals and gases (including CO_2).

The disasters occur when something disturbs the layers. An earthquake, a landslide, or even winds can trigger the phenomenon. As waves form and move across the lake, the layers within the lake are mixed. When the deep water containing the CO_2 rises, the dissolved CO_2 is released from the solution (similar to the bubbles released on opening a can of soda).

At Lake Nyos, where 1700 people died, the cloud of CO_2 spilled over the edge of the crater and traveled down a river valley. Since CO_2 is denser than air, the cloud stayed near the ground. It traveled at an amazing speed of 45 mph and killed people as far away as 25 miles.

Although scientists don't know precisely what causes the water layers to turn over, they have succeeded in measuring the rate at which gas seeps into the lake bottoms. The rate is so fast that some scientists think the bottom waters of Lake Nyos could be saturated in less than 20 years, and Lake Monoun could be saturated in less than 10 years.

Scientists and engineers are working to lower gas concentrations in both lakes. In Lake Monoun, water is being pumped through pipes from the lake bottom to the surface to release the gas slowly. Lake Nyos is a larger lake and represents a more difficult problem. One end of the lake is supported by a weak natural dam. If the dam were to break, the water from the lake would spill into a valley with about 10,000 residents and could trigger a CO_2 release.

Solutions also function as diluting agents in reactions in which the undiluted reactants would combine with each other too violently. Moreover, a solution of known concentration provides a convenient method for delivering specific amounts of reactants.

14.6 Concentration of Solutions

The concentration of a solution expresses the amount of solute dissolved in a given quantity of solvent or solution. Because reactions are often conducted in solution, it's important to understand the methods of expressing concentration and to know how to prepare solutions of particular concentrations. The concentration of a solution may be expressed qualitatively or quantitatively. Let's begin with a look at the qualitative methods of expressing concentration.

Dilute and Concentrated Solutions

When we say a solution is *dilute* or *concentrated,* we are expressing, in a relative way, the amount of solute present. One gram of a compound and 2 g of a compound

Getting Clothes CO₂ Clean!

Dry cleaning has been used for many years to clean clothes that can't be placed in water (such as silk, rayon, wool) without severe damage or shrinking. The dry-cleaning process involves no water. Instead clothes are treated for stains and then washed in perchloroethylene (perc), an organic liquid. Unfortunately perc may be a health hazard and is classified as both an air pollutant and environmental contaminant. This means big trouble for small dry cleaners who now must dispose of it as hazardous waste and also worry about health risks to employees. Consequently researchers are looking for a solvent that can replace perc for dry cleaning.

One interesting alternative is to use carbon dioxide for dry cleaning. You may think of CO_2 as a greenhouse gas, the gas you breathe in and out, or the solid called dry ice. One thing's for sure—we don't often consider CO_2 as a liquid. Scientists

have already started to consider liquid CO_2 as a dry cleaning alternative. Why? There is no shortage of CO_2 in the world. It can be collected from the waste generated in industry. No special disposal procedures are needed since it can be recycled and leaks would be nontoxic to humans and the environment in small quantities.

The largest obstacle for the dry-cleaning process is that CO_2 doesn't dissolve most polymers, oils, waxes, and substances that get caught in clothes. Of course, neither does water; it requires the help of detergents. So scientists have now developed polymers that can be used as detergents in liquid CO_2. Joseph DeSimone (University of North Carolina at Chapel Hill) made detergent polymers called copolymers, which are really just two polymers joined together. One end is soluble in CO_2; the other attracts oils and waxes. The new detergent polymers form

micelles that trap the oils and waxes within the micelle and dissolve in CO_2.

Now machines that look like bank vaults can clean 50–70 pounds of clothes per load with CO_2 and about 5 ounces of polymer detergent. One extra bonus—since the machine operates at room temperature, stains don't require pretreating! Leather and suede can also be cleaned using CO_2. Someday soon you may walk into your local dry-cleaning shop and get your clothes CO_2 cleaned!

---

in solution are both dilute solutions when compared with the same volume of a solution containing 20 g of a compound. Ordinary concentrated hydrochloric acid contains 12 mol of HCl per liter of solution. In some laboratories the dilute acid is made by mixing equal volumes of water and the concentrated acid. In other laboratories the concentrated acid is diluted with two or three volumes of water, depending on its use. The term **dilute solution,** then, describes a solution that contains a relatively small amount of dissolved solute. Conversely, a **concentrated solution** contains a relatively large amount of dissolved solute.

dilute solution
concentrated solution

Mass Percent Solution

The mass percent method expresses the concentration of the solution as the percent of solute in a given mass of solution. It says that for a given mass of solution a certain percent of that mass is solute. Suppose we take a bottle from the reagent shelf that reads "sodium hydroxide, NaOH, 10%." This statement means that for every 100 g of this solution, 10 g will be NaOH and 90 g will be water. (Note that this amount of solution is 100 g and not 100 mL.) We could also make this same concentration of solution by dissolving 2.0 g of NaOH in 18 g of water. Mass percent concentrations are most generally used for solids dissolved in liquids:

$$\text{mass percent} = \frac{\text{g solute}}{\text{g solute + g solvent}} \times 100 = \frac{\text{g solute}}{\text{g solution}} \times 100$$

Note that mass percent is independent of the formula for the solute.

As instrumentation advances are made in chemistry, our ability to measure the concentration of dilute solutions is increasing as well. In addition to mass percent, chemists now commonly use **parts per million (ppm)**:

parts per million (ppm)

$$\text{parts per million} = \frac{\text{g solute}}{\text{g solute} + \text{g solvent}} \times 1{,}000{,}000$$

Currently, air and water contaminants, drugs in the human body, and pesticide residues are measured in parts per million.

Example 14.2   What is the mass percent of sodium hydroxide in a solution that is made by dissolving 8.00 g NaOH in 50.0 g H_2O?

Solution

grams of solute (NaOH) =   8.00 g

grams of solvent (H_2O) = 50.0 g

$$\left(\frac{8.00 \text{ g NaOH}}{8.00 \text{ g NaOH} + 50.0 \text{ g } H_2O} \right)100 = 13.8\% \text{ NaOH solution}$$

Example 14.3   What masses of potassium chloride and water are needed to make 250. g of 5.00% solution?

Solution   The percent expresses the mass of the solute:

250. g = total mass of solution

5.00% of 250. g = (0.0500)(250. g) = 12.5 g KCl (solute)

250. g − 12.5 g = 238 H_2O

Dissolving 12.5 g KCl in 238 g H_2O gives a 5.00% KCl solution.

Example 14.4   A 34.0% sulfuric acid solution has a density of 1.25 g/mL. How many grams of H_2SO_4 are contained in 1.00 L of this solution?

Solution   Since H_2SO_4 is the solute, we first solve the mass percent equation for grams of solute:

$$\text{mass percent} = \frac{\text{g solute}}{\text{g solution}} \times 100$$

$$\text{g solute} = \frac{\text{mass percent} \times \text{g solution}}{100}$$

The mass percent is given so we need to determine the grams of solution. The mass of the solution can be calculated from the density data. Convert density (g/mL) to grams:

$$1.00 \text{ L} = 1.00 \times 10^3 \text{ mL}$$

$$\left(\frac{1.25 \text{ g}}{\text{mL}} \right)(1.00 \times 10^3 \text{ mL}) = 1250 \text{ g} \quad \text{(mass of solution)}$$

Now we have all the figures to calculate the grams of solute:

$$\text{g solute} = \frac{(34.0 \text{ g } H_2SO_4)(1250 \text{ g})}{100 \text{ g}} = 425 \text{ g } H_2SO_4$$

Thus 1.00 L of 34.0% H_2SO_4 solution contains 425 g H_2SO_4.

Practice 14.2

What is the mass percent of Na_2SO_4 in a solution made by dissolving 25.0 g Na_2SO_4 in 225.0 g H_2O?

Mass/Volume Percent (m/v)

This method expresses concentration as grams of solute per 100 mL of solution. With this system, a 10.0% (m/v) glucose solution is made by dissolving 10.0 g of glucose in water, diluting to 100 mL, and mixing. The 10.0% (m/v) solution could also be made by diluting 20.0 g to 200 mL, 50.0 g to 500 mL, and so on. Of course, any other appropriate dilution ratio may be used:

$$\text{mass/volume percent} = \frac{\text{g solute}}{\text{mL solution}} \times 100$$

Volume Percent

Solutions that are formulated from two liquids are often expressed as *volume percent* with respect to the solute. The volume percent is the volume of a liquid in 100 mL of solution. The label on a bottle of ordinary rubbing alcohol reads "isopropyl alcohol, 70% by volume." Such a solution could be made by mixing 70 mL of alcohol with water to make a total volume of 100 mL, but we cannot use 30 mL of water because the two volumes are not necessarily additive:

$$\text{volume percent} = \frac{\text{volume of liquid in question}}{\text{total volume of solution}} \times 100$$

Volume percent is used to express the concentration of alcohol in beverages. Wines generally contain 12% alcohol by volume. This translates into 12 mL of alcohol in each 100 mL of wine. The beverage industry also uses the concentration unit of *proof* (twice the volume percent). Pure alcohol is 100%, therefore 200 proof. Scotch whiskey is 86 proof or 43% alcohol.

Molarity

Mass percent solutions do not equate or express the molar masses of the solute in solution. For example, 1000. g of 10.0% NaOH solution contains 100. g NaOH;

FIGURE 14.7 ▶
Preparation of a 1 M solution.

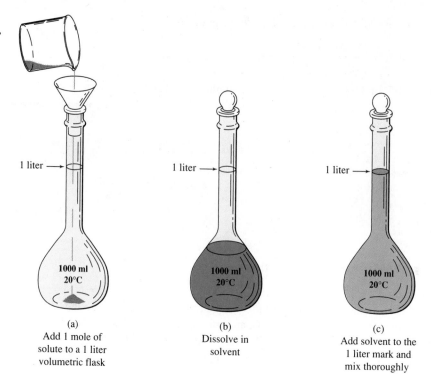

(a)
Add 1 mole of
solute to a 1 liter
volumetric flask

(b)
Dissolve in
solvent

(c)
Add solvent to the
1 liter mark and
mix thoroughly

1000. g of 10.0% KOH solution contains 100. g KOH. In terms of moles of NaOH and KOH, these solutions contain

$$\text{mol NaOH} = (100.\ \text{g NaOH})\left(\frac{1\ \text{mol NaOH}}{40.00\ \text{g NaOH}}\right) = 2.50\ \text{mol NaOH}$$

$$\text{mol KOH} = (100.\ \text{g KOH})\left(\frac{1\ \text{mol KOH}}{56.11\ \text{g KOH}}\right) = 1.78\ \text{mol KOH}$$

From these figures we see that the two 10.0% solutions do not contain the same number of moles of NaOH and KOH. As a result we find that a 10.0% NaOH solution contains more reactive base than a 10.0% KOH solution.

We need a method of expressing concentration that will easily indicate how many moles of solute are present per unit volume of solution. For this purpose the concentration known as molarity is used.

A 1-molar solution contains 1 mol of solute per liter of solution. For example, to make a 1-molar solution of sodium hydroxide, NaOH, we dissolve 40 g NaOH (1 mol) in water and dilute the solution with more water to a volume of 1 L. The solution contains 1 mol of the solute in 1 L of solution and is said to be 1 molar in concentration. Figure 14.7 illustrates the preparation of a 1-molar solution. Note that the volume of the solute and the solvent together is 1 L.

The concentration of a solution can, of course, be varied by using more or less

molarity (M)   solute or solvent; but in any case the **molarity** of a solution is the number of moles of solute per liter of solution. The abbreviation for molarity is *M*. The units of molarity are moles per liter. The expression "2.0 *M* NaOH" means a 2.0 molar solution of NaOH (2.0 mol, or 80 g, of NaOH dissolved to make 1.0 L of solution).

$$\text{molarity} = M = \frac{\text{number of moles of solute}}{\text{liter of solution}} = \frac{\text{moles}}{\text{liter}}$$

Flasks that are calibrated to contain specific volumes at a particular temperature are used to prepare solutions of a desired concentration. These *volumetric flasks* have a calibration mark on the neck that accurately indicate the measured volume. Molarity is based on a specific volume of solution and therefore will vary slightly with temperature because volume varies with temperature:

(1000 mL H_2O at 20°C = 1001 mL at 25°C)

Suppose we want to make 500 mL of 1 *M* solution. This solution can be prepared by determining the mass of 0.5 mol of the solute and diluting with water in a 500-mL (0.5-L) volumetric flask. The molarity will be

$$M = \frac{0.5 \text{ mol solute}}{0.5 \text{ L solution}} = 1 \text{ molar}$$

You can see that it isn't necessary to have a liter of solution to express molarity. All we need to know is the number of moles of dissolved solute and the volume of solution. Thus 0.001 mol NaOH in 10 mL of solution is 0.1 *M*:

$$\left(\frac{0.001 \text{ mol}}{10 \text{ mL}}\right)\left(\frac{1000 \text{ mL}}{1 \text{ L}}\right) = 0.1 \ M$$

When we stop to think that a balance is not calibrated in moles but in grams, we can incorporate grams into the molarity formula. We do so by using the relationship

$$\text{moles} = \frac{\text{grams of solute}}{\text{molar mass}}$$

Substituting this relationship into our expression for molarity, we get

$$M = \frac{\text{mol}}{\text{L}} = \frac{\text{g solute}}{\text{molar mass solute} \times \text{L solution}}$$

We can now determine the mass of any amount of a solute that has a known formula, dilute it to any volume, and calculate the molarity of the solution using this formula.

Molarities of concentrated acids commonly used in the laboratory:

HCl	12 M
$HC_2H_3O_2$	17 M
HNO_3	16 M
H_2SO_4	18 M

What is the molarity of a solution containing 1.4 mol of acetic acid, $HC_2H_3O_2$, in 250. mL of solution?

Example 14.5

By the unit conversion method we note that the concentration given in the problem statement is 1.4 mol per 250. mL (mol/mL). Since molarity = mol/L, the needed conversion is

Solution

$$\frac{\text{mol}}{\text{mL}} \longrightarrow \frac{\text{mol}}{\text{L}} = M$$

$$\left(\frac{1.4 \text{ mol}}{250. \text{ mL}}\right)\left(\frac{1000 \text{ mL}}{\text{L}}\right) = \frac{5.6 \text{ mol}}{\text{L}} = 5.6 \ M$$

Example 14.6   What is the molarity of a solution made by dissolving 2.00 g of potassium chlorate in enough water to make 150. mL of solution?

Solution   We use the unit conversion method. The steps in the conversions must lead to units of moles/liter:

$$\frac{g\ KClO_3}{mL} \longrightarrow \frac{g\ KClO_3}{L} \longrightarrow \frac{mol\ KClO_3}{L} = M$$

The data are

$$\text{mass } KClO_3 = 2.00\ g \qquad \text{molar mass } KClO_3 = 122.6\ g/mol \qquad \text{volume} = 150.\ mL$$

$$\left(\frac{2.00\ g\ \cancel{KClO_3}}{150.\ \cancel{mL}}\right)\left(\frac{1000\ \cancel{mL}}{L}\right)\left(\frac{1\ mol\ KClO_3}{122.6\ g\ \cancel{KClO_3}}\right) = \frac{0.109\ mol}{L} = 0.109\ M\ KClO_3$$

Example 14.7   How many grams of potassium hydroxide are required to prepare 600. mL of 0.450 M KOH solution?

Solution   The conversion is

$$\text{milliliters} \longrightarrow \text{liters} \longrightarrow \text{moles} \longrightarrow \text{grams}$$

The data are

$$\text{volume} = 600.\ mL \quad M = \frac{0.450\ mol}{L} \quad \text{molar mass } KOH = \frac{56.11\ g}{mol}$$

The calculation is

$$(600.\ \cancel{mL})\left(\frac{1\ \cancel{L}}{1000\ \cancel{mL}}\right)\left(\frac{0.450\ \cancel{mol}}{\cancel{L}}\right)\left(\frac{56.11\ g\ KOH}{\cancel{mol}}\right) = 15.1\ g\ KOH$$

Practice 14.3

What is the molarity of a solution made by dissolving 7.50 g of magnesium nitrate, $Mg(NO_3)_2$, in enough water to make 25.0 mL of solution?

Practice 14.4

How many grams of sodium chloride are needed to prepare 125 mL of a 0.037 M NaCl solution?

Example 14.8   Calculate the number of moles of nitric acid in 325 mL of 16 M HNO_3 solution.

Solution   Use the equation

$$\text{moles} = \text{liters} \times M$$

Substitute the data given in the problem and solve:

$$\text{moles} = (0.325\ \cancel{L})\left(\frac{16\ \text{mol HNO}_3}{1\ \cancel{L}}\right) = 5.2\ \text{mol HNO}_3$$

What volume of 0.250 M solution can be prepared from 16.0 g of potassium carbonate?   **Example 14.9**

Solution

We start with 16.0 g K_2CO_3; we need to find the volume of 0.250 M solution that can be prepared from this amount of K_2CO_3. The conversion therefore is

$$\text{g } K_2CO_3 \longrightarrow \text{mol } K_2CO_3 \longrightarrow \text{L solution}$$

The data are

$$\text{mass } K_2CO_3 = 16.0\ \text{g} \quad M = \frac{0.250\ \text{mol}}{1\ \text{L}} \quad \text{molar mass } K_2CO_3 = \frac{138.2\ \text{g}}{1\ \text{mol}}$$

$$(16.0\ \cancel{\text{g } K_2CO_3})\left(\frac{1\ \cancel{\text{mol } K_2CO_3}}{138.2\ \cancel{\text{g } K_2CO_3}}\right)\left(\frac{1\ \text{L}}{0.250\ \cancel{\text{mol } K_2CO_3}}\right) = 0.463\ \text{L (463 mL)}$$

Thus 463 mL of 0.250 M solution can be made from 16.0 g K_2CO_3.

How many milliliters of 2.00 M HCl will react with 28.0 g NaOH?   **Example 14.10**

Solution

Step 1. Write and balance the equation for the reaction:

$$HCl(aq) + NaOH(aq) \longrightarrow NaCl(aq) + H_2O(aq)$$

The equation states that 1 mol of HCl reacts with 1 mol of NaOH.

Step 2. Find the number of moles NaOH in 28.0 g NaOH:

$$\text{g NaOH} \longrightarrow \text{mol NaOH}$$

$$(28.0\ \text{g NaOH})\left(\frac{1\ \text{mol}}{40.00\ \text{g}}\right) = 0.700\ \text{mol NaOH}$$

$$28.0\ \text{g NaOH} = 0.700\ \text{mol NaOH}$$

Step 3. Solve for moles and volume of HCl needed. From Steps 1 and 2 we see that 0.700 mol HCl will react with 0.700 mol NaOH, because the ratio of moles reacting is 1:1. We know that 2.00 M HCl contains 2.00 mol HCl per liter, and so the volume that contains 0.700 mol HCl will be less than 1 L:

$$\text{mol NaOH} \longrightarrow \text{mol HCl} \longrightarrow \text{L HCl} \longrightarrow \text{mL HCl}$$

$$(0.700\ \cancel{\text{mol NaOH}})\left(\frac{1\ \cancel{\text{mol HCl}}}{1\ \cancel{\text{mol NaOH}}}\right)\left(\frac{1\ \text{L HCl}}{2.00\ \cancel{\text{mol HCl}}}\right) = 0.350\ \text{L HCl}$$

$$(0.350\ \cancel{\text{L}}\ \text{HCl})\left(\frac{1000\ \text{mL}}{1\ \cancel{\text{L}}}\right) = 350.\ \text{mL HCl}$$

Therefore 350. mL of 2.00 M HCl contains 0.700 mol HCl and will react with 0.700 mol, or 28.0 g, of NaOH.

TABLE 14.4	Concentration Units for Solutions	
Units	**Symbol**	**Definition**
Mass percent	% m/m	$\dfrac{\text{Mass solute}}{\text{Mass solution}} \times 100$
Parts per million	ppm	$\dfrac{\text{Mass solute}}{\text{Mass solution}} \times 1{,}000{,}000$
Mass/volume percent	% m/v	$\dfrac{\text{Mass solute}}{\text{mL solution}} \times 100$
Volume percent	% v/v	$\dfrac{\text{mL solute}}{\text{mL solution}} \times 100$
Molarity	M	$\dfrac{\text{Moles solute}}{\text{L solution}}$
Molality	m	$\dfrac{\text{Moles solute}}{\text{kg solvent}}$

Practice 14.5

What volume of 0.035 M AgNO$_3$ can be made from 5.0 g of AgNO$_3$?

Practice 14.6

How many milliliters of 0.50 M NaOH are required to react completely with 25.00 mL of 1.5 M HCl?

We've now examined several ways to measure concentration of solutions quantitatively. A summary of these concentration units is found in Table 14.4.

Dilution Problems

Chemists often find it necessary to dilute solutions from one concentration to another by adding more solvent to the solution. If a solution is diluted by adding pure solvent, the volume of the solution increases, but the number of moles of solute in the solution remains the same. Thus the moles/liter (molarity) of the solution decreases. Always read a problem carefully to distinguish between (1) how much solvent must be added to dilute a solution to a particular concentration, and (2) to what volume a solution must be diluted to prepare a solution of a particular concentration.

Example 14.11 Calculate the molarity of a sodium hydroxide solution that is prepared by mixing 100. mL of 0.20 M NaOH with 150. mL of water. Assume the volumes are additive.

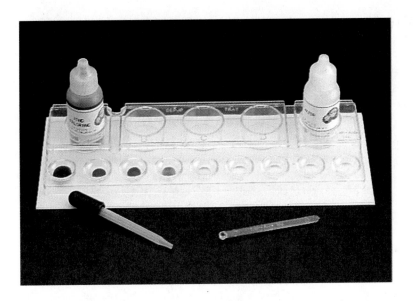

◀ **A serial dilution. The concentration of food coloring in well 1 (far left) is 1 part per 10 (by weight), well 2 is 1 part per 100, well 3 is 1 part per 1000, and so on. The concentration in well 6 is 1 part per million (ppm).**

Solution

This problem is a dilution problem. If we double the volume of a solution by adding water, we cut the concentration in half. Therefore, the concentration of the above solution should be less than 0.10 M. In the dilution, the moles of NaOH remain constant; the molarity and volume change. The final volume is (100. mL + 150. mL) or 250. mL.

To solve this problem, (1) calculate the moles of NaOH in the original solution, and (2) divide the moles of NaOH by the final volume of the solution to obtain the new molarity.

Step 1. Calculate the moles of NaOH in the original solution:

$$M = \frac{mol}{L} \qquad mol = L \times M$$

$$(0.100 \ \cancel{L})\left(\frac{0.20 \ mol \ NaOH}{1 \ \cancel{L}}\right) = 0.020 \ mol \ NaOH$$

Step 2. Solve for the new molarity, taking into account that the total volume of the solution after dilution is 250. mL (0.250 L):

$$M = \frac{0.020 \ mol \ NaOH}{0.250 \ L} = 0.080 \ M \ NaOH$$

Alternative Solution

When the moles of solute in a solution before and after dilution are the same, then the moles before and after dilution may be set equal to each other:

$$mol_1 = mol_2$$

where mol_1 = moles before dilution and mol_2 = moles after dilution. Then

$$mol_1 = L_1 \times M_1 \qquad mol_2 = L_2 \times M_2$$

$$L_1 \times M_1 = L_2 \times M_2$$

When both volumes are in the same units, a more general statement can be made:

$$V_1 \times M_1 = V_2 \times M_2$$

For this problem

$V_1 = 100.\ \text{mL}$ $\qquad\qquad$ $M_1 = 0.20\ M$

$V_2 = 150.\ \text{mL} + 100.\ \text{mL}$ $\qquad$ $M_2 = (\text{unknown})$

Then

$$(100.\ \text{mL})(0.20\ M) = (250.\ \text{mL})M_2$$

Solving for M_2, we get

$$M_2 = \frac{(100.\ \text{mL})(0.20\ M)}{250.\ \text{mL}} = 0.080\ M\ \text{NaOH}$$

Practice 14.7

Calculate the molarity of a solution prepared by diluting 125 mL of 0.400 M $K_2Cr_2O_7$ with 875 mL of water.

Example 14.12 How many grams of silver chloride will be precipitated by adding sufficient silver nitrate to react with 1500. mL of 0.400 M barium chloride solution?

$$2\ AgNO_3(aq) + BaCl_2(aq) \longrightarrow 2\ AgCl(s) + Ba(NO_3)_2(aq)$$

Solution

This problem is a stoichiometry problem. The fact that $BaCl_2$ is in solution means that we need to consider the volume and concentration of the solution in order to determine the number of moles of $BaCl_2$ reacting.

Step 1. Determine the number of moles of $BaCl_2$ in 1500. mL of 0.400 M solution:

$$M = \frac{\text{mol}}{\text{L}} \qquad \text{mol} = \text{L} \times M \qquad 1500.\ \text{mL} = 1.500\ \text{L}$$

$$(1.500\ \text{L})\left(\frac{0.400\ \text{mol } BaCl_2}{\text{L}}\right) = 0.600\ \text{mol } BaCl_2$$

Step 2. Use the mole-ratio method to calculate the moles and grams of AgCl:

$$\text{mol } BaCl_2 \longrightarrow \text{mol AgCl} \longrightarrow \text{g AgCl}$$

$$(0.600\ \text{mol } BaCl_2)\left(\frac{2\ \text{mol AgCl}}{1\ \text{mol } BaCl_2}\right)\left(\frac{143.4\ \text{g AgCl}}{\text{mol AgCl}}\right) = 172\ \text{g AgCl}$$

Practice 14.8

How many grams of lead(II) iodide will be precipitated by adding sufficient $Pb(NO_3)_2$ to react with 750 mL of 0.250 M KI solution?

$$2\ KI(aq) + Pb(NO_3)_2(aq) \longrightarrow PbI_2(s) + 2\ KNO_3(aq)$$

14.7 Colligative Properties of Solutions

Two solutions—one containing 1 mol (60.06 g) of urea, NH_2CONH_2, and the other containing 1 mol (342.3 g) of sucrose, $C_{12}H_{22}O_{11}$, in 1 kg of water—both have a freezing point of $-1.86°C$, not $0°C$ as for pure water. Urea and sucrose are distinctly different substances, yet they lower the freezing point of the water by the same amount. The only thing apparently common to these two solutions is that each contains 1 mol (6.022×10^{23} molecules) of solute and 1 kg of solvent. In fact, when we dissolve 1 mol of any nonionizable solute in 1 kg of water, the freezing point of the resulting solution is $-1.86°C$.

These results lead us to conclude that the freezing-point depression for a solution containing 6.022×10^{23} solute molecules (particles) and 1 kg of water is a constant, namely, 1.86°C. Freezing-point depression is a general property of solutions. Furthermore, the amount by which the freezing point is depressed is the same for all solutions made with a given solvent; that is, each solvent shows a characteristic *freezing-point depression constant*. Freezing-point depression constants for several solvents are given in Table 14.5.

The solution formed by the addition of a nonvolatile solute to a solvent has a lower freezing point, a higher boiling point, and a lower vapor pressure than that of the pure solvent. These effects are related and are known as colligative properties. The **colligative properties** are properties that depend only on the number of solute particles in a solution and not on the nature of those particles. Freezing-point depression, boiling-point elevation, and vapor-pressure lowering are colligative properties of solutions.

colligative properties

TABLE 14.5 **Freezing-Point Depression and Boiling-Point Elevation Constants of Selected Solvents**

Solvent	Freezing point of pure solvent (°C)	Freezing-point depression constant, K_f $\left(\dfrac{°C\ kg\ solvent}{mol\ solute}\right)$	Boiling point of pure solvent (°C)	Boiling-point elevation constant, K_b $\left(\dfrac{°C\ kg\ solvent}{mol\ solute}\right)$
Water	0.00	1.86	100.0	0.512
Acetic acid	16.6	3.90	118.5	3.07
Benzene	5.5	5.1	80.1	2.53
Camphor	178	40	208.2	5.95

Microencapsulation

Producing chemical reactions so that they occur at precisely the correct moment is a common task for chemists. For this to happen, one or more of the reactants must be stored separately and released under controlled conditions precisely when the reaction is desired. One technique that accomplishes this is microencapsulation, in which reactive chemicals—solids, liquids, or gases—are sealed in tiny capsules. The material forming the wall of the capsule is carefully chosen so that the encapsulated chemicals can be released at the appropriate time. This release can be accomplished in a variety of ways—dissolving the capsules; diffusion through the capsule walls; and by mechanical, thermal, electrical, or chemical disruption of the capsules.

In one type of microencapsulation, water diffuses into the capsule and forms a solution that then diffuses into the surroundings at a constant rate. Some types of capsules contain materials that dissolve at a certain level of acidity and form pores in the capsule through which the encapsulated materials escape. Still other types of capsules dissolve completely over a given period of time.

Applications of microencapsulation are found everywhere. Carbonless paper, often used in receipts, makes use of pressure-sensitive microcapsules containing colorless dye precursors. Another reactive substance is present and converts the precursor to the colored form when pressure is applied by a pen or printer.

Microencapsulated products are found in our kitchens. Flavorings are encapsulated to make them easier to store in a powdered state, cut evaporation, and reduce reactions with the air. These advantages increase shelf life. Flavoring microcapsules may also be heat sensitive and release their contents during cooking or pressure sensitive (as in chewing gum) and release their contents upon chewing.

Still other encapsulated products are found in our bathrooms. Time-release microencapsulation is used in deodorants, moisturizers, colognes, and perfumes. The encapsulation process prevents evaporation, decomposition, and unwanted reactions with the air and other ingredients. Drugs and medications are frequently encapsulated to dissolve slowly over a long period of time in the body. These medications generally work in the intestinal tract, but some may also be given by injection to work within other tissues.

Fragrances have undergone microencapsulation in such products as cosmetics, health care products, detergents, and even foods. Encapsulated fragrances are responsible for the ever-present scratch-and-sniff labels found in children's books and fashion magazines. When the paper is scratched or pulled open, the fragrance is released into the air.

Other applications of microencapsulation include time-release pesticides and neutralizer for contact lenses, as well as special additives in detergents, cleaners, and paints.

◄ **Many items we come across in our daily lives are microencapsulated. These are just a few examples.**

The colligative properties of a solution can be considered in terms of vapor pressure. The vapor pressure of a pure liquid depends on the tendency of molecules to escape from its surface. If 10% of the molecules in a solution are nonvolatile solute molecules, the vapor pressure of the solution is 10% lower than that of the pure solvent. The vapor pressure is lower because the surface of the solution contains 10% nonvolatile molecules and 90% of the volatile solvent molecules. A liquid

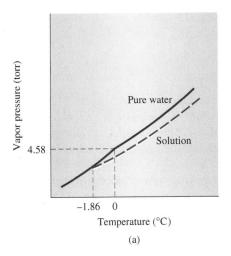

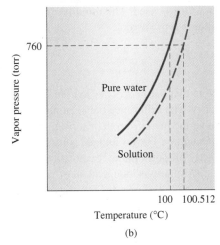

boils when its vapor pressure equals the pressure of the atmosphere. We can thus see that the solution just described as having a lower vapor pressure will have a higher boiling point than the pure solvent. The solution with a lowered vapor pressure doesn't boil until it has been heated above the boiling point of the solvent (see Figure 14.8a). Each solvent has its own characteristic boiling-point elevation constant (Table 14.5). The boiling-point elevation constant is based on a solution that contains 1 mol of solute particles per kilogram of solvent. For example, the boiling-point elevation constant for a solution containing 1 mol of solute particles per kilogram of water is 0.512°C, which means that this water solution will boil at 100.512°C.

The freezing behavior of a solution can also be considered in terms of lowered vapor pressure. Figure 14.8b shows the vapor-pressure relationships of ice, water, and a solution containing 1 mol of solute per kilogram of water. The freezing point of water is at the intersection of the liquid and solid vapor-pressure curves (i.e., at the point where water and ice have the same vapor pressure). Because the vapor pressure of the liquid is lowered by the solute, the vapor-pressure curve of the solution does not intersect the vapor-pressure curve of the solid until the solution has been cooled below the freezing point of pure water. So the solution must be cooled below 0°C in order for it to freeze.

The foregoing discussion dealing with freezing-point depressions is restricted to *un-ionized* substances. The discussion of boiling-point elevations is restricted to *nonvolatile* and un-ionized substances. The colligative properties of ionized substances are not under consideration at this point; we will discuss them in Chapter 15.

Some practical applications involving colligative properties are (1) use of salt–ice mixtures to provide low freezing temperatures for homemade ice cream, (2) use of sodium chloride or calcium chloride to melt ice from streets, and (3) use of ethylene glycol–water mixtures as antifreeze in automobile radiators (ethylene glycol also raises the boiling point of radiator fluid, thus allowing the engine to operate at a higher temperature).

Both the freezing-point depression and the boiling-point elevation are directly proportional to the number of moles of solute per kilogram of solvent. When we deal with the colligative properties of solutions, another concentration expression,

▲
Engine coolant is one application of colligative properties. The addition of coolant to the water in a radiator raises its boiling point and lowers its freezing point.

molality (m)   *molality,* is used. The **molality (*m*)** of a solute is the number of moles of solute per kilogram of solvent:

$$m = \frac{\text{mol solute}}{\text{kg solvent}}$$

Note that a lowercase *m* is used for molality concentrations and a capital *M* for molarity. The difference between molality and molarity is that molality refers to moles of solute *per kilogram of solvent,* whereas molarity refers to moles of solute *per liter of solution.* For un-ionized substances, the colligative properties of a solution are directly proportional to its molality.

Molality is independent of volume. It is a mass-to-mass relationship of solute to solvent and allows for experiments, such as freezing-point depression and boiling-point elevation, to be conducted at variable temperatures.

The following equations are used in calculations involving colligative properties and molality:

$$\Delta t_f = mK_f \qquad \Delta t_b = mK_b \qquad m = \frac{\text{mol solute}}{\text{kg solvent}}$$

m = molality; mol solute/kg solvent

Δt_f = freezing-point depression; °C

Δt_b = boiling-point elevation; °C

K_f = freezing-point depression constant; °C kg solvent/mol solute

K_b = boiling-point elevation constant; °C kg solvent/mol solute

Sodium chloride or calcium ▶ chloride is used to melt ice on snowy streets and highways.

What is the molality (m) of a solution prepared by dissolving 2.70 g CH_3OH in 25.0 g H_2O?

Example 14.13

Solution

Since $m = \dfrac{\text{mol solute}}{\text{kg solvent}}$,   the conversion is

$$\dfrac{2.70 \text{ g } CH_3OH}{25.0 \text{ g } H_2O} \longrightarrow \dfrac{\text{mol } CH_3OH}{25.0 \text{ g } H_2O} \longrightarrow \dfrac{\text{mol } CH_3OH}{1 \text{ kg } H_2O}$$

The molar mass of CH_3OH is $(12.01 + 4.032 + 16.00)$ or 32.04 g/mol:

$$\left(\dfrac{2.70 \text{ g } CH_3OH}{25.0 \text{ g } H_2O}\right)\left(\dfrac{1 \text{ mol } CH_3OH}{32.04 \text{ g } CH_3OH}\right)\left(\dfrac{1000 \text{ g } H_2O}{1 \text{ kg } H_2O}\right) = \dfrac{3.37 \text{ mol } CH_3OH}{1 \text{ kg } H_2O}$$

The molality is 3.37 m.

Practice 14.9

What is the molality of a solution prepared by dissolving 150.0 g $C_6H_{12}O_6$ in 600.0 g H_2O?

A solution is made by dissolving 100. g of ethylene glycol, $C_2H_6O_2$, in 200. g of water. What is the freezing point of this solution?

Example 14.14

Solution

To calculate the freezing point of the solution, we first need to calculate Δt_f, the change in freezing point. Use the equation

$$\Delta t_f = mK_f = \dfrac{\text{mol solute}}{\text{kg solvent}} \times K_f$$

K_f (for water): $\dfrac{1.86°C \text{ kg solvent}}{\text{mol solute}}$ (from Table 14.5)

mol solute: $(100. \text{ g } C_2H_6O_2)\left(\dfrac{1 \text{ mol } C_2H_6O_2}{62.07 \text{ g } C_2H_6O_2}\right) = 1.61 \text{ mol } C_2H_6O_2$

kg solvent: $(200. \text{ g } H_2O)\left(\dfrac{1 \text{ kg}}{1000 \text{ g}}\right) = 0.200 \text{ kg } H_2O$

$$\Delta t_f = \left(\dfrac{1.61 \text{ mol } C_2H_6O_2}{0.200 \text{ kg } H_2O}\right)\left(\dfrac{1.86°C \text{ kg } H_2O}{1 \text{ mol } C_2H_6O_2}\right) = 15.0°C$$

The freezing-point depression, 15.0°C, must be subtracted from 0°C, the freezing point of the pure solvent (water):

freezing point of solution = freezing point of solvent $- \Delta t_f$

$$= 0.0°C - 15.0°C = -15.0°C$$

Therefore the freezing point of the solution is $-15.0°C$.

Example 14.15 A solution made by dissolving 4.71 g of a compound of unknown molar mass in 100.0 g of water has a freezing point of $-1.46°C$. What is the molar mass of the compound?

Solution   First substitute the data in $\Delta t_f = mK_f$ and solve for m:

$$\Delta t_f = +1.46 \text{ (since the solvent, water, freezes at } 0°C)$$

$$K_f = \frac{1.86°C \text{ kg } H_2O}{\text{mol solute}}$$

$$1.46°C = mK_f = m \times \frac{1.86°C \text{ kg } H_2O}{\text{mol solute}}$$

$$m = \frac{1.46°C \times \text{mol solute}}{1.86°C \times \text{kg } H_2O} = \frac{0.785 \text{ mol solute}}{\text{kg } H_2O}$$

Now convert the data, 4.71 g solute/100.0 g H_2O, to g/mol:

$$\left(\frac{4.71 \text{ g solute}}{100.0 \text{ g } H_2O}\right)\left(\frac{1000 \text{ g } H_2O}{1 \text{ kg } H_2O}\right)\left(\frac{1 \text{ kg } H_2O}{0.785 \text{ mol solute}}\right) = 60.0 \text{ g/mol}$$

The molar mass of the compound is 60.0 g/mol.

Practice 14.10

What is the freezing point of the solution in Practice Exercise 14.9? What is the boiling point?

14.8   Osmosis and Osmotic Pressure

When red blood cells are put into distilled water, they gradually swell and in time may burst. If red blood cells are put in a 5% urea (or a 5% salt) solution, they gradu-

semipermeable membrane   ally shrink and take on a wrinkled appearance. The cells behave in this fashion because they are enclosed in semipermeable membranes. A **semipermeable membrane** allows the passage of water (solvent) molecules through it in either direction but prevents the passage of larger solute molecules or ions. When two solutions of different concentrations (or water and a water solution) are separated by a semipermeable membrane, water diffuses through the membrane from the solution of lower concentration into the solution of higher concentration. The diffusion of water, either from a dilute solution or from pure water, through a semipermeable membrane into a

osmosis   solution of higher concentration is called **osmosis.**

A 0.90% (0.15 M) sodium chloride solution is known as a *physiological saline solution* because it is *isotonic* with blood plasma; that is, it has the same osmotic pressure as blood plasma. Because each mole of NaCl yields about 2 mol of ions when in solution, the solute particle concentration in physiological saline solution is nearly 0.30 M. Five percent glucose solution (0.28 M) is also approximately isotonic with blood plasma. Blood cells neither swell nor shrink in an isotonic solution.

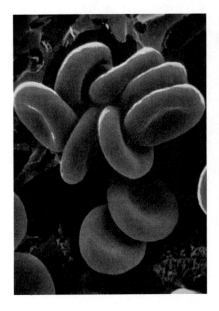
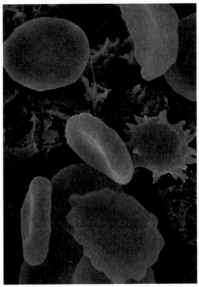
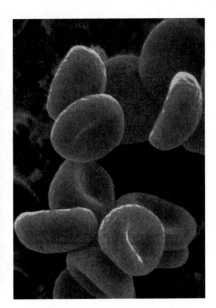

▲
Human red blood cells. Left: In an isotonic solution the concentration is the same inside and outside the cell (0.9% saline). Center: In a hypertonic solution (1.6% saline) water leaves the cells causing them to crenate (shrink). Right: In a hypotonic solution (0.2% saline) the cells swell as water moves into the cell center. Cells do not change in size. Magnification is 260,000×.

The cells described in the preceding paragraph swell in water because water is *hypotonic* to cell plasma. The cells shrink in 5% urea solution because the urea solution is *hypertonic* to the cell plasma. To prevent possible injury to blood cells by osmosis, fluids for intravenous use are usually made up at approximately isotonic concentration.

All solutions exhibit *osmotic pressure,* which is another colligative property. Osmotic pressure is dependent only on the concentration of the solute particles and is independent of their nature. The osmotic pressure of a solution can be measured by determining the amount of counterpressure needed to prevent osmosis; this pressure can be very large. The osmotic pressure of a solution containing 1 mol of solute particles in 1 kg of water is about 22.4 atm, which is about the same as the pressure exerted by 1 mol of a gas confined in a volume of 1 L at 0°C.

Osmosis has a role in many biological processes, and semipermeable membranes occur commonly in living organisms. An example is the roots of plants, which are covered with tiny structures called root hairs; soil water enters the plant by osmosis, passing through the semipermeable membranes covering the root hairs. Artificial or synthetic membranes can also be made.

Osmosis can be demonstrated with the simple laboratory setup shown in Figure 14.9. As a result of osmotic pressure, water passes through the cellophane membrane into the thistle tube, causing the solution level to rise. In osmosis the net transfer of water is always from a less concentrated to a more concentrated solution; that is, the effect is toward equalization of the concentration on both sides of the membrane. Note that the effective movement of water in osmosis is always from the region of *higher water concentration* to the region of *lower water concentration.*

Osmosis can be explained by assuming that a semipermeable membrane has passages that permit water molecules and other small molecules to pass in either direction. Both sides of the membrane are constantly being struck by water molecules in random motion. The number of water molecules crossing the membrane is proportional to the number of water molecule-to-membrane impacts per unit of time. Because the solute molecules or ions reduce the concentration of water, there are

FIGURE 14.9 ▶
Laboratory demonstration of osmosis: As a result of osmosis, water passes through the membrane causing the solution to rise in the thistle tube.

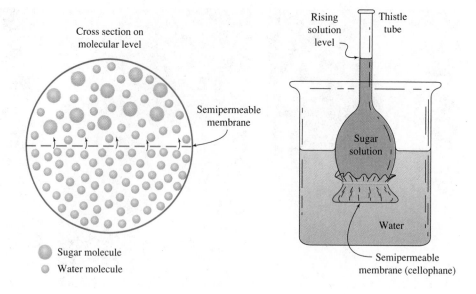

Cross section on molecular level

Semipermeable membrane

Sugar molecule

Water molecule

Rising solution level

Thistle tube

Sugar solution

Water

Semipermeable membrane (cellophane)

more water molecules and thus more water molecule impacts on the side with the lower solute concentration (more dilute solution). The greater number of water molecule-to-membrane impacts on the dilute side thus causes a net transfer of water to the more concentrated solution. Again, note that the overall process involves the net transfer, by diffusion through the membrane, of water molecules from a region of higher water concentration (dilute solution) to one of lower water concentration (more concentrated solution).

This is a simplified picture of osmosis. No one has ever seen the hypothetical passages that allow water molecules and other small molecules or ions to pass through them. Alternative explanations have been proposed, but our discussion has been confined to water solutions. Osmotic pressure is a general colligative property, however, and is known to occur in nonaqueous systems.

Concepts in Review

1. Describe the types of solutions.
2. List the general properties of solutions.
3. Describe and illustrate the process by which an ionic substance dissolves in water.
4. Indicate the effects of temperature and pressure on the solubility of solids and gases in liquids.
5. Identify and explain the factors affecting the rate at which a solid dissolves in a liquid.
6. Use a solubility table or graph to determine whether a solution is saturated, unsaturated, or supersaturated at a given temperature.
7. Calculate the mass percent or volume percent for a solution.
8. Calculate the amount of solute in a given quantity of a solution when given the mass percent or volume percent of a solution.
9. Calculate the molarity of a solution from the volume and the mass, or moles, of solute.
10. Calculate the mass of a substance necessary to prepare a solution of specified volume and molarity.
11. Determine the resulting molarity in a typical dilution problem.
12. Apply stoichiometry to chemical reactions involving solutions.
13. Explain the effect of a solute on the vapor pressure of a solvent.

14. Explain the effect of a solute on boiling point and freezing point of a solution.

15. Calculate the boiling and freezing points of a solution from concentration data.

16. Calculate molality and molar mass of a solute from boiling/freezing point data.

17. Explain the process of osmosis.

Key Terms

colligative properties (14.7)
concentrated solution (14.3)
concentration of a solution (14.2)
dilute solution (14.6)
immiscible (14.2)
miscible (14.2)

molality (m) (14.7)
molarity (M) (14.6)
osmosis (14.8)
parts per million (ppm) (14.6)
saturated solution (14.3)
semipermeable membrane (14.8)

solubility (14.2)
solute (14.1)
solution (14.1)
solvent (14.1)
supersaturated solution (14.3)
unsaturated solution (14.6)

Questions

1. Sketch the orientation of water molecules (a) about a single sodium ion and (b) about a single chloride ion in solution.

2. Estimate the number of grams of sodium fluoride that would dissolve in 100 g of water at 50°C. (Table 14.2)

3. What is the solubility at 25°C of these substances? (Figure 14.4)
   (a) potassium chloride   (c) potassium nitrate
   (b) potassium chlorate

4. What is different in the solubility trend of the potassium halides compared with that of the lithium halides and the sodium halides? (Table 14.2)

5. What is the solubility, in grams of solute per 100 g of H_2O, of (a) $KClO_3$ at 60°C; (b) HCl at 20°C; (c) Li_2SO_4 at 80°C; and (d) KNO_3 at 0°C? (Figure 14.4)

6. Which substance, KNO_3 or NH_4Cl, shows the greater increase in solubility with increased temperature? (Figure 14.4)

7. Does a 2-molal solution in benzene or a 1-molal solution in camphor show the greater freezing-point depression? (Table 14.5)

8. What would be the total surface area if the 1-cm cube in Figure 14.5 were cut into cubes 0.01 cm on a side?

9. At which temperatures—10°C, 20°C, 30°C, 40°C, or 50°C—would you expect a solution made from 63 g of ammonium chloride and 150 g of water to be unsaturated? (Figure 14.4)

10. Explain why the rate of dissolving decreases. (Figure 14.6)

11. Explain how a supersaturated solution of $NaC_2H_3O_2$ can be prepared and proven to be supersaturated.

12. Assume that the thistle tube in Figure 14.9 contains 1.0 M sugar solution and that the water in the beaker has just been replaced by a 2.0 M solution of urea. Would the solution level in the thistle tube continue to rise, remain constant, or fall? Explain.

13. Name and distinguish between the two components of a solution.

14. Is it always apparent in a solution which component is the solute, for example, in a solution of a liquid in a liquid?

15. Explain why the solute does not settle out of a solution.

16. Is it possible to have one solid dissolved in another? Explain.

17. An aqueous solution of KCl is colorless, $KMnO_4$ is purple, and $K_2Cr_2O_7$ is orange. What color would you expect of an aqueous solution of $Na_2Cr_2O_7$? Explain.

18. Explain why hexane will dissolve benzene but will not dissolve sodium chloride.

19. Some drinks like tea are consumed either hot or cold, whereas others like Coca Cola are drunk only cold. Why?

20. Why is air considered to be a solution?

21. In which will a teaspoonful of sugar dissolve more rapidly, 200 mL of iced tea or 200 mL of hot coffee? Explain in terms of the KMT.

22. What is the effect of pressure on the solubility of gases in liquids? Of solids in liquids?

23. Why do smaller particles dissolve faster than large ones?

24. In a saturated solution containing undissolved solute, solute is continuously dissolving, but the concentration of the solution remains unchanged. Explain.

25. Explain why there is no apparent reaction when crystals of $AgNO_3$ and NaCl are mixed, but a reaction is apparent immediately when solutions of $AgNO_3$ and NaCl are mixed.

26. What do we mean when we say that concentrated nitric acid, HNO_3, is 16 molar?

27. Will 1 L of 1 M NaCl contain more chloride ions than 0.5 L of 1 M $MgCl_2$? Explain.

28. Champagne is usually cooled in a refrigerator prior to opening. It's also opened very carefully. What would happen if a warm bottle of champagne is shaken and opened quickly and forcefully?

29. Describe how you would prepare 750 mL of 5 M NaCl solution.

30. Explain in terms of the KMT how a semipermeable membrane functions when placed between pure water and a 10% sugar solution.

31. Which has the higher osmotic pressure, a solution containing 100 g of urea, NH_2CONH_2, in 1 kg H_2O or a solution containing 150 g of glucose, $C_6H_{12}O_6$, in 1 kg H_2O?

32. Explain why a lettuce leaf in contact with salad dressing containing salt and vinegar soon becomes wilted and limp whereas another lettuce leaf in contact with plain water remains crisp.

33. A group of shipwreck survivors floated for several days on a life raft before being rescued. Those who had drunk some seawater were found to be suffering the most from dehydration. Explain.

34. Arrange the following bases (in descending order) according to the volume of each that will react with 1 L of 1 M HCl:
    (a) 1 M NaOH
    (b) 1.5 M $Ca(OH)_2$
    (c) 2 M KOH
    (d) 0.6 M $Ba(OH)_2$

*35. Explain in terms of vapor pressure why the boiling point of a solution containing a nonvolatile solute is higher than that of the pure solvent.

36. Explain why the freezing point of a solution is lower than the freezing point of the pure solvent.

37. Which would be colder, a glass of water and crushed ice or a glass of Seven-Up and crushed ice? Explain.

38. When water and ice are mixed, the temperature of the mixture is 0°C. But, if methyl alcohol and ice are mixed, a temperature of $-10°C$ is readily attained. Explain why the two mixtures show such different temperature behavior.

39. Which would be more effective in lowering the freezing point of 500. g of water?
    (a) 100. g of sucrose, $C_{12}H_{22}O_{11}$, or 100. g of ethyl alcohol, C_2H_5OH
    (b) 100. g of sucrose or 20.0 g of ethyl alcohol
    (c) 20.0 g of ethyl alcohol or 20.0 g of methyl alcohol, CH_3OH

40. Is the molarity of a 5 m aqueous solution of NaCl greater or less than 5 M? Explain.

Paired Exercises

41. Which of the substances listed below are reasonably soluble and which are insoluble in water? (See Figure 14.2 or Appendix V.)
    (a) KOH
    (b) $NiCl_2$
    (c) ZnS
    (d) $AgC_2H_3O_2$
    (e) Na_2CrO_4

42. Which of the substances listed below are reasonably soluble and which are insoluble in water? (See Figure 14.2 or Appendix V.)
    (a) PbI_2
    (b) $MgCO_3$
    (c) $CaCl_2$
    (d) $Fe(NO_3)_3$
    (e) $BaSO_4$

Percent Solutions

43. Calculate the mass percent of the following solutions:
    (a) 25.0 g NaBr + 100.0 g H_2O
    (b) 1.20 g K_2SO_4 + 10.0 g H_2O

44. Calculate the mass percent of the following solutions:
    (a) 40.0 g $Mg(NO_3)_2$ + 500.0 g H_2O
    (b) 17.5 g $NaNO_3$ + 250.0 g H_2O

45. How many grams of a solution that is 12.5% by mass $AgNO_3$ would contain 30.0 g of $AgNO_3$?

46. How many grams of a solution that is 12.5% by mass $AgNO_3$ would contain 0.400 mol of $AgNO_3$?

47. Calculate the mass percent of the following solutions:
    (a) 60.0 g NaCl + 200.0 g H_2O
    (b) 0.25 mol $HC_2H_3O_2$ + 3.0 mol H_2O

48. Calculate the mass percent of the following solutions:
    (a) 145.0 g NaOH in 1.5 kg H_2O
    (b) 1.0 m solution of $C_6H_{12}O_6$ in water

49. How much solute is present in 65 g of 5.0% KCl solution?

50. How much solute is present in 250. g of 15.0% K_2CrO_4 solution?

51. Calculate the mass/volume percent of a solution made by dissolving 22.0 g of CH_3OH (methanol) in C_2H_5OH (ethanol) to make 100. mL of solution.

52. Calculate the mass/volume percent of a solution made by dissolving 4.20 g of NaCl in H_2O to make 12.5 mL of solution.

53. What is the volume percent of 10.0 mL of CH_3OH (methanol) dissolved in water to a volume of 40.0 mL?

54. What is the volume percent of 2.0 mL of hexane, C_6H_{14}, dissolved in benzene, C_6H_6, to a volume of 9.0 mL?

Molarity Problems

55. Calculate the molarity of the following solutions:
    (a) 0.10 mol of solute in 250 mL of solution
    (b) 2.5 mol of NaCl in 0.650 L of solution
    (c) 53.0 g of Na_2CrO_4 in 1.00 L of solution
    (d) 260 g of $C_6H_{12}O_6$ in 800. mL of solution

56. Calculate the molarity of the following solutions:
    (a) 0.025 mol of HCl in 10. mL of solution
    (b) 0.35 mol $BaCl_2 \cdot 2\ H_2O$ in 593 mL of solution
    (c) 1.50 g of $Al_2(SO_4)_3$ in 2.00 L of solution
    (d) 0.0282 g of $Ca(NO_3)_2$ in 1.00 mL of solution

57. Calculate the number of moles of solute in each of the following solutions:
    (a) 40.0 L of 1.0 M LiCl
    (b) 25.0 mL of 3.00 M H_2SO_4

58. Calculate the number of moles of solute in each of the following solutions:
    (a) 349 mL of 0.0010 M NaOH
    (b) 5000. mL of 3.1 M $CoCl_2$

59. Calculate the grams of solute in each of the following solutions:
    (a) 150 L of 1.0 M NaCl
    (b) 260 mL of 18 M H_2SO_4

60. Calculate the grams of solute in each of the following solutions:
    (a) 0.035 L of 10.0 M HCl
    (b) 8.00 mL of 8.00 M $Na_2C_2O_4$

61. How many milliliters of 0.256 M KCl solution will contain the following?
    (a) 0.430 mol of KCl
    (b) 20.0 g of KCl

62. How many milliliters of 0.256 M KCl solution will contain the following?
    (a) 10.0 mol of KCl
    *(b) 71.0 g of chloride ion, Cl^-

Dilution Problems

63. What will be the molarity of the resulting solutions made by mixing the following? Assume volumes are additive.
    (a) 100. mL 1.0 M HCl + 150 mL 2.0 M HCl
    (b) 25.0 mL 12.5 M NaCl + 75.0 mL 2.00 M NaCl

64. What will be the molarity of the resulting solutions made by mixing the following? Assume volumes are additive.
    (a) 200. mL of 12 M HCl + 200.0 mL H_2O
    (b) 60.0 mL of 0.60 M $ZnSO_4$ + 500. mL H_2O

65. Calculate the volume of concentrated reagent required to prepare the diluted solutions indicated:
    (a) 15 M NH_3 to prepare 50. mL of 6.0 M NH_3
    (b) 18 M H_2SO_4 to prepare 250 mL of 10.0 M H_2SO_4

66. Calculate the volume of concentrated reagent required to prepare the diluted solutions indicated:
    (a) 12 M HCl to prepare 400. mL of 6.0 M HCl
    (b) 16 M HNO_3 to prepare 100. mL of 2.5 M HNO_3

67. Calculate the molarity of the solutions made by mixing 250 mL of 0.75 M H_2SO_4 with
    (a) 150 mL of H_2O
    (b) 250 mL of 0.70 M H_2SO_4

68. Calculate the molarity of the solutions made by mixing 250 mL of 0.75 M H_2SO_4 with
    (a) 400. mL of 2.50 M H_2SO_4
    (b) 375 mL of H_2O

Stoichiometry Problems

69. Given the equation
$BaCl_2(aq) + K_2CrO_4(aq) \longrightarrow BaCrO_4(s) + 2\ KCl(aq)$,
calculate
  (a) The grams of $BaCrO_4$ that can be obtained from 100.0 mL of 0.300 M $BaCl_2$
  (b) The volume of 1.0 M $BaCl_2$ solution needed to react with 50.0 mL of 0.300 M K_2CrO_4 solution

71. Given the equation
$6\ FeCl_2(aq) + K_2Cr_2O_7(aq) + 14\ HCl(aq) \longrightarrow$
$\quad 6\ FeCl_3(aq) + 2\ CrCl_3(aq) + 2\ KCl(aq) + 7\ H_2O(l)$,
calculate
  (a) moles KCl produced from 2.0 mol $FeCl_2$
  (b) moles $CrCl_3$ produced from 1.0 mol $FeCl_2$
  (c) moles $FeCl_2$ required to react with 0.050 mol $K_2Cr_2O_7$
  (d) milliliters of 0.060 M $K_2Cr_2O_7$ required to react with 0.025 mol $FeCl_2$
  (e) milliliters of 6.0 M HCl required to react with 15.0 mL 6.0 M $FeCl_2$

70. Given the equation
$3\ MgCl_2(aq) + 2\ Na_3PO_4(aq) \longrightarrow$
$\quad\quad\quad\quad\quad Mg_3(PO_4)_2(s) + 6\ NaCl(aq)$,
calculate
  (a) milliliters of 0.250 M Na_3PO_4 that will react with 50.0 mL of 0.250 M $MgCl_2$
  (b) grams of $Mg_3(PO_4)_2$ that will be formed from 50.0 mL of 0.250 M $MgCl_2$

72. Given the equation
$2\ KMnO_4(aq) + 16\ HCl(aq) \longrightarrow$
$\quad 2\ MnCl_2(aq) + 5\ Cl_2(g) + 8\ H_2O(l) + 2\ KCl(aq)$,
calculate
  (a) moles Cl_2 produced from 0.050 mol $KMnO_4$
  (b) moles of HCl required to react with 1.0 L of 2.0 M $KMnO_4$
  (c) milliliters of 6.0 M HCl required to react with 200. mL of 0.50 M $KMnO_4$
  (d) liters of Cl_2 gas at STP produced by the reaction of 75.0 mL of 6.0 M HCl

Molality and Colligative Properties Problems

73. Calculate the molality of these solutions:
  (a) 14.0 g CH_3OH in 100.0 g of H_2O
  (b) 2.50 mol of benzene, C_6H_6, in 250 g of hexane, C_6H_{14}

75. What is the (a) molality, (b) freezing point, and (c) boiling point of a solution containing 2.68 g of naphthalene, $C_{10}H_8$, in 38.4 g of benzene, C_6H_6?

***77.** The freezing point of a solution of 8.00 g of an unknown compound dissolved in 60.0 g of acetic acid is 13.2°C. Calculate the molar mass of the compound.

74. Calculate the molality of these solutions:
  (a) 1.0 g $C_6H_{12}O_6$ in 1.0 g H_2O
  (b) 0.250 mol iodine in 1.0 kg H_2O

76. What is the (a) molality, (b) freezing point, and (c) boiling point of a solution containing 100.0 g of ethylene glycol, $C_2H_6O_2$, in 150.0 g of water?

***78.** What is the molar mass of a compound if 4.80 g of the compound dissolved in 22.0 g of H_2O gives a solution that freezes at −2.50°C?

Additional Exercises

79. How many grams of solution, 10.0% NaOH by mass, are required to neutralize 150 mL of a 1.0 M HCl solution?

***80.** How many grams of solution, 10.0% NaOH by mass, are required to neutralize 250.0 g of a 1.0 m solution of HCl?

***81.** A sugar syrup solution contains 15.0% sugar, $C_{12}H_{22}O_{11}$, by mass and has a density of 1.06 g/mL.
  (a) How many grams of sugar are in 1.0 L of this syrup?
  (b) What is the molarity of this solution?
  (c) What is the molality of this solution?

***82.** A solution of 3.84 g C_4H_2N (empirical formula) in 250.0 g of benzene depresses the freezing point of benzene 0.614°C. What is the molecular formula for the compound?

***83.** Hydrochloric acid, HCl, is sold as a concentrated aqueous solution (12.0 mol/L). If the density of the solution is 1.18 g/mL, determine the molality of the solution.

***84.** How many grams of KNO_3 are needed to make 450 mL of a solution that is to contain 5.5 mg/mL of potassium ion? Calculate the molarity of the solution.

85. What mass of 5.50% solution can be prepared from 25.0 g KCl?

86. Physiological saline, NaCl, solutions used in intravenous injections have a concentration of 0.90% NaCl (mass/volume).
    (a) How many grams of NaCl are needed to prepare 500.0 mL of this solution?
    *(b) How much water must evaporate from this solution to give a solution that is 9.0% NaCl (mass/volume)?

*87. A solution is made from 50.0 g KNO_3 and 175 g H_2O. How many grams of water must evaporate to give a saturated solution of KNO_3 in water at 20°C? (See Figure 14.4.)

88. What volume of 70.0% rubbing alcohol can you prepare if you have only 150 mL of pure isopropyl alcohol on hand?

89. At 20°C, an aqueous solution of HNO_3 that is 35.0% HNO_3 by mass has a density of 1.21 g/mL.
    (a) How many grams of HNO_3 are present in 1.00 L of this solution?
    (b) What volume of this solution will contain 500. g HNO_3?

*90. What is the molarity of a nitric acid solution, if the solution is 35.0% HNO_3 by mass and has a density of 1.21 g/mL?

91. To what volume must a solution of 80.0 g H_2SO_4 in 500.0 mL of solution be diluted to give a 0.10 M solution?

92. Calculate the milliliters of water that must be added to 300.0 mL of 1.40 M HCl to make a solution that is 0.500 M HCl?

93. A 10.0-mL sample of 16 M HNO_3 is diluted to 500.0 mL. What is the molarity of the final solution?

94. Given a 5.00 M KOH solution, how would you prepare 250.0 mL of 0.625 M KOH?

95. (a) How many moles of hydrogen will be liberated from 200.0 mL of 3.00 M HCl reacting with an excess of magnesium? The equation is

    $$Mg(s) + 2 HCl(aq) \longrightarrow MgCl_2(aq) + H_2(g)$$

    (b) How many liters of hydrogen gas, H_2, measured at 27°C and 720 torr, will be obtained? (*Hint*: Use the ideal gas equation.)

*96. What is the molarity of an HCl solution, 150.0 mL of which, when treated with excess magnesium, liberates 3.50 L of H_2 gas measured at STP?

97. Suppose you start with 100. mL of distilled water, then you add one drop (20 drops/mL) of concentrated acetic acid, $HC_2H_3O_2$ (17.8 M). What is the molarity of the resulting solution?

98. Which will be more effective in neutralizing stomach acid, HCl: a tablet containing 12.0 g $Mg(OH)_2$ or a tablet containing 10.0 g $Al(OH)_3$? Show evidence for your answer.

99. Which would be more effective as an antifreeze in an automobile radiator? A solution containing
    (a) 10 kg of methyl alcohol, CH_3OH, or 10 kg of ethyl alcohol, C_2H_5OH?
    (b) 10 m solution of methyl alcohol or 10 m solution of ethyl alcohol?

100. Automobile battery acid is 38% H_2SO_4 and has a density of 1.29 g/mL. Calculate the molality and the molarity of this solution.

101. A sugar solution made to feed hummingbirds contains 1.00 lb of sugar, $C_{12}H_{22}O_{11}$, to 4.00 lb of water. Can this solution be put outside without freezing where the temperature falls to 20.0°F at night? Show evidence for your answer.

*102. What is the (a) molality and (b) boiling point of an aqueous sugar, $C_{12}H_{22}O_{11}$, solution that freezes at −5.4°C?

103. A solution of 6.20 g $C_2H_6O_2$ in water has a freezing point of −0.372°C. How many grams of H_2O are in the solution?

104. What (a) mass and (b) volume of ethylene glycol ($C_2H_6O_2$, density = 1.11 g/mL) should be added to 12.0 L of water in an automobile radiator to protect it from freezing at −20°C? (c) To what temperature Fahrenheit will the radiator be protected?

105. Can a saturated solution ever be a dilute solution? Explain.

106. What volume of 0.65 M HCl is needed to completely neutralize 12 g NaOH?

*107. If 150 mL of 0.055 M HNO_3 are needed to completely neutralize 1.48 g of an *impure* sample of sodium hydrogen carbonate (baking soda), what percent of the sample is baking soda?

108. (a) How much water must be added to concentrated sulfuric acid, H_2SO_4 (17.8 M), to prepare 8.4 L of 1.5 M sulfuric acid solution?
     (b) How many moles of H_2SO_4 are in each milliliter of the original concentrate?
     (c) How many moles are in each milliliter of the diluted solution?

109. An aqueous solution freezes at −3.6°C. What is its boiling temperature?

*110. How would you prepare a 6.00 M HNO$_3$ solution if only 3.00 M and 12.0 M solutions of the acid are available for mixing?

*111. A 20.0-mL portion of an HBr solution of unknown strength is diluted to exactly 240 mL. If 100.0 mL of this diluted solution requires 88.4 mL of 0.37 M NaOH to achieve complete neutralization, what was the strength of the original HBr solution?

112. When 80.5 mL of 0.642 M Ba(NO$_3$)$_2$ is mixed with 44.5 mL of 0.743 M KOH, a precipitate of Ba(OH)$_2$ forms. How many grams of Ba(OH)$_2$ do you expect?

113. Exactly 300. g of a 5.0% sucrose solution is to be prepared. How many grams of a 2.0% solution of sucrose would contain the same number of grams of sugar?

114. A 0.25 M solution of lithium carbonate, Li$_2$CO$_3$, a drug used to treat manic depression, is prepared.
  (a) How many moles of Li$_2$CO$_3$ are present in 45.8 mL of the solution?
  (b) How many grams of Li$_2$CO$_3$ are in 750 mL of the same solution?
  (c) How many milliliters of the solution would be needed to supply 6.0 g of the solute?
  (d) If the solution has a density of 1.22 g/mL, what is its mass percent?

115. If a student accidentally mixed 400.0 mL of 0.35 M HCl with 1100 mL of 0.65 M HCl, what would be the final molarity of the hydrochloric acid solution?

Answers to Practice Exercises

14.1	unsaturated	**14.5**	0.84 L (840 mL)	**14.9**	1.387 m
14.2	10.0% Na$_2$SO$_4$ solution	**14.6**	75 mL NaOH	**14.10**	freezing point $= -2.58°$C, boiling
14.3	2.02 M	**14.7**	5.00×10^{-2} M		point $= 100.71°$C
14.4	0.27 g NaCl	**14.8**	43 g		

PUTTING IT TOGETHER
Review for Chapters 12–14

Multiple Choice: *Choose the correct answer to each of the following.*

1. Which of these statements is *not* one of the principal assumptions of the kinetic-molecular theory for an ideal gas?
   (a) All collisions of gaseous molecules are perfectly elastic.
   (b) A mole of any gas occupies 22.4 L at STP.
   (c) Gas molecules have no attraction for one another.
   (d) The average kinetic energy for molecules is the same for all gases at the same temperature.

2. Which of the following is not equal to 1.00 atm?
   (a) 760. cm Hg   (c) 760. mm Hg
   (b) 29.9 in. Hg   (d) 760. torr

3. If the pressure on 45 mL of gas is changed from 600. torr to 800. torr, the new volume will be
   (a) 60 mL   (c) 0.045 L
   (b) 34 mL   (d) 22.4 L

4. The volume of a gas is 300. mL at 740. torr and 25°C. If the pressure remains constant and the temperature is raised to 100.°C, the new volume will be
   (a) 240. mL   (c) 376 mL
   (b) 1.20 L   (d) 75.0 mL

5. The volume of a dry gas is 4.00 L at 15.0°C and 745 torr. What volume will the gas occupy at 40.0°C and 700. torr?
   (a) 4.63 L   (b) 3.46 L   (c) 3.92 L   (d) 4.08 L

6. A sample of Cl_2 occupies 8.50 L at 80.0°C and 740. mm Hg. What volume will the Cl_2 occupy at STP?
   (a) 10.7 L   (b) 6.75 L   (c) 11.3 L   (d) 6.40 L

7. What volume will 8.00 g O_2 occupy at 45°C and 2.00 atm?
   (a) 0.462 L   (b) 104 L   (c) 9.62 L   (d) 3.26 L

8. The density of NH_3 gas at STP is
   (a) 0.760 g/mL   (c) 1.32 g/mL
   (b) 0.760 g/L   (d) 1.32 g/L

9. The ratio of the relative rate of effusion of methane, CH_4, to sulfur dioxide, SO_2, is
   (a) $^{64}/_{16}$ (b) $^{16}/_{64}$ (c) $^{1}/_4$ (d) $^{2}/_1$

10. Measured at 65°C and 500. torr, the mass of 3.21 L of a gas is 3.5 g. The molar mass of this gas is
    (a) 21 g/mole   (c) 24 g/mole
    (b) 46 g/mole   (d) 130 g/mole

11. Box A contains O_2 (molar mass = 32.0) at a pressure of 200 torr. Box B, which is identical to box A in volume, contains twice as many molecules of CH_4 (molar mass = 16.0) as the molecules of O_2 in box A. The temperatures of the gases are identical. The pressure in box B is
    (a) 100 torr   (c) 400 torr
    (b) 200 torr   (d) 800 torr

12. A 300.-mL sample of oxygen, O_2, is collected over water at 23°C and 725 torr. If the vapor pressure of water at 23°C is 21.0 torr, the volume of dry O_2 at STP is
    (a) 256 mL   (c) 341 mL
    (b) 351 mL   (d) 264 mL

13. A tank containing 0.01 mol of neon and 0.04 mol of helium shows a pressure of 1 atm. What is the partial pressure of neon in the tank?
    (a) 0.8 atm   (c) 0.2 atm
    (b) 0.01 atm   (d) 0.5 atm

14. How many liters of NO_2 (at STP) can be produced from 25.0 g Cu reacting with concentrated nitric acid?

$$Cu(s) + 4\ HNO_3(aq) \longrightarrow$$
$$Cu(NO_3)_2(aq) + 2\ H_2O(l) + 2\ NO_2(g)$$

   (a) 4.41 L   (b) 8.82 L   (c) 17.6 L   (d) 44.8 L

15. How many liters of butane vapor are required to produce 2.0 L CO_2 at STP?

$$2\ C_4H_{10}(g) + 13\ O_2(g) \longrightarrow 8\ CO_2(g) + 10\ H_2O(g)$$
   butane

   (a) 2.0 L   (b) 4.0 L   (c) 0.80 L   (d) 0.50 L

16. What volume of CO_2 (at STP) can be produced when 15.0 g C_2H_6 and 50.0 g O_2 are reacted?

$$2\ C_2H_6(g) + 7\ O_2(g) \longrightarrow 4\ CO_2(g) + 6\ H_2O(g)$$

   (a) 20.0 L   (b) 22.4 L   (c) 35.0 L   (d) 5.6 L

17. Which of these gases has the highest density at STP?
    (a) N_2O (b) NO_2 (c) Cl_2 (d) SO_2

18. What is the density of CO_2 at 25°C and 0.954 atm?
    (a) 1.72 g/L   (c) 0.985 g/L
    (b) 2.04 g/L   (d) 1.52 g/L

19. How many molecules are present in 0.025 mol of H_2 gas?
    (a) 1.5×10^{22} molecules
    (b) 3.37×10^{23} molecules
    (c) 2.40×10^{25} molecules
    (d) 1.50×10^{22} molecules

20. 5.60 L of a gas at STP has a mass of 13.0 g. What is the molar mass of the gas?
    (a) 33.2 g/mol   (c) 66.4 g/mol
    (b) 26.0 g/mol   (d) 52.0 g/mol

21. The heat of fusion of water is
    (a) 4.184 J/g   (c) 2.26 kJ/g
    (b) 335 J/g   (d) 2.26 kJ/mol

22. The heat of vaporization of water is
    (a) 4.184 J/g   (c) 2.26 kJ/g
    (b) 335 J/g   (d) 2.26 kJ/mol

23. The specific heat of water is
    (a) 4.184 J/g°C   (c) 2.26 kJ/g°C
    (b) 335 J/g°C   (d) 18 J/g°C

24. The density of water at 4°C is
    (a) 1.0 g/mL   (c) 18.0 g/mL
    (b) 80 g/mL   (d) 14.7 lb/in.³

25. SO_2 can be properly classified as a(n)
    (a) basic anhydride   (c) anhydrous salt
    (b) hydrate   (d) acid anhydride

26. When compared to H_2S, H_2Se, and H_2Te, water is found to have the highest boiling point because it
    (a) has the lowest molar mass
    (b) is the smallest molecule
    (c) has the highest bonding
    (d) forms hydrogen bonds better than the others

27. In which of the following molecules will hydrogen bonding be important?
    (a) H—F   (c) H—Br

    (b) S—H   (d) H—C—O—C—H
        |
        H

28. Which of the following is an incorrect equation?
    (a) $H_2SO_4 + 2\,NaOH \longrightarrow Na_2SO_4 + 2\,H_2O$
    (b) $C_2H_6 + O_2 \longrightarrow 2\,CO_2 + 3\,H_2$
    (c) $2\,H_2O \xrightarrow[H_2SO_4]{\text{electrolysis}} 2\,H_2 + O_2$
    (d) $Ca + 2\,H_2O \longrightarrow H_2 + Ca(OH)_2$

29. Which of the following is an incorrect equation?
    (a) $C + H_2O(g) \xrightarrow{1000°C} CO(g) + H_2(g)$
    (b) $CaO + H_2O \longrightarrow Ca(OH)_2$
    (c) $2\,NO_2 + H_2O \longrightarrow 2\,HNO_3$
    (d) $Cl_2 + H_2O \longrightarrow HCl + HOCl$

30. How many kilojoules are required to change 85 g of water at 25°C to steam at 100.°C?
    (a) 219 kJ (b) 27 kJ (c) 590 kJ (d) 192 kJ

31. A chunk of 0°C ice, mass 145 g, is dropped into 75 g of water at 62°C. The heat of fusion of water is 335 J/g. The result, after thermal equilibrium is attained, will be
    (a) 87 g ice and 133 g liquid water, all at 0°C
    (b) 58 g ice and 162 g liquid water, all at 0°C
    (c) 220 g water at 7°C
    (d) 220 g water at 17°C

32. The formula for iron(II) sulfate heptahydrate is
    (a) $Fe_2SO_4 \cdot 7\,H_2O$ (c) $FeSO_4 \cdot 7\,H_2O$
    (b) $Fe(SO_4)_2 \cdot 6\,H_2O$ (d) $Fe_2(SO_4)_3 \cdot 7\,H_2O$

33. The process by which a solid changes directly to a vapor is called
    (a) vaporization   (c) sublimation
    (b) evaporation   (d) condensation

34. Hydrogen bonding
    (a) occurs only between water molecules
    (b) is stronger than covalent bonding
    (c) can occur between NH_3 and H_2O
    (d) results from strong attractive forces in ionic compounds

35. A liquid boils when
    (a) the vapor pressure of the liquid equals the external pressure above the liquid
    (b) the heat of vaporization exceeds the vapor pressure
    (c) the vapor pressure equals 1 atm
    (d) the normal boiling temperature is reached

36. Consider two beakers, one containing 50 mL of liquid A and the other 50 mL of liquid B. The boiling point of A is 90°C and that of B is 72°C. Which of these statements is correct?
    (a) A will evaporate faster than B.
    (b) B will evaporate faster than A.
    (c) Both A and B evaporate at the same rate.
    (d) Insufficient data to answer the question.

37. 95.0 g of 0.0°C ice is added to exactly 100. g of water at 60.0°C. When the temperature of the mixture first reaches 0.0°C, the mass of ice still present is
    (a) 0.0 g (c) 10.0 g
    (b) 20.0 g (d) 75.0 g

38. Which of the following is not a general property of solutions?
    (a) a homogeneous mixture of two or more substances
    (b) variable composition
    (c) dissolved solute breaks down to individual molecules
    (d) the same chemical composition, the same chemical properties, and the same physical properties in every part

39. If NaCl is soluble in water to the extent of 36.0 g NaCl/100 g H_2O at 20°C, then a solution at 20°C containing 45 g NaCl/150 g H_2O would be
    (a) dilute   (c) supersaturated
    (b) saturated   (d) unsaturated

40. If 5.00 g NaCl is dissolved in 25.0 g of water, the percent of NaCl by mass is
    (a) 16.7 (c) 0.20
    (b) 20.0 (d) no correct answer given

41. How many grams of 9.0% $AgNO_3$ solution will contain 5.3 g $AgNO_3$?
    (a) 47.7 (c) 59
    (b) 0.58 (d) no correct answer given

42. The molarity of a solution containing 2.5 mol of acetic acid, $HC_2H_3O_2$, in 400. mL of solution is
    (a) 0.063 M (c) 0.103 M
    (b) 1.0 M (d) 6.3 M

43. What volume of 0.300 M KCl will contain 15.3 g KCl?
(a) 1.46 L   (c) 61.5 mL
(b) 683 mL   (d) 4.60 L

44. What mass of $BaCl_2$ will be required to prepare 200. mL of 0.150 M solution?
(a) 0.750 g   (b) 156 g   (c) 6.25 g   (d) 31.2 g

Problems 45–47 relate to the reaction

$$CaCO_3 + 2\ HCl \longrightarrow CaCl_2 + H_2O + CO_2$$

45. What volume of 6.0 M HCl will be needed to react with 0.350 mol of $CaCO_3$?
(a) 42.0 mL   (c) 117 mL
(b) 1.17 L   (d) 583 mL

46. If 400. mL of 2.0 M HCl reacts with excess $CaCO_3$, the volume of CO_2 produced, measured at STP, is
(a) 18 L   (b) 5.6 L   (c) 9.0 L   (d) 56 L

47. If 5.3 g $CaCl_2$ is produced in the reaction, what is the molarity of the HCl used if 25 mL of it reacted with excess $CaCO_3$?
(a) 3.8 M (b) 0.19 M (c) 0.38 M (d) 0.42 M

48. If 20.0 g of the nonelectrolyte urea, $CO(NH_2)_2$, is dissolved in 25.0 g of water, the freezing point of the solution will be
(a) $-2.47°C$ (c) $-24.7°C$
(b) $-1.40°C$ (d) $-3.72°C$

49. When 256 g of a nonvolatile, nonelectrolyte unknown were dissolved in 500. g H_2O, the freezing point was found to be $-2.79°C$. The molar mass of the unknown solute is
(a) 357   (b) 62.0   (c) 768   (d) 341

50. How many milliliters of 6.0 M H_2SO_4 must you use to prepare 500. mL of 0.20 M sulfuric acid solution?
(a) 30   (b) 17   (c) 12   (d) 100

51. How many milliliters of water must be added to 200. mL of 1.40 M HCl to make a solution that is 0.500 M HCl?
(a) 360. mL   (c) 140. mL
(b) 560. mL   (d) 280. mL

52. Which procedure is most likely to increase the solubility of most solids in liquids?
(a) stirring
(b) pulverizing the solid
(c) heating the solution
(d) increasing the pressure

53. The addition of a crystal of $NaClO_3$ to a solution of $NaClO_3$ causes additional crystals to precipitate. The original solution was
(a) unsaturated   (c) saturated
(b) dilute   (d) supersaturated

54. Which of these anions will not form a precipitate with silver ions, Ag^+?
(a) Cl^- (b) NO_3^- (c) Br^- (d) CO_3^{2-}

55. Which of these salts are considered to be soluble in water?
(a) $BaSO_4$ (b) NH_4Cl (c) AgI   (d) PbS

56. A solution of ethyl alcohol and benzene is 40% alcohol by volume. Which statement is correct?
(a) The solution contains 40 mL of alcohol in 100 mL of solution.
(b) The solution contains 60 mL of benzene in 100 mL of solution.
(c) The solution contains 40 mL of alcohol in 100 g of solution.
(d) The solution is made by dissolving 40 mL of alcohol in 60 mL of benzene.

Free Response Questions

1. Which solution should have a higher boiling point: 215 mL of a 10.0% (m/v) aqueous KCl solution or 224 mL of a 1.10 M aqueous NaCl solution?

2. A glass containing 345 mL of a soft drink (a carbonated beverage) was left sitting out on a kitchen counter. If the CO_2 released at room temperature (25°C) and pressure (1 atm) occupies 1.40 L, at a minimum, what is the concentration (in ppm) of the CO_2 in the original soft drink (assume the density of the original soft drink is 0.965 g/mL).

3. Dina and Murphy were trying to react 100. mL of a 0.10 M HCl solution with KOH. The procedure called for a 10% KOH solution. Dina made a 10% mass/v solution while Murphy made a 10% by mass solution. (Assume there is no volume change upon dissolving KOH.) Which solution required less volume to fully react with 100. mL of the HCl solution?

4. A flask containing 825 mL of a solution containing 0.355 moles of $CuSO_4$ was left open overnight. The next morning the flask only contained 755 mL of solution.
(a) Which of the pathways shown below best represents evaporation and why are the others wrong?
(b) What is the concentration (molarity) of the $CuSO_4$ solution remaining in the flask?

5. Three students at Jamston High, Zack, Gaye, and Lamont, each had the opportunity to travel over spring break. As part of a project, each of them measured the boiling point of pure water at their vacation spot. Zack found a boiling point of 93.9°C, Gaye measured 101.1°C, and Lamont read 100.°C. Which student most likely went snow skiing near Ely, Nevada, and which student most likely went water skiing in Honolulu? From the boiling point information, what can you surmise about the Dead Sea region, the location of the third student's vacation? Explain.

6. Why does a change in pressure of a gas significantly affect its volume whereas a change in pressure on a solid or liquid has negligible effect on their respective volumes? If the picture below represents a liquid at the molecular level, draw what you might expect a solid and a gas to look like.

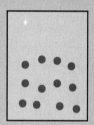

7. (a) If you filled up three balloons with equal volumes of hydrogen, argon, and carbon dioxide gas, all at the same temperature and pressure, which balloon would weigh the most? the least? Explain.

(b) If you filled up three balloons with equal masses of nitrogen, oxygen, and neon, all to the same volume at the same temperature, which would have the lowest pressure?

8. Ray ran a double-displacement reaction using 0.050 mol $CuCl_2$ and 0.10 mol $AgNO_3$. The resulting white precipitate was removed by filtration. The filtrate was accidentally left open on the lab bench for over a week and when Ray returned the flask contained solid, blue crystals. Ray weighed the crystals and found they had a mass of 14.775 g. Was Ray expecting this number? If not, what did he expect?

9. Why is it often advantageous or even necessary to run reactions in solution rather than mixing two solids? Would you expect reactions run in the gas phase to be more similar to solutions or to solids? Why?

CHAPTER

15

Acids, Bases, and Salts

▲ Maintaining the correct water acidity allows us to safely enjoy the fun of a swimming pool.

369

Acids are important chemicals. They are used in cooking to produce the surprise of tartness (from lemons) and to release CO_2 bubbles from leavening agents in baking. Vitamin C is an acid that is an essential nutrient in our diet. Our stomachs release acid to aid in digestion. Excess stomach acid can produce heartburn and indigestion. Bacteria in our mouths produce acids that can dissolve tooth enamel to form cavities. In our recreational activities we are concerned about acidity levels in swimming pools and spas. Acids are essential in the manufacture of detergents, plastics, and storage batteries. The acid base properties of substances are found in all areas of our lives. In this chapter we consider the properties of acids, bases, and salts.

15.1 Acids and Bases

The word *acid* is derived from the Latin *acidus,* meaning "sour" or "tart," and is also related to the Latin word *acetum,* meaning "vinegar." Vinegar has been around since antiquity as a product of the fermentation of wine and apple cider. The sour constituent of vinegar is acetic acid, $HC_2H_3O_2$. Characteristic properties commonly associated with acids include the following:

1. sour taste
2. change the color of litmus, a vegetable dye, from blue to red
3. react with
   - metals such as zinc and magnesium to produce hydrogen gas
   - hydroxide bases to produce water and an ionic compound (salt)
   - carbonates to produce carbon dioxide

These properties are due to the hydrogen ions, H^+, released by acids in a water solution.

Classically, a *base* is a substance capable of liberating hydroxide ions, OH^-, in water solution. Hydroxides of the alkali metals (Group IA) and alkaline earth metals (Group IIA), such as LiOH, NaOH, KOH, $Ca(OH)_2$, and $Ba(OH)_2$, are the most common inorganic bases. Water solutions of bases are called *alkaline solutions* or *base solutions.* Some of the characteristic properties commonly associated with bases include the following:

1. bitter or caustic taste
2. a slippery, soapy feeling
3. the ability to change litmus from red to blue
4. the ability to interact with acids

Several theories have been proposed to answer the question "What is an acid and a base?" One of the earliest, most significant of these theories was advanced in 1884 by Svante Arrhenius (1859–1927), a Swedish scientist, who stated that "an acid is a hydrogen-containing substance that dissociates to produce hydrogen ions, and a base is a hydroxide-containing substance that dissociates to produce hydroxide ions in aqueous solutions." Arrhenius postulated that the hydrogen ions are produced by the dissociation of acids in water, and that the hydroxide ions are produced by the dissociation of bases in water:

$$HA \longrightarrow H^+(aq) + A^-(aq)$$
acid

$$MOH \longrightarrow M^+(aq) + OH^-(aq)$$
base

An Arrhenius acid solution contains an excess of H^+ ions.
An Arrhenius base solution contains an excess of OH^- ions.

In 1923, the Brønsted–Lowry proton transfer theory was introduced by J. N. Brønsted (1897–1947), a Danish chemist, and T. M. Lowry (1847–1936), an English chemist. This theory states that an acid is a proton donor and a base is a proton acceptor.

A Brønsted–Lowry acid is a proton (H^+) donor.
A Brønsted–Lowry base is a proton (H^+) acceptor.

Consider the reaction of hydrogen chloride gas with water to form hydrochloric acid:

$$HCl(g) + H_2O(l) \longrightarrow H_3O^+(aq) + Cl^-(aq) \qquad (1)$$

In the course of the reaction, HCl donates, or gives up, a proton to form a Cl^- ion, and H_2O accepts a proton to form the H_3O^+ ion. Thus, HCl is an acid and H_2O is a base, according to the Brønsted–Lowry theory.

A hydrogen ion, H^+, is nothing more than a bare proton and does not exist by itself in an aqueous solution. In water H^+ combines with a polar water molecule to form a hydrated hydrogen ion, H_3O^+, commonly called a **hydronium ion.** The H^+ is attracted to a polar water molecule, forming a bond with one of the two pairs of unshared electrons:

hydronium ion

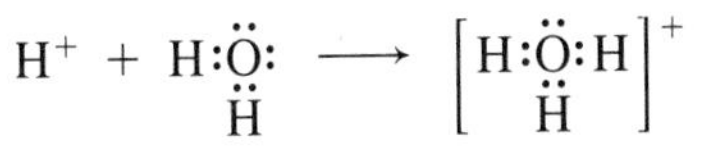

$$H^+ + H\text{:}\overset{\cdot\cdot}{\underset{\cdot\cdot}{O}}\text{:} \longrightarrow \left[H\text{:}\overset{\cdot\cdot}{\underset{\cdot\cdot}{O}}\text{:}H \right]^+$$
$$\phantom{H^+ + H\text{:}}\underset{H}{|} \phantom{\longrightarrow \left[H\text{:}} \underset{H}{|}$$
hydronium ion

Note the electron structure of the hydronium ion. For simplicity we often use H^+ instead of H_3O^+ in equations, with the explicit understanding that H^+ is always hydrated in solution.

When a Brønsted–Lowry acid donates a proton, as illustrated in equation (1), it forms the conjugate base of that acid. When a base accepts a proton, it forms the conjugate acid of that base. A conjugate acid and base are produced as products. The formulas of a conjugate acid–base pair differ by one proton (H^+). Consider what happens when $HCl(g)$ is bubbled through water, as shown by this equation:

conjugate acid–base pair

$$HCl(g) + H_2O(l) \longrightarrow Cl^-(aq) + H_3O^+(aq)$$

conjugate acid–base pair

acid   base   base   acid

The conjugate acid–base pairs are $HCl-Cl^-$ and $H_3O^+-H_2O$. The conjugate base of HCl is Cl^-, and the conjugate acid of Cl^- is HCl. The conjugate base of H_3O^+ is H_2O, and the conjugate acid of H_2O is H_3O^+.

Another example of conjugate acid–base pairs can be seen in this equation:

$$NH_4^+ + H_2O \longrightarrow H_3O^+ + NH_3$$

acid   base   acid   base

Here the conjugate acid–base pairs are $NH_4^+-NH_3$ and $H_3O^+-H_2O$

Example 15.1   Write the formula for (a) the conjugate base of H_2O and of HNO_3, and (b) the conjugate acid of SO_4^{2-} and of $C_2H_3O_2^-$.

Solution   (a) To write the conjugate base of an acid, remove one proton from the acid formula:

Remember: The difference between an acid or a base and its conjugate is one proton, H^+.

$$H_2O \xrightarrow{-H^+} OH^- \quad \text{(conjugate base)}$$

$$HNO_3 \xrightarrow{-H^+} NO_3^- \quad \text{(conjugate base)}$$

Note that, by removing an H^+, the conjugate base becomes more negative than the acid by one minus charge.

(b) To write the conjugate acid of a base, add one proton to the formula of the base:

$$SO_4^{2-} \xrightarrow{+H^+} HSO_4^- \quad \text{(conjugate acid)}$$

$$C_2H_3O_2^- \xrightarrow{+H^+} HC_2H_3O_2 \quad \text{(conjugate acid)}$$

In each case the conjugate acid becomes more positive than the base by a $+1$ charge due to the addition of H^+.

Practice 15.1

Indicate the conjugate base for these acids:
(a) H_2CO_3 (b) HNO_2 (c) $HC_2H_3O_2$

Practice 15.2

Indicate the conjugate acid for these bases:
(a) HSO_4^- (b) NH_3 (c) OH^-

A more general concept of acids and bases was introduced by Gilbert N. Lewis. The Lewis theory deals with the way in which a substance with an unshared pair of electrons reacts in an acid–base type of reaction. According to this theory a base is any substance that has an unshared pair of electrons (electron-pair donor), and an acid is any substance that will attach itself to or accept a pair of electrons.

> **A Lewis acid is an electron-pair acceptor.**
> **A Lewis base is an electron-pair donor.**

In the reaction

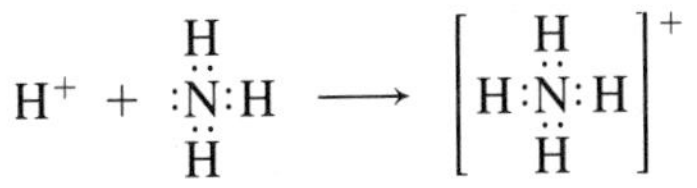

$$
H^+ \; + \; :\!\!\underset{\displaystyle H}{\overset{\displaystyle H}{N}}\!\!:\!H \; \longrightarrow \; \left[H\!:\!\!\underset{\displaystyle H}{\overset{\displaystyle H}{N}}\!\!:\!H \right]^+
$$

acid   base

The H^+ is a Lewis acid and $:NH_3$ is a Lewis base. According to the Lewis theory, substances other than proton donors (e.g., BF_3) behave as acids:

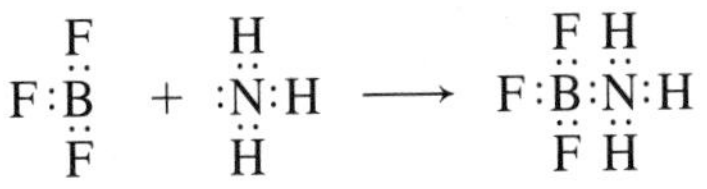

$$
F\!:\!\!\underset{\displaystyle F}{\overset{\displaystyle F}{B}} \; + \; :\!\!\underset{\displaystyle H}{\overset{\displaystyle H}{N}}\!\!:\!H \; \longrightarrow \; F\!:\!\!\underset{\displaystyle F}{\overset{\displaystyle F}{B}}\!:\!\!\underset{\displaystyle H}{\overset{\displaystyle H}{N}}\!\!:\!H
$$

acid   base

These three theories, which explain how acid–base reactions occur, are summarized in Table 15.1. We will generally use the theory that best explains the reaction under consideration. Most of our examples will refer to aqueous solutions. Note that in an aqueous acidic solution the H^+ ion concentration is always greater than OH^- ion concentration. And vice versa—in an aqueous basic solution the OH^- ion concentration is always greater than the H^+ ion concentration. When the H^+ and OH^- ion concentrations in a solution are equal, the solution is neutral; that is, it is neither acidic nor basic.

TABLE 15.1	Summary of Acid–Base Definitions	
Theory	**Acid**	**Base**
Arrhenius	A hydrogen-containing substance that produces hydrogen ions in aqueous solution	A hydroxide-containing substance that produces hydroxide ions in aqueous solution
Brønsted–Lowry	A proton (H^+) donor	A proton (H^+) acceptor
Lewis	Any species that will bond to an unshared pair of electrons (electron-pair acceptor)	Any species that has an unshared pair of electrons (electron-pair donor)

15.2 Reactions of Acids

In aqueous solutions the H^+ or H_3O^+ ions are responsible for the characteristic reactions of acids. The following reactions are in an aqueous medium.

Reaction with Metals Acids react with metals that lie above hydrogen in the activity series of elements to produce hydrogen and an ionic compound (salt) (see Section 17.5):

$$acid + metal \longrightarrow hydrogen + ionic\ compound$$

$$2\ HCl(aq) + Ca(s) \longrightarrow H_2(g) + CaCl_2(aq)$$

$$H_2SO_4(aq) + Mg(s) \longrightarrow H_2(g) + MgSO_4(aq)$$

$$6\ HC_2H_3O_2(aq) + 2\ Al(s) \longrightarrow 3\ H_2(g) + 2\ Al(C_2H_3O_2)_3(aq)$$

Acids such as nitric acid (HNO_3) are oxidizing substances (see Chapter 17) and react with metals to produce water instead of hydrogen. For example,

$$3\ Zn(s) + 8\ HNO_3(dilute) \longrightarrow 3\ Zn(NO_3)_2(aq) + 2\ NO(g) + 4\ H_2O(l)$$

Reaction with Bases The interaction of an acid and a base is called a *neutralization reaction.* In aqueous solutions the products of this reaction are a salt and water:

$$acid + base \longrightarrow salt + water$$

$$HBr(aq) + KOH(aq) \longrightarrow KBr(aq) + H_2O(l)$$

$$2\ HNO_3(aq) + Ca(OH)_2(aq) \longrightarrow Ca(NO_3)_2(aq) + 2\ H_2O(l)$$

$$2\ H_3PO_4(aq) + 3\ Ba(OH)_2(aq) \longrightarrow Ba_3(PO_4)_2(s) + 6\ H_2O(l)$$

Reaction with Metal Oxides This reaction is closely related to that of an acid with a base. With an aqueous acid solution, the products are a salt and water:

$$acid + metal\ oxide \longrightarrow salt + water$$

$$2\ HCl(aq) + Na_2O(s) \longrightarrow 2\ NaCl(aq) + H_2O(l)$$

$$H_2SO_4(aq) + MgO(s) \longrightarrow MgSO_4(aq) + H_2O(l)$$

$$6\ HCl(aq) + Fe_2O_3(s) \longrightarrow 2\ FeCl_3(aq) + 3\ H_2O(l)$$

Reaction with Carbonates Many acids react with carbonates to produce carbon dioxide, water, and an ionic compound:

Carbonic acid (H_2CO_3) is not the product because it is unstable and spontaneously decomposes into water and carbon dioxide.

$$H_2CO_3(aq) \longrightarrow CO_2(g) + H_2O(l)$$

$$acid + carbonate \longrightarrow salt + water + carbon\ dioxide$$

$$2\ HCl(aq) + Na_2CO_3(aq) \longrightarrow 2\ NaCl(aq) + H_2O(l) + CO_2(g)$$

$$H_2SO_4(aq) + MgCO_3(s) \longrightarrow MgSO_4(aq) + H_2O(l) + CO_2(g)$$

Pucker Power

The candy counter is filled with treats that cause your lips to pucker and stimulate your tongue to send signals to your brain saying "SOUR." The human tongue has four types of taste receptors (known as "taste buds"). Sweet, bitter, salty, and sour taste buds are each concentrated in a different area of the tongue. These receptors are molecules that fit together with molecules in the candy (or other food), sending a signal to the brain, which is interpreted as a taste.

Substances that taste sour are acids. The substances that produce this sour taste in candy and other confections are often malic acid and/or citric acid.

citric acid

malic acid

Citric acid tastes more sour than malic acid. When these substances are mixed with other ingredients such as sugar, corn syrup, flavorings, and preservatives, the result is a candy both sweet and tart. The same acids can be mixed with synthetic rubber or chicle (dried latex from the sapodilla tree) to produce bubble gum with real pucker power. The

Sugar, food coloring, flavors

$NaHCO_3$

Gum

Sodium citrate and sodium malate

▲
Cross section of a Mad Dawg™ gum ball.

most sour of this type of gum is called Face Slammers™. Try some on your favorite 12-year-old.

Gum manufacturers have gone a step further in incorporating acid–base chemistry into their products for an even greater surprise. In Mad Dawg™ gum, the initial taste is sour. But after a couple of minutes of chewing, brightly colored foam begins to accumulate in your mouth and ooze out over your lips. What's going on here?

The foam is a mixture of sugar and saliva mixed with carbon dioxide bubbles released when several of the gum's ingredients are mixed in the watery environment of your mouth. The citric and malic acids dissociate to form hydrogen ions while sodium hydrogen carbonate (baking soda) dissolves into sodium and hydrogen carbonate ions:

$$NaHCO_3(s) \longrightarrow Na^+(aq) + HCO_3^-(aq)$$

The hydrogen ions from the acids mix with the hydrogen carbonate ions from the baking soda to produce water and carbon dioxide gas:

$$H^+(aq) + HCO_3^-(aq) \longrightarrow H_2O(l) + CO_2(g)$$

The acids stimulate production of saliva, and the food coloring adds color to the foamy mess. The major problem for chemists in creating this treat was keeping the reactants apart until the consumer pops the gum into his or her mouth. As solids, sodium hydrogen carbonate and citric acid do not react. But introduce the slightest amount of water and the process begins. When a tablet such as Alka Seltzer™ is manufactured, the ingredients (solid citric acid, sodium hydrogen carbonate, aspirin, and flavoring) are compressed into a tablet that is sealed in a dry foil packet. When opened and dropped into water, the reaction (and the relief) begins and the bubbles are released.

In Mad Dawg™ gum balls, the center core is moist gum and the coating is applied in solution. Some of the early versions of these gum balls actually exploded as they were removed from the candy machine. To eliminate this problem, multiple coatings are now used to keep the acid in one layer (on the outside to give the first sour taste) and the sodium hydrogen carbonate in an inner layer (see diagram). When you crunch on the gum ball, the layers begin to mix in your saliva and the fun begins!

citric acid → citrate ion + $H^+(aq)$

15.3   Reactions of Bases

The OH⁻ ions are responsible for the characteristic reactions of bases. The following reactions are in an aqueous medium.

Reaction with Acids Bases react with acids to produce a salt and water. See reaction of acids with bases in Section 15.2.

Amphoteric Hydroxides Hydroxides of certain metals, such as zinc, aluminum, and chromium, are **amphoteric**; that is, they are capable of reacting as either an acid or a base. When treated with a strong acid, they behave like bases; when reacted with a strong base, they behave like acids:

amphoteric

$$Zn(OH)_2(s) + 2\ HCl(aq) \longrightarrow ZnCl_2(aq) + 2\ H_2O(l)$$

$$Zn(OH)_2(s) + 2\ NaOH(aq) \longrightarrow Na_2Zn(OH)_4(aq)$$

Reaction of NaOH and KOH with Certain Metals Some amphoteric metals react directly with the strong bases sodium hydroxide and potassium hydroxide to produce hydrogen:

$$base + metal + water \longrightarrow salt + hydrogen$$

$$2\ NaOH(aq) + Zn(s) + 2\ H_2O(l) \longrightarrow Na_2Zn(OH)_4(aq) + H_2(g)$$

$$2\ KOH(aq) + 2\ Al(s) + 6\ H_2O(l) \longrightarrow 2\ KAl(OH)_4(aq) + 3\ H_2(g)$$

15.4   Salts

Chemists use the terms ionic compound and salt interchangeably.

Salts are very abundant in nature. Most of the rocks and minerals of Earth's mantle are salts of one kind or another. Huge quantities of dissolved salts also exist in the oceans. Salts can be considered compounds derived from acids and bases. They consist of positive metal or ammonium ions combined with negative nonmetal ions (OH⁻ and O²⁻ excluded). The positive ion is the base counterpart and the nonmetal ion is the acid counterpart:

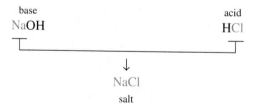

Salts are usually crystalline and have high melting and boiling points.

You may want to review Chapter 6 for nomenclature of acids, bases, and salts.

From a single acid such as hydrochloric acid (HCl), we can produce many chloride compounds by replacing the hydrogen with metal ions (e.g., NaCl, KCl, RbCl, CaCl₂, NiCl₂). Hence the number of known salts greatly exceeds the number of known acids and bases. If the hydrogen atoms of a binary acid are replaced by a nonmetal, the resulting compound has covalent bonding and is therefore not considered to be ionic (e.g., PCl₃, S₂Cl₂, Cl₂O, NCl₃, ICl).

These strange mineral formations called "tufa" exist at Mono Lake, California. Tufa is formed by water bubbling through sand saturated with NaCl, Na_2CO_3, and Na_2SO_4.

15.5 Electrolytes and Nonelectrolytes

We can show that solutions of certain substances are conductors of electricity with a simple conductivity apparatus, which consists of a pair of electrodes connected to a voltage source through a light bulb and switch (see Figure 15.1). If the medium between the electrodes is a conductor of electricity, the light bulb will glow when the switch is closed. When chemically pure water is placed in the beaker and the switch is closed, the light does not glow, indicating that water is a virtual nonconductor. When we dissolve a small amount of sugar in the water and test the solution, the light still does not glow, showing that a sugar solution is also a nonconductor. But, when a small amount of salt, NaCl, is dissolved in water and this solution is tested, the light glows brightly. Thus the salt solution conducts electricity. A fundamental difference exists between the chemical bonding in sugar and that in salt. Sugar is a covalently bonded (molecular) substance; common salt is a substance with ionic bonds.

FIGURE 15.1
A conductivity apparatus shows the difference in conductivity of solutions. (a) Distilled water does not conduct electricity. (b) Sugar water is a nonelectrolyte. (c) Salt water is a strong electrolyte and conducts electricity.
▼

(a)

(b)

(c)

TABLE 15.2	Representative Electrolytes and Nonelectrolytes		
Electrolytes		**Nonelectrolytes**	
H_2SO_4 $HC_2H_3O_2$		$C_{12}H_{22}O_{11}$ (sugar)	CH_3OH (methyl alcohol)
HCl NH_3		C_2H_5OH (ethyl alcohol)	$CO(NH_2)_2$ (urea)
HNO_3 K_2SO_4		$C_2H_4(OH)_2$ (ethylene glycol)	O_2
NaOH $NaNO_3$		$C_3H_5(OH)_3$ (glycerol)	H_2O

electrolyte
nonelectrolyte

Substances whose aqueous solutions are conductors of electricity are called **electrolytes.** Substances whose solutions are nonconductors are known as **nonelectrolytes.** The classes of compounds that are electrolytes are acids, bases, and other ionic compounds (salts). Solutions of certain oxides also are conductors because the oxides form an acid or a base when dissolved in water. One major difference between electrolytes and nonelectrolytes is that electrolytes are capable of producing ions in solution, whereas nonelectrolytes do not have this property. Solutions that contain a sufficient number of ions will conduct an electric current. Although pure water is essentially a nonconductor, many city water supplies contain enough dissolved ionic matter to cause the light to glow dimly when the water is tested in a conductivity apparatus. Table 15.2 lists some common electrolytes and nonelectrolytes.

Acids, bases, and salts are electrolytes.

 ### 15.6   Dissociation and Ionization of Electrolytes

Arrhenius received the 1903 Nobel Prize in chemistry for his work on electrolytes. He found that a solution conducts electricity because the solute dissociates immediately upon dissolving into electrically charged particles (ions). The movement of these ions toward oppositely charged electrodes causes the solution to be a conductor. According to his theory, solutions that are relatively poor conductors contain electrolytes that are only partly dissociated. Arrhenius also believed that ions exist in solution whether or not an electric current is present. In other words the electric current does not cause the formation of ions. Remember that positive ions are cations; negative ions are anions.

dissociation

We have seen that sodium chloride crystals consist of sodium and chloride ions held together by ionic bonds. **Dissociation** is the process by which the ions of a salt separate as the salt dissolves. When placed in water, the sodium and chloride ions are attracted by the polar water molecules, which surround each ion as it dissolves. In water, the salt dissociates, forming hydrated sodium and chloride ions (see Figure 15.2). The sodium and chloride ions in solution are surrounded by a specific number

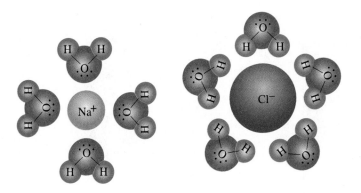

◀ FIGURE 15.2
Hydrated sodium and chloride ions. When sodium chloride dissolves in water, each Na⁺ and Cl⁻ ion becomes surrounded by water molecules. The negative end of the water dipole is attracted to the Na⁺ ion, and the positive end is attracted to the Cl⁻ ion.

of water dipoles and have less attraction for each other than they had in the crystalline state. The equation representing this dissociation is

$$NaCl(s) + (x + y)\ H_2O \longrightarrow Na^+(H_2O)_x + Cl^-(H_2O)_y$$

A simplified dissociation equation in which the water is omitted but understood to be present is

$$NaCl(s) \longrightarrow Na^+(aq) + Cl^-(aq)$$

Remember that sodium chloride exists in an aqueous solution as hydrated ions and not as NaCl units, even though the formula NaCl (or $Na^+ + Cl^-$) is often used in equations.

The chemical reactions of salts in solution are the reactions of their ions. For example, when sodium chloride and silver nitrate react and form a precipitate of silver chloride, only the Ag^+ and Cl^- ions participate in the reaction. The Na^+ and NO_3^- remain as ions in solution:

$$Ag^+(aq) + Cl^-(aq) \longrightarrow AgCl(s)$$

Ionization is the formation of ions; it occurs as a result of a chemical reaction of certain substances with water. Glacial acetic acid (100% $HC_2H_3O_2$) is a liquid that behaves as a nonelectrolyte when tested by the method described in Section 15.5. But a water solution of acetic acid conducts an electric current (as indicated by the dull-glowing light of the conductivity apparatus). The equation for the reaction with water, which forms hydronium and acetate ions, is

ionization

$$\underset{\text{acid}}{HC_2H_3O_2} + \underset{\text{base}}{H_2O} \rightleftharpoons \underset{\text{acid}}{H_3O^+} + \underset{\text{base}}{C_2H_3O_2^-}$$

or, in the simplified equation,

$$HC_2H_3O_2 \rightleftharpoons H^+ + C_2H_3O_2^-$$

In this ionization reaction, water serves not only as a solvent but also as a base according to the Brønsted–Lowry theory.

Hydrogen chloride is predominantly covalently bonded, but when dissolved in water, it reacts to form hydronium and chloride ions:

$$HCl(g) + H_2O(l) \longrightarrow H_3O^+(aq) + Cl^-(aq)$$

FIGURE 15.3 ▶
HCl solution (left) is 100% ionized, while in HC$_2$H$_3$O$_2$ solution (right) almost all of the solute is in molecular form. HCl is a strong acid, while HC$_2$H$_3$O$_2$ is a weak acid. Note: The water molecules in the solution are not shown in this figure.

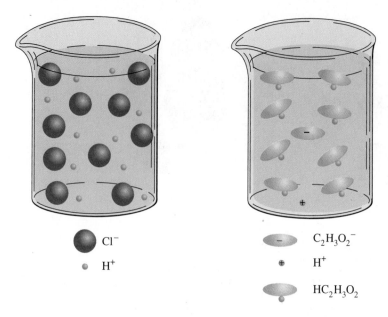

Cl⁻ — Cl^-
H⁺   — H^+

$C_2H_3O_2^-$
H^+
$HC_2H_3O_2$

When a hydrogen chloride solution is tested for conductivity, the light glows brilliantly, indicating many ions in the solution.

Ionization occurs in each of the preceding two reactions with water, producing ions in solution. The necessity for water in the ionization process can be demonstrated by dissolving hydrogen chloride in a nonpolar solvent such as hexane, and testing the solution for conductivity. The solution fails to conduct electricity, indicating that no ions are produced.

The terms *dissociation* and *ionization* are often used interchangeably to describe processes taking place in water. But, strictly speaking, the two are different. In the dissociation of a salt, the salt already exists as ions; when it dissolves in water, the ions separate, or dissociate, and increase in mobility. In the ionization process, ions are produced by the reaction of a compound with water.

15.7   Strong and Weak Electrolytes

strong electrolyte
weak electrolyte

Electrolytes are classified as strong or weak depending on the degree, or extent, of dissociation or ionization. **Strong electrolytes** are essentially 100% ionized in solution; **weak electrolytes** are much less ionized (based on comparing 0.1 *M* solutions). Most electrolytes are either strong or weak, with a few classified as moderately strong or weak. Most salts are strong electrolytes. Acids and bases that are strong electrolytes (highly ionized) are called *strong acids* and *strong bases*. Acids and bases that are weak electrolytes (slightly ionized) are called *weak acids* and *weak bases*.

For equivalent concentrations, solutions of strong electrolytes contain many more ions than do solutions of weak electrolytes. As a result, solutions of strong electrolytes are better conductors of electricity. Consider the two solutions, 1 *M* HCl and 1 *M* HC$_2$H$_3$O$_2$. Hydrochloric acid is almost 100% ionized; acetic acid is about 1% ionized. (See Figure 15.3.) Thus HCl is a strong acid and HC$_2$H$_3$O$_2$ is a weak acid. Hydrochloric acid has about 100 times as many hydronium ions in solution as acetic acid, making the HCl solution much more acidic.

We can distinguish between strong and weak electrolytes experimentally using the apparatus described in Section 15.5. A 1 M HCl solution causes the light to glow brilliantly, but a 1 M $HC_2H_3O_2$ solution causes only a dim glow. The strong base sodium hydroxide, NaOH, can be distinguished in a similar fashion from the weak base ammonia, NH_3. The ionization of a weak electrolyte in water is represented by an equilibrium equation showing that both the un-ionized and ionized forms are present in solution. In the equilibrium equation of $HC_2H_3O_2$ and its ions, we say that the equilibrium lies "far to the left" because relatively few hydrogen and acetate ions are present in solution:

$$HC_2H_3O_2(aq) \rightleftharpoons H^+(aq) + C_2H_3O_2^-(aq)$$

We have previously used a double arrow in an equation to represent reversible processes in the equilibrium between dissolved and undissolved solute in a saturated solution. A double arrow ($\rightleftharpoons$) is also used in the ionization equation of soluble weak electrolytes to indicate that the solution contains a considerable amount of the un-ionized compound in equilibrium with its ions in solution. (See Section 16.1 for a discussion of reversible reactions.) A single arrow is used to indicate that the electrolyte is essentially all in the ionic form in the solution. For example, nitric acid is a strong acid; nitrous acid is a weak acid. Their ionization equations in water may be indicated as

$$HNO_3(aq) \xrightarrow{H_2O} H^+(aq) + NO_3^-(aq)$$
$$HNO_2(aq) \overset{H_2O}{\rightleftharpoons} H^+(aq) + NO_2^-(aq)$$

Practically all soluble salts, acids (such as sulfuric, nitric, and hydrochloric acids), and bases (such as sodium, potassium, calcium, and barium hydroxides) are strong electrolytes. Weak electrolytes include numerous other acids and bases such as acetic acid, nitrous acid, carbonic acid, and ammonia. The terms *strong acid, strong base, weak acid,* and *weak base* refer to whether an acid or base is a strong or weak electrolyte. A brief list of strong and weak electrolytes is given in Table 15.3.

Electrolytes yield two or more ions per formula unit upon dissociation—the actual number being dependent on the compound. Dissociation is complete or nearly complete for nearly all soluble ionic compounds and for certain other strong elec-

TABLE 15.3	Strong and Weak Electrolytes		
Strong electrolytes		**Weak electrolytes**	
Most soluble salts	$HClO_4$	$HC_2H_3O_2$	$H_2C_2O_4$
H_2SO_4	NaOH	H_2CO_3	H_3BO_3
HNO_3	KOH	HNO_2	HClO
HCl	$Ca(OH)_2$	H_2SO_3	NH_3
HBr	$Ba(OH)_2$	H_2S	HF

trolytes, such as those given in Table 15.3. The following are dissociation equations for several strong electrolytes. In all cases the ions are actually hydrated:

$$NaOH \xrightarrow{H_2O} Na^+(aq) + OH^-(aq) \qquad \text{2 ions in solution per formula unit}$$

$$Na_2SO_4 \xrightarrow{H_2O} 2\ Na^+(aq) + SO_4^{2-}(aq) \qquad \text{3 ions in solution per formula unit}$$

$$Fe_2(SO_4)_3 \xrightarrow{H_2O} 2\ Fe^{3+}(aq) + 3\ SO_4^{2-}(aq)$$

5 ions in solution per formula unit

One mole of NaCl will give 1 mol of Na^+ ions and 1 mol of Cl^- ions in solution, assuming complete dissociation of the salt. One mole of $CaCl_2$ will give 1 mol of Ca^{2+} ions and 2 mol of Cl^- ions in solution:

$$NaCl \xrightarrow{H_2O} Na^+(aq) + Cl^-(aq)$$
1 mol   1 mol   1 mol

$$CaCl_2 \xrightarrow{H_2O} Ca^{2+}(aq) + 2\ Cl^-(aq)$$
1 mol   1 mol   2 mol

Example 15.2   What is the molarity of each ion in a solution of (a) 2.0 M NaCl, and (b) 0.40 M K_2SO_4? Assume complete dissociation.

Solution   (a) According to the dissociation equation,

$$NaCl \xrightarrow{H_2O} Na^+(aq) + Cl^-(aq)$$
1 mol   1 mol   1 mol

the concentration of Na^+ is equal to that of NaCl (1 mol NaCl $\longrightarrow$ 1 mol Na^+), and the concentration of Cl^- is also equal to that of NaCl. Therefore the concentrations of the ions in 2.0 M NaCl are 2.0 M Na^+ and 2.0 M Cl^-.

(b) According to the dissociation equation,

$$K_2SO_4 \xrightarrow{H_2O} 2\ K^+(aq) + SO_4^{2-}(aq)$$
1 mol   2 mol   1 mol

the concentration of K^+ is twice that of K_2SO_4 and the concentration of SO_4^{2-} is equal to that of K_2SO_4. Therefore the concentrations of the ions in 0.40 M K_2SO_4 are 0.80 M K^+ and 0.40 M SO_4^{2-}.

Practice 15.3

What is the molarity of each ion in a solution of (a) 0.050 M $MgCl_2$, and (b) 0.070 M $AlCl_3$?

Colligative Properties of Electrolyte Solutions

We have learned that when 1 mol of sucrose, a nonelectrolyte, is dissolved in 1000 g of water, the solution freezes at $-1.86°C$. When 1 mol NaCl is dissolved in 1000 g of water, the freezing point of the solution is not $-1.86°C$, as might be expected, but is closer to $-3.72°C$ (-1.86×2). The reason for the lower freezing point is that 1 mol NaCl in solution produces 2 mol of particles ($2 \times 6.022 \times 10^{23}$ ions) in solution. Thus the freezing-point depression produced by 1 mol NaCl is essentially equivalent to that produced by 2 mol of a nonelectrolyte. An electrolyte such as $CaCl_2$, which yields three ions in water, gives a freezing-point depression of about three times that of a nonelectrolyte. These freezing-point data provide additional evidence that electrolytes dissociate when dissolved in water. The other colligative properties are similarly affected by substances that yield ions in aqueous solutions.

15.8 Ionization of Water

The more we study chemistry, the more intriguing the water molecule becomes. Two equations commonly used to show how water ionizes are

$$H_2O + H_2O \rightleftharpoons H_3O^+ + OH^-$$
$$\text{acid} \quad \text{base} \quad\quad \text{acid} \quad\quad \text{base}$$

and

$$H_2O \rightleftharpoons H^+ + OH^-$$

The first equation represents the Brønsted–Lowry concept, with water reacting as both an acid and a base, forming a hydronium ion and a hydroxide ion. The second equation is a simplified version, indicating that water ionizes to give a hydrogen and a hydroxide ion. Actually, the proton, H^+, is hydrated and exists as a hydronium ion. In either case equal molar amounts of acid and base are produced so that water is neutral, having neither H^+ nor OH^- ions in excess. The ionization of water at 25°C produces an H^+ ion concentration of 1.0×10^{-7} mol/L and an OH^- ion concentration of 1.0×10^{-7} mol/L. Square brackets, [ ], are used to indicate that the concentration is in moles per liter. Thus $[H^+]$ means the concentration of H^+ is in moles per liter. These concentrations are usually expressed as

$$[H^+] \text{ or } [H_3O^+] = 1.0 \times 10^{-7} \text{ mol/L}$$
$$[OH^-] = 1.0 \times 10^{-7} \text{ mol/L}$$

These figures mean that about two out of every billion water molecules are ionized. This amount of ionization, small as it is, is a significant factor in the behavior of water in many chemical reactions.

15.9 Introduction to pH

The acidity of an aqueous solution depends on the concentration of hydrogen or hydronium ions. The pH scale of acidity gives us a simple, convenient numerical way to state the acidity of a solution. Values on the pH scale are obtained by mathematical conversion of H^+ ion concentrations to pH by the expression:

$$pH = -\log[H^+]$$

where $[H^+] = H^+$ or H_3O^+ ion concentration in moles per liter. The **pH** is defined as the *negative* logarithm of the H^+ or H_3O^+ concentration in moles per liter:

pH

$$pH = -\log[H^+] = -\log(1 \times 10^{-7}) = -(-7) = 7$$

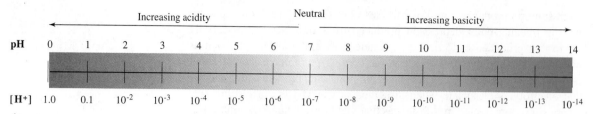

FIGURE 15.4
The pH scale of acidity and basicity.

For example, the pH of pure water at 25°C is 7 and is said to be neutral; that is, it is neither acidic nor basic, because the concentrations of H^+ and OH^- are equal. Solutions that contain more H^+ ions than OH^- ions have pH values less than 7, and solutions that contain less H^+ ions than OH^- ions have values greater than 7.

> **pH < 7.00 is an acidic solution**
> **pH = 7.00 is a neutral solution**
> **pH > 7.00 is a basic solution**

When $[H^+] = 1 \times 10^{-5}$ mol/L, pH = 5 (acidic)

When $[H^+] = 1 \times 10^{-9}$ mol/L, pH = 9 (basic)

Instead of saying that the hydrogen ion concentration in the solution is 1×10^{-5} mol/L, it's customary to say that the pH of the solution is 5. The smaller the pH value, the more acidic the solution (see Figure 15.4).

The pH scale, along with its interpretation, is given in Table 15.4, and Table 15.5 lists the pH of some common solutions. Note that a change of only one pH unit means a tenfold increase or decrease in H^+ ion concentration. For example, a solution with a pH of 3.0 is ten times more acidic than a solution with a pH of 4.0. A simplified method of determining pH from $[H^+]$ follows:

$$[H^+] = 1 \times 10^{-5} \quad \longleftarrow \quad \text{pH = this number (5)}$$
$$\text{pH = 5}$$

when this number
is exactly 1

$$[H^+] = 2 \times 10^{-5} \quad \longleftarrow \quad \text{pH is between this number and}$$
$$\text{next lower number (4 and 5)}$$
$$\text{pH = 4.7}$$

when this number
is between 1 and 10

logarithm

Calculating the pH value for H^+ ion concentrations requires the use of logarithms, which are exponents. The **logarithm** (log) of a number is simply the power to which 10 must be raised to give that number. Thus the log of 100 is 2 ($100 = 10^2$), and the log of 1000 is 3 ($1000 = 10^3$). The log of 500 is 2.70, but you can't determine this value easily without a scientific calculator.

Help on using calculators is found in Appendix II.

Let's determine the pH of a solution with $[H^+] = 2 \times 10^{-5}$ using a calculator. Enter 2×10^{-5} into your calculator and press the log key. The number $-4.69 \ldots$ will be displayed. The pH is then

Remember: Change the sign on your calculator since pH = −log[H⁺].

$$\text{pH} = -\log[H^+] = -(-4.69 \ldots) = 4.7$$

TABLE 15.4	pH Scale for Expressing Acidity	
$[H^+]$ (mol/L)	pH	
1×10^{-14}	14	
1×10^{-13}	13	
1×10^{-12}	12	Increasing
1×10^{-11}	11	basicity
1×10^{-10}	10	
1×10^{-9}	9	
1×10^{-8}	8	
1×10^{-7}	7	Neutral
1×10^{-6}	6	
1×10^{-5}	5	
1×10^{-4}	4	
1×10^{-3}	3	Increasing
1×10^{-2}	2	acidity
1×10^{-1}	1	
1×10^{0}	0	

TABLE 15.5	The pH of Common Solutions
Solution	pH
Gastric juice	1.0
0.1 M HCl	1.0
Lemon juice	2.3
Vinegar	2.8
0.1 M $HC_2H_3O_2$	2.9
Orange juice	3.7
Tomato juice	4.1
Coffee, black	5.0
Urine	6.0
Milk	6.6
Pure water (25°C)	7.0
Blood	7.4
Household ammonia	11.0
1 M NaOH	14.0

Next we must determine the correct number of significant figures in the logarithm. The rules for logs are different from those we use in other math operations. The number of decimal places for a log must equal the number of significant figures in the original number. Since 2×10^{-5} has one significant figure, we should round the log to one decimal place (4.69 . . .) = 4.7.

What is the pH of a solution with an $[H^+]$ of (a) 1.0×10^{-11}, (b) 6.0×10^{-4}, and (c) 5.47×10^{-8}?

Example 15.3

Solution

(a) $[H^+] = 1.0 \times 10^{-11}$
(2 significant figures)
$pH = -\log(1.0 \times 10^{-11})$
$pH = 11.00$
(2 decimal places)

(b) $[H^+] = 6.0 \times 10^{-4}$
(2 significant figures)
$\log 6.0 \times 10^{-4} = -3.22$
$pH = -\log[H^+]$
$pH = -(-3.22) = 3.22$
(2 decimal places)

(c) $[H^+] = 5.47 \times 10^{-8}$
(3 significant figures)
$\log 5.47 \times 10^{-8} = -7.262$
$pH = -\log[H^+]$
$pH = -(-7.262) = 7.262$
(3 decimal places)

Practice 15.4

What is the pH of a solution with $[H^+]$ of (a) 3.9×10^{-12} M, (b) 1.3×10^{-3} M, and (c) 3.72×10^{-6} M?

In the Pink—A Sign of Corrosion

As our fleet of commercial airplanes ages, scientists are looking for new ways to detect corrosion and stop it *before* it becomes a safety issue. The problem is that planes are subjected to heat, rain, and wind—all factors that can contribute to metal corrosion. How can a maintenance crew detect corrosion on a huge aircraft since many imaging techniques work on only small areas at a time? Imagine a plane that could tell the maintenance crew it needed repair.

Gerald Frankel and Jim Zhang from Ohio State University have developed a paint that detects changes in pH. Corrosion of metals (in contact with air and water) results in a chemical reaction that produces hydroxide ions, which increase

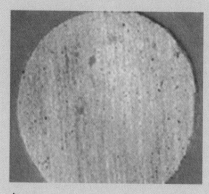

▲ **Phenolphthalein paint after 8 days.**

pH. Frankel and Zhang made a clear acrylic coating that they mixed with phenolphthalein. This acid–base indica-

tor turns bright fuschia when the pH increases above 8.0. The scientists can show the "paint" turns pink at corrosion sites as small as 15 μm deep.

Technicians at Wright Air Force Base in Ohio think this could lead to a whole new way to detect corrosion. William Mullins says, "You could walk down the vehicle and see there's a pink spot." This method would be especially good at detecting corrosion concealed around rivets and where metals overlap. The only limitation appears to be that the coating must be clear.

The next time your airline maintenance crew sees pink spots, you may be in for a delay or a change of aircraft!

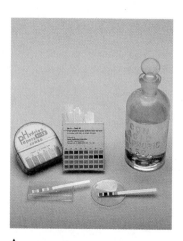

▲
FIGURE 15.5
pH test paper for determining the approximate acidity of solutions.

The measurement and control of pH is extremely important in many fields. Proper soil pH is necessary to grow certain types of plants successfully. The pH of certain foods is too acidic for some diets. Many biological processes are delicately controlled pH systems. The pH of human blood is regulated to very close tolerances through the uptake or release of H^+ by mineral ions, such as HCO_3^-, HPO_4^{2-}, and $H_2PO_4^-$. Changes in the pH of the blood by as little as 0.4 pH unit result in death.

Compounds with colors that change at particular pH values are used as indicators in acid–base reactions. For example, phenolphthalein, an organic compound, is colorless in acid solution and changes to pink at a pH of 8.3. When a solution of sodium hydroxide is added to a hydrochloric acid solution containing phenolphthalein, the change in color (from colorless to pink) indicates that all the acid is neutralized. Commercially available pH test paper, such as that shown in Figure 15.5, contains chemical indicators. The indicator in the paper takes on different colors when wetted with solutions of different pH. Thus the pH of a solution can be estimated by placing a drop on the test paper and comparing the color of the test paper with a color chart calibrated at different pH values. Common applications of pH test indicators are the kits used to measure and adjust the pH of swimming pools, hot tubs, and saltwater aquariums. Electronic pH meters are used for making rapid and precise pH determinations.

15.10 Neutralization

neutralization

The reaction of an acid and a base to form a salt and water is known as **neutralization.** We've seen this reaction before, but now with our knowledge about ions and ionization, let's reexamine the process of neutralization.

◀ The progress of a titration can be monitored by computer graphing.

Consider the reaction that occurs when solutions of sodium hydroxide and hydrochloric acid are mixed. The ions present initially are Na^+ and OH^- from the base and H^+ and Cl^- from the acid. The products, sodium chloride and water, exist as Na^+ and Cl^- ions and H_2O molecules. A chemical equation representing this reaction is

$$HCl(aq) + NaOH(aq) \longrightarrow NaCl(aq) + H_2O(l)$$

This equation, however, does not show that HCl, NaOH, and NaCl exist as ions in solution. The following total ionic equation gives a better representation of the reaction:

$$(H^+ + Cl^-) + (Na^+ + OH^-) \longrightarrow Na^+ + Cl^- + H_2O(l)$$

This equation shows that the Na^+ and Cl^- ions did not react. These ions are called **spectator ions** because they were present but did not take part in the reaction. The only reaction that occurred was that between the H^+ and OH^- ions. Therefore the equation for the neutralization can be written as this net ionic equation:

spectator ion

$$\underset{\text{acid}}{H^+(aq)} + \underset{\text{base}}{OH^-(aq)} \longrightarrow \underset{\text{water}}{H_2O(l)}$$

This simple net ionic equation represents not only the reaction of sodium hydroxide and hydrochloric acid, but also the reaction of any strong acid with any water-soluble hydroxide base in an aqueous solution. The driving force of a neutralization reaction is the ability of an H^+ ion and an OH^- ion to react and form a molecule of un-ionized water.

This meter measures the ▶
pH of garden soil.

The amount of acid, base, or other species in a sample can be determined by
titration   **titration,** which measures the volume of one reagent required to react with a measured mass or volume of another reagent.

Consider the titration of an acid with a base. A measured volume of acid of
unknown concentration is placed in a flask, and a few drops of an indicator solution
are added. Base solution of known concentration is slowly added from a buret to the
acid until the indicator changes color. The indicator selected is one that changes
color when the stoichiometric quantity (according to the equation) of base has been
added to the acid. At this point, known as the *end point of the titration,* the titration
is complete, and the volume of base used to neutralize the acid is read from the buret.
The concentration or amount of acid in solution can be calculated from the titration
data and the chemical equation for the reaction. Let's look at some examples.

Example 15.4   Suppose that 42.00 mL of 0.150 M NaOH solution is required to neutralize 50.00 mL
of hydrochloric acid solution. What is the molarity of the acid solution?

Solution   The equation for the reaction is

$$NaOH(aq) + HCl(aq) \longrightarrow NaCl(aq) + H_2O(l)$$

In this neutralization NaOH and HCl react in a 1:1 mole ratio. Therefore the moles of
HCl in solution are equal to the moles of NaOH required to react with it. First we calculate the moles of NaOH used, and from this value we determine the moles of HCl:

> *Data*: 42.00 mL of 0.150 M NaOH   50.00 mL HCl
>   Molarity of acid = M (unknown)

Determine the moles of NaOH:

$$M = mol/L \quad 42.00 \text{ mL} = 0.04200 \text{ L}$$

$$(0.04200 \text{ L})\left(\frac{0.150 \text{ mol NaOH}}{1 \text{ L}}\right) = 0.00630 \text{ mol NaOH}$$

Since NaOH and HCl react in a 1:1 ratio, 0.00630 mol HCl was present in the
50.00 mL of HCl solution. Therefore the molarity of the HCl is

$$M = \frac{mol}{L} = \frac{0.00630 \text{ mol HCl}}{0.05000 \text{ L}} = 0.126 \text{ } M \text{ HCl}$$

Suppose that 42.00 mL of 0.150 M NaOH solution is required to neutralize 50.00 mL of H_2SO_4 solution. What is the molarity of the acid solution?

Example 15.5

Solution

The equation for the reaction is

$$2 \text{ NaOH}(aq) + \text{H}_2\text{SO}_4(aq) \longrightarrow \text{Na}_2\text{SO}_4(aq) + 2 \text{ H}_2\text{O}(l)$$

The same amount of base (0.00630 mol NaOH) is used in this titration as in Example 15.4, but the mole ratio of acid to base in the reaction is 1:2. The moles of H_2SO_4 reacted can be calculated using the mole-ratio method:

Data: 42.00 mL of 0.150 M NaOH = 0.00630 mol NaOH

$$(0.00630 \text{ mol NaOH})\left(\frac{1 \text{ mol H}_2\text{SO}_4}{2 \text{ mol NaOH}}\right) = 0.00315 \text{ mol H}_2\text{SO}_4$$

Therefore 0.00315 mol H_2SO_4 was present in 50.00 mL of H_2SO_4 solution. The molarity of the H_2SO_4 is

$$M = \frac{\text{mol}}{\text{L}} = \frac{0.00315 \text{ mol H}_2\text{SO}_4}{0.05000 \text{ L}} = 0.0630 \text{ } M \text{ H}_2\text{SO}_4$$

A 25.00-mL sample of H_2SO_4 solution required 14.26 mL 0.2240 M NaOH for complete neutralization. What is the molarity of the sulfuric acid?

Example 15.6

Solution

The equation for the reaction is

$$2 \text{ NaOH}(aq) + \text{H}_2\text{SO}_4(aq) \longrightarrow \text{Na}_2\text{SO}_4(aq) + 2 \text{ H}_2\text{O}(l)$$

The moles of NaOH needed are

$$\text{moles NaOH} = (V_{\text{NaOH}})(M_{\text{NaOH}})$$

$$= (0.01426 \text{ L})\left(0.2240 \frac{\text{mol}}{\text{L}}\right)$$

$$= 0.003194 \text{ mol NaOH}$$

Since the mole ratio of acid to base is $\dfrac{1 \text{ H}_2\text{SO}_4}{2 \text{ NaOH}}$ the moles of acid in the sample are

$$(0.003914 \text{ mol NaOH})\left(\frac{1 \text{ H}_2\text{SO}_4}{2 \text{ NaOH}}\right) = 0.001597 \text{ mol H}_2\text{SO}_4$$

Now to find the molarity of the sample we divide the moles of acid by its original volume:

$$\left(\frac{0.001597 \text{ mol H}_2\text{SO}_4}{0.02500 \text{ L}}\right) = 0.06388 \text{ } M \text{ H}_2\text{SO}_4$$

Practice 15.5

A 50.0-mL sample of HCl required 24.81 mL of 0.1250 M NaOH for neutralization. What is the molarity of the acid?

15.11 Writing Net Ionic Equations

In Section 15.10, we wrote the reaction of hydrochloric acid and sodium hydroxide in three different equations:

(1) $HCl(aq) + NaOH(aq) \longrightarrow NaCl(aq) + H_2O(l)$

(2) $(H^+ + Cl^-) + (Na^+ + OH^-) \longrightarrow Na^+ + Cl^- + H_2O(l)$

(3) $H^+ + OH^- \longrightarrow H_2O$

un-ionized equation

total ionic equation

net ionic equation

In the **un-ionized equation** (1), compounds are written in their molecular, or formula expressions. In the **total ionic equation** (2), compounds are written to show the form in which they are predominantly present: strong electrolytes as ions in solution; and nonelectrolytes, weak electrolytes, precipitates, and gases in their molecular (or un-ionized) forms. In the **net ionic equation** (3), only those molecules or ions that have changed are included in the equation; ions or molecules that do not change (the spectators) are omitted.

When balancing equations thus far, we've been concerned only with the atoms of the individual elements. Because ions are electrically charged, ionic equations often end up with a net electrical charge. A balanced equation must have the same net charge on each side, whether that charge is positive, negative, or zero. Therefore when balancing ionic equations, we must make sure that both the same number of each kind of atom and the same net electrical charge are present on each side.

Here is a list of rules for writing ionic equations:

1. Strong electrolytes in solution are written in their ionic form.
2. Weak electrolytes are written in their molecular (un-ionized) form.
3. Nonelectrolytes are written in their molecular form.
4. Insoluble substances, precipitates, and gases are written in their molecular forms.
5. The net ionic equation should include only substances that have undergone a chemical change. Spectator ions are omitted from the net ionic equation.
6. Equations must be balanced, both in atoms and in electrical charge.

Study the following examples. In each one the un-ionized equation is given. Write the total ionic equation and the net ionic equation for each.

Example 15.7

$HNO_3(aq) + KOH(aq) \longrightarrow KNO_3(aq) + H_2O(l)$

un-ionized equation

Solution

$(H^+ + NO_3^-) + (K^+ + OH^-) \longrightarrow (K^+ + NO_3^-) + H_2O$

total ionic equation

$H^+ + OH^- \longrightarrow H_2O$

net ionic equation

The HNO_3, KOH, and KNO_3 are soluble, strong electrolytes. The K^+ and NO_3^- ions are spectator ions, have not changed, and are not included in the net ionic equation. Water is a nonelectrolyte and is written in the molecular form.

Example 15.8

$$2\,AgNO_3(aq) + BaCl_2(aq) \longrightarrow 2\,AgCl(s) + Ba(NO_3)_2(aq)$$
un-ionized equation

$$(2\,Ag^+ + 2\,NO_3^-) + (Ba^{2+} + 2\,Cl^-) \longrightarrow 2\,AgCl(s) + (Ba^{2+} + 2\,NO_3^-)$$
total ionic equation

$$Ag^+ + Cl^- \longrightarrow AgCl(s)$$
net ionic equation

Solution

Although AgCl is an ionic compound, it is written in the un-ionized form on the right side of the ionic equations because most of the Ag^+ and Cl^- ions are no longer in solution but have formed a precipitate of AgCl. The Ba^{2+} and NO_3^- ions are spectator ions.

Example 15.9

$$Na_2CO_3(aq) + H_2SO_4(aq) \longrightarrow Na_2SO_4(aq) + H_2O(l) + CO_2(g)$$
un-ionized equation

$$(2\,Na^+ + CO_3^{2-}) + (2\,H^+ + SO_4^{2-}) \longrightarrow (2\,Na^+ + SO_4^{2-}) + H_2O(l) + CO_2(g)$$
total ionic equation

$$CO_3^{2-} + 2\,H^+ \longrightarrow H_2O(l) + CO_2(g)$$
net ionic equation

Solution

Carbon dioxide, CO_2, is a gas and evolves from the solution; Na^+ and SO_4^{2-} are spectator ions.

Example 15.10

$$HC_2H_3O_2(aq) + NaOH(aq) \longrightarrow NaC_2H_3O_2(aq) + H_2O(l)$$
un-ionized equation

$$HC_2H_3O_2 + (Na^+ + OH^-) \longrightarrow (Na^+ + C_2H_3O_2^-) + H_2O$$
total ionic equation

$$HC_2H_3O_2 + OH^- \longrightarrow C_2H_3O_2^- + H_2O$$
net ionic equation

Solution

Acetic acid, $HC_2H_3O_2$, a weak acid, is written in the molecular form, but sodium acetate, $NaC_2H_3O_2$, a soluble salt, is written in the ionic form. The Na^+ ion is the only spectator ion in this reaction. Both sides of the net ionic equation have a -1 electrical charge.

Example 15.11

$$Mg(s) + 2\,HCl(aq) \longrightarrow MgCl_2(aq) + H_2(g)$$
un-ionized equation

$$Mg + (2\,H^+ + 2\,Cl^-) \longrightarrow (Mg^{2+} + 2\,Cl^-) + H_2(g)$$
total ionic equation

$$Mg + 2 H^+ \longrightarrow Mg^{2+} + H_2(g)$$
net ionic equation

Solution   The net electrical charge on both sides of the equation is +2.

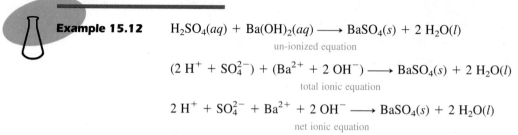

Example 15.12 $$H_2SO_4(aq) + Ba(OH)_2(aq) \longrightarrow BaSO_4(s) + 2 H_2O(l)$$
un-ionized equation

$$(2 H^+ + SO_4^{2-}) + (Ba^{2+} + 2 OH^-) \longrightarrow BaSO_4(s) + 2 H_2O(l)$$
total ionic equation

$$2 H^+ + SO_4^{2-} + Ba^{2+} + 2 OH^- \longrightarrow BaSO_4(s) + 2 H_2O(l)$$
net ionic equation

Solution   Barium sulfate, $BaSO_4$, is a highly insoluble salt. If we conduct this reaction using the conductivity apparatus described in Section 15.5, the light glows brightly at first but goes out when the reaction is complete because almost no ions are left in solution. The $BaSO_4$ precipitates out of solution, and water is a nonconductor of electricity.

Practice 15.6

Write the net ionic equation for

$$3 H_2S(aq) + 2 Bi(NO_3)_3(aq) \longrightarrow Bi_2S_3(s) + 6 HNO_3(aq)$$

15.12 Acid Rain

Acid rain is defined as any atmospheric precipitation that is more acidic than usual. The increase in acidity might be from natural or industrial sources. Rain acidity varies throughout the world and across the United States. The pH of rain is generally lower in the eastern United States and higher in the west. Unpolluted rain has a pH of 5.6, and so is slightly acidic. This acidity results from the dissolution of carbon dioxide in the water producing carbonic acid:

$$CO_2(g) + H_2O(l) \longrightarrow H_2CO_3(aq) \rightleftharpoons H^+(aq) + HCO_3^-(aq)$$

Although the details of acid rain formation are not yet fully understood, chemists know the general process involves the following steps:

1. emission of nitrogen and sulfur oxides into the air
2. transportation of these oxides throughout the atmosphere
3. chemical reactions between the oxides and water forming sulfuric acid, H_2SO_4, and nitric acid, HNO_3
4. rain or snow, which carries the acids to the ground

The oxides may also be deposited directly on a dry surface and become acidic when normal rain falls on them.

◀ **Marble masterpieces, sculpted to last forever, are slowly disappearing as acid rain dissolves the calcium carbonate (CaCO₃).**

Acid rain is not a new phenomenon. Rain was probably acidic in the early days of our planet as volcanic eruptions, fires, and decomposition of organic matter released large volumes of nitrogen and sulfur oxides into the atmosphere. Use of fossil fuels, especially since the industrial revolution about 250 years ago, has made significant changes in the amounts of pollutants being released into the atmosphere. As increasing amounts of fossil fuels have been burned, more and more sulfur and nitrogen oxides have poured into the atmosphere, thus increasing the acidity of rain.

Acid rain affects a variety of factors in our environment. For example, freshwater plants and animals decline significantly when rain is acidic; large numbers of fish and plants die when acidic water from spring thaws enters the lakes. Aluminum is leached from the soil into lakes by acidic rainwater where the aluminum compounds adversely affect the gills of fish. In addition to leaching aluminum from the soil, acid rain also causes other valuable minerals, such as magnesium and calcium, to dissolve and run into lakes and streams. It can also dissolve the waxy protective coat on plant leaves making them vulnerable to attack by bacteria and fungi.

In our cities, acid rain is responsible for extensive and continuing damage to buildings, monuments, and statues. It reduces the durability of paint and promotes the deterioration of paper, leather, and cloth. In short, we are just beginning to explore the effects of acid rain on human beings and on our food chain.

15.13 Colloids: An Introduction

When we add sugar to a flask of water and shake it, the sugar dissolves and forms a clear homogeneous *solution.* When we do the same experiment with very fine sand and water, the sand particles form a *suspension,* which settles when the shaking stops. When we repeat the experiment again using ordinary cornstarch, we find that

Foam Cars—Wave of the Future?

Metal foams are a new class of materials that may revolutionize the car industry. Automotive makers know that one key to making cars that are fuel efficient is to decrease weight. Up till now that weight reduction meant higher cost (in materials like titanium and aluminum) and problems in crash testing (since light vehicles don't absorb energy well).

German automotive supplier Willhelm Karmann (whose company manufactured the Volkswagen Karmann-Ghia) developed an aluminum foam composite material with some amazing properties. Parts made of this foam composite weigh 30–50% less than an equivalent steel part and are ten times stiffer. The material is so light it floats in water and although it costs 20–25% more than steel it could be used for as much as 20% of a compact car. Since the surface of the new material is not smooth it would likely be used in structural areas of the car (firewalls, roof panels, luggage compartment walls, etc.)

How is aluminum foam made? Two layers of aluminum sheet and a middle powder layer (made of titanium metal hydride and aluminum powder) are rolled together under very high pressure

to make a single flat sheet. This sheet metal is then processed in traditional ways to make a variety of 3-D shapes. Then the sheet is placed in a 1148°F oven for two minutes. This quick bake allows the Al metal to melt and mix with H_2 gas (released from the titanium hydride) making foam. The sheet rises just like a cake (increasing five or seven

times in thickness). When the foam cools it is a rigidly formed 3-D structure between two aluminum skins.

Not only is the new aluminum foam part much lighter than its steel counterpart; it also performs well in crash tests. One of these days you may climb into a car that is really a foam colloid—with great fuel economy and crash resistance.

the starch does not dissolve in cold water. But if the mixture is heated and stirred, the starch forms a cloudy, opalescent *dispersion*. This dispersion does not appear to be clear and homogeneous like the sugar solution, yet it is not obviously heterogeneous and does not settle like the sand suspension. In short, its properties are intermediate between those of the sugar solution and those of the sand suspension. The starch dispersion is actually a *colloid,* a name derived from the Greek *kolla,* meaning "glue," and was coined by the English scientist Thomas Graham in 1861.

colloid

As it is now used, the word **colloid** means a dispersion in which the dispersed particles are larger than the solute ions or molecules of a true solution and smaller than the particles of a mechanical suspension. The term does not imply a gluelike quality, although most glues are colloidal materials. The size of colloidal particles ranges from a lower limit of about 1 nm (10^{-7} cm) to an upper limit of about 1000 nm (10^{-4} cm). There are eight types of colloids, which are summarized in Table 15.6.

TABLE 15.6	Types of Colloidal Dispersions	
Type	**Name**	**Examples**
Gas in liquid	foam	whipped cream, soapsuds
Gas in solid	solid foam	Styrofoam, foam rubber, pumice
Liquid in gas	liquid aerosol	fog, clouds
Liquid in liquid	emulsion	milk, vinegar in oil salad dressing, mayonnaise
Liquid in solid	solid emulsion	cheese, opals, jellies
Solid in gas	solid aerosol	smoke, dust in air
Solid in liquid	sol	india ink, gold sol
Solid in solid	solid sol	tire rubber, certain gems (e.g., rubies)

The fundamental difference between a colloidal dispersion and a true solution is the size, not the nature, of the particles. The solute particles in a solution are usually single ions or molecules that may be hydrated to varying degrees. Colloidal particles are usually aggregations of ions or molecules. However, the molecules of some polymers, such as proteins, are large enough to be classified as colloidal particles when in solution. To fully appreciate the differences in relative sizes, the volumes (not just the linear dimensions) of colloidal particles and solute particles must be compared. The difference in volumes can be approximated by assuming that the particles are spheres. A large colloidal particle has a diameter of about 500 nm, whereas a fair-sized ion or molecule has a diameter of about 0.5 nm. Thus the diameter of the colloidal particle is about 1000 times that of the solute particle. Because the volumes of spheres are proportional to the cubes of their diameters, we can calculate that the volume of a colloidal particle can be up to a billion ($10^3 \times 10^3 \times 10^3 = 10^9$) times greater than that of a solution particle.

15.14 Properties of Colloids

In 1827, while observing a strongly illuminated aqueous suspension of pollen under a high-powered microscope, Robert Brown (1773–1858) noted that the pollen grains appeared to have a trembling, erratic motion. He later determined that this erratic motion is not confined to pollen but is characteristic of colloidal particles in general. This random motion of colloidal particles is called **Brownian movement.** We can readily observe such movement by confining cigarette smoke in a small transparent chamber and illuminating it with a strong beam of light at right angles to the optical axis of the microscope. The smoke particles appear as tiny randomly moving lights because the light is reflected from their surfaces. This motion is due to the continual bombardment of the smoke particles by air molecules. Since Brownian movement can be seen when colloidal particles are dispersed in either a gaseous or a liquid medium, it affords nearly direct visual proof that matter at the molecular level is moving randomly, as postulated by the kinetic-molecular theory.

Brownian movement

FIGURE 15.6 ▶
Tyndall effect. A beam of
light is visible in a colloidal
suspension (left) but not in a
true solution (right).

When an intense beam of light is passed through an ordinary solution and viewed at an angle, the beam passing through the solution is hardly visible. A beam of light, however, is clearly visible and sharply outlined when it is passed through a colloidal dispersion (see Figure 15.6). This phenomenon is known as the **Tyndall effect.** The Tyndall effect, like the Brownian movement, can be observed in nearly all colloidal dispersions. It occurs because the colloidal particles are large enough to scatter the rays of visible light. The ions or molecules of true solutions are too small to scatter light and therefore do not exhibit a noticeable Tyndall effect.

Tyndall effect

Another important characteristic of colloids is that the particles have relatively huge surface areas. We saw in Section 14.6 that the surface area is increased tenfold when a 1-cm cube is divided into 1000 cubes with sides of 0.1 cm. When a 1-cm cube is divided into colloidal-size cubes measuring 10^{-6} cm, the combined surface area of all the particles becomes a million times greater than that of the original cube.

Colloidal particles become electrically charged when they adsorb ions on their surfaces. *Adsorption* should not be confused with *absorption.* Adsorption refers to the adhesion of molecules or ions to a surface, whereas absorption refers to the taking in of one material by another material. Adsorption occurs because the atoms or ions at the surface of a particle are not completely surrounded by other atoms or ions as are those in the interior. Consequently these surface atoms or ions attract and adsorb ions or polar molecules from the dispersion medium onto the surfaces of the colloidal particles. This property is directly related to the large surface area presented by the many tiny particles.

15.15   Applications of Colloidal Properties

Activated charcoal has an enormous surface area, approximately 1 million square centimeters per gram in some samples. Hence, charcoal is very effective in selectively adsorbing the polar molecules of some poisonous gases and is therefore used in gas masks. Charcoal can be used to adsorb impurities from liquids as well as from gases, and large amounts are used to remove substances that have objectionable tastes and odors from water supplies. In sugar refineries activated charcoal is used to adsorb colored impurities from the raw sugar solutions.

A process widely used for dust and smoke control in many urban and industrial areas was devised by an American, Frederick Cottrell (1877–1948). The Cottrell process takes advantage of the fact that the particulate matter in dust and smoke is electrically charged. Air to be cleaned of dust or smoke is passed between electrode plates charged with a high voltage. Positively charged particles are attracted to, neutralized, and thereby precipitated at the negative electrodes. Negatively charged particles are removed in the same fashion at the positive electrodes. Large Cottrell units are fitted with devices for automatic removal of precipitated material. In some installations, particularly at cement mills and smelters, the value of the dust collected may be sufficient to pay for the precipitation equipment. Small units, designed for removing dust and pollen from air in the home, are now on the market. Unfortunately, Cottrell units remove only particulate matter; they cannot remove gaseous pollutants such as carbon monoxide, sulfur dioxide, and nitrogen oxides.

Colloidal particles become electrically charged when they adsorb ions on their surface.

Thomas Graham found that a parchment membrane would allow the passage of true solutions but would prevent the passage of colloidal dispersions. Dissolved solutes can be removed from colloidal dispersions through the use of such a membrane by a process called **dialysis.** The membrane itself is called a *dialyzing membrane.* Artificial membranes are made from such materials as parchment paper, collodion, or certain kinds of cellophane. Dialysis can be demonstrated by putting a colloidal starch dispersion and some copper(II) sulfate solution in a parchment paper bag and suspending it in running water. In a few hours the blue color of the copper(II) sulfate has disappeared, and only the starch dispersion remains in the bag.

dialysis

A life-saving application of dialysis has been the development of artificial kidneys. The blood of a patient suffering from partial kidney failure is passed through the artificial kidney machine for several hours, during which time the soluble waste products are removed by dialysis.

Concepts in Review

1. State the general characteristics of acids and bases.
2. Define an acid and base in terms of the Arrhenius, Brønsted–Lowry, and Lewis theories.
3. Identify acid–base conjugate pairs in a reaction.
4. When given the reactants, complete and balance equations for the reactions of acids with bases, metals, metal oxides, and carbonates.
5. When given the reactants, complete and balance equations of the reaction of an amphoteric hydroxide with either a strong acid or a strong base.
6. Write balanced equations for the reaction of sodium hydroxide or potassium hydroxide with zinc and with aluminum.
7. Classify common compounds as electrolytes or nonelectrolytes.
8. Distinguish between strong and weak electrolytes.

9. Explain the process of dissociation and ionization. Indicate how they differ.
10. Write equations for the dissociation and/or ionization of acids, bases, and salts in water.
11. Describe and write equations for the ionization of water.
12. Explain how pH expresses hydrogen ion concentration or hydronium ion concentration.
13. Given pH as an integer, calculate the H^+ molarity, and vice versa.
14. Use a calculator to calculate pH values from corresponding H^+ molarities.
15. Explain the process of acid–base neutralization.
16. Calculate the molarity or volume of an acid or base solution from appropriate titration data.
17. Write un-ionized, total ionic, and net ionic equations for chemical equations.

Key Terms

amphoteric (15.3)
Brownian movement (15.14)
colloid (15.13)
dialysis (15.15)
dissociation (15.6)
electrolyte (15.5)
hydronium ion (15.1)

ionization (15.6)
logarithm (15.9)
net ionic equation (15.11)
neutralization (15.10)
nonelectrolyte (15.5)
pH (15.9)
spectator ion (15.10)

strong electrolyte (15.7)
titration (15.10)
total ionic equation (15.11)
Tyndall effect (15.14)
un-ionized equation (15.11)
weak electrolyte (15.7)

Questions

1. Since a hydrogen ion and a proton are identical, what differences exist between the Arrhenius and Brønsted–Lowry definitions of an acid? (Table 15.1)

2. According to Figure 15.1, what type of substance must be in solution for the bulb to light?

3. Which of the following classes of compounds are electrolytes: acids, alcohols, bases, salts? (Table 15.2)

4. What two differences are apparent in the arrangement of water molecules about the hydrated ions as depicted in Figure 15.2?

5. The pH of a solution with a hydrogen ion concentration of 0.003 M is between what two whole numbers? (Table 15.4)

6. Which is more acidic, tomato juice or blood? (Table 15.5)

7. Use the three acid–base theories (Arrhenius, Brønsted–Lowry, and Lewis) to define an acid and a base.

8. For each acid–base theory referred to in Question 7, write an equation illustrating the neutralization of an acid with a base.

9. Write the Lewis structure for the (a) bromide ion, (b) hydroxide ion, and (c) cyanide ion. Why are these ions considered to be bases according to the Brønsted–Lowry and Lewis acid–base theories?

10. Into what three classes of compounds do electrolytes generally fall?

11. Name each compound listed in Table 15.3.

12. A solution of HCl in water conducts an electric current, but a solution of HCl in hexane does not. Explain this behavior in terms of ionization and chemical bonding.

13. How do ionic compounds exist in their crystalline structure? What occurs when they are dissolved in water?

14. An aqueous methyl alcohol, CH_3OH, solution does not conduct an electric current, but a solution of sodium hydroxide, NaOH, does. What does this information tell us about the OH group in the alcohol?

15. Why does molten NaCl conduct electricity?

16. Explain the difference between dissociation of ionic compounds and ionization of molecular compounds.

17. Distinguish between strong and weak electrolytes.

18. Explain why ions are hydrated in aqueous solutions.

19. What is the main distinction between water solutions of strong and weak electrolytes?

20. What are the relative concentrations of $H^+(aq)$ and $OH^-(aq)$ in (a) a neutral solution, (b) an acid solution, and (c) a basic solution?

21. Write the net ionic equation for the reaction of a strong acid with a water-soluble hydroxide base in an aqueous solution.

22. The solubility of HCl gas in water, a polar solvent, is much greater than its solubility in hexane, a nonpolar solvent. How can you account for this difference?

23. Pure water, containing equal concentrations of both acid and base ions, is neutral. Why?

24. Indicate the fundamental difference between a colloidal dispersion and a true solution.

25. Explain the Tyndall effect and how it may be used to distinguish between a colloidal dispersion and a true solution.

26. Explain the process of dialysis, giving a practical application in society.

Paired Exercises

27. Identify the conjugate acid–base pairs in the following equations:
(a) $HC_2H_3O_2 + H_2SO_4 \rightleftharpoons H_2C_2H_3O_2^+ + HSO_4^-$
(b) The two-step ionization of sulfuric acid,
$$H_2SO_4 + H_2O \longrightarrow H_3O^+ + HSO_4^-$$
$$HSO_4^- + H_2O \rightleftharpoons H_3O^+ + SO_4^{2-}$$
(c) $HClO_4 + H_2O \longrightarrow H_3O^+ + ClO_4^-$
(d) $CH_3O^- + H_3O^+ \longrightarrow CH_3OH + H_2O$

29. Complete and balance these equations:
(a) $Mg(s) + HCl(aq) \longrightarrow$
(b) $BaO(s) + HBr(aq) \longrightarrow$
(c) $Al(s) + H_2SO_4(aq) \longrightarrow$
(d) $Na_2CO_3(aq) + HCl(aq) \longrightarrow$
(e) $Fe_2O_3(s) + HBr(aq) \longrightarrow$
(f) $Ca(OH)_2(aq) + H_2CO_3(aq) \longrightarrow$

31. Which of these compounds are electrolytes? Consider each substance to be mixed with water.
(a) HCl
(b) CO_2
(c) $CaCl_2$
(d) $C_{12}H_{22}O_{11}$ (sugar)
(e) C_3H_7OH (rubbing alcohol)
(f) CCl_4 (insoluble)

33. Calculate the molarity of the ions present in these salt solutions. Assume each salt to be 100% dissociated:
(a) $0.015\ M$ NaCl
(b) $4.25\ M$ NaKSO₄
(c) $0.20\ M$ CaCl₂
(d) 22.0 g KI in 500. mL of solution

35. In Exercise 33, how many grams of each ion would be present in 100. mL of each solution?

37. What is the molar concentration of all ions present in a solution prepared by mixing the following? Neglect the concentration of H^+ and OH^- from water. Also, assume volumes of solutions are additive.
(a) 30.0 mL of 1.0 M NaCl and 40.0 mL of 1.0 M NaCl
(b) 30.0 mL of 1.0 M HCl and 30.0 mL of 1.0 M NaOH
*(c) 100.0 mL of 0.40 M KOH and 100.0 mL of 0.80 M HCl

39. Given the data for the following separate titrations, calculate the molarity of the HCl:

	mL HCl	Molarity HCl	mL NaOH	Molarity NaOH
(a)	40.13	M	37.70	0.728
(b)	19.00	M	33.66	0.306
(c)	27.25	M	18.00	0.555

28. Identify the conjugate acid–base pairs in the following equations:
(a) $HCl + NH_3 \longrightarrow NH_4^+ + Cl^-$
(b) $HCO_3^- + OH^- \rightleftharpoons CO_3^{2-} + H_2O$
(c) $HCO_3^- + H_3O^+ \rightleftharpoons H_2CO_3 + H_2O$
(d) $HC_2H_3O_2 + H_2O \rightleftharpoons H_3O^+ + C_2H_3O_2^-$

30. Complete and balance these equations:
(a) $NaOH(aq) + HBr(aq) \longrightarrow$
(b) $KOH(aq) + HCl(aq) \longrightarrow$
(c) $Ca(OH)_2(aq) + HI(aq) \longrightarrow$
(d) $Al(OH)_3(s) + HBr(aq) \longrightarrow$
(e) $Na_2O(s) + HClO_4(aq) \longrightarrow$
(f) $LiOH(aq) + FeCl_3(aq) \longrightarrow$

32. Which of these compounds are electrolytes? Consider each substance to be mixed with water.
(a) $NaHCO_3$ (baking soda)
(b) N_2 (insoluble gas)
(c) $AgNO_3$
(d) $HCOOH$ (formic acid)
(e) $RbOH$
(f) K_2CrO_4

34. Calculate the molarity of the ions present in these salt solutions. Assume each salt to be 100% dissociated:
(a) $0.75\ M$ ZnBr₂
(b) $1.65\ M$ Al₂(SO₄)₃
(c) 900. g $(NH_4)_2SO_4$ in 20.0 L of solution
(d) 0.0120 g $Mg(ClO_3)_2$ in 1.00 mL of solution

36. In Exercise 34, how many grams of each ion would be present in 100. mL of each solution?

38. What is the molar concentration of all ions present in a solution prepared by mixing the following? Neglect the concentration of H^+ and OH^- from water. Also, assume volumes of solutions are additive.
(a) 100.0 mL of 2.0 M KCl and 100.0 mL of 1.0 M CaCl₂
(b) 35.0 mL of 0.20 M Ba(OH)₂ and 35.0 mL of 0.20 M H₂SO₄
(c) 1.00 L of 1.0 M AgNO₃ and 500. mL of 2.0 M NaCl

40. Given the data for the following separate titrations, calculate the molarity of the NaOH:

	mL HCl	Molarity HCl	mL NaOH	Molarity NaOH
(a)	37.19	0.126	31.91	M
(b)	48.04	0.482	24.02	M
(c)	13.13	1.425	39.39	M

41. Rewrite the following unbalanced equations, changing them into balanced net ionic equations. All reactions are in water solution.
(a) $K_2SO_4(aq) + Ba(NO_3)_2(aq) \longrightarrow$
$$KNO_3(aq) + BaSO_4(s)$$
(b) $CaCO_3(s) + HCl(aq) \longrightarrow$
$$CaCl_2(aq) + CO_2(g) + H_2O(l)$$
(c) $Mg(s) + HC_2H_3O_2(aq) \longrightarrow$
$$Mg(C_2H_3O_2)_2(aq) + H_2(g)$$

42. Rewrite the following unbalanced equations, changing them into balanced net ionic equations. All reactions are in water solution.
(a) $H_2S(g) + CdCl_2(aq) \longrightarrow CdS(s) + HCl(aq)$
(b) $Zn(s) + H_2SO_4(aq) \longrightarrow ZnSO_4(aq) + H_2(g)$
(c) $AlCl_3(aq) + Na_3PO_4(aq) \longrightarrow$
$$AlPO_4(s) + NaCl(aq)$$

43. In the following pairs which solution is more acidic? All are water solutions. Explain your answer.
(a) $1\ M$ HCl or $1\ M\ H_2SO_4$?
(b) $1\ M$ HCl or $1\ M\ HC_2H_3O_2$?

44. In the following pairs which solution is more acidic? All are water solutions. Explain your answer.
(a) $1\ M$ HCl or $2\ M$ HCl?
(b) $1\ M\ HNO_3$ or $1\ M\ H_2SO_4$?

45. What volume (in milliliters) of $0.245\ M$ HCl will neutralize 10.0 g $Al(OH)_3$? The equation is
$$3\ HCl(aq) + Al(OH)_3(s) \longrightarrow AlCl_3(aq) + 3\ H_2O(l)$$

46. What volume (in milliliters) of $0.245\ M$ HCl will neutralize 50.0 mL of $0.100\ M\ Ca(OH)_2$? The equation is
$$2\ HCl(aq) + Ca(OH)_2(aq) \longrightarrow CaCl_2(aq) + 2\ H_2O(l)$$

*47. A 0.200-g sample of impure NaOH requires 18.25 mL of $0.2406\ M$ HCl for neutralization. What is the percent of NaOH in the sample?

*48. A batch of sodium hydroxide was found to contain sodium chloride as an impurity. To determine the amount of impurity, a 1.00-g sample was analyzed and found to require 49.90 mL of $0.466\ M$ HCl for neutralization. What is the percent of NaCl in the sample?

*49. What volume of H_2 gas, measured at $27°C$ and $700.$ torr, can be obtained by reacting 5.00 g of zinc metal with $100.$ mL of $0.350\ M$ HCl? The equation is
$$Zn(s) + 2\ HCl(aq) \longrightarrow ZnCl_2(aq) + H_2(g)$$

*50. What volume of H_2 gas, measured at $27°C$ and $700.$ torr, can be obtained by reacting 5.00 g of zinc metal with $200.$ mL of $0.350\ M$ HCl? The equation is
$$Zn(s) + 2\ HCl(aq) \longrightarrow ZnCl_2(aq) + H_2(g)$$

51. Calculate the pH of solutions having these H^+ ion concentrations:
(a) $0.01\ M$
(b) $1.0\ M$
(c) $6.5 \times 10^{-9}\ M$

52. Calculate the pH of solutions having these H^+ ion concentrations:
(a) $1 \times 10^{-7}\ M$
(b) $0.50\ M$
(c) $0.00010\ M$

53. Calculate the pH of
(a) orange juice, $3.7 \times 10^{-4}\ M\ H^+$
(b) vinegar, $2.8 \times 10^{-3}\ M\ H^+$

54. Calculate the pH of
(a) black coffee, $5.0 \times 10^{-5}\ M\ H^+$
(b) limewater, $3.4 \times 10^{-11}\ M\ H^+$

Additional Exercises

55. What is the concentration of Ca^{2+} ions in a solution of CaI_2 having an I^- ion concentration of $0.520\ M$?

56. How many milliliters of $0.40\ M$ HCl can be made by diluting $100.$ mL of $12\ M$ HCl with water?

57. If 29.26 mL of $0.430\ M$ HCl neutralizes 20.40 mL of $Ba(OH)_2$ solution, what is the molarity of the $Ba(OH)_2$ solution? The reaction is
$$Ba(OH)_2(aq) + 2\ HCl(aq) \longrightarrow BaCl_2(aq) + 2\ H_2O(l)$$

58. A 1 m solution of acetic acid, $HC_2H_3O_2$, in water freezes at a lower temperature than a 1 m solution of ethyl alcohol, C_2H_5OH, in water. Explain.

59. At the same cost per pound, which alcohol, CH_3OH or C_2H_5OH, would be more economical to purchase as an antifreeze for your car? Why?

60. How does a hydronium ion differ from a hydrogen ion?

61. Arrange, in decreasing order of freezing points, 1 m aqueous solutions of HCl, $HC_2H_3O_2$, $C_{12}H_{22}O_{11}$ (sucrose), and $CaCl_2$. (List the one with the highest freezing point first.)

62. At $100°C$ the H^+ concentration in water is about 1×10^{-6} mol/L, about ten times that of water at $25°C$. At which of these temperatures is
(a) the pH of water the greater?
(b) the hydrogen ion (hydronium ion) concentration the higher?
(c) the water neutral?

63. What is the relative difference in H^+ concentration in solutions that differ by one pH unit?

64. A sample of pure sodium carbonate with a mass of 0.452 g was dissolved in water and neutralized with 42.4 mL of hydrochloric acid. Calculate the molarity of the acid:
$$Na_2CO_3(aq) + 2\,HCl(aq) \longrightarrow$$
$$2\,NaCl(aq) + CO_2(g) + H_2O(l)$$

65. What volume (mL) of 0.1234 M HCl is needed to neutralize 2.00 g $Ca(OH)_2$?

66. How many grams of KOH are required to neutralize 50.00 mL of 0.240 M HNO_3?

67. Two drops (0.1 mL) of 1.0 M HCl are added to water to make 1.0 L of solution. What is the pH of this solution if the HCl is 100% ionized?

68. What volume of concentrated (18.0 M) sulfuric acid must be used to prepare 50.0 L of 5.00 M solution?

69. If 3.0 g NaOH is added to 500. mL of 0.10 M HCl, will the resulting solution be acidic or basic? Show evidence for your answer.

70. If 380 mL of 0.35 M $Ba(OH)_2$ is added to 500.0 mL of 0.65 M HCl, will the mixture be acidic or basic? Find the pH of the resulting solution.

71. If 50.00 mL of 0.2000 M HCl is titrated with 0.2000 M NaOH, find the pH of the solution after the following amounts of base have been added:
(a) 0.000 mL   **(e)** 49.90 mL
(b) 10.00 mL   **(f)** 49.99 mL
(c) 25.00 mL   **(g)** 50.00 mL
(d) 49.00 mL
Plot your answers on a graph with pH on the y-axis and mL NaOH on the x-axis.

72. Sulfuric acid reacts with NaOH:
(a) Write a balanced equation for the reaction producing Na_2SO_4.
(b) How many milliliters of 0.10 M NaOH are needed to react with 0.0050 mol H_2SO_4?
(c) How many grams of Na_2SO_4 will also form?

*73. Lactic acid (found in sour milk) has an empirical formula of $HC_3H_5O_3$. A 1.0-g sample of lactic acid required 17.0 mL of 0.65 M NaOH to reach the end point of a titration. What is the molecular formula for lactic acid?

74. A 10.0-mL sample of HNO_3 was diluted to a volume of 100.00 mL. Then, 25 mL of that diluted solution was needed to neutralize 50.0 mL of 0.60 M KOH. What was the concentration of the original nitric acid?

75. The pH of a solution of a strong acid was determined to be 3. If water is then added to dilute this solution, would the pH change? Why or why not? Could enough water ever be added to raise the pH of an acid solution above 7?

76. Solution X has a pH of 2. Solution Y has a pH of 4. On the basis of this information, which of the following is true?
(a) the $[H^+]$ of X is one-half that of Y
(b) the $[H^+]$ of X is twice that of Y
(c) the $[H^+]$ of X is 100 times that of Y
(d) the $[H^+]$ of X is $1/100$ that of Y

77. A student is given three solutions: an acid, a base, and one that is neither acidic nor basic. The student performs tests on these solutions and records their properties. For each result, tell whether it is the property of an acid, a base, or whether you cannot decide.
(a) the solution has $[H^+] = 1 \times 10^{-7}\,M$
(b) the solution has $[OH^-] = 1 \times 10^{-2}\,M$
(c) the solution turns litmus red
(d) the solution is a good conductor of electricity

Answers to Practice Exercises

15.1 (a) HCO_3^-, (b) NO_2^-, (c) $C_2H_3O_2^-$
15.2 (a) H_2SO_4, (b) NH_4^+, (c) H_2O
15.3 (a) 0.050 M Mg^{2+}, 0.10 M Cl^-,
    (b) 0.070 M Al^{3+}, 0.21 M Cl^-
15.4 (a) 11.41, (b) 2.89, (c) 5.429
15.5 (c) 0.0620 M HCl
15.6 $3\,H_2S(aq) + 2\,Bi^{3+}(aq) \longrightarrow Bi_2S_3(s) + 6\,H^+(aq)$

CHAPTER

16

Chemical Equilibrium

▲ A coral reef is a system in ecological equilibrium with the ocean surrounding it.

CHAPTER 16 / OUTLINE

Thus far we've considered chemical change as proceeding from reactants to products. Does that mean that the change then stops? No, but often it appears to be the case at the macroscopic level. A solute dissolves until the solution becomes saturated. Once a solid remains undissolved in a container, the system appears to be at rest. The human body is a marvelous chemical factory, yet from day to day it appears to be quite the same. For example, the blood remains at a constant pH, even though all sorts of chemical reactions are taking place. Another example is a terrarium, which can be watered and sealed for long periods of time with no ill effects. Or an antacid, which absorbs excess stomach acid and does *not* change the pH of the stomach. In all of these cases reactions are proceeding, even though visible signs of chemical change are absent. Similarly when a system is at equilibrium, chemical reactions are dynamic at the molecular level. In this chapter we will consider chemical systems as they approach equilibrium conditions.

16.1 Reversible Reactions

In the preceding chapters we treated chemical reactions mainly as reactants changing to products. However, many reactions do not go to completion. Some reactions do not go to completion because they are reversible; that is, when the products are formed, they react to produce the starting reactants.

We've encountered reversible systems before. One is the vaporization of a liquid by heating and its subsequent condensation by cooling:

liquid + heat $\longrightarrow$ vapor

vapor + cooling $\longrightarrow$ liquid

The conversion between nitrogen dioxide, NO_2, and dinitrogen tetroxide, N_2O_4, shows us visible evidence of the reversibility of a reaction. The NO_2 is a reddish-brown gas that changes with cooling to N_2O_4, a colorless gas. The reaction is reversible by heating N_2O_4:

$$2\,NO_2(g) \xrightarrow{\text{cooling}} N_2O_4(g)$$

$$N_2O_4(g) \xrightarrow{\text{heating}} 2\,NO_2(g)$$

These two reactions may be represented by a single equation with a double arrow, $\rightleftharpoons$, to indicate that the reactions are taking place in both directions at the same time:

$$2\,NO_2(g) \rightleftharpoons N_2O_4(g)$$

This reversible reaction can be demonstrated by sealing samples of NO_2 in two tubes and placing one tube in warm water and the other in ice water (see Figure 16.1).

A **reversible chemical reaction** is one in which the products formed react to produce the original reactants. Both the forward and reverse reactions occur

reversible chemical reaction

FIGURE 16.1 ▶
Reversible reaction of NO₂
and N₂O₄. More of the
reddish-brown NO₂ molecules
are visible in the tube that is
heated (right, 80°C) than in
the tube that is cooled (left,
0°C).

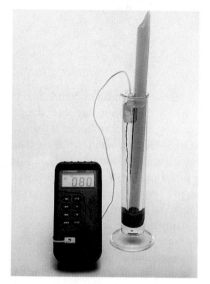

simultaneously. The forward reaction is called *the reaction to the right,* and the reverse reaction is called *the reaction to the left.* A double arrow is used in the equation to indicate that the reaction is reversible.

16.2   Rates of Reaction

chemical kinetics

Every reaction has a rate, or speed, at which it proceeds. Some are fast and some are extremely slow. The study of reaction rates and reaction mechanisms is known as **chemical kinetics.**

The rate of a reaction is variable and depends on the concentration of the reacting species, the temperature, the presence of catalysts, and the nature of the reactants. Consider the hypothetical reaction

$$A + B \longrightarrow C + D \quad \text{(forward reaction)}$$

$$C + D \longrightarrow A + B \quad \text{(reverse reaction)}$$

in which a collision between A and B is necessary for a reaction to occur. The rate at which A and B react depends on the concentration or the number of A and B molecules present; it will be fastest, for a fixed set of conditions, when they are first mixed (as shown by the height of the red line in Figure 16.2). As the reaction proceeds, the number of A and B molecules available for reaction decreases, and the rate of reaction slows down (seen as the red line flattens in Figure 16.2). If the reaction is reversible, the speed of the reverse reaction is zero at first (blue line in Figure 16.2) and gradually increases as the concentrations of C and D increase. As the number of A and B molecules decreases, the forward rate slows down because A and B cannot find one another as often in order to accomplish a reaction. To counteract this diminishing rate of reaction, an excess of one reagent is often used to keep the reaction from becoming impractically slow. Collisions between molecules may be compared to video games. When many objects are on the screen, collisions occur frequently; but if only a few objects are present, collisions can usually be avoided.

FIGURE 16.2
The rates of the forward and reverse reactions become equal at some point in time. The forward reaction rate (red) decreases as a result of decreasing amounts of reactants. The reverse reaction rate (blue) starts at zero and increases as the amount of product increases. When the two rates become equal (purple), a state of chemical equilibrium has been reached.

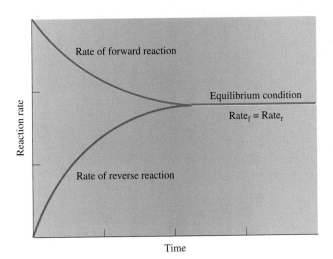

16.3 Chemical Equilibrium

equilibrium

Any system at **equilibrium** represents a dynamic state in which two or more opposing processes are taking place at the same time and at the same rate. A chemical equilibrium is a dynamic system in which two or more opposing chemical reactions are going on at the same time and at the same rate. When the rate of the forward reaction is exactly equal to the rate of the reverse reaction, a condition of **chemical equilibrium** exists (see purple line in Figure 16.2). The concentrations of the products and the reactants are not changing, and the system appears to be at a standstill because the products are reacting at the same rate at which they are being formed.

chemical equilibrium

Chemical equilibrium:
rate of forward reaction = rate of reverse reaction

A saturated salt solution is in a condition of equilibrium:

$$NaCl(s) \rightleftharpoons Na^+(aq) + Cl^-(aq)$$

At equilibrium, salt crystals are continuously dissolving, and Na^+ and Cl^- ions are continuously crystallizing. Both processes are occurring at the same rate.

The ionization of weak electrolytes is another chemical equilibrium system:

$$HC_2H_3O_2(aq) + H_2O(l) \rightleftharpoons H_3O^+(aq) + C_2H_3O_2^-(aq)$$

In this reaction the equilibrium is established in a 1 *M* solution when the forward reaction has gone about 1%—that is, when only 1% of the acetic acid molecules in solution have ionized. Therefore only a relatively few ions are present, and the acid behaves as a weak electrolyte.

The reaction represented by

$$H_2(g) + I_2(g) \underset{}{\overset{700 \text{ K}}{\rightleftharpoons}} 2 \text{ HI}(g)$$

provides another example of chemical equilibrium. Theoretically, 1.00 mol of hydrogen should react with 1.00 mol of iodine to yield 2.00 mol of hydrogen iodide. Actually, when 1.00 mol H_2 and 1.00 mol I_2 are reacted at 700 K, only 1.58 mol HI is present when equilibrium is attained. Since 1.58 is 79% of the theoretical yield of 2.00 mol HI, the forward reaction is only 79% complete at equilibrium. The equilibrium mixture will also contain 0.21 mol each of unreacted H_2 and I_2 (1.00 mol − 0.79 mol = 0.21 mol):

$$H_2(g) + I_2(g) \overset{700 \text{ K}}{\longrightarrow} 2 \text{ HI}(g)$$

This equation represents the condition if the reaction were 100% complete; 2.00 mol HI would be formed and no H_2 and I_2 would be left unreacted.

$$\underset{\substack{0.21 \\ \text{mol}}}{H_2(g)} + \underset{\substack{0.21 \\ \text{mol}}}{I_2(g)} \overset{700 \text{ K}}{\rightleftharpoons} \underset{\substack{1.58 \\ \text{mol}}}{2 \text{ HI}(g)}$$

This equation represents the actual equilibrium attained starting with 1.00 mol each of H_2 and I_2. It shows that the forward reaction is only 79% complete.

16.4   Le Chatelier's Principle

Le Chatelier's principle

In 1888, the French chemist Henri Le Chatelier (1850–1936) set forth a simple, far-reaching generalization on the behavior of equilibrium systems. This generalization, known as **Le Chatelier's principle,** states

> **If a stress is applied to a system in equilibrium, the system will respond in such a way as to relieve that stress and restore equilibrium under a new set of conditions.**

The application of Le Chatelier's principle helps us predict the effect of changing conditions in chemical reactions. We will examine the effect of changes in concentration, temperature, and volume.

16.5   Effect of Concentration on Equilibrium

The manner in which the rate of a chemical reaction depends on the concentration of the reactants must be determined experimentally. Many simple, one-step reactions result from a collision between two molecules or ions. The rate of such one-step reactions can be altered by changing the concentration of the reactants or products. An increase in concentration of the reactants provides more individual reacting species for collisions and results in an increase in the rate of reaction.

    An equilibrium is disturbed when the concentration of one or more of its components is changed. As a result the concentration of all species will change, and a new equilibrium mixture will be established. Consider the hypothetical equilibrium represented by the equation

$$A + B \rightleftharpoons C + D$$

where A and B react in one step to form C and D. When the concentration of B is increased, the following occurs:

1. The rate of the reaction to the right (forward) increases. This rate is proportional to the concentration of A times the concentration of B.
2. The rate to the right becomes greater than the rate to the left.
3. Reactants A and B are used faster than they are produced; C and D are produced faster than they are used.
4. After a period of time, rates to the right and left become equal, and the system is again in equilibrium.
5. In the new equilibrium the concentration of A is less, and the concentrations of B, C, and D are greater than in the original equilibrium.

Conclusion: The equilibrium has shifted to the right.

Applying this change in concentration to the equilibrium mixture of 1.00 mol of hydrogen and 1.00 mol of iodine from Section 16.3, we find that, when an additional 0.20 mol I_2 is added, the yield of HI (based on H_2) is 85% (1.70 mol) instead of 79%. Here is how the two systems compare after the new equilibrium mixture is reached:

Concentration	Change
$[H_2]$	?
$[I_2]$	*increase*
$[HI]$	?

Original equilibrium	**New equilibrium**
1.00 mol H_2 + 1.00 mol I_2	1.00 mol H_2 + 1.20 mol I_2
Yield: 79% HI	Yield: 85% HI (based on H_2)
Equilibrium mixture contains:	Equilibrium mixture contains:
1.58 mol HI	1.70 mol HI
0.21 mol H_2	0.15 mol H_2
0.21 mol I_2	0.35 mol I_2

Analyzing this new system, we see that, when 0.20 mol I_2 is added, the equilibrium shifts to the right to counteract the increase in I_2 concentration. Some of the H_2 reacts with added I_2 and produces more HI, until an equilibrium mixture is established again. When I_2 is added, the concentration of I_2 increases, the concentration of H_2 decreases, and the concentration of HI increases.

Concentration	Change
$[H_2]$	*decrease*
$[I_2]$	*increase*
$[HI]$	*increase*

Practice 16.1

Use a chart like those in the margin to show what would happen to the concentrations of each substance in the system

$$H_2(g) + I_2(g) \rightleftharpoons 2 \, HI(g)$$

upon adding (a) more H_2 and (b) more HI.

The equation

$$Fe^{3+}(aq) + SCN^-(aq) \rightleftharpoons Fe(SCN)^{2+}(aq)$$

pale yellow   colorless   red

represents an equilibrium that is used in certain analytical procedures as an indicator because of the readily visible, intense red color of the complex $Fe(SCN)^{2+}$ ion. A very dilute solution of iron(III), Fe^{3+}, and thiocyanate, SCN^-, is light red. When the concentration of either Fe^{3+} or SCN^- is increased, the equilibrium shift to the right is observed by an increase in the intensity of the color, resulting from the formation of additional $Fe(SCN)^{2+}$.

If either Fe^{3+} or SCN^- is removed from solution, the equilibrium will shift to the left, and the solution will become lighter in color. When Ag^+ is added to the solution, a white precipitate of silver thiocyanate (AgSCN) is formed, thus removing SCN^- ion from the equilibrium:

$$Ag^+(aq) + SCN^-(aq) \rightleftharpoons AgSCN(s)$$

The system accordingly responds to counteract the change in SCN^- concentration by shifting the equilibrium to the left. This shift is evident by a decrease in the intensity of the red color due to a decreased concentration of $Fe(SCN)^{2+}$.

Now consider the effect of changing the concentrations in the equilibrium mixture of chlorine water. The equilibrium equation is

$$Cl_2(aq) + 2 H_2O(l) \rightleftharpoons HOCl(aq) + H_3O^+(aq) + Cl^-(aq)$$

The variation in concentrations and the equilibrium shifts are tabulated in the following table. An X in the second or third column indicates that the reagent is increased or decreased. The fourth column indicates the direction of the equilibrium shift.

Reagent	Concentration		Equilibrium shift
	Increase	Decrease	
Cl_2	—	X	Left
H_2O	X	—	Right
HOCl	X	—	Left
H_3O^+	—	X	Right
Cl^-	X	—	Left

Consider the equilibrium in a 0.100 M acetic acid solution:

$$HC_2H_3O_2(aq) + H_2O(l) \rightleftharpoons H_3O^+(aq) + C_2H_3O_2^-(aq)$$

In this solution the concentration of the hydronium ion, H_3O^+, which is a measure of the acidity, is 1.34×10^{-3} mol/L, corresponding to a pH of 2.87. What will happen to the acidity when 0.100 mol of sodium acetate, $NaC_2H_3O_2$, is added to 1 L of 0.100 M acetic acid, $HC_2H_3O_2$? When $NaC_2H_3O_2$ dissolves, it dissociates into sodium ions, Na^+, and acetate ions, $C_2H_3O_2^-$. The acetate ion from the salt is a common ion to the acetic acid equilibrium system and increases the total acetate ion concentration in the solution. As a result the equilibrium shifts to the left, decreasing the hydronium ion concentration and lowering the acidity of the solution. Evidence of this decrease in acidity is shown by the fact that the pH of a solution that is 0.100 M in $HC_2H_3O_2$ and 0.100 M in $NaC_2H_3O_2$ is 4.74. The pH of several different solutions of $HC_2H_3O_2$ and $NaC_2H_3O_2$ is shown in the table that follows. Each time the acetate ion is increased, the pH increases, indicating a further shift in the equilibrium toward un-ionized acetic acid.

Concentration	Change
$[HC_2H_3O_2]$	increase
$[H_3O^+]$	decrease
$[C_2H_3O_2^-]$	**increase**

Solution	pH
1 L 0.100 M $HC_2H_3O_2$	2.87
1 L 0.100 M $HC_2H_3O_2$ + 0.100 mol $NaC_2H_3O_2$	4.74
1 L 0.100 M $HC_2H_3O_2$ + 0.200 mol $NaC_2H_3O_2$	5.05
1 L 0.100 M $HC_2H_3O_2$ + 0.300 mol $NaC_2H_3O_2$	5.23

In summary we can say that when the concentration of a reagent on the left side of an equation is increased, the equilibrium shifts to the right. When the concentration of a reagent on the right side of an equation is increased, the equilibrium shifts to the left. In accordance with Le Chatelier's principle the equilibrium always shifts in the direction that tends to reduce the concentration of the added reactant.

Practice 16.2

Aqueous chromate ion, CrO_4^{2-}, exists in equilibrium with aqueous dichromate ion, $Cr_2O_7^{2-}$ in an acidic solution. What effect will (a) increasing the dichromate ion, and (b) adding HCl have on the equilibrium?

$$2\,CrO_4^{2-}(aq) + 2\,H^+(aq) \rightleftharpoons Cr_2O_7^{2-}(aq) + H_2O(l)$$

16.6   Effect of Volume on Equilibrium

Changes in volume significantly affect the reaction rate only when one or more of the reactants or products is a gas, and the reaction is run in a closed container. In these cases the effect of decreasing the volume of the reacting gases is equivalent to increasing their concentrations. In the reaction

$$CaCO_3(s) \overset{\Delta}{\rightleftharpoons} CaO(s) + CO_2(g)$$

calcium carbonate decomposes into calcium oxide and carbon dioxide when heated about 825°C. Decreasing the volume of the container speeds up the reverse reaction and causes the equilibrium to shift to the left. Decreasing the volume increases the concentration of CO_2, the only gaseous substance in the reaction.

If the volume of the container is decreased, the pressure of the gas will increase. In a system composed entirely of gases, this decrease in the volume of the container will cause the reaction and the equilibrium to shift to the side that contains the smaller number of molecules. To clarify your thinking: When the container volume is decreased, the pressure in the container is increased. The system tries to lower this

Gaseous ammonia is often used to add nitrogen to the fields before planting and during early growth.

Haber received the Nobel prize in chemistry for this process in 1918.

pressure by reducing the number of molecules. Let's consider an example that shows these effects.

Prior to World War I, Fritz Haber (1868–1934) invented the first major process for the fixation of nitrogen. In this process nitrogen and hydrogen are reacted together in the presence of a catalyst at moderately high temperature and pressure to produce ammonia:

$$N_2(g) + 3 H_2(g) \rightleftharpoons 2 NH_3(g) + 92.5 \text{ kJ}$$

1 mol	3 mol	2 mol
1 volume	3 volumes	2 volumes

The left side of the equation in the Haber process represents 4 mol of gas combining to give 2 mol of gas on the right side of the equation. A decrease in the volume of the container shifts the equilibrium to the right. This decrease in volume results in a higher concentration of both reactants and products. The equilibrium shifts to the right toward fewer molecules.

When the total number of gaseous molecules on both sides of an equation is the same, a change in volume does not cause an equilibrium shift. The following reaction is an example:

$$N_2(g) + O_2(g) \rightleftharpoons 2 NO(g)$$

1 mol	1 mol	2 mol
1 volume	1 volume	2 volumes
6.022×10^{23} molecules	6.022×10^{23} molecules	$2(6.022 \times 10^{23})$ molecules

When the volume of the container is decreased, the rate of both the forward and the reverse reactions will increase because of the higher concentrations of N_2, O_2, and NO. But the equilibrium will not shift because the number of molecules is the same on both sides of the equation and the effects on concentration are the same on both forward and reverse rates.

Example 16.1   What effect would a decrease in volume of the container have on the position of equilibrium in these reactions?

(a) $2 SO_2(g) + O_2(g) \rightleftharpoons 2 SO_3(g)$

(b) $H_2(g) + Cl_2(g) \rightleftharpoons 2 HCl(g)$

(c) $N_2O_4(g) \rightleftharpoons 2 NO_2(g)$

(a) The equilibrium will shift to the right because the substance on the right has a smaller number of moles than those on the left.

(b) The equilibrium position will be unaffected because the moles of gases on both sides of the equation are the same.

(c) The equilibrium will shift to the left because $N_2O_4(g)$ represents the smaller number of molecules.

Solution

Practice 16.3

What effect would a decrease in the container's volume have on the position of the equilibrium in these reactions?

(a) $2\ NO(g) + Cl_2(g) \rightleftharpoons 2\ NOCl(g)$

(b) $COBr_2(g) \rightleftharpoons CO(g) + Br_2(g)$

16.7 Effect of Temperature on Equilibrium

When the temperature of a system is raised, the rate of reaction increases because of increased kinetic energy and more frequent collisions of the reacting species. In a reversible reaction the rate of both the forward and the reverse reactions is increased by an increase in temperature; however, the reaction that absorbs heat increases to a greater extent, and the equilibrium shifts to favor that reaction.

High temperatures can cause the destruction or decomposition of the reactants or products.

An increase in temperature generally increases the rate of reaction. Molecules at elevated temperatures have more kinetic energy; their collisions are thus more likely to result in a reaction.

When heat is applied to a system in equilibrium, the reaction that absorbs heat is favored. When the process, as written, is endothermic, the forward reaction is increased. When the reaction is exothermic, the reverse reaction is favored. In this sense heat may be treated as a reactant in endothermic reactions or as a product in exothermic reactions. Therefore temperature is analogous to concentration when applying Le Chatelier's principle to heat effects on a chemical reaction.

Hot coke (C) is a very reactive element. In the reaction

$$C(s) + CO_2(g) + heat \rightleftharpoons 2\ CO(g)$$

very little if any CO is formed at room temperature. At 1000°C, the equilibrium mixture contains about an equal number of moles of CO and CO_2. Since the reaction is endothermic, the equilibrium is shifted to the right at higher temperatures.

When phosphorus trichloride reacts with dry chlorine gas to form phosphorus pentachloride, the reaction is exothermic:

$$PCl_3(l) + Cl_2(g) \rightleftharpoons PCl_5(s) + 88\ kJ$$

Heat must continuously be removed during the reaction to obtain a good yield of the product. According to Le Chatelier's principle, heat will cause the product, PCl_5, to decompose, re-forming PCl_3 and Cl_2. The equilibrium mixture at 200°C contains 52% PCl_5, and at 300°C it contains 3% PCl_5, verifying that heat causes the equilibrium to shift to the left.

▲
Light sticks. The chemical reaction that produces light in these light sticks is endothermic. Placing the light stick in hot water (right) favors this reaction, producing a brighter light than when the light stick is in ice water (left).

Example 16.2   What effect would an increase in temperature have on the position of the equilibrium in these reactions?

$$4\ HCl(g) + O_2(g) \rightleftharpoons 2\ H_2O(g) + 2\ Cl_2(g) + 95.4\ kJ \qquad (1)$$

$$H_2(g) + Cl_2(g) \rightleftharpoons 2\ HCl(g) + 185\ kJ \qquad (2)$$

$$CH_4(g) + 2\ O_2(g) \rightleftharpoons CO_2(g) + 2\ H_2O(g) + 890\ kJ \qquad (3)$$

$$N_2O_4(g) + 58.6\ kJ \rightleftharpoons 2\ NO_2(g) \qquad (4)$$

$$2\ CO_2(g) + 566\ kJ \rightleftharpoons 2\ CO(g) + O_2(g) \qquad (5)$$

$$H_2(g) + I_2(g) + 51.9\ kJ \rightleftharpoons 2\ HI(g) \qquad (6)$$

Solution   Reactions (1), (2), and (3) are exothermic; an increase in temperature will cause the equilibrium to shift to the left. Reactions (4), (5), and (6) are endothermic; an increase in temperature will cause the equilibrium to shift to the right.

Practice 16.4

What effect would an increase in temperature have on the position of the equilibrium in these reactions?

(a) $2\ SO_2(g) + O_2(g) \rightleftharpoons 2\ SO_3(g) + 198\ kJ$

(b) $H_2(g) + CO_2(g) + 41\ kJ \rightleftharpoons H_2O(g) + CO(g)$

16.8   Effect of Catalysts on Equilibrium

catalyst   A **catalyst** is a substance that influences the rate of a reaction and can be recovered essentially unchanged at the end of the reaction. A catalyst does not shift the equilibrium of a reaction; it affects only the speed at which the equilibrium is reached. It does this by lowering the activation energy for the reaction (see Figure 16.3). **Acti-**

activation energy   **vation energy** is the minimum energy required for the reaction to occur. A catalyst speeds up a reaction by lowering the activation energy while not changing the energies of reactants or products. If a catalyst does not affect the equilibrium, then it follows that it must affect the rate of both the forward and the reverse reactions equally.

The reaction between phosphorus trichloride and sulfur is highly exothermic, but it's so slow that very little product, thiophosphoryl chloride, is obtained, even after prolonged heating. When a catalyst, such as aluminum chloride is added, the reaction is complete in a few seconds:

$$PCl_3(l) + S(s) \xrightarrow{\ AlCl_3\ } PSCl_3(l)$$

The lab preparation of oxygen uses manganese dioxide as a catalyst to increase the rates of decomposition of both potassium chlorate and hydrogen peroxide:

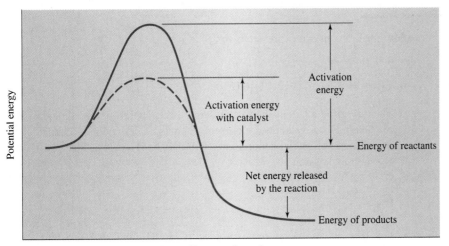

◀ FIGURE 16.3
Energy diagram for an exothermic reaction. Energy is put into the reaction (activation energy) to initiate the process. In the reaction shown, all of the activation energy and the net energy are released as the reaction proceeds to products. Note that the presence of a catalyst lowers the activation energy but does not change the energies of the reactants or the products.

Potential energy

Activation energy

Activation energy with catalyst

Energy of reactants

Net energy released by the reaction

Energy of products

Progress of reaction

$$2 \text{ KClO}_3(s) \xrightarrow[\Delta]{\text{MnO}_2} 2 \text{ KCl}(s) + 3 \text{ O}_2(g)$$

$$2 \text{ H}_2\text{O}_2(aq) \xrightarrow{\text{MnO}_2} 2 \text{ H}_2\text{O}(l) + \text{O}_2(g)$$

Catalysts are extremely important to industrial chemistry. Hundreds of chemical reactions that are otherwise too slow to be of practical value have been put to commercial use once a suitable catalyst was found. And in the area of biochemistry, catalysts are of supreme importance because nearly all chemical reactions in all forms of life are completely dependent on biochemical catalysts known as *enzymes*.

16.9 Equilibrium Constants

In a reversible chemical reaction at equilibrium, the concentrations of the reactants and products are constant. At equilibrium the rates of the forward and reverse reactions are equal, and an equilibrium constant expression can be written relating the products to the reactants. For the general reaction

$$a\text{A} + b\text{B} \rightleftharpoons c\text{C} + d\text{D}$$

at a given temperature, the following equilibrium constant expression can be written:

$$K_{eq} = \frac{[\text{C}]^c[\text{D}]^d}{[\text{A}]^a[\text{B}]^b}$$

where K_{eq}, the **equilibrium constant**, is constant at a particular temperature. The quantities in brackets are the concentrations of each substance in moles per liter. The superscript letters *a, b, c,* and *d* are the coefficients of the substances in the balanced equation. According to convention, we place the concentrations of the products (the substances on the right side of the equation as written) in the numerator and the concentrations of the reactants in the denominator.

equilibrium constant, K_{eq}

Note: The exponents are the same as the coefficients in the balanced equation.

Example 16.3   Write equilibrium constant expressions for

(a) $3 H_2(g) + N_2(g) \rightleftharpoons 2 NH_3(g)$

(b) $CO(g) + 2 H_2(g) \rightleftharpoons CH_3OH(g)$

Solution

(a) The only product, NH_3, has a coefficient of 2. Therefore the numerator will be $[NH_3]^2$. Two reactants are present, H_2 with a coefficient of 3 and N_2 with a coefficient of 1. The denominator will thus be $[H_2]^3[N_2]$. The equilibrium constant expression is

$$K_{eq} = \frac{[NH_3]^2}{[H_2]^3[N_2]}$$

(b) For this equation the numerator is $[CH_3OH]$ and the denominator is $[CO][H_2]^2$. The equilibrium constant expression is

$$K_{eq} = \frac{[CH_3OH]}{[CO][H_2]^2}$$

Practice 16.5

Write equilibrium constant expressions for

(a) $2 N_2O_5(g) \rightleftharpoons 4 NO_2(g) + O_2(g)$

(b) $4 NH_3(g) + 3 O_2(g) \rightleftharpoons 2 N_2(g) + 6 H_2O(g)$

The magnitude of an equilibrium constant indicates the extent to which the forward and reverse reactions take place. When K_{eq} is greater than 1, the amount of products at equilibrium is greater than the amount of reactants. When K_{eq} is less than 1, the amount of reactants at equilibrium is greater than the amount of products. A very large value for K_{eq} indicates that the forward reaction goes essentially to completion. A very small K_{eq} means that the reverse reaction goes nearly to completion and that the equilibrium is far to the left (toward the reactants). Two examples follow:

$$H_2(g) + I_2(g) \rightleftharpoons 2 HI(g) \qquad K_{eq} = 54.8 \text{ at } 425°C$$

This K_{eq} indicates that more product than reactant exists at equilibrium.

$$COCl_2(g) \rightleftharpoons CO(g) + Cl_2(g) \qquad K_{eq} = 7.6 \times 10^{-4} \text{ at } 400°C$$

This K_{eq} indicates that $COCl_2$ is stable and that very little decomposition to CO and Cl_2 occurs at 400°C. The equilibrium is far to the left.

Units are generally not included in values of K_{eq} for reasons beyond the scope of this book.

When the molar concentrations of all species in an equilibrium reaction are known, the K_{eq} can be calculated by substituting the concentrations into the equilibrium constant expression.

Calculate the K_{eq} for the following reaction based on concentrations of $PCl_5 = 0.030$ mol/L, $PCl_3 = 0.97$ mol/L, and $Cl_2 = 0.97$ mol/L at 300°C.

$$PCl_5(g) \rightleftharpoons PCl_3(g) + Cl_2(g)$$

Example 16.4

First write the K_{eq} expression; then substitute the respective concentrations into this equation and solve:

$$K_{eq} = \frac{[PCl_3][Cl_2]}{[PCl_5]} = \frac{(0.97)(0.97)}{(0.030)} = 31$$

Solution

Remember: Units are not included for K$_{eq}$.

This K_{eq} is considered to be a fairly large value, indicating that at 300°C the decomposition of PCl_5 proceeds far to the right.

Practice 16.6

Calculate the K_{eq} for this reaction. Is the forward or the reverse reaction favored?

$$2 NO(g) + O_2(g) \rightleftharpoons 2 NO_2(g)$$

when $[NO] = 0.050\ M$, $[O_2] = 0.75\ M$, and $[NO_2] = 0.25\ M$.

16.10 Ion Product Constant for Water

We've seen that water ionizes to a slight degree. This ionization is represented by these equilibrium equations:

$$H_2O + H_2O \rightleftharpoons H_3O^+ + OH^- \qquad (1)$$

$$H_2O \rightleftharpoons H^+ + OH^- \qquad (2)$$

Equation (1) is the more accurate representation of the equilibrium because free protons (H^+) do not exist in water. Equation (2) is a simplified and often-used representation of the water equilibrium. The actual concentration of H^+ produced in pure water is minute and amounts to only 1.00×10^{-7} mol/L at 25°C. In pure water,

$$[H^+] = [OH^-] = 1.00 \times 10^{-7}\ mol/L$$

since both ions are produced in equal molar amounts, as shown in equation (2).

The $H_2O \rightleftharpoons H^+ + OH^-$ equilibrium exists in water and in all water solutions. A special equilibrium constant called the **ion product constant for water, K_w,** applies to this equilibrium. The constant K_w is defined as the product of the H^+ ion concentration and the OH^- ion concentration, each in moles per liter:

ion product constant for water, K$_w$

$$K_w = [H^+][OH^-]$$

The numerical value of K_w is 1.00×10^{-14}, since for pure water at 25°C,

$$K_w = [H^+][OH^-] = (1.00 \times 10^{-7})(1.00 \times 10^{-7}) = 1.00 \times 10^{-14}$$

TABLE 16.1	Relationship of H^+ and OH^- Concentrations in Water Solutions			
$[H^+]$	$[OH^-]$	K_w	pH	pOH
1.00×10^{-2}	1.00×10^{-12}	1.00×10^{-14}	2.00	12.00
1.00×10^{-4}	1.00×10^{-10}	1.00×10^{-14}	4.00	10.00
2.00×10^{-6}	5.00×10^{-9}	1.00×10^{-14}	5.70	8.30
1.00×10^{-7}	1.00×10^{-7}	1.00×10^{-14}	7.00	7.00
1.00×10^{-9}	1.00×10^{-5}	1.00×10^{-14}	9.00	5.00

The value of K_w for all water solutions at 25°C is the constant 1.00×10^{-14}. It is important to realize that as the concentration of one of these ions, H^+ or OH^-, increases, the other decreases. However, the product of $[H^+]$ and $[OH^-]$ always equals 1.00×10^{-14}. This relationship can be seen in the examples shown in Table 16.1. If the concentration of one ion is known, the concentration of the other can be calculated from the K_w expression.

$$K_w = [H^+][OH^-] \qquad [H^+] = \frac{K_w}{[OH^-]} \qquad [OH^-] = \frac{K_w}{[H^+]}$$

Example 16.5   What is the concentration of (a) H^+, and (b) OH^- in a 0.001 M HCl solution? Remember that HCl is 100% ionized.

Solution   (a) Since all the HCl is ionized, H^+ = 0.001 mol/L or 1×10^{-3} mol/L:

$$HCl \longrightarrow H^+ + Cl^-$$
$$ 0.001\ M \quad 0.001\ M$$

$$[H^+] = 1 \times 10^{-3} \text{ mol/L}$$

(b) To calculate the $[OH^-]$ in this solution, use the following equation and substitute the values for K_w and $[H^+]$:

$$[OH^-] = \frac{K_w}{[H^+]}$$

$$[OH^-] = \frac{1.00 \times 10^{-14}}{1 \times 10^{-3}} = 1 \times 10^{-11} \text{ mol/L}$$

Practice 16.7

Determine the $[H^+]$ and $[OH^-]$ in
(a) $5.0 \times 10^{-5}\ M$ HNO_3 (b) $2.0 \times 10^{-6}\ M$ KOH

What is the pH of a 0.010 M NaOH solution? Assume that NaOH is 100% ionized. **Example 16.6**

Since all the NaOH is ionized, $[OH^-] = 0.010$ mol/L or 1.0×10^{-2} mol/L. **Solution**

$$NaOH \longrightarrow Na^+ + OH^-$$
$$0.010\ M \quad 0.010\ M$$

To find the pH of the solution, we first calculate the H^+ ion concentration. Use the following equation and substitute the values for K_w and $[OH^-]$:

$$[H^+] = \frac{K_w}{[OH^-]} = \frac{1.00 \times 10^{-14}}{1.0 \times 10^{-2}} = 1.0 \times 10^{-12}\ \text{mol/L}$$

$$pH = -\log[H^+] = -\log(1.0 \times 10^{-12}) = 12.00$$

Practice 16.8

Determine the pH for the following solutions:
(a) $5.0 \times 10^{-5}\ M$ HNO$_3$ (b) $2.0 \times 10^{-6}\ M$ KOH

Just as pH is used to express the acidity of a solution, pOH is used to express the basicity of an aqueous solution. The pOH is related to the OH^- ion concentration in the same way that the pH is related to the H^+ ion concentration.

$$pOH = -\log[OH^-]$$

Thus a solution in which $[OH^-] = 1.0 \times 10^{-2}$, as in Example 16.6, will have pOH = 2.00.

In pure water, where $[H^+] = 1.00 \times 10^{-7}$ and $[OH^-] = 1.00 \times 10^{-7}$, the pH is 7.0, and the pOH is 7.0. The sum of the pH and pOH is always 14.0:

$$pH + pOH = 14.00$$

In Example 16.6, the pH can also be found by first calculating the pOH from the OH^- ion concentration and then subtracting from 14.00.

$$pH = 14.00 - pOH = 14.00 - 2.00 = 12.00$$

Table 16.1 summarizes the relationship between $[H^+]$ and $[OH^-]$ in water solutions.

16.11 Ionization Constants

In addition to K_w, several other equilibrium constants are commonly used. Let's consider the equilibrium constant for acetic acid in solution. Because it is a weak acid, an equilibrium is established between molecular $HC_2H_3O_2$ and its ions in solution:

$$HC_2H_3O_2(aq) \rightleftharpoons H^+(aq) + C_2H_3O_2^-(aq)$$

The ionization constant expression is the concentration of the products divided by the concentration of the reactants:

$$K_a = \frac{[H^+][C_2H_3O_2^-]}{[HC_2H_3O_2]}$$

acid ionization constant, K_a

The constant is called the **acid ionization constant, K_a,** a special type of equilibrium constant.

The concentration of water in the solution is large compared to other concentrations and does not change appreciably.

At 25°C, a 0.100 M $HC_2H_3O_2$ solution is 1.34% ionized and has an $[H^+]$ of 1.34×10^{-3} mol/L. From this information we can calculate the ionization constant for acetic acid.

A 0.100 M solution initially contains 0.100 mol of acetic acid per liter. Of this, 0.100 mol, only 1.34%, or 1.34×10^{-3} mol, is ionized, which gives an $[H^+] = 1.34 \times 10^{-3}$ mol/L. Because each molecule of acid that ionizes yields one H^+ and one $C_2H_3O_2^-$, the concentration of $C_2H_3O_2^-$ ions is also 1.34×10^{-3} mol/L. This ionization leaves $0.100 - 0.00134 = 0.099$ mol/L of un-ionized acetic acid.

Acid	Initial concentration (mol/L)	Equilibrium concentration (mol/L)
$[HC_2H_3O_2]$	0.100	0.099
$[H^+]$	0	0.00134
$[C_2H_3O_2^-]$	0	0.00134

Substituting these concentrations in the equilibrium expression, we obtain the value for K_a:

$$K_a = \frac{[H^+][C_2H_3O_2^-]}{[HC_2H_3O_2]} = \frac{(1.34 \times 10^{-3})(1.34 \times 10^{-3})}{(0.099)} = 1.8 \times 10^{-5}$$

The K_a for acetic acid, 1.8×10^{-5}, is small and indicates that the position of the equilibrium is far toward the un-ionized acetic acid. In fact, a 0.100 M acetic acid solution is 99% un-ionized.

Once the K_a for acetic acid is established, it can be used to describe other systems containing H^+, $C_2H_3O_2^-$, and $HC_2H_3O_2$ in equilibrium at 25°C. The ionization constants for several other weak acids are listed in Table 16.2.

Example 16.7   What is the $[H^+]$ in a 0.50 M $HC_2H_3O_2$ solution? The ionization constant, K_a, for $HC_2H_3O_2$ is 1.8×10^{-5}.

Solution   To solve this problem, first write the equilibrium equation and the K_a expression:

$$HC_2H_3O_2 \rightleftharpoons H^+ + C_2H_3O_2^- \qquad K_a = \frac{[H^+][C_2H_3O_2^-]}{[HC_2H_3O_2]} = 1.8 \times 10^{-5}$$

We know that the initial concentration of $HC_2H_3O_2$ is 0.50 M. We also know from the ionization equation that one $C_2H_3O_2^-$ is produced for every H^+ produced; that is,

| TABLE 16.2 | Ionization Constants (K_a) of Weak Acids at 25°C | | | | |

Acid	Formula	K_a	Acid	Formula	K_a
Acetic	$HC_2H_3O_2$	1.8×10^{-5}	Hydrocyanic	HCN	4.0×10^{-10}
Benzoic	$HC_7H_5O_2$	6.3×10^{-5}	Hypochlorous	HClO	3.5×10^{-8}
Carbolic (phenol)	HC_6H_5O	1.3×10^{-10}	Nitrous	HNO_2	4.5×10^{-4}
Cyanic	HCNO	2.0×10^{-4}	Hydrofluoric	HF	6.5×10^{-4}
Formic	$HCHO_2$	1.8×10^{-4}			

the $[H^+]$ and the $[C_2H_3O_2^-]$ are equal. To solve, let $Y = [H^+]$, which also equals the $[C_2H_3O_2^-]$. The un-ionized $[HC_2H_3O_2]$ remaining will then be $0.50 - Y$, the starting concentration minus the amount that ionized:

$$[H^+] = [C_2H_3O_2^-] = Y \qquad [HC_2H_3O_2] = 0.50 - Y$$

	Initial	Equilibrium
$[H^+]$	0	Y
$[C_2H_3O_2]$	0	Y
$[HC_2H_3O_2]$	0.5	0.5 − Y

Substituting these values into the K_a expression, we obtain

$$K_a = \frac{(Y)(Y)}{0.50 - Y} = \frac{Y^2}{0.50 - Y} = 1.8 \times 10^{-5}$$

An exact solution of this equation for Y requires the use of a mathematical equation known as the *quadratic equation*. However, an approximate solution is obtained if we assume that Y is small and can be neglected compared with 0.50. Then $0.50 - Y$ will be equal to approximately 0.50. The equation now becomes

The quadratic equation is
$$y = \frac{-b \pm \sqrt{b^2 - 4ac}}{2a}$$
for the equation
$$ay^2 + by + c = 0$$

$$\frac{Y^2}{0.50} = 1.8 \times 10^{-5}$$

$$Y^2 = 0.50 \times 1.8 \times 10^{-5} = 0.90 \times 10^{-5} = 9.0 \times 10^{-6}$$

Taking the square root of both sides of the equation, we obtain

$$Y = \sqrt{9.0 \times 10^{-6}} = 3.0 \times 10^{-3} \text{ mol/L}$$

Thus the $[H^+]$ is approximately 3.0×10^{-3} mol/L in a 0.50 M $HC_2H_3O_2$ solution. The exact solution to this problem, using the quadratic equation, gives a value of 2.99×10^{-3} mol/L for $[H^+]$, showing that we were justified in neglecting Y compared with 0.50.

Practice 16.9

Calculate the hydrogen ion concentration in (a) 0.100 M hydrocyanic acid (HCN) solution, and (b) 0.0250 M carbolic acid (HC_6H_5O) solution.

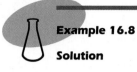

Example 16.8

Solution

Calculate the percent ionization in a 0.50 M $HC_2H_3O_2$ solution.

The percent ionization of a weak acid, $HA(aq) \rightleftharpoons H^+(aq) + A^-(aq)$, is found by dividing the concentration of the H^+ or A^- ions at equilibrium by the initial concentration of HA. For acetic acid,

> *Percent ionization of an acid, HA, is determined by dividing the concentration of H^+ or A^- ions at equilibrium by the initial concentration of HA and multiplying by 100.*

$$\frac{\text{concentration of } [H^+] \text{ or } [C_2H_3O_2^-]}{\text{initial concentration of } [HC_2H_3O_2]} \times 100 = \text{percent ionized}$$

To solve this problem, we first need to calculate the $[H^+]$. This calculation has already been done for a 0.50 M solution:

$$[H^+] = 3.0 \times 10^{-3} \text{ mol/L in a } 0.50 \ M \text{ solution} \qquad \text{(from Example 16.7)}$$

This $[H^+]$ represents a fractional amount of the initial 0.50 M $HC_2H_3O_2$. Therefore

$$\frac{3.0 \times 10^{-3} \text{ mol/L}}{0.50 \text{ mol/L}} \times 100 = 0.60\% \text{ ionized}$$

A 0.50 M $HC_2H_3O_2$ solution is 0.60% ionized.

Practice 16.10

Calculate the percent ionization for
(a) 0.100 M hydrocyanic acid (HCN)
(b) 0.0250 M carbolic acid (HC_6H_5O)

16.12 Solubility Product Constant

solubility product constant, K_{sp}

The **solubility product constant, K_{sp},** is the equilibrium constant of a slightly soluble salt. To evaluate K_{sp}, consider this example. The solubility of AgCl in water is 1.3×10^{-5} mol/L at 25°C. The equation for the equilibrium between AgCl and its ions in solution is

$$AgCl(s) \rightleftharpoons Ag^+(aq) + Cl^-(aq)$$

The equilibrium constant expression is

$$K_{eq} = \frac{[Ag^+][Cl^-]}{[AgCl(s)]}$$

The amount of solid AgCl does not affect the equilibrium system provided that some is present. In other words the concentration of solid AgCl is constant whether 1 mg or 10 g of the salt is present. Therefore the product obtained by multiplying the two constants K_{eq} and $[AgCl(s)]$ is also a constant. This is the solubility product constant, K_{sp}:

TABLE 16.3 Solubility Product Constants (K_{sp}) at 25°C

Compound	K_{sp}	Compound	K_{sp}
AgCl	1.7×10^{-10}	CaF_2	3.9×10^{-11}
AgBr	5×10^{-13}	CuS	9×10^{-45}
AgI	8.5×10^{-17}	$Fe(OH)_3$	6×10^{-38}
$AgC_2H_3O_2$	2×10^{-3}	PbS	7×10^{-29}
Ag_2CrO_4	1.9×10^{-12}	$PbSO_4$	1.3×10^{-8}
$BaCrO_4$	8.5×10^{-11}	$Mn(OH)_2$	2.0×10^{-13}
$BaSO_4$	1.5×10^{-9}		

$$K_{eq} \times [AgCl(s)] = [Ag^+][Cl^-] = K_{sp}$$

$$K_{sp} = [Ag^+][Cl^-]$$

The K_{sp} is equal to the product of the $[Ag^+]$ and the $[Cl^-]$, each in moles per liter. When 1.3×10^{-5} mol/L of AgCl dissolves, it produces 1.3×10^{-5} mol/L each of Ag^+ and Cl^-. From these concentrations the K_{sp} can be calculated:

$$[Ag^+] = 1.3 \times 10^{-5} \text{ mol/L} \qquad [Cl^-] = 1.3 \times 10^{-5} \text{ mol/L}$$

$$K_{sp} = [Ag^+][Cl^-] = (1.3 \times 10^{-5})(1.3 \times 10^{-5}) = 1.7 \times 10^{-10}$$

Once the K_{sp} value for AgCl is established, it can be used to describe other systems containing Ag^+ and Cl^-.

The K_{sp} expression does not have a denominator. It consists only of the concentrations (mol/L) of the ions in solution. As in other equilibrium expressions, each of these concentrations is raised to a power that is the same number as its coefficient in the balanced equation. Here are equilibrium equations and the K_{sp} expressions for several other substances:

$$AgBr(s) \rightleftharpoons Ag^+(aq) + Br^-(aq) \qquad K_{sp} = [Ag^+][Br^-]$$

$$BaSO_4(s) \rightleftharpoons Ba^{2+}(aq) + SO_4^{2-}(aq) \qquad K_{sp} = [Ba^{2+}][SO_4^{2-}]$$

$$Ag_2CrO_4(s) \rightleftharpoons 2Ag^+(aq) + CrO_4^{2-}(aq) \qquad K_{sp} = [Ag^+]^2[CrO_4^{2-}]$$

$$CuS(s) \rightleftharpoons Cu^{2+}(aq) + S^{2-}(aq) \qquad K_{sp} = [Cu^{2+}][S^{2-}]$$

$$Mn(OH)_2(s) \rightleftharpoons Mn^{2+}(aq) + 2\,OH^-(aq) \qquad K_{sp} = [Mn^{2+}][OH^-]^2$$

$$Fe(OH)_3(s) \rightleftharpoons Fe^{3+}(aq) + 3\,OH^-(aq) \qquad K_{sp} = [Fe^{3+}][OH^-]^3$$

Table 16.3 lists K_{sp} values for these and several other substances.

When the product of the molar concentration of the ions in solution (each raised to its proper power) is greater than the K_{sp} for that substance, precipitation should occur. If the ion product is less than the K_{sp} value, no precipitation will occur.

Example 16.9   Write K_{sp} expressions for AgI and PbI$_2$, both of which are slightly soluble salts.

Solution   First write the equilibrium equations:

$$AgI(s) \rightleftharpoons Ag^+(aq) + I^-(aq)$$

$$PbI_2(s) \rightleftharpoons Pb^{2+}(aq) + 2\,I^-(aq)$$

Since the concentration of the solid crystals is constant, the K_{sp} equals the product of the molar concentrations of the ions in solution. In the case of PbI$_2$, the [I$^-$] must be squared:

$$K_{sp} = [Ag^+][I^-] \qquad K_{sp} = [Pb^{2+}][I^-]^2$$

Example 16.10 The K_{sp} value for lead sulfate is 1.3×10^{-8}. Calculate the solubility of PbSO$_4$ in grams per liter.

Solution   First write the equilibrium equation and the K_{sp} expression:

$$PbSO_4 \rightleftharpoons Pb^{2+}(aq) + SO_4^{2-}(aq)$$

$$K_{sp} = [Pb^{2+}][SO_4^{2-}] = 1.3 \times 10^{-8}$$

Since the lead sulfate that is in solution is completely dissociated, the [Pb^{2+}] or [SO$_4^{2-}$] is equal to the solubility of PbSO$_4$ in moles per liter. Let

$$Y = [Pb^{2+}] = [SO_4^{2-}]$$

Substitute Y into the K_{sp} equation and solve:

$$[Pb^{2+}][SO_4^{2-}] = (Y)(Y) = 1.3 \times 10^{-8}$$

$$Y^2 = 1.3 \times 10^{-8}$$

$$Y = 1.1 \times 10^{-4} \text{ mol/L}$$

The solubility of PbSO$_4$ therefore is 1.1×10^{-4} mol/L. Now convert mol/L to g/L:

1 mol of PbSO$_4$ has a mass of (207.2 g + 32.07 g + 64.00 g) or 303.3 g

$$\left(\frac{1.1 \times 10^{-4} \text{ mol}}{L}\right)\left(\frac{303.3 \text{ g}}{\text{mol}}\right) = 3.3 \times 10^{-2} \text{ g/L}$$

The solubility of PbSO$_4$ is 3.3×10^{-2} g/L.

Practice 16.11

Write the K_{sp} expression for
(a) Cr(OH)$_3$ (b) Cu$_3$(PO$_4$)$_2$

Practice 16.12

The K_{sp} value for CuS is 9.0×10^{-45}. Calculate the solubility of CuS in grams per liter.

An ion added to a solution already containing that ion is called a *common ion*. When a common ion is added to an equilibrium solution of a weak electrolyte or a slightly soluble salt, the equilibrium shifts according to Le Chatelier's principle. For example, when silver nitrate, $AgNO_3$, is added to a saturated solution of AgCl, ($AgCl(s) \rightleftharpoons Ag^+ + Cl^-$), the equilibrium shifts to the left due to the increase in the $[Ag^+]$. As a result, the $[Cl^-]$ and the solubility of AgCl decreases. The AgCl and $AgNO_3$ have the common ion Ag^+. A shift in the equilibrium position upon addition of an ion already contained in the solution is known as the **common ion effect.**

common ion effect

Silver nitrate is added to a saturated AgCl solution until the $[Ag^+]$ is 0.10 *M*. What will be the $[Cl^-]$ remaining in solution?   **Example 16.11**

This is an example of the common ion effect. The addition of $AgNO_3$ puts more Ag^+ in solution; the Ag^+ combines with Cl^- and causes the equilibrium to shift to the left, reducing the $[Cl^-]$ in solution. After the addition of Ag^+ to the mixture, the $[Ag^+]$ and $[Cl^-]$ in solution are no longer equal.   **Solution**

We use the K_{sp} to calculate the $[Cl^-]$ remaining in solution. The K_{sp} is constant at a particular temperature and remains the same no matter how we change the concentration of the species involved:

$$K_{sp} = [Ag^+][Cl^-] = 1.7 \times 10^{-10} \qquad [Ag^+] = 0.10 \text{ mol/L}$$

We then substitute the $[Ag^+]$ into the K_{sp} expression and calculate the $[Cl^-]$:

$$[0.10][Cl^-] = 1.7 \times 10^{-10}$$

$$[Cl^-] = \frac{1.7 \times 10^{-10}}{0.10} = 1.7 \times 10^{-9} \text{ mol/L}$$

This calculation shows a 10,000-fold reduction of Cl^- ions in solution. It illustrates that Cl^- ions may be quantitatively removed from solution with an excess of Ag^+ ions.

Practice 16.13

Sodium sulfate, Na_2SO_4, is added to a saturated solution of $BaSO_4$ until the concentration of the sulfate ion is 2.0×10^{-2} *M*. What will be the concentration of the Ba^{2+} ions remaining in solution?

16.13 **Hydrolysis**

Hydrolysis is the term used for the general reaction in which a water molecule is split. For example, the net ionic hydrolysis reaction for a sodium acetate solution is   *hydrolysis*

$$C_2H_3O_2^-(aq) + H_2O(l) \rightleftharpoons HC_2H_3O_2(aq) + OH^-(aq)$$

TABLE 16.4	Ionic Composition of Salts and the Nature of the Aqueous Solutions They Form		

Type of salt	Nature of aqueous solution	Examples
Weak base–strong acid	Acidic	NH_4Cl, NH_4NO_3
Strong base–weak acid	Basic	$NaC_2H_3O_2$, K_2CO_3
Weak base–weak acid	Depends on the salt	$NH_4C_2H_3O_2$, NH_4NO_2
Strong base–strong acid	Neutral	$NaCl$, KBr

In this reaction the water molecule is split, with the H^+ combining with $C_2H_3O_2^-$ to give the weak acid $HC_2H_3O_2$ and the OH^- going into solution, making the solution more basic.

Salts that contain an ion of a weak acid undergo hydrolysis. For example, a $0.10\,M$ NaCN solution has a pH of 11.1. The hydrolysis reaction that causes this solution to be basic is

$$CN^-(aq) + H_2O(l) \rightleftharpoons HCN(aq) + OH^-(aq)$$

If a salt contains the ion of a weak base, the ion hydrolyzes to produce an acidic solution. An example is ammonium chloride, which produces the NH_4^+ and Cl^- in solution. The NH_4^+ hydrolyzes to produce an acidic solution:

$$NH_4^+(aq) + H_2O(l) \longrightarrow NH_3(aq) + H_3O^+(aq)$$

The ions of a salt derived from a strong acid and a strong base, such as NaCl, do not undergo hydrolysis and thus form neutral solutions. Table 16.4 lists the ionic composition of various salts and the nature of the aqueous solutions that they form.

Practice 16.14

Indicate whether these salts would produce an acidic, basic, or neutral aqueous solution:
(a) KCN (b) $NaNO_3$ (c) NH_4Br

16.14 Buffer Solutions: The Control of pH

The control of pH within narrow limits is critically important in many chemical applications and vitally important in many biological systems. For example, human blood must be maintained between pH 7.35 and 7.45 for the efficient transport of oxygen from the lungs to the cells. This narrow pH range is maintained by buffer systems in the blood.

buffer solution   A **buffer solution** resists changes in pH when diluted or when small amounts of acid or base are added. Two common types of buffer solutions are (1) a weak acid mixed with its conjugate base, and (2) a weak base mixed with its conjugate acid.

Exchange of Oxygen and Carbon Dioxide in the Blood

The transport of oxygen and carbon dioxide between the lungs and tissues is a complex process that involves several reversible reactions, each of which behaves in accordance with Le Chatelier's principle.

The binding of oxygen to hemoglobin is a reversible reaction. The oxygen molecule must attach to the hemoglobin (Hb) and then later detach. The equilibrium equation for this reaction can be written:

$$Hb + O_2 \rightleftharpoons HbO_2$$

In the lungs the concentration of oxygen is high and favors the forward reaction. Oxygen quickly binds to the hemoglobin until it is saturated with oxygen.

In the tissues the concentration of oxygen is lower and in accordance with Le Chatelier's principle: The equilibrium position shifts to the left and the hemoglobin releases oxygen to the tissues. Approximately 45% of the oxygen diffuses out of the capillaries into the tissues, where it may be picked up by *myoglobin,* another carrier molecule.

Myoglobin functions as an oxygen-storage molecule, holding the oxygen until it is required in the energy-producing portions of the cell. The reaction between myoglobin (Mb) and oxygen can be written as an equilibrium reaction:

$$Mb + O_2 \rightleftharpoons MbO_2$$

The hemoglobin and myoglobin equations are very similar, so what accounts for the transfer of the oxygen from the hemoglobin to the myoglobin? Although both equilibria involve similar interactions, the affinity between oxygen and

▲ **Oxygen and carbon dioxide are exchanged in the red blood cells when they are in capillaries.**

hemoglobin is different from the affinity between myoglobin and oxygen. In the tissues the position of the hemoglobin equilibrium is such that it is 55% saturated with oxygen, whereas the myoglobin is at 90% oxygen saturation. Under these conditions hemoglobin will release oxygen, while myoglobin will bind oxygen. Thus oxygen is loaded onto hemoglobin in the lungs and unloaded in the tissue's cells.

Carbon dioxide produced in the cells must be removed from the tissues. Oxygen-depleted hemoglobin molecules accomplish this by becoming carriers of carbon dioxide. The carbon dioxide does not bind at the heme site as the oxygen does, but rather at one end of the protein chain. When carbon dioxide dissolves in water, some of the CO_2 reacts to release hydrogen ions:

$$CO_2 + H_2O \rightleftharpoons HCO_3^- + H^+$$

To facilitate the removal of CO_2 from the tissues, this equilibrium needs to be moved toward the right. This shift is accomplished by the removal of H^+ from the tissues by the hemoglobin molecule. The deoxygenated hemoglobin molecule can bind H^+ ions as well as CO_2. In the lungs this whole process is reversed so the CO_2 is removed from the hemoglobin and exhaled.

Molecules that are similar in structure to the oxygen molecule can become involved in competing equilibria. Hemoglobin is capable of binding with carbon monoxide, CO, nitrogen monoxide, NO, and cyanide, CN^-. The extent of the competition depends on the affinity. Since these molecules have a greater affinity for hemoglobin than oxygen, they will effectively displace oxygen from hemoglobin. For example,

$$HbO_2 + CO \rightleftharpoons HbCO + O_2$$

Since the affinity of hemoglobin for CO is 150 times stronger than its affinity for oxygen, the equilibrium position lies far to the right. This explains why CO is a poisonous substance and why oxygen is administered to victims of CO poisoning. The hemoglobin molecules can only transport oxygen if the CO is released and the oxygen shifts the equilibrium toward the left.

A saltwater aquarium is a ▶ buffer system.

The action of a buffer system can be understood by considering a solution of acetic acid and sodium acetate. The weak acid, $HC_2H_3O_2$, is mostly un-ionized and is in equilibrium with its ions in solution. The sodium acetate is completely ionized:

$$HC_2H_3O_2(aq) \rightleftharpoons H^+(aq) + C_2H_3O_2^-(aq)$$

$$NaC_2H_3O_2(aq) \longrightarrow Na^+(aq) + C_2H_3O_2^-(aq)$$

Because the sodium acetate is completely ionized, the solution contains a much higher concentration of acetate ions than would be present if only acetic acid were in solution. The acetate ion represses the ionization of acetic acid and also reacts with water, causing the solution to have a higher pH (be more basic) than an acetic acid solution (see Section 16.5). Thus a 0.1 M acetic acid solution has a pH of 2.87, but a solution that is 0.1 M in acetic acid and 0.1 M in sodium acetate has a pH of 4.74. This difference in pH is the result of the common ion effect.

A buffer solution has a built-in mechanism that counteracts the effect of adding acid or base. Consider the effect of adding HCl or NaOH to an acetic acid–sodium acetate buffer. When a small amount of HCl is added, the acetate ions of the buffer combine with the H^+ ions from HCl to form un-ionized acetic acid, thus neutralizing the added acid and maintaining the approximate pH of the solution. When NaOH is added, the OH^- ions react with acetic acid to neutralize the added base and thus maintain the approximate pH. The equations for these reactions are

$$H^+(aq) + C_2H_3O_2^-(aq) \rightleftharpoons HC_2H_3O_2(aq)$$

$$OH^-(aq) + HC_2H_3O_2(aq) \rightleftharpoons H_2O(l) + C_2H_3O_2^-(aq)$$

Data comparing the changes in pH caused by adding HCl and NaOH to pure water and to an acetic acid–sodium acetate buffer solution are shown in Table 16.5.

The human body has a number of buffer systems. One of these, the hydrogen carbonate–carbonic acid buffer, $HCO_3^- - H_2CO_3$, maintains the blood plasma at a pH of 7.4. The phosphate system, $HPO_4^{2-} - H_2PO_4^-$, is an important buffer in the red blood cells as well as in other places in the body.

TABLE 16.5	Changes in pH Caused by the Addition of HCl and NaOH		
Solution		pH	Change in pH
H_2O (1000 mL)		7	—
H_2O + 0.010 mol HCl		2	5
H_2O + 0.010 mol NaOH		12	5
Buffer solution (1000 mL)			
0.10 M $HC_2H_3O_2$ + 0.10 M $NaC_2H_3O_2$		4.74	—
Buffer + 0.010 mol HCl		4.66	0.08
Buffer + 0.010 mol NaOH		4.83	0.09

Concepts in Review

1. Describe a reversible reaction.

2. Explain why the rate of the forward reaction decreases and the rate of the reverse reaction increases as a chemical reaction approaches equilibrium.

3. Describe the qualitative effect of Le Chatelier's principle.

4. Predict how the rate of a chemical reaction is affected by (a) changes in the concentration of reactants, (b) changes in volume of gaseous reactants, (c) changes in temperature, and (d) the presence of a catalyst.

5. Write the equilibrium constant expression for a chemical reaction from a balanced chemical equation.

6. Explain the meaning of the numerical constant, K_{eq}, when given the concentration of the reactants and products in equilibrium.

7. Calculate the concentration of one substance in equilibrium when given the equilibrium constant and the concentrations of all the other substances.

8. Calculate the equilibrium constant, K_{eq}, when given the concentration of reactants and products in equilibrium.

9. Calculate the ionization constant for a weak acid from appropriate data.

10. Calculate the concentrations of all the chemical species in a solution of a weak acid when given the percent ionization or the ionization constant.

11. Compare the relative strengths of acids by using their ionization constants.

12. Use the ion product constant for water, K_w, to calculate $[H^+]$, $[OH^-]$, pH, and pOH when given any one of these quantities.

13. Calculate the solubility product constant, K_{sp}, of a slightly soluble salt when given its solubility, or vice versa.

14. Compare relative solubilities of salts if solubility products are known.

15. Discuss the common ion effect on a system at equilibrium.

16. Explain hydrolysis and why some salts form acidic or basic aqueous solutions.

17. Explain how a buffer solution is able to counteract the addition of small amounts of either H^+ or OH^- ions.

Key Terms

acid ionization constant, K_a (16.11)
activation energy (16.8)
buffer solution (16.14)
catalyst (16.8)
chemical equilibrium (16.3)

chemical kinetics (16.2)
common ion effect (16.12)
equilibrium (16.3)
equilibrium constant, K_{eq} (16.9)
hydrolysis (16.13)

ion product constant for water, K_w (16.10)
Le Chatelier's principle (16.4)
reversible chemical reaction (16.1)
solubility product constant, K_{sp} (16.12)

Questions

1. How would you expect the two tubes in Figure 16.1 to appear if both are at 25°C?

2. Is the reaction $N_2O_4 \rightleftharpoons 2\ NO_2$ exothermic or endothermic? (Figure 16.1)

3. At equilibrium how do the forward and reverse reaction rates compare? (Figure 16.2)

4. For each solution in Table 16.1, what is the sum of the pH plus the pOH? What would be the pOH of a solution whose pH was -1?

5. Of the acids listed in Table 16.2, which ones are stronger than acetic acid and which are weaker?

6. Tabulate the relative order of molar solubilities of AgCl, AgBr, AgI, $AgC_2H_3O_2$, $PbSO_4$, $BaSO_4$, $BaCrO_4$, and PbS. List the most soluble first. (Table 16.3)

7. Which compound in the following pairs has the greater molar solubility? (Table 16.3)
   (a) $Mn(OH)_2$ or Ag_2CrO_4
   (b) $BaCrO_4$ or Ag_2CrO_4

8. Explain how the acetic acid–sodium acetate buffer system maintains its pH when 0.010 mol of HCl is added to 1 L of the buffer solution. (Table 16.5)

9. Explain why a precipitate of NaCl forms when HCl gas is passed into a saturated aqueous solution of NaCl.

10. Why does the rate of a reaction usually increase when the concentration of one of the reactants is increased?

11. If pure hydrogen iodide, HI, is placed in a vessel at 700 K, will it decompose? Explain.

12. Why does an increase in temperature cause the rate of reaction to increase?

13. Describe how equilibrium is reached when the substances A and B are first mixed and react as

$$A + B \rightleftharpoons C + D$$

14. With dilution, aqueous solutions of acetic acid, $HC_2H_3O_2$, show increased ionization. For example, a 1.0 M solution of acetic acid is 0.42% ionized, whereas a 0.10 M solution is 1.34% ionized. Explain the behavior using the ionization equation and equilibrium principles.

15. A 1.0 M solution of acetic acid ionizes less and has a higher concentration of H^+ ions than a 0.10 M acetic acid solution. Explain this behavior. (See Question 14 for data.)

16. What would cause two separate samples of pure water to have slightly different pH values?

17. Why are the pH and pOH equal in pure water?

18. Explain why silver acetate is more soluble in nitric acid than in water. (*Hint*: Write the equilibrium equation first and then consider the effect of the acid on the acetate ion.) What would happen if hydrochloric acid were used in place of nitric acid?

19. Dissolution of sodium acetate, $NaC_2H_3O_2$, in pure water gives a basic solution. Why? (*Hint*: A small amount of $HC_2H_3O_2$ is formed.)

20. Describe why the pH of a buffer solution remains almost constant when a small amount of acid or base is added to it.

Paired Exercises

21. Express these reversible systems in equation form:
    (a) a mixture of ice and liquid water at 0°C
    (b) crystals of Na_2SO_4 in a saturated aqueous solution of Na_2SO_4

22. Express these reversible systems in equation form:
    (a) liquid water and vapor at 100°C in a pressure cooker
    (b) a closed system containing boiling sulfur dioxide, SO_2

23. Consider this system at equilibrium:

$$4 NH_3(g) + 3 O_2(g) \rightleftharpoons$$
$$2 N_2(g) + 6 H_2O(g) + 1531 \text{ kJ}$$

(a) Is the reaction exothermic or endothermic?

(b) If the system's state of equilibrium is disturbed by the addition of O_2, in which direction, left or right, must the reaction occur to reestablish equilibrium? After the new equilibrium has been established, how will the final molar concentrations of NH_3, O_2, N_2, and H_2O compare (increase or decrease) with their concentrations before the addition of the O_2?

24. Consider this system at equilibrium:

$$4 NH_3(g) + 3 O_2(g) \rightleftharpoons$$
$$2 N_2(g) + 6 H_2O(g) + 1531 \text{ kJ}$$

(a) If the system's state of equilibrium is disturbed by the addition of N_2, in which direction, left or right, must the reaction occur to reestablish equilibrium? After the new equilibrium has been established, how will the final molar concentrations of NH_3, O_2, N_2, and H_2O compare (increase or decrease) with their concentrations before the addition of the N_2?

(b) If the system's state of equilibrium is disturbed by the addition of heat, in which direction will the reaction occur, left or right, to reestablish equilibrium?

25. Consider this system at equilibrium:

$$N_2(g) + 3 H_2(g) \rightleftharpoons 2 NH_3(g) + 92.5 \text{ kJ}$$

Complete the following table. Indicate changes in moles by entering I, D, N, or ? in the table. (I = increase, D = decrease, N = no change, ? = insufficient information to determine.)

Change of stress imposed on the system at equilibrium	Direction of reaction, left or right, to reestablish equilibrium	Change in number of moles		
		N_2	H_2	NH_3
(a) Add N_2				
(b) Remove H_2				
(c) Decrease volume of reaction vessel				
(d) Increase temperature				

26. Consider this system at equilibrium:

$$N_2(g) + 3 H_2(g) \rightleftharpoons 2 NH_3(g) + 92.5 \text{ kJ}$$

Complete the following table. Indicate changes in moles by entering I, D, N, or ? in the table. (I = increase, D = decrease, N = no change, ? = insufficient information to determine.)

Change of stress imposed on the system at equilibrium	Direction of reaction, left or right, to reestablish equilibrium	Change in number of moles		
		N_2	H_2	NH_3
(a) Add NH_3				
(b) Increase volume of reaction vessel				
(c) Add catalyst				
(d) Add both H_2 and NH_3				

27. For the following equations, tell in which direction, left or right, the equilibrium will shift when these changes are made: The temperature is increased, the pressure is increased by decreasing the volume of the reaction vessel, and a catalyst is added.

(a) $3 O_2(g) + 271 \text{ kJ} \rightleftharpoons 2 O_3(g)$

(b) $CH_4(g) + Cl_2(g) \rightleftharpoons CH_3Cl(g) + HCl(g) + 110 \text{ kJ}$

(c) $2 NO(g) + 2 H_2(g) \rightleftharpoons N_2(g) + 2 H_2O(g) + 665 \text{ kJ}$

28. For the following equations, tell in which direction, left or right, the equilibrium will shift when these changes are made: The temperature is increased, the pressure is increased by decreasing the volume of the reaction vessel, and a catalyst is added.

(a) $2 SO_3(g) + 197 \text{ kJ} \rightleftharpoons 2 SO_2(g) + O_2(g)$

(b) $4 NH_3(g) + 3 O_2(g) \rightleftharpoons$
$$2 N_2(g) + 6 H_2O(g) + 1531 \text{ kJ}$$

(c) $OF_2(g) + H_2O(g) \rightleftharpoons O_2(g) + 2 HF(g) + 318 \text{ kJ}$

29. Utilizing Le Chatelier's principle, indicate the shift (if any) that would occur to

$$C_2H_6(g) + \text{heat} \rightleftharpoons C_2H_4(g) + H_2(g)$$

(a) if the concentration of hydrogen gas is decreased

(b) if the temperature is lowered

(c) if a catalyst is added

30. Utilizing Le Chatelier's principle, indicate the shift (if any) that would occur to

$$C_2H_6(g) + \text{heat} \rightleftharpoons C_2H_4(g) + H_2(g)$$

(a) if C_2H_6 is removed from the system

(b) if the volume of the container is increased

(c) if the temperature is raised

31. Write the equilibrium constant expression for these reactions:
(a) $4 HCl(g) + O_2(g) \rightleftharpoons 2 Cl_2(g) + 2 H_2O(g)$
(b) $N_2(g) + 3 H_2(g) \rightleftharpoons 2 NH_3(g)$
(c) $PCl_5(g) \rightleftharpoons PCl_3(g) + Cl_2(g)$

33. Write the solubility product expression, K_{sp}, for these substances:
(a) CuS   (c) $PbBr_2$
(b) $BaSO_4$ (d) Ag_3AsO_4

35. What effect will decreasing the $[H^+]$ of a solution have on (a) pH, (b) pOH, (c) $[OH^-]$, and (d) K_w?

37. Decide whether these salts form an acidic, basic, or neutral solution when dissolved in water:
(a) KCl   (c) K_2SO_4
(b) Na_2CO_3 (d) $(NH_4)_2SO_4$

39. Write hydrolysis equations for aqueous solutions of these salts:
(a) KNO_2
(b) $Mg(C_2H_3O_2)_2$

41. Write hydrolysis equations for these ions:
(a) HCO_3^-
(b) NH_4^+

43. One of the important pH-regulating systems in the blood consists of a carbonic acid–sodium hydrogen carbonate buffer:

$H_2CO_3(aq) \rightleftharpoons H^+(aq) + HCO_3^-(aq)$
$NaHCO_3(aq) \longrightarrow Na^+(aq) + HCO_3^-(aq)$

Explain how this buffer resists changes in pH when excess acid, H^+, gets into the bloodstream.

45. Calculate (a) the $[H^+]$, (b) the pH, and (c) the percent ionization of a 0.25 M solution of $HC_2H_3O_2$. ($K_a = 1.8 \times 10^{-5}$)

47. A 1.000 M solution of a weak acid, HA, is 0.52% ionized. Calculate the ionization constant, K_a, for the acid.

49. Calculate the percent ionization and pH of solutions of $HC_2H_3O_2$ ($K_a = 1.8 \times 10^{-5}$) having the following molarities: (a) 1.0 M, (b) 0.10 M, and (c) 0.010 M.

*51. A 0.37 M solution of a weak acid, HA, has a pH of 3.7. What is the K_a for this acid?

53. A common laboratory reagent is 6.0 M HCl. Calculate the $[H^+]$, $[OH^-]$, pH, and pOH of this solution.

55. Calculate the pH and the pOH of these solutions:
(a) 0.00010 M HCl
(b) 0.010 M NaOH
*(c) saturated $Fe(OH)_3$ solution ($K_{sp} = 6.0 \times 10^{-38}$)

57. Calculate the $[OH^-]$ in these solutions:
(a) $[H^+] = 1.0 \times 10^{-4}$
(b) $[H^+] = 2.8 \times 10^{-6}$

32. Write the equilibrium constant expression for these reactions:
(a) $HClO_2(aq) \rightleftharpoons H^+(aq) + ClO_2^-(aq)$
(b) $HC_2H_3O_2(aq) \rightleftharpoons H^+(aq) + C_2H_3O_2^-(aq)$
(c) $4 NH_3(g) + 5 O_2(g) \rightleftharpoons 4 NO(g) + 6 H_2O(g)$

34. Write the solubility product expression, K_{sp}, for these substances:
(a) $Fe(OH)_3$ (c) CaF_2
(b) Sb_2S_5 (d) $Ba_3(PO_4)_2$

36. What effect will increasing the $[H^+]$ of a solution have on (a) pH, (b) pOH, (c) $[OH^-]$, and (d) K_w?

38. Decide whether these salts form an acidic, basic, or neutral solution when dissolved in water:
(a) $Ca(CN)_2$ (c) $NaNO_2$
(b) $BaBr_2$ (d) NaF

40. Write hydrolysis equations for aqueous solutions of these salts:
(a) NH_4NO_3
(b) Na_2SO_3

42. Write hydrolysis equations for these ions:
(a) OCl^-
(b) ClO_2^-

44. One of the important pH-regulating systems in the blood consists of a carbonic acid–sodium hydrogen carbonate buffer:

$H_2CO_3(aq) \rightleftharpoons H^+(aq) + HCO_3^-(aq)$
$NaHCO_3(aq) \longrightarrow Na^+(aq) + HCO_3^-(aq)$

Explain how this buffer resists changes in pH when excess base, OH^-, gets into the bloodstream.

46. Calculate (a) the $[H^+]$, (b) the pH, and (c) the percent ionization of a 0.25 M solution of phenol, HC_6H_5O. ($K_a = 1.3 \times 10^{-10}$)

48. A 0.15 M solution of a weak acid, HA, has a pH of 5. Calculate the ionization constant, K_a, for the acid.

50. Calculate the percent ionization and pH of solutions of HClO ($K_a = 3.5 \times 10^{-8}$) having the following molarities: (a) 1.0 M, (b) 0.10 M, and (c) 0.010 M.

*52. A 0.23 M solution of a weak acid, HA, has a pH of 2.89. What is the K_a for this acid?

54. A common laboratory reagent is 1.0 M NaOH. Calculate the $[H^+]$, $[OH^-]$, pH, and pOH of this solution.

56. Calculate the pH and the pOH of these solutions:
(a) 0.0025 M NaOH
(b) 0.10 M HClO ($K_a = 3.5 \times 10^{-8}$)
*(c) saturated $Fe(OH)_2$ solution ($K_{sp} = 8.0 \times 10^{-16}$)

58. Calculate the $[OH^-]$ in these solutions:
(a) $[H^+] = 4.0 \times 10^{-9}$
(b) $[H^+] = 8.9 \times 10^{-2}$

59. Calculate the $[H^+]$ in these solutions:
    (a) $[OH^-] = 6.0 \times 10^{-7}$
    (b) $[OH^-] = 1 \times 10^{-8}$

61. Given the following solubility data, calculate the solubility product constant for each substance:
    (a) $BaSO_4$, 3.9×10^{-5} mol/L
    (b) Ag_2CrO_4, 7.8×10^{-5} mol/L
    (c) $CaSO_4$, 0.67 g/L
    (d) $AgCl$, 0.0019 g/L

63. Calculate the molar solubility for these substances:
    (a) Ag_2SO_4, $K_{sp} = 1.5 \times 10^{-5}$
    (b) $Mg(OH)_2$, $K_{sp} = 7.1 \times 10^{-12}$

65. For each substance in Question 63 calculate the solubility in grams per 100. mL of solution.

*67. Solutions containing 100. mL of 0.010 M Na_2SO_4 and 100. mL of 0.001 M $Pb(NO_3)_2$ are mixed. Show by calculation whether or not a precipitate will form. Assume the volumes are additive. (K_{sp} $PbSO_4 = 1.3 \times 10^{-8}$)

69. How many moles of AgBr will dissolve in 1.0 L of 0.10 M NaBr? ($K_{sp} = 5.0 \times 10^{-13}$ for AgBr)

71. Calculate the $[H^+]$ and the pH of a buffer solution that is 0.20 M in $HC_2H_3O_2$ and contains sufficient sodium acetate to make the $[C_2H_3O_2^-]$ equal to 0.10 M. (K_a for $HC_2H_3O_2 = 1.8 \times 10^{-5}$)

73. When 1.0 mL of 1.0 M HCl is added to 50. mL of 1.0 M NaCl, the $[H^+]$ changes from 1×10^{-7} M to 2.0×10^{-2} M. Calculate the initial pH and the pH change in the solution.

60. Calculate the $[H^+]$ in these solutions:
    (a) $[OH^-] = 4.5 \times 10^{-6}$
    (b) $[OH^-] = 7.3 \times 10^{-4}$

62. Given the following solubility data, calculate the solubility product constant for each substance:
    (a) ZnS, 3.5×10^{-12} mol/L
    (b) $Pb(IO_3)_2$, 4.0×10^{-5} mol/L
    (c) Ag_3PO_4, 6.73×10^{-3} g/L
    (d) $Zn(OH)_2$, 2.33×10^{-4} g/L

64. Calculate the molar solubility for these substances:
    (a) $BaCO_3$, $K_{sp} = 2.0 \times 10^{-9}$
    (b) $AlPO_4$, $K_{sp} = 5.8 \times 10^{-19}$

66. For each substance in Question 64 calculate the solubility in grams per 100. mL of solution.

*68. Solutions containing 50.0 mL of 1.0×10^{-4} M $AgNO_3$ and 100. mL of 1.0×10^{-4} M NaCl are mixed. Show by calculation whether or not a precipitate will form. Assume the volumes are additive. (K_{sp} $AgCl = 1.7 \times 10^{-10}$)

70. How many moles of AgBr will dissolve in 1.0 L of 0.10 M $MgBr_2$? ($K_{sp} = 5.0 \times 10^{-13}$ for AgBr)

72. Calculate the $[H^+]$ and the pH of a buffer solution that is 0.20 M in $HC_2H_3O_2$ and contains sufficient sodium acetate to make the $[C_2H_3O_2^-]$ equal to 0.20 M. (K_a for $HC_2H_3O_2 = 1.8 \times 10^{-5}$)

74. When 1.0 mL of 1.0 M HCl is added to 50. mL of a buffer solution that is 1.0 M in $HC_2H_3O_2$ and 1.0 M in $NaC_2H_3O_2$, the $[H^+]$ changes from 1.8×10^{-5} M to 1.9×10^{-5} M. Calculate the initial pH and the pH change in the solution.

Additional Exercises

75. What is the maximum number of moles of HI that can be obtained from a reaction mixture containing 2.30 mol I_2 and 2.10 mol H_2?

76. (a) How many moles of HI are produced when 2.00 mol H_2 and 2.00 mol I_2 are reacted at 700 K? (Reaction is 79% complete.)
    (b) Addition of 0.27 mol I_2 to the system increases the yield of HI to 85%. How many moles of H_2, I_2, and HI are now present?
    (c) From the data in part (a), calculate K_{eq} for the reaction at 700 K.

*77. After equilibrium is reached in the reaction of 6.00 g H_2 with 200. g I_2 at 500. K, analysis shows that the flask contains 64.0 g of HI. How many moles of H_2, I_2, and HI are present in this equilibrium mixture?

78. What is the equilibrium constant of the reaction

    $$PCl_3(g) + Cl_2(g) \rightleftharpoons PCl_5(g)$$

    if a 20. L flask contains 0.10 mol PCl_3, 1.50 mol Cl_2, and 0.22 mol PCl_5?

79. If the rate of a reaction doubles for every 10°C rise in temperature, how much faster will the reaction go at 100°C than at 30°C?

80. Calculate the ionization constant for the following acids. Each acid ionizes as follows: $HA \rightleftharpoons H^+ + A^-$.

Acid	Acid concentration	$[H^+]$
Hypochlorous, HOCl	0.10 M	5.9×10^{-5} mol/L
Propanoic, $HC_3H_5O_2$	0.15 M	1.4×10^{-3} mol/L
Hydrocyanic, HCN	0.20 M	8.9×10^{-6} mol/L

81. The K_{sp} of CaF_2 is 3.9×10^{-11}. Calculate (a) the molar concentrations of Ca^{2+} and F^- in a saturated solution, and (b) the grams of CaF_2 that will dissolve in 500. mL of water.

82. The following solutions are mixed. Calculate whether or not a precipitate will form:
   (a) 100 mL of 0.010 M Na_2SO_4 and 100 mL of 0.001 M $Pb(NO_3)_2$
   (b) 50.0 mL of 1.0×10^{-4} M $AgNO_3$ and 100. mL of 1.0×10^{-4} M NaCl
   (c) 1.0 g $Ca(NO_3)_2$ in 150 mL H_2O and 250 mL of 0.01 M NaOH
   K_{sp} $PbSO_4 = 1.3 \times 10^{-8}$
   K_{sp} $AgCl = 1.7 \times 10^{-10}$
   K_{sp} $Ca(OH)_2 = 1.3 \times 10^{-6}$

83. If $BaCl_2$ is added to a saturated $BaSO_4$ solution until the $[Ba^{2+}]$ is 0.050 M,
   (a) what concentration of SO_4^{2-} remains in solution?
   (b) how much $BaSO_4$ remains dissolved in 100. mL of the solution? ($K_{sp} = 1.5 \times 10^{-9}$ for $BaSO_4$)

84. The K_{sp} for $PbCl_2$ is 2.0×10^{-5}. Will a precipitate form when 0.050 mol $Pb(NO_3)_2$ and 0.010 mol NaCl are dissolved in 1.0 L H_2O? Show evidence for your answer.

85. Suppose the concentration of a solution is 0.10 M Ba^{2+} and 0.10 M Sr^{2+}. Which sulfate, $BaSO_4$ or $SrSO_4$, will precipitate first when a dilute solution of H_2SO_4 is added dropwise to the solution? Show evidence for your answer. ($K_{sp} = 1.5 \times 10^{-9}$ for $BaSO_4$ and $K_{sp} = 3.5 \times 10^{-7}$ for $SrSO_4$)

86. Calculate the K_{eq} for the reaction

$$SO_2(g) + O_2(g) \rightleftharpoons SO_3(g)$$

when the equilibrium concentrations of the gases at 530°C are $[SO_3] = 11.0$ M, $[SO_2] = 4.20$ M, and $[O_2] = 0.60 \times 10^{-3}$ M.

87. If it takes 0.048 g BaF_2 to saturate 15.0 mL of water, what is the K_{sp} of BaF_2?

88. The K_{eq} for the formation of ammonia gas from its elements is 4.0. If the equilibrium concentrations of nitrogen gas and hydrogen gas are both 2.0 M, what is the equilibrium concentration of the ammonia gas?

89. The K_{sp} of $SrSO_4$ is 7.6×10^{-7}. Should precipitation occur when 25.0 mL of 1.0×10^{-3} M $SrCl_2$ solution is mixed with 15.0 mL of 2.0×10^{-3} M Na_2SO_4? Show proof.

***90.** The solubility of Hg_2I_2 in H_2O is 3.04×10^{-7} g/L. The reaction $Hg_2I_2 \rightleftharpoons Hg_2^{2+} + 2I^-$ represents the equilibrium. Calculate the K_{sp}.

91. Under certain circumstances, when oxygen gas is heated, it can be converted into ozone according to the following reaction equation:

$$3 O_2(g) + heat \rightleftharpoons 2 O_3(g)$$

Name three different ways that you could increase the production of the ozone.

92. One day in a laboratory, some water spilled on a table. In just a few minutes the water had evaporated. Some days later, a similar amount of water spilled again. This time, the water remained on the table after 7 or 8 hours. Name three conditions that could have changed in the lab to cause this difference.

***93.** All a snowbound skier had to eat were walnuts! He was carrying a bag holding 12 dozen nuts. With his mittened hands, he cracked open the shells. Each nut that was opened resulted in one kernel and two shell halves. When he tired of the cracking and got ready to do some eating, he discovered he had 194 total pieces (whole nuts, shell halves, and kernels). What is the K_{eq} for this reaction?

94. For the reaction $CO(g) + H_2O(g) \rightleftharpoons CO_2(g) + H_2(g)$ at a certain temperature, K_{eq} is 1. At equilibrium would you expect to find
   (a) only CO and H_2
   (b) mostly CO_2 and H_2
   (c) about equal concentrations of CO and H_2O, compared to CO_2 and H_2
   (d) mostly CO and H_2O
   (e) only CO and H_2O
   Explain your answer briefly.

95. Write the equilibrium constant expressions for these reactions:
   (a) $3 O_2(g) \rightleftharpoons 2 O_3(g)$
   (b) $H_2O(g) \rightleftharpoons H_2O(l)$
   (c) $MgCO_3(s) \rightleftharpoons MgO(s) + CO_2(g)$
   (d) $2 Bi^{3+}(aq) + 3 H_2S(aq) \rightleftharpoons Bi_2S_3(s) + 6 H^+(aq)$

96. Reactants A and B are mixed, each initially at a concentration of 1.0 M. They react to produce C according to this equation:

$$2 A + B \rightleftharpoons C$$

When equilibrium is established, the concentration of C is found to be 0.30 M. Calculate the value of K_{eq}.

97. At a certain temperature, K_{eq} is 2.2×10^{-3} for the reaction

$$2 \text{ICl}(g) \rightleftharpoons \text{I}_2(g) + \text{Cl}_2(g)$$

Now calculate the K_{eq} value for the reaction

$$\text{I}_2(g) + \text{Cl}_2(g) \rightleftharpoons 2 \text{ICl}(g)$$

98. One drop of $1 M$ OH^- ion is added to a $1 M$ solution of HNO_2. What will be the effect of this addition on the equilibrium concentration of the following?

$$\text{HNO}_2(aq) \rightleftharpoons \text{H}^+(aq) + \text{NO}_2^-(aq)$$

(a) $[\text{OH}^-]$
(b) $[\text{H}^+]$
(c) $[\text{NO}_2^-]$
(d) $[\text{HNO}_2]$

*99. At 500°C, the reaction

$$\text{SO}_2(g) + \text{NO}_2(g) \rightleftharpoons \text{NO}(g) + \text{SO}_3(g)$$

has $K_{eq} = 90$. What will be the equilibrium concentrations of the four gases if the two reactants begin with equal concentrations of $0.50 M$?

100. How many grams of CaSO_4 will dissolve in 600. mL of water? ($K_{sp} = 2.0 \times 10^{-4}$ for CaSO_4)

101. A student found that 0.098 g of PbF_2 was dissolved in 400. mL of saturated PbF_2. What is the K_{sp} for the lead(II) fluoride?

Answers to Practice Exercises

16.1 (a)

Concentration	Change
$[\text{H}_2]$	increase
$[\text{I}_2]$	decrease
$[\text{HI}]$	increase

(b)

Concentration	Change
$[\text{H}_2]$	increase
$[\text{I}_2]$	increase
$[\text{HI}]$	increase

16.2 (a) Equilibrium shifts left; (b) equilibrium shifts right
16.3 (a) Equilibrium shifts right; (b) equilibrium shifts left

16.4 (a) Equilibrium shifts left; (b) equilibrium shifts right
16.5 (a) $K_{eq} = \dfrac{[\text{NO}_2]^4[\text{O}_2]}{[\text{N}_2\text{O}_5]^2}$; (b) $K_{eq} = \dfrac{[\text{N}_2]^2[\text{H}_2\text{O}]^6}{[\text{NH}_3]^4[\text{O}_2]^3}$
16.6 $K_{eq} = 33$; the forward reaction is favored
16.7 (a) $[\text{H}^+] = 5.0 \times 10^{-5}$ $[\text{OH}^-] = 2.0 \times 10^{-10}$
 (b) $[\text{H}^+] = 5.0 \times 10^{-9}$ $[\text{OH}^-] = 2.0 \times 10^{-6}$
16.8 (a) 4.30; (b) 8.30
16.9 (a) 6.3×10^{-6}; (b) 1.8×10^{-6}
16.10 (a) 6.3×10^{-3}% ionized; (b) 7.2×10^{-3}% ionized
16.11 (a) $K_{sp} = [\text{Cr}^{3+}][\text{OH}^-]^3$;
 (b) $K_{sp} = [\text{Cu}^{2+}]^3[\text{PO}_4^{3-}]^2$
16.12 9.1×10^{-21} g/L
16.13 7.5×10^{-8} mol/L
16.14 (a) basic; (b) neutral; (c) acidic

Oxidation–Reduction

▲
Rust is being removed from
this iron sculpture.

The variety of oxidation–reduction reactions that affect us every day is amazing. Our society runs on batteries—in our calculators, laptop computers, cars, toys, radios, televisions, and more. We paint iron railings and galvanize nails to combat corrosion. We electroplate jewelry and computer chips with very thin coatings of gold or silver. We bleach our clothes and develop our photographs in solutions using chemical reactions that involve electron transfer. We test for glucose in urine or alcohol in the breath with reactions that show vivid color changes. Plants turn energy into chemical compounds through a series of reactions called photosynthesis. These reactions all involve the transfer of electrons between substances in a chemical process called *oxidation–reaction.*

17.1 Oxidation Number

The oxidation number of an atom (sometimes called its *oxidation state*) represents the number of electrons lost, gained, or unequally shared by an atom. Oxidation numbers can be zero, positive, or negative. An oxidation number of zero means the atom has the same number of electrons assigned to it as there are in the free neutral atom. A positive oxidation number means the atom has fewer electrons assigned to it than in the neutral atom, and a negative oxidation number means the atom has more electrons assigned to it than in the neutral atom.

The oxidation number of an atom that has lost or gained electrons to form an ion is the same as the positive or negative charge of the ion. (See Table 17.1.) In the ionic compound NaCl, the oxidation numbers are clearly $+1$ for the Na^+ ion and -1 for the Cl^- ion. The Na^+ ion has one less electron than the neutral Na atom, and the Cl^- ion has one more electron than the neutral Cl atom. In $MgCl_2$, two electrons have transferred from the Mg atom to the two Cl atoms; the oxidation number of Mg is $+2$.

In covalently bonded substances where electrons are shared between two atoms, oxidation numbers are assigned by an arbitrary system based on relative electronegativities. For symmetrical covalent molecules such as H_2 and Cl_2, each atom is assigned an oxidation number of zero because the bonding pair of electrons is shared equally between two like atoms, neither of which is more electronegative than the other:

$$H\!:\!H \qquad :\ddot{C}l\!:\!\ddot{C}l:$$

When the covalent bond is between two unlike atoms, the bonding electrons are shared unequally because the more electronegative element has a greater attraction for them. In this case the oxidation numbers are determined by assigning both electrons to the more electronegative element.

Thus in compounds with covalent bonds such as NH_3 and H_2O,

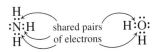

TABLE 17.1	Oxidation Numbers for Common Ions	
Ion	**Oxidation number**
H^+	$+1$
Na^+	$+1$
K^+	$+1$
Li^+	$+1$
Ag^+	$+1$
Cu^{2+}	$+2$
Ca^{2+}	$+2$
Ba^{2+}	$+2$
Fe^{2+}	$+2$
Mg^{2+}	$+2$
Zn^{2+}	$+2$
Al^{3+}	$+3$
Fe^{3+}	$+3$
Cl^-	-1
Br^-	-1
F^-	-1
I^-	-1
S^{2-}	-2
O^{2-}	-2

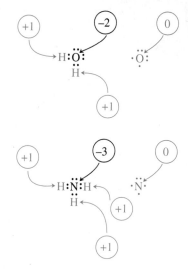

oxidation number
oxidation state

The oxidation number for Cu metal is 0 while the oxidation number for Cu^{2+} ions in the crystal is 2+.

the pairs of electrons are unequally shared between the atoms and are attracted toward the more electronegative elements, N and O. This causes the N and O atoms to be relatively negative with respect to the H atoms. At the same time it causes the H atoms to be relatively positive with respect to the N and O atoms. In H_2O, both pairs of shared electrons are assigned to the O atom, giving it two electrons more than the neutral O atom, and each H atom is assigned one electron less than the neutral H atom. Therefore the oxidation number of the O atom is -2, and the oxidation number of each H atom is $+1$. In NH_3, the three pairs of shared electrons are assigned to the N atom, giving it three electrons more than the neutral N atom, and each H atom has one electron less than the neutral atom. Therefore the oxidation number of the N atom is -3, and the oxidation number of each H atom is $+1$.

Assigning correct oxidation numbers to elements is essential for balancing oxidation–reduction equations.

The **oxidation number** or **oxidation state** of an element is an integer value assigned to each element in a compound or ion that allows us to keep track of electrons associated with each atom. Oxidation numbers have a variety of uses in chemistry—from writing formulas to predicting properties of compounds and assisting in the balancing of oxidation–reduction reactions in which electrons are transferred.

As a starting point, the oxidation number of an uncombined element, regardless of whether it is monatomic or diatomic, is zero. Rules for assigning oxidation numbers are summarized in Table 17.2.

Use the following steps to find the oxidation number for an element within a compound.

Step 1. Write the oxidation number of each known atom below the atom in the formula.

Step 2. Multiply each oxidation number by the number of atoms of that element in the compound.

Step 3. Write an expression indicating the sum of all the oxidation numbers in the compound. Remember: The sum of the oxidation numbers in a compound must equal zero.

TABLE 17.2	Rules for Assigning Oxidation Number

1. All elements in their free state (uncombined with other elements) have an oxidation number of zero (e.g., Na, Cu, Mg, H_2, O_2, Cl_2, N_2).
2. H is $+1$, except in metal hydrides, where it is -1 (e.g., NaH, CaH_2).
3. O is -2, except in peroxides, where it is -1, and in OF_2, where it is $+2$.
4. The metallic element in an ionic compound has a positive oxidation number.
5. In covalent compounds the negative oxidation number is assigned to the most electronegative atom.
6. The algebraic sum of the oxidation numbers of the elements in a compound is zero.
7. The algebraic sum of the oxidation numbers of the elements in a polyatomic ion is equal to the charge of the ion.

Determine the oxidation number for carbon in carbon dioxide:

Example 17.1

$$CO_2$$
Step 1 -2
Step 2 $(-2)2$
Step 3 $C + (-4) = 0$
Step 3 $C = +4$ (oxidation number for carbon)

Determine the oxidation number for sulfur in sulfuric acid:

Example 17.2

$$H_2SO_4$$
Step 1 $+1$ -2
Step 2 $2(+1) = +2$ $4(-2) = -8$
Step 3 $+2 + S + (-8) = 0$
              $S = +6$ (oxidation number for sulfur)

Practice 17.1

Determine the oxidation number of (a) S in Na_2SO_4, (b) As in K_3AsO_4, and (c) C in $CaCO_3$.

Oxidation numbers in a polyatomic ion (ions containing more than one atom) are determined in a similar fashion, except that in a polyatomic ion the sum of the oxidation numbers must equal the charge on the ion instead of zero.

Determine the oxidation number for manganese in the permanganate ion MnO_4^-:

Example 17.3

$$MnO_4^-$$
Step 1 -2
Step 2 $(-2)4$
Step 3 $Mn + (-8) = -1$ (the charge on the ion)
              $Mn = +7$ (oxidation number for manganese)

Example 17.4   Determine the oxidation number for carbon in the oxalate ion $C_2O_4^{2-}$:

$$C_2O_4^{2-}$$

Step 1 $(-2)4$
Step 2 $2C + (-8) = -2$ (the charge on the ion)
Step 3 $2C = +6$
                $C = +3$ (oxidation number for C)

Practice 17.2

Determine the oxidation numbers of (a) N in NH_4^+, (b) Cr in $Cr_2O_7^{2-}$, and (c) P in PO_4^{3-}.

Example 17.5   Determine the oxidation number of each element in (a) KNO_3, and (b) SO_4^{2-}.

Solution

(a) Potassium is a Group IA metal; therefore it has an oxidation number of $+1$. The oxidation number of each O atom is -2 (Table 17.2, Rule 3). Using these values and the fact that the sum of the oxidation numbers of all the atoms in a compound is zero, we can determine the oxidation number of N:

$$KNO_3$$
$$+1 + N + 3(-2) = 0$$
$$N = +6 - 1 = +5$$

The oxidation numbers are K, $+1$; N, $+5$; O, -2.

(b) Because SO_4^{2-} is an ion, the sum of oxidation numbers of the S and the O atoms must be -2, the charge of the ion. The oxidation number of each O atom is -2 (Table 17.2, Rule 3). Then

$$SO_4^{2-}$$
$$S + 4(-2) = -2, \quad S - 8 = -2$$
$$S = -2 + 8 = +6$$

The oxidation numbers are S, $+6$; O, -2.

Practice 17.3

Determine the oxidation number of each element in these species:
(a) $BeCl_2$ (b) $HClO$ (c) H_2O_2 (d) NH_4^+ (e) BrO_3^-

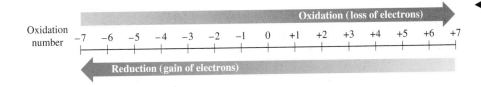

►FIGURE 17.1
Oxidation and reduction.
Oxidation results in an increase in the oxidation number, and reduction results in a decrease in the oxidation number.

17.2 Oxidation–Reduction

Oxidation–reduction, also known as **redox,** is a chemical process in which the oxidation number of an element is changed. The process may involve the complete transfer of electrons to form ionic bonds or only a partial transfer or shift of electrons to form covalent bonds.

oxidation–reduction
redox

   Oxidation occurs whenever the oxidation number of an element increases as a result of losing electrons. Conversely, **reduction** occurs whenever the oxidation number of an element decreases as a result of gaining electrons. For example, a change in oxidation number from $+2$ to $+3$ or from -1 to 0 is oxidation; a change from $+5$ to $+2$ or from -2 to -4 is reduction (see Figure 17.1). Oxidation and reduction occur simultaneously in a chemical reaction; one cannot take place without the other.

oxidation
reduction

   Many combination, decomposition, and single-displacement reactions involve oxidation–reduction. Let's examine the combustion of hydrogen and oxygen from this point of view:

$$2\ H_2 + O_2 \longrightarrow 2\ H_2O$$

Both reactants, hydrogen and oxygen, are elements in the free state and have an oxidation number of zero. In the product (water), hydrogen has been oxidized to $+1$ and oxygen reduced to -2. The substance that causes an increase in the oxidation state of another substance is called an **oxidizing agent.** The substance that causes a decrease in the oxidation state of another substance is called a **reducing agent.** In this reaction the oxidizing agent is free oxygen, and the reducing agent is free hydrogen. In the reaction

oxidizing agent

reducing agent

$$Zn(s) + H_2SO_4(aq) \longrightarrow ZnSO_4(aq) + H_2(g)$$

metallic zinc is oxidized, and hydrogen ions are reduced. Zinc is the reducing agent, and hydrogen ions, the oxidizing agent. Electrons are transferred from the zinc metal to the hydrogen ions. The reaction is better expressed as

$$Zn^0 + 2\ H^+ + SO_4^{2-} \longrightarrow Zn^{2+} + SO_4^{2-} + H_2^0$$

> **Oxidation:** **Increase in oxidation number**
>   **Loss of electrons**
> **Reduction:** **Decrease in oxidation number**
>   **Gain of electrons**

▲
In the reaction between Zn and H_2SO_4 the Zn is oxidized while hydrogen is reduced.

The oxidizing agent is reduced and gains electrons. The reducing agent is oxidized and loses electrons. The transfer of electrons is characteristic of all redox reactions.

17.3   Balancing Oxidation–Reduction Equations

Many simple redox equations can be balanced readily by inspection, or by trial and error:

$$Na + Cl_2 \longrightarrow NaCl \qquad \text{(unbalanced)}$$

$$2\,Na + Cl_2 \longrightarrow 2\,NaCl \qquad \text{(balanced)}$$

Balancing this equation is certainly not complicated. But as we study more complex reactions and equations such as

$$P + HNO_3 + H_2O \longrightarrow NO + H_3PO_4 \qquad \text{(unbalanced)}$$

$$3\,P + 5\,HNO_3 + 2\,H_2O \longrightarrow 5\,NO + 3\,H_3PO_4 \qquad \text{(balanced)}$$

the trial-and-error method of balancing equations takes an unnecessarily long time.

One systematic method for balancing oxidation–reduction equations is based on the transfer of electrons between the oxidizing and reducing agents. Consider the first equation again:

$$Na^0 + Cl_2^0 \longrightarrow Na^+Cl^- \qquad \text{(unbalanced)}$$

The superscript 0 shows the oxidation number is 0 for elements in their uncombined state

In this reaction sodium metal loses one electron per atom when it changes to a sodium ion. At the same time chlorine gains one electron per atom. Because chlorine is diatomic, two electrons per molecule are needed to form a chloride ion from each atom. These electrons are furnished by two sodium atoms. Stepwise, the reaction may be written as two half-reactions, the oxidation half-reaction and the reduction half-reaction:

$$
\begin{array}{ll}
2\,Na^0 \longrightarrow 2\,Na^+ + 2\,e^- & \text{oxidation half-reaction} \\
\underline{Cl_2^0 + 2\,e^- \longrightarrow \qquad\qquad 2\,Cl^-} & \text{reduction half-reaction} \\
Cl_2^0 + 2\,Na^0 \longrightarrow 2\,Na^+Cl^- &
\end{array}
$$

When the two half-reactions, each containing the same number of electrons, are added together algebraically, the electrons cancel out. In this reaction there are no excess electrons; the two electrons lost by the two sodium atoms are utilized by chlorine. In all redox reactions the loss of electrons by the reducing agent must equal the gain of electrons by the oxidizing agent. Here, sodium is oxidized and chlorine is reduced. Chlorine is the oxidizing agent; sodium is the reducing agent.

In the following examples we use the change-in-oxidation-number method, a system for balancing more complicated redox equations.

Example 17.6   Balance the equation

$$Sn + HNO_3 \longrightarrow SnO_2 + NO_2 + H_2O \qquad \text{(unbalanced)}$$

Solution   **Step 1.** Assign oxidation numbers to each element to identify the elements being oxidized and those being reduced. Write the oxidation numbers below each element to avoid confusing them with ionic charge:

$$Sn + HNO_3 \longrightarrow SnO_2 + NO_2 + H_2O$$

$$\begin{array}{cccccc}
0 & +1 & +5 & -2 & +4 & -2 & +4 & -2 & +1 & -2
\end{array}$$

Note that the oxidation numbers of Sn and N have changed.

Step 2. Now write two new equations, using only the elements that change in oxidation number. Then add electrons to bring the equations into electrical balance. One equation represents the oxidation step; the other represents the reduction step. Remember: Oxidation produces electrons; reduction uses electrons.

$$Sn^0 \longrightarrow Sn^{4+} + 4\ e^- \qquad \text{oxidation}$$
$$\text{Sn}^0 \text{ loses 4 electrons}$$

$$N^{5+} + 1\ e^- \longrightarrow N^{4+} \qquad \text{reduction}$$
$$\text{N}^{5+} \text{ gains 1 electron}$$

Step 3. Multiply the two equations by the smallest whole numbers that will make the electrons lost by oxidation equal to the number of electrons gained by reduction. In this reaction the oxidation step is multiplied by 1 and the reduction step by 4. The equations become

$$Sn^0 \longrightarrow Sn^{4+} + 4\ e^- \qquad \text{oxidation}$$
$$\text{Sn}^0 \text{ loses 4 electrons}$$

$$4\ N^{5+} + 4\ e^- \longrightarrow 4\ N^{4+} \qquad \text{reduction}$$
$$4\ \text{N}^{5+} \text{ gains 4 electrons}$$

We have now established the ratio of the oxidizing to the reducing agent as being four atoms of N to one atom of Sn.

Step 4. Transfer the coefficient in front of each substance in the balanced oxidation–reduction equations to the corresponding substance in the original equation. We need to use 1 Sn, 1 SnO$_2$, 4 HNO$_3$, and 4 NO$_2$:

$$Sn + 4\ HNO_3 \longrightarrow SnO_2 + 4\ NO_2 + H_2O \quad \text{(unbalanced)}$$

Step 5. In the usual manner, balance the remaining elements that are not oxidized or reduced to give the final balanced equation:

$$Sn + 4\ HNO_3 \longrightarrow SnO_2 + 4\ NO_2 + 2\ H_2O \quad \text{(balanced)}$$

In balancing the final elements, we must not change the ratio of the elements that were oxidized and reduced.

Finally, check to ensure that both sides of the equation have the same number of atoms of each element. The final balanced equation contains 1 atom of Sn, 4 atoms of N, 4 atoms of H, and 12 atoms of O on each side.

Because each new equation presents a slightly different problem and because proficiency in balancing equations requires practice, let's work through two more examples.

Balance the equation

$$I_2 + Cl_2 + H_2O \longrightarrow HIO_3 + HCl \quad \text{(unbalanced)}$$

Example 17.7

Solution

Step 1. Assign oxidation numbers:

$$I_2 + Cl_2 + H_2O \longrightarrow HIO_3 + HCl$$

$$
\begin{array}{cccccc}
\underset{0}{I_2} & \underset{0}{Cl_2} & \underset{+1}{H_2}\underset{-2}{O} & \underset{+1\ +5\ -2}{HIO_3} & \underset{+1\ -1}{HCl}
\end{array}
$$

The oxidation numbers of I_2 and Cl_2 have changed, I_2 from 0 to +5, and Cl_2 from 0 to −1.

Step 2. Write the oxidation and reduction steps. Balance the number of atoms and then balance the electrical charge using electrons:

$$I_2 \longrightarrow 2\ I^{5+} + 10\ e^-$$ oxidation   (10 e^- are needed to balance the +10 charge)

I₂ loses 10 electrons

$$Cl_2 + 2\ e^- \longrightarrow 2\ Cl^-$$ reduction   (2 e^- are needed to balance the −2 charge)

Cl₂ gains 2 electrons

Step 3. Adjust loss and gain of electrons so that they are equal. Multiply the oxidation step by 1 and the reduction step by 5:

$$I_2 \longrightarrow 2\ I^{5+} + 10\ e^-$$ oxidation

I₂ loses 10 electrons

$$5\ Cl_2 + 10\ e^- \longrightarrow 10\ Cl^-$$ reduction

5 Cl₂ gain 10 electrons

Step 4. Transfer the coefficients from the balanced redox equations into the original equation. We need to use 1 I_2, 2 HIO_3, 5 Cl_2, and 10 HCl:

$$I_2 + 5\ Cl_2 + H_2O \longrightarrow 2\ HIO_3 + 10\ HCl \quad \text{(unbalanced)}$$

Step 5. Balance the remaining elements, H and O:

$$I_2 + 5\ Cl_2 + 6\ H_2O \longrightarrow 2\ HIO_3 + 10\ HCl \quad \text{(balanced)}$$

Check: The final balanced equation contains 2 atoms of I, 10 atoms of Cl, 12 atoms of H, and 6 atoms of O on each side.

Example 17.8   Balance the equation

$$K_2Cr_2O_7 + FeCl_2 + HCl \longrightarrow CrCl_3 + KCl + FeCl_3 + H_2O \quad \text{(unbalanced)}$$

Solution

Step 1. Assign oxidation numbers (Cr and Fe have changed):

$$K_2Cr_2O_7 + FeCl_2 + HCl \longrightarrow CrCl_3 + KCl + FeCl_3 + H_2O$$

$$
\begin{array}{ccccccc}
\underset{+1\ +6\ -2}{K_2Cr_2O_7} & \underset{+2\ -1}{FeCl_2} & \underset{+1\ -1}{HCl} & \underset{+3\ -1}{CrCl_3} & \underset{+1\ -1}{KCl} & \underset{+3\ -1}{FeCl_3} & \underset{+1\ -2}{H_2O}
\end{array}
$$

Step 2. Write the oxidation and reduction steps. Balance the number of atoms and then balance the electrical charge using electrons:

$$Fe^{2+} \longrightarrow Fe^{3+} + 1\ e^-$$ oxidation

Fe²⁺ loses 1 electron

$$2\ Cr^{6+} + 6\ e^- \longrightarrow 2\ Cr^{3+}$$ reduction

2 Cr⁶⁺ gain 6 electrons

Step 3. Balance the loss and gain of electrons. Multiply the oxidation step by 6 and the reduction step by 1 to equalize the transfer of electrons.

$$6\ Fe^{2+} \longrightarrow 6\ Fe^{3+} + 6\ e^{-} \qquad \text{oxidation}$$

6 Fe^{2+} lose 6 electrons

$$2\ Cr^{6+} + 6\ e^{-} \longrightarrow 2\ Cr^{3+} \qquad \text{reduction}$$

2 Cr^{6+} gain 6 electrons

Step 4. Transfer the coefficients from the balanced redox equations into the original equation. (Note that one formula unit of $K_2Cr_2O_7$ contains two Cr atoms.) We need to use 1 $K_2Cr_2O_7$, 2 $CrCl_3$, 6 $FeCl_2$, and 6 $FeCl_3$:

$$K_2Cr_2O_7 + 6\ FeCl_2 + HCl \longrightarrow$$
$$2\ CrCl_3 + KCl + 6\ FeCl_3 + H_2O \quad \text{(unbalanced)}$$

Step 5. Balance the remaining elements in this order: K, Cl, H, O.

$$K_2Cr_2O_7 + 6\ FeCl_2 + 14\ HCl \longrightarrow$$
$$2\ CrCl_3 + 2\ KCl + 6\ FeCl_3 + 7\ H_2O \quad \text{(balanced)}$$

Check: The final balanced equation contains 2 K atoms, 2 Cr atoms, 7 O atoms, 6 Fe atoms, 26 Cl atoms, and 14 H atoms on each side.

Practice 17.4

Balance these equations using the change-in-oxidation-number method:
(a) $HNO_3 + S \longrightarrow NO_2 + H_2SO_4 + H_2O$
(b) $CrCl_3 + MnO_2 + H_2O \longrightarrow MnCl_2 + H_2CrO_4$
(c) $KMnO_4 + HCl + H_2S \longrightarrow KCl + MnCl_2 + S + H_2O$

17.4 Balancing Ionic Redox Equations

The main difference between balancing ionic redox equations and molecular redox equations is in how we handle the ions. In the ionic redox equations, besides having the same number of atoms of each element on both sides of the final equation, we must also have equal net charges. In assigning oxidation numbers, we must therefore remember to consider the ionic charge.

Several methods are used to balance ionic redox equations, including, with slight modification, the oxidation-number method just shown for molecular equations. But the most popular method is probably the ion–electron method.

The ion–electron method uses ionic charges and electrons to balance ionic redox equations. Oxidation numbers are not formally used, but it is necessary to determine what is being oxidized and what is being reduced. The method is as follows:

1. Write the two half-reactions that contain the elements being oxidized and reduced using the entire formula of the ion or molecule.
2. Balance the elements other than oxygen and hydrogen.

Oxidation–reduction reactions are the basis for many interesting applications. Consider photochromic glass, which is used for lenses in light-sensitive glasses. These lenses, manufactured by the Corning Glass Company, can change from transmitting 85% of light to only transmitting 22% of light when exposed to bright sunlight.

Photochromic glass is composed of linked tetrahedrons of silicon and oxygen atoms jumbled in a disorderly array, with crystals of silver chloride caught between the silica tetrahedrons. When the glass is clear, the visible light passes right through the molecules. The glass absorbs ultraviolet light, however, and this energy triggers an oxidation–reduction reaction between Ag^+ and Cl^-:

$$Ag^+ + Cl^- \xrightarrow{\text{UV light}} Ag^0 + Cl^0$$

◀ **An oxidation–reduction reaction causes these photochromic glasses to change from light to dark in bright sunlight.**

To prevent the reaction from reversing itself immediately, a few ions of Cu^+ are incorporated into the silver chloride crystal. These Cu^+ ions react with the newly formed chlorine atoms:

$$Cu^+ + Cl^0 \longrightarrow Cu^{2+} + Cl^-$$

The silver atoms move to the surface of the crystal and form small colloidal clusters of silver metal. This metallic silver absorbs visible light, making the lens appear dark (colored).

As the glass is removed from the light, the Cu^{2+} ions slowly move to the surface of the crystal where they interact with the silver metal:

$$Cu^{2+} + Ag^0 \longrightarrow Cu^+ + Ag^+$$

The glass clears as the silver ions rejoin chloride ions in the crystals.

3. Balance oxygen and hydrogen.
   Acidic solution: For reactions in acidic solution, use H^+ and H_2O to balance oxygen and hydrogen. For each oxygen, use one H_2O. Then add H^+ as needed to balance the hydrogen atoms.
   Basic solution: For reactions in alkaline solutions, first balance as though the reaction were in an acidic solution, using Steps 1–3. Then add as many OH^- ions to each side of the equation as there are H^+ ions in the equation. Now combine the H^+ and OH^- ions into water (for example, 4 H^+ and 4 OH^- give 4 H_2O). Rewrite the equation, canceling equal numbers of water molecules that appear on opposite sides of the equation.
4. Add electrons (e^-) to each half-reaction to bring them into electrical balance.
5. Since the loss and gain of electrons must be equal, multiply each half-reaction by the appropriate number to make the number of electrons the same in each half-reaction.
6. Add the two half-reactions together, canceling electrons and any other identical substances that appear on opposite sides of the equation.

Example 17.9 Balance this equation using the ion–electron method:

$$MnO_4^- + S^{2-} \longrightarrow Mn^{2+} + S^0 \quad \text{(acidic solution)}$$

Solution

Step 1. Write two half-reactions, one containing the element being oxidized and the other the element being reduced (use the entire molecule or ion):

$$S^{2-} \longrightarrow S^0 \qquad \text{oxidation}$$

$$MnO_4^- \longrightarrow Mn^{2+} \quad \text{reduction}$$

Step 2. Balance elements other than oxygen and hydrogen (accomplished in Step 1: 1 S and 1 Mn on each side).

Step 3. Balance O and H. Remember the solution is acidic. The oxidation requires neither O nor H, but the reduction equation needs $4 \, H_2O$ on the right and $8 \, H^+$ on the left.

$$S^{2-} \longrightarrow S^0$$

$$8 \, H^+ + MnO_4^- \longrightarrow Mn^{2+} + 4 \, H_2O$$

Step 4. Balance each half-reaction electrically with electrons:

$$S^{2-} \longrightarrow S^0 + 2 \, e^-$$
net charge $= -2$ on each side

$$5 \, e^- + 8 \, H^+ + MnO_4^- \longrightarrow Mn^{2+} + 4 \, H_2O$$
net charge $= +2$ on each side

Step 5. Equalize loss and gain of electrons. In this case multiply the oxidation equation by 5 and the reduction equation by 2:

$$5 \, S^{2-} \longrightarrow 5 \, S^0 + 10 \, e^-$$

$$10 \, e^- + 16 \, H^+ + 2 \, MnO_4^- \longrightarrow 2 \, Mn^{2+} + 8 \, H_2O$$

Step 6. Add the two half-reactions together, canceling the $10 \, e^-$ from each side, to obtain the balanced equation:

$$5 \, S^{2-} \longrightarrow 5 \, S^0 + \cancel{10 \, e^-}$$
$$\underline{\cancel{10 \, e^-} + 16 \, H^+ + 2 \, MnO_4^- \longrightarrow 2 \, Mn^{2+} + 8 \, H_2O}$$
$$16 \, H^+ + 2 \, MnO_4^- + 5 \, S^{2-} \longrightarrow 2 \, Mn^{2+} + 5 \, S^0 + 8 \, H_2O \quad \text{(balanced)}$$

Check: Both sides of the equation have a charge of $+4$ and contain the same number of atoms of each element.

Example 17.10

Balance this equation:

$$CrO_4^{2-} + Fe(OH)_2 \longrightarrow Cr(OH)_3 + Fe(OH)_3 \quad \text{(basic solution)}$$

Solution

Step 1. Write the two half-reactions:

$$Fe(OH)_2 \longrightarrow Fe(OH)_3 \quad \text{oxidation}$$

$$CrO_4^{2-} \longrightarrow Cr(OH)_3 \quad \text{reduction}$$

Step 2. Balance elements other than H and O (accomplished in Step 1).

Step 3. Remember the solution is basic. Balance O and H as though the solution were acidic. Use H_2O and H^+. To balance O and H in the oxidation equation, add $1 \, H_2O$ on the left and $1 \, H^+$ on the right side:

$$Fe(OH)_2 + H_2O \longrightarrow Fe(OH)_3 + H^+$$

Add 1 OH$^-$ to each side:

$$Fe(OH)_2 + H_2O + OH^- \longrightarrow Fe(OH)_3 + H^+ + OH^-$$

Combine H$^+$ and OH$^-$ as H$_2$O and rewrite, canceling H$_2$O on each side:

$$Fe(OH)_2 + \cancel{H_2O} + OH^- \longrightarrow Fe(OH)_3 + \cancel{H_2O}$$

$$\boxed{Fe(OH)_2 + OH^- \qquad Fe(OH)_3} \qquad \text{(oxidation)}$$

To balance O and H in the reduction equation, add 1 H$_2$O on the right and 5 H$^+$ on the left:

$$CrO_4^{2-} + 5\,H^+ \longrightarrow Cr(OH)_3 + H_2O$$

Add 5 OH$^-$ to each side:

$$CrO_4^{2-} + 5\,H^+ + 5\,OH^- \longrightarrow Cr(OH)_3 + H_2O + 5\,OH^-$$

Combine 5 H$^+$ + 5 OH$^- \longrightarrow$ 5 H$_2$O:

$$CrO_4^{2-} + 5\,H_2O \longrightarrow Cr(OH)_3 + H_2O + 5\,OH^-$$

Rewrite, canceling 1 H$_2$O from each side:

$$\boxed{CrO_4^{2-} + 4\,H_2O \qquad Cr(OH)_3 + 5\,OH^-} \qquad \text{(reduction)}$$

Step 4. Balance each half-reaction electrically with electrons:

$$Fe(OH)_2 + OH^- \longrightarrow Fe(OH)_3 + e^-$$
$$\text{(balanced oxidation equation)}$$

$$CrO_4^{2-} + 4\,H_2O + 3\,e^- \longrightarrow Cr(OH)_3 + 5\,OH^-$$
$$\text{(balanced reduction equation)}$$

Step 5. Equalize the loss and gain of electrons. Multiply the oxidation reaction by 3:

$$3\,Fe(OH)_2 + 3\,OH^- \longrightarrow 3\,Fe(OH)_3 + 3\,e^-$$
$$CrO_4^{2-} + 4\,H_2O + 3\,e^- \longrightarrow Cr(OH)_3 + 5\,OH^-$$

Step 6. Add the two half-reactions together, canceling the 3 e$^-$ and 3 OH$^-$ from each side of the equation:

$$3\,Fe(OH)_2 + 3\,OH^- \longrightarrow 3\,Fe(OH)_3 + 3\cancel{e^-}$$
$$\underline{CrO_4^{2-} + 4\,H_2O + 3\cancel{e^-} \longrightarrow Cr(OH)_3 + 5\,OH^-}$$
$$CrO_4^{2-} + 3\,Fe(OH)_2 + 4\,H_2O \longrightarrow Cr(OH)_3 + 3\,Fe(OH)_3 + 2\,OH^- \quad \text{(balanced)}$$

Check: Each side of the equation has a charge of -2 and contains the same number of atoms of each element.

Practice 17.5

Balance these equations using the ion–electron method:
(a) $I^- + NO_2^- \longrightarrow I_2 + NO$ (acidic solution)
(b) $Cl_2 + IO_3^- \longrightarrow IO_4^- + Cl^-$ (basic solution)
(c) $AuCl_4^- + Sn^{2+} \longrightarrow Sn^{4+} + AuCl + Cl^-$

Ionic equations can also be balanced using the change-in-oxidation-number method shown in Example 17.6. To illustrate this method, let's use the equation from Example 17.10.

Balance this equation using the change-in-oxidation-number method:

Example 17.11

$$CrO_4^{2-} + Fe(OH)_2 \longrightarrow Cr(OH)_3 + Fe(OH)_3 \text{ (basic solution)}$$

Solution

Steps 1 and 2. Assign oxidation numbers and balance the charges with electrons:

$$Cr^{6+} + 3\,e^- \longrightarrow Cr^{3+} \quad \text{reduction}$$
$$Cr^{6+} \text{ gains 3 } e^-$$

$$Fe^{2+} \longrightarrow Fe^{3+} + e^- \quad \text{oxidation}$$
$$Fe^{2+} \text{ loses 1 } e^- \longrightarrow$$

Step 3. Equalize the loss and gain of electrons, and then multiply the oxidation step by 3:

$$Cr^{6+} + 3\,e^- \longrightarrow Cr^{3+}$$
$$Cr^{6+} \text{ gains 3 } e^-$$

$$3\,Fe^{2+} \longrightarrow 3\,Fe^{3+} + 3\,e^-$$
$$3\,Fe^{2+} \text{ lose 3 } e^-$$

Step 4. Transfer coefficients back to the original equation:

$$CrO_4^{2-} + 3\,Fe(OH)_2 \longrightarrow Cr(OH)_3 + 3\,Fe(OH)_3$$

Step 5. Balance electrically. Because the solution is basic, use OH^- to balance charges. The charge on the left side is -2, and on the right side is 0. Add 2 OH^- ions to the right side of the equation:

$$CrO_4^{2-} + 3\,Fe(OH)_2 \longrightarrow Cr(OH)_3 + 3\,Fe(OH)_3 + 2\,OH^-$$

Adding 4 H_2O to the left side balances the equation:

$$CrO_4^{2-} + 3\,Fe(OH)_2 + 4\,H_2O \longrightarrow$$
$$Cr(OH)_3 + 3\,Fe(OH)_3 + 2\,OH^- \quad \text{(balanced)}$$

Check: Each side of the equation has a charge of -2 and contains the same number of atoms of each element.

▲ FIGURE 17.2
A coil of copper placed in a silver nitrate solution forms silver crystals on the wire. The pale blue of the solution indicates the presence of copper ions.

Practice 17.6

Balance these equations using the change-in-oxidation-number method:
(a) $Zn \longrightarrow Zn(OH)_4^{2-} + H_2$ (basic solution)
(b) $H_2O_2 + Sn^{2+} \longrightarrow Sn^{4+}$ (acidic solution)
(c) $Cu + Cu^{2+} \longrightarrow Cu_2O$ (basic solution)

17.5 Activity Series of Metals

Knowledge of the relative chemical reactivities of the elements helps us predict the course of many chemical reactions. For example, calcium reacts with cold water to produce hydrogen, and magnesium reacts with steam to produce hydrogen. Therefore calcium is considered a more reactive metal than magnesium:

$$Ca(s) + 2 H_2O(l) \longrightarrow Ca(OH)_2(aq) + H_2(g)$$
$$Mg(s) + H_2O(g) \longrightarrow MgO(s) + H_2(g)$$
$$\text{steam}$$

The difference in their activity is attributed to the fact that calcium loses its two valence electrons more easily than magnesium and is therefore more reactive and/or more readily oxidized than magnesium.

When a coil of copper is placed in a solution of silver nitrate ($AgNO_3$), free silver begins to plate out on the copper. (See Figure 17.2.) After the reaction has continued for some time, we can observe a blue color in the solution, indicating the presence of copper(II) ions. The equations are

$Cu^0(s) + 2 AgNO_3(aq) \longrightarrow 2 Ag^0(s) + Cu(NO_3)_2(aq)$	
$Cu^0(s) + 2 Ag^+(aq) \longrightarrow 2 Ag^0(s) + Cu^{2+}(aq)$	net ionic equation
$Cu^0(s) \longrightarrow Cu^{2+}(aq) + 2 e^-$	oxidation of Cu^0
$Ag^+(aq) + e^- \longrightarrow Ag^0(s)$	reduction of Ag^+

If a coil of silver is placed in a solution of copper(II) nitrate, $Cu(NO_3)_2$, no reaction is visible.

$$Ag^0(s) + Cu(NO_3)_2(aq) \longrightarrow \text{no reaction}$$

In the reaction between Cu and $AgNO_3$, electrons are transferred from Cu^0 atoms to Ag^+ ions in solution. Copper has a greater tendency than silver to lose electrons, so an electrochemical force is exerted upon silver ions to accept electrons from copper atoms. When an Ag^+ ion accepts an electron, it is reduced to an Ag^0 atom and is no longer soluble in solution. At the same time, Cu^0 is oxidized and goes into solution as Cu^{2+} ions. From this reaction we can conclude that copper is more reactive than silver.

Metals such as sodium, magnesium, zinc, and iron that react with solutions of acids to liberate hydrogen are more reactive than hydrogen. Metals such as copper, silver, and mercury that do not react with solutions of acids to liberate hydrogen are less reactive than hydrogen. By studying a series of reactions such as these, we can list metals according to their chemical activity, placing the most active at the top and the least active at the bottom. This list is called the **activity series of metals.** Table 17.3 lists some of the common metals in the series. The arrangement corresponds to

TABLE 17.3 Activity Series of Metals

Ease of oxidation ↑

K	$\longrightarrow K^+$	$+ e^-$
Ba	$\longrightarrow Ba^{2+}$	$+ 2 e^-$
Ca	$\longrightarrow Ca^{2+}$	$+ 2 e^-$
Na	$\longrightarrow Na^+$	$+ e^-$
Mg	$\longrightarrow Mg^{2+}$	$+ 2 e^-$
Al	$\longrightarrow Al^{3+}$	$+ 3 e^-$
Zn	$\longrightarrow Zn^{2+}$	$+ 2 e^-$
Cr	$\longrightarrow Cr^{3+}$	$+ 3 e^-$
Fe	$\longrightarrow Fe^{2+}$	$+ 2 e^-$
Ni	$\longrightarrow Ni^{2+}$	$+ 2 e^-$
Sn	$\longrightarrow Sn^{2+}$	$+ 2 e^-$
Pb	$\longrightarrow Pb^{2+}$	$+ 2 e^-$
H_2	$\longrightarrow$ **$2 H^+$**	**$+ 2 e^-$**
Cu	$\longrightarrow Cu^{2+}$	$+ 2 e^-$
As	$\longrightarrow As^{3+}$	$+ 3 e^-$
Ag	$\longrightarrow Ag^+$	$+ e^-$
Hg	$\longrightarrow Hg^{2+}$	$+ 2 e^-$
Au	$\longrightarrow Au^{3+}$	$+ 3 e^-$

activity series of metals

the ease with which the elements are oxidized or lose electrons, with the most easily oxidizable element listed first. More extensive tables are available in chemistry reference books.

The general principles governing the arrangement and use of the activity series are as follows:

1. The reactivity of the metals listed decreases from top to bottom.
2. A free metal can displace the ion of a second metal from solution, provided that the free metal is above the second metal in the activity series.
3. Free metals above hydrogen react with nonoxidizing acids in solution to liberate hydrogen gas.
4. Free metals below hydrogen do not liberate hydrogen from acids.
5. Conditions such as temperature and concentration may affect the relative position of some of these elements.

Here are two examples using the activity series of metals.

Example 17.12

Solution

Will zinc metal react with dilute sulfuric acid?

From Table 17.3, we see that zinc is above hydrogen; therefore zinc atoms will lose electrons more readily than hydrogen atoms. Hence zinc atoms will reduce hydrogen ions from the acid to form hydrogen gas and zinc ions. In fact, these reagents are commonly used for the laboratory preparation of hydrogen. The equation is

$$Zn(s) + H_2SO_4(aq) \longrightarrow ZnSO_4(aq) + H_2(g)$$

$$Zn(s) + 2\,H^+(aq) \longrightarrow Zn^{2+}(aq) + H_2(g) \qquad \text{(net ionic equation)}$$

Example 17.13

Solution

Will a reaction occur when copper metal is placed in an iron(II) sulfate solution?

No, copper lies below iron in the series, loses electrons less easily than iron, and therefore will not displace iron(II) ions from solution. In fact, the reverse is true. When an iron nail is dipped into a copper(II) sulfate solution, it becomes coated with free copper. The equations are

$$Cu(s) + FeSO_4(aq) \longrightarrow \text{no reaction}$$

$$Fe(s) + CuSO_4(aq) \longrightarrow FeSO_4(aq) + Cu(s)$$

From Table 17.3, we may abstract the following pair in their relative position to each other:

$$Fe \longrightarrow Fe^{2+} + 2\,e^-$$

$$Cu \longrightarrow Cu^{2+} + 2\,e^-$$

According to Principle 2 on the use of the activity series, we can predict that free iron will react with copper(II) ions in solution to form free copper metal and iron(II) ions in solution:

$$Fe(s) + Cu^{2+}(aq) \longrightarrow Fe^{2+}(aq) + Cu(s) \quad \text{(net ionic equation)}$$

Practice 17.7

Indicate whether these reactions will occur:
(a) Sodium metal is placed in dilute hydrochloric acid.
(b) A piece of lead is placed in magnesium nitrate solution.
(c) Mercury is placed in a solution of silver nitrate.

17.6   Electrolytic and Voltaic Cells

electrolysis
electrolytic cell

The process in which electrical energy is used to bring about chemical change is known as **electrolysis.** An **electrolytic cell** uses electrical energy to produce a chemical reaction. The use of electrical energy has many applications in industry—for example, in the production of sodium, sodium hydroxide, chlorine, fluorine, magnesium, aluminum, and pure hydrogen and oxygen, and in the purification and electroplating of metals.

What happens when an electric current is passed through a solution? Let's consider a hydrochloric acid solution in a simple electrolytic cell, as shown in Figure 17.3. The cell consists of a source of direct current (a battery) connected to two electrodes that are immersed in a solution of hydrochloric acid. The negative electrode is called the **cathode** because cations are attracted to it. The positive electrode is called the **anode** because anions are attracted to it. The cathode is attached to the negative pole and the anode to the positive pole of the battery. The battery supplies electrons to the cathode.

cathode
anode

When the electric circuit is completed, positive hydronium ions (H_3O^+) migrate to the cathode where they pick up electrons and evolve as hydrogen gas. At the same time the negative chloride ions (Cl^-) migrate to the anode, where they lose electrons and evolve as chlorine gas.

Reaction at the cathode:

$$H_3O^+ + 1\ e^- \longrightarrow H^0 + H_2O \qquad \text{(reduction)}$$

$$H^0 + H^0 \longrightarrow H_2$$

FIGURE 17.3▶
During the electrolysis of a hydrochloric acid solution, positive hydronium ions are attracted to the cathode, where they gain electrons and form hydrogen gas. Chloride ions migrate to the anode, where they lose electrons and form chlorine gas. The equation for this process is
2 HCl(aq) ⟶ H₂(g) + Cl₂(g).

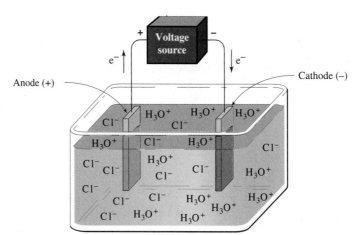

Reaction at the anode:

$$Cl^- \longrightarrow Cl^0 + 1\,e^- \qquad \text{(oxidation)}$$

$$Cl^0 + Cl^0 \longrightarrow Cl_2$$

$$2\,HCl(aq) \xrightarrow{\text{electrolysis}} H_2(g) + Cl_2(g) \qquad \text{net reaction}$$

Note that oxidation–reduction has taken place. Chloride ions lost electrons (were oxidized) at the anode, and hydronium ions gained electrons (were reduced) at the cathode.

Oxidation always occurs at the anode and reduction at the cathode.

When concentrated sodium chloride solutions (brines) are electrolyzed, the products are sodium hydroxide, hydrogen, and chlorine. The overall reaction is

$$2\,Na^+(aq) + 2\,Cl^-(aq) + 2\,H_2O(l) \xrightarrow{\text{electrolysis}}$$
$$2\,Na^+(aq) + 2\,OH^-(aq) + H_2(g) + Cl_2(g)$$

The net ionic equation is

$$2\,Cl^-(aq) + 2\,H_2O(l) \longrightarrow 2\,OH^-(aq) + H_2(g) + Cl_2(g)$$

During electrolysis, Na^+ ions move toward the cathode and Cl^- ions move toward the anode. The anode reaction is similar to that of hydrochloric acid; the chlorine is liberated:

$$2\,Cl^-(aq) \longrightarrow Cl_2(g) + 2\,e^-$$

Even though Na^+ ions are attracted by the cathode, the facts show that hydrogen is liberated there. No evidence of metallic sodium is found, but the area around the cathode tests alkaline from the accumulated OH^- ions. The reaction at the cathode is

$$2\,H_2O(l) + 2\,e^- \longrightarrow H_2(g) + 2\,OH^-(aq)$$

If electrolysis is allowed to continue until all the chloride is reacted, the solution remaining will contain only sodium hydroxide, which on evaporation yields solid NaOH. Large amounts of sodium hydroxide and chlorine are made by this process.

When molten sodium chloride (without water) is subjected to electrolysis, metallic sodium and chlorine gas are formed:

$$2\,Na^+(l) + 2\,Cl^-(l) \xrightarrow{\text{electrolysis}} 2\,Na(l) + Cl_2(g)$$

An important electrochemical application is the electroplating of metals. Electroplating is the art of covering a surface or an object with a thin adherent electrodeposited metal coating. Electroplating is done for protection of the surface of the base metal or for a purely decorative effect. The layer deposited is surprisingly thin, varying from as little as 5×10^{-5} cm to 2×10^{-3} cm, depending on the metal and the intended use. The object to be plated is set up as the cathode and is immersed in a solution containing ions of the plating metal. When an electric current passes through the solution, metal ions that migrate to the cathode are reduced, depositing on the object as the free metal. In most cases the metal deposited on the object is replaced

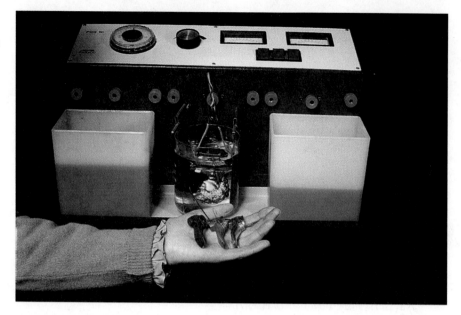

in the solution by using an anode of the same metal. The following equations show
the chemical changes in the electroplating of nickel:

Reaction at the cathode: $Ni^{2+}(aq) + 2\ e^- \longrightarrow Ni(s)$ Ni plated out
on an object

Reaction at the anode: $Ni(s) \longrightarrow Ni^{2+}(aq) + 2\ e^-$ Ni replenished
in solution

Metals commonly used in commercial electroplating are copper, nickel, zinc, lead,
cadmium, chromium, tin, gold, and silver.

In the electrolytic cell shown in Figure 17.3, electrical energy from the voltage
source is used to bring about nonspontaneous redox reactions. The hydrogen and
chlorine produced have more potential energy than was present in the hydrochloric
acid before electrolysis.

Conversely, some spontaneous redox reactions can be made to supply useful
amounts of electrical energy. When a piece of zinc is put in a copper(II) sulfate solution, the zinc quickly becomes coated with metallic copper. We expect this coating to
happen because zinc is above copper in the activity series; copper(II) ions are therefore reduced by zinc atoms:

$$Zn^0(s) + Cu^{2+}(aq) \longrightarrow Zn^{2+}(aq) + Cu^0(s)$$

This reaction is clearly a spontaneous redox reaction, but simply dipping a zinc rod
into a copper(II) sulfate solution will not produce useful electric current. However,
when we carry out this reaction in the cell shown in Figure 17.4, an electric current
is produced. The cell consists of a piece of zinc immersed in a zinc sulfate solution
and connected by a wire through a voltmeter to a piece of copper immersed in
copper(II) sulfate solution. The two solutions are connected by a salt bridge. Such
a cell produces an electric current and a potential of about 1.1 volts when both

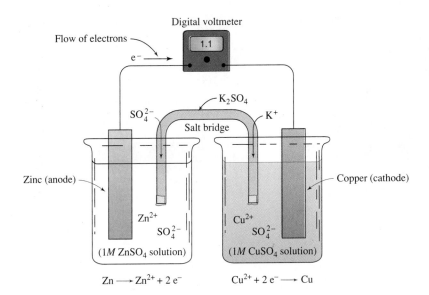

◄ **FIGURE 17.4**
Zinc–copper voltaic cell.
The cell has a potential of
1.1 volts when ZnSO₄ and
CuSO₄ solutions are 1.0 M.
The salt bridge provides
electrical contact between
the two half-cells.

solutions are 1.0 M in concentration. A cell that produces electric current from a spontaneous chemical reaction is called a **voltaic cell.** A voltaic cell is also known as a *galvanic cell.*

voltaic cell

The driving force responsible for the electric current in the zinc–copper cell originates in the great tendency of zinc atoms to lose electrons relative to the tendency of copper(II) ions to gain electrons. In the cell shown in Figure 17.4, zinc atoms lose electrons and are converted to zinc ions at the zinc electrode surface; the electrons flow through the wire (external circuit) to the copper electrode. Here copper(II) ions pick up electrons and are reduced to copper atoms, which plate out on the copper electrode. Sulfate ions flow from the CuSO₄ solution via the salt bridge into the ZnSO₄ solution (internal circuit) to complete the circuit. The equations for the reactions of this cell are

anode $Zn^0(s) \longrightarrow Zn^{2+}(aq) + 2\ e^-$ (oxidation)

cathode $Cu^{2+}(aq) + 2\ e^- \longrightarrow Cu^0(s)$ (reduction)

net ionic $Zn^0(s) + Cu^{2+}(aq) \longrightarrow Zn^{2+}(aq) + Cu^0(s)$

overall $Zn(s) + CuSO_4(aq) \longrightarrow ZnSO_4(aq) + Cu(s)$

The redox reaction, the movement of electrons in the metallic or external part of the circuit, and the movement of ions in the solution or internal part of the circuit of the copper–zinc cell are very similar to the actions that occur in the electrolytic cell of Figure 17.3. The only important difference is that the reactions of the zinc–copper cell are spontaneous. This spontaneity is the crucial difference between all voltaic and electrolytic cells.

Voltaic cells use chemical reactions to produce electrical energy, and electrolytic cells use electrical energy to produce chemical reactions.

FIGURE 17.5 ▶
(a) A common acid-type dry
cell. (b) Diagram of an alkaline zinc–mercury cell.

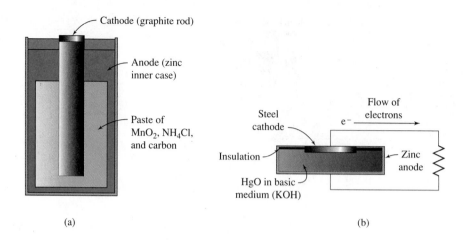

(a)   (b)

Calculators, watches, radios, walkmen, and portable CD players are powered by small efficient voltaic cells called *dry cell batteries.* Called dry cells because they do not contain a liquid electrolyte (like the voltaic cells discussed earlier), dry cell batteries are found in several different versions.

The *acid*-type dry cell battery contains a zinc inner case that functions as the anode. A carbon (graphite) rod runs through the center and is in contact with the zinc case at one end and a moist paste of solid MnO_2, NH_4Cl, and carbon that functions as the cathode. (See Figure 17.5a.) The cell produces about 1.5 volts. The *alkaline*-type dry cell battery is the same as the acid type except the NH_4Cl is replaced by either KOH or NaOH. These dry cells typically last longer because the zinc anode corrodes more slowly in basic conditions. A third type of dry cell is the *zinc–mercury* cell shown in Figure 17.5b. The reactions occurring in this cell are

anode	$Zn^0 + 2\ OH^- \longrightarrow ZnO + H_2O + 2\ e^-$	(oxidation)
cathode	$HgO + H_2O + 2\ e^- \longrightarrow Hg^0 + 2\ OH^-$	(reduction)
net ionic	$Zn^0 + Hg^{2+} \longrightarrow Zn^{2+} + Hg^0$	
overall	$Zn^0 + HgO \longrightarrow ZnO + Hg^0$	

To offset the relatively high initial cost, this cell (a) provides current at a very steady potential of about 1.5 volts; (b) has an exceptionally long service life—that is, high energy output to weight ratio; (c) is completely self-contained; and (d) can be stored for relatively long periods of time when not in use.

An automobile storage battery is an energy reservoir. The charged battery acts as a voltaic cell and through chemical reactions furnishes electrical energy to operate the starter, lights, radio, and so on. When the engine is running, a generator or alternator produces and forces an electric current through the battery and, by electrolytic chemical action, restores it to the charged condition.

The cell unit consists of a lead plate filled with spongy lead and a lead dioxide plate, both immersed in dilute sulfuric acid solution, which serves as the electrolyte (see Figure 17.6). When the cell is discharging, or acting as a voltaic cell, these reactions occur:

Pb plate (anode):	$Pb^0 \longrightarrow Pb^{2+} + 2\ e^-$	(oxidation)
PbO_2 plate (cathode):	$PbO_2 + 4\ H^+ + 2\ e^- \longrightarrow Pb^{2+} + 2\ H_2O$	(reduction)

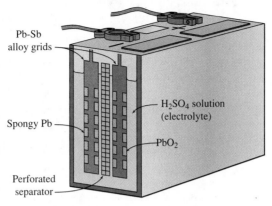

Pb-Sb alloy grids

H$_2$SO$_4$ solution (electrolyte)

Spongy Pb

PbO$_2$

Perforated separator

Figure 17.6
Cross-sectional diagram of a lead storage battery cell.

Net ionic redox reaction: $Pb^0 + PbO_2 + 4\,H^+ \longrightarrow 2\,Pb^{2+} + 2\,H_2O$

Precipitation reaction on plates: $Pb^{2+}(aq) + SO_4^{2-}(aq) \longrightarrow PbSO_4(s)$

Because lead(II) sulfate is insoluble, the Pb^{2+} ions combine with SO_4^{2-} ions to form a coating of $PbSO_4$ on each plate. The overall chemical reaction of the cell is

$$Pb(s) + PbO_2(s) + 2\,H_2SO_4(aq) \xrightarrow[\text{cycle}]{\text{discharge}} 2\,PbSO_4(s) + 2\,H_2O(l)$$

The cell can be recharged by reversing the chemical reaction. This reversal is accomplished by forcing an electric current through the cell in the opposite direction. Lead sulfate and water are reconverted to lead, lead (IV) oxide, and sulfuric acid:

$$2\,PbSO_4(s) + 2\,H_2O(l) \xrightarrow[\text{cycle}]{\text{charge}} Pb(s) + PbO_2(s) + 2\,H_2SO_4(aq)$$

The electrolyte in a lead storage battery is a 38% by mass sulfuric acid solution having a density of 1.29 g/mL. As the battery is discharged, sulfuric acid is removed, thereby decreasing the density of the electrolyte solution. The state of charge or discharge of the battery can be estimated by measuring the density (or specific gravity) of the electrolyte solution with a hydrometer. When the density has dropped to about 1.05 g/mL, the battery needs recharging.

In a commercial battery, each cell consists of a series of cell units of alternating lead–lead(IV) oxide plates separated and supported by wood, glass wool, or fiberglass. The energy storage capacity of a single cell is limited, and its electrical potential is only about 2 volts. Therefore a bank of six cells is connected in series to provide the 12-volt output of the usual automobile battery.

A New Look for a Great Lady

Corrosion has been a problem for the world since the earliest use of metals. The complexities of corrosion are exemplified by our famous Statue of Liberty. The beautiful lady was given to the United States by France in 1886. She is constructed of a skeleton of iron bars (ribs) that were bent to precisely conform to her shape and covered with a thin skin of copper. Her skin is connected to the skeleton with special copper "saddles" or bands that overlap the iron bars and are riveted in place.

Iron will corrode (oxidize) in moist air and, if in contact with copper and an electrolyte (seawater), the corrosion rate can increase 100-fold. The harbor environment of New York fosters accelerated corrosion, and within less than a century the iron ribs had corroded to a point where the lady was in serious danger of collapse.

To repair and reinforce the statue, each of the iron bars (over 1300 in all) were replaced one at a time with stainless

▲ **The Statue of Liberty was in serious danger of collapse before renovation.**

steel, which corrodes much less easily. At the same time, the copper saddles were coated with Teflon to insulate them from contact with the iron and further reduce future corrosion.

The external appearance of the statue was left essentially unchanged. The bluish green patina is a natural product formed as copper undergoes atmospheric corrosion. Patina may take several forms and exists as a protective film that adheres tightly to the surface of the copper and reduces the rate of corrosion underneath.

The patina of the Statue of Liberty is primarily $CuSO_4 \cdot 3\,Cu(OH)_2$. It has protected the lady well—during her hundred years, the copper skin has thinned only about 4%. Unfortunately, acid rain appears to be changing the patina to $CuSO_4 \cdot 2\,Cu(OH)_2$, which doesn't bond as closely to the copper surface. Chemists are concerned that this change will result in the loss of patina and increased corrosion of the copper skin.

Concepts in Review

1. Assign oxidation numbers to all the elements in a compound or ion.

2. Determine which element is being oxidized and which element is being reduced in an oxidation–reduction reaction.

3. Identify the oxidizing agent and the reducing agent in an oxidation–reduction reaction.

4. Balance oxidation–reduction equations in molecular and ionic forms.

5. Outline the general principles concerning the activity series of the metals.

6. Use the activity series to determine whether a proposed single-displacement reaction will occur.

7. Distinguish between an electrolytic and a voltaic cell.

8. Draw a voltaic cell that will produce electric current from an oxidation–reduction reaction involving two metals and their salts.

9. Identify the anode reaction and the cathode reaction in an electrolytic or voltaic cell.

Key Terms

activity series of metals (17.5)
anode (17.6)
cathode (17.6)
electrolysis (17.6)

electrolytic cell (17.6)
oxidation (17.2)
oxidation number (17.1)
oxidation–reduction (17.2)

oxidation state (17.1)
oxidizing agent (17.2)
redox (17.2)

reducing agent (17.2)
reduction (17.2)
voltaic cell (17.6)

Questions

1. In the equation

$$I_2 + 5 Cl_2 + 6 H_2O \longrightarrow 2 HIO_3 + 10 HCl$$

   (a) has iodine been oxidized or has it been reduced?
   (b) has chlorine been oxidized or has it been reduced? (Figure 17.1)

2. Which element of each pair is more active? (Table 17.3)
   (a) Ag or Al
   (b) Na or Ba
   (c) Ni or Cu

3. Will the following combinations react in aqueous solution? (Table 17.3)
   (a) $Zn + Cu^{2+}$ (e) $Ba + FeCl_2$
   (b) $Ag + H^+$ (f) $Pb + NaCl$
   (c) $Sn + Ag^+$ (g) $Ni + Hg(NO_3)_2$
   (d) $As + Mg^{2+}$ (h) $Al + CuSO_4$

4. The reaction between powdered aluminum and iron(III) oxide (in the thermite process) producing molten iron is very exothermic.
   (a) Write the equation for the chemical reaction that occurs.
   (b) Explain in terms of Table 17.3 why a reaction occurs.
   (c) Would you expect a reaction between powdered iron and aluminum oxide?
   (d) Would you expect a reaction between powdered aluminum and chromium(III) oxide?

5. Write equations for the chemical reaction of aluminum, chromium, gold, iron, copper, magnesium, mercury, and zinc with dilute solutions of (a) hydrochloric acid and (b) sulfuric acid. If a reaction will not occur, write "no reaction" as the product. (Table 17.3)

6. An $NiCl_2$ solution is placed in the apparatus shown in Figure 17.3, instead of the HCl solution shown. Write equations for
   (a) the anode reaction
   (b) the cathode reaction
   (c) the net electrochemical reaction

7. What is the major distinction between the reactions occurring in Figures 17.3 and 17.4?

8. In the cell shown in Figure 17.4,
   (a) what would be the effect of removing the voltmeter and connecting the wires shown coming to the voltmeter?
   (b) what would be the effect of removing the salt bridge?

9. Why are oxidation and reduction said to be complementary processes?

10. When molten $CaBr_2$ is electrolyzed, calcium metal and bromine are produced. Write equations for the two half-reactions that occur at the electrodes. Label the anode half-reaction and the cathode half-reaction.

11. Why is direct current used instead of alternating current in the electroplating of metals?

12. What property of lead(IV) oxide and lead(II) sulfate makes it unnecessary to have salt bridges in the cells of a lead storage battery?

13. Explain why the density of the electrolyte in a lead storage battery decreases during the discharge cycle.

14. In one type of alkaline cell used to power devices such as portable radios, Hg^{2+} ions are reduced to metallic mercury when the cell is being discharged. Does this reduction occur at the anode or the cathode? Explain.

15. Differentiate between an electrolytic cell and a voltaic cell.

16. Why is a porous barrier or a salt bridge necessary in some voltaic cells?

Paired Exercises

17. What is the oxidation number of the underlined element in each compound?
   (a) <u>N</u>aCl
   (b) Fe<u>Cl</u>$_3$
   (c) <u>Pb</u>O$_2$
   (d) Na<u>N</u>O$_3$
   (e) H$_2$<u>S</u>O$_3$
   (f) <u>N</u>H$_4$Cl

18. What is the oxidation number of the underlined element in each compound?
   (a) K<u>Mn</u>O$_4$
   (b) <u>I</u>$_2$
   (c) <u>N</u>H$_3$
   (d) K<u>Cl</u>O$_3$
   (e) K$_2$<u>Cr</u>O$_4$
   (f) K$_2$<u>Cr</u>$_2$O$_7$

19. What is the oxidation number of the underlined elements?
    (a) $\underline{S}^{2-}$
    (b) $\underline{N}O_2^-$
    (c) $Na_2\underline{O}_2$
    (d) $\underline{Bi}^{3+}$

20. What is the oxidation number of the underlined elements?
    (a) $\underline{O}_2$
    (b) $\underline{As}O_4^{3-}$
    (c) $Fe(\underline{O}H)_3$
    (d) $\underline{I}O_3^-$

21. In the following half-reactions, which element is changing oxidation state? Is the half-reaction an oxidation or a reduction? Supply the proper number of electrons to each side to balance each equation.
    (a) $Zn^{2+} \longrightarrow Zn$
    (b) $2\ Br^- \longrightarrow Br_2$
    (c) $MnO_4^- + 8\ H^+ \longrightarrow Mn^{2+} + 4\ H_2O$
    (d) $Ni \longrightarrow Ni^{2+}$

22. In the following half-reactions, which element is changing oxidation state? Is the half-reaction an oxidation or a reduction? Supply the proper number of electrons to each side to balance each equation:
    (a) $SO_3^{2-} + H_2O \longrightarrow SO_4^{2-} + 2\ H^+$
    (b) $NO_3^- + 4\ H^+ \longrightarrow NO + 2\ H_2O$
    (c) $S_2O_4^{2-} + 2\ H_2O \longrightarrow 2\ SO_3^{2-} + 4\ H^+$
    (d) $Fe^{2+} \longrightarrow Fe^{3+}$

23. In the following unbalanced equations, identify
    (a) the oxidized element and the reduced element
    (b) the oxidizing agent and the reducing agent
        (1) $Cr + HCl \longrightarrow CrCl_3 + H_2$
        (2) $SO_4^{2-} + I^- + H^+ \longrightarrow H_2S + I_2 + H_2O$

24. In the following unbalanced equations, identify
    (a) the oxidized element and the reduced element
    (b) the oxidizing agent and the reducing agent
        (1) $AsH_3 + Ag^+ + H_2O \longrightarrow H_3AsO_4 + Ag + H^+$
        (2) $Cl_2 + NaBr \longrightarrow NaCl + Br_2$

25. Balance these equations using the change-in-oxidation-number method:
    (a) $Zn + S \longrightarrow ZnS$
    (b) $AgNO_3 + Pb \longrightarrow Pb(NO_3)_2 + Ag$
    (c) $Fe_2O_3 + CO \longrightarrow Fe + CO_2$
    (d) $H_2S + HNO_3 \longrightarrow S + NO + H_2O$
    (e) $MnO_2 + HBr \longrightarrow MnBr_2 + Br_2 + H_2O$

26. Balance these equations using the change-in-oxidation-number method:
    (a) $Cl_2 + KOH \longrightarrow KCl + KClO_3 + H_2O$
    (b) $Ag + HNO_3 \longrightarrow AgNO_3 + NO + H_2O$
    (c) $CuO + NH_3 \longrightarrow N_2 + Cu + H_2O$
    (d) $PbO_2 + Sb + NaOH \longrightarrow PbO + NaSbO_2 + H_2O$
    (e) $H_2O_2 + KMnO_4 + H_2SO_4 \longrightarrow$
        $O_2 + MnSO_4 + K_2SO_4 + H_2O$

27. Balance these ionic redox equations using the ion–electron method. These reactions occur in acidic solution.
    (a) $Zn + NO_3^- \longrightarrow Zn^{2+} + NH_4^+$
    (b) $NO_3^- + S \longrightarrow NO_2 + SO_4^{2-}$
    (c) $PH_3 + I_2 \longrightarrow H_3PO_2 + I^-$
    (d) $Cu + NO_3^- \longrightarrow Cu^{2+} + NO$
    *(e) $ClO_3^- + Cl^- \longrightarrow Cl_2$

28. Balance these ionic redox equations using the ion–electron method. These reactions occur in acidic solution.
    (a) $ClO_3^- + I^- \longrightarrow I_2 + Cl^-$
    (b) $Cr_2O_7^{2-} + Fe^{2+} \longrightarrow Cr^{3+} + Fe^{3+}$
    (c) $MnO_4^- + SO_2 \longrightarrow Mn^{2+} + SO_4^{2-}$
    (d) $H_3AsO_3 + MnO_4^- \longrightarrow H_3AsO_4 + Mn^{2+}$
    *(e) $Cr_2O_7^{2-} + H_3AsO_3 \longrightarrow Cr^{3+} + H_3AsO_4$

29. Balance these ionic redox equations using the ion–electron method. These reactions occur in basic solutions.
    (a) $Cl_2 + IO_3^- \longrightarrow Cl^- + IO_4^-$
    (b) $MnO_4^- + ClO_2^- \longrightarrow MnO_2 + ClO_4^-$
    (c) $Se \longrightarrow Se^{2-} + SeO_3^{2-}$
    *(d) $Fe_3O_4 + MnO_4^- \longrightarrow Fe_2O_3 + MnO_2$
    *(e) $BrO^- + Cr(OH)_4^- \longrightarrow Br^- + CrO_4^{2-}$

30. Balance these ionic redox equations using the ion–electron method. These reactions occur in basic solutions.
    (a) $MnO_4^- + SO_3^{2-} \longrightarrow MnO_2 + SO_4^{2-}$
    (b) $ClO_2 + SbO_2^- \longrightarrow ClO_2^- + Sb(OH)_6^-$
    (c) $Al + NO_3^- \longrightarrow NH_3 + Al(OH)_4^-$
    *(d) $P_4 \longrightarrow HPO_3^{2-} + PH_3$
    *(e) $Al + OH^- \longrightarrow Al(OH)_4^- + H_2$

Additional Exercises

31. The chemical reactions taking place during discharge in a lead storage battery are

    $Pb + SO_4^{2-} \longrightarrow PbSO_4$
    $PbO_2 + SO_4^{2-} + 4\ H^+ \longrightarrow PbSO_4 + 2\ H_2O$

    (a) Complete each half-reaction by supplying electrons.
    (b) Which reaction is oxidation and which is reduction?
    (c) Which reaction occurs at the anode of the battery?

32. Use this unbalanced redox equation

    $KMnO_4 + HCl \longrightarrow KCl + MnCl_2 + H_2O + Cl_2$

    to indicate
    (a) the oxidizing agent
    (b) the reducing agent
    (c) the number of electrons that are transferred per mole of oxidizing agent

33. How many moles of NO gas will be formed by the reaction of 25.0 g of silver with nitric acid?

$$Ag + HNO_3 \longrightarrow AgNO_3 + NO + H_2O \text{ (acid solution)}$$

34. What volume of chlorine gas, measured at STP, is required to react with excess KOH to form 0.300 mol $KClO_3$?

$$Cl_2 + KOH \longrightarrow KCl + KClO_3 + H_2O \text{ (acid solution)}$$

35. What mass of $KMnO_4$ is needed to react with 100. mL H_2O_2 solution? ($d = 1.031$ g/mL, 9.0% H_2O_2 by mass)

$$H_2O_2 + KMnO_4 + H_2SO_4 \longrightarrow$$
$$O_2 + MnSO_4 + K_2SO_4 + H_2O \qquad \text{(acid solution)}$$

*36. What volume of 0.200 M $K_2Cr_2O_7$ will be required to oxidize 5.00 g H_3AsO_3?

$$Cr_2O_7^{2-} + H_3AsO_3 \longrightarrow Cr^{3+} + H_3AsO_4 \text{ (acid solution)}$$

*37. What volume of 0.200 M $K_2Cr_2O_7$ will be required to oxidize the Fe^{2+} ion in 60.0 mL of 0.200 M $FeSO_4$ solution?

$$Cr_2O_7^{2-} + Fe^{2+} \longrightarrow Cr^{3+} + Fe^{3+} \qquad \text{(acid solution)}$$

*38. A sample of crude potassium iodide was analyzed using this reaction (not balanced):

$$I^- + SO_4^{2-} \longrightarrow I_2 + H_2S \quad \text{(acid solution)}$$

If a 4.00-g sample of crude KI produced 2.79 g of iodine, what is the percent purity of the KI?

*39. What volume of NO gas, measured at 28°C and 744 torr, will be formed by the reaction of 0.500 mol Ag reacting with excess nitric acid?

$$Ag + HNO_3 \longrightarrow AgNO_3 + NO + H_2O \text{ (acid solution)}$$

40. How many moles of H_2 can be produced from 100.0 g Al according to this reaction?

$$Al + OH^- \longrightarrow Al(OH)_4^- + H_2 \quad \text{(basic solution)}$$

41. There is something incorrect about these half-reactions:
(a) $Cu^+ + e^- \longrightarrow Cu^{2+}$
(b) $Pb^{2+} + e^{2-} \longrightarrow Pb$
Identify what is wrong, and correct it.

42. Why can oxidation *never* occur without reduction?

43. The following observations were made concerning metals A, B, C, and D.
(a) When a strip of metal A is placed in a solution of B^{2+} ions, no reaction is observed.
(b) Similarly, A in a solution containing C^+ ions produces no reaction.
(c) When a strip of metal D is placed in a solution of C^+ ions, black metallic C deposits on the surface of D, and the solution tests positively for D^{2+} ions.

(d) When a piece of metallic B is placed in a solution of D^{2+} ions, metallic D appears on the surface of B and B^{2+} ions are found in the solution.
Arrange the ions, A^+, B^{2+}, C^+, and D^{2+}, in order of their ability to attract electrons. List them in order of increasing ability.

44. Tin normally has oxidation numbers of 0, +2, and +4. Which of these species can be an oxidizing agent, which can be a reducing agent, and which can be both? In each case what product would you expect as the tin reacts?

45. Manganese is an element that can exist in numerous oxidation states. In each of these compounds identify the oxidation number of the manganese. Which compound would you expect to be the best oxidizing agent and why?
(a) $Mn(OH)_2$
(b) MnF_3
(c) MnO_2
(d) K_2MnO_4
(e) $KMnO_4$

46. Which equations represent oxidations?
(a) $Mg \longrightarrow Mg^{2+}$
(b) $SO_2 \longrightarrow SO_3$
(c) $KMnO_4 \longrightarrow MnO_2$
(d) $Cl_2O_3 \longrightarrow Cl^-$

*47. In the following equation, note the reaction between manganese(IV) oxide and bromide ions:

$$MnO_2 + Br^- \longrightarrow Br_2 + Mn^{2+}$$

(a) Balance this redox reaction in acidic solution.
(b) How many grams of MnO_2 would be needed to produce 100.0 mL of 0.05 M Mn^{2+}?
(c) How many liters of bromine vapor at 50°C and 1.4 atm would also result?

48. Use the table shown to complete the following reactions. If no reaction occurs, write NR:
(a) $F_2 + Cl^- \longrightarrow$
(b) $Br_2 + Cl^- \longrightarrow$
(c) $I_2 + Cl^- \longrightarrow$
(d) $Br_2 + I^- \longrightarrow$

Activity
↑ F_2
Cl_2
Br_2
I_2

(ease of reduction)

49. Manganese metal reacts with HCl to give hydrogen gas and the Mn^{2+} ion in solution. Write a balanced equation for the reaction.

50. If zinc is allowed to react with dilute nitric acid, zinc is oxidized to the +2 ion, while the nitrate ion can be reduced to ammonium, NH_4^+. Write a balanced equation for the reaction in acidic solution.

51. In the following equations, identify the
   (a) atom or ion oxidized
   (b) atom or ion reduced
   (c) oxidizing agent
   (d) reducing agent
   (e) change in oxidation number associated with each oxidizing process
   (f) change in oxidation number associated with each reducing process
   (1) $C_3H_8 + O_2 \longrightarrow CO_2 + H_2O$
   (2) $HNO_3 + H_2S \longrightarrow NO + S + H_2O$
   (3) $CuO + NH_3 \longrightarrow N_2 + H_2O + Cu$
   (4) $H_2O_2 + Na_2SO_3 \longrightarrow Na_2SO_4 + H_2O$
   (5) $H_2O_2 \longrightarrow H_2O + O_2$

52. In the galvanic cell shown in the diagram, a strip of silver is placed in a solution of silver nitrate, and a strip of lead is placed in a solution of lead(II) nitrate. The two beakers are connected with a salt bridge. Determine
   (a) the anode
   (b) the cathode
   (c) where oxidation occurs
   (d) where reduction occurs
   (e) which direction electrons flow through the wire
   (f) which direction ions flow through the solution

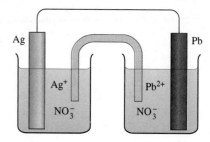

Answers to Practice Exercises

17.1 (a) $S = +6$, (b) $As = +5$, (c) $C = +4$
17.2 (*Note*: $H = +1$ even though it comes second in the formula; N is a nonmetal.)
   (a) $N = -3$, (b) $Cr = +6$, (c) $P = +5$
17.3 (a) $Be = +2$; $Cl = -1$, (b) $H = +1$; $Cl = +1$; $O = -2$,
   (c) $H = +1$; $O = -1$, (d) $N = -3$; $H = +1$,
   (e) $Br = +5$; $O = -2$
17.4 (a) $6\,HNO_3 + S \longrightarrow 6\,NO_2 + H_2SO_4 + 2\,H_2O$
   (b) $2\,CrCl_3 + 3\,MnO_2 + 2\,H_2O \longrightarrow 3\,MnCl_2 + 2\,H_2CrO_4$,
   (c) $2\,KMnO_4 + 6\,HCl + 5\,H_2S \longrightarrow$
        $2\,KCl + 2\,MnCl_2 + 5\,S + 8\,H_2O$

17.5 (a) $4\,H^+ + 2\,I^- + 2\,NO_2^- \longrightarrow I_2 + 2\,NO + 2\,H_2O$
   (b) $2\,OH^- + Cl_2 + IO_3^- \longrightarrow IO_4^- + H_2O + 2\,Cl^-$
   (c) $AuCl_4^- + Sn^{2+} \longrightarrow Sn^{4+} + AuCl + 3\,Cl^-$
17.6 (a) $Zn + 2\,H_2O + 2\,OH^- \longrightarrow Zn(OH)_4^{2-} + H_2$
   (b) $H_2O_2 + Sn^{2+} + 2\,H^+ \longrightarrow Sn^{4+} + 2\,H_2O$
   (c) $Cu + Cu^{2+} + 2\,OH^- \longrightarrow Cu_2O + H_2O$
17.7 (a) yes, (b) no, (c) no

PUTTING IT TOGETHER
Review for Chapters 15–17

Multiple Choice: *Choose the correct answer to each of the following.*

1. When the reaction

   $Al + HCl \longrightarrow$

   is completed and balanced, this term appears in the balanced equation:
   (a) $3\,HCl$ (b) $AlCl_2$ (c) $3\,H_2$ (d) $4\,Al$

2. When the reaction

   $CaO + HNO_3 \longrightarrow$

   is completed and balanced, this term appears in the balanced equation:
   (a) H_2 (b) $2\,H_2$ (c) $2\,CaNO_3$ (d) H_2O

3. When the reaction

   $H_3PO_4 + KOH \longrightarrow$

   is completed and balanced, this term appears in the balanced equation:
   (a) H_3PO_4 (c) KPO_4
   (b) $6\,H_2O$ (d) $3\,KOH$

4. When the reaction

   $HCl + Cr_2(CO_3)_3 \longrightarrow$

   is completed and balanced, this term appears in the balanced equation:
   (a) Cr_2Cl (b) $3\,HCl$ (c) $3\,CO_2$ (d) H_2O

5. Which of these is not a salt?
   (a) $K_2Cr_2O_7$ (c) $Ca(OH)_2$
   (b) $NaHCO_3$ (d) $Na_2C_2O_4$

6. Which of these is not an acid?
   (a) H_3PO_4 (b) H_2S (c) H_2SO_4 (d) NH_3

7. Which of these is a weak electrolyte?
   (a) NH_4OH (c) K_3PO_4
   (b) $Ni(NO_3)_2$ (d) $NaBr$

8. Which of these is a nonelectrolyte?
   (a) $HC_2H_3O_2$ (c) $KMnO_4$
   (b) $MgSO_4$ (d) CCl_4

9. Which of these is a strong electrolyte?
   (a) H_2CO_3 (c) NH_4OH
   (b) HNO_3 (d) H_3BO_3

10. Which of these is a weak electrolyte?
    (a) $NaOH$ (c) $HC_2H_3O_2$
    (b) $NaCl$ (d) H_2SO_4

11. A solution has an H^+ concentration of $3.4 \times 10^{-5}\,M$. The pH is
    (a) 4.47 (b) 5.53 (c) 3.53 (d) 5.47

12. A solution with a pH of 5.85 has an H^+ concentration of
    (a) $7.1 \times 10^{-5}\,M$ (c) $3.8 \times 10^{-4}\,M$
    (b) $7.1 \times 10^{-6}\,M$ (d) $1.4 \times 10^{-6}\,M$

13. If 16.55 mL of 0.844 M NaOH is required to titrate 10.00 mL of a hydrochloric acid solution, the molarity of the acid solution is
    (a) 0.700 M (c) 1.40 M
    (b) 0.510 M (d) 0.255 M

14. What volume of 0.462 M NaOH is required to neutralize 20.00 mL of 0.391 M HNO_3?
    (a) 23.6 mL (c) 9.03 mL
    (b) 16.9 mL (d) 11.8 mL

15. 25.00 mL of H_2SO_4 solution requires 18.92 mL of 0.1024 M NaOH for complete neutralization. The molarity of the acid is
    (a) 0.1550 M (c) 0.07750 M
    (b) 0.03875 M (d) 0.06765 M

16. Dilute hydrochloric acid is a typical acid, as shown by its
    (a) color (b) odor (c) solubility (d) taste

17. What is the pH of a 0.00015 M HCl solution?
    (a) 4.0 (c) between 3 and 4
    (b) 2.82 (d) no correct answer given

18. The chloride ion concentration in 300. mL of 0.10 M $AlCl_3$ is
    (a) 0.30 M (c) 0.030 M
    (b) 0.10 M (d) 0.90 M

19. The amount of $BaSO_4$ that will precipitate when 100. mL of 0.10 M $BaCl_2$ and 100. mL of 0.10 M Na_2SO_4 are mixed is
    (a) 0.010 mol (c) 23 g
    (b) 0.10 mol (d) no correct answer given

20. The freezing point of a 0.50 m NaCl aqueous solution will be about
    (a) $-1.86°C$ (c) $-2.79°C$
    (b) $-0.93°C$ (d) no correct answer given

21. The equation

    $HC_2H_3O_2 + H_2O \rightleftharpoons H_3O^+ + C_2H_3O_2^-$ implies that

    (a) If you start with 1.0 mol $HC_2H_3O_2$, 1.0 mol H_3O^+ and 1.0 mol $C_2H_3O_2^-$ will be produced.
    (b) An equilibrium exists between the forward reaction and the reverse reaction.
    (c) At equilibrium, equal molar amounts of all four substances will exist.
    (d) The reaction proceeds all the way to the products, then reverses, going all the way back to the reactants.

22. If the reaction A + B $\rightleftharpoons$ C + D is initially at equilibrium, and then more A is added, which of the following is not true?
(a) More collisions of A and B will occur; the rate of the forward reaction will thus be increased.
(b) The equilibrium will shift toward the right.
(c) The moles of B will be increased.
(d) The moles of D will be increased.

23. What will be the H^+ concentration in a 1.0 M HCN solution? ($K_a = 4.0 \times 10^{-10}$)
(a) $2.0 \times 10^{-5} M$ (c) $4.0 \times 10^{-10} M$
(b) $1.0 M$ (d) $2.0 \times 10^{-10} M$

24. What is the percent ionization of HCN in Exercise 23?
(a) 100%   (c) $2.0 \times 10^{-3}\%$
(b) $2.0 \times 10^{-8}\%$ (d) $4.0 \times 10^{-8}\%$

25. If $[H^+] = 1 \times 10^{-5} M$, which of the following is not true?
(a) pH = 5   (c) $[OH^-] = 1 \times 10^{-5} M$
(b) pOH = 9   (d) The solution is acidic.

26. If $[H^+] = 2.0 \times 10^{-4} M$, then $[OH^-]$ will be
(a) $5.0 \times 10^{-9} M$ (c) $2.0 \times 10^{-4} M$
(b) 3.70   (d) $5.0 \times 10^{-11} M$

27. The solubility product of $PbCrO_4$ is 2.8×10^{-13}. The solubility of $PbCrO_4$ is
(a) $5.3 \times 10^{-7} M$ (c) $7.8 \times 10^{-14} M$
(b) $2.8 \times 10^{-13} M$ (d) $1.0 M$

28. The solubility of AgBr is $6.3 \times 10^{-7} M$. The value of the solubility product is
(a) 6.3×10^{-7} (c) 4.0×10^{-48}
(b) 4.0×10^{-13} (d) 4.0×10^{-15}

29. Which of these solutions would be the best buffer solution?
(a) $0.10 M$ $HC_2H_3O_2$ + $0.10 M$ $NaC_2H_3O_2$
(b) $0.10 M$ HCl
(c) $0.10 M$ HCl + $0.10 M$ NaCl
(d) pure water

30. For the reaction $H_2(g) + I_2(g) \rightleftharpoons 2$ HI(g), at 700 K, $K_{eq} = 56.6$. If an equilibrium mixture at 700 K was found to contain 0.55 M HI and 0.21 M H_2, the I_2 concentration must be
(a) $0.046 M$ (c) $22 M$
(b) $0.025 M$ (d) $0.21 M$

31. The equilibrium constant for the reaction
2 A + B $\rightleftharpoons$ 3 C + D is
(a) $\dfrac{[C]^3[D]}{[A]^2[B]}$ (c) $\dfrac{[3C][D]}{[2A][B]}$
(b) $\dfrac{[2A][B]}{[3C][D]}$ (d) $\dfrac{[A]^2[B]}{[C]^3[D]}$

32. In the equilibrium represented by
$$N_2(g) + O_2(g) \rightleftharpoons 2\ NO_2(g)$$
as the pressure is increased, the amount of NO_2 formed
(a) increases   (b) decreases
(c) remains the same
(d) increases and decreases irregularly

33. Which factor will not increase the concentration of ammonia as represented by this equation?
$$3\ H_2(g) + N_2(g) \rightleftharpoons 2\ NH_3(g) + 92.5\ kJ$$
(a) increasing the temperature
(b) increasing the concentration of N_2
(c) increasing the concentration of H_2
(d) increasing the pressure

34. If HCl(g) is added to a saturated solution of AgCl, the concentration of Ag^+ in solution
(a) increases
(b) decreases
(c) remains the same
(d) increases and decreases irregularly

35. The solubility of $CaCO_3$ at 20°C is 0.013 g/L. What is the K_{sp} for $CaCO_3$?
(a) 1.3×10^{-8} (c) 1.7×10^{-8}
(b) 1.3×10^{-4} (d) 1.7×10^{-4}

36. The K_{sp} for $BaCrO_4$ is 8.5×10^{-11}. What is the solubility of $BaCrO_4$ in grams per liter?
(a) 9.2×10^{-6} (c) 2.3×10^{-3}
(b) 0.073   (d) 8.5×10^{-11}

37. What will be the $[Ba^{2+}]$ when 0.010 mol Na_2CrO_4 is added to 1.0 L of saturated $BaCrO_4$ solution? See Exercise 36 for K_{sp}.
(a) $8.5 \times 10^{-11} M$ (c) $9.2 \times 10^{-6} M$
(b) $8.5 \times 10^{-9} M$ (d) $9.2 \times 10^{-4} M$

38. Which would occur if a small amount of sodium acetate crystals, $NaC_2H_3O_2$, were added to 100 mL of 0.1 M $HC_2H_3O_2$ at constant temperature?
(a) The number of acetate ions in the solution would decrease.
(b) The number of acetic acid molecules would decrease.
(c) The number of sodium ions in solution would decrease.
(d) The H^+ concentration in the solution would decrease.

39. If the temperature is decreased for the endothermic reaction
$$A + B \rightleftharpoons C + D$$
which of the following is true?
(a) The concentration of A will increase.
(b) No change will occur.
(c) The concentration of B will decrease.
(d) The concentration of D will increase.

40. In K_2SO_4, the oxidation number of sulfur is
    (a) +2   (b) +4   (c) +6   (d) −2

41. In $Ba(NO_3)_2$, the oxidation number of N is
    (a) +5   (b) −3   (c) +4   (d) −1

42. In the reaction

    $$H_2S + 4\,Br_2 + 4\,H_2O \longrightarrow H_2SO_4 + 8\,HBr$$

    the oxidizing agent is
    (a) H_2S (b) Br_2 (c) H_2O (d) H_2SO_4

43. In the reaction

    $$VO_3^- + Fe^{2+} + 4\,H^+ \longrightarrow VO^{2+} + Fe^{3+} + 2\,H_2O$$

    the element reduced is
    (a) V   (b) Fe   (c) O   (d) H

Questions 44–46 pertain to the activity series.

   K   Ca   Mg   Al   Zn   Fe   H   Cu   Ag

44. Which of these pairs will not react in water solution?
    (a) Zn, $CuSO_4$ (c) Fe, $AgNO_3$
    (b) Cu, $Al_2(SO_4)_3$ (d) Mg, $Al_2(SO_4)_3$

45. Which element is the most easily oxidized?
    (a) K   (b) Mg   (c) Zn   (d) Cu

46. Which element will reduce Cu^{2+} to Cu but will not reduce Zn^{2+} to Zn?
    (a) Fe   (b) Ca   (c) Ag   (d) Mg

47. In the electrolysis of fused (molten) $CaCl_2$, the product at the negative electrode is
    (a) Ca^{2+} (b) Cl^- (c) Cl_2 (d) Ca

48. In its reactions, a free element from Group IIA in the periodic table is most likely to
    (a) be oxidized   (c) be unreactive
    (b) be reduced   (d) gain electrons

49. In the partially balanced redox equation

    $$3\,Cu + HNO_3 \longrightarrow 3\,Cu(NO_3)_2 + 2\,NO + H_2O$$

    the coefficient needed to balance H_2O is
    (a) 8   (b) 6   (c) 4   (d) 3

50. Which reaction does not involve oxidation–reduction?
    (a) burning sodium in chlorine
    (b) chemical union of Fe and S
    (c) decomposition of $KClO_3$
    (d) neutralization of NaOH with H_2SO_4

51. How many moles of Fe^{2+} can be oxidized to Fe^{3+} by 2.50 mol Cl_2 according to this equation?

    $$Fe^{2+} + Cl_2 \longrightarrow Fe^{3+} + Cl^-$$

    (a) 2.50 mol   (c) 1.00 mol
    (b) 5.00 mol   (d) 22.4 mol

52. How many grams of sulfur can be produced in this reaction from 100 mL of 6.00 M HNO_3?

    $$HNO_3 + H_2S \longrightarrow S + NO + H_2O$$

    (a) 28.9 g   (b) 19.3 g   (c) 32.1 g   (d) 289 g

53. Which of these ions can be reduced by H_2?
    (a) Hg^{2+} (b) Sn^{2+} (c) Zn^{2+} (d) K^+

54. Which of the following is *not* true of a zinc–mercury cell?
    (a) It provides current at a steady potential.
    (b) It has a short service life.
    (c) It is self-contained.
    (d) It can be stored for long periods of time.

Balancing Oxidation–Reduction Equations
Balance each equation.

55. $P + HNO_3 \longrightarrow HPO_3 + NO + H_2O$

56. $MnSO_4 + PbO_2 + H_2SO_4 \longrightarrow$
    $$HMnO_4 + PbSO_4 + H_2O$$

57. $Cr_2O_7^{2-} + Cl^- \longrightarrow Cr^{3+} + Cl_2$ (acidic solution)

58. $MnO_4^- + AsO_3^{3-} \longrightarrow Mn^{2+} + AsO_4^{3-}$ (acidic solution)

59. $S^{2-} + Cl_2 \longrightarrow SO_4^{2-} + Cl^-$ (basic solution)

60. $Zn + NO_3^- \longrightarrow Zn(OH)_4^{2-} + NH_3$ (basic solution)

61. $KOH + Cl_2 \longrightarrow KCl + KClO + H_2O$ (basic solution)

62. $As + ClO_3^- \longrightarrow H_3AsO_3 + HClO$ (acidic solution)

63. $MnO_4^- + Cl^- \longrightarrow Mn^{2+} + Cl_2$ (acidic solution)

64. $H_2O_2 + Cl_2O_7 \longrightarrow ClO_2^- + O_2$ (basic solution)

Free Response Questions

1. You are investigating the properties of two new metallic elements found on Pluto, Bz and Yz. Bz reacts with aqueous HCl to produce hydrogen gas and $BzCl_3$. Yz has no reaction with aqueous HCl. However, the formula for the compound it forms with chlorine is $YzCl_2$. Write the balanced reaction that should occur if a galvanic cell is set up with Bz and Yz electrodes in solutions containing the metallic ions.

2. Suppose that 25 mL of an iron(II) nitrate solution is added to a beaker containing aluminum metal. Assume the reaction went to completion with no excess reagents. The solid iron produced was removed by filtration.
   (a) Write a balanced redox equation for the reaction.
   (b) For which solution, the initial iron(II) nitrate or the solution after the solid iron was filtered out, would the freezing point be lower? Explain your answer.

3. 50. mL of a 0.10 M HCl solution is poured equally into two flasks.
   (a) What is the pH of the HCl solution?
   (b) Next, 0.050 mol Zn is added to Flask A and 0.050 mol Cu is added to Flask B. Determine the pH of each solution after approximately 20 minutes.

4. (a) Write a balanced acid–base reaction that produces Na_2S.
   (b) If Na_2S is added to an aqueous solution of H_2S ($K_a = 9.1 \times 10^{-8}$), will the pH of the solution rise or fall? Explain.

5. (a) Would you expect a reaction to take place between HCN (aq) and $AgNO_3$ (aq) (K_{sp} for AgCN = 5.97×10^{-17})? Explain, and if a reaction occurs, write the net ionic equation.
   (b) If NaCN is added to distilled water, would you expect the solution to be acidic, basic, or neutral? Explain using any chemical equations that may be appropriate.

6. For each set of beakers below, draw a picture you might expect to see when the contents of the beakers are mixed together and allowed to react.

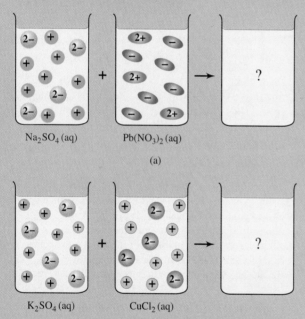

Na_2SO_4 (aq) $Pb(NO_3)_2$ (aq)

(a)

K_2SO_4 (aq) $CuCl_2$ (aq)

7. The picture below represents the equilibrium condition of
$$2A_3X \rightleftarrows 2A_2X + A_2$$

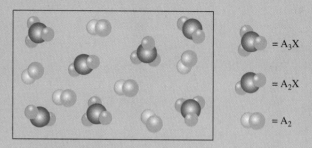

$= A_3X$
$= A_2X$
$= A_2$

   (a) What is the equilibrium constant?
   (b) Does the equilibrium lie to the left or to the right?
   (c) Do you think the reaction is a redox reaction? Explain your answer.

8. The picture below represents the equilibrium condition of the reaction $X_2 + 2G \rightleftarrows X_2G_2$.

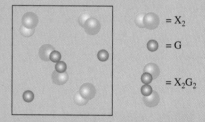

$= X_2$
$= G$
$= X_2G_2$

   (a) What is the equilibrium constant?
   (b) If the ratio of reactants to products increased when the temperature was raised, was the reaction exothermic or endothermic?
   (c) Provide a logical explanation for why the equilibrium shifts to the right when the pressure is increased at constant temperature.

9. The hydroxide ion concentration of a solution is 3.4×10^{-10} M. Balance the following equation:
$$Fe^{2+}(aq) + MnO_4^-(aq) \longrightarrow Fe^{3+}(aq) + Mn^{2+}(aq)$$

CHAPTER 18 / OUTLINE

The nucleus of the atom is a source of tremendous energy. Harnessing this energy has enabled us to fuel power stations, treat cancer, and preserve food. We use isotopes in medicine to diagnose illness and to detect minute quantities of drugs or hormones. Researchers use radioactive tracers to sequence the human genome. We also use nuclear processes to detect explosives in luggage and to establish the age of objects such as human artifacts and rocks. In this chapter we'll consider properties of nuclei and their applications in our lives.

18.1   Discovery of Radioactivity

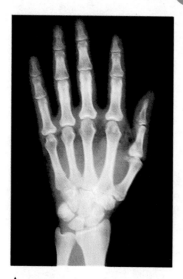

▲
X-ray technology opened the door for a new world of diagnosis and treatment.

In 1895, Wilhelm Konrad Roentgen (1845–1923) made an important breakthrough that eventually led to the discovery of radioactivity. Roentgen discovered X-rays when he observed that a vacuum discharge tube enclosed in a thin, black cardboard box caused a nearby piece of paper coated with the salt barium platinocyanide to glow with a brilliant phosphorescence. From this and other experiments he concluded that certain rays, which he called X-rays, were emitted from the discharge tube, penetrated the box, and caused the salt to glow. Roentgen's observations that X-rays could penetrate other bodies and affect photographic plates led to the development of X-ray photography.

Shortly after this discovery, Antoine Henri Becquerel (1852–1908) attempted to show a relationship between X-rays and the phosphorescence of uranium salts. In one experiment he wrapped a photographic plate in black paper, placed a sample of uranium salt on it, and exposed it to sunlight. The developed photographic plate showed that rays emitted from the salt had penetrated the paper. When Becquerel attempted to repeat the experiment, the sunlight was intermittent, so he placed the entire setup in a drawer. Several days later he developed the photographic plate, expecting to find it only slightly affected. To his amazement he found an intense image on the plate. He repeated the experiment in total darkness and obtained the same results, proving that the uranium salt emitted rays that affected the photographic plate without its being exposed to sunlight. Thus did the discovery of radioactivity come about, a combination of numerous experiments by the finest minds of the day—and serendipity. Becquerel later showed that the rays coming from uranium are able to ionize air and are also capable of penetrating thin sheets of metal.

radioactivity   The name *radioactivity* was coined two years later (in 1898) by Marie Curie. **Radioactivity** is the spontaneous emission of particles and/or rays from the nucleus of an atom. Elements having this property are said to be radioactive.

In 1898, Marie Sklodowska Curie (1867–1934) and her husband Pierre Curie (1859–1906) turned their research interests to radioactivity. In a short time the Curies discovered two new elements, polonium and radium, both of which are radioactive. To confirm their work on radium, they processed 1 ton of pitchblende residue ore to obtain 0.1 g of pure radium chloride, which they used to make further studies on the properties of radium and to determine its atomic mass.

In 1899, Ernest Rutherford began to investigate the nature of the rays emitted from uranium. He found two particles, which he called *alpha* and *beta particles*.

TABLE 18.1	Isotopic Notation for Several Particles (and Small Isotopes) Associated with Nuclear Chemistry		
Particle	Symbol	Atomic number Z	Mass number A
Neutron	$_0^1 n$	0	1
Proton	$_1^1 H$	1	1
Beta particle (electron)	$_{-1}^0 e$	−1	0
Positron (positive electron)	$_{+1}^0 e$	1	0
Alpha particle (helium nucleus)	$_2^4 He$	2	4
Deuteron (heavy hydrogen nucleus)	$_1^2 H$	1	2

Soon he realized that uranium, while emitting these particles, was changing into another element. By 1912, over 30 radioactive isotopes were known, and many more are known today. The *gamma ray,* a third type of emission from radioactive materials similar to an X-ray, was discovered by Paul Villard (1860–1934) in 1900. Rutherford's description of the nuclear atom led scientists to attribute the phenomenon of radioactivity to reactions taking place in the nuclei of atoms.

The symbolism and notation we described for isotopes in Chapter 5 is also very useful in nuclear chemistry:

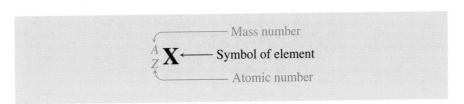

$$_Z^A X$$

— Mass number
— Symbol of element
— Atomic number

For example, $_{92}^{238} U$ represents a uranium isotope with an atomic number of 92 and a mass number of 238. This isotope is also designated as U-238 or uranium-238 and contains 92 protons and 146 neutrons. The protons and neutrons collectively are known as **nucleons.** The mass number is the total number of nucleons in the nucleus. Table 18.1 shows the isotopic notations for several particles associated with nuclear chemistry.

nucleon

When we speak of isotopes, we mean atoms of the same element with different masses, such as $_8^{16}O$, $_8^{17}O$, $_8^{18}O$. In nuclear chemistry we use the term **nuclide** to mean any isotope of any atom. Thus $_8^{16}O$ and $_{92}^{235}U$ are referred to as nuclides. Nuclides that spontaneously emit radiation are referred to as *radionuclides.*

nuclide

18.2 Natural Radioactivity

Radioactive elements continuously undergo **radioactive decay,** or disintegration, to form different elements. The chemical properties of an element are associated with its electronic structure, but radioactivity is a property of the nucleus. Therefore

radioactive decay

neither ordinary changes of temperature and pressure nor the chemical or physical state of an element has any effect on its radioactivity.

The principal emissions from the nuclei of radionuclides are known as alpha rays (or particles), beta rays (or particles), and gamma rays. Upon losing an alpha or beta particle, the radioactive element changes into a different element. We will explain this process in detail later.

half-life

Each radioactive nuclide disintegrates at a specific and constant rate, which is expressed in units of half-life. The **half-life** ($t_{1/2}$) is the time required for one-half of a specific amount of a radioactive nuclide to disintegrate. The half-lives of the elements range from a fraction of a second to billions of years. To illustrate, suppose we start with 1.0 g of $^{226}_{88}$Ra, ($t_{1/2}$ = 1620 years):

Element	$t_{1/2}$
$^{238}_{92}$U	4.5×10^9 years
$^{226}_{88}$Ra	1620 years
$^{15}_{6}$C	2.4 seconds

$$1.0 \text{ g } ^{226}_{88}\text{Ra} \xrightarrow[1620 \text{ years}]{t_{1/2}} 0.50 \text{ g } ^{226}_{88}\text{Ra} \xrightarrow[1620 \text{ years}]{t_{1/2}} 0.25 \text{ g } ^{226}_{88}\text{Ra}$$

The half-lives of the various radioisotopes of the same element differ dramatically. Half-lives for certain isotopes of radium, carbon, and uranium are listed in Table 18.2.

Example 18.1   The half-life of $^{131}_{53}$I is 8 days. How much $^{131}_{53}$I from a 32-g sample remains after five half-lives?

Solution   Using the following graph, we can find the number of grams of ^{131}I remaining after one half-life:

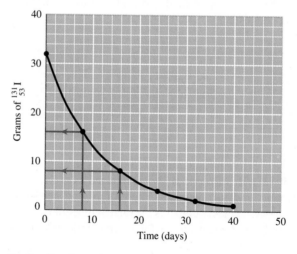

Trace a perpendicular line from 8 days on the x-axis to the line on the graph. Now trace a horizontal line from this point on the plotted line to the y-axis and read the corresponding grams of ^{131}I. Continue this process for each half-life, adding 8 days to the previous value on the x-axis:

Half-lives	0	1	2	3	4	5
Number of days		8	16	24	32	40
Amount remaining	32 g	16 g	8 g	4 g	2 g	1 g

Starting with 32 g, 1 g ^{131}I remains after five half-lives (40 days).

TABLE 18.2	Half-Lives for Radium, Carbon, and Uranium Isotopes		
Isotope	Half-life	Isotope	Half-life
Ra-223	11.7 days	C-14	5668 years
Ra-224	3.64 days	C-15	2.4 seconds
Ra-225	14.8 days	U-235	7.1×10^8 years
Ra-226	1620 years	U-238	4.5×10^9 years
Ra-228	6.7 years		

Example 18.2

In how many half-lives will 10.0 g of a radioactive nuclide decay to less than 10% of its original value?

Solution

We know that 10% of the original amount is 1.0 g. After the first half-life, half the original material remains and half has decayed (5.00 g). After the second half-life, one-fourth of the original material remains (i.e., one-half of 5.00 g). This progression continues, reducing the quantity remaining by half for each half-life that passes.

Half-lives	0	1	2	3	4
Percent remaining	100%	50%	25%	12.5%	6.25%
Amount remaining	10.0 g	5.00 g	2.50 g	1.25 g	0.625 g

Therefore the amount remaining will be less than 10% sometime between the third and the fourth half-lives.

Practice 18.1

The half-life of $^{14}_6C$ is 5668 years. How much $^{14}_6C$ will remain after six half-lives in a sample that initially contains 25.0 g?

Nuclides are said to be either *stable* (nonradioactive) or *unstable* (radioactive). Elements that have atomic numbers greater than 83 (bismuth) are naturally radioactive, although some of the nuclides have extremely long half-lives. Some of the naturally occurring nuclides of elements 81, 82, and 83 are radioactive, and some are stable. Only a few naturally occurring elements that have atomic numbers less than 81 are radioactive. However, no stable isotopes of element 43 (technetium) or of element 61 (promethium) are known.

Radioactivity is believed to be a result of an unstable ratio of neutrons to protons in the nucleus. Stable nuclides of elements up to about atomic number 20 generally have about a 1:1 neutron-to-proton ratio. In elements above number 20, the neutron-to-proton ratio in the stable nuclides gradually increases to about 1.5:1 in element number 83 (bismuth). When the neutron-to-proton ratio is too high or too low, alpha, beta, or other particles are emitted to achieve a more stable nucleus.

FIGURE 18.1 ▶
**The effect of an electromag-
netic field on alpha particles,
beta particles, and gamma
rays. Lighter beta particles
are deflected considerably
more than alpha particles.
Alpha and beta particles are
deflected in opposite directions. Gamma radiation is not
affected by the electromagnetic field.**

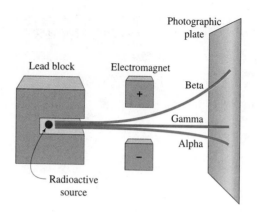

18.3   Alpha Particles, Beta Particles, and Gamma Rays

The classical experiment proving that alpha and beta particles are oppositely charged was performed by Marie Curie (see Figure 18.1). She placed a radioactive source in a hole in a lead block and positioned two poles of a strong electromagnet so that the radiations that were given off passed between them. The paths of three different kinds of radiation were detected by means of a photographic plate placed some distance beyond the electromagnet. The lighter beta particles were strongly deflected toward the positive pole of the electromagnet; the heavier alpha particles were less strongly deflected and in the opposite direction. The uncharged gamma rays were not affected by the electromagnet and struck the photographic plates after traveling along a path straight out of the lead block.

Alpha Particles

alpha particle

An **alpha particle** (α) consists of two protons and two neutrons, has a mass of about 4 amu, and a charge of $+2$. It is a helium nucleus which is usually given one of the following symbols: α or ^4_2He. When an alpha particle is emitted from the nucleus, a different element is formed. The atomic number of the new element is 2 less, and the mass is 4 amu less, than that of the starting element.

> **Loss of an alpha particle from the nucleus results in**
>   **loss of 4 in the mass number (A)**
>   **loss of 2 in the atomic number (Z)**

For example, when $^{238}_{92}\text{U}$ loses an alpha particle, $^{234}_{90}\text{Th}$ is formed, because two neutrons and two protons are lost from the uranium nucleus. This disintegration may be written as a nuclear equation:

$$^{238}_{92}\text{U} \longrightarrow {}^{234}_{90}\text{Th} + \alpha \qquad \text{or} \qquad ^{238}_{92}\text{U} \longrightarrow {}^{234}_{90}\text{Th} + {}^4_2\text{He}$$

For the loss of an alpha particle from $^{226}_{88}\text{Ra}$, the equation is

$$^{226}_{88}\text{Ra} \longrightarrow \ ^{222}_{86}\text{Rn} + \ ^{4}_{2}\text{He} \quad \text{or} \quad ^{226}_{88}\text{Ra} \longrightarrow \ ^{222}_{86}\text{Rn} + \alpha$$

A nuclear equation, like a chemical equation, consists of reactants and products and must be balanced. To have a balanced nuclear equation, the sum of the mass numbers (superscripts) on both sides of the equation must be equal, and the sum of the atomic numbers (subscripts) on both sides of the equation must be equal:

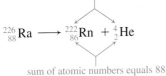

What new nuclide will be formed when $^{230}_{90}\text{Th}$ loses an alpha particle? The new nuclide will have a mass of 226 amu and will contain 88 protons, so its atomic number is 88. Locate the corresponding element on the periodic chart—in this case, $^{226}_{88}\text{Ra}$ or radium-226.

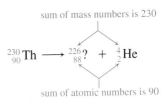

Beta Particles

The **beta particle** (β) is identical in mass and charge to an electron; its charge is -1. Both a beta particle and a proton are produced by the decomposition of a neutron:

beta particle

$$^{1}_{0}\text{n} \longrightarrow \ ^{1}_{1}\text{p} + \ ^{0}_{-1}\text{e}$$

The beta particle leaves, and the proton remains in the nucleus. When an atom loses a beta particle from its nucleus, a different element is formed that has essentially the same mass but an atomic number that is 1 greater than that of the starting element. The beta particle is written as β or $^{0}_{-1}\text{e}$.

> **Loss of a beta particle from the nucleus results in**
> **no change in the mass number (A)**
> **increase of 1 in the atomic number (Z)**

Examples of equations in which a beta particle is lost are

$$^{234}_{90}\text{Th} \longrightarrow \ ^{234}_{91}\text{Pa} + \beta$$

$$^{234}_{91}\text{Pa} \longrightarrow \ ^{234}_{92}\text{U} + \ ^{0}_{-1}\text{e}$$

$$^{210}_{82}\text{Pb} \longrightarrow \ ^{210}_{83}\text{Bi} + \beta$$

Gamma Rays

Gamma rays (γ) are photons of energy. A gamma ray is similar to an X-ray but is more energetic. They have no electrical charge and no measurable mass. Gamma rays are released from the nucleus in many radioactive changes along with either

gamma ray

alpha or beta particles. Gamma radiation does not result in a change of atomic number or the mass of an element.

> **Loss of a gamma ray from the nucleus results in**
> **no change in mass number (A) or atomic number (Z)**

Example 18.3   (a) Write an equation for the loss of an alpha particle from the nuclide $^{194}_{78}\text{Pt}$.
(b) What nuclide is formed when $^{228}_{88}\text{Ra}$ loses a beta particle from its nucleus?

Solution   (a) Loss of an alpha particle, ^4_2He, results in a decrease of 4 in the mass number and a decrease of 2 in the atomic number:

Mass of new nuclide: $A - 4$ or $194 - 4 = 190$

Atomic number of new nuclide: $Z - 2$ or $78 - 2 = 76$

Looking up element number 76 on the periodic table, we find it to be osmium, Os. The equation then is

$$^{194}_{78}\text{Pt} \longrightarrow \; ^{190}_{76}\text{Os} + \, ^4_2\text{He}$$

(b) The loss of a beta particle from a $^{228}_{88}\text{Ra}$ nucleus means a gain of 1 in the atomic number with no essential change in mass. The new nuclide will have an atomic number of $(Z + 1)$ or 89, which is actinium, Ac:

$$^{228}_{88}\text{Ra} \longrightarrow \; ^{228}_{89}\text{Ac} + \, ^{\;0}_{-1}\text{e}$$

Example 18.4   What nuclide will be formed when $^{214}_{82}\text{Pb}$ successively emits two beta particles, then one alpha particle from its nucleus? Write successive equations showing these changes.

Solution   The changes brought about in the three steps outlined are as follows:

     β loss: Increase of 1 in the atomic number; no change in mass

     β loss: Increase of 1 in the atomic number; no change in mass

     α loss: Decrease of 2 in the atomic number; decrease of 4 in the mass

The equations are

$$^{214}_{82}\text{Pb} \longrightarrow \; ^{214}_{83}\text{X} + \beta \longrightarrow \; ^{214}_{84}\text{X} + \beta \longrightarrow \; ^{210}_{82}\text{X} + \alpha$$

where X stands for the new nuclide formed. Looking up each of these elements by their atomic numbers, we rewrite the equations

$$^{214}_{82}\text{Pb} \xrightarrow{\;\;\beta\;\;} \; ^{214}_{83}\text{Bi} \xrightarrow{\;\;\beta\;\;} \; ^{214}_{84}\text{Po} \xrightarrow{\;\;\alpha\;\;} \; ^{210}_{82}\text{Pb}$$

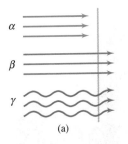
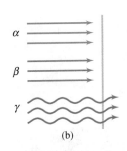
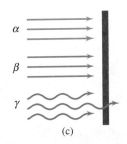

(a)   (b)   (c)

◀ FIGURE 18.2
Relative penetrating ability of alpha, beta, and gamma radiation. (a) Thin sheet of paper; (b) thin sheet of aluminum; (c) 5-cm lead block.

TABLE 18.3   *Characteristics of Nuclear Radiation*

Radiation	Symbol	Mass (amu)	Electrical charge	Velocity	Composition	Ionizing power
Alpha	$\alpha, {}_{2}^{4}\text{He}$	4	+2	Variable, less than 10% the speed of light	He nucleus	High
Beta	$\beta, {}_{-1}^{0}e$	$\dfrac{1}{1837}$	−1	Variable, up to 90% the speed of light	Identical to an electron	Moderate
Gamma	γ	0	0	Speed of light	Photons or electromagnetic waves of energy	Almost none

Practice 18.2

What nuclide will be formed when ${}_{86}^{222}\text{Rn}$ emits an alpha particle?

The ability of radioactive rays to pass through various objects is in proportion to the speed at which they leave the nucleus. Gamma rays travel at the velocity of light (186,000 miles per second) and are capable of penetrating several inches of lead. The velocities of beta particles are variable, the fastest being about nine-tenths the velocity of light. Alpha particles have velocities less than one-tenth the velocity of light. Figure 18.2 illustrates the relative penetrating power of these rays. A few sheets of paper will stop alpha particles; a thin sheet of aluminum will stop both alpha and beta particles; and a 5-cm block of lead will reduce, but not completely stop, gamma radiation. In fact, it is difficult to stop all gamma radiation. Table 18.3 summarizes the properties of alpha, beta, and gamma radiation.

18.4   Radioactive Disintegration Series

The naturally occurring radioactive elements with a higher atomic number than lead (Pb) fall into three orderly disintegration series. Each series proceeds from one

These three series begin with the elements uranium, thorium, and actinium.

▲
The age of geologic formations like these found in Petrified Forest National Park can be determined using disintegration series.

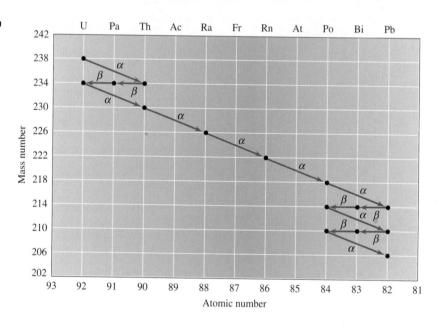

FIGURE 18.3 ▶
The uranium disintegration series. $^{238}_{92}U$ **decays by a series of alpha (α) and beta (β) emissions to the stable nuclide** $^{206}_{82}Pb$.

element to the next by the loss of either an alpha or a beta particle, finally ending in a nonradioactive nuclide. The uranium series starts with $^{238}_{92}U$ and ends with $^{206}_{82}Pb$. The thorium series starts with $^{232}_{90}Th$ and ends with $^{208}_{82}Pb$. The actinium series starts with $^{235}_{92}U$ and ends with $^{207}_{82}Pb$. A fourth series, the neptunium series, starts with the synthetic element $^{241}_{94}Pu$ and ends with the stable bismuth nuclide $^{209}_{83}Bi$. The uranium series is shown in Figure 18.3. Gamma radiation, which accompanies alpha and beta radiation, is not shown.

By using these a series and the half-lives of its members, scientists have been able to approximate the age of certain geologic deposits. This approximation is done by comparing the amount of $^{238}_{92}U$ with the amount of $^{206}_{82}Pb$ and other nuclides in the series that are present in a particular geologic formation. Rocks found in Canada and Finland have been calculated to be about 3.0×10^9 (3 billion) years old. Some meteorites have been determined to be 4.5×10^9 years old.

Practice 18.3

What nuclides are formed when $^{238}_{92}U$ undergoes the following decays?
(a) an alpha particle and a beta particle
(b) three alpha particles and two beta particles

18.5 Transmutation of Elements

Transmutation is the conversion of one element into another by either natural or artificial means. Transmutation occurs spontaneously in natural radioactive disintegrations. Alchemists tried for centuries to convert lead and mercury into gold by artificial means, but transmutation by artificial means was not achieved until 1919, when Ernest Rutherford succeeded in bombarding the nuclei of nitrogen atoms with alpha particles and produced oxygen nuclides and protons. The nuclear equation for this transmutation can be written as

transmutation

$$^{14}_{7}N + \alpha \longrightarrow ^{17}_{8}O + ^{1}_{1}H \quad \text{or} \quad ^{14}_{7}N + ^{4}_{2}He \longrightarrow ^{17}_{8}O + ^{1}_{1}H$$

It is believed that the alpha particle enters the nitrogen nucleus, forming $^{18}_{9}F$ as an intermediate, which then decomposes into the products.

Rutherford's experiments opened the door to nuclear transmutations of all kinds. Atoms were bombarded by alpha particles, neutrons, protons, deuterons ($^{2}_{1}H$), electrons, and so forth. Massive instruments were developed for accelerating these particles to very high speeds and energies to aid their penetration of the nucleus. The famous cyclotron was developed by E. O. Lawrence (1901–1958) at the University of California; later instruments include the Van de Graaf electrostatic generator, the betatron, and the electron and proton synchrotrons. With these instruments many nuclear transmutations became possible. Equations for a few of these are as follows:

$$^{7}_{3}Li + ^{1}_{1}H \longrightarrow 2\,^{4}_{2}He$$

$$^{40}_{18}Ar + ^{1}_{1}H \longrightarrow ^{40}_{19}K + ^{1}_{0}n$$

$$^{23}_{11}Na + ^{1}_{1}H \longrightarrow ^{23}_{12}Mg + ^{1}_{0}n$$

$$^{114}_{48}Cd + ^{2}_{1}H \longrightarrow ^{115}_{48}Cd + ^{1}_{1}H$$

$$^{2}_{1}H + ^{2}_{1}H \longrightarrow ^{3}_{1}H + ^{1}_{1}H$$

$$^{209}_{83}Bi + ^{2}_{1}H \longrightarrow ^{210}_{84}Po + ^{1}_{0}n$$

$$^{16}_{8}O + ^{1}_{0}n \longrightarrow ^{13}_{6}C + ^{4}_{2}He$$

$$^{238}_{92}U + ^{12}_{6}C \longrightarrow ^{244}_{98}Cf + 6\,^{1}_{0}n$$

18.6 Artificial Radioactivity

Irene Joliot-Curie (daughter of Pierre and Marie Curie) and her husband Frederic Joliot-Curie observed that when aluminum-27 is bombarded with alpha particles, neutrons and positrons (positive electrons) are emitted as part of the products. When

▲
Aerial view of Stanford University's Linear Accelerator.

CHEMISTRY IN ACTION How Old Is This Object?

When archaeologists discover a fossil or artifact, how do they determine its age? They couldn't count rings in fossils (as we do with trees), so an American chemist developed a technique called radiocarbon dating. W. F. Libby, who received the Nobel Prize in chemistry in 1960 for his work, based his method on the decay rate of C-14.

Radiocarbon dating works this way: Carbon dioxide in the atmosphere contains a fixed ratio of radioactive C-14 to ordinary C-12.

Plants that consume carbon dioxide during photosynthesis and animals that eat the plants contain the same proportion of C-14 to C-12 as long as they are alive. When an organism dies, the amount of C-12 remains fixed, but the C-14 content diminishes according to its half-life (5668 years). By comparing the ratio of C-14 to C-12 in an object to the same ratio in living plants, we can estimate the age of the object being evaluated. In 5668 years, one-half of the radiocarbon initially present will have undergone decomposition. In 11,336 years, one-fourth of the original C-14 will be left. The age of fossil material, archaeological specimens, and old wood is then determined by measuring the remaining C-14.

◀ Radiocarbon dating can be used to verify the age of artifacts from Aztec civilizations.

The age of specimens from ancient Egyptian tombs calculated by radiocarbon dating correlates closely with the chronological age established by Egyptologists. Charcoal samples obtained at Darrington Walls, a woodhenge in Great Britain, were determined to be about 4000 years old. Radiocarbon dating instruments currently in use enable researchers to date specimens back as far as 70,000 years.

Radioactive decay has also been used to date samples other than those containing carbon. For example, the age of rock formations containing uranium has been approximated by determining the ratio of U-238 to Pb-206. Lead-206 is the last isotope formed in the U-238 disintegration series. Thus a geologic deposit containing a 1:1 ratio of U-238 to Pb-206 corresponds to a time lapse of one half-life of U-238, which is 4.5×10^9 years, assuming that all the lead came from the decay of U-238. The age of Moon rocks returned to Earth by the *Apollo* missions were calculated by similar techniques.

---

the source of alpha particles is removed, neutrons cease to be produced, but positrons continue to be emitted. This observation suggested that the neutrons and positrons come from two separate reactions. It also indicated that a product of the first reaction is radioactive. After further investigation they discovered that, when aluminum-27 is bombarded with alpha particles, phosphorus-30 and neutrons are produced. Phosphorus-30 is radioactive, has a half-life of 2.5 minutes, and decays to silicon-30 with the emission of a positron. The equations for these reactions are

$$^{27}_{13}\text{Al} + {}^{4}_{2}\text{He} \longrightarrow {}^{30}_{15}\text{P} + {}^{1}_{0}\text{n}$$

$$^{30}_{15}\text{P} \longrightarrow {}^{30}_{14}\text{Si} + {}^{0}_{+1}\text{e}$$

artificial radioactivity
induced radioactivity

The radioactivity of nuclides produced in this manner is known as **artificial radioactivity** or **induced radioactivity.** Artificial radionuclides behave like natural radioactive elements in two ways: They disintegrate in a definite fashion and they have a specific half-life. The Joliot-Curies received the Nobel Prize in chemistry in 1935 for the discovery of artificial, or induced, radioactivity.

18.7   Measurement of Radioactivity

Radiation from radioactive sources is so energetic that it is called *ionizing radiation*. When it strikes an atom or a molecule, one or more electrons are knocked off, and an ion is created. The Geiger counter, an instrument commonly used to detect and measure radioactivity, depends on this fact. The instrument consists of a Geiger–Müller detecting tube and a counting device. The detector tube is a pair of oppositely charged electrodes in an argon gas-filled chamber fitted with a thin window. When radiation, such as a beta particle, passes through the window into the tube, some argon is ionized, and a momentary pulse of current (discharge) flows between the electrodes. These current pulses are electronically amplified in the counter and appear as signals in the form of audible clicks, flashing lights, meter de-fections, or numerical readouts (Figure 18.4).

▲ **FIGURE 18.4**
Geiger–Müller survey meter.

The amount of radiation that an individual encounters can be measured by a film badge. This badge contains a piece of photographic film in a light-proof holder and is worn in areas where radiation might be encountered. The silver grains in the film will darken when exposed to radiation. The badges are processed after a predeter-mined time interval to determine the amount of radiation the wearer has been exposed to.

A scintillation counter is used to measure radioactivity for biomedical applica-tions. A scintillator is composed of molecules that emit light when exposed to ioniz-ing radiation. A light-sensitive detector counts the flashes and converts them to a nu-merical readout.

The *curie* is the unit used to express the amount of radioactivity produced by an element. One **curie** (Ci) is defined as the quantity of radioactive material giving 3.7×10^{10} disintegrations per second. The basis for this figure is pure radium, which has an activity of 1 Ci/g. Because the curie is such a large quantity, the millicurie and microcurie, representing one-thousandth and one-millionth of a curie, respectively, are more practical and more commonly used.

curie (Ci)

The curie only measures radioactivity emitted by a radionuclide. Different units are required to measure exposure to radiation. The **roentgen (R)** quantifies exposure to gamma or X-rays; 1 roentgen is defined as the amount of radiation required to produce 2.1×10^9 ions/cm^3 of dry air. The **rad (radiation *absorbed dose*)** is defined as the amount of radiation that provides 0.01 J of energy per kilogram of matter. The amount of radiation absorbed will change depending on the type of matter. The roentgen and the rad are numerically similar; 1 roentgen of gamma radiation pro-vides 0.92 rad in bone tissue.

roentgen (R)

rad (radiation absorbed dose)

Neither the rad nor the roentgen indicates the biological damage caused by radi-ation. One rad of alpha particles has the ability to cause ten times more damage than 1 rad of gamma rays or beta particles. Another unit, **rem (*roentgen* equivalent to *man*)** takes into account the degree of biological effect caused by the type of radia-tion exposure; 1 rem is equal to the dose in rads multiplied by a factor specific to the form of radiation. The factor is 10 for alpha particles and 1 for both beta particles and gamma rays. Units of radiation are summarized in Table 18.4.

rem (roentgen equivalent to man)

TABLE 18.4	Radiation Units	
Unit	Measure	Equivalent
curie (Ci)	rate of decay of a radioactive substance	1 Ci = 3.7×10^{10} disintegrations/sec
roentgen (R)	exposure based on the quantity of ionization produced in air	1 R = 2.1×10^9 ions/cm^3
rad	absorbed dose of radiation	1 rad = 0.01 J/kg matter
rem	radiation dose equivalent	1 rem = 1 rad $\times$ factor
gray (Gy) (SI unit)	energy absorbed by tissue	1 Gy = 1 J/kg tissue (1 Gy = 100 rad)

18.8   Nuclear Fission

nuclear fission

In **nuclear fission** a heavy nuclide splits into two or more intermediate-sized fragments when struck in a particular way by a neutron. The fragments are called *fission products*. As the atom splits, it releases energy and two or three neutrons, each of which can cause another nuclear fission. The first instance of nuclear fission was reported in January 1939 by the German scientists Otto Hahn (1879–1968) and Fritz Strassmann (1902–1980). Detecting isotopes of barium, krypton, cerium, and lanthanum after bombarding uranium with neutrons led scientists to believe that the uranium nucleus had been split.

Characteristics of nuclear fission are as follows:

1. Upon absorption of a neutron, a heavy nuclide splits into two or more smaller nuclides (fission products).
2. The mass of the nuclides formed ranges from about 70–160 amu.
3. Two or more neutrons are produced from the fission of each atom.
4. Large quantities of energy are produced as a result of the conversion of a small amount of mass into energy.
5. Most nuclides produced are radioactive and continue to decay until they reach a stable nucleus.

One process by which this fission takes place is illustrated in Figure 18.5. When a heavy nucleus captures a neutron, the energy increase may be sufficient to cause deformation of the nucleus until the mass finally splits into two fragments, releasing energy and usually two or more neutrons.

In a typical fission reaction, a $^{235}_{92}$U nucleus captures a neutron and forms unstable $^{236}_{92}$U. This $^{236}_{92}$U nucleus undergoes fission, quickly disintegrating into two fragments, such as $^{139}_{56}$Ba and $^{94}_{36}$Kr, and three neutrons. The three neutrons in turn may be captured by three other $^{235}_{92}$U atoms, each of which undergoes fission, producing nine neutrons, and so on. A reaction of this kind, in which the products cause the reaction to continue or magnify, is known as a **chain reaction.** For a chain reaction to continue, enough fissionable material must be present so that each atomic fission causes, on average, at least one additional fission. The minimum quantity of an element needed to support a self-sustaining chain reaction is called

chain reaction

CHEMISTRY IN ACTION
Does Your Food Glow in the Dark?

Over the past 25 years the Food and Drug Administration (FDA) has approved irradiation to delay ripening or kill microbes and insects in wheat, potatoes, fresh fruits, and poultry as well as spices. But the only regular use of irradiation on food in the United States has been on spices.

Now the FDA has approved irradiation of red meat and new legislation allows the labels about irradiated food to be in much smaller type. The general public has not yet accepted the use of radiation to reduce microbes and help preserve food. Why? Many people believe that irradiating food makes it radioactive. This is not true; in fact, the most that irradiation of food does is to produce compounds similar to those created by cooking and also to reduce the vitamin content of some food.

How does irradiation of food work? The radioactive source currently used is cobalt-60. It is contained in a concrete cell (see diagram) with 6-foot-thick walls. Inside the cell is a pool of water

with racks of thin Co-60 rods suspended above. When the rods are not being used they are submerged in the water, which absorbs the gamma radiation. When food to be irradiated moves into the cells the Co-60 rods are lifted from the water, and the boxes of food move among them on a conveyor being irradiated from all sides.

Scientists who have investigated the process say that food irradiation is safe. The Centers for Disease Control and Pre-

vention in Atlanta estimates that food-borne illness causes as many as 9,000 deaths per year. Irradiation provides one way to reduce microbial contamination in food. It does not solve the problem of careless handling of food by processors, and long-term studies on humans have not yet been concluded regarding irradiated food supplies. Once again we face the issue of balancing benefit and risk.

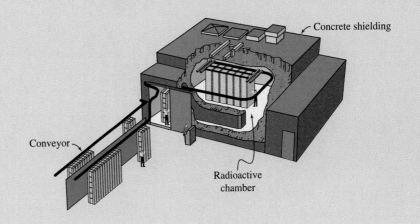

the **critical mass.** Since energy is released in each atomic fission, chain reactions provide a steady supply of energy. (A chain reaction is illustrated in Figure 18.6.) Two of the many possible ways in which $^{235}_{92}U$ may fission are shown by:

critical mass

$$^{235}_{92}U + ^{1}_{0}n \longrightarrow ^{139}_{56}Ba + ^{94}_{36}Kr + 3\,^{1}_{0}n$$

$$^{235}_{92}U + ^{1}_{0}n \longrightarrow ^{144}_{54}Xe + ^{90}_{38}Sr + 2\,^{1}_{0}n$$

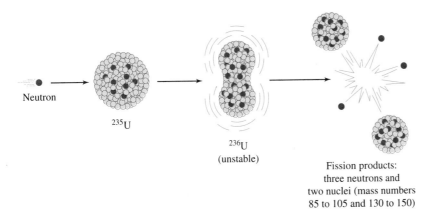

◄ **FIGURE 18.5**
The fission process. When a neutron is captured by a heavy nucleus, the nucleus becomes more unstable. The more energetic nucleus begins to deform, resulting in fission. Two nuclear fragments and three neutrons are produced by this fission process.

479

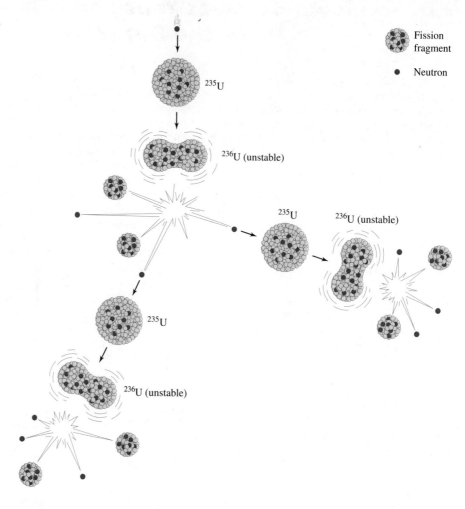

18.9   Nuclear Power

Nearly all electricity for commercial use is produced by machines consisting of a turbine linked by a drive shaft to an electrical generator. The energy required to run the turbine can be supplied by falling water, as in hydroelectric power plants, or by steam as in thermal power plants.

The world's demands for energy, largely from fossil fuels, is heavy. At the present rates of consumption, the estimated world supply of fossil fuels is sufficient for only a few centuries. Although the United States has large coal and oil shale deposits, it currently imports over 40% of its oil supply. We clearly need to develop alternative energy sources. At present, uranium is the most productive alternative energy source, and about 17% of the electrical energy used in the United States is generated from power plants using uranium fuel.

A nuclear power plant is a thermal power plant in which heat is produced by a nuclear reactor instead of by combustion of fossil fuel. The major components of a nuclear reactor are

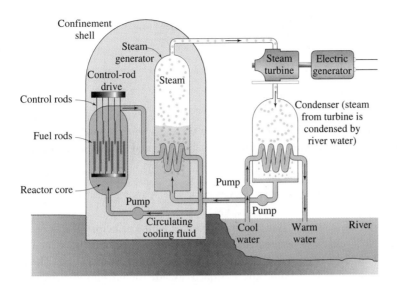

1. an arrangement of nuclear fuel, called the reactor core
2. a control system, which regulates the rate of fission and thereby the rate of heat generation
3. a cooling system, which removes the heat from the reactor and also keeps the core at the proper temperature

One type of reactor uses metal slugs containing uranium enriched from the normal 0.7% U-235 to about 3% U-235. The self-sustaining fission reaction is moderated, or controlled, by adjustable control rods containing substances that slow down and capture some of the neutrons produced. Ordinary water, heavy water, and molten sodium are typical coolants used. Energy obtained from nuclear reactions in the form of heat is used in the production of steam to drive turbines for generating electricity. (See Figure 18.7.)

The potential dangers of nuclear power were tragically demonstrated by the accidents at Three Mile Island, Pennsylvania (1979), and Chernobyl in the former U.S.S.R. (1986). Both accidents resulted from the loss of coolant to the reactor core. The reactors at Three Mile Island were covered by concrete containment buildings and therefore released a relatively small amount of radioactive material into the atmosphere. But because the Soviet Union did not require containment structures on nuclear power plants, the Chernobyl accident resulted in 31 deaths and the resettlement of 135,000 people. The release of large quantities of I-131, Cs-134, and Cs-137 appears to be causing long-term health problems in that exposed population.

Another major disadvantage of nuclear power is its highly radioactive waste products, some of which have half-lives of thousands of years. As yet, no technology has been developed that disposes of these dangerous wastes in complete safety.

In the United States, reactors designed for commercial power production use uranium oxide, U_3O_8, that is enriched with the relatively scarce fissionable U-235 isotope. Because the supply of U-235 is limited, a new type of reactor known as the *breeder reactor* has been developed. Breeder reactors produce additional fissionable material at the same time that the fission reaction is occurring. In a breeder reactor, excess neutrons convert nonfissionable isotopes, such as U-238 or Th-232, to fissionable isotopes, Pu-239 or U-233 shown on the next page.

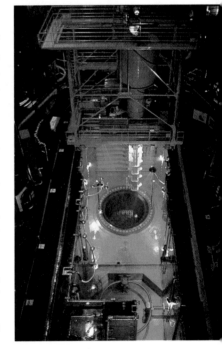

▲
Interior view of a reactor hall at a nuclear power plant in France. The core of the reactor is situated in the water pool (blue area).

$$^{238}_{92}\text{U} + ^{1}_{0}\text{n} \longrightarrow ^{239}_{92}\text{U} \xrightarrow{\beta} ^{239}_{93}\text{Np} \xrightarrow{\beta} ^{239}_{94}\text{Pu}$$

$$^{232}_{90}\text{Th} + ^{1}_{0}\text{n} \longrightarrow ^{233}_{90}\text{Th} \xrightarrow{\beta} ^{233}_{91}\text{Pa} \xrightarrow{\beta} ^{233}_{92}\text{U}$$

These transmutations make it possible to greatly extend the supply of fuel for nuclear reactors. No breeder reactors are presently in commercial operation in the United States, but a number of them are being operated in Europe and Great Britain.

18.10   The Atomic Bomb

The atomic bomb is a fission bomb; it operates on the principle of a very fast chain reaction that releases a tremendous amount of energy. An atomic bomb and a nuclear reactor both depend on self-sustaining nuclear fission chain reactions. The essential difference is that in a bomb the fission is "wild," or uncontrolled, whereas in a nuclear reactor the fission is moderated and carefully controlled. A minimum critical mass of fissionable material is needed for a bomb, or a major explosion will not occur. When a quantity smaller than the critical mass is used, too many neutrons formed in the fission step escape without combining with another nucleus, and a chain reaction does not occur. Therefore the fissionable material of an atomic bomb must be stored as two or more subcritical masses and brought together to form the critical mass at the desired time of explosion. The temperature developed in an atomic bomb is believed to be about 10 million degrees Celsius.

The nuclides used in atomic bombs are U-235 and Pu-239. Uranium deposits contain about 0.7% of the U-235 isotope, the remainder being U-238. Uranium-238 does not undergo fission except with very high energy neutrons. It was discovered, however, that U-238 captures a low energy neutron without undergoing fission and that the product, U-239, changes to Pu-239 (plutonium) by a beta-decay process. Plutonium-239 readily undergoes fission upon capture of a neutron and is therefore useful for nuclear weapons. The equations for the nuclear transformations are

$$^{238}_{92}\text{U} + ^{1}_{0}\text{n} \longrightarrow ^{239}_{92}\text{U} \xrightarrow{\beta} ^{239}_{93}\text{Np} \xrightarrow{\beta} ^{239}_{94}\text{Pu}$$

The hazards of an atomic bomb explosion include not only shock waves from the explosive pressure and tremendous heat, but also intense radiation in the form of alpha particles, beta particles, gamma rays, and ultraviolet rays. Gamma rays and X-rays can penetrate deeply into the body, causing burns, sterilization, and gene mutation, which can adversely affect future generations. Both radioactive fission products and unfissioned material are present after the explosion. If the bomb explodes near the ground, many tons of dust are lifted into the air. Radioactive material adhering to this dust, known as *fallout,* is spread by air currents over wide areas of the land and constitutes a lingering source of radiation hazard.

Today nuclear war is probably the most awesome threat facing civilization. Only two rather primitive fission-type atom bombs were used to destroy the Japanese cities of Hiroshima and Nagasaki and bring World War II to an early end. The threat of nuclear war is increased by the fact that the number of nations possessing nuclear weapons is steadily increasing.

◀ **The mushroom cloud is a signature of uncontrolled fission in an atomic bomb.**

18.11   Nuclear Fusion

The process of uniting the nuclei of two light elements to form one heavier nucleus is known as **nuclear fusion.** Such reactions can be used for producing energy, because the masses of the two nuclei that fuse into a single nucleus are greater than the mass of the nucleus formed by their fusion. The mass differential is liberated in the form of energy. Fusion reactions are responsible for the tremendous energy output of the sun. Thus aside from relatively small amounts from nuclear fission and radioactivity, fusion reactions are the ultimate source of our energy, even the energy from fossil fuels. They are also responsible for the devastating power of the thermonuclear, or hydrogen, bomb.

nuclear fusion

Fusion reactions require temperatures on the order of tens of millions of degrees for initiation. Such temperatures are present in the Sun but have been produced only momentarily on Earth. For example, the hydrogen, or fusion, bomb is triggered by the temperature of an exploding fission bomb. Two typical fusion reactions are

$$\underset{\text{tritium}}{{}_{1}^{3}\text{H}} + \underset{\text{deuterium}}{{}_{1}^{2}\text{H}} \longrightarrow {}_{2}^{4}\text{He} + {}_{0}^{1}\text{n} + \text{energy}$$

$$\underset{\substack{3.0150 \\ \text{amu}}}{{}_{1}^{3}\text{H}} + \underset{\substack{1.0079 \\ \text{amu}}}{{}_{1}^{1}\text{H}} \longrightarrow \underset{\substack{4.0026 \\ \text{amu}}}{{}_{2}^{4}\text{He}} + \text{energy}$$

The total mass of the reactants in the second equation is 4.0229 amu, which is 0.0203 amu greater than the mass of the product. This difference in mass is manifested in the great amount of energy liberated.

During the past 50–55 years, a great deal of research in the United States and in other countries, especially the former Soviet Union, has focused on controlled nuclear fusion reactions. The goal of controlled nuclear fusion has not yet been attained, although the required ignition temperature has been reached in several devices. Evidence to date leads us to believe that we can develop a practical fusion power reactor. Fusion power, if we can develop it, will be far superior to fission power for the following reasons:

1. Virtually infinite amounts of energy are possible from fusion. Uranium supplies for fission power are limited, but heavy hydrogen, or deuterium (the most likely fusion fuel), is abundant. It is estimated that the deuterium in a cubic mile of seawater used as fusion fuel can provide more energy than the petroleum reserves of the entire world.

▲
Solar flares such as these are indications of fusion reactions occurring at temperatures of millions of degrees.

2. From an environmental viewpoint, fusion power is much "cleaner" than fission power because fusion reactions (in contrast to uranium and plutonium fission reactions) do not produce large amounts of long-lived and dangerously radioactive isotopes.

18.12 Mass–Energy Relationship in Nuclear Reactions

Large amounts of energy are released in nuclear reactions; thus significant amounts of mass are converted to energy. We stated earlier that the amount of mass converted to energy in chemical changes is insignificant compared to the amount of energy released in a nuclear reaction. In fission reactions about 0.1% of the mass is converted into energy. In fusion reactions as much as 0.5% of the mass may be changed into energy. The Einstein equation, $E = mc^2$, can be used to calculate the energy liberated, or available, when the mass loss is known. For example, in the reaction

$$\underset{7.016 \text{ g}}{^{7}_{3}\text{Li}} + \underset{1.008 \text{ g}}{^{1}_{1}\text{H}} \longrightarrow \underset{4.003 \text{ g}}{^{4}_{2}\text{He}} + \underset{4.003 \text{ g}}{^{4}_{2}\text{He}} + \text{energy}$$

the mass difference between the reactants and products (8.024 g − 8.006 g) is 0.018 g. The energy equivalent to this amount of mass is 1.62×10^{12} J. By comparison, this is more than 4 million times greater than the 3.9×10^5 J of energy obtained from the complete combustion of 12.01 g (1 mol) of carbon.

The mass of a nucleus is actually less than the sum of the masses of the protons and neutrons that make up that nucleus. The difference between the mass of the protons and the neutrons in a nucleus and the mass of the nucleus is known as the **mass defect.** The energy equivalent to this difference in mass is known as the **nuclear binding energy.** This energy is the amount that would be required to break a nucleus into its individual protons and neutrons. The higher the binding energy, the more stable the nucleus. Elements of intermediate atomic masses have high binding energies. For example, iron (element number 26) has a very high binding energy and therefore a very stable nucleus. Just as electrons attain less energetic and more stable arrangements through ordinary chemical reactions, neutrons and protons attain less energetic and more stable arrangements through nuclear fission or fusion reactions. Thus when uranium undergoes fission, the products have less mass (and greater binding energy) than the original uranium. In like manner, when hydrogen and lithium fuse to form helium, the helium has less mass (and greater binding energy) than the hydrogen and lithium. It is this conversion of mass to energy that accounts for the very large amounts of energy associated with both fission and fusion reactions.

mass defect
nuclear binding energy

18.13 Transuranium Elements

The elements that follow uranium on the periodic table and that have atomic numbers greater than 92 are known as the **transuranium elements.** They are synthetic radioactive elements; none of them occur naturally.

The first transuranium element, number 93, was discovered in 1939 by Edwin M. McMillan (1907–1991) at the University of California while he was investigating

transuranium element

the fission of uranium. He named it neptunium for the planet Neptune. In 1941, element 94, plutonium, was identified as a beta-decay product of neptunium:

$$^{238}_{93}\text{Np} \longrightarrow {}^{238}_{94}\text{Pu} + {}^{0}_{-1}\text{e}$$

$$^{239}_{93}\text{Np} \longrightarrow {}^{239}_{94}\text{Pu} + {}^{0}_{-1}\text{e}$$

Plutonium is one of the most important fissionable elements known today.

Since 1964, the discoveries of eight new transuranium elements, numbers 104–112, have been announced. These elements have been produced in minute quantities by high energy particle accelerators.

18.14 Biological Effects of Radiation

Radiation with energy to dislocate bonding electrons and create ions when passing through matter is classified as **ionizing radiation.** Alpha particles, beta particles, gamma rays, and X-rays fall into this classification. Ionizing radiation can damage or kill living cells and can be particularly devastating when it strikes the cell nuclei and affects molecules involved in cell reproduction. The effects of radiation on living organisms fall into these general categories: (1) acute or short-term effects, (2) long-term effects, and (3) genetic effects.

ionizing radiation

Acute Radiation Damage

High levels of radiation, especially from gamma rays or X-rays, produce nausea, vomiting, and diarrhea. The effect has been likened to a sunburn throughout the body. If the dosage is high enough, death will occur in a few days. The damaging effects of radiation appear to be centered in the nuclei of the cells, and cells that are undergoing rapid cell division are most susceptible to damage. It is for this reason that cancers are often treated with gamma radiation from a Co-60 source. Cancerous cells multiply rapidly and are destroyed by a level of radiation that does not seriously damage normal cells.

Long-Term Radiation Damage

Protracted exposure to low levels of any form of ionizing radiation can weaken an organism and lead to the onset of malignant tumors, even after fairly long time delays. The largest exposure to synthetic sources of radiation is from X-rays. Evidence suggests that the lives of early workers in radioactivity and X-ray technology may have been shortened by long-term radiation damage.

Strontium-90 isotopes are present in the fallout from atmospheric testing of nuclear weapons. Strontium is in the same periodic-table group as calcium, and its chemical behavior is similar to that of calcium. Hence when foods contaminated with Sr-90 are eaten, Sr-90 ions are laid down in the bone tissue along with ordinary calcium ions. Strontium-90 is a beta emitter with a half-life of 28 years. Blood cells manufactured in bone marrow are affected by the radiation from Sr-90. Hence there is concern that Sr-90 accumulation in the environment may cause an increase in the incidence of leukemia and bone cancers.

A Window into Living Organisms

Imagine viewing a living process as it is occurring. This was the dream of many scientists in the past as they tried to extract this knowledge from dead tissue. Today, because of innovations in nuclear chemistry, this dream is a common, everyday occurrence.

Compounds containing a radionuclide are described as being *labeled* or *tagged*. These compounds undergo their normal chemical reactions, but their location can be detected because of their radioactivity. When such compounds are given to a plant or an animal, the movement of the nuclide can be traced through the organism by the use of a Geiger counter or other detecting device.

In an early use of the tracer technique, the pathway by which CO_2 becomes fixed into carbohydrate ($C_6H_{12}O_6$) during photosynthesis was determined. The net equation for photosynthesis is

$$6\ CO_2 + 6\ H_2O \longrightarrow C_6H_{12}O_6 + 6\ O_2$$

Radioactive $^{14}CO_2$ was injected into a colony of green algae, and the algae was then placed in the dark and killed at selected time intervals. When the radioactive compounds were separated by paper chromatography and analyzed, the results elucidated a series of light-independent photosynthetic reactions.

Biological research using tracer techniques have determined

1. the rate of phosphate uptake by plants, using radiophosphorus
2. the flow of nutrients in the digestive tract using radioactive barium compounds
3. the accumulation of iodine in the thyroid gland, using radioactive iodine
4. the absorption of iron by the hemoglobin of the blood, using radioactive iron

▲ A radioactive tracer is injected into this patient and absorbed by the brain. The PET scanner detects photons emitted by the tracer and produces an image used in medical diagnosis and research.

In chemistry, uses for tracers are unlimited. The study of reaction mechanisms, the measurement of the rates of chemical reactions, and the determination of physical constants are just a few of the areas of application.

Radioactive tracers are commonly used in medical diagnosis. The radionuclide must be effective at a low concentration and have a short half-life to reduce the possibility of damage to the patient.

Radioactive iodine (I-131) is used to determine thyroid function, where the body concentrates iodine. In this process a small amount of radioactive potassium or sodium iodide is ingested. A detector is focused on the thyroid gland and measures the amount of iodine in the gland. This picture is then compared to that of a normal thyroid to detect any differences.

Doctors examine the heart's pumping performance and check for evidence of obstruction in coronary arteries by *nuclear scanning*. The radionuclide Tl-201, when injected into the bloodstream, lodges in healthy heart muscle. Thallium-201 emits gamma radiation, which is detected by a special imaging device called a *scintillation camera*. The data obtained are simultaneously translated into pictures by a computer. With this technique doctors can observe whether heart tissue has died after a heart attack and whether blood is flowing freely through the coronary passages.

One of the most recent applications of nuclear chemistry is the use of positron emission tomography (PET) in the measurement of dynamic processes in the body, such as oxygen use or blood flow. In this application a compound is made that contains a positron-emitting nuclide such as C-11, O-15, or N-13. The compound is injected into the body, and the patient is placed in an instrument that detects the positron emission. A computer produces a three-dimensional image of the area.

PET scans have been used to locate the areas of the brain involved with epileptic seizures. Glucose tagged with C-11 is injected, and an image of the brain is produced. Since the brain uses glucose almost exclusively for energy, diseased areas that use glucose at a rate different than normal tissue can then be identified.

Genetic Effects

The information needed to create an individual of a particular species, be it a bacterial cell or a human being, is contained within the nucleus of a cell. This genetic information is encoded in the structure of DNA (deoxyribonucleic acid) molecules, which make up genes. The DNA molecules form precise duplicates of themselves when cells divide, thus passing genetic information from one generation to the next. Radiation can damage DNA molecules. If the damage is not severe enough to prevent the individual from reproducing, a mutation may result. Most mutation-induced traits are undesirable. Unfortunately, if the bearer of the altered genes survives to reproduce, these traits are passed along to succeeding generations. In other words the genetic effects of increased radiation exposure are found in future generations, not in the present generation.

Because radioactive rays are hazardous to health and living tissue, special precautions must be taken in designing laboratories and nuclear reactors, in disposing of waste materials, and in monitoring the radiation exposure of people working in this field.

Concepts in Review

1. Outline the historical development of nuclear chemistry, including the major contributions of Henri Becquerel, Marie Curie, Ernest Rutherford, Irene Joliet-Curie, Otto Hahn, Fritz Strassmann, and Edwin McMillen.

2. Write balanced nuclear chemical equations using isotopic notation.

3. Determine the amount of radionuclide remaining after a given period of time when the starting amount and half-life are given.

4. List the characteristics that distinguish alpha particles, beta particles, and gamma rays from the standpoint of mass, charge, relative velocities, and penetrating power.

5. Describe the effect of a magnetic field on alpha particles, beta particles, and gamma rays.

6. Describe a radioactive disintegration series, and predict which isotope would be formed by the loss of specified numbers of alpha and beta particles from a given radionuclide.

7. Discuss the transmutation of elements.

8. Indicate the methods used for the detection of radiation.

9. Distinguish between radioactive disintegration and nuclear fission reactions.

10. Explain how the fission of U-235 can lead to a chain reaction and why a critical mass is necessary.

11. Explain how the energy from nuclear fission is converted to electrical energy.

12. Explain the difference between fission reactions in a nuclear reactor and those of an atomic bomb.

13. Explain what is meant by the term *nuclear fusion* and why a massive effort to develop controlled nuclear fusion is in progress.

14. Indicate the significance of mass defect and nuclear binding energy.

15. Indicate the major effects of radiation on living organisms.

16. Explain how the age of objects can be determined using radioactivity.

17. Indicate several current uses for radioactive tracers.

Key Terms

alpha particle (18.3)
artificial radioactivity (18.6)
beta particle (18.3)
chain reaction (18.8)
critical mass (18.8)
curie (Ci) (18.7)
gamma ray (18.3)
half-life (18.2)

induced radioactivity (18.6)
ionizing radiation (18.14)
mass defect (18.12)
nuclear binding energy (18.12)
nuclear fission (18.8)
nuclear fusion (18.11)
nucleon (18.1)
nuclide (18.1)

rad (radiation absorbed dose) (18.7)
radioactive decay (18.2)
radioactivity (18.1)
rem (roentgen equivalent to man) (18.7)
roentgen (R) (18.7)
transmutation (18.5)
transuranium element (18.13)

Questions

Questions refer to tables, figures, and key words and concepts defined in the chapter. A particularly challenging question or exercise is indicated with an asterisk.

1. To afford protection from radiation injury, which kind of radiation requires (a) the most shielding? (b) the least shielding?

2. Why is an alpha particle deflected less than a beta particle in passing through an electromagnetic field?

3. Name three pairs of nuclides that might be obtained by fissioning U-235 atoms.

4. Identify these people and their associations with the early history of radioactivity:
   (a) Antoine Henri Becquerel
   (b) Marie and Pierre Curie
   (c) Wilhelm Roentgen
   (d) Ernest Rutherford
   (e) Otto Hahn and Fritz Strassmann

5. Why is the radioactivity of an element unaffected by the usual factors that affect the rate of chemical reactions, such as ordinary changes of temperature and concentration?

6. Distinguish between the terms *isotope* and *nuclide*.

7. The half-life of Pu-244 is 76 million years. If Earth's age is about 5 billion years, discuss the feasibility of finding this nuclide as a naturally occurring nuclide.

8. Tell how alpha, beta, and gamma radiation are distinguished from the standpoint of
   (a) charge
   (b) relative mass
   (c) nature of particle or ray
   (d) relative penetrating power

9. Distinguish between natural and artificial radioactivity.

10. What is a radioactive disintegration series?

11. Briefly discuss the transmutation of elements.

12. Stable Pb-208 is formed from Th-232 in the thorium disintegration series by successive, α, β, β, α, α, α, α, β, β, α particle emissions. Write the symbol (including mass and atomic number) for each nuclide formed in this series.

13. The nuclide Np-237 loses a total of seven alpha particles and four beta particles. What nuclide remains after these losses?

14. Bismuth-211 decays by alpha emission to give a nuclide that in turn decays by beta emission to yield a stable nuclide. Show these two steps with nuclear equations.

15. What was Otto Hahn and Fritz Strassmann's contribution to nuclear physics?

16. What is a breeder reactor? Explain how it accomplishes the "breeding."

17. What is the essential difference between the nuclear reactions in a nuclear reactor and those in an atomic bomb?

18. Why must a certain minimum amount of fissionable material be present before a self-supporting chain reaction can occur?

19. What is mass defect and nuclear binding energy?

20. Explain why radioactive rays are classified as ionizing radiation.

21. Give a brief description of the biological hazards associated with radioactivity.

22. Strontium-90 has been found to occur in radioactive fallout. Why is there so much concern about this radionuclide being found in cow's milk? (Half-life of Sr-90 is 28 years.)

23. What is a radioactive tracer? How is it used?

24. Describe the radiocarbon method for dating archaeological artifacts.

25. How might radioactivity be used to locate a leak in an underground pipe?

26. Anthropologists have found bones whose age suggests that the human line may have emerged in Africa as much as 4 million years ago. If wood or charcoal were found with such bones, would C-14 dating be useful in dating the bones? Explain.

Paired Exercises

27. Indicate the number of protons, neutrons, and nucleons in these nuclei:
    (a) $^{35}_{17}Cl$ (b) $^{226}_{88}Ra$

28. Indicate the number of protons, neutrons, and nucleons in these nuclei:
    (a) $^{235}_{92}U$ (b) $^{82}_{35}Br$

29. How are the mass and the atomic number of a nucleus affected by the loss of an alpha particle?

31. Write nuclear equations for the alpha decay of
    (a) $^{218}_{85}\text{At}$ (b) $^{221}_{87}\text{Fr}$

33. Write nuclear equations for the beta decay of
    (a) $^{14}_{6}\text{C}$ (b) $^{137}_{55}\text{Cs}$

35. Write nuclear equations for the conversion of $^{13}_{6}\text{C}$ to $^{14}_{6}\text{C}$.

37. Complete and balance these nuclear equations by supplying the missing particles:
    (a) $^{27}_{13}\text{Al} + ^{4}_{2}\text{He} \longrightarrow ^{30}_{15}\text{P} + \underline{\quad}$
    (b) $^{27}_{14}\text{Si} \longrightarrow ^{0}_{+1}\text{e} + \underline{\quad}$
    (c) $\underline{\quad} + ^{2}_{1}\text{H} \longrightarrow ^{13}_{7}\text{N} + ^{1}_{0}\text{n}$
    (d) $\underline{\quad} \longrightarrow ^{82}_{36}\text{Kr} + ^{0}_{-1}\text{e}$

39. Strontium-90 has a half-life of 28 years. If a 1.00-mg sample were stored for 112 years, what mass of Sr-90 would remain?

*41. Consider the fission reaction

$$^{235}_{92}\text{U} + ^{1}_{0}\text{n} \longrightarrow ^{94}_{38}\text{Sr} + ^{139}_{54}\text{Xe} + 3\,^{1}_{0}\text{n} + \text{energy}$$

Calculate the following using this mass data (1.0 g is equivalent to 9.0×10^{13} J):

U-235 = 235.0439 amu   Sr-94 = 93.9154 amu
Xe-139 = 138.9179 amu   n = 1.0087 amu

    (a) the energy released in joules for a single event (one uranium atom splitting)
    (b) the energy released in joules per mole of uranium splitting
    (c) the percentage of mass lost in the reaction

30. How are the mass and the atomic number of a nucleus affected by the loss of a beta particle?

32. Write nuclear equations for the alpha decay of
    (a) $^{192}_{78}\text{Pt}$ (b) $^{210}_{84}\text{Po}$

34. Write nuclear equations for the beta decay of
    (a) $^{239}_{93}\text{Np}$ (b) $^{90}_{38}\text{Sr}$

36. Write nuclear equations for the conversion of $^{30}_{15}\text{P}$ to $^{30}_{14}\text{Si}$.

38. Complete and balance these nuclear equations by supplying the missing particles:
    (a) $^{66}_{29}\text{Cu} \longrightarrow ^{66}_{30}\text{Zn} + \underline{\quad}$
    (b) $^{0}_{-1}\text{e} + \underline{\quad} \longrightarrow ^{7}_{3}\text{Li}$
    (c) $^{27}_{13}\text{Al} + ^{4}_{2}\text{He} \longrightarrow ^{30}_{14}\text{Si} + \underline{\quad}$
    (d) $^{85}_{37}\text{Rb} + \underline{\quad} \longrightarrow ^{82}_{35}\text{Br} + ^{4}_{2}\text{He}$

40. Strontium-90 has a half-life of 28 years. If a sample was tested in 1980 and found to be emitting 240 counts/min, in what year would the same sample be found to be emitting 30 counts/min? How much of the original Sr-90 would be left?

42. Consider the fusion reaction

$$^{1}_{1}\text{H} + ^{2}_{1}\text{H} \longrightarrow ^{3}_{2}\text{He} + \text{energy}$$

Calculate the following using this mass data (1.0 g is equivalent to 9.0×10^{13} J):

$^{1}_{1}\text{H}$ = 1.00794 amu
$^{2}_{1}\text{H}$ = 2.01410 amu
$^{3}_{2}\text{He}$ = 3.01603 amu

    (a) the energy released in joules per mole of He-3 formed
    (b) the percentage of mass lost in the reaction

Additional Exercises

43. If radium costs $50,000 a gram, how much will 0.0100 g of $^{226}\text{RaCl}_2$ cost if the price is based only on the radium content?

44. An archaeological specimen was analyzed and found to be emitting only 25% as much C-14 radiation per gram of carbon as newly cut wood. How old is this specimen?

45. Barium-141 is a beta emitter. What is the half-life if a 16.0-g sample of the nuclide decays to 0.500 g in 90 minutes?

*46. Calculate (a) the mass defect and (b) the binding energy of $^{7}_{3}\text{Li}$ using the mass data:
    $^{7}_{3}\text{Li}$ = 7.0160 g   n = 1.0087 g
    p = 1.0073 g e^{-} = 0.00055 g
    1.0 g $\equiv 9.0 \times 10^{13}$ J (from $E = mc^2$)

*47. In the disintegration series $^{235}_{92}\text{U} \longrightarrow ^{207}_{82}\text{Pb}$, how many alpha and beta particles are emitted?

48. List three devices used for radiation detection and explain their operation.

49. The half-life of I-123 is 13 hours. If 10 mg of I-123 is administered to a patient, how much I-123 remains after 3 days and 6 hours?

50. Clearly distinguish between fission and fusion. Give an example of each.

51. Starting with 1 g of a radioactive isotope whose half-life is 10 days, sketch a graph showing the pattern of decay for that material. On the x-axis, plot time (you may want to simply show multiples of the half-life), and on the y-axis, plot mass of material remaining. Then after completing the graph, explain why a sample never really gets to the point where *all* of its radioactivity is considered to be gone.

52. Identify each missing product (name the element and give its atomic number and mass number) by balancing the following nuclear equations:
   (a) $^{235}U + {}^{1}_{0}n \longrightarrow {}^{143}Xe + 3 {}^{1}_{0}n +$ _____
   (b) $^{235}U + {}^{1}_{0}n \longrightarrow {}^{102}Y + 3 {}^{1}_{0}n +$ _____
   (c) $^{14}N + {}^{1}_{0}n \longrightarrow {}^{1}H +$ _____

53. Consider these reactions:
   (a) $H_2O(l) \longrightarrow H_2O(g)$
   (b) $2 H_2(g) + O_2(g) \longrightarrow 2 H_2O(g)$
   (c) ${}^{2}_{1}H + {}^{2}_{1}H \longrightarrow {}^{3}_{1}H + {}^{1}_{1}H$
   The following energy values belong to one of these equations:

   energy₁   115.6 kcal released
   energy₂   10.5 kcal absorbed
   energy₃ 7.5×10^7 kcal released

   Match the equation to the energy value and briefly explain your choices.

54. When ${}^{235}_{92}U$ is struck by a neutron, the unstable isotope ${}^{236}_{92}U$ results. When that daughter isotope undergoes fission, there are numerous possible products. If strontium-90 and three neutrons are the results of one such fission, what is the other product?

55. Write balanced nuclear equations for
   (a) beta emission by ${}^{29}_{12}Mg$
   (b) alpha emission by ${}^{150}_{60}Nd$
   (c) positron emission by ${}^{72}_{33}As$

56. Rubidium-87, a beta emitter, is the product of positron emission. Identify
   (a) the product of rubidium-87 decay
   (b) the precursor of rubidium-87

57. Potassium-42 is used to locate brain tumors. Its half-life is 12.5 hours. Starting with 15.4 mg, what fraction will remain after 100 hours? If it was necessary to have at least 1 μg for a particular procedure, could you hold the original sample for 200 hours before using it?

58. How much of a sample of cesium-137 ($t_{1/2}$ = 30 years) must have been present originally if, after 270 years, 15.0 g remains?

59. Suppose that the existence of element 114 were confirmed and reported. What element would this new substance fall beneath in the periodic table? Would it be a metal? What typical ion might you expect it to form in solution?

60. Cobalt-60 has a half-life of 5.26 years. If 1.00 g of ^{60}Co was allowed to decay, how many grams would be left after
   (a) one half-life?
   (b) two half-lives?
   (c) four half-lives?
   (d) ten half-lives?

61. Write balanced equations to show these changes:
   (a) alpha emission by boron-11
   (b) beta emission by strontium-88
   (c) neutron absorption by silver-107
   (d) proton emission by potassium-41
   (e) electron absorption by antimony-116

62. The ${}^{14}_{6}C$ content of an ancient piece of wood was found to be one-sixteenth of that in living trees. How many years old is this piece of wood, if the half-life of carbon-14 is 5668 years?

63. The curie is equal to 3.7×10^{10} disintegrations/sec, and the becquerel is equivalent to just 1 disintegration/sec. Suppose a hospital has a 150-g radioactive source with an activity of 1.24 Ci. What is its activity in becquerels?

Answers to Practice Exercises

18.1 0.391 g
18.2 ${}^{218}_{84}Po$

18.3 (a) ${}^{234}_{91}Pa$
      (b) ${}^{226}_{88}Ra$

CHAPTER 19

Introduction to Organic Chemistry

The nylon used in windsurfing sails is formed from a polymer containing carbon.

Numerous substances throughout nature incorporate silicon or carbon in their structures. Silicon is the staple of the geologist—it combines with oxygen in various ways to produce silica and a family of compounds known as the silicates. These compounds form the chemical foundation of most sand, rocks, and soil. In the living world, carbon combines with hydrogen, oxygen, nitrogen, and sulfur to form millions of compounds.

The petroleum industry and the myriad of polymer products we find indispensable are two of the many industries that depend on carbon chemistry. Synthetic fibers (clothing and carpeting) and plastics (containers, compact discs, computer terminals, and pens) are made from carbon compounds. Cold remedies, cleaning products, nutrients for space travel, convenience foods, and countless drugs, both legal and illegal, have all come about from our understanding of the chemistry of carbon.

19.1   The Beginnings of Organic Chemistry

During the late 18th and the early 19th centuries, the fact that compounds obtained from animal and vegetable sources defied the established rules for inorganic compounds baffled chemists. They knew that compound formation in inorganic compounds is due to a simple attraction between positively and negatively charged elements, and usually only one, or at most a few, compounds could be made from a given group of two or three elements. But one group of four elements—carbon, hydrogen, oxygen, and nitrogen—gave rise to a large number of remarkably stable compounds.

vital-force theory

No organic compounds had ever been synthesized from inorganic substances and chemists had no other explanation for the complexities of organic compounds, so they believed these compounds were formed by a "vital force." The **vital-force theory** held that organic substances could originate only from living material. In 1828, the results of a simple experiment by German chemist Friedrich Wöhler (1800–1882) proved to be the end of this theory. As he attempted to prepare ammonium cyanate, NH_4CNO, by heating cyanic acid, $HCNO$, and ammonia, NH_3, Wöhler obtained a white crystalline substance he identified as urea, $H_2N-CO-NH_2$. Wöhler knew urea to be an authentic organic substance because it had been isolated from urine. The implications of Wöhler's results were not immediately recognized, but the fact that one organic compound had been isolated from inorganic compounds changed the face of chemistry forever.

organic chemistry

With Wöhler's work, it was apparent that no vital force other than skill and knowledge was needed to make organic chemicals in the laboratory. Today the branch of chemistry that deals with carbon compounds, **organic chemistry,** does not imply that these compounds must originate from living matter. A few special kinds of carbon compounds (e.g., carbon oxides, metal carbides, and metal carbonates) are excluded from the organic classification because their chemistry is more closely related to that of inorganic substances.

The field of organic chemistry is vast; it includes not only all living organisms but also a great many other materials that we use daily. Foodstuffs (fats, proteins, carbohydrates); fuels; fabrics; wood and paper products; paints and varnishes; plastics; dyes; soaps and detergents; cosmetics; medicinals; and rubber products—all are organic materials.

The sources of organic compounds are carbon-containing raw materials—petroleum and natural gas, coal, carbohydrates, fats, and oils. In the United States we produce about 250 billion pounds of organic chemicals from these sources, which amounts to more than 1100 pounds per year for every man, woman, and child. About 90% of this 250 billion pounds comes from petroleum and natural gas. Because world reserves of petroleum and natural gas are finite, we will sometime have to rely on other sources to make the vast amount of organic substances that we depend on. Fortunately we know how to synthesize many organic compounds from sources other than petroleum, although at a much greater expense.

The colors in fabric dyes come from organic materials.

19.2 The Carbon Atom

The carbon atom is central to all organic compounds. The atomic number of carbon is 6, and its electron structure is $1s^2 2s^2 2p^2$. Two stable isotopes of carbon exist, C-12 and C-13. In addition, carbon has several radioactive isotopes, C-14 being the most widely known of these because of its use in radiocarbon dating.

A carbon atom usually forms four covalent bonds. The most common geometric arrangement of these bonds is tetrahedral (see Figure 19.1). In this structure the four covalent bonds are not planar about the carbon atom but are directed toward the corners of a regular tetrahedron. (A tetrahedron is a solid figure with four sides.) The angle between these tetrahedral bonds is 109.5°.

With four valence electrons, the carbon atom ($\cdot\overset{\cdot}{C}\cdot$) forms four single covalent bonds by sharing electrons with other atoms. The structures of methane and carbon tetrachloride illustrate this point:

methane   carbon tetrachloride

Actually, these compounds have a tetrahedral shape with bond angles of 109.5° (see Figure 19.2), but the bonds are often drawn at right angles. In methane each bond is formed by the sharing of electrons between a carbon and a hydrogen atom.

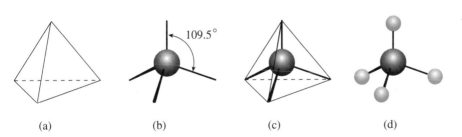

(a)   (b)   (c)   (d)

◄ FIGURE 19.1
Tetrahedral structure of carbon. (a) A regular tetrahedron; (b) a carbon atom with tetrahedral bonds; (c) a carbon atom within a regular tetrahedron; (d) a methane molecule, CH_4.

CH₄

Cl
Cl   Cl
Cl

CCl₄

▲
FIGURE 19.2
Space-filling models of CH₄ and CCl₄.

Carbon–carbon bonds are formed because carbon atoms can share electrons with other carbon atoms. One, two, or three pairs of electrons can be shared between two carbon atoms, forming a single, double, or triple bond, respectively:

$$\cdot\overset{\cdot}{C}:\overset{\cdot}{C}\cdot \qquad \cdot\overset{\cdot}{C}::\overset{\cdot}{C}\cdot \qquad \cdot C:::C\cdot$$

$$\cdot\overset{\cdot}{C}{-}\overset{\cdot}{C}\cdot \qquad \cdot\overset{\cdot}{C}{=}\overset{\cdot}{C}\cdot \qquad \cdot C{\equiv}C\cdot$$

single bond   double bond   triple bond

Each dash represents a covalent bond. Carbon, more than any other element, has the ability to form chains of covalently bonded atoms. This bonding ability is the main reason for the large number of organic compounds. Three examples are shown here. It's easy to see how, through this bonding ability, long chains of carbon atoms form by linking one carbon atom to another through covalent bonds:

$$\cdot\overset{\cdot}{C}:\overset{\cdot}{C}:\overset{\cdot}{C}\cdot \qquad \cdot\overset{\cdot}{C}{-}\overset{\cdot}{C}{-}\overset{\cdot}{C}\cdot \qquad$$

three carbon atoms bonded by single bonds

seven-carbon chain

ten carbon atoms bonded together

Carbon forms so many different compounds that a system for grouping the molecules is necessary. Organic molecules are classified according to structural features. The members of each class of compounds contain a characteristic atom or group of atoms called a **functional group**. Molecules with the same functional group share similarities in structure that result in similar chemical properties. Thus we need study only a few members of a particular class of compounds to be able to predict the behavior of other molecules in that class. In this brief introduction to organic chemistry we will consider two categories of compounds: hydrocarbons and hydrocarbon derivatives.

functional group

19.3   Hydrocarbons

hydrocarbon

saturated hydrocarbon
unsaturated hydrocarbon

Hydrocarbons are compounds composed entirely of carbon and hydrogen atoms bonded to each other by covalent bonds. These molecules are further classified as saturated or unsaturated. **Saturated hydrocarbons** have only single bonds between carbon atoms. These hydrocarbons are classified as *alkanes*. **Unsaturated hydrocarbons** contain a double or triple bond between two carbon atoms and include *alkenes, alkynes,* and *aromatic* compounds. These classifications are summarized in Figure 19.3.

Fossil fuels—natural gas, petroleum, and coal—are the principal sources of hydrocarbons. Natural gas is primarily methane with small amounts of ethane, propane, and butane. Petroleum is a mixture of hydrocarbons from which gasoline, kerosene, fuel oil, lubricating oil, paraffin wax, and petrolatum (themselves mixtures of hydrocarbons) are separated. Coal tar, a volatile by-product of the steel industry's

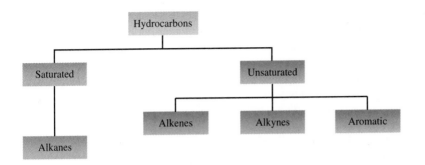

◀ FIGURE 19.3
General classification of hydrocarbons.

process of making coke from coal, is the source of many valuable chemicals, including the aromatic hydrocarbons benzene, toluene, and naphthalene.

19.4 Alkanes

The **alkanes,** also known as *paraffins* or *saturated hydrocarbons,* are straight- or branched-chain hydrocarbons with only single covalent bonds between the carbon atoms. We will study the alkanes in some detail because many other classes of organic compounds are derivatives of these substances. Be sure to learn the names of the first ten members of the alkane series, because they are the basis for naming other classes of compounds.

Methane, CH_4, is the first member of the alkane series. Members having two, three, and four carbon atoms are ethane, propane, and butane, respectively. The first four alkanes have common names and must be memorized, but the next six names are derived from Greek numbers. The names and formulas of the first ten alkanes are given in Table 19.1.

Successive compounds in the alkane series differ from each other in composition by one carbon and two hydrogen atoms. When each member of a series differs

alkane

Learn the names of the alkanes in Table 19.1; they form the stem of many other names.

TABLE 19.1 **Names, Formulas, and Physical Properties of Straight-Chain Alkanes**

Name	Molecular formula C_nH_{2n+2}	Condensed structural formula	Boiling point (°C)	Melting point (°C)
Methane	CH_4	CH_4	−161	−183
Ethane	C_2H_6	CH_3CH_3	−88	−172
Propane	C_3H_8	$CH_3CH_2CH_3$	−45	−187
Butane	C_4H_{10}	$CH_3CH_2CH_2CH_3$	−0.5	−138
Pentane	C_5H_{12}	$CH_3CH_2CH_2CH_2CH_3$	36	−130
Hexane	C_6H_{14}	$CH_3CH_2CH_2CH_2CH_2CH_3$	69	−95
Heptane	C_7H_{16}	$CH_3CH_2CH_2CH_2CH_2CH_2CH_3$	98	−90
Octane	C_8H_{18}	$CH_3CH_2CH_2CH_2CH_2CH_2CH_2CH_3$	125	−57
Nonane	C_9H_{20}	$CH_3CH_2CH_2CH_2CH_2CH_2CH_2CH_2CH_3$	151	−54
Decane	$C_{10}H_{22}$	$CH_3CH_2CH_2CH_2CH_2CH_2CH_2CH_2CH_2CH_3$	174	−30

Oil contains alkanes as well
as other hydrocarbons.

▲
The combustion of alkanes
forms the basis for the gas
and petroleum industry.

from the next member by a CH_2 group, the series is called a **homologous series.** The members of a homologous series are similar in structure but differ in formula. All common classes of organic compounds exist in homologous series, which can be represented by a general formula. For open-chain alkanes the general formula is C_nH_{2n+2}, where n corresponds to the number of carbon atoms in the molecule. The formulas of specific alkanes are easily determined from this general formula. Thus for pentane, $n = 5$ and $2n + 2 = 12$, so its formula is C_5H_{12}. For hexadecane, a 16-carbon alkane, the formula is $C_{16}H_{34}$.

One single reaction of alkanes has inspired people to explore equatorial jungles, to endure the heat and sandstorms of the deserts of Africa and the Middle East, to mush across the frozen Arctic, and to drill holes in Earth more than 30,000 feet deep! The substance is oil and the reaction is combustion with oxygen to produce heat energy. Combustion reactions overshadow all other reactions of alkanes in economic importance. For example, note the heat generated when methane reacts with oxygen:

$$CH_4(g) + 2\ O_2(g) \longrightarrow CO_2(g) + 2\ H_2O(g) + 802.5\ kJ\ (191.8\ kcal)$$

Thermal energy is converted to mechanical and electrical energy all over the world. But combustion reactions are not usually of great interest to organic chemists, because carbon dioxide and water are the only chemical products of complete combustion. Aside from their combustibility, alkanes are limited in reactivity.

19.5   Structural Formulas and Isomerism

The properties of an organic substance are dependent on its molecular structure. By structure we mean the way in which the atoms bond within the molecule. The majority of organic compounds are made from relatively few elements—carbon, hydrogen, oxygen, nitrogen, and the halogens. In these compounds carbon has four bonds to each atom, nitrogen three bonds, oxygen two bonds, and hydrogen and the halogens one bond to each atom:

$$-\overset{|}{\underset{|}{C}}- \qquad H- \qquad -O- \qquad -\overset{|}{\underset{|}{N}}- \qquad Cl- \qquad Br- \qquad I- \qquad F-$$

Alkane molecules contain only carbon–carbon and carbon–hydrogen bonds. Each carbon atom is joined to four other atoms by four single covalent bonds. These bonds are separated by angles of 109.5° (corresponding to those angles formed by lines drawn from the center of a regular tetrahedron to its corners). Alkane molecules are essentially nonpolar. Because of this low polarity, these molecules have very little intermolecular attraction and therefore relatively low boiling points compared with other organic compounds of similar molar mass.

The three-dimensional character of atoms and molecules is difficult to portray without models or computer-generated drawings. Methane and ethane are shown here in Lewis structure and line structure form:

Lewis diagrams can be drawn
for alkanes, but these
molecules usually are
represented by replacing
electron pairs with single
lines.

$$H:\overset{H}{\underset{H}{C}}:H \qquad H-\overset{\overset{H}{|}}{\underset{\underset{H}{|}}{C}}-H$$

methane

$$H:\overset{\overset{H}{\cdot\cdot}}{\underset{\underset{H}{\cdot\cdot}}{C}}:\overset{\overset{H}{\cdot\cdot}}{\underset{\underset{H}{\cdot\cdot}}{C}}:H \qquad H-\overset{\overset{H\ \ H}{|\ \ \ |}}{\underset{\underset{H\ \ H}{|\ \ \ |}}{C-C}}-H$$

ethane

To write the correct structural formula for propane, C_3H_8, the next member of the alkane series, we need to place each atom in the molecule. An alkane contains only single bonds, so each carbon atom must be bonded to four other atoms by either C—C or C—H bonds. Hydrogen must be bonded to only one carbon atom by a C—H bond, since C—H—C bonds do not occur, and an H—H bond would simply represent a hydrogen molecule. Thus the only possible structure for propane is

$$
\begin{array}{ccccccc}
 & H & & H & & H & \\
 & | & & | & & | & \\
H & - & C & - & C & - & C & - & H \\
 & | & & | & & | & \\
 & H & & H & & H &
\end{array}
$$

propane

However, it's possible to write two structural formulas corresponding to the molecular formula C_4H_{10} (butane). Two C_4H_{10} compounds with these structural formulas actually exist:

$$
\begin{array}{ccccccccc}
H & H & H & H \\
| & | & | & | \\
H-C-C-C-C-H \\
| & | & | & | \\
H & H & H & H
\end{array}
\quad \text{and} \quad
\begin{array}{c}
H \\
H \diagdown \; | \; \diagup H \\
C \\
H \quad | \quad H \\
| \quad | \quad | \\
H-C-C-C-H \\
| \quad | \quad | \\
H \quad H \quad H
\end{array}
$$

normal butane   2-methylpropane

The butane with the unbranched carbon chain is called *normal butane* (abbreviated *n*-butane); it boils at 0.5°C and melts at −138.3°C. The branched-chain butane is called 2-methylpropane; it boils at −11.7°C and melts at −159.5°C. These differences in physical properties are sufficient to establish that the two compounds, though they have the same molecular formula, are different substances. The structural arrangements of the atoms in methane, ethane, propane, butane, and 2-methylpropane are shown in Figure 19.4.

This phenomenon of two or more compounds having the same molecular formula but different structural arrangements is called **isomerism.** The individual compounds are called **isomers.** Isomerism is common among organic compounds and is another reason for the large number of known compounds. There are 3 isomers of pentane, 5 isomers of hexane, 9 isomers of heptane, 18 isomers of octane, 35 isomers of nonane, and 75 isomers of decane. The phenomenon of isomerism is a compelling reason for using structural formulas.

isomerism

isomers

> **Isomers are compounds with the same molecular formula but different structural formulas.**

To save time and space, condensed structural formulas, in which the atoms and groups attached to a carbon atom are written to the right of that carbon atom, are often used. For example, the condensed structural formula for pentane is $CH_3CH_2CH_2CH_2CH_3$ or $CH_3(CH_2)_3CH_3$. Some condensed structural formulas are shown in Figure 19.4.

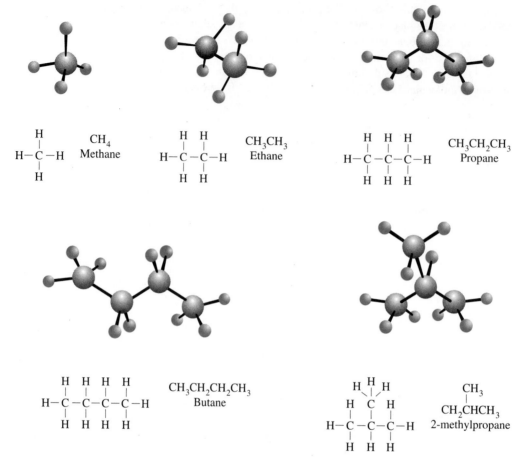

▲
FIGURE 19.4
Ball-and-stick models illustrating structural formulas of methane, ethane, propane, butane, and 2-methylpropane. Condensed structural formulas are shown above the names.

Let's interpret the condensed structural formula for propane:

$$\overset{1}{CH_3}\overset{2}{CH_2}\overset{3}{CH_3}$$

Carbon-1 has three hydrogen atoms attached to it and is bonded to C-2, which has two hydrogen atoms on it and which is bonded to C-3. Carbon-3 has three hydrogen atoms bonded to it.

Example 19.1   Pentane, C_5H_{12}, has three isomers. Write their structural formulas and their condensed structural formulas.

Solution   In a problem of this kind, it's best to first write the carbon skeleton with the longest continuous carbon chain—in this case, five carbon atoms. We complete the structure by attaching hydrogen atoms around each carbon atom so that each carbon atom has four bonds. The carbon atoms at the ends of the chain need three hydrogen atoms. The three inner carbon atoms each need two hydrogen atoms to give them four bonds:

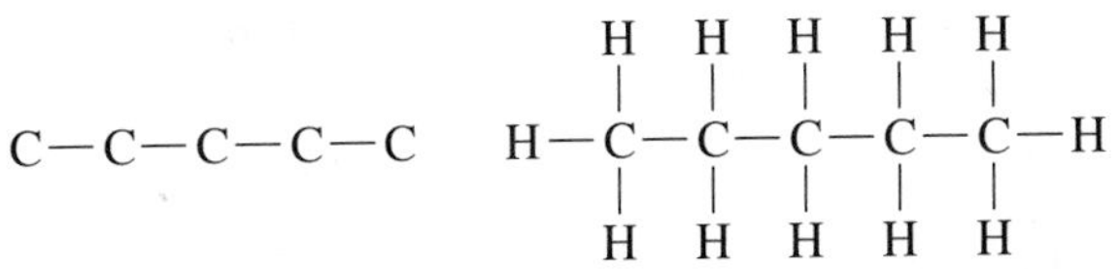

C—C—C—C—C

$$H-\overset{\displaystyle H}{\underset{\displaystyle H}{\overset{\displaystyle |}{\underset{\displaystyle |}{C}}}}-\overset{\displaystyle H}{\underset{\displaystyle H}{\overset{\displaystyle |}{\underset{\displaystyle |}{C}}}}-\overset{\displaystyle H}{\underset{\displaystyle H}{\overset{\displaystyle |}{\underset{\displaystyle |}{C}}}}-\overset{\displaystyle H}{\underset{\displaystyle H}{\overset{\displaystyle |}{\underset{\displaystyle |}{C}}}}-\overset{\displaystyle H}{\underset{\displaystyle H}{\overset{\displaystyle |}{\underset{\displaystyle |}{C}}}}-H$$

$CH_3CH_2CH_2CH_2CH_3$

For the next isomer, we write a four-carbon chain and attach the fifth carbon atom to either of the middle carbon atoms (don't use the end ones):

$$\overset{\displaystyle C}{\underset{\displaystyle |}{}}$$
C—C—C—C

$$\overset{\displaystyle C}{\underset{\displaystyle |}{}}$$
C—C—C—C

These structures represent the same compound.

Now add the 12 hydrogen atoms to complete the structure:

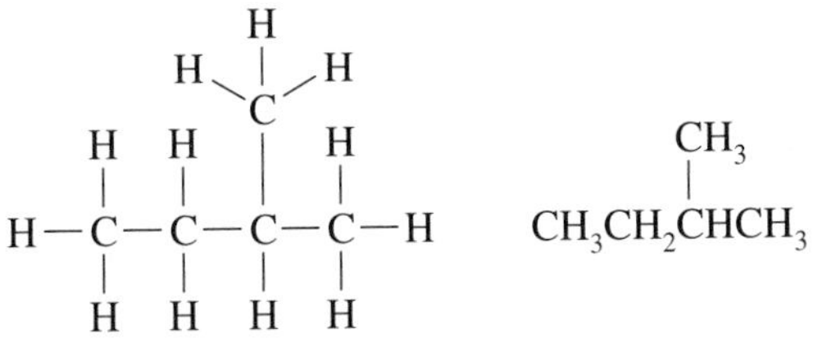

$$H-C-C-C-C-H$$

$CH_3CH_2CHCH_3$ or $CH_3CH_2CH(CH_3)_2$

with CH_3 branch

For the third isomer, write a three-carbon chain, attach the remaining two carbon atoms to the central carbon atom, and complete the structure by adding the 12 hydrogen atoms:

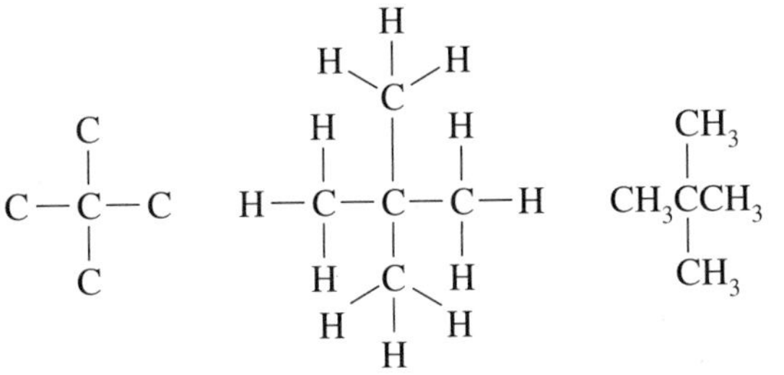

$$\overset{\displaystyle C}{\underset{\displaystyle C}{C-C-C}}$$

$$H-C-C-C-H$$

CH_3CCH_3 or $C(CH_3)_4$
with CH_3 above and CH_3 below

19.6 Naming Alkanes

In the early years of organic chemistry, each new compound was given a name, usually by the person who had isolated or synthesized it. Names were not systematic but did carry some information—often about the origin of the substance. Wood alcohol (methanol), for example, was so named because it was obtained by destructive distillation or pyrolysis of wood.

It soon became apparent that a naming system was needed and, in 1892, such a system was proposed and adopted. In its present form the International Union of Pure and Applied Chemistry (IUPAC) system is generally unambiguous and internationally accepted. However, a great many well-established common names and abbreviations (such as TNT and DDT) have continued to be used because of their brevity and/or convenience. So you need a knowledge of both the IUPAC system and the common names.

To name organic compounds using the IUPAC system, you must recognize certain common alkyl groups. **Alkyl groups** have the general formula C_nH_{2n+1} (one less hydrogen atom than the corresponding alkane). The name of the group is formed from the name of the corresponding alkane by simply dropping -*ane* and substituting a -*yl* ending. The names and formulas of selected alkyl groups are given in Table 19.2. The letter "R" is often used in formulas to represent any of the possible alkyl groups:

alkyl group

$$R = C_nH_{2n+1} \text{ (any alkyl group)}$$

The following IUPAC rules are all that are needed to name a great many alkanes. In later sections these rules will be extended to cover other classes of compounds, but advanced texts or references must be consulted for the complete system.

Rule 1. Select the longest continuous chain of carbon atoms as the parent compound, and consider all alkyl groups attached to it as branch chains or substituents that have replaced hydrogen atoms of the parent hydrocarbon. If two chains of equal length are found, use the chain that has the larger number of substituents attached to it. The alkane's name consists of the parent compound's name prefixed by the names of the alkyl groups attached to it.

Rule 2. Number the carbon atoms in the parent carbon chain starting from the end closest to the first carbon atom that has an alkyl or other group substituted for a hydrogen atom. If the first substituent from each end is on the same-numbered carbon, go to the next substituent to determine which end of the chain to start numbering.

Rule 3. Name each alkyl group and designate its position on the parent carbon chain by a number (e.g., 2-methyl means a methyl group attached to C-2).

TABLE 19.2 Names and Formulas of Selected Alkyl Groups

Formula	Name	Formula	Name
CH_3-	methyl	CH_3CH- with CH_3 above	isopropyl
CH_3CH_2-	ethyl		
$CH_3CH_2CH_2-$	propyl		
$CH_3CH_2CH_2CH_2-$	butyl	CH_3CHCH_2- with CH_3 above	isobutyl
$CH_3(CH_2)_3CH_2-$	pentyl		
$CH_3(CH_2)_4CH_2-$	hexyl	CH_3CH_2CH- with CH_3 above	sec-butyl (secondary butyl)
$CH_3(CH_2)_5CH_2-$	heptyl		
$CH_3(CH_2)_6CH_2-$	octyl	CH_3C- with CH_3 above and CH_3 below	tert-butyl (tertiary butyl)
$CH_3(CH_2)_7CH_2-$	nonyl		
$CH_3(CH_2)_8CH_2-$	decyl		

Rule 4. When the same alkyl-group branch chain occurs more than once, indicate this repetition by a prefix (*di-*, *tri-*, *tetra-*, and so forth) written in front of the alkyl-group name (e.g., *dimethyl* indicates two methyl groups). The numbers indicating the alkyl-group positions are separated by a comma and followed by a hyphen and are placed in front of the name (e.g., 2,3-dimethyl).

Rule 5. When several different alkyl groups are attached to the parent compound, list them in alphabetical order (e.g., ethyl before methyl in 3-ethyl-4-methyloctane). Prefixes are not included in alphabetical ordering (ethyl comes before dimethyl).

Let's use the IUPAC system to name this compound:

$$\overset{4}{C}H_3-\overset{3}{C}H_2-\overset{2}{C}H-\overset{1}{C}H_3 \quad \text{or} \quad \overset{1}{C}H_3-\overset{2}{C}H-\overset{3}{C}H_2-\overset{4}{C}H_3$$
$$\underset{CH_3}{|} \qquad\qquad\qquad \underset{CH_3}{|}$$

2-methylbutane

The longest continuous chain contains four carbon atoms. Therefore we use the parent compound name, butane. The methyl group, CH_3-, attached to C-2 is named as a prefix to butane, the "2-" indicating the point of its attachment on the butane chain.

How would we write the structural formula for 2-methylpentane? Its name tells us how. The parent compound, pentane, contains five carbons. We write and number the five-carbon skeleton of pentane, put a methyl group on C-2 (because of the "2-methyl" in the name), and add hydrogens to give each carbon four bonds:

$$\overset{5}{C}-\overset{4}{C}-\overset{3}{C}-\overset{2}{C}-\overset{1}{C} \qquad \overset{5}{C}-\overset{4}{C}-\overset{3}{C}-\overset{2}{C}-\overset{1}{C} \qquad CH_3-CH_2-CH_2-CH-CH_3$$
$$\underset{CH_3}{|} \qquad\qquad\qquad\qquad \underset{CH_3}{|}$$

2-methylpentane

Could this compound be called 4-methylpentane? No, in the IUPAC system the parent carbon chain is numbered starting from the end *nearest* the branch chain.

It is very important to understand that the *sequence* of atoms and groups—not the way the sequence is written—determines the name of a compound. These formulas all represent 2-methylpentane; note that carbon numbering does not have to follow a straight line:

$$\overset{1}{C}H_3-\overset{2}{C}H-\overset{3}{C}H_2-\overset{4}{C}H_2-\overset{5}{C}H_3 \qquad\qquad \overset{5}{C}H_3-\overset{4}{C}H_2-\overset{3}{C}H_2-\overset{2}{C}H-\overset{1}{C}H_3$$
$$\underset{CH_3}{|} \qquad\qquad\qquad\qquad\qquad \overset{CH_3}{\overset{|}{}}$$

$$\overset{CH_3}{\overset{2|}{\overset{}{}}}$$
$$\overset{1}{C}H_3-\overset{2}{C}H-\overset{3}{C}H_2 \qquad\qquad \overset{CH_3}{\overset{2|}{CH}}$$
$$\underset{4|}{}\qquad \overset{1}{CH_3}\diagdown\overset{3}{}\diagup\overset{4}{CH_2}$$
$$\overset{5}{CH_2-CH_3} \qquad\qquad \underset{CH_2}{}\underset{\overset{5}{CH_3}}{}$$

The following formulas and names demonstrate other aspects of the IUPAC nomenclature system:

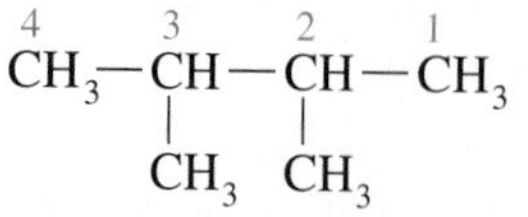

$$\overset{4}{C}H_3-\overset{3}{C}H-\overset{2}{C}H-\overset{1}{C}H_3$$
with CH_3 CH_3 below
2,3-dimethylbutane

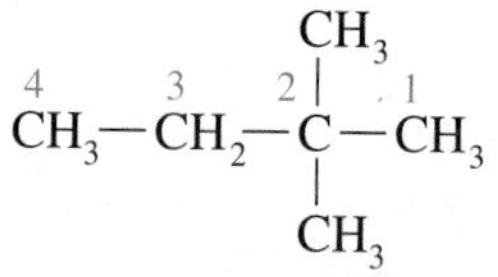

$$\overset{4}{C}H_3-\overset{3}{C}H_2-\overset{2}{C}-\overset{1}{C}H_3$$
with CH_3 above and CH_3 below
2,2-dimethylbutane

In 2,3-dimethylbutane, the longest carbon chain is four, indicating a butane; "dimethyl" indicates two methyl groups; "2,3-" means that one CH_3 is on C-2 and one is on C-3. In 2,2-dimethylbutane, both methyl groups are on the same carbon atom; both numbers are required.

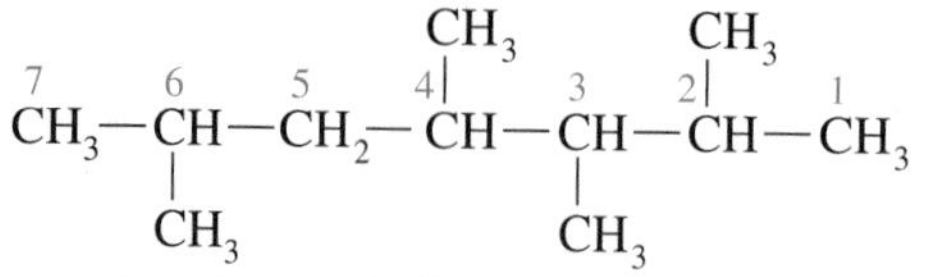

$$\overset{7}{C}H_3-\overset{6}{C}H-\overset{5}{C}H_2-\overset{4}{C}H-\overset{3}{C}H-\overset{2}{C}H-\overset{1}{C}H_3$$
with CH_3 groups
2,3,4,6-tetramethylheptane (not 2,4,5,6-)

Note that this molecule is numbered from right to left.

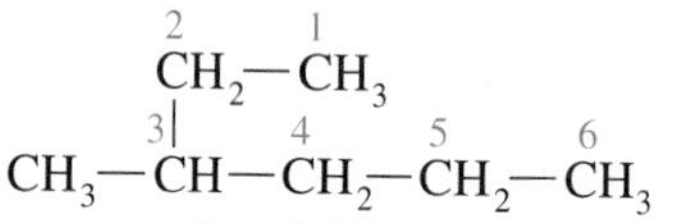

$$\overset{2}{C}H_2-\overset{1}{C}H_3$$
$$\overset{3}{C}H_3-\overset{}{C}H-\overset{4}{C}H_2-\overset{5}{C}H_2-\overset{6}{C}H_3$$
3-methylhexane

The longest continuous chain in 3-methylhexane has six carbons.

In the next structure, the longest carbon chain is eight. The groups attached or substituted for hydrogen on the octane chain are named in alphabetical order.

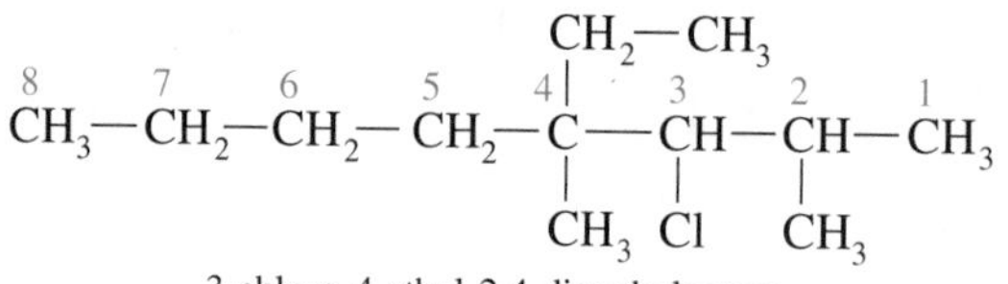

$$\overset{8}{C}H_3-\overset{7}{C}H_2-\overset{6}{C}H_2-\overset{5}{C}H_2-\overset{4}{C}-\overset{3}{C}H-\overset{2}{C}H-\overset{1}{C}H_3$$
with CH_2-CH_3 above and CH_3 Cl CH_3 below
3-chloro-4-ethyl-2,4-dimethyloctane

Example 19.2 Write the formulas for (a) 3-ethylpentane, and (b) 2,2,4-trimethylpentane.

Solution

(a) The name *pentane* indicates a five-carbon chain. Write five connecting carbon atoms and number them. Attach an ethyl group, CH_3CH_2-, to C-3. Now add hydrogen atoms to give each carbon atom four bonds: C-1 and C-5 each need three hydrogen atoms; C-2 and C-4 each need two hydrogen atoms; and C-3 needs one hydrogen atom.

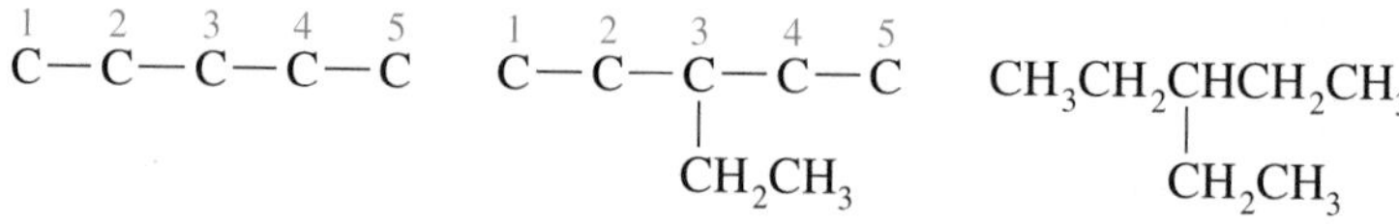

$$\overset{1}{C}-\overset{2}{C}-\overset{3}{C}-\overset{4}{C}-\overset{5}{C} \qquad \overset{1}{C}-\overset{2}{C}-\overset{3}{C}-\overset{4}{C}-\overset{5}{C} \qquad CH_3CH_2CHCH_2CH_3$$
with CH_2CH_3 and CH_2CH_3 below
3-ethylpentane

(b) Pentane indicates a five-carbon chain. Write five connecting carbon atoms and number them. There are three methyl groups, CH_3-, in the compound (trimethyl), two attached to C-2 and one attached to C-4. Attach these three methyl groups to their respective carbon atoms. Now add hydrogen atoms to give each carbon atom four bonds. Thus C-1 and C-5 each need three hydrogen atoms; C-2 does not need any hydrogen atoms; C-3 needs two hydrogen atoms; and C-4 needs one hydrogen atom. The formula is complete:

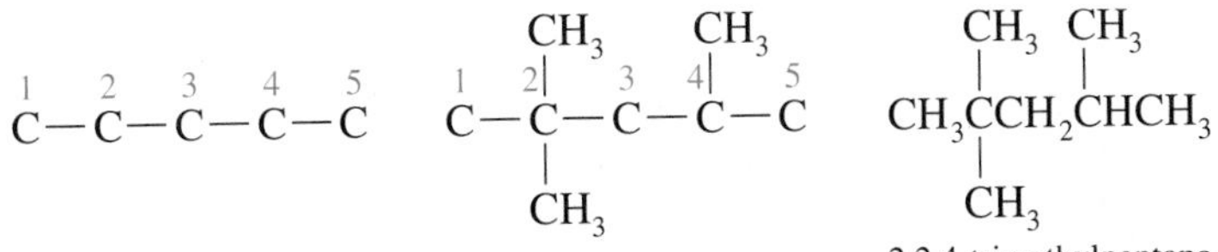

$$
\overset{1}{C}-\overset{2}{C}-\overset{3}{C}-\overset{4}{C}-\overset{5}{C}
$$

$$
\overset{1}{C}-\overset{2}{\underset{|}{C}}-\overset{3}{C}-\overset{4}{\underset{|}{C}}-\overset{5}{C}
$$
with CH$_3$ on C-2 (two) and CH$_3$ on C-4

$$CH_3CCH_2CHCH_3$$
with CH$_3$, CH$_3$ above and CH$_3$ below

2,2,4-trimethylpentane

Example 19.3

Name these compounds:

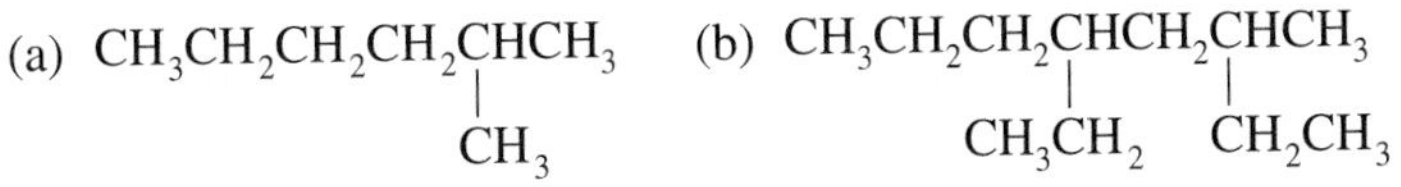

(a) $CH_3CH_2CH_2CH_2CHCH_3$ with CH$_3$ below

(b) $CH_3CH_2CH_2CHCH_2CHCH_3$ with CH$_3$CH$_2$ and CH$_2$CH$_3$ below

Solution

(a) The longest continuous carbon chain contains six carbon atoms (Rule 1). Thus the parent name is hexane. Number the carbon chain from right to left so that the methyl group attached to C-2 is given the lowest possible number (Rule 2). With a methyl group on C-2, the name of the compound is 2-methylhexane (Rule 3).

(b) The longest continuous carbon chain contains eight carbon atoms:

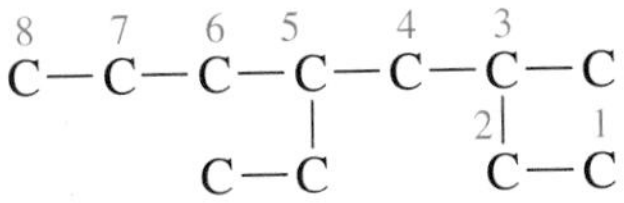

$$
\overset{8}{C}-\overset{7}{C}-\overset{6}{C}-\overset{5}{C}-\overset{4}{C}-\overset{3}{C}-C
$$
with C—C below C-5 and C—C (2,1) below C-3

The parent name is octane. As the chain is numbered, a methyl group is on C-3 and an ethyl group is on C-5. The name of the compound is 5-ethyl-3-methyloctane. Note that ethyl is named before methyl (alphabetical order) (Rule 5).

Practice 19.1

Name these alkanes:

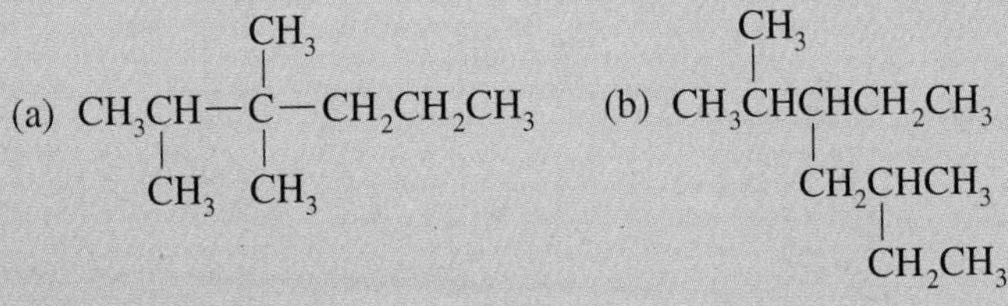

(a) $CH_3CH-\underset{|}{\overset{|}{C}}-CH_2CH_2CH_3$ with CH$_3$ above and CH$_3$, CH$_3$ below

(b) $CH_3CHCHCH_2CH_3$ with CH$_3$ above and CH$_2$CHCH$_3$ then CH$_2$CH$_3$ below

---

19.7 Alkenes and Alkynes

Alkenes and alkynes are classified as unsaturated hydrocarbons. They are said to be unsaturated because, unlike alkanes, their molecules do not contain the maximum

alkene
alkyne

possible number of hydrogen atoms. **Alkenes** have two less hydrogen atoms, and **alkynes** have four less hydrogen atoms than alkanes with a comparable number of carbon atoms. Alkenes contain at least one double bond between adjacent carbon atoms, while alkynes contain at least one triple bond between adjacent carbon atoms.

> **Alkenes contain a carbon–carbon double bond.**
> **Alkynes contain a carbon–carbon triple bond.**

Remember: In a homologous series the formulas of successive members differ by increments of CH_2.

The simplest alkene is ethylene (or ethene), $CH_2{=}CH_2$, and the simplest alkyne is acetylene (or ethyne), $CH{\equiv}CH$ (Figure 19.5). Ethylene and acetylene are the first members of a homologous series (e.g., $CH_2{=}CH_2$, $CH_3CH{=}CH_2$, and $CH_3CH_2CH{=}CH_2$). Huge quantities of alkenes are made by cracking and dehydrogenating alkanes during the processing of crude oils. These alkenes are used to manufacture motor fuels, polymers, and petrochemicals. Alkene molecules, like those of alkanes, have very little polarity. Hence the physical properties of alkenes are similar to those of the corresponding saturated hydrocarbons.

> **General formula for alkenes: C_nH_{2n}**
> **General formula for alkynes: C_nH_{2n-2}**

Table 19.3 gives the names and formulas for several alkenes and alkynes.

▲
Acetylene is used for welding steel girders and car exhaust systems.

TABLE 19.3 Names and Formulas for Several Alkenes and Alkynes	
Formula	**IUPAC name**
$CH_2{=}CH_2$	ethene
$CH_3CH{=}CH_2$	propene
$CH_3CH_2CH{=}CH_2$	1-butene
$CH_3CH{=}CHCH_3$	2-butene
$CH_3C{=}CH_2$ 　$|$ 　CH_3	2-methylpropene
$CH{\equiv}CH$	ethyne
$CH_3C{\equiv}CH$	propyne
$CH_3CH_2C{\equiv}CH$	1-butyne
$CH_3C{\equiv}CCH_3$	2-butyne

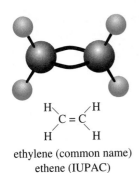

◀ FIGURE 19.5
**Ball-and-stick models for
ethylene and acetylene.**

H–C≡C–H
acetylene (common name)
ethyne (IUPAC)

ethylene (common name)
ethene (IUPAC)

19.8 Naming Alkenes and Alkynes

The names of alkenes and alkynes are derived from the corresponding alkanes. To name an alkene (or alkyne) by the IUPAC system,

1. Select the longest carbon–carbon chain that has a double or triple bond.
2. Name this parent compound as you would an alkane but change the *-ane* ending to *-ene* for an alkene or to *-yne* for an alkyne; thus, propane changes to propene or propyne:

$CH_3CH_2CH_3$ $CH_3CH=CH_2$ $CH_3C\equiv CH$
propane   propene   propyne

3. Number the carbon chain of the parent compound starting with the end nearer to the double or triple bond. Use the smaller of the two numbers on the double- or triple-bonded carbon atoms to indicate the position of the double or triple bond. Place this number in front of the alkene or alkyne name; 2-butene means that the carbon–carbon double bond is between C-2 and C-3.
4. Side chains and other groups are treated as in naming alkanes. Name the substituent group, and designate its position on the parent chain with a number.

Study the following examples of named alkenes and alkynes:

$\overset{4}{C}H_3\overset{3}{C}H_2\overset{2}{C}H=\overset{1}{C}H_2$ $\overset{1}{C}H_3\overset{2}{C}H=\overset{3}{C}HCH_3^{4}$
1-butene   2-butene

$$\overset{6}{C}H_3\overset{5}{C}H_2\overset{4}{C}H_2$$
$$|3 \quad 2 \qquad 1$$
$$CHCH=CH_2$$
$$|$$
$$CH_3CH_2CH_2$$
3-propyl-1-hexene

$\overset{4}{C}H_3\overset{3}{C}H_2\overset{2}{C}\equiv\overset{1}{C}H$
1-butyne

$$\qquad\qquad\qquad CH_3$$
$$\overset{1}{C}H_3-\overset{2}{C}\equiv\overset{3}{C}-\overset{4}{C}H-\overset{5}{C}H-\overset{6}{C}H_3$$
$$|$$
$$CH_3$$
4,5-dimethyl-2-hexyne

To write a structural formula from a systematic name, the naming process is reversed. For example, how would we write the structural formula for 4-methyl-2-pentene? The name indicates (1) five carbons in the longest chain, (2) a double bond between C-2 and C-3, and (3) a methyl group on C-4.

Write five carbon atoms in a row. Place a double bond between C-2 and C-3, and place a methyl group on C-4. Now add hydrogen atoms to give each carbon atom four bonds. Carbons 1 and 5 each need three hydrogen atoms; C-2, C-3, and C-4 each need one hydrogen atom.

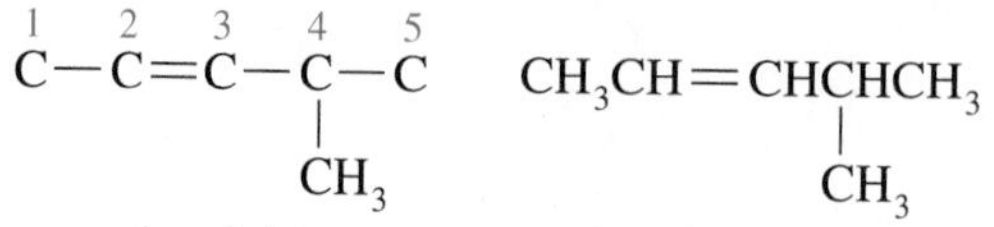

$$\overset{1}{C}-\overset{2}{C}=\overset{3}{C}-\overset{4}{C}-\overset{5}{C} \qquad CH_3CH=CHCHCH_3$$
$$\hspace{3.2cm} | \hspace{3.8cm} |$$
$$\hspace{3.2cm} CH_3 \hspace{3.5cm} CH_3$$

carbon skeleton   4-methyl-2-pentene

Example 19.4   Write structural formulas for (a) 7-methyl-2-octene, and (b) 3-hexyne.

Solution   (a) Octene, like octane, indicates an eight-carbon chain. The chain contains a double bond between C-2 and C-3 and a methyl group on C-7. Write eight carbon atoms in a row, place a double bond between C-2 and C-3, and place a methyl group on C-7. Now add hydrogen atoms to give each carbon atom four bonds:

$$\overset{1}{C}-\overset{2}{C}=\overset{3}{C}-\overset{4}{C}-\overset{5}{C}-\overset{6}{C}-\overset{7}{C}-\overset{8}{C} \qquad CH_3CH=CHCH_2CH_2CH_2CHCH_3$$
$$\hspace{4.3cm} | \hspace{5.3cm} |$$
$$\hspace{4.3cm} CH_3 \hspace{5cm} CH_3$$

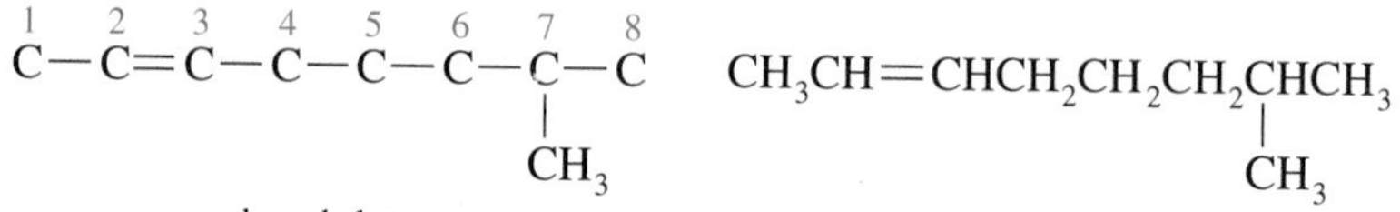

carbon skeleton   7-methyl-2-octene

(b) The stem *hex-* indicates a six-carbon chain; the suffix *-yne* indicates a carbon–carbon triple bond; the number 3 locates the triple bond between C-3 and C-4. Write six carbon atoms in a row and place a triple bond between C-3 and C-4. Now add hydrogen atoms to give each carbon atom four bonds; C-3 and C-4 don't need hydrogen atoms:

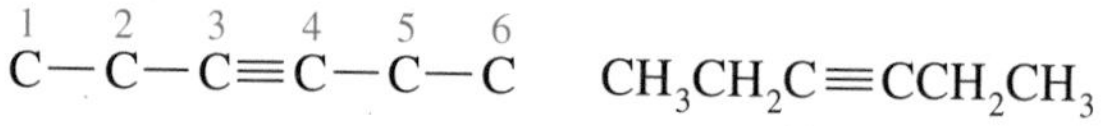

$$\overset{1}{C}-\overset{2}{C}-\overset{3}{C}\equiv\overset{4}{C}-\overset{5}{C}-\overset{6}{C} \qquad CH_3CH_2C\equiv CCH_2CH_3$$
3-hexyne

Practice 19.2

Name these compounds:

(a) 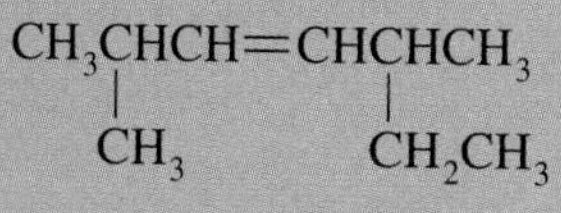 $CH_3CHCH=CHCHCH_3$ (b) 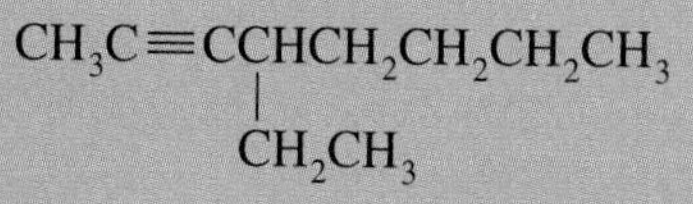 $CH_3C\equiv CCHCH_2CH_2CH_2CH_3$
$\hspace{0.8cm} | \hspace{1.8cm} |$ $\hspace{4.5cm} |$
$\hspace{0.8cm} CH_3 \hspace{1.3cm} CH_2CH_3$ $\hspace{3.8cm} CH_2CH_3$

19.9   Reactions of Alkenes

The alkenes are much more reactive than the corresponding alkanes. This greater reactivity is due to the carbon–carbon double bonds. In our brief overview we will limit our discussion to the most common reaction of alkenes.

Addition

In organic chemistry a reaction in which two substances join to produce one compound is called an **addition reaction.** Addition at the carbon–carbon double bond is the most common reaction of alkenes. Hydrogen, halogens (Cl_2 or Br_2), hydrogen halides, and water are some of the reagents that can be added to unsaturated hydrocarbons. Ethylene, for example, reacts in this fashion:

$$CH_2{=}CH_2 \ + \ H_2 \ \xrightarrow[\text{1 atm}]{\text{Pt} \ \ 25°C} \ CH_3{-}CH_3$$
ethylene   ethane

Pt 25° C indicates the catalyst, **Pt,** and other necessary conditions for the reaction.

$$CH_2{=}CH_2 \ + \ Br{-}Br \ \longrightarrow \ CH_2Br{-}CH_2Br$$
1,2-dibromoethane

Visible evidence of the Br_2 addition is the disappearance of the reddish brown color of bromine as it reacts.

$$CH_2{=}CH_2 \ + \ HCl \ \longrightarrow \ CH_3CH_2Cl$$
chloroethane

$$CH_2{=}CH_2 \ + \ HOH \ \xrightarrow{H^+} \ CH_3CH_2OH$$
ethanol
(ethyl alcohol)

The H^+ indicates that the reaction is carried out under acidic conditions.

Note that the double bond is broken and the unsaturated alkene molecules become saturated by an addition reaction. Reactions of this kind can occur with almost any molecule that contains a carbon–carbon double bond.

19.10 Aromatic Hydrocarbons

Benzene and all substances with structures and chemical properties resembling benzene are classified as **aromatic compounds.** The word *aromatic* originally referred to the rather pleasant odor many of these substances possess, but this meaning has been dropped. Benzene, the parent substance of the aromatic hydrocarbons, was first isolated by Michael Faraday in 1825. Its correct molecular formula, C_6H_6, was established a few years later, but finding a reasonable structural formula that would account for the properties of benzene was difficult.

Finally, in 1865, August Kekulé proposed that the carbon atoms in a benzene molecule are arranged in a six-membered ring with one hydrogen atom bonded to each carbon atom and with three carbon–carbon double bonds:

A turning point in the history of chemistry, Kekulé's structures mark the beginning of our understanding of structure in aromatic compounds. His formulas do have one serious shortcoming, though: They represent benzene and related substances as highly unsaturated compounds. Yet benzene does not readily undergo addition reactions like a typical alkene. For example, benzene does not decolorize bromine solutions rapidly. Instead, the chemical behavior of benzene resembles that of an alkane. Its typical reactions are the substitution type, where a hydrogen atom is replaced by some other group:

$$C_6H_6 + Cl_2 \xrightarrow{\text{Fe}} C_6H_5Cl + HCl$$

Modern theory suggests that the benzene molecule is a hybrid of the two Kekulé structures shown earlier.

For convenience, chemists usually write the structure of benzene as one or the other of these abbreviated forms:

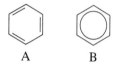

A   B

In both representations, it is understood that there is a carbon atom and a hydrogen atom at each corner of the hexagon. The classical Kekulé structure is represented by A; the modern molecular orbital structure is represented by B. These hexagonal structures are used to represent the structural formulas of benzene derivatives—that is, substances in which one or more hydrogen atoms in the ring have been replaced by other atoms or groups. Chlorobenzene, C_6H_5Cl, for example, is written in this fashion:

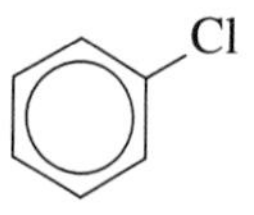

chlorobenzene, C_6H_5Cl

This notation indicates that the chlorine atom has replaced a hydrogen atom and is bonded directly to a carbon atom in the ring. Thus the correct formula for chlorobenzene is C_6H_5Cl, not C_6H_6Cl.

19.11   Naming Aromatic Compounds

A substituted benzene is derived by replacing one or more of benzene's hydrogen atoms with another atom or group of atoms. Thus a monosubstituted benzene has the formula C_6H_5G, where G is the group replacing a hydrogen atom. Because all the hydrogen atoms in benzene are equivalent, it doesn't matter which hydrogen is replaced by the monosubstituted group.

Monosubstituted Benzenes

Some monosubstituted benzenes are named by adding the name of the substituent group as a prefix to the word *benzene*. The name is written as one word. Here are some examples:

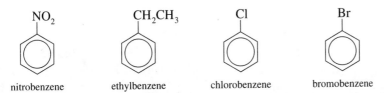

nitrobenzene   ethylbenzene   chlorobenzene   bromobenzene

Certain monosubstituted benzenes have special names that should be learned. These are parent names for further substituted compounds:

toluene
(methylbenzene)

phenol
(hydroxybenzene)

benzoic acid

aniline
(aminobenzene)

The C_6H_5— group is known as the phenyl group (pronounced *fen-ill*). The name *phenyl* is used for compounds that cannot be easily named as benzene derivatives. For example, the following compounds are named as derivatives of alkanes:

3-chloro-2-phenylpentane

diphenylmethane

Disubstituted Benzenes

When two substituent groups replace two hydrogen atoms in a benzene molecule, three isomers are possible. The prefixes *ortho-, meta-,* and *para-* (abbreviated *o-, m-,* and *p-*) are used to name these disubstituted benzenes. In the ortho-compound the substituents are located on adjacent carbon atoms, in the meta-compound they are one carbon apart, and in the para-compound they are on opposite sides of the ring. When the two groups are different, name them alphabetically followed by the word *benzene*.

The dichlorobenzenes, $C_6H_4Cl_2$, illustrate this method of naming. The three isomers have different physical properties, indicating that they are different substances. Note that the para-isomer is a solid and the other two are liquids at room temperature.

▲
Moth crystals or mothballs used to be made of naphthalene. Today many are para-dichlorobenzene.

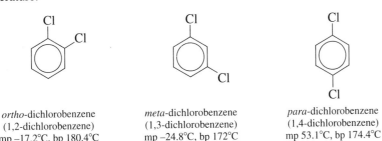

ortho-dichlorobenzene
(1,2-dichlorobenzene)
mp −17.2°C, bp 180.4°C

meta-dichlorobenzene
(1,3-dichlorobenzene)
mp −24.8°C, bp 172°C

para-dichlorobenzene
(1,4-dichlorobenzene)
mp 53.1°C, bp 174.4°C

When one substituent corresponds to a monosubstituted benzene with a special name, the monosubstituted compound becomes the parent name for the disubstituted

compound. In the following examples the parent compounds are phenol, aniline, toluene, and benzoic acid:

o-nitrophenol   p-bromoaniline   m-nitrotoluene   p-aminobenzoic acid

Tri- and Polysubstituted Benzenes

When a benzene ring has more than two substituents, the carbon atoms in the ring are numbered. Numbering starts at one of the substituted groups and goes either clockwise or counterclockwise, but must be done in the direction that gives the lowest possible numbers to the substituent groups. When the compound is named as a derivative of the special parent compound, the substituent of the parent compound is considered to be on C-1 of the ring (the CH_3 group is on C-1 in 2,4,6-tribromo-toluene). The following examples illustrate this system:

1,3,5-trinitrobenzene   1,2,4-tribromobenzene (not 1,4,6-)   2,4,6-trinitrotoluene (TNT)   5-bromo-2-chlorophenol

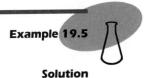

Write formulas and names for all possible isomers of (a) chloronitrobenzene, **Example 19.5**
$C_6H_4Cl(NO_2)$, and (b) tribromobenzene, $C_6H_3Br_3$.

Solution

(a) The name and formula indicate a chloro-group, Cl, and a nitro-group, NO_2, attached to a benzene ring. There are six positions in which to place these two groups. They can be ortho-, meta-, or para- to each other.

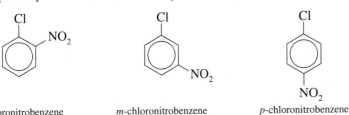

o-chloronitrobenzene   m-chloronitrobenzene   p-chloronitrobenzene

(b) For tribromobenzene, start by placing the three bromo-groups in the 1, 2, and 3 positions, then the 1, 2, and 4 positions, and so on, until all the possible isomers are formed. Name each isomer to check that no duplicate formulas have been written:

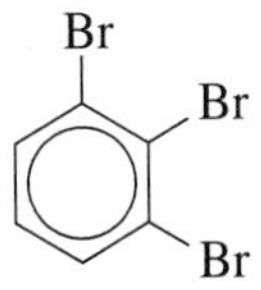

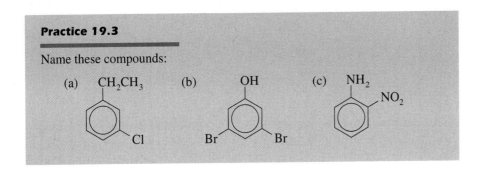

1,2,3-tribromobenzene 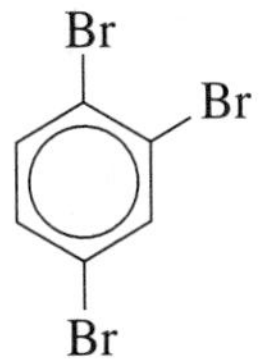   1,2,4-tribromobenzene 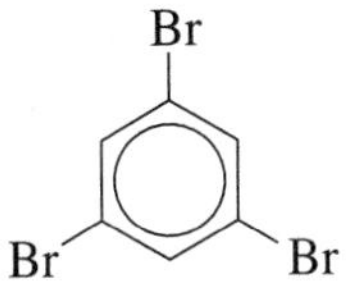   1,3,5-tribromobenzene

Tribromobenzene has only three isomers. However, if one Br is replaced by a Cl, six isomers are possible.

Practice 19.3

Name these compounds:

(a) CH_2CH_3 (b) OH   (c) NH_2 NO_2

Cl   Br   Br

19.12 Hydrocarbon Derivatives

Hydrocarbon derivatives are compounds that can be synthesized from a hydrocarbon. These derivatives contain not only carbon and hydrogen, but such additional elements as oxygen, nitrogen, or a halogen. The compounds in each class have

TABLE 19.4	Classes of Hydrocarbon Derivatives

Class	General formula	Structure of functional group	Sample structural formula	Name IUPAC	Name Common
Alkyl halide	R*X	$-X$ X = F, Cl, Br, I	CH_3Cl CH_3CH_2Cl	chloromethane chloroethane	methyl chloride ethyl chloride
Alcohol	ROH	$-OH$	CH_3OH CH_3CH_2OH	methanol ethanol	methyl alcohol ethyl alcohol
Ether	R—O—R	R—O—R	CH_3-O-CH_3 $CH_3CH_2-O-CH_2CH_3$	methoxymethane ethoxyethane	dimethyl ether diethyl ether
Aldehyde	R—C=O with H above C	—C=O with H above C	H—C=O with H above; $CH_3-C=O$ with H below	methanal ethanal	formaldehyde acetaldehyde
Ketone	R—C—R with O double bonded below	R—C—R with O double bonded below	CH_3-C-CH_3 with O double bonded below $CH_3-C-CH_2CH_3$ with O double bonded below	propanone 2-butanone	acetone methyl ethyl ketone
Carboxylic acid	R—C with =O and OH	—C with =O and OH	HCOOH CH_3COOH	methanoic acid ethanoic acid	formic acid acetic acid
Ester	R—C with =O and OR	—C with =O and OR	$HCOOCH_3$ CH_3COOCH_3	methyl methanoate methyl ethanoate	methyl formate methyl acetate

*The letter "R" is used to indicate any of the many possible alkyl groups.

similarities in structure and properties. We will consider the classes of hydrocarbon derivatives shown in Table 19.4, which is divided into two sections of different color. The compounds in the first section don't contain a C=O group in their molecules, while those in the second section all contain a C=O group. A carbon atom double bonded to an oxygen atom is called a *carbonyl group.*

19.13 Alkyl Halides

Alkanes react with a halogen in ultraviolet light to produce an alkyl halide, R—X. *Halogenation* is a general term for the substitution of a halogen atom for a hydrogen atom.

CHEMISTRY IN ACTION The End for Graffiti?

Imagine a clear, hard coating that repels water, doesn't hold paint or dirt, and even repels adhesive tape. In a world filled with graffiti and grime, a nonstick coating would go far in cleaning up our cities. Donald Schmidt, a Dow polymer chemist, developed just such a water-

based cross-linked fluorochemical coating during the late 1980s.

The coating is a mixture of two different polymers: a reactive surfactant and a cross-linking agent. The polymers dissolve in water and can be brushed or sprayed on any surface to form a thin layer of twisted molecules. When the coating is heated, the spaghettilike molecules link, preventing the coating from redissolving in water.

To enhance the nonstick properties of the coating, other solvents with high boiling points are added, which makes the coating dry more slowly, allowing the strands of polymer to get closer. The coating is very smooth with no tiny crevices where objects can stick.

The 3M Corporation is investigating potential uses for this coating material. Initial plans include hard surfaces such as kitchen counters, bathroom tile, automobile finishes, and even wallpaper. In addition, the coating may be useful on airplane wings to facilitate deicing, and the medical profession may use the coating to prevent blood cells from clogging an artificial heart.

Since the material resists paint and markers, it's being considered as a way to combat graffiti. The greatest difficulty in this application is the tremendous variety of surfaces requiring treatment. Scientists are working to develop the proper formulation for the various commercial uses.

$$RH + X_2 \xrightarrow[\text{light}]{UV} RX + HX \qquad (X = Cl \text{ or } Br)$$

When a specific halogen is used, the name reflects this. For example, when chlorine is the halogen, the process is called *chlorination*.

$$CH_3CH_3 + Cl_2 \xrightarrow[\text{light}]{UV} CH_3CH_2Cl + HCl$$
chloroethane

The reaction yields alkyl halides, RX, which are useful as intermediates for the manufacture of other substances.

A well-known reaction of methane and chlorine is shown by the equation

$$CH_4 + Cl_2 \xrightarrow[\text{light}]{UV} CH_3Cl + HCl$$
chloromethane

According to IUPAC rules, organic halides are named by giving the halogen substituent as a prefix of the parent alkane (e.g., CH_3Cl is chloromethane).

Alkyl halides are primarily used as industrial solvents. Carbon tetrachloride, CCl_4, was once used extensively in the dry-cleaning process but it has been replaced because of its toxicity and carcinogenic effects. Common dry-cleaning solvents are now dichloromethane, CH_2Cl_2, and 1,1,1-trichloroethane, CH_3CCl_3.

Organic halides are also used as anesthetics. Chloroform, $CHCl_3$, was once a popular anesthetic. It's no longer used, however, because it is harmful to the respiratory system. Instead, such compounds as halothane, $CF_3CHClBr$, are being used. (See the Chemistry in Action on p. 520.)

Composed of carbon, hydrogen, fluorine, and chlorine, chlorinated fluorocarbons (CFCs) are alkyl halides used in aerosol propellants and refrigerants. CFCs have been replaced by other compounds because they generate chlorine atoms in the upper atmosphere, which deplete the ozone layer (see Section 12.17). Many aerosols are now propelled by carbon dioxide. New refrigerant molecules (called hydrofluorocarbons, HFCs) are being used in refrigerators and air conditioners.

19.14   Alcohols

alcohol   **Alcohols** are organic compounds whose molecules contain a hydroxyl ($-OH$) group bonded to a saturated carbon atom. Thus if we substitute an $-OH$ for an H in CH_4, we get CH_3OH, methyl alcohol. The functional group of the alcohols is $-OH$. The general formula for alcohols is ROH, where R is an alkyl or a substituted alkyl group.

Alcohols differ from metal hydroxides in that they do not dissociate or ionize in water. The $-OH$ group is attached to the carbon atom by a covalent bond and not by an ionic bond. The alcohols form a homologous series, with methanol, CH_3OH, as the first member of the series. Models of the structural arrangements of the atoms in methanol and ethanol are shown in Figure 19.6.

primary alcohol
secondary alcohol
tertiary alcohol
Alcohols are classified as **primary** (1°), **secondary** (2°), or **tertiary** (3°), depending on whether the carbon atom to which the $-OH$ group is attached is bonded to one, two, or three other carbon atoms, respectively. Generalized formulas for 1°, 2°, and 3° alcohols are

$$
\begin{array}{ccc}
\quad H \quad & \quad R \quad & \quad R \quad \\
\quad | \quad & \quad | \quad & \quad | \quad \\
R-C-OH & R-C-OH & R-C-OH \\
\quad | \quad & \quad | \quad & \quad | \quad \\
\quad H \quad & \quad H \quad & \quad R \quad \\
\text{primary alcohol} & \text{secondary alcohol} & \text{tertiary alcohol}
\end{array}
$$

Examples of these alcohols are listed in Table 19.5. Methanol, CH_3OH, is grouped with the primary alcohols.

Molecular structures with more than one $-OH$ group attached to a single carbon atom are generally not stable. But an alcohol molecule can contain two or

FIGURE 19.6 ▶
Ball-and-stick models illustrating structural formulas of methanol and ethanol.

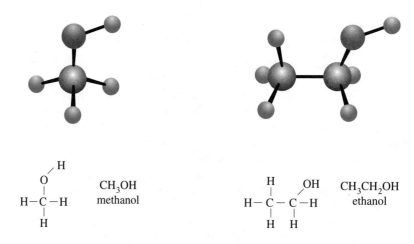

$$
\begin{array}{l}
\quad \diagup H \\
O \\
| \\
H-C-H \\
| \\
H
\end{array}
\quad
\begin{array}{l}
CH_3OH \\
\text{methanol}
\end{array}
$$

$$
\begin{array}{l}
\quad H \quad\quad OH \\
\quad | \quad\quad | \\
H-C-C-H \\
\quad | \quad\quad | \\
\quad H \quad\quad H
\end{array}
\quad
\begin{array}{l}
CH_3CH_2OH \\
\text{ethanol}
\end{array}
$$

TABLE 19.5   Names and Classifications of Alcohols

Class	Formula	IUPAC name	Common name*	Boiling point (°C)
Primary	CH_3OH	methanol	Methyl alcohol	65.0
Primary	CH_3CH_2OH	ethanol	Ethyl alcohol	78.5
Primary	$CH_3CH_2CH_2OH$	1-propanol	n-propyl alcohol	97.4
Primary	$CH_3CH_2CH_2CH_2OH$	1-butanol	n-butyl alcohol	118
Primary	$CH_3(CH_2)_3CH_2OH$	1-pentanol	n-pentyl alcohol	138
Primary	$CH_3(CH_2)_6CH_2OH$	1-octanol	n-octyl alcohol	195
Primary	CH_3CHCH_2OH | CH_3	2-methyl-1-propanol	isobutyl alcohol	108
Secondary	CH_3CHCH_3 | OH	2-propanol	isopropyl alcohol	82.5
Secondary	$CH_3CH_2CHCH_3$ | OH	2-butanol	sec-butyl alcohol	91.5
Tertiary	CH_3 | CH_3-C-OH | CH_3	2-methyl-2-propanol	tert-butyl alcohol	82.9
Dihydroxy	$HOCH_2CH_2OH$	1,2-ethanediol	ethylene glycol	197
Trihydroxy	$HOCH_2CHCH_2OH$ | OH	1,2,3-propanetriol	glycerol or glycerine	290

*The abbreviations n, sec, and tert stand for normal, secondary, and tertiary, respectively.

more —OH groups if each —OH is attached to a different carbon atom. Accordingly, alcohols are also classified as monohydroxy, dihydroxy, trihydroxy, and so on, on the basis of the number of hydroxyl groups per molecule. **Polyhydroxy alcohol** is a general term for an alcohol that has more than one —OH group per molecule.

polyhydroxy alcohol

Methanol

Methanol is a volatile (bp 65°C), highly flammable liquid. It is poisonous and capable of causing blindness or death if taken internally. Exposure to methanol vapors for even short periods of time is dangerous. Despite the hazards, over 8 billion pounds of methanol (3.6×10^9 kg) are manufactured annually and used

1. in the conversion of methanol to formaldehyde, which is primarily used to make polymers
2. in the manufacture of other chemicals, especially various kinds of esters
3. to denature ethyl alcohol (rendering it unfit as a beverage)
4. as an industrial solvent
5. as an inexpensive and temporary antifreeze for radiators (it isn't a satisfactory permanent antifreeze because its boiling point is lower than that of water)

▲
This vintner is checking alcohol content, color, and sediment before wine is bottled.

Ethanol

Ethanol is without doubt the most widely known alcohol. Huge quantities of this substance are prepared by fermentation using starch and sugar as the raw materials. Conversion of simple sugars to ethanol is accomplished by the yeast enzyme zymase:

$$C_6H_{12}O_6 \xrightarrow{\text{zymase}} 2\ CH_3CH_2OH\ +\ 2\ CO_2$$
$$\text{glucose} \qquad\qquad\qquad \text{ethanol}$$

Ethanol is economically significant as

1. an intermediate in the manufacture of other chemicals such as acetaldehyde, acetic acid, ethyl acetate, and diethyl ether
2. a solvent for many organic substances (e.g., pharmaceutical, perfumes, flavorings)
3. an ingredient in alcoholic beverages

Ethanol acts physiologically as a food, a drug, and a poison. It's a food in the limited sense that the body metabolizes small amounts of it to carbon dioxide and water with the production of energy. In moderate quantities ethanol causes drowsiness and depresses brain functions so that activities requiring skill and judgment (such as automobile driving) are impaired. In larger quantities it causes nausea, vomiting, impaired perception, and lack of coordination. If a very large amount is consumed, unconsciousness and ultimately death may occur.

For industrial use ethanol is often denatured (rendered unfit for drinking) by adding small amounts of methanol and other denaturants that are extremely difficult to remove. Denaturing is required by the federal government to protect the beverage alcohol tax source. Special tax-free use permits are issued to scientific and industrial users who require pure ethanol for nonbeverage uses.

Other widely used alcohols include the following:

1. Isopropyl alcohol (2-propanol) is the principal ingredient in rubbing alcohol formulations.
2. Ethylene glycol is the main compound in antifreezes; it is also used in the manufacture of synthetic fibers and in the paint industry.
3. Glycerol, also known as glycerine, is a syrupy, sweet-tasting liquid used in the manufacture of polymers and explosives, as an emollient in cosmetics, as a humectant in tobacco, and as a sweetener.

19.15 Naming Alcohols

If you know how to name alkanes, it's easy to name alcohols by the IUPAC system. Unfortunately, several alcohols are also known by common names, so it's often necessary to know both names for a given alcohol. The common name is usually formed from the name of the alkyl group that is attached to the —OH group, followed by the word *alcohol*. (See examples given below and in Table 19.5.) To name an alcohol by the IUPAC system,

1. Select the longest continuous chain of carbon atoms containing the hydroxyl group.
2. Number the carbon atoms in this chain so that the one bearing the —OH group has the lowest possible number.
3. Form the parent alcohol name by replacing the final -e of the corresponding alkane name by -ol. When isomers are possible, locate the position of the —OH by placing the number (hyphenated) of the carbon atom to which the —OH is bonded immediately before the parent alcohol name.
4. Name each alkyl side chain (or other group) and designate its position by number.

Study the following examples of this naming system and also those shown in Table 19.5.

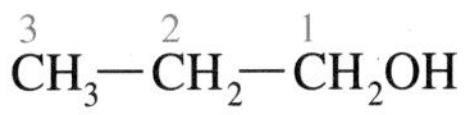

$$\overset{3}{CH_3}-\overset{2}{CH_2}-\overset{1}{CH_2OH}$$

1-propanol

$$\overset{1}{CH_3}-\overset{2}{CH}-\overset{3}{CH_3}$$
$$\qquad\quad |$$
$$\qquad\quad OH$$

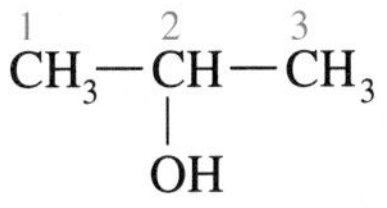

2-propanol
(isopropyl alcohol)

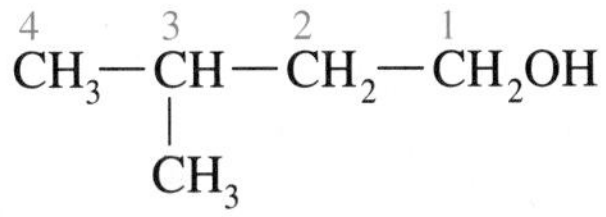

$$\overset{4}{CH_3}-\overset{3}{CH}-\overset{2}{CH_2}-\overset{1}{CH_2OH}$$
$$\qquad\quad |$$
$$\qquad\quad CH_3$$

3-methyl-1-butanol

$$\overset{2}{HOCH_2}-\overset{1}{CH_2OH}$$

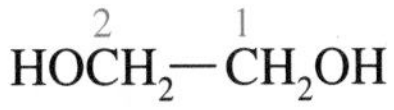

1,2-ethanediol
(ethylene glycol)

Name this alcohol by the IUPAC system:

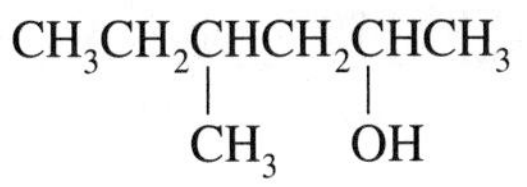

Example 19.6

$$CH_3CH_2CHCH_2CHCH_3$$
$$\qquad\quad |\qquad\quad |$$
$$\qquad\quad CH_3\quad OH$$

Step 1. The longest continuous carbon chain containing the —OH group has six carbon atoms.

Solution

Step 2. This carbon chain is numbered from right to left so that the —OH group has the smallest possible number. In this case the —OH is on C-2:

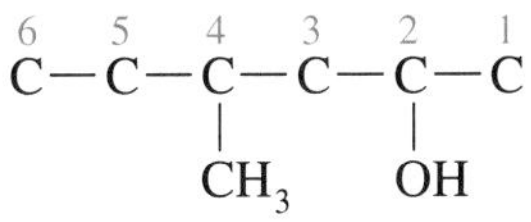

$$\overset{6}{C}-\overset{5}{C}-\overset{4}{C}-\overset{3}{C}-\overset{2}{C}-\overset{1}{C}$$
$$\qquad\qquad |\qquad\quad |$$
$$\qquad\qquad CH_3\quad OH$$

Step 3. The name of the six-carbon alkane is hexane. Replace the final e in hexane by -ol, forming the name hexanol. Since the —OH is on C-2, place a 2- before hexanol to give it the parent alcohol name of 2-hexanol.

Step 4. A methyl group, —CH$_3$, is located on C-4. Therefore the full name of the compound is 4-methyl-2-hexanol.

Example 19.7   Write the structural formula of 3,3-dimethyl-2-hexanol.

Solution

Step 1. The 2-hexanol refers to a six-carbon chain with an —OH group on C-2.
Write the skeleton structure with an —OH on C-2. Now place the two
methyl groups (3,3-dimethyl) on C-3:

$$
\begin{array}{c}
\overset{1}{C}-\overset{2}{C}-\overset{3}{C}-\overset{4}{C}-\overset{5}{C}-\overset{6}{C} \\
\;\;\;\;\;\; | \\
\;\;\;\;\;\; OH
\end{array}
\qquad
\begin{array}{c}
\;\;\;\;\;\;\;\;\;\;\; CH_3 \\
\;\;\;\;\;\;\;\;\;\;\; | \\
\overset{1}{C}-\overset{2}{C}-\overset{3}{C}-\overset{4}{C}-\overset{5}{C}-\overset{6}{C} \\
\;\;\; | \;\;\;\; | \\
\;\;\; HO \;\; CH_3
\end{array}
$$

Step 2. Finally, add hydrogen atoms to give each carbon atom four bonds:

$$
\begin{array}{c}
\;\;\;\;\;\;\;\;\;\;\;\; CH_3 \\
\;\;\;\;\;\;\;\;\;\;\;\; | \\
CH_3CH - C - CH_2CH_2CH_3 \\
\;\;\;\; | \;\;\;\; | \\
\;\;\; HO \;\;\;\; CH_3
\end{array}
$$

3,3-dimethyl-2-hexanol

Practice 19.4

Name these alcohols and classify each as primary, secondary, or tertiary:

(a) $CH_3CH_2CHCH_2CH_2CHCH_3$
    $\quad\quad\;\; | \quad\quad\quad\quad |$
    $\quad\quad\; OH \quad\quad\quad CH_3$

(b) $\quad\quad OH$
    $\quad\quad |$
    $CH_3CCH_2CHCH_3$
    $\quad | \quad\quad\; |$
    $\; CH_3 \; CH_3$

19.16 Ethers

ether   **Ethers** have the general formula ROR′. The two groups, R and R′, may be derived
from saturated, unsaturated, or aromatic hydrocarbons and, for a given ether, may be
alike or different. Table 19.6 lists structural formulas and names for different ethers.

Saturated ethers have little chemical reactivity, but because they readily dissolve
a great many organic substances, ethers are often used as solvents in both laboratory
and manufacturing operations.

Alcohols (ROH) and ethers (ROR′) are isomeric, having the same molecular
formula but different structural formulas. For example, the molecular formula for
ethanol and dimethyl ether is C_2H_6O, but the structural formulas are

$\quad\quad\quad CH_3CH_2OH \quad\quad CH_3-O-CH_3$
$\quad\quad\quad\quad$ ethanol $\quad\quad\quad\quad$ dimethyl ether

These two molecules are extremely different in both physical and chemical properties. Ethanol boils at 78.3°C, and dimethyl ether boils at −23.7°C. In addition,
ethanol is capable of intermolecular hydrogen bonding and therefore has a much
higher boiling point. It also has a greater solubility in water than dimethyl ether.

TABLE 19.6	Names and Structural Formulas of Ethers	

Name*	Formula	Boiling point (°C)
Dimethyl ether (Methoxymethane)	CH_3-O-CH_3	−24
Methyl ethyl ether (Methoxyethane)	$CH_3CH_2-O-CH_3$	8
Diethyl ether (Ethoxyethane)	$CH_3CH_2-O-CH_2CH_3$	35
Ethyl isopropyl ether (2-Ethoxypropane)	$CH_3CH_2-O-\overset{\textstyle\,}{C}HCH_3$ $\quad\quad\quad\quad\ \ \mid$ $\quad\quad\quad\quad CH_3$	54
Divinyl ether	$CH_2=CH-O-CH=CH_2$	39
Anisole (Methoxybenzene)	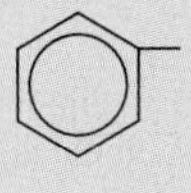 ⬡— OCH_3	154

*The IUPAC name is in parentheses.

Naming Ethers

Like alcohols, individual ethers can have several names. Once widely used as an anesthetic, the ether with the formula $CH_3CH_2-O-CH_2CH_3$ is called diethyl ether, ethyl ether, ethoxyethane, or simply ether. Common names of ethers are formed from the names of the groups attached to the oxygen atom followed by the word *ether*:

$$CH_3 - \boxed{O} - CH_3$$

methyl ether methyl

dimethyl ether

$$CH_3 - \boxed{O} - CH_2CH_3$$

methyl ether ethyl

methyl ethyl ether

In the IUPAC System, ethers are named as alkoxy, $RO-$, derivatives of the alkane corresponding to the longest carbon–carbon chain in the molecule. To name an ether by this system, use these steps:

1. Select the longest carbon–carbon chain and label it with the name of the corresponding alkane.
2. Change the *-yl* ending of the other hydrocarbon group to *-oxy* to obtain the alkoxy group name. For example, CH_3O- is called *methoxy*.
3. Combine the two names from Steps 1 and 2, giving the alkoxy name first, to form the ether name:

$$CH_3 - O - CH_2CH_3$$

This is the longest carbon–carbon chain, so call it *ethane*.

This is the other group; modify its name to *methoxy* and combine with ethane to obtain the name of the ether, *methoxyethane*.

CHEMISTRY IN ACTION Ethers

One of the great advances of modern medicine, ethyl ether is most widely known as a general anesthetic for surgery. Two Americans, Crawford W. Long and William T. Morton, were the first to use ether in this manner. Long, a physician, used ether in a surgical operation as early as 1842 but didn't publish his discovery until 1849. Morton, a dentist, used ether as an anesthetic for dental work in 1846. He publicly demonstrated its effectiveness by administering ether to a patient undergoing surgery at the Massachusetts General Hospital during that same year.

The word *anesthesia* is from the Greek, meaning insensibility, and was suggested to Morton by the poet and physician Oliver Wendell Holmes. A *general anesthetic* is a substance or combination of substances that produces both unconsciousness and insensitivity to pain. Many other substances, including other ethers such as divinyl ether (Vinethene) and methoxyflurane (Penthrane), are general inhalation anesthetics:

$$CH_2=CH-O-CH=CH_2$$
divinyl ether

$$CHCl_2CF_2-O-CH_3$$
methoxyflurane

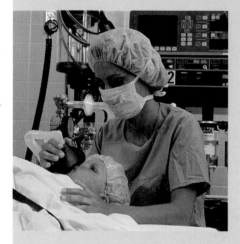

◀ **Anesthetist giving anesthetic to a child.**

These substances are superior in some respects, and have replaced ethyl ether as a general anesthetic.

Ether produces unconsciousness by depressing the activity of the central nervous system. The major disadvantages of ether include flammability, irritation of respiratory passages, and occurrence of nausea and vomiting after its use. These hazards have resulted in a change toward the use of other substances (such as nitrous oxide, N_2O, or halogenated compounds like halothane, $CF_3CHClBr$) as general anesthetics. These substitute compounds also pose hazards and must be used with caution.

Ether is an excellent extracting medium. It acts as a good solvent for separating lipids (ether soluble) from carbohydrates and proteins (ether insoluble). Ether is also the solvent of choice for cocaine users; the technique known as "free-basing" separates cocaine from other substances by extracting it into ether. One method used to apprehend cocaine manufacturers involves tracking large ether shipments.

Thus

$CH_3CH_2-O-CH_2CH_3$ is ethoxyethane
$CH_3CH_2CH_2-O-CH_2CH_2CH_2CH_3$ is propoxybutane

19.17 Aldehydes and Ketones

The aldehydes and ketones are closely related classes of compounds. Their structures contain the **carbonyl group**, $>C=O$, a carbon double bonded to oxygen. **Aldehydes** have at least one hydrogen atom bonded to the carbonyl group, while **ketones** have two alkyl (R) or aromatic (Ar) groups bonded to the carbonyl group:

carbonyl group
aldehyde
ketone

$$\underset{\text{aldehydes}}{R-\overset{\overset{\textstyle O}{\|}}{C}-H \quad Ar-\overset{\overset{\textstyle O}{\|}}{C}-H} \qquad \underset{\text{ketones}}{R-\overset{\overset{\textstyle O}{\|}}{C}-R \quad R-\overset{\overset{\textstyle O}{\|}}{C}-Ar \quad Ar-\overset{\overset{\textstyle O}{\|}}{C}-Ar}$$

520

In a linear expression the aldehyde group is often written as CHO or CH=O. For example,

$$CH_3CHO \quad \text{is equivalent to} \quad CH_3\overset{\displaystyle O}{\overset{\displaystyle \|}{C}}-H$$

In the linear expression of a ketone, the carbonyl group is written as CO; thus

$$CH_3COCH_3 \quad \text{is equivalent to} \quad CH_3\overset{\displaystyle O}{\overset{\displaystyle \|}{C}}CH_3$$

Formaldehyde is the most widely used aldehyde. It is a poisonous, irritating gas that is very soluble in water. Marketed as a 40% aqueous solution called *formalin,* the largest use of this chemical is in the manufacture of polymers. About 2.1 billion pounds (9.6×10^8 kg) of formaldehyde are manufactured annually in the United States. Formaldehyde vapors are intensely irritating to the mucous membranes, and ingestion may cause severe abdominal pains, leading to coma and death.

Acetone and methyl ethyl ketone are widely used organic solvents. Acetone, in particular, is used in very large quantities for this purpose. The U.S. production of acetone is about 1.9 billion pounds (8.7×10^8 kg) annually. It is used as a solvent in the manufacture of drugs, chemicals, and explosives; for removal of paints, varnishes, and fingernail polish; and as a solvent in the plastics industry. Methyl ethyl ketone (MEK) is also widely used as a solvent, especially for lacquers.

19.18 Naming Aldehydes and Ketones

Aldehydes

The IUPAC names of aldehydes are obtained by dropping the final *-e* and adding *-al* to the name of the parent hydrocarbon (i.e., the longest carbon–carbon chain carrying the −CHO group). The aldehyde carbon is always at the end of the carbon chain, is understood to be carbon number 1, and does not need to be numbered. The first member of the homologous series, $H_2C=O$, is methanal. The name *methanal* is derived from the hydrocarbon methane, which contains one carbon atom. The second member of the series is ethanal, the third member of the series is propanal, and so on:

$$CH_4 \qquad H-\overset{\displaystyle O}{\overset{\displaystyle \|}{C}}-H \qquad CH_3CH_3 \qquad CH_3\overset{\displaystyle O}{\overset{\displaystyle \|}{C}}-H$$

methane   methanal   ethane   ethanal

(from methane + *al*)   (from ethane + *al*)

The longest carbon chain containing the aldehyde group is the parent compound. Other groups attached to this chain are numbered and named as we have done previously. For example,

$$CH_3CH_2\overset{\displaystyle CH_3}{\overset{\displaystyle |}{C}}HCH_2CH_2\overset{\displaystyle O}{\overset{\displaystyle \|}{C}}-H$$

4-methylhexanal

Common names for some aldehydes are widely used. The common names for the aldehydes are derived from the common names of the carboxylic acids (see Table 19.7). The *-ic acid* or *-oic acid* ending of the acid name is dropped and is replaced with the suffix *-aldehyde*. Thus the name of the one-carbon acid, formic acid, becomes formaldehyde for the one-carbon aldehyde:

$$
\underset{\text{formic acid}}{\text{H}-\overset{\displaystyle\text{O}}{\overset{\displaystyle\|}{\text{C}}}-\text{OH}}
\qquad
\underset{\text{formaldehyde}}{\text{H}-\overset{\displaystyle\text{O}}{\overset{\displaystyle\|}{\text{C}}}-\text{H}}
\qquad
\underset{\text{acetic acid}}{\text{CH}_3\overset{\displaystyle\text{O}}{\overset{\displaystyle\|}{\text{C}}}-\text{OH}}
\qquad
\underset{\text{acetaldehyde}}{\text{CH}_3\overset{\displaystyle\text{O}}{\overset{\displaystyle\|}{\text{C}}}-\text{H}}
$$

The IUPAC name of a ketone is derived from the name of the alkane corresponding to the longest carbon chain that contains the ketone–carbonyl group. The parent name is formed by changing the *-e* ending of the alkane to *-one*. If the chain is longer than four carbons, it's numbered so that the carbonyl carbon has the smallest number possible, and this number is prefixed to the name of the ketone. Other groups bonded to the parent chain are named and numbered as previously indicated for hydrocarbons and alcohols. For example,

$$
\underset{\text{propanone}}{\text{CH}_3-\overset{\displaystyle\text{O}}{\overset{\displaystyle\|}{\text{C}}}-\text{CH}_3}
\qquad
\underset{\text{2-pentanone}}{\overset{5}{\text{CH}_3}\overset{4}{\text{CH}_2}\overset{3}{\text{CH}_2}-\overset{2}{\overset{\displaystyle\text{O}}{\overset{\displaystyle\|}{\text{C}}}}-\overset{1}{\text{CH}_3}}
\qquad
\underset{\text{4-methyl-3-hexanone}}{\overset{1}{\text{CH}_3}\overset{2}{\text{CH}_2}-\overset{3}{\overset{\displaystyle\text{O}}{\overset{\displaystyle\|}{\text{C}}}}-\underset{\underset{\text{CH}_3}{|}}{\overset{4}{\text{C}}\text{H}}\overset{5}{\text{CH}_2}\overset{6}{\text{CH}_3}}
$$

Note that in 4-methyl-3-hexanone the carbon chain is numbered from left to right to give the ketone group the lowest possible number.

An alternative non-IUPAC method commonly used to name simple ketones is to list the names of the alkyl or aromatic groups attached to the carbonyl carbon together with the word *ketone*.

$$
\underset{(\text{methyl ethyl ketone})}{\underset{\substack{\uparrow\qquad\uparrow\qquad\quad\uparrow\\ \text{methyl \; ketone}\qquad\text{ethyl}}}{\text{CH}_3-\overset{\displaystyle\text{O}}{\overset{\displaystyle\|}{\text{C}}}-\text{CH}_2\text{CH}_3}}
\qquad\qquad
\underset{(\text{methyl isopropyl ketone})}{\underset{\substack{\uparrow\qquad\uparrow\qquad\quad\uparrow\\ \text{methyl \; ketone}\qquad\text{isopropyl}}}{\text{CH}_3-\overset{\displaystyle\text{O}}{\overset{\displaystyle\|}{\text{C}}}-\underset{\overset{|}{\text{CH}_3}}{\overset{\overset{\text{CH}_3}{|}}{\text{C}}\text{H}}\text{CH}_3}}
$$

Two of the most widely used ketones are commonly known by non-IUPAC names: propanone is called acetone, and butanone is known as methyl ethyl ketone, or MEK.

Example 19.8   Write the formulas and the names for the straight-chain five- and six-carbon aldehydes.

Solution   The IUPAC names are based on the five- and six-carbon alkanes. Drop the *-e* of the alkane name and add the suffix *-al*. Pentane, C_5, becomes pentanal and hexane, C_6, becomes hexanal.

$$
\underset{\text{pentanal}}{\text{CH}_3\text{CH}_2\text{CH}_2\text{CH}_2\overset{\displaystyle\text{O}}{\overset{\displaystyle\|}{\text{C}}}-\text{H}}
\qquad\qquad
\underset{\text{hexanal}}{\text{CH}_3\text{CH}_2\text{CH}_2\text{CH}_2\text{CH}_2\overset{\displaystyle\text{O}}{\overset{\displaystyle\|}{\text{C}}}-\text{H}}
$$

Give two names for these ketones:

Example 19.9

(a) CH₃CH₂CCH₂CHCH₃
    (O double bond on C, CH₃ branch)

(b) CH₃CH₂CH₂C—⟨benzene ring⟩
    (O double bond on C)

Solution

(a) The parent carbon chain that contains the carbonyl group has six carbons. Number this chain from the end nearest to the carbonyl group. The ketone group is on C-3, and a methyl group is on C-5. The six-carbon alkane is hexane. Drop the -e from hexane and add -one to give the parent name hexanone. Prefix the name hexanone with 3- to locate the ketone group and with 5-methyl- to locate the methyl group. The name is 5-methyl-3-hexanone. The common name is ethyl isobutyl ketone since the C=O has an ethyl group and an isobutyl group bonded to it.

(b) The longest chain has four carbons. The parent ketone name is butanone, derived by dropping the -e of butane and adding -one. The butanone has a phenyl group attached to C-1. The IUPAC name is therefore 1-phenyl-1-butanone. The common name is phenyl *n*-propyl ketone, since the C=O group has a phenyl and an *n*-propyl group bonded to it.

Practice 19.5

Write the IUPAC names for these molecules:

(a) H—C—CHCH₂CH₂
        | |
       CH₃   CH₂CH₃
    (O double bond on C)

(b) CH₃CH₂CH₂CHCH₂CHCCH₃
              | | ‖
             CH₃   CH₃   O

19.19 Carboxylic Acids

Organic acids, known as **carboxylic acids,** are characterized by the functional group called a **carboxyl group.** The carboxyl group is represented in the following ways:

carboxylic acid
carboxyl group

—C—OH   or   —COOH   or   —CO₂H
 ‖
 O

Open-chain carboxylic acids form a homologous series. The carboxyl group is always at the end of a carbon chain, and the carbon atom in this group is understood to be C-1 in naming the compound.

To name a carboxylic acid by the IUPAC system, first identify the longest carbon chain including the carboxyl group. Then form the acid name by dropping the -e from the corresponding parent hydrocarbon name and adding -oic acid. Thus the names corresponding to the C-1, C-2, and C-3 acids are methanoic acid, ethanoic acid, and propanoic acid. These names are, of course, derived from methane, ethane, and propane:

CH_4	methane	HCOOH	methanoic acid
CH_3CH_3	ethane	CH_3COOH	ethanoic acid
$CH_3CH_2CH_3$	propane	CH_3CH_2COOH	propanoic acid

The IUPAC system is neither the only nor the most generally used method of naming acids. Organic acids are usually known by common names. Methanoic, ethanoic, and propanoic acids are commonly called formic, acetic, and propionic acids, respectively. These names usually refer to a natural source of the acid and are not really systematic. Formic acid was named from the Latin word *formica,* meaning "ant." This acid contributes to the stinging sensation of ant bites. Acetic acid is found in vinegar and is so named from the Latin word for vinegar. The name of butyric acid is derived from the Latin term for butter, since it is a constituent of butterfat. Many of the carboxylic acids, principally those having even numbers of carbon atoms ranging from 4 to about 20, exist in combined form in plant and animal fats. These acids are called *saturated fatty acids.* Table 19.7 lists the IUPAC and common names of the more important saturated acids.

The simplest aromatic acid is benzoic acid. *Ortho*-hydroxybenzoic acid is known as salicylic acid, the basis for many salicylate drugs such as aspirin. There are three methylbenzoic acids, known as *o-, m-,* and *p*-toluic acids.

benzoic acid

salicylic acid
(*o*-hydroxybenzoic acid)

acetylsalicylic acid
(aspirin)

p-toluic acid

The bark of a willow tree is a natural source of salicylic acid.

TABLE
19.7   **Formulas and Names of Saturated Carboxylic Acids**

Formula	IUPAC name	Common name
HCOOH	methanoic acid	formic acid
CH_3COOH	ethanoic acid	acetic acid
CH_3CH_2COOH	propanoic acid	propionic acid
$CH_3(CH_2)_2COOH$	butanoic acid	butyric acid
$CH_3(CH_2)_3COOH$	pentanoic acid	valeric acid
$CH_3(CH_2)_4COOH$	hexanoic acid	caproic acid
$CH_3(CH_2)_6COOH$	octanoic acid	caprylic acid
$CH_3(CH_2)_8COOH$	decanoic acid	capric acid
$CH_3(CH_2)_{10}COOH$	dodecanoic acid	lauric acid
$CH_3(CH_2)_{12}COOH$	tetradecanoic acid	myristic acid
$CH_3(CH_2)_{14}COOH$	hexadecanoic acid	palmitic acid
$CH_3(CH_2)_{16}COOH$	octadecanoic acid	stearic acid
$CH_3(CH_2)_{18}COOH$	eicosanoic acid	arachidic acid

19.20 Esters

Carboxylic acids react with alcohols in an acidic medium to form esters. **Esters** have the general formula RCOOR′ where R′ can be any type of saturated, unsaturated, or aromatic hydrocarbon group. The functional group of the ester is —COOR′:

ester

$$RC\overset{\displaystyle O}{\underset{\displaystyle OR'}{\diagup}} \qquad -C\overset{\displaystyle O}{\underset{\displaystyle OR'}{\diagup}} \quad \text{or} \quad -COOR'$$

ester   functional group of an ester

The reaction of acetic acid and ethyl alcohol is shown as an example of the group of reactions called *esterification*. In addition to the ester, a molecule of water is formed as a product.

$$\underset{\substack{\text{acetic acid}\\ \text{(ethanoic acid)}}}{CH_3C}\overset{\displaystyle O}{\|}\!\!-\!\!OH \; + \; H\!\!-\!\!\underset{\substack{\text{ethyl alcohol}\\ \text{(ethanol)}}}{O\!-\!CH_2CH_3} \;\underset{}{\overset{H^+}{\rightleftharpoons}}\; \underset{\substack{\text{ethyl acetate}\\ \text{(ethyl ethanoate)}}}{CH_3C}\overset{\displaystyle O}{\|}\!\!-\!\!OCH_2CH_3 \; + \; H_2O$$

Esters are alcohol derivatives of carboxylic acids. They are named in much the same way as salts. The alcohol part (R′ in OR′) is named first, followed by the name of the acid modified to end in *-ate*. The *-ic* ending of the organic acid name is replaced by the ending *-ate*. Thus in the IUPAC system, ethanoic acid becomes ethanoate. In the common names, acetic acid becomes acetate. To name an ester, be sure to recognize the portion of the ester molecule that comes from the acid and the portion that comes from the alcohol. In the general formula for an ester, the RC=O comes from the acid and the R′O comes from the alcohol:

$$\underset{\substack{\text{acid} \qquad \text{alcohol}}}{R-C}\overset{\displaystyle O}{\|}\!\!-\!\!O\!-\!R' \qquad \underset{\substack{\text{acetic}\qquad\text{methyl}\\ \text{acid}\qquad\text{alcohol}}}{CH_3C}\overset{\displaystyle O}{\|}\!\!-\!\!OCH_3$$

methyl acetate
(methyl ethanoate)

Esters occur naturally in many varieties of plant life. They often have pleasant, fragrant, fruity odors and are used as flavoring and scenting agents. Esters are insoluble in water but soluble in alcohol. Table 19.8 lists selected esters.

19.21 Polymers—Macromolecules

Some very large molecules (macromolecules) exist in nature, containing tens of thousands of atoms. Some of these, such as starch, glycogen, cellulose, proteins, and DNA, have molar masses in the millions and are central to many of our life processes. Synthetic macromolecules touch every phase of modern living. Today it's hard to imagine a world without polymers. Textiles for clothing, carpeting, and draperies; shoes; toys; automobile parts; construction materials; synthetic rubber; chemical equipment; medical supplies; cooking utensils; synthetic leather; recre-

TABLE 19.8	Odors and Flavors of Selected Esters		
Formula	**IUPAC name**	**Common name**	**Odor or flavor**
$CH_3COOCH_2CH_2CHCH_3$ CH_3	isopentyl ethanoate	isoamyl acetate	banana, pear
$CH_3CH_2CH_2COOCH_2CH_3$	ethyl butanoate	ethyl butyrate	pineapple
$HCOOCH_2CHCH_3$ CH_3	isobutyl methanoate	isobutyl formate	raspberry
$CH_3COOCH_2(CH_2)_6CH_3$	octyl ethanoate	*n*-octyl acetate	orange
⌬—$COOCH_3$ ⌬—OH	methyl-2-hydroxy benzoate	methyl salicylate	wintergreen

Synthetic polymers are used ▶ to form commercial products, such as curtains, carpets, and disposable diapers.

ational equipment—the list goes on. The vast majority of these polymeric materials are based on petroleum. Petroleum is nonreplaceable; therefore our dependence on polymers is another good reason for not squandering our limited world supply of petroleum.

    The process of forming very large, high-molar-mass molecules from smaller units is called **polymerization.** The large molecule, or unit, is called the **polymer** and the small repeating unit, the **monomer.** Polymers containing more than one kind of monomer are called **copolymers.** The term *polymer* is derived from the Greek word *polumerēs,* meaning "having many parts." Ethylene is a monomer and polyethylene is a polymer. Because of their large size, polymers are often called *macromole-*

polymerization
polymer
monomer
copolymer

cules. Some synthetic polymers are called *plastics.* The word *plastic* means "capable of being molded, or pliable."

Polyethylene is an example of a synthetic polymer. Ethylene, derived from petroleum, is made to react with itself to form polyethylene (or polythene).

Polyethylene is a long-chain hydrocarbon made from many ethylene units:

$$n \; CH_2{=}CH_2 \quad {-}CH_2CH_2[CH_2CH_2]_n CH_2CH_2CH_2CH_2{-}$$

ethylene   polyethylene $\longrightarrow$

TABLE 19.9	Polymers Derived from Modified Ethylene Monomers	
Monomer	**Polymer**	**Uses**
$CH_2{=}CH_2$ Ethylene	$-(CH_2{-}CH_2)_n$ Polyethylene	Packing material, molded articles, containers, toys
$CH_2{=}CH$ $\;\;\vert$ $\;\;CH_3$ Propylene	$\left(CH_2{-}CH\right)$ $\qquad\;\;\vert$ $\qquad CH_3{}_n$ Polypropylene	Textile fibers, molded articles, lightweight ropes, autoclavable biological equipment
$CH_2{=}CH$ $\;\;\vert$ $\;\;Cl$ Vinyl chloride	$\left(CH_2{-}CH\right)$ $\qquad\;\;\vert$ $\qquad Cl\;_n$ Polyvinyl chloride	Garden hoses, pipes, molded articles, floor tile, electrical insulation, vinyl leather
$CH_2{=}CH$ $\;\;\vert$ $\;\;CN$ Acrylonitrile	$\left(CH_2{-}CH\right)$ $\qquad\;\;\vert$ $\qquad CN\;_n$ Orlon, Acrilan	Textile fibers
$CF_2{=}CF_2$ Tetrafluoroethylene	$-(CF_2{-}CF_2)_n$ Teflon	Gaskets, valves, insulation, heat-resistant and chemical-resistant coatings, linings for pots and pans
$CH_2{=}CH$ (benzene ring) Styrene	$\left(CH_2{-}CH\right)_n$ (benzene ring) Polystyrene	Molded articles, styrofoam, insulation, toys, disposable food containers
$CH_2{=}C{-}CH_3$ $\qquad\;\vert$ $\qquad C{-}O{-}CH_3$ $\qquad\;\Vert$ $\qquad O$ Methylmethacrylate	$\left(CH_2{-}C\;\;{\overset{CH_3}{\underset{\underset{OCH_3}{O{=}C}}{\vert}}}\right)_n$ Lucite, Plexiglas (acrylic resins)	Contact lenses, clear sheets for windows and optical uses, molded articles, automobile finishes

A typical polyethylene molecule is made up of about 2500–25,000 ethylene molecules joined in a continuous structure. Over 12 billion pounds of polyethylene are produced annually in the United States. Its uses are as varied as any single substance known and include chemical equipment, packaging material, electrical insulation, films, industrial protective clothing, and toys.

Ethylene derivatives, in which one or more hydrogen atoms have been replaced by other atoms or groups, can also be polymerized. Many of our commercial synthetic polymers are made from such modified ethylene monomers. For example, $CH_2\!=\!CH-$ is a vinyl group and $CH_2\!=\!CHCl$ is vinyl chloride. Vinyl chloride can be polymerized to polyvinylchloride (PVC). The names, structures, and some uses for several of these polymers are given in Table 19.9.

Concepts in Review

1. Describe the tetrahedral nature of the carbon atom.
2. Identify the different types of bonding between carbon atoms.
3. Explain what is meant by a homologous series.
4. Describe the phenomenon of isomerism.
5. Write the names and formulas for the first ten normal alkanes.
6. Write the names and structural formulas for the common alkyl groups, C_nH_{2n+1}.
7. Write common or IUPAC names for each class of compounds discussed in this chapter.
8. Write structural formulas from compound names using the classifications given in this chapter.
9. Write structural formulas for the isomers of a compound when given a name or molecular formula.
10. Write equations for the combustion of alkanes.
11. List the major classes of hydrocarbon derivatives and their functional groups.
12. Distinguish by structure an alkane, an alkene, and an alkyne.
13. Write equations for addition reactions of alkenes.
14. Describe the nature of benzene and how its properties differ from open-chain unsaturated hydrocarbons.
15. Recognize and write formulas for benzene compounds.
16. Recognize and identify primary, secondary, and tertiary alcohols.
17. Write the equations for the preparation of esters from carboxylic acids and alcohols.
18. Identify the alcohol and the carboxylic acid needed to prepare a given ester.
19. Write structural formulas for polymers derived from modified ethylene monomers when given a monomer, or vice versa.

Key Terms

addition reaction (19.9)
alcohol (19.14)
aldehyde (19.17)
alkane (19.4)
alkene (19.7)
alkyl group (19.6)
alkyne (19.7)
aromatic compound (19.10)
carbonyl group (19.17)
carboxyl group (19.19)
carboxylic acid (19.19)

copolymer (19.21)
ester (19.20)
ether (19.16)
functional group (19.2)
homologous series (19.4)
hydrocarbon (19.3)
isomerism (19.5)
isomers (19.5)
ketone (19.17)
monomer (19.21)
organic chemistry (19.1)

polyhydroxy alcohol (19.14)
polymer (19.21)
polymerization (19.21)
primary alcohol (19.14)
saturated hydrocarbon (19.3)
secondary alcohol (19.14)
tertiary alcohol (19.14)
unsaturated hydrocarbon (19.3)
vital-force theory (19.1)

Questions

1. What bonding characteristic of carbon is primarily responsible for the existence of so many organic compounds?

2. What is the most common geometric arrangement of covalent carbon bonds? Illustrate this structure.

3. In addition to single bonds, what other types of bonds can carbon atoms form? Give examples.

4. Draw a Lewis structure for
   (a) a single carbon atom
   (b) molecules of methane, ethene, and ethyne

5. Write the names and draw the structural formulas for the first ten normal alkanes.

6. Write the names and draw the structural formulas for all possible alkyl groups, C_nH_{2n+1}, containing from one to four carbon atoms.

7. Which one of these compounds belongs to a different homologous series than the others?
   (a) C_2H_4 (c) C_6H_{14}
   (b) CH_4 (d) C_5H_{12}

8. Which word does not belong with the others?
   (a) alkane   (d) ethane
   (b) paraffin   (e) ethylene
   (c) saturated   (f) pentane

9. What is the single most important reaction of alkanes?

10. Draw the structure for vinyl acetylene, which has the formula C_4H_4 and contains one double bond and one triple bond.

11. An open-chain hydrocarbon has the formula C_6H_8. What possible combinations of carbon–carbon double bonds and/or carbon–carbon triple bonds can be in this compound?

12. Why is ethylene glycol (1,2-ethanediol) superior to methyl alcohol (methanol) as an antifreeze for automobile radiators?

13. Alcohols are considered to be toxic to the human body—with ethanol, the alcohol in alcoholic beverages, being the least toxic one. What are the hazards of ingesting (a) methanol and (b) ethanol?

14. What is wrong with the name 1-methylpentane?

Paired Exercises

15. Name these normal alkyl groups:
    (a) $C_5H_{11}-$
    (b) $C_7H_{15}-$

16. Name these normal alkyl groups:
    (a) $C_8H_{17}-$
    (b) $C_{10}H_{21}-$

17. Write condensed structural formulas for the five isomers of hexane.

18. Write condensed structural formulas for the nine isomers of heptane.

19. Give IUPAC names for the following:
    (a) CH_3CH_2Cl
    (b) $CH_3CHClCH_3$
    (c) $(CH_3)_2CHCH_2Cl$

20. Give IUPAC names for the following:
    (a) $CH_3CH_2CH_2Cl$
    (b) $(CH_3)_3CCl$
    (c) $CH_3CHClCH_2CH_3$

21. Give IUPAC names for these compounds:
    (a) $CH_3CHCH_2CHCH_2CH_2CH_3$
         $\quad\;\;|\qquad\quad|$
         $\quad CH_3\quad CH_2CH_3$
    (b) $CH_3CHCH_2CHCH_2-CHCH_2CH_3$
         $\quad\;\;|\qquad\;\;|\qquad\quad|$
         $\quad CH_3\quad CH_2CH_3\;\; CH_2CH_3$

22. Give IUPAC names for these compounds:
    (a) $CH_3CHCH_2CH_2CHCH_3$
         $\quad\;\;|\qquad\qquad\;\;|$
         $\quad CH_2\qquad\quad CH_2$
         $\quad\;\;|\qquad\qquad\;\;|$
         $\quad CH_3\qquad\quad CH_3$
    (b) $CH_3CH_2CH_2CHCH_2CH_3$
         $\qquad\qquad\qquad|$
         $\qquad\qquad\; CH_3CHCH_3$

23. Draw structural formulas of these compounds:
    (a) 2,4-dimethylpentane
    (b) 2,2-dimethylpentane
    (c) 3-isopropyloctane

24. Draw structural formulas of these compounds:
    (a) 4-ethyl-2-methylhexane
    (b) 4-*tert*-butylheptane
    (c) 4-ethyl-7-isopropyl-2,4,8-trimethyldecane

25. One name in each pair is incorrect. Draw structures corresponding to each name and indicate which name is incorrect.
  (a) 2-methylbutane and 3-methylbutane
  (b) 2-ethylbutane and 3-methylpentane

27. Draw structural formulas for all the isomers of
  (a) CH_3Br
  (b) CH_2Cl_2
  (c) C_2H_5Cl
  (d) C_3H_7Br

29. Using alkenes and any other necessary inorganic reagents, write reactions showing the formation of
  (a) $CH_3CH_2CHClCH_2Cl$
  (b) CH_3CH_2Br

31. Draw structures containing two carbon atoms for the following classes of compounds:
  (a) alkene
  (b) alkyne
  (c) alkyl halide
  (d) alcohol

33. Give the IUPAC name for each compound in Exercise 31.

35. Draw structural formulas for the following:
  (a) chloromethane
  (b) vinyl chloride
  (c) chloroform
  (d) 1,1-dibromoethene

37. Draw structural formulas for the following:
  (a) 2,5-dimethyl-3-hexene
  (b) 2-ethyl-3-methyl-1-pentene
  (c) 4-methyl-2-pentene

39. Name these compounds:
  (a) $CH_3CH{=}CCH_2CH_2CH_3$
  $\qquad\qquad\quad |$
  $\qquad\qquad\ CH_3$
  (b) $CH_3C{=}C-CH_3$
  $\qquad\ \ |\quad\ \ |$
  $\qquad\ CH_3\ \ CH_3$

41. Complete these reactions and name the products:
  (a) $CH_2{=}CHCH_3 + Br_2 \longrightarrow$
  (b) $CH_2{=}CH_2 + HBr \longrightarrow$
  (c) $CH_3CH{=}CHCH_3 + H_2 \xrightarrow[\text{1 atm}]{\text{Pt 25°C}}$

26. One name in each pair is incorrect. Draw structures corresponding to each name and indicate which name is incorrect.
  (a) 2-dimethylbutane and 2,2-dimethylbutane
  (b) 2,4-dimethylhexane and 2-ethyl-4-methyl-pentane

28. Draw structural formulas for all the isomers of
  (a) C_4H_9I
  (b) $C_3H_6Cl_2$
  (c) C_3H_6BrCl
  (d) $C_4H_8Cl_2$

30. Using alkenes and any other necessary inorganic reagents, write reactions showing the formation of
  (a) $CH_3CHBrCH_2Br$
  (b) $CH_3CHBrCH_3$

32. Draw structures containing two carbon atoms for the following classes of compounds:
  (a) ether
  (b) aldehyde
  (c) carboxylic acid
  (d) ester

34. Give the IUPAC name for each compound in Exercise 32.

36. Draw structural formulas for the following:
  (a) hexachloroethane
  (b) iodoethyne
  (c) 6-bromo-3-methyl-3-hexene-1-yne
  (d) 1,2-dibromoethene

38. Draw structural formulas for the following:
  (a) 1,2-diphenylethene
  (b) 3-penten-1-yne
  (c) 3-phenyl-1-butyne

40. Name these compounds:
  (a) $CH_3CH_2CHCH{=}CH_2$
  $\qquad\qquad\ |$
  $\qquad\qquad CH$
  $\qquad H_3C\diagup\ \diagdown CH_3$
  (b) $CH_3CH_2CH{=}CCH_2CH_3$
  $\qquad\qquad\qquad\ \ |$
  $\qquad\qquad\qquad\ CH_3$

42. Complete these reactions and name the products:
  (a) $CH_2{=}CH_2 + H_2O \xrightarrow{\text{H+}}$
  (b) $CH{\equiv}CH + 2\,Br_2 \longrightarrow$
  (c) $CH_2{=}CH_2 + H_2 \xrightarrow[\text{1 atm}]{\text{Pt 25°C}}$

43. Name these compounds:

(a) 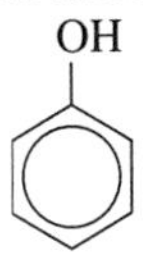 OH

(b) 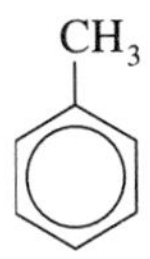 CH$_3$

(c) 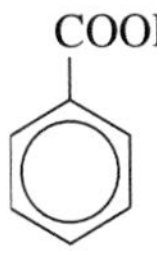 COOH

(d) 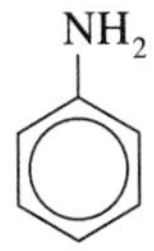 NH$_2$

(e) 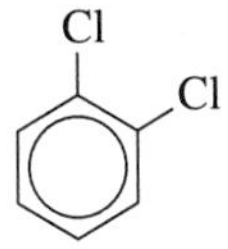 Cl
Cl

(f) 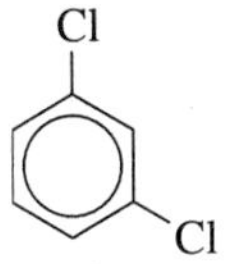 Cl
Cl

44. Name these compounds:

(a) 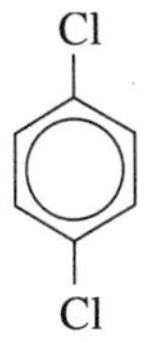 Cl
Cl

(b) 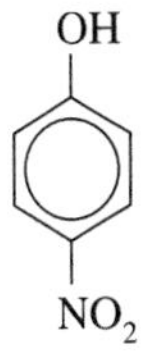 OH
NO$_2$

(c) 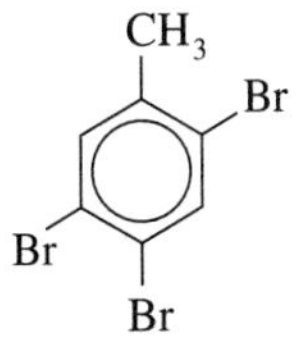 CH$_3$
Br
Br
Br

(d) 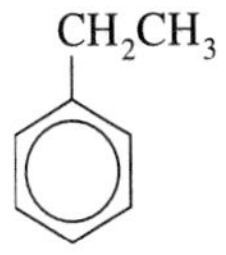 CH$_2$CH$_3$

(e) 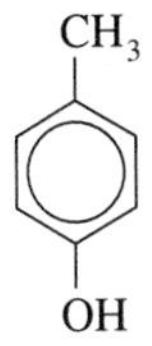 CH$_3$
OH

(f) 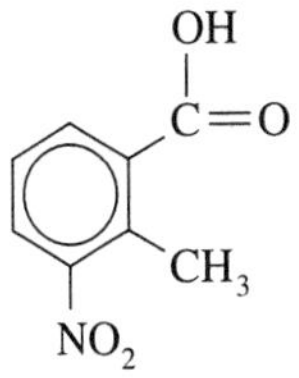 OH
C=O
CH$_3$
NO$_2$

45. Draw structural formulas for
(a) benzene
(b) toluene
(c) benzoic acid
(d) aniline

46. Draw structural formulas for
(a) phenol
(b) *o*-bromochlorobenzene
(c) 1,3-dichloro-5-nitrobenzene
(d) *m*-dinitrobenzene

47. Draw structural formulas for
(a) ethylbenzene
(b) 1,3,5-tribromobenzene

48. Draw structural formulas for
(a) *tert*-butylbenzene
(b) 1,1-diphenylethane

49. Draw structural formulas and write the IUPAC names for all isomers of trichlorobenzene, C$_6$H$_3$Cl$_3$.

50. Draw structural formulas and write the IUPAC names for all isomers of dichlorobromobenzene, C$_6$H$_3$Cl$_2$Br.

51. Write the IUPAC name for these alcohols. Classify each as primary, secondary, or tertiary alcohols.
(a) 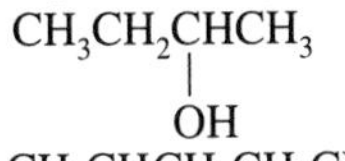 CH$_3$CH$_2$CHCH$_3$
OH
(b) 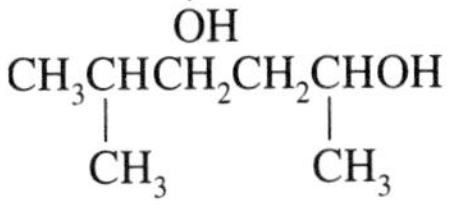 CH$_3$CHCH$_2$CH$_2$CHOH
CH$_3$ CH$_3$

52. Write the IUPAC name for these alcohols. Classify each as primary, secondary, or tertiary alcohols.
OH
(a) 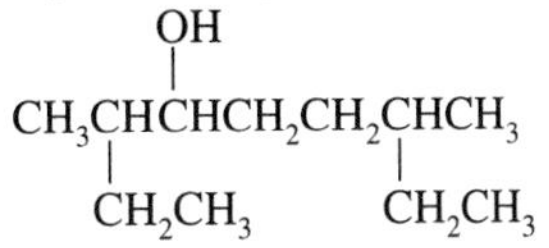 CH$_3$CHCHCH$_2$CH$_2$CHCH$_3$
CH$_2$CH$_3$ CH$_2$CH$_3$
(b) 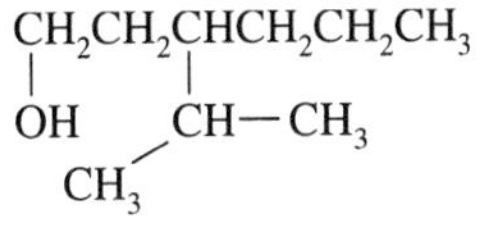 CH$_2$CH$_2$CHCH$_2$CH$_2$CH$_3$
OH   CH—CH$_3$
CH$_3$

53. Draw structural formulas for
(a) 2-pentanol
(b) isopropyl alcohol

54. Draw structural formulas for
(a) 2,2-dimethyl-1-heptanol
(b) 1,3-propanediol

55. Name these aldehydes:
(a) CH$_2$=O
O
(b) 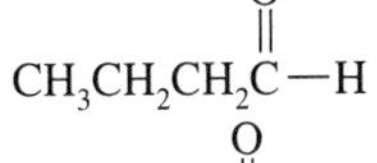 CH$_3$CH$_2$CH$_2$C—H
O
(c) CH$_3$CHCH$_2$C—H
CH$_3$

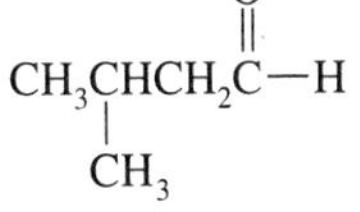

56. Name these aldehydes:
(a) 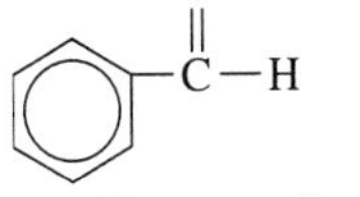
O
C—H
H   O
(b) O=CCH$_2$CH$_2$C—H
O
(c) CH$_3$CHCH$_2$C—H
OH

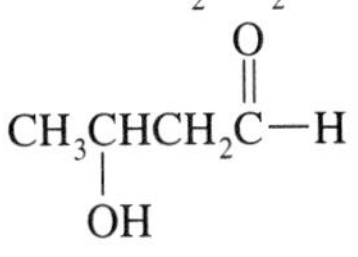

57. Name these ketones:
(a) CH_3COCH_3
(b) $CH_3CH_2COCH_3$
(c)

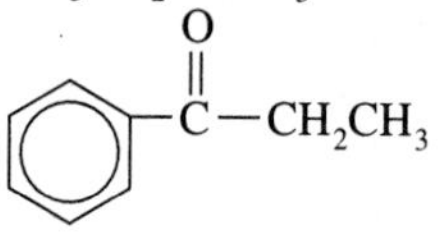

58. Name these ketones:

(a) 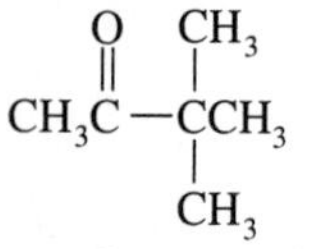 $CH_3\overset{\overset{O}{\|}}{C}-\overset{\overset{CH_3}{|}}{\underset{\underset{CH_3}{|}}{C}}CH_3$

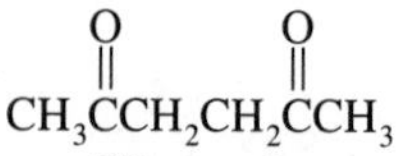

(b) $CH_3\overset{\overset{O}{\|}}{C}CH_2CH_2\overset{\overset{O}{\|}}{C}CH_3$

(c) $CH_3\overset{\overset{CH_3}{|}}{\underset{\underset{OH}{|}}{C}}-CH_2\overset{\overset{O}{\|}}{C}CH_3$

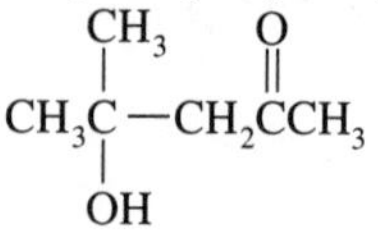

59. Name these acids:
(a) $CH_3CHBrCOOH$
(b) $CH_2{=}CHCH_2COOH$
(c) $CH_3CH_2CH_2COOH$
(d)

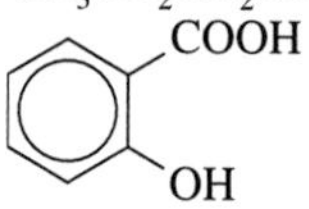

60. Name these acids:
(a) 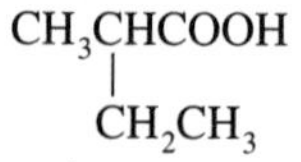 $CH_3\underset{\underset{CH_2CH_3}{|}}{CH}COOH$

(b)

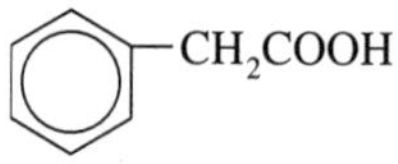

(c) $CH_3CH_2CH_2CH_2COOH$
(d)

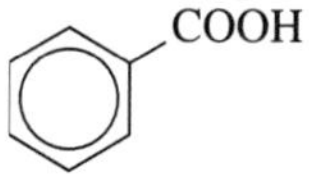

61. Draw structural formulas for these esters:
(a) ethyl formate
(b) methyl ethanoate
(c) isopropyl propanoate

62. Draw structural formulas for these esters:
(a) *n*-nonyl acetate
(b) ethyl benzoate
(c) methyl salicylate

63. Write IUPAC names for these esters:
(a)  $CH_3\underset{\underset{O}{\|}}{C}OCH_2CH_3$

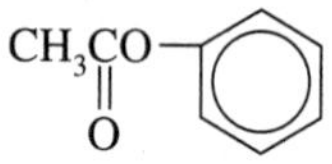

(b) $CH_3\underset{\underset{O}{\|}}{C}O-$

(c) $CH_3CH_2CH_2\overset{\overset{O}{\|}}{C}-O-\overset{\overset{CH_3}{|}}{C}HCH_3$

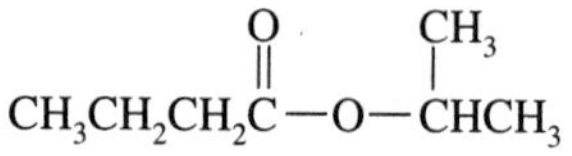

64. Write IUPAC names for these esters:
(a) 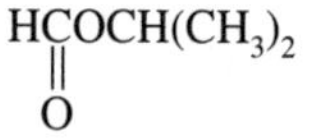 $H\underset{\underset{O}{\|}}{C}OCH(CH_3)_2$

(b)

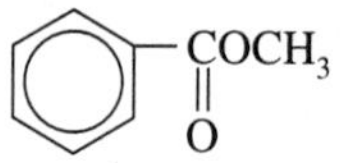

(c) 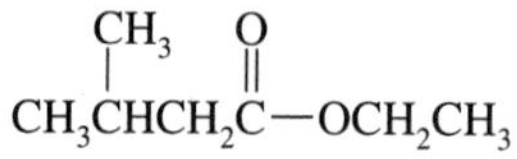 $CH_3\overset{\overset{CH_3}{|}}{C}HCH_2\overset{\overset{O}{\|}}{C}-OCH_2CH_3$

65. Complete these equations:
(a) $CH_3COOH + NaOH \longrightarrow$
(b) 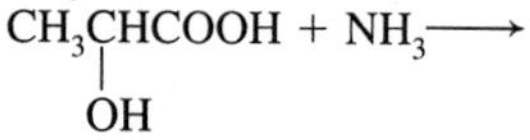 $CH_3\underset{\underset{OH}{|}}{C}HCOOH + NH_3\longrightarrow$

66. Complete these equations:
(a) $CH_3COOH + KOH \longrightarrow$
(b)

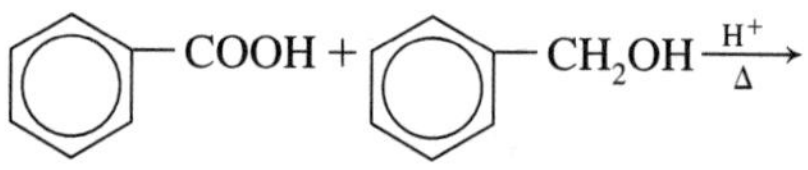

67. Ethylene and its derivatives are the most common monomers for polymers. Write formulas for these ethylene-based polymers:
(a) polyethylene
(b) polyvinyl chloride

68. Ethylene and its derivatives are the most common monomers for polymers. Write formulas for these ethylene-based polymers:
(a) polyacrylonitrile
(b) Teflon

Additional Exercises

69. Draw the structural formulas and write the IUPAC names for all isomers of (a) pentyne, C_5H_8, and (b) hexyne, C_6H_{10}.

70. Eight open-chain isomeric alcohols have the formula $C_5H_{11}OH$.
(a) Draw the structural formula and write the IUPAC name for each of these alcohols.
(b) Indicate which of these isomers are primary, secondary, and tertiary alcohols.

71. What is the molar mass of an open-chain saturated alcohol containing 30 carbon atoms? This alcohol is present in beeswax as an ester.

72. Write the common name and structure of the ether that is isomeric with (a) 1-propanol, (b) ethanol, and (c) isopropyl alcohol.

73. Give the IUPAC and common names and structures of all ethers having the molecular formula $C_5H_{12}O$.

74. Draw structural formulas for propanal and propanone. From these formulas, do you think that aldehydes and ketones are isomeric with each other? Show evidence and substantiate your answer by testing with a four-carbon aldehyde and ketone.

75. Give the common and IUPAC names for the first five straight-chain carboxylic acids.

76. Draw the structural formulas and write the IUPAC names for all the isomers of hexanoic acid, $CH_3(CH_2)_4COOH$.

77. Write equations for the preparation of these esters:
(a) ethyl formate
(b) methyl propanoate
(c) *n*-propyl benzoate

78. (a) Draw a structural formula showing the polymer that can be formed from the following monomers (show four units): (1) propylene, (2) 1-butene, (3) 2-butene
(b) How many ethylene units are in a polyethylene molecule that has a molar mass of 35,000?

79. How many different dibromobenzenes are there? How many tribromobenzenes? Show structures.

*80. What is the oxidation number of carbon in these compounds?
(a) CH_3OH
(b) CO_2
(c) C_6H_6

81. A compound has the following composition:
$C = 24.3\%$
$H = 4.1\%$
$Cl = 71.7\%$
If 140.3 mL of the vapor at 100°C and 740 mm Hg pressure has a mass of 0.442 g, what is the molecular formula of the compound? Draw possible structures of the compound.

82. A compound was found to contain 24 g of carbon, 4 g of hydrogen, and 32 g of oxygen. The molar mass of the compound is 60. Which of these structures is correct for that compound?
(a) C_2H_5OH (d) COH
(b) CH_3CHO (e) CH_3COOH
(c) CH_3OCH_3

83. Write formulas for
(a) ethyl alcohol   (d) decane
(b) iodomethane   (e) *tert*-butyl alcohol
(c) 2-chloropentane

84. Write equations for the reactions in acid solution between
(a) methanol and formic acid
(b) butanol and butanoic acid
(c) hexanol and hexanoic acid
Name the esters produced in each case.

85. Draw the structural formula for
(a) a six-carbon alkane with methyl branches on two adjacent carbons
(b) a six-carbon alcohol with an —OH group not at the end of the chain
(c) a three-carbon organic (carboxylic) acid
(d) a four-carbon aldehyde
(e) the ester formed when a four-carbon alcohol reacts with a three-carbon acid

Answers to Practice Exercises

19.1 (a) 2,3,3-trimethylhexane,
(b) 3-ethyl-2,5-dimethyl-heptane
19.2 (a) 2,5-dimethyl-3-heptene,
(b) 4-ethyl-2-octyne

19.3 (a) *m*-chloroethylbenzene,
(b) 3,5-dibromophenol,
(c) *o*-nitroaniline
19.4 (a) secondary, 6-methyl-3-heptanol;
(b) tertiary, 2,4-dimethyl-2-pentanol

19.5 (a) 2-methylhexanal,
(b) 3,5-dimethyl-2-octanone

CHAPTER

20

Introduction to Biochemistry

▲
The energy to climb mountains, breathe, and live our busy daily lives comes from chemical reactions performed by biomolecules.

CHAPTER 20 / OUTLINE

The study of life has long fascinated us—it's probably the most intriguing of all scientific studies, but the answer to our question, "What is life?" still eludes us. The chemical bases for certain fundamental biological processes are well understood; we know how sunlight and carbon dioxide are converted into food. We can pinpoint specific genes and identify their function. We even use DNA to "fingerprint" suspects at crime scenes. But, still we search for an answer to "what is life?"

Common molecular principles underscore the diversity of life, which enable us to produce essential substances, such as insulin for diabetics, using bacteria as the manufacturing machine. Other substances, such as the human growth factor, are being genetically engineered as well.

Biochemistry is more important than ever to the practice of medicine and the study of nutrition. Today we routinely measure the severity of a heart attack by the levels of certain enzymes in the blood, and the role of cholesterol, fats, and trace elements in our diet are current topics of great interest.

20.1 Chemistry in Living Organisms

Chemical substances present in living organisms—from microbes to humans—range in complexity from water and simple salts to DNA (deoxyribonucleic acid) molecules containing tens of thousands of atoms. Four of the chemical elements, hydrogen, carbon, nitrogen, and oxygen, make up approximately 95% of the mass of living matter. Small amounts of sulfur, phosphorus, calcium, sodium, potassium, chlorine, magnesium, and iron, together with trace amounts of many other elements such as copper, manganese, zinc, cobalt, and iodine, are also found in living organisms. The human body consists of about 60% water with some tissues having a water content as high as 80%.

Biochemistry is the branch of chemistry concerned with the chemical reactions occurring in living organisms. Its scope includes such processes as growth, respiration, digestion, metabolism, and reproduction.

The four major classes of biomolecules upon which all life depends are carbohydrates, lipids, proteins, and nucleic acids. Each kind of living organism has an amazing ability to select and synthesize a large portion of the many complicated molecules needed for its existence. In fact, the processes carried out in a living organism are similar to those of a highly automated, smoothly running chemical factory. But unlike chemical factories, living organisms are able to expand (grow), repair damage (if not too severe), and, finally, reproduce themselves.

biochemistry

20.2 Carbohydrates

Chemically, **carbohydrates** are polyhydroxy aldehydes or polyhydroxy ketones or substances that yield these compounds when hydrolyzed. The name *carbohydrate* was given to this class of compounds many years ago by French scientists who

carbohydrate

▲
The carbohydrates in these hay rolls are storage molecules for energy from the Sun.

called them *hydrates de carbone* because their empirical formulas approximated $(C \cdot H_2O)_n$. However, the hydrogen and oxygen do not actually exist as water or in hydrate form as we have seen in such compounds as $BaCl_2 \cdot 2\ H_2O$. Empirical formulas used to represent carbohydrates are $C_x(H_2O)_y$ and $(CH_2O)_n$.

Carbohydrates, also known as saccharides, occur naturally in plants and are one of the three principal classes of animal food. The other two classes of foods are lipids (fats) and proteins. Plants are able to synthesize carbohydrates by the photosynthetic process. Animals are incapable of this synthesis and are dependent on the plant kingdom for their source of carbohydrates. Animals are, however, capable of converting plant carbohydrate into glycogen (animal carbohydrate) and storing this glycogen throughout the body as a reserve energy source. The amount of energy available from carbohydrates is about 17 kJ/g (4 Cal/g).

Carbohydrates exist as sugars, starches, and cellulose. The simplest of these are the sugars. The names for sugars end in *-ose* (e.g., glucose, sucrose, maltose). Carbohydrates are classified as monosaccharides, disaccharides, oligosaccharides, and polysaccharides according to the number of monosaccharide units linked in a molecule.

Monosaccharides

monosaccharide

Monosaccharides are carbohydrates that cannot be hydrolyzed to simpler carbohydrate units. They are often called simple sugars, the most common of which is glucose. Monosaccharides containing three-, four-, five-, and six-carbon atoms are called *trioses, tetroses, pentoses,* and *hexoses,* respectively. Monosaccharides that contain an aldehyde group on one carbon atom and a hydroxyl group on each of the other carbon atoms are called *aldoses*. Ketoses are monosaccharides that contain a ketone group on one carbon atom and a hydroxyl group on each of the other carbons.

$$H-\overset{\displaystyle H}{\underset{\displaystyle \|}{C}}=O \qquad -OH \qquad \overset{\displaystyle O}{\underset{\displaystyle \|}{\diagup C \diagdown}}$$

aldehyde   hydroxyl   ketone

We can write structural formulas for many monosaccharides, but only a limited number are of biological importance. Sixteen different isomeric aldohexoses of formula $C_6H_{12}O_6$ are known; glucose and galactose are the most important of these. Most sugars exist predominantly in a cyclic structure, forming a five- or six-membered ring. For example, glucose exists as a ring in which C-1 is bonded through an oxygen atom to C-5:

H—C=O
H—C—OH
H—C—OH
CH$_2$OH

erythrose
(an aldotetrose)

open-chain form
of glucose

cyclic form of glucose

Glucose Glucose, $C_6H_{12}O_6$, is the most important of the monosaccharides. An aldohexose found in the free state in plants and animal tissues, glucose is commonly known as *dextrose* or *grape sugar* and is a component of the disaccharides sucrose, maltose, and lactose; it is also the monomer of the polysaccharides starch, cellulose, and glycogen. Among the common sugars, glucose is of intermediate sweetness (see Table 20.1).

Glucose is the key sugar of the body and is carried by the bloodstream to all body parts. The concentration of glucose in the blood is normally 80–100 mg per 100 mL of blood. Because glucose is the most abundant carbohydrate in the blood, it is also sometimes known as *blood sugar*. Glucose requires no digestion and therefore may be given intravenously to patients who cannot take food by mouth. Glucose is found in the urine of those who have diabetes mellitus (sugar diabetes), a condition called *glycosuria*.

TABLE 20.1	Relative Sweetness of Sugars
Fructose	100
Sucrose	58
Glucose	43
Maltose	19
Galactose	19
Lactose	9.2
Invert sugar	75

Galactose Galactose, $C_6H_{12}O_6$, is also an aldohexose and occurs along with glucose in the disaccharide lactose and in many oligo- and polysaccharides, such as pectin, gums, and mucilages. It is an isomer of glucose, differing only in the spatial arrangement of the $-H$ and $-OH$ groups around C-4. Galactose is synthesized in the mammary glands to make the lactose of milk, and is less than half as sweet as glucose.

galactose
(an aldohexose)

cyclic form of galactose

A severe inherited disease, called galactosemia, is the inability of infants to metabolize galactose. The galactose concentration increases markedly in the blood and also appears in the urine. Galactosemia causes vomiting, diarrhea, enlargement of the liver, and often mental retardation. If not recognized within a few days after birth, it can lead to death. But if diagnosis is made early and lactose is excluded from the diet, the symptoms disappear and normal growth is resumed.

Fructose Fructose, $C_6H_{12}O_6$, also known as levulose, is a ketohexose and occurs in fruit juices, honey, and (along with glucose) as a constituent of the disaccharide sucrose. Fructose is the major constituent of the polysaccharide inulin, a starchlike substance present in many plants, such as dahlia tubers, chicory roots, and Jerusalem artichokes. Fructose is the sweetest of all the sugars, being about twice as sweet as glucose. This sweetness accounts for the sweet taste of honey; the enzyme invertase, which is present in bees, splits sucrose into glucose and fructose. Fructose is metabolized directly but is also readily converted to glucose in the liver.

$$
\begin{array}{c}
\overset{1}{C}H_2OH \\
|\ 2 \\
C=O \\
|\ 3 \\
HO-C-H \\
|\ 4 \\
H-C-OH \\
|\ 5 \\
H-C-OH \\
|\ 6 \\
CH_2OH
\end{array}
$$

fructose
(a ketohexose)

cyclic form of fructose

Ribose Ribose, $C_5H_{10}O_5$, is an aldopentose and is present in adenosine triphosphate (ATP), one of the chemical energy carriers in the body. Ribose and one of its derivatives, deoxyribose, are also important components of the nucleic acids RNA and DNA, respectively—the genetic information carriers in the body.

$$
\begin{array}{c}
\overset{1}{H}-C=O \\
|\ 2 \\
H-C-OH \\
|\ 3 \\
H-C-OH \\
|\ 4 \\
H-C-OH \\
|\ 5 \\
CH_2OH
\end{array}
$$

ribose
(an aldopentose)

cyclic form of ribose

Disaccharides

Lactose is a disaccharide found in breast milk.
▼

Disaccharides are carbohydrates whose molecules yield two molecules of the same or of different monosaccharides when hydrolyzed. The three disaccharides that are especially important from a biological viewpoint are sucrose, lactose, and maltose. Sucrose, $C_{12}H_{22}O_{11}$, which is commonly known as *table sugar,* is found in the free state throughout the plant kingdom. Sugar cane contains 15–20% sucrose, and sugar beets contain 10–17%. Maple syrup and sorghum are also good sources of sucrose.

Lactose, $C_{12}H_{22}O_{11}$, or *milk sugar,* is found free in nature mainly in the milk of mammals. Human milk contains about 6.7% lactose, and cow's milk contains about 4.5%.

Maltose, $C_{12}H_{22}O_{11}$, is found in sprouting grain but occurs much less commonly (in nature) than either sucrose or lactose. Maltose is prepared commercially by the partial hydrolysis of starch, catalyzed either by enzymes or by dilute acids.

Disaccharides are not used directly in the body but are first hydrolyzed to monosaccharides. Disaccharides yield two monosaccharide molecules when hydrolyzed in the laboratory at elevated temperatures in the presence of hydrogen ions (acids) as catalysts. In biological systems, enzymes (biochemical catalysts) carry out the reaction. A different enzyme is required for the hydrolysis of each of the three disaccharides:

$$\text{Sucrose + Water} \xrightarrow{\text{H}^+ \text{ or sucrase}} \text{Glucose + Fructose}$$

$$\text{Lactose + Water} \xrightarrow{\text{H}^+ \text{ or lactase}} \text{Galactose + Glucose}$$

$$\text{Maltose + Water} \xrightarrow{\text{H}^+ \text{ or maltase}} \text{Glucose + Glucose}$$

The structure of a disaccharide is derived from two monosaccharide molecules by the elimination of a water molecule between them. In maltose, for example, the two monosaccharides are glucose. The water molecule is eliminated between the OH group on C-1 of one glucose unit and the OH group on C-4 of the other glucose unit. Thus the two glucose units are joined through an oxygen atom at C-1 and C-4.

▲
Cotton is a complex carbohydrate made up of almost pure cellulose.

Sucrose consists of a glucose unit and a fructose unit linked through an oxygen atom from C-1 on glucose to C-2 on fructose. In lactose, the linkage is from C-1 of galactose through an oxygen atom to C-4 of glucose.

Polysaccharides

Polysaccharides are also called *complex carbohydrates* and can be hydrolyzed to a large number of monosaccharide units. The molar masses of polysaccharides range up to 1 million or more. Three of the most important polysaccharides are starch, glycogen, and cellulose.

polysaccharide

Starch is a polymer of glucose and is found mainly in the seeds, roots, and tubers of plants. Corn, wheat, potatoes, rice, and cassava are the chief sources of starch whose principal use is for food.

Glycogen is the reserve carbohydrate of the animal kingdom and is often called *animal starch*. Glycogen is formed in the body by polymerization of glucose and is stored especially in the liver and in muscle tissue. Glycogen also occurs in some insects and lower plants including fungi and yeasts.

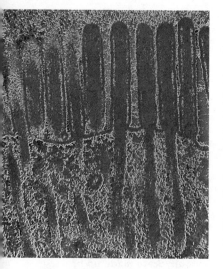

Microvilli (red) in the epithelium of the human duodenum are responsible for the absorption of food. The finger-like projections maximize absorption capacity.

Cellulose, like starch and glycogen, is also a polymer of glucose. It differs from starch and glycogen in the manner in which the cyclic glucose units are linked to form chains. Cellulose is the most abundant organic substance found in nature. It is the chief structural component of plants and wood. Cotton fibers are almost pure cellulose, and wood, after removal of moisture, consists of about 50% cellulose. An important substance in the textile and paper industries, cellulose is also used to make rayon fibers, photographic film, guncotton, and cellophane. Humans cannot utilize cellulose as food because they lack the necessary enzymes to hydrolyze it to glucose. However, cellulose is an important source of bulk in the diet.

The digestion and metabolism of carbohydrates is a complex biochemical process. It starts in the mouth where the enzyme amylase in the saliva begins the hydrolysis of starch to maltose and temporarily stops in the stomach where hydrochloric acid deactivates the enzyme. Digestion continues in the intestines where the hydrochloric acid is neutralized and pancreatic enzymes complete the hydrolysis to maltose. The enzyme maltase then catalyzes the digestion of maltose to glucose:

$$\text{Starch} \xrightarrow{\text{amylase}} \text{Dextrins} \xrightarrow{\text{amylase}} \text{Maltose} \xrightarrow{\text{maltase}} \text{Glucose}$$

Other specific enzymes in the intestines convert sucrose and lactose to monosaccharides. Glucose is absorbed through the intestinal walls into the bloodstream where it is transported to the cells to be used for energy. Excess glucose is rapidly removed by the liver and muscle tissue, where it is polymerized and stored as glycogen. As the body calls for it, glycogen is converted back to glucose, which is ultimately oxidized to carbon dioxide and water with the release of energy. This energy is used by the body for maintenance, growth, and other normal functions.

20.3 Lipids

lipids

Lipids are a group of oily, greasy organic substances found in living organisms that are water insoluble but soluble in organic solvents, such as diethyl ether, benzene, and chloroform. Unlike carbohydrates, lipids share no common chemical structure. The most abundant lipids are the fats and oils, which make up one of the three important classes of foods.

fats
oils
triacylglycerols or triglycerides

Fats and **oils** are esters of glycerol and predominantly long-chain fatty acids (carboxylic acids). Fats and oils are also called **triacylglycerols** or **triglycerides,** since each molecule is derived from one molecule of glycerol and three molecules of fatty acid:

$$
\begin{array}{ccc}
 & & \text{O} \\
 & & \| \\
\text{CH}_2-\text{O}-\!\!\!&\text{C}-\text{R} \\
 & & \text{O} \\
 & & \| \\
\text{CH}-\text{O}-\!\!\!&\text{C}-\text{R}' \\
 & & \text{O} \\
 & & \| \\
\text{CH}_2-\text{O}-\!\!\!&\text{C}-\text{R}''
\end{array}
$$

glycerol portion →

general formula
for a triacylglycerol

$$
\begin{array}{cc}
 & \text{O} \\
 & \| \\
\text{CH}_2-\text{O}-\text{C}-\text{C}_{17}\text{H}_{35} \\
 & \text{O} \\
 & \| \\
\text{CH}-\text{O}-\text{C}-\text{C}_{15}\text{H}_{31} \\
 & \text{O} \\
 & \| \\
\text{CH}_2-\text{O}-\text{C}-\text{C}_{11}\text{H}_{23}
\end{array}
$$

typical triacylglycerol
containing three different
fatty acids

The formulas of triacylglycerol molecules vary for the following reasons:

1. The length of the fatty acid chain may vary from 4 to 20 carbons, but the number of carbon atoms in the chain is nearly always even.
2. Each fatty acid may be saturated, or it may be unsaturated and contain one, two, or three carbon–carbon double bonds.
3. An individual triacylglycerol may, and frequently does, contain three different fatty acids.

The most abundant saturated fatty acids in fats and oils are lauric, myristic, palmitic, and stearic acids (see Table 19.7). The most abundant unsaturated acids in fats and oils contain 18 carbon atoms and have one, two, or three carbon–carbon double bonds. Their formulas are

$$CH_3(CH_2)_7CH=CH(CH_2)_7COOH$$
oleic acid

$$CH_3(CH_2)_4CH=CHCH_2CH=CH(CH_2)_7COOH$$
linoleic acid

$$CH_3CH_2CH=CHCH_2CH=CHCH_2CH=CH(CH_2)_7COOH$$
linolenic acid

Other significant unsaturated fatty acids are palmitoleic acid (with 16 carbons) and arachidonic acid (with 20 carbons).

$$CH_3(CH_2)_5CH=CH(CH_2)_7COOH \qquad CH_3(CH_2)_4(CH=CHCH_2)_4CH_2CH_2COOH$$
palmitoleic acid $\qquad\qquad\qquad\qquad$ arachidonic acid

Three unsaturated fatty acids—linoleic, linolenic, and arachidonic—are essential for animal nutrition and must be supplied in the diet. Diets lacking these fatty acids lead to impaired growth and reproduction, and such skin disorders as eczema and dermatitis. We don't require fats in our diet except as a source of these three fatty acids.

The major physical difference between fats and oils is that fats are solid and oils are liquid at room temperature. Since the glycerol part of the structure is the same for a fat and an oil, the difference must be due to the fatty acid end of the molecule. Fats contain a higher proportion of saturated fatty acids, whereas oils contain higher amounts of unsaturated fatty acids. The term *polyunsaturated* means that the molecules of a particular product each contain several double bonds.

Fats and oils are obtained from natural sources. In general, fats come from animal sources and oils from vegetable sources. Olive, cottonseed, corn, soybean, linseed, and other oils are obtained from the fruit or seed of their respective vegetable sources. Table 20.2 shows the major constituents of several fats and oils.

Fats are an important energy source for humans and normally account for about 25–50% of caloric intake. When oxidized to carbon dioxide and water, fats supply about 39 kJ/g (9.3 Cal/g), which is more than twice the amount obtained from carbohydrates and proteins.

Fats are digested in the small intestine where they are first emulsified by the bile salts and then hydrolyzed to di- and monoglycerides, fatty acids, and glycerol. The fatty acids pass through the intestinal wall and are coated with a protein to increase solubility in the blood. They are then transported to various parts of the body where they are broken down in a series of enzyme-catalyzed reactions for the production of energy. Part of the hydrolyzed fat is converted back into fat in the adipose tissue and

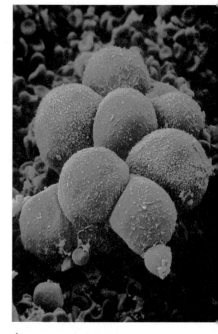

▲
Fat cells can result in clogged arteries, but they also provide a major reserve of potential energy.

TABLE 20.2	Fatty Acid Composition of Selected Fats and Oils				
	Fatty acid (%)				
Fat or oil	Myristic acid	Palmitic acid	Stearic acid	Oleic acid	Linoleic acid
Animal fat					
Butter[a]	7–10	23–26	10–13	30–40	4–5
Lard	1–2	28–30	12–18	41–48	6–7
Vegetable oil					
Olive	0–1	5–15	1–4	49–84	4–12
Peanut	—	6–9	2–6	50–70	13–26
Corn	0–2	7–11	3–4	43–49	34–42
Cottonseed	0–2	19–24	1–2	23–33	40–48
Soybean	0–2	6–10	2–4	21–29	50–59
Linseed[b]	—	4–7	2–5	9–38	3–43

[a]Butyric acid, 3–4%
[b]Linolenic acid, 25–58%

stored as a reserve source of energy. These fat deposits also function to insulate against loss of heat as well as to protect vital organs against mechanical injury.

Solid fats are preferable to oils for the manufacture of soaps and for use as certain food products. Hydrogenation of oils to make them solid is carried out on a large commercial scale. In this process hydrogen, bubbled through hot oil containing a finely dispersed nickel catalyst, adds to the carbon–carbon double bonds of the oil to saturate the double bonds and form fats. In practice, only some of the double bonds are allowed to become saturated. The product that is marketed as solid "shortening" is used for cooking and baking. Oils and fats are also partially hydrogenated to improve their shelf life. Rancidity in fats and oils results from air oxidation at points of unsaturation, producing low-molar-mass aldehydes and acids of disagreeable odor and flavor.

Soap is made by hydrolyzing fats or oils with aqueous NaOH. This hydrolysis process, called *saponification,* requires 3 mol NaOH per mole of fat:

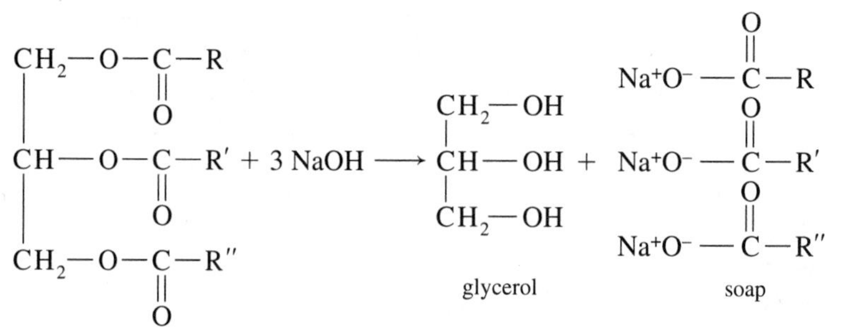

The most common soaps are the sodium salts of long-chain fatty acids, such as sodium stearate, $NaOOCC_{17}H_{35}$; sodium palmitate, $NaOOCC_{15}H_{31}$; and sodium oleate, $NaOOCC_{17}H_{33}$.

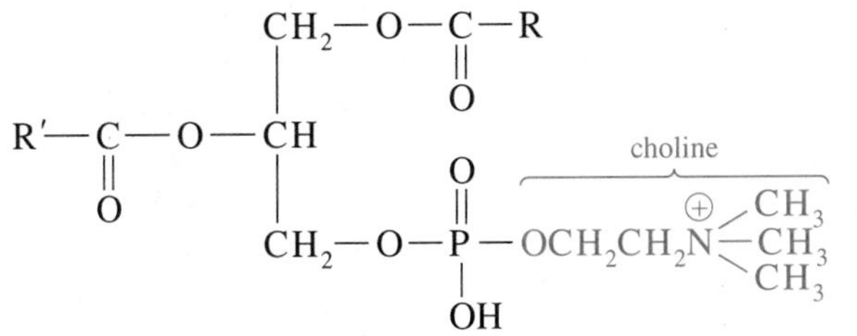

◀ FIGURE 20.1
Formulas for a
phospholipid, a glycolipid,
and steroids.

a lecithin (a phospholipid)

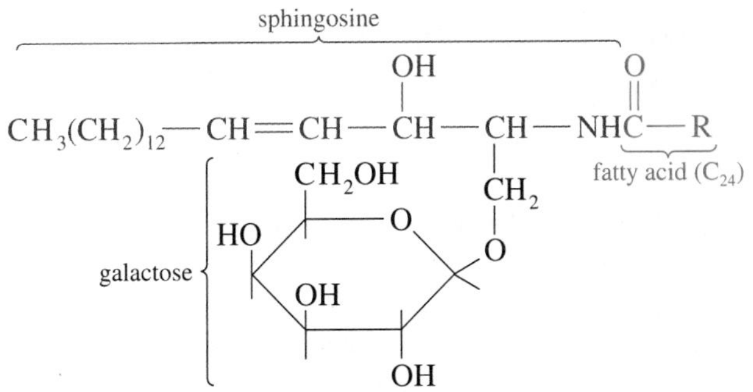

a cerebroside (a glycolipid)

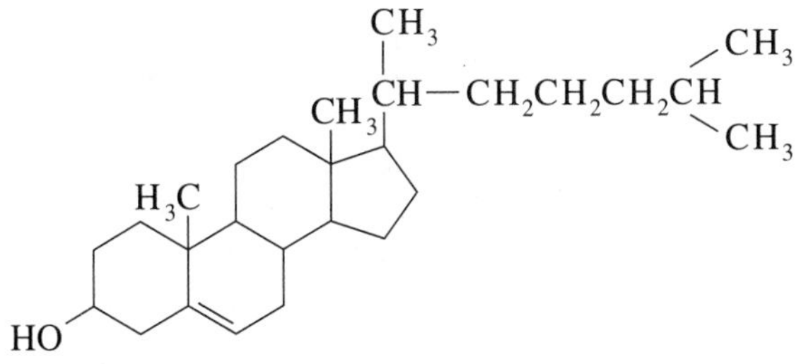

cholesterol (a steroid)

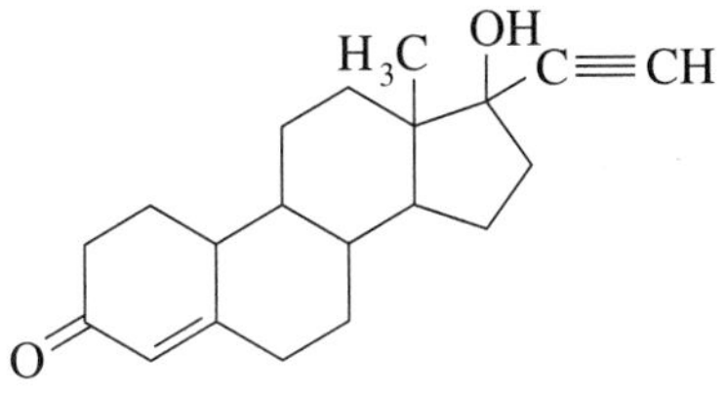

Norlutin
(a birth control pill)

Other principal classes of lipids, besides fats and oils, are phospholipids, glyco-lipids, and steroids (see Figure 20.1). The phospholipids are found in all animal and vegetable cells and are abundant in the brain, the spinal cord, egg yolk, and liver. Glycolipids (cerebrosides) contain a long-chain alcohol called *sphingosine*. They contain no glycerol but do contain a monosaccharide (usually galactose). Glyco-lipids are found in many different tissues but, as the name *cerebroside* indicates, occur in large quantities in brain tissue.

Steroids all have a four-fused carbocyclic-ring system (as in cholesterol) with various side groups attached to the rings. Cholesterol is the most abundant steroid in the body. It occurs in the brain, the spinal column, and nervous tissue, and it is the principal constituent of gallstones. The body synthesizes about 1 g of cholesterol per day, while about 0.3 g per day is ingested in the average diet. The major sources of cholesterol in the diet are meat, liver, and egg yolk. The cholesterol level in the blood generally rises with a person's age and body weight. High blood-level cholesterol is associated with atherosclerosis (hardening of the arteries), which results in reduced flow of blood and high blood pressure. Cholesterol is needed by the body to synthesize other steroids, some of which regulate male and female sexual characteristics. Many synthetic birth control pills are modified steroids that interfere with a woman's normal conception cycle.

20.4   Amino Acids and Proteins

Proteins are the third important class of foodstuffs. Some common foods with high (over 10%) protein content are gelatin, fish, beans, nuts, cheese, eggs, poultry, and meat. Proteins are present in all body tissue. They can form structural elements, such as hair, fingernails, wool, and silk. Proteins also function as enzymes that regulate the countless chemical reactions taking place in every living organism. About 15% of the human body weight is protein. Chemically, proteins are polymers of amino acids with molar masses ranging up to more than 50 million.

amino acid   **Amino acids** are carboxylic acids that contain an amino ($-NH_2$) group attached to C-2 (the alpha carbon), and are thus called α-*amino acids.* They also contain another variable group, R. The R group represents any of the various groups that make up the specific amino acids. For example, when R is H$-$, the amino acid is glycine; when R is CH_3-, the amino acid is alanine; when R is $CH_3SCH_2CH_2-$, the amino acid is methionine.

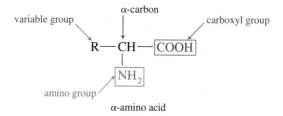

α-amino acid

Some amino acids have two amino groups and some contain two acid groups.

There are approximately 200 different known amino acids in nature. Some are found in only one particular species of plant or animal, others in only a few life-forms. But 20 of these amino acids are found in almost all proteins. Furthermore, these same 20 amino acids are used by all forms of life in the synthesis of proteins; their names, formulas, and abbreviations are given in Table 20.3. Eight are considered essential amino acids, since the human body is *not* capable of synthesizing them. Therefore they must be supplied in our diets if we are to enjoy normal health.

protein   **Proteins** are polymeric substances that yield primarily amino acids on hydrolysis. The bond connecting the amino acids in a protein is commonly called a

peptide linkage   **peptide linkage** or peptide bond. If we combine two glycine molecules by eliminating a water molecule between the amino group of one and the carboxyl group of the second glycine, we form a compound containing the amide structure and the peptide linkage. The compound containing the two amino acid groups is called a *dipeptide:*

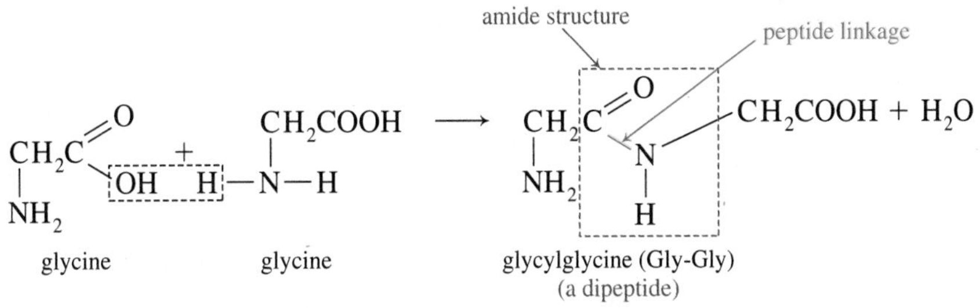

glycine   glycine   glycylglycine (Gly-Gly)
(a dipeptide)

The product formed from two glycine molecules is called *glycylglycine* (abbreviated Gly-Gly). Note that the molecule still has a free amino group at one end and a free carboxyl group at the other end. The formation of Gly-Gly is considered the first step in the synthesis of a protein, since each end of the molecule is capable of joining to another amino acid. We can thus visualize the formation of a protein by joining a great many amino acids in this fashion. Another example showing a tripeptide (three amino acids linked together) follows. This compound contains two peptide linkages:

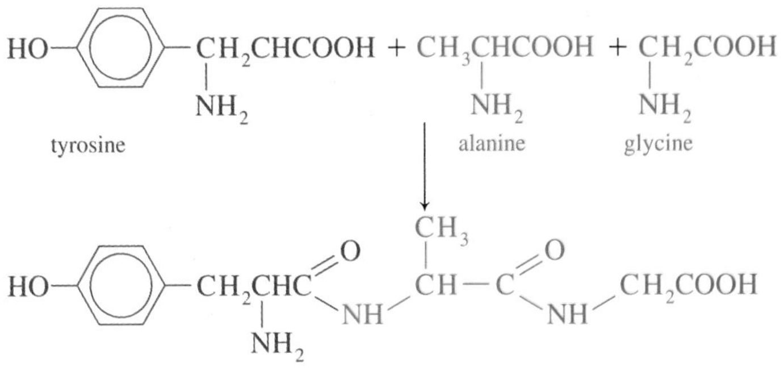

tyrosine   alanine   glycine

tyrosylalanylglycine (a tripeptide) (Tyr-Ala-Gly)

There are five other tripeptide combinations of these three amino acids using only one unit of each amino acid. Peptides containing up to about 40–50 amino acid units in a chain are **polypeptides.** Still longer chains of amino acids are proteins.

The amino acid units in a peptide are called *amino acid residues* or simply residues. (They no longer are amino acids because they have lost an H atom from their amino groups and an OH from their carboxyl groups.) In linear peptides, one end of the chain has a free amino group and the other end a free carboxyl group. The amino-group end is called the *N-terminal residue* and the other end the *C-terminal residue:*

polypeptide

$$\overset{1\quad 2\quad 3\quad 4\quad 5\quad 6\quad 7}{\text{Ala-Pro-Tyr-Met-Gly-Lys-Gly}}$$

N-terminal residue   *C*-terminal residue

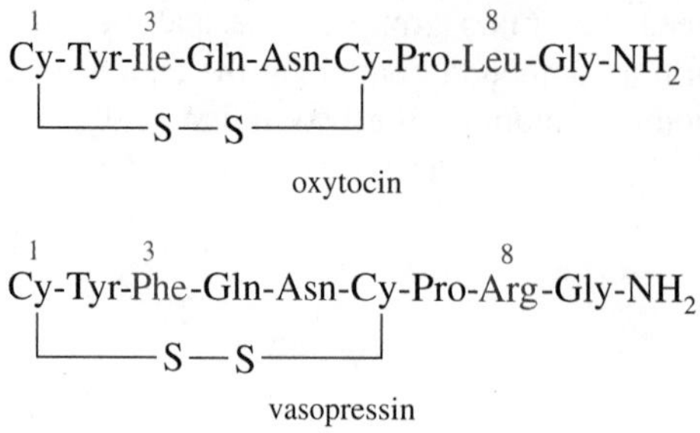

FIGURE 20.2 ▶
Amino acid sequences of oxytocin and vasopressin. The difference in only two amino acids (shown in color) in these two compounds results in very different physiological activity.

$$\overset{1}{\text{Cy}}-\text{Tyr}-\overset{3}{\text{Ile}}-\text{Gln}-\text{Asn}-\text{Cy}-\text{Pro}-\overset{8}{\text{Leu}}-\text{Gly}-\text{NH}_2$$

oxytocin

$$\overset{1}{\text{Cy}}-\text{Tyr}-\overset{3}{\text{Phe}}-\text{Gln}-\text{Asn}-\text{Cy}-\text{Pro}-\overset{8}{\text{Arg}}-\text{Gly}-\text{NH}_2$$

vasopressin

The sequence of amino acids in a chain is numbered starting with the *N*-terminal residue, which is usually written to the left with the *C*-terminal residue at the right.

Peptides are named as acyl derivatives of the *C*-terminal amino acid with the *C*-terminal unit keeping its complete name. The *-ine* ending of all but the *C*-terminal amino acid is changed to *-yl,* and these are listed in the order in which they appear, starting with the *N*-terminal amino acid:

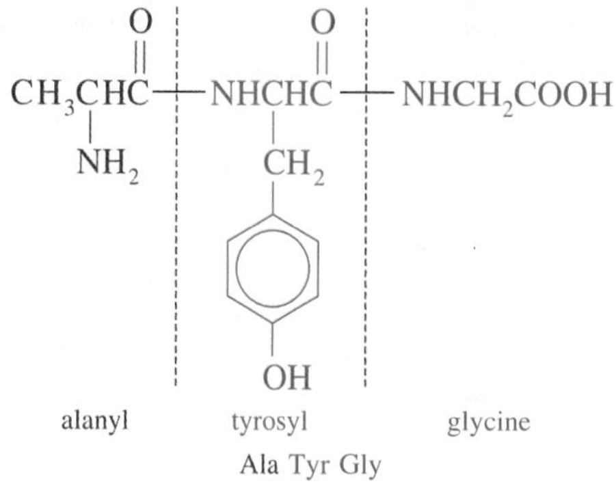

alanyl   tyrosyl   glycine

Ala Tyr Gly

Thus Ala-Tyr-Gly is called *alanyltyrosylglycine,* and Arg-Gln-His-Ala is arginylglutamylhistidylalanine.

Many small, naturally occurring polypeptides have significant biochemical functions. The amino acid sequences of two of these, oxytocin and vasopressin, are shown in Figure 20.2. Oxytocin controls uterine contractions during labor in childbirth and also causes contraction of the smooth muscles of the mammary gland, resulting in milk secretion. Vasopressin in high concentration raises the blood pressure and has been used in surgical shock treatment for this purpose. Vasopressin is also an antidiuretic, regulating the excretion of fluid by the kidneys. The absence of vasopressin leads to diabetes insipidus. This condition, which is characterized by excretion of up to 30 L of urine per day, may be controlled by administering vasopressin or its derivatives. Oxytocin and vasopressin are similar nonapeptides, differing only at positions 3 and 8.

Determining the sequence of the amino acids in even one protein molecule was once a formidable task. The amino acid sequence of beef insulin was announced in 1955 by the British biochemist Frederick Sanger (b. 1918). Finding the sequence of this structure required several years of effort by Sanger's team. He was awarded the 1958 Nobel Prize in chemistry for this work. As a result of this work, automated amino acid sequencers have been developed that can determine the sequence of an

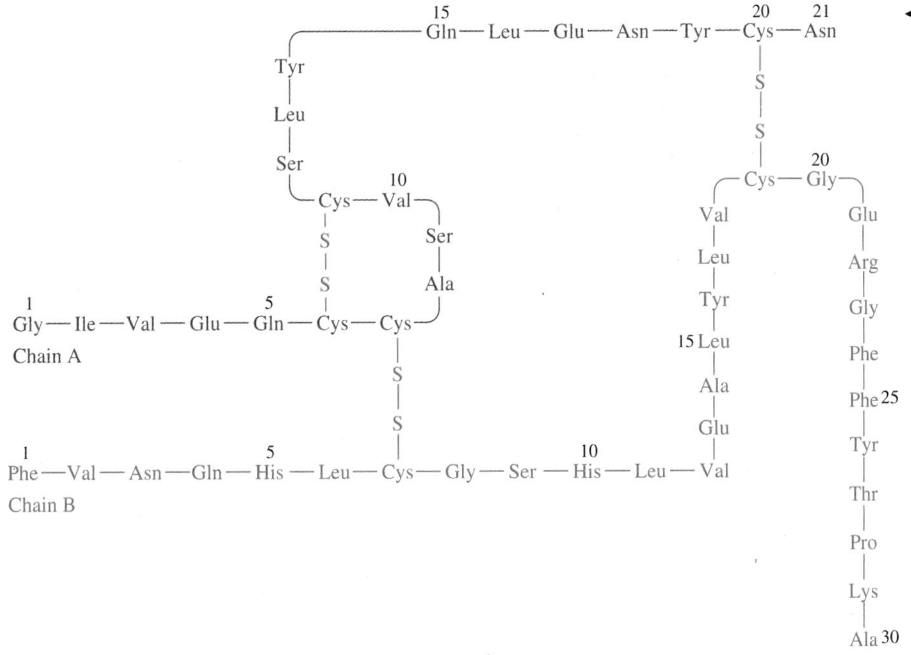

◀ **FIGURE 20.3**
Amino acid sequence of beef insulin.

average-sized protein in a few days. Beef insulin consists of 51 amino acid units in two polypeptide chains. The two chains are connected by disulfide linkages ($-S-S-$) of two cysteine residues at two different sites. The structure is shown in Figure 20.3. Insulins from other animals, including humans, differ slightly by one, two, or three amino acid residues.

Protein digestion takes place in the stomach and the small intestine. Here, digestive enzymes hydrolyze proteins to smaller peptides and amino acids, which pass through the walls of the intestines, are absorbed by the blood, and are transported to the liver and other tissues of the body. The body does not store free amino acids. They are used to synthesize

1. proteins to replace and repair body tissues
2. other nitrogen-containing substances, such as certain hormones and heme
3. nucleic acids
4. enzymes that control the synthesis of other necessary products, such as carbohydrates and fats

Proteins are catabolized (degraded) to carbon dioxide, water, and urea. Urea, containing the protein nitrogen, is eliminated from the body in the urine.

Carbohydrates and fats are used primarily to supply heat and energy to the body. Proteins, on the other hand, are used mainly to repair and replace worn-out tissue. Tissue proteins are continuously being broken down and resynthesized. Therefore protein must be continually supplied to the body in the diet. It is nothing short of amazing how organisms pick out the desired amino acids from the bloodstream and put them together in proper order to synthesize a needed protein. Now let's look at nucleic acids, which control the synthesis of proteins.

TABLE 20.3	Common Amino Acids Derived from Proteins	
Name	**Abbreviation**	**Formula**
Alanine	Ala	$CH_3CHCOOH$ 	 NH_2

Arginine	Arg	$NH_2-C-NH-CH_2CH_2CH_2CHCOOH$
		      ||
		     NH NH_2

Asparagine	Asn	$NH_2C-CH_2CHCOOH$
		    || |
		     O NH_2

Aspartic acid	Asp	$HOOCCH_2\ CHCOOH$
		             |
		           NH_2

Cysteine	Cys	$HSCH_2CHCOOH$
		         |
		       NH_2

Glutamic acid	Glu	$HOOCCH_2\ CH_2CHCOOH$
		                |
		             NH_2

Glutamine	Gln	$NH_2\ CCH_2CH_2CHCOOH$
		      || |
		      O NH_2

Glycine	Gly	$HCHCOOH$
		      |
		     NH_2

Histidine	His	$N-CH$
		 || ||
		 $HC\ \ \ \ C-CH_2CHCOOH$
		   $\diagdown N \diagup$ |
		     | NH_2
		     H

Isoleucine*	Ile	$CH_3CH_2CH-CHCOOH$
		           | |
		         CH_3 NH_2

Leucine*	Leu	$(CH_3)_2CHCH_2CHCOOH$
		                |
		             NH_2

20.5   Nucleic Acids

Explaining how hereditary material duplicates itself was a baffling problem for biologists. This explanation and the answer to the question, "Why are the offspring of a given species undeniably of that species?" eluded biologists for many years. Many thought the chemical basis for heredity lay in the structure of proteins. But they couldn't find how protein reproduced itself. The answer to the hereditary problem was finally found in the structure of the nucleic acids.

Name	Abbreviation	Formula
Lysine*	Lys	$NH_2CH_2CH_2CH_2CH_2CHCOOH$ $\quad\quad\quad\quad\quad\quad\quad	$ $\quad\quad\quad\quad\quad\quad\quad NH_2$
Methionine*	Met	$CH_3SCH_2CH_2CHCOOH$ $\quad\quad\quad\quad\quad\quad	$ $\quad\quad\quad\quad\quad\quad NH_2$
Phenylalanine*	Phe	$\text{⬡}-CH_2CHCOOH$ $\quad\quad\quad\quad	$ $\quad\quad\quad\quad NH_2$
Proline	Pro	⬠ ring with N–H; $-COOH$
Serine	Ser	$HOCH_2CHCOOH$ $\quad\quad\quad\quad	$ $\quad\quad\quad\quad NH_2$
Threonine*	Thr	$CH_3CH-CHCOOH$ $\quad\quad\quad	\quad\quad	$ $\quad\quad\quad OH \quad NH_2$
Tryptophan*	Trp	indole ring $-C-CH_2CHCOOH$ $\quad\quad\quad\quad\quad || \quad\quad	$ $\quad\quad\quad\quad\quad CH \quad NH_2$
Tyrosine	Tyr	$HO-\text{⬡}-CH_2CHCOOH$ $\quad\quad\quad\quad\quad\quad	$ $\quad\quad\quad\quad\quad\quad NH_2$
Valine*	Val	$(CH_3)_2CHCHCOOH$ $\quad\quad\quad\quad\quad	$ $\quad\quad\quad\quad\quad NH_2$

*Amino acids essential in human nutrition.

The unit structure of all living things is the cell. Suspended in the nucleus of cells are chromosomes, which consist largely of proteins and nucleic acids. The nucleic acids and proteins are intimately associated in complexes called *nucleoproteins*. Nucleic acids either contain the sugar deoxyribose or they contain the sugar ribose. Accordingly, they are called deoxyribonucleic acid (DNA) and ribonucleic acid (RNA). DNA was discovered in 1869 by Swiss physiologist Friedrich Miescher (1844–1895), who extracted it from the nuclei of cells.

DNA is a polymeric substance made up of thousands of units called nucleotides. DNA The fundamental components of the nucleotides in DNA are phosphoric acid, deoxyribose (a pentose sugar), and the four nitrogen-containing bases, adenine, thymine, guanine, and cytosine (abbreviated as A, T, G, and C). Phosphoric acid is

FIGURE 20.4 ▶
Fundamental components of
nucleotides in DNA.

phosphoric acid

thymine (T)

cytosine (C)

2-deoxyribose

adenine (A)

guanine (G)

nucleotide

obtained from minerals in the diet; deoxyribose is synthesized in the body from glucose; and the four nitrogen bases are made in the body from amino acids. The formulas for these compounds are given in Figure 20.4.

A **nucleotide** in DNA consists of one of the four bases linked to a deoxyribose sugar, which in turn is linked to a phosphate group. Each nucleotide has the following sequence:

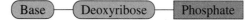

The structures for a single nucleotide and a segment of a polynucleotide (DNA) are shown in Figure 20.5.

In 1953, the American biologist James D. Watson (b. 1928) and the British physicist Francis H. C. Crick (b. 1916) announced their double-stranded helix structure for DNA. This concept was a milestone in the history of biology, and in 1962 Watson and Crick, along with Maurice H. F. Wilkin (b. 1916), who did the brilliant X-ray diffraction studies on DNA, were awarded the Nobel Prize in medicine and physiology.

The structure of DNA, according to Watson and Crick, consists of two polymeric strands of nucleotides in the form of a double helix, with both nucleotide strands coiled around the same axis (see Figure 20.6). Along each strand are alternate phosphate and deoxyribose units with one of the four bases adenine, guanine, cytosine, or thymine attached to deoxyribose as a side group. The double helix is held together by hydrogen bonds extending from the base on one strand of the double helix to a complementary base on the other strand. Watson and Crick furthermore ascertained that adenine is always hydrogen bonded to thymine, and guanine is always hydrogen bonded to cytosine. Previous analytical work by others substantiated this concept of complementary bases by showing that the molar ratio of adenine to thymine in DNA is approximately $1:1$ and that of guanine to cytosine is also approximately $1:1$.

The structure of DNA has been compared to a ladder twisted into a double helix, with the rungs of the ladder perpendicular to the twisted railings. The phosphate and

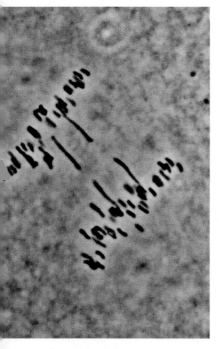

The chromosomes in this mammal cell are in the final stage of cell division. They contain nucleic acids and proteins replicated for the new cell.

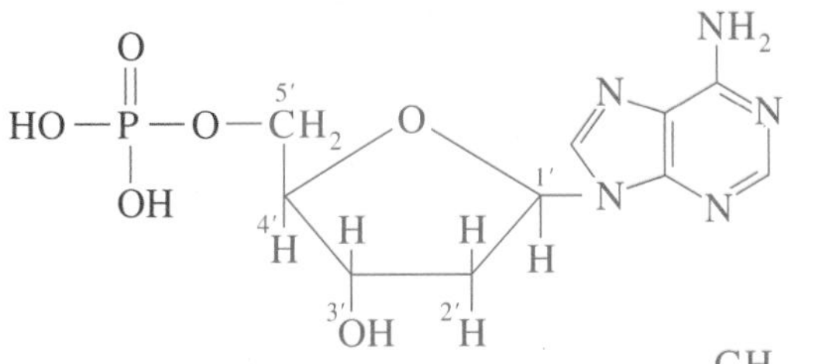

(a) **Deoxyadenosine-5'-monophosphate**

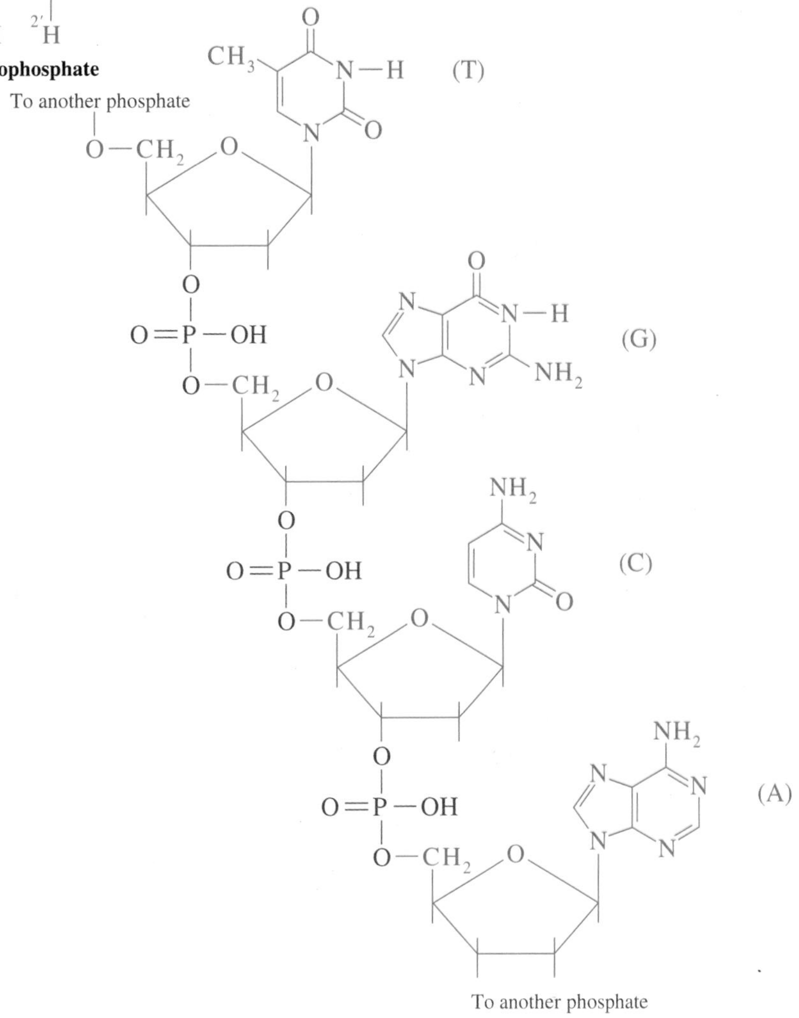

(b) **Four nucleotide units of a DNA strand**

▲
FIGURE 20.5
(a) A single nucleotide, adenine deoxyribonucleotide. (b) A segment of one strand of deoxyribonucleic acid (DNA) showing four nucleotides, including those of adenine (A), cytosine (C), guanine (G), and thymine (T).

deoxyribose units alternate along the two railings of the ladder, and two nitrogen bases form each rung of the ladder. The DNA structure is illustrated in Figure 20.6.

For any individual of any species, the sequence of base pairs and the length of the nucleotide chains in DNA molecules contain the coded messages that determine all of the characteristics of that individual. In this sense the DNA molecule is a template that stores information for recall as needed. DNA contains the genetic code of life, which is passed on from one generation to another.

RNA is a polymer of nucleotides, but it differs from DNA in that (1) it is single-stranded; (2) it contains the pentose sugar ribose instead of deoxyribose;

RNA

FIGURE 20.6 ▶
Double-stranded helix structure of DNA.

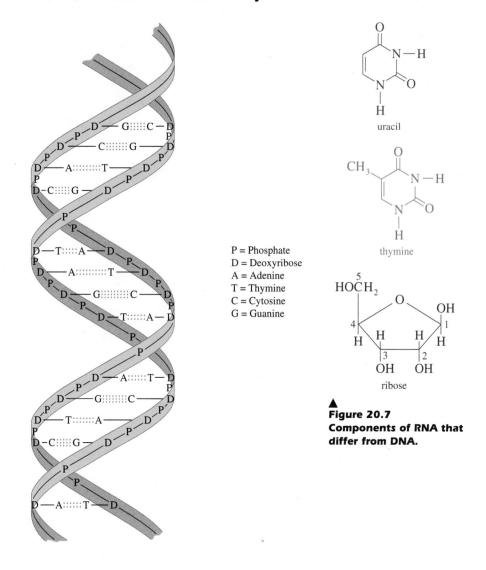

P = Phosphate
D = Deoxyribose
A = Adenine
T = Thymine
C = Cytosine
G = Guanine

uracil

thymine

ribose

▲
Figure 20.7
Components of RNA that differ from DNA.

and (3) it contains uracil instead of thymine as one of its four nitrogen bases (see Figure 20.7).

transcription   **Transcription** is the process by which DNA directs the synthesis of RNA. The nucleotide sequence of only one strand of DNA is transcribed into a single strand of RNA. This transcription occurs in a complementary fashion. Where there is a guanine base in DNA, a cytosine base will occur in RNA. Cytosine is transcribed to guanine, thymine to adenine, and adenine to uracil.

The main function of RNA is to direct the synthesis of proteins. RNA is produced in the cell nucleus but performs its function outside of the nucleus. Three kinds of RNA are produced directly from DNA: messenger RNA (mRNA), transfer RNA (tRNA), and ribosomal RNA (rRNA). Messenger RNA contains bases in the exact order transcribed from a strip of the DNA master code. The base sequence on the mRNA in turn establishes the sequence of amino acids that comprise a specific protein. The relatively small transfer RNA molecules bring specific amino acids to the site of protein synthesis. There is at least one different tRNA for each amino acid.

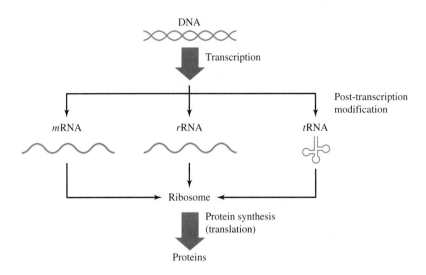

The actual site of protein synthesis is a ribosome, which is composed of rRNA and protein. The function of the rRNA is not completely understood. However, the ribosome is believed to move along the mRNA chain and to aid in the polymerization of amino acids in the order prescribed by the base sequence of the mRNA chain. The flow of genetic information usually is in one direction, from DNA to RNA to proteins (Figure 20.8).

20.6   DNA and Genetics

Heredity is the process by which the physical and mental characteristics of parents are transferred to their offspring. For this process to occur, the material responsible for genetic transfer must be able to make exact copies of itself. The design for replication is built into Watson and Crick's DNA structure, first by the nature of its double helical structure and second by the complementary nature of its nitrogen bases, where adenine will bond only to thymine and guanine only to cytosine. The DNA double helix unwinds, or "unzips," into two separate strands at the hydrogen bonds between the bases. Each strand then serves as a template combining only with the proper free nucleotides to produce two identical replicas of itself. This replication of DNA occurs in the cell just before the cell divides, thereby giving each daughter cell the full genetic code of the original cell. This process is illustrated in Figure 20.9.

As we have indicated, DNA is an integral part of the chromosomes. Each species carries a specific number of chromosomes in the nucleus of each of its cells. The number of chromosomes varies with different species. Humans have 23 pairs, or 46 chromosomes. Chromosomes are long, threadlike bodies composed of nucleic acids and proteins that contain the basic units of heredity called genes. **Genes** are segments of the DNA chain that contain the codes for the formation of polypeptides and RNAs. Hundreds of genes can exist along a DNA chain.

genes

In ordinary cell division, known as **mitosis,** each DNA molecule forms a duplicate by uncoiling to single strands. Each strand then assembles the complementary portion from available free nucleotides to form a duplicate of the original DNA

mitosis

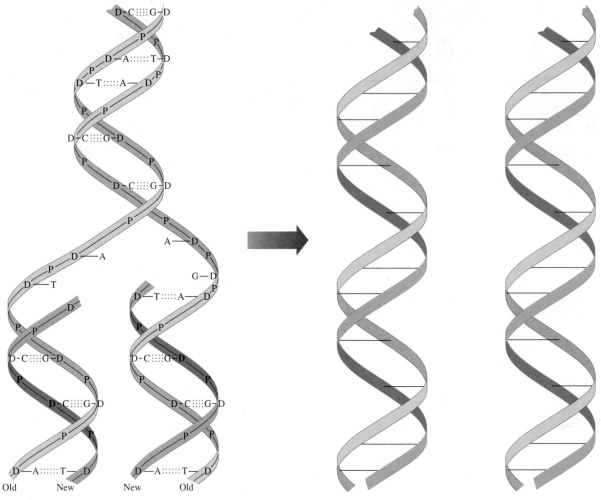

▲
FIGURE 20.9
Replication of DNA. The two helices unwind, separating at the point of the hydrogen bonds. Each strand then serves as a template, recombining with the proper nucleotides to duplicate itself as a double-stranded helix.

molecule. After cell division is completed, each daughter gene contains the genetic material that corresponds exactly to those present in the original cell before division.

However, in almost all higher forms of life, reproduction takes place by union of the male sperm with the female egg. Cell splitting to form the sperm cell and the egg cell occurs by a different and more complicated process called *meiosis*. In **meiosis,** the genetic material is divided so that each daughter cell receives one chromosome from each pair. After meiosis, the egg cell and the sperm cell each carry only half of the chromosomes from its original cell. Between them they form a new cell that once again contains the correct number of chromosomes and all the hereditary characteristics of the species. Thus the offspring derives half of its genetic characteristics from the father and half from the mother.

meiosis

Nature is not 100% perfect. Occasionally, DNA replication is not perfect, or a section of the DNA molecule is damaged by X-rays, radioactivity, or drugs, and a mutant organism is produced. For example, in the disease of sickle-cell anemia, a large proportion of the red blood cells form into sickle shapes instead of the usual globular shape. This irregularity limits the ability of the blood to carry oxygen and causes the person to be weak and unable to fight infection, leading to early death. Sickle-cell anemia is due to one misplaced amino acid in the structure of hemoglobin. The sickle-cell-producing hemoglobin has a valine residue where a glutamic acid residue should be located. Sickle-cell anemia is an inherited disease indicating a fault in the DNA coding transmitted from parent to child. Many biological disorders and ailments have been traced directly to a deficiency in the genetic information of DNA.

20.7 Enzymes

Enzymes are the catalysts of biochemical reactions. Most enzymes are proteins, and   *enzyme*
they catalyze nearly all of the reactions that occur in living cells. Uncatalyzed reactions that may require hours of boiling in the presence of a strong acid or a strong base can occur in a fraction of a second in the presence of the proper enzyme at room temperature and nearly neutral pH. This process is all the more remarkable when we realize that enzymes do not actually cause chemical reactions. They act as catalysts by greatly lowering the activation energy of specific biochemical reactions. The lowered activation energy permits these reactions to proceed at high speed at body temperature.

Louis Pasteur (1822–1895) was one of the first scientists to study enzyme-catalyzed reactions. He believed that living yeasts or bacteria were required for these reactions, which he called fermentations—for example, the conversion of glucose to alcohol by yeasts. In 1897, Eduard Büchner (1860–1917) made a cell-free filtrate that contained enzymes prepared by grinding yeast cells with very fine sand. The enzymes in this filtrate converted glucose to alcohol, thus proving that the presence of living cells is not required for enzyme activity. For this work Büchner received the Nobel Prize in chemistry in 1907.

Each organism contains thousands of enzymes. Some enzymes are simple proteins consisting only of amino acid units. Others are conjugated and consist of a protein part, or *apoenzyme,* and a nonprotein part, or *coenzyme.* Both parts are essential, and a functioning enzyme consisting of both the protein and nonprotein parts is called a *holoenzyme.*

Apoenzyme + Coenzyme = Holoenzyme

Often the coenzyme is a vitamin, and the same coenzyme may be associated with many different enzymes.

For some enzymes, an inorganic component, such as a metal ion (e.g., Ca^{2+}, Mg^{2+}, or Zn^{2+}), is required. This inorganic component is an *activator.* From the standpoint of function, an activator is analogous to a coenzyme, but inorganic components are not called coenzymes.

Another remarkable property of enzymes is their specificity of reaction; that is, a certain enzyme will catalyze the reaction of a specific type of substance. For example, the enzyme maltase catalyzes the reaction of maltose and water to form glucose. Maltase has no effect on the other two common disaccharides sucrose and lactose. Each of these sugars requires a specific enzyme—sucrase to hydrolyze sucrose, lactase to hydrolyze lactose. (See the hydrolysis equations in Section 20.2.)

The substance acted on by an enzyme is called the **substrate.** Sucrose is the sub-   *substrate*
strate of the enzyme sucrase. Enzymes are named by adding the suffix *-ase* to the

Industrial Strength Enzymes

Not only are enzymes important in biology, they are important in industry. Enzymes offer two major advantages to manufacturing processes and in commercial products: first, they cause very large increases in reaction rates even at room temperature; second, they are relatively specific and can be used to target selected reactants. Perhaps the biggest disadvantage to industrial enzymes is their relative short supply (and therefore higher cost as compared to traditional chemical treatments).

About 25% of all industrial enzymes are used to convert cornstarch into syrups that are equivalent in sweetness and in calories to ordinary table sugar. More than 5 billion pounds of such syrups are produced annually. The product is a high-fructose syrup equivalent in sweetness to sucrose. One of these syrups, sold commercially since 1968, contains by dry weight about 42% fructose, 50% glucose, and 8% other carbohydrates.

Industrial enzymes also offer solutions to environmental pollution. For example, the paper industry, like other industries that use chemicals, is concerned with minimizing processes that produce potentially hazardous waste. Paper is produced from wood chips by first digesting the cellulose structure with calcium sulfite and then bleaching the pulp with chlorine to obtain a bright white paper. An excess of chlorine must be used because the pulp is not completely broken down. This excess creates a significant disposal problem as chlorine is environmentally hazardous. Developments in biotechnology offer a solution. The enzymes needed to complete wood fiber digestion (cellulase and hemicellulase) have been produced in larger quantities via genetic engineering. With such enzymes to finish degrading the wood pulp, paper manufacturers can markedly decrease the amount of chlorine used as bleach.

◀ **These jeans have been treated with enzymes to fade and soften them.**

Consumer goods are increasingly impacted by enzyme technology. Many detergents are better cleansing agents because they contain enzymes; fully 40% of all industrially produced enzymes are used in detergents. Meat tenderizers often contain papain, an enzyme that breaks down protein molecules.

Even clothing manufacturers are finding uses for enzymes. Many denim products are enzyme-treated to replace stone-washing, a process in which the material is washed with pumice to soften the fabric's appearance and remove some of the dye. Because this abrasion weakens the fabric, some manufacturers now use "biostoning." The denim is treated with the enzyme, cellulase, which changes the fabric's appearance without weakening the fabric structure.

Purified enzymes have a number of medical applications. For genetic diseases characterized by the loss of a specific enzyme, a treatment known as enzyme-replacement therapy has been developed. For example, Tay–Sachs disease leads to the accumulation of excess intracellular polysaccharides because specific digestive enzymes are unavailable. Polysaccharide buildup can lead to mental retardation, paralysis, blindness, and death. Current research is aimed at developing an appropriate microcapsule package that will transport additional digestive enzymes to the affected cells. Such enzyme-replacement therapy has also been proposed for removing toxic substances from the bloodstream.

root of the substrate name. Note, for example, the derivations of maltase, sucrase, and lactase from maltose, sucrose, and lactose. Many enzymes, especially digestive enzymes, have common names such as pepsin, rennin, trypsin, and so on. These names have no systematic significance.

Enzymes act according to the following general sequence. Enzyme (E) and substrate (S) combine to form an enzyme–substrate intermediate (E–S). This intermediate decomposes to give the product (P) and regenerate the enzyme:

$$E + S \rightleftharpoons E\text{–}S \longrightarrow E + P$$

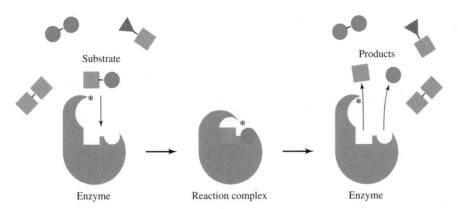

◀ FIGURE 20.10
Enzyme–substrate
interaction illustrating
specificity of an enzyme by
the lock-and-key model.

For the hydrolysis of maltose, the sequence is

$$\underset{\text{E}}{\text{Maltase}} + \underset{\text{S}}{\text{Maltose}} \rightleftharpoons \underset{\text{E–S}}{\text{Maltase–Maltose}}$$

$$\underset{\text{E–S}}{\text{Maltase–Maltose}} + H_2O \longrightarrow \underset{\text{E}}{\text{Maltase}} + \underset{\text{P}}{\text{2 Glucose}}$$

Enzyme specificity is believed to be due to the particular shape of a small part of the enzyme, its active site, which exactly fits a complementary-shaped part of the substrate (see Figure 20.10). This interaction is analogous to a lock and key; the substrate is the lock and the enzyme, the key. Just as a key opens only the lock it fits, the enzyme acts only on a molecule that fits its particular shape. When the substrate and the enzyme come together, they form a substrate–enzyme complex unit. The substrate, activated by the enzyme in the complex, reacts to form the products, regenerating the enzyme.

A more recent model of the enzyme–substrate catalytic site, known as the "induced-fit" model, visualizes a flexible site of enzyme–substrate attachment, with the substrate inducing a change in the enzyme shape to fit the shape of the substrate. This model allows for the possibility that in some cases the enzyme might wrap itself around the substrate and so form the correct shape of lock and key. Thus the enzyme does not need to have an exact preformed catalytic site to match the substrate.

Concepts in Review

1. Classify carbohydrates as mono-, di-, or polysaccharides.

2. Draw structural formulas in the open-chain and cyclic forms for glucose, fructose, galactose, and ribose.

3. Draw structural formulas for maltose and sucrose.

4. State the properties and general occurrence of glucose, galactose, fructose, and ribose.

5. Understand the manner in which monosaccharides are linked in maltose, lactose, and sucrose.

6. Write equations for the enzyme-catalyzed hydrolysis of maltose, lactose, and sucrose.

7. Give the monosaccharide composition of maltose, lactose, sucrose, starch, cellulose, and glycogen.

8. Discuss the similarities and differences between starch and cellulose.

9. Discuss, in simple terms, the metabolism of carbohydrates in the human body.

10. Rate the relative sweetness of the common mono- and disaccharides.

11. Give the general formula for fats and oils.

12. Write the names and formulas of the fatty acids that most commonly occur in fats and oils.

13. State which fatty acids are essential to human diets.

14. Write the structure of a triacylglycerol when given the fatty acid composition.

15. Write an equation for the saponification of a fat or an oil with caustic soda, NaOH.

16. Tell how a fat differs from an oil.

17. Explain the "hydrogenation" of vegetable oils, and the purposes for this process.

18. Draw the structural formula for cholesterol and the structural feature that is common to all steroids.

19. Distinguish among these three lipids: fats, phospholipids, and glycolipids.

20. List five foods that are major sources of proteins.

21. Explain the meaning of α-amino acids and the significance of these compounds in naturally occurring protein material.

22. Show the structural formula of a di-, tri-, or polypeptide that will be formed by combining amino acids.

23. Describe the functions and metabolic fate of amino acids and proteins.

24. Name and write the formulas of the six fundamental components of DNA.

25. Write the structure for a segment of a polynucleotide that contains up to four nucleotides.

26. Explain the three structural differences between DNA and RNA.

27. Describe the double-helix structure of DNA according to Watson and Crick.

28. Explain the concept of complementary bases and how it relates to DNA.

29. Discuss the role of DNA in genetics.

30. Distinguish between cell division in mitosis and meiosis.

31. Discuss the role of enzymes in the body and the theory of how they function.

32. Tell what is meant by the specificity of an enzyme.

Key Terms

amino acid (20.4)	enzyme (20.7)	mitosis (20.6)	polypeptide (20.4)	substrate (20.7)
biochemistry (20.1)	fats (20.3)	monosaccharide (20.2)	polysaccharide (20.2)	transcription (20.5)
carbohydrate (20.2)	genes (20.6)	nucleotide (20.5)	protein (20.4)	triacylglycerols or triglycerides (20.3)
disaccharide (20.2)	lipids (20.3)	oils (20.3)	RNA (20.5)	
DNA (20.5)	meiosis (20.6)	peptide linkage (20.4)		

Questions

1. Of the sugars listed in Table 20.1, which is the sweetest disaccharide? Which is the sweetest monosaccharide? (Invert sugar is a mixture of glucose and fructose, so it should not be considered.)

2. Are the fatty acids in vegetable oils more saturated or unsaturated than those in animal fats? Explain your answer. (Table 20.2)

3. Which of the common amino acids have more than one carboxyl group? Which have more than one amino group? (Table 20.3)

4. How many disulfide linkages are there in each molecule of beef insulin? (Figure 20.3)

5. In the four nucleotide units of DNA shown in Figure 20.5, which of these components are part of the backbone chain, and which are off to the side: the nitrogen bases, the deoxyribose, and the phosphoric acid?

6. In the double-stranded helix structure of DNA, which nitrogen bases are always hydrogen-bonded to the following: cytosine, thymine, adenine, and guanine? (Figure 20.6)

7. Life is dependent on four major classes of biomolecules. What are they?

8. Indicate the three types of carbohydrates. Which is the simplest?

9. What is an aldose, an aldotetrose, a ketose, a ketohexose? Give an example of each.

10. Classify the following as a monosaccharide, disaccharide, or polysaccharide: glucose, sucrose, maltose, fructose, cellulose, lactose, glycogen, galactose, starch, and ribose.

11. Draw structural formulas in the open-chain form for ribose, glucose, fructose, and galactose.

12. Draw structural formulas in the cyclic form for ribose, glucose, fructose, and galactose.

13. State the properties and the sources of ribose, glucose, fructose, and galactose.

14. The molecular formula for lactic acid is $C_3H_6O_3$, and its structural formula is $CH_3CH(OH)COOH$. Is this compound a carbohydrate? Explain.

15. What is the monosaccharide composition of
    (a) sucrose
    (b) maltose
    (c) lactose
    (d) starch
    (e) cellulose
    (f) glycogen?

16. Draw structural formulas in the cyclic form for sucrose and maltose.

17. If the most common monosaccharides have the formula $C_6H_{12}O_6$, why do the resulting disaccharides have the formula $C_{12}H_{22}O_{11}$, rather than $C_{12}H_{24}O_{12}$?

18. Write equations, using structural formulas in the cyclic form, for the hydrolysis of
    (a) sucrose
    (b) maltose
    What enzymes catalyze these reactions?

19. Discuss the similarities and differences between starch and cellulose.

20. In what form is carbohydrate stored in the body?

21. Discuss, in simple terms, the metabolism of carbohydrates in the human body.

22. State the natural sources of sucrose, maltose, lactose, and starch.

23. Invert sugar, obtained by the hydrolysis of sucrose to an equal-molar mixture of fructose and glucose, is commonly used as a sweetener in commercial food preparations. Why is invert sugar sweeter than the original sucrose?

24. What properties of molecules cause them to be classified as lipids?

25. Write structural formulas for glycerol, stearic acid, palmitic acid, oleic acid, and linoleic acid.

26. Distinguish both chemically and physically between a fat and a vegetable oil.

27. What is a triacylglycerol? Give an example.

28. Write the structure for tristearin, a fat in which all the fatty acid units are stearic acid.

29. Write the structure of a triacylglycerol that contains one unit each of linoleic, stearic, and oleic acids. How many other formulas are possible in which the triacylglycerol contains one unit each of these acids?

30. Write equations for the saponification of
    (a) tripalmitin (a fat in which all the fatty acids are palmitic acid)
    (b) the triacylglycerol of Question 29.
    Which product(s) are soaps?

31. How can vegetable oils be solidified? What is the advantage of solidifying these oils?

32. What functions do fats have in the human body?

33. Which fatty acids are essential to human diets?

34. Draw the structural formula of cholesterol.

35. Draw the ring structure that is common to all steroids.

36. List six foods that are major sources of proteins.

37. What functional groups are present in amino acids?

38. Why are the amino acids of proteins called α-amino acids?

39. Write the full structure for the two possible dipeptides containing glycine and phenylalanine.

40. Write structures for
    (a) glycylglycine
    (b) glycylglycylalanine
    (c) leucylmethionylglycylserine.

41. Using amino acid abbreviations, write all the possible tripeptides containing one unit each of glycine, phenylalanine, and leucine.

42. What are essential amino acids? Write the names of the amino acids that are essential for humans.

43. When proteins are eaten by a human, what are the metabolic fates of the protein material?

44. Why should protein be continually included in a balanced diet?

45. Write structural formulas for the compounds that make up DNA.

46. (a) What are the three units that make up a nucleotide?
    (b) List the components of the four types of nucleotides found in DNA.
    (c) Write the structure and name of one of these nucleotides.

47. Briefly describe the structure of DNA as proposed by Watson and Crick.

48. What is the role of hydrogen bonding in the structure of DNA?

49. Explain the concept of complementary bases and how it relates to DNA.

50. A segment of a DNA strand has a base sequence of C-G-A-T-T-G-C-A. What is the base sequence of the other complementary strand of the double helix?

51. Explain the replication process of DNA.

52. Briefly discuss the relationship of DNA to genetics.

53. What are the three differences between DNA and RNA in terms of structure?

54. Distinguish between cell division in mitosis and in meiosis.

55. What are enzymes and what is their role in the body?

56. What is meant by specificity of an enzyme?

57. In the polypeptide Tyr-Gly-His-Phe-Val, identify the N-terminal and the C-terminal residues.

58. How are polypeptides numbered? Number the polypeptide in Question 57.

59. What is the name of the bond that bonds one α-amino acid to another in a protein?

60. Differentiate between the lock-and-key and induced-fit models for enzyme function.

61. What is meant by enzyme specificity?

62. What is the function of enzymes in the body?

63. Use molecular structure to explain the following:
    (a) Fructose is a ketone.
    (b) Glucose is an aldehyde.

64. What is the simplest empirical formula for a carbohydrate?

65. How can you discern the difference between an amino acid and an ordinary carboxylic acid?

*66. What change in oxidation state is experienced by the carbon in the combustion of sugar, $C_{12}H_{22}O_{11}$, to CO_2?

67. Write the formula for the triacylglycerol formed from
    (a) glycerol and butanoic acid
    (b) glycerol and one molecule each of stearic ($C_{17}H_{35}COOH$), palmitic ($C_{15}H_{31}COOH$), and myristic ($C_{13}H_{27}COOH$) acids

68. The molar mass of cellulose is approximately 6.0×10^5 and the molar mass of a soluble starch is on the order of 4.0×10^3. The monomer unit in both of these molecules has the empirical formula, $C_6H_{10}O_5$. The units are about 5.0×10^{-10} m long. About how many units occur in each molecule, and how long are the molecules of cellulose and starch as a result?

PUTTING IT TOGETHER
Review for Chapters 18–20

Multiple Choice: *Choose the correct answer to each of the following.*

1. If $^{238}_{92}U$ loses an alpha particle, the resulting nuclide is
   (a) $^{237}_{92}U$ (b) $^{234}_{90}Th$ (c) $^{238}_{93}Np$ (d) $^{236}_{90}Th$

2. If $^{210}_{82}Pb$ loses a beta particle, the resulting nuclide is
   (a) $^{209}_{83}Bi$ (b) $^{210}_{81}Ti$ (c) $^{206}_{80}Hg$ (d) $^{210}_{83}Bi$

3. In the equation $^{209}_{83}Bi + ? \longrightarrow {}^{210}_{84}Po + {}^{1}_{0}n$, the missing bombarding particle would be
   (a) $^{2}_{1}H$ (b) $^{1}_{0}n$ (c) $^{4}_{2}He$ (d) $^{0}_{-1}e$

4. Which of the following is not a characteristic of nuclear fission?
   (a) Upon absorption of a proton, a heavy nucleus splits into two or more smaller nuclei.
   (b) Two or more neutrons are produced from the fission of each atom.
   (c) Large quantities of energy are produced.
   (d) Most nuclei formed are radioactive.

5. The half-life of Sn-121 is 10 days. If you started with 40 g of this isotope, how much would you have left 30 days later?
   (a) 10 g (b) none (c) 15 g (d) 5 g

6. $^{241}_{94}Pu$ successively emits β, α, α, β, α, α. At that point, the nuclide has become
   (a) $^{225}_{94}Pu$ (b) $^{225}_{88}Ra$ (c) $^{207}_{84}Po$ (d) $^{219}_{84}Po$

7. Calculate the nuclear binding energy of $^{56}_{26}Fe$.
   Mass data:
   $^{56}_{26}Fe = 55.9349$ g/mol
   n = 1.0087 g/mol $e^- = 0.00055$ g/mol
   p = 1.0073 g/mol 1.0 g = 9.0×10^{13} J
   (a) 4.8×10^{13} J/mol (c) 0.5302 g/mol
   (b) 56.4651 g/mol (d) 4.9×10^{15} J/mol

8. The radioactivity ray with the greatest penetrating ability is
   (a) alpha (c) gamma
   (b) beta (d) proton

9. In a nuclear reaction,
   (a) mass is lost
   (b) mass is gained
   (c) mass is converted into energy
   (d) energy is converted into mass

10. As the temperature of a radionuclide increases, its half-life
    (a) increases (c) remains the same
    (b) decreases (d) fluctuates

11. The nuclide that has the longest half-life is
    (a) $^{238}_{92}U$ (b) $^{210}_{82}Pb$ (c) $^{234}_{90}Th$ (d) $^{222}_{88}Ra$

12. Which of the following is not a unit of radiation?
    (a) curie (b) roentgen (c) rod (d) rem

13. When $^{235}_{92}U$ is bombarded by a neutron, the atom can fission into
    (a) $^{124}_{53}I + {}^{109}_{47}Ag + 2\,{}^{1}_{0}n$
    (b) $^{123}_{50}Sn + {}^{110}_{42}Mo + 2\,{}^{1}_{0}n$
    (c) $^{134}_{56}Ba + {}^{128}_{36}Xe + 2\,{}^{1}_{0}n$
    (d) $^{90}_{38}Sr + {}^{143}_{58}Ce + 2\,{}^{1}_{0}n$

14. In the nuclear equation
    $$^{45}_{21}Sc + {}^{1}_{0}n \longrightarrow X + {}^{1}_{1}H$$
    the nuclide X that is formed is
    (a) $^{45}_{22}Ti$ (b) $^{45}_{20}Ca$ (c) $^{46}_{22}Ti$ (d) $^{45}_{20}K$

15. What type of radiation is a very energetic form of photon?
    (a) alpha (c) gamma
    (b) beta (d) positron

16. When $^{239}_{92}U$ decays to $^{239}_{93}Np$, what particle is emitted?
    (a) positron (c) alpha particle
    (b) neutron (d) beta particle

17. The roentgen is the unit of radiation that measures
    (a) an absorbed dose of radiation
    (b) exposure to X-rays
    (c) the dose from a different type of radiation
    (d) the rate of decay of a radioactive substance

18. Which of the following is not a correct name for the alkane shown with it?
    (a) C_2H_6, ethane (c) C_7H_{16}, heptane
    (b) C_5H_{12}, propane (d) $C_{10}H_{22}$, decane

19. The structural formula of *o*-xylene is

20. The ester

$$CH_3\overset{\displaystyle O}{\overset{\displaystyle \|}{C}}-O-\overset{\displaystyle CH_3}{\overset{\displaystyle |}{C}}HCH_3$$

can be made from which alcohol and carboxylic acid?

(a) $CH_3\overset{\displaystyle O}{\overset{\displaystyle \|}{C}}-OH$, $CH_3CH_2CH_2OH$

(b) $CH_3\overset{\displaystyle O}{\overset{\displaystyle \|}{C}}-OH$, $HO-\overset{\displaystyle CH_3}{\overset{\displaystyle |}{C}}HCH_2OH$

(c) $H-\overset{\displaystyle O}{\overset{\displaystyle \|}{C}}-OH$, $CH_3\overset{\displaystyle OH}{\overset{\displaystyle |}{C}}HCH_3$

(d) $CH_3\overset{\displaystyle O}{\overset{\displaystyle \|}{C}}-OH$, $CH_3\overset{\displaystyle OH}{\overset{\displaystyle |}{C}}HCH_3$

21. The product of the reaction

$$CH_3CH{=}CH_2 + H_2O + H^+ \longrightarrow$$

is a(n)

(a) alcohol   (c) alkyne

(b) aldehyde (d) carboxylic acid

22. The number of isomers of butyl alcohol, C_4H_9OH, is

(a) 2 (b) 3 (c) 4 (d) 6

23. What is the correct name for this compound?

$$CH_3-\overset{\displaystyle \overset{\displaystyle CH_3}{\displaystyle |}}{C}H-CH-\overset{\displaystyle \overset{\displaystyle CH_3}{\displaystyle |}}{C}H-CH{=}CH_2$$
$$\overset{\displaystyle |}{CH_3-CH-CH_3}$$

(a) isobutane

(b) 2,4-methyl-3-propyl-5-hexene

(c) 3,5-dimethyl-4-isopropyl-1-hexene

(d) 4,4-diisopropylhexene

24. Which of these acids is named incorrectly?

(a) $CH_3CH_2CH_2COOH$, butyric acid

(b) $HCOOH$, formic acid

(c) CH_3CH_2COOH, propic acid

(d) CH_3COOH, acetic acid

25. With acid as a catalyst, ethanol and formic acid will react to form

(a) $CH_3\overset{\displaystyle O}{\overset{\displaystyle \|}{C}}-OH$

(b) $H-\overset{\displaystyle O}{\overset{\displaystyle \|}{C}}-O-CH_2CH_3$

(c) $CH_3\overset{\displaystyle O}{\overset{\displaystyle \|}{C}}-O-CH_3$

(d) $CH_3\overset{\displaystyle O}{\overset{\displaystyle \|}{C}}-O-\overset{\displaystyle O}{\overset{\displaystyle \|}{C}}CH_3$

26. The number of isomers of C_6H_{14} is

(a) 3 (b) 5 (c) 6 (d) 8

27. An open-chain hydrocarbon of formula C_6H_8 can have in its formula

(a) one carbon–carbon double bond

(b) two carbon–carbon double bonds

(c) one carbon–carbon triple bond

(d) one carbon–carbon double bond and one carbon–carbon triple bond

28. The reaction

$$CH_2{=}CH_2 + Br_2 \longrightarrow CH_2BrCH_2Br$$

represents

(a) dehalogenation (c) addition

(b) substitution   (d) dehydration

29. The general formula for a ketone is

(a) RCHO (c) RCOOR

(b) ROR   (d) R_2CO

30. Which of the following cannot be an aromatic compound?

(a) C_6H_5OH (c) C_6H_{14}

(b) C_6H_6 (d) $C_6H_5CH_3$

31. What is the correct name for this compound?

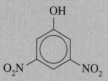

(a) m-dinitrophenol   (c) 3,5-dinitrophenol

(b) 2,4-dinitrophenol (d) 1,3-dinitrophenol

32. Which of these pairs are not isomers?

(a) CH_3OCH_2Cl and CH_2ClCH_2OH

(b) CH_3CH_2CHO and $CH_3OCH_2CH_3$

(c) $CH_3OCH_2OCH_3$ and $CH_3CH(OH)CH_2OH$

(d) $C_6H_4(CH_3)_2$ and $C_6H_5CH_2CH_3$

33. The reaction of $CH_3CH{=}CHCH_3 + HBr$ produces

(a) $CH_3CH_2CH_2CH_3 + Br_2$

(b) $CH_3CHBrCHBrCH_3 + H_2$

(c) $CH_3CHBrCH_2CH_3$

(d) $CH_3CH_2CH_2CH_2Br$

34. Polyvinyl chloride is a polymer of

(a) $CH_2{=}CCl_2$

(b) $CF_2{=}CF_2$

(c) $CH_2{=}CHCl$

(d) $C_6H_5CH{=}CHCl$

35. What is the correct name for this compound?

$$CH_3\overset{\displaystyle O}{\overset{\displaystyle \|}{C}}-O-\overset{\displaystyle \overset{\displaystyle CH_3}{\displaystyle |}}{\underset{\displaystyle \underset{\displaystyle H}{\displaystyle |}}{C}}-CH_3$$

(a) acetyl-2-propanoate

(b) propyl acetate

(c) ethyl isopropylate

(d) isopropyl ethanoate

36. Teflon is a polymer of

(a) $CH_2{=}CCl_2$ (c) $CH_2{=}CHCl$

(b) $CF_2{=}CF_2$ (d) $C_6H_5CH{=}CHCl$

37. Sugars are members of a group of compounds with the general name
    (a) carbohydrates (c) proteins
    (b) lipids   (d) steroids

38. The products formed when maltose is hydrolyzed are
    (a) glucose and fructose
    (b) glucose and galactose
    (c) glucose and glucose
    (d) galactose and fructose

39. Which is *not* true of starch?
    (a) It is a polysaccharide.
    (b) It is hydrolyzed to maltose.
    (c) It is composed of glucose units.
    (d) It is not digestible by humans.

40. Lactose is
    (a) a monosaccharide
    (b) a disaccharide composed of galactose and glucose
    (c) a disaccharide composed of two glucose units
    (d) a decomposition product of starch

41. Which is *not* true of glucose?
    (a) It is a monosaccharide.
    (b) It is a component of sucrose, maltose, lactose, starch, glycogen, and cellulose.
    (c) It is a ketohexose.
    (d) It is the main source of energy for the body.

42. The sweetest of the common sugars is
    (a) fructose (c) glucose
    (b) sucrose   (d) maltose

43. Which of the following is *not* true?
    (a) Lactose is sweeter than sucrose.
    (b) Glycogen is known as animal starch.
    (c) Cellulose is the most abundant organic substance in nature.
    (d) Lactose is a disaccharide known as milk sugar.

44. Which is *not* formed in the saponification of a fat?
    (a) glycerol
    (b) amino acids
    (c) soap
    (d) a metal salt of a long-chain fatty acid

45. Which of these lipids does *not* contain a glycerol unit as part of its structure?
    (a) a fat   (c) a glycolipid
    (b) a phospholipid (d) an oil

46. An α-amino acid always contains
    (a) an amino group on the carbon atom adjacent to the carboxyl group
    (b) a carboxyl group at each end of the molecule
    (c) two amino groups
    (d) alternating amino and carboxyl groups

47. Which of these amino acids contains sulfur?
    (a) alanine   (c) cysteine
    (b) histidine (d) glycine

48. A compound containing ten amino acid molecules linked together is called a
    (a) protein   (c) deca-amino acid
    (b) polypeptide   (d) nucleotide

49. Which of the following is *not* a correct statement about DNA and RNA?
    (a) DNA contains deoxyribose, whereas RNA contains ribose.
    (b) Both DNA and RNA are polymers made up of nucleotides.
    (c) DNA directs the synthesis of proteins and RNA contains the genetic code of life.
    (d) DNA exists as a double helix, whereas RNA exists as a single helix.

50. Which of these bases is found in RNA, but not in DNA?
    (a) thymine (c) guanine
    (b) adenine (d) uracil

51. Complementary base pairs in DNA are linked through the formation of
    (a) phosphate ester bonds
    (b) peptide linkages
    (c) hydrogen bonds
    (d) ionic bonds

52. In a DNA double helix, hydrogen bonding occurs between
    (a) adenine and thymine
    (b) thymine and guanine
    (c) adenine and uracil
    (d) cytosine and thymine

53. Which of these scientists did *not* receive the Nobel Prize for the structure of DNA?
    (a) Crick   (c) Sanger
    (b) Watson (d) Wilkins

54. The substance acted on by an enzyme is called a(n)
    (a) catalyst   (c) coenzyme
    (b) apoenzyme   (d) substrate

55. The process during which a cell splits to form a sperm or an egg cell is called
    (a) mitosis (c) translation
    (b) meiosis (d) transcription

Free Response Questions

1. The nuclide $^{223}_{87}$Fr emits three radioactive particles losing 8 mass units and 3 atomic number units. Propose a radioactive decay series and write the symbol for the resulting nuclide.

2. Biological molecules such as methionine enkephalin, leucine enkephalin, and β-endorphin begin with a common tetrapeptide tyrosyl-glycyl-glycyl-phenylalanyl-. Draw the tetrapeptide sequence and write its abbreviated form.

3. What functional groups are in prostaglandin A_2?

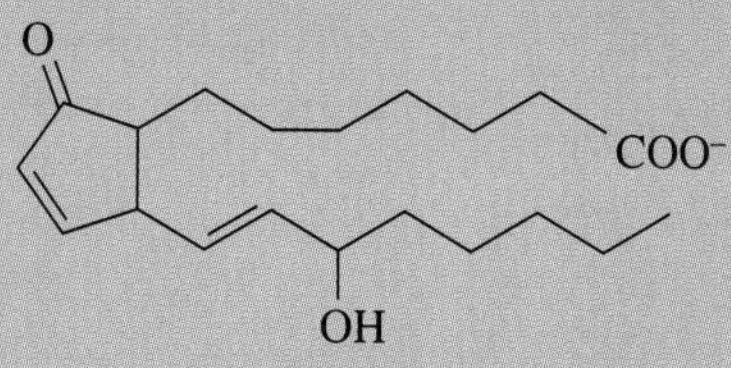

prostaglandin A_2

4. Which of the following molecules would you expect to be significantly soluble in water? Why?
$CH_3CH_2CH_2OH$, $CH_3CH_2CH_2CHO$, $CH_3CH_2COCH_3$, $CH_3CH_2CH_2CH_3$
Hint: Remember "like dissolves like."

5. Which of these compounds are structural isomers of each other?
methyl propyl ether, butanoic acid, acetone, propene, 2-butanol, phenol, methyl propanoate

6. (a) Indicate which two amino acids could form an ester bond using their R group side chains.
(b) Draw the product of cholesterol reacting with bromine.

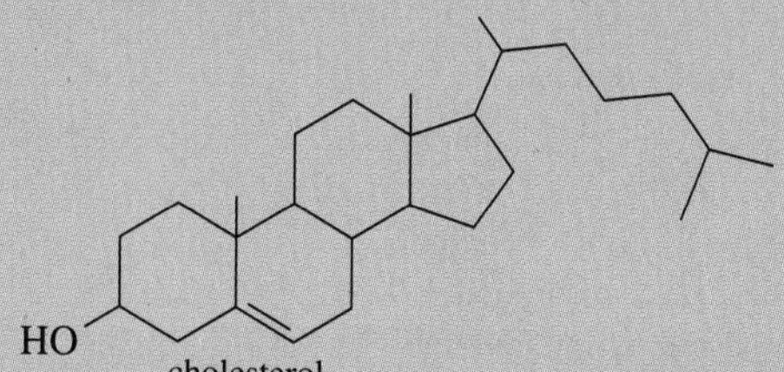

cholesterol

(c) Is cholesterol soluble in water? How do you know?

7. (a) Carboxypeptidase A is a digestive enzyme. A zinc ion is necessary for enzyme activity. What is the zinc ion called?
(b) Some RNA strands also show enzymatic activity but DNA does not. What are two structural differences between RNA and DNA?
(c) Do you think we have radioactive isotopes in our bodies? Explain.

8. (a) For each OH group in open chain glucose, indicate if it is primary, secondary, or tertiary.
(b) Does tyrosine contain an ortho, meta, or parasubstituted aromatic ring?
(c) How many amino acids are in a pentapeptide? How many molecules of water were lost when it was formed?
(d) Based on what you know about the hydrolysis of disaccharides, if a person has high blood sugar levels, do you think it would be better for them to cut down on lactose or maltose? Why?

9. How many half-lives are required to change 96 g of a sample of a radioactive isotope to 1.5 g over approximately 24 days? What is the approximate half-life of this isotope?

10. (a) What type of process is $^6_3Li + ^1_0n \longrightarrow ^3_1H + ^4_2He$?
(b) How is this process different from radioactive decay?
(c) Could nuclear fission be classified as the same type process?

Mathematical Review

Multiplication   Multiplication is a process of adding any given number or quantity to itself a certain number of times. Thus, 4 times 2 means 4 added two times, or 2 added together four times, to give the product 8. Various ways of expressing multiplication are

$$ab \qquad a \times b \qquad a \cdot b \qquad a(b) \qquad (a)(b)$$

Each of these expressions means a times b, or a multiplied by b, or b times a.

When $a = 16$ and $b = 24$,   we have $16 \times 24 = 384$.

The expression $°F = (1.8 \times °C) + 32$ means that we are to multiply 1.8 times the Celsius degrees and add 32 to the product. When $°C$ equals 50,

$$°F = (1.8 \times 50) + 32 = 90 + 32 = 122°F$$

The result of multiplying two or more numbers together is known as the *product.*

Division   The word *division* has several meanings. As a mathematical expression, it is the process of finding how many times one number or quantity is contained in another. Various ways of expressing division are

$$a \div b \qquad \frac{a}{b} \qquad a/b$$

Each of these expressions means a divided by b.

When $a = 15$ and $b = 3$, $\dfrac{15}{3} = 5.$

The number above the line is called the *numerator;* the number below the line is the *denominator.* Both the horizontal and the slanted (/) division signs also mean "per." For example, in the expression for density, the mass per unit volume:

$$\text{density} = \text{mass/volume} = \frac{\text{mass}}{\text{volume}} = \text{g/mL}$$

The diagonal line still refers to a division of grams by the number of milliliters occupied by that mass. The result of dividing one number into another, is called the *quotient.*

Proper fraction	Decimal fraction	Proper fraction
$\dfrac{1}{8}$	= 0.125 =	$\dfrac{125}{1000}$
$\dfrac{1}{10}$	= 0.1 =	$\dfrac{1}{10}$
$\dfrac{3}{4}$	= 0.75 =	$\dfrac{75}{100}$
$\dfrac{1}{100}$	= 0.01 =	$\dfrac{1}{100}$
$\dfrac{1}{4}$	= 0.25 =	$\dfrac{25}{100}$

Fractions and Decimals   A fraction is an expression of division, showing that the numerator is divided by the denominator. A *proper fraction* is one in which the numerator is smaller than the denominator. In an *improper fraction,* the numerator is the larger number. A decimal or a decimal fraction is a proper fraction in which the denominator is some power of 10. The decimal fraction is determined by carrying out the division of the proper fraction. Examples of proper fractions and their decimal fraction equivalents are shown in the accompanying table.

Addition of Numbers with Decimals   To add numbers with decimals, we use the same procedure as that used when adding whole numbers, but we always line up the decimal points in the same column. For example, add 8.21 + 143.1 + 0.325:

```
    8.21
 +143.1
 +  0.325
 --------
  151.635
```

When adding numbers that express units of measurement, we must be certain that the numbers added together all have the same units. For example, what is the total length of three pieces of glass tubing: 10.0 cm, 125 mm, and 8.4 cm? If we simply add the numbers, we obtain a value of 143.4, but we are not certain what the unit of measurement is. To add these lengths correctly, first change 125 mm to 12.5 cm. Now all the lengths are expressed in the same units and can be added:

```
 10.0 cm
 12.5 cm
  8.4 cm
 -------
 30.9 cm
```

Subtraction of Numbers with Decimals   To subtract numbers containing decimals, we use the same procedure as for subtracting whole numbers, but we always line up the decimal points in the same column. For example, subtract 20.60 from 182.49:

```
 182.49
 - 20.60
 -------
 161.89
```

Multiplication of Numbers with Decimals   To multiply two or more numbers together that contain decimals, we first multiply as if they were whole numbers. Then, to locate the decimal point in the product, we add together the number of digits to the right of the decimal in all the numbers multiplied together. The product should have this same number of digits to the right of the decimal point.

Multiply $2.05 \times 2.05 = 4.2025$ (total of four digits to the right of the decimal). Here are more examples:

$14.25 \times 6.01 \times 0.75 = 64.231875$ (six digits to the right of the decimal)
$39.26 \times 60 = 1255.60$ (two digits to the right of the decimal)

If a number is a measurement, the answer must be adjusted to the correct number of significant figures.

Division of Numbers with Decimals To divide numbers containing decimals, we first relocate the decimal points of the numerator and denominator by moving them to the right as many places as needed to make the denominator a whole number. (Move the decimal of both the numerator and the denominator the same amount and in the same direction.) For example,

$$\frac{136.94}{4.1} = \frac{1369.4}{41}$$

The decimal point adjustment in this example is equivalent to multiplying both numerator and denominator by 10. Now we carry out the division normally, locating the decimal point immediately above its position in the dividend:

$$41\overline{)1269.4} \quad \frac{0.441}{26.25} = \frac{44.1}{2625} = 2625\overline{)44.1000} $$

$$\overset{33.4}{41\overline{)1269.4}} \qquad \frac{0.441}{26.25} = \frac{44.1}{2625} = \overset{0.0168}{2625\overline{)44.1000}}$$

These examples are guides to the principles used in performing the various mathematical operations illustrated. Every student of chemistry should learn to use a calculator for solving mathematical problems (see Appendix II). The use of a calculator will save many hours of doing tedious calculations. After solving a problem, the student should check for errors and evaluate the answer to see if it is logical and consistent with the data given.

Algebraic Equations Many mathematical problems that are encountered in chemistry fall into the following algebraic forms. Solutions to these problems are simplified by first isolating the desired term on one side of the equation. This rearrangement is accomplished by treating both sides of the equation in an identical manner until the desired term is isolated.

(a) $a = \dfrac{b}{c}$

To solve for b, multiply both sides of the equation by c:

$$a \times c = \frac{b}{c} \times c$$

$$b = a \times c$$

To solve for c, multiply both sides of the equation by $\dfrac{c}{a}$:

$$a \times \frac{c}{a} = \frac{b}{c} \times \frac{c}{a}$$

$$c = \frac{b}{a}$$

(b) $\dfrac{a}{b} = \dfrac{c}{d}$

To solve for a, multiply both sides of the equation by b:

$$\dfrac{a}{\not{b}} \times \not{b} = \dfrac{c}{d} \times b$$

$$a = \dfrac{c \times b}{d}$$

To solve for b, multiply both sides of the equation by $\dfrac{b \times d}{c}$:

$$\dfrac{a}{\not{b}} \times \dfrac{\not{b} \times d}{c} = \dfrac{\not{c}}{\not{d}} \times \dfrac{b \times \not{d}}{\not{c}}$$

$$\dfrac{a \times d}{c} = b$$

(c) $a \times b = c \times d$

To solve for a, divide both sides of the equation by b:

$$\dfrac{a \times \not{b}}{\not{b}} = \dfrac{c \times d}{b}$$

$$a = \dfrac{c \times d}{b}$$

(d) $\dfrac{b - c}{a} = d$

To solve for b, first multiply both sides of the equation by a:

$$\dfrac{\not{a}(b - c)}{\not{a}} = d \times a$$

$$b - c = d \times a$$

Then add c to both sides of the equation:

$$b - \not{c} + \not{c} = d \times a + c$$

$$b = (d \times a) + c$$

When $a = 1.8$, $c = 32$, and $d = 35$,

$$b = (35 \times 1.8) + 32 = 63 + 32 = 95$$

Expression of Large and Small Numbers In scientific measurement and calculations, we often encounter very large and very small numbers—for example, 0.00000384 and 602,000,000,000,000,000,000,000. These numbers are troublesome to write and awkward to work with, especially in calculations. A convenient method of expressing these large and small numbers in a simplified form is by means of exponents or powers of 10. This method of expressing numbers is known as **scientific or exponential notation.**

An **exponent** is a number written as a superscript following another number. Exponents are often called *powers* of numbers. The term *power* indicates how many times the number is used as a factor. In the number 10^2, 2 is the exponent, and the number means 10 squared, or 10 to the second power, or $10 \times 10 = 100$. Three other examples are

scientific or exponential notation

exponent

$$3^2 = 3 \times 3 = 9$$
$$3^4 = 3 \times 3 \times 3 \times 3 = 81$$
$$10^3 = 10 \times 10 \times 10 = 1000$$

For ease of handling, large and small numbers are expressed in powers of 10. Powers of 10 are used because multiplying or dividing by 10 coincides with moving the decimal point in a number by one place. Thus, a number multiplied by 10^1 would move the decimal point one place to the right; 10^2, two places to the right; 10^{-2}, two places to the left. To express a number in powers of 10, we move the decimal point in the original number to a new position, placing it so that the number is a value between 1 and 10. This new decimal number is multiplied by 10 raised to the proper power. For example, to write the number 42,389 in exponential form, the decimal point is placed between the 4 and the 2 (4.2389), and the number is multiplied by 10^4; thus, the number is 4.2389×10^4:

$$42,389 = 4.2389 \times 10^4$$
$$4\,3\,2\,1$$

The exponent of 10 (4) tells us the number of places that the decimal point has been moved from its original position. If the decimal point is moved to the left, the exponent is a positive number; if it is moved to the right, the exponent is a negative number. To express the number 0.00248 in exponential notation (as a power of 10), the decimal point is moved three places to the right; the exponent of 10 is -3, and the number is 2.48×10^{-3}.

$$0.00248 = 2.48 \times 10^{-3}$$
$$1\,2\,3$$

Study the following examples.

$$1237 = 1.237 \times 10^3$$
$$988 = 9.88 \times 10^2$$
$$147.2 = 1.472 \times 10^2$$
$$2{,}200{,}000 = 2.2 \times 10^6$$
$$0.0123 = 1.23 \times 10^{-2}$$
$$0.00005 = 5 \times 10^{-5}$$
$$0.000368 = 3.68 \times 10^{-4}$$

Exponents in multiplication and division The use of powers of 10 in multiplication and division greatly simplifies locating the decimal point in the answer. In multiplication, first change all numbers to powers of 10, then multiply the numerical portion in the usual manner, and finally add the exponents of 10 algebraically, expressing them as a power of 10 in the product. In multiplication, the exponents (powers of 10) are added algebraically.

$$10^2 \times 10^3 = 10^{(2+3)} = 10^5$$
$$10^2 \times 10^2 \times 10^{-1} = 10^{(2+2-1)} = 10^3$$

Multiply: $\qquad\qquad$ (40,000)(4200)

Change to powers of 10: $\quad$ $(4 \times 10^4)(4.2 \times 10^3)$

Rearrange: $\qquad\qquad$ $(4 \times 4.2)(10^4 \times 10^3)$

$\qquad\qquad\qquad$ $16.8 \times 10^{(4+3)}$

$\qquad\qquad\qquad$ 16.8×10^7 $\quad$ or $\quad$ 1.68×10^8 $\quad$ (Answer)

Multiply: $\quad$ (380)(0.00020)

$\qquad\qquad$ $(3.80 \times 10^2)(2.0 \times 10^{-4})$

$\qquad\qquad$ $(3.80 \times 2.0)(10^2 \times 10^{-4})$

$\qquad\qquad$ $7.6 \times 10^{(2-4)}$

$\qquad\qquad$ 7.6×10^{-2} $\quad$ or $\quad$ 0.076 $\quad$ (Answer)

Multiply: $\quad$ (125)(284)(0.150)

$\qquad\qquad$ $(1.25 \times 10^2)(2.84 \times 10^2)(1.50 \times 10^{-1})$

$\qquad\qquad$ $(1.25)(2.84)(1.50)(10^2 \times 10^2 \times 10^{-1})$

$\qquad\qquad$ $5.325 \times 10^{(2+2-1)}$

$\qquad\qquad$ 5.33×10^3 $\quad$ (Answer)

In division, after changing the numbers to powers of 10, move the 10 and its exponent from the denominator to the numerator, changing the sign of the exponent. Carry out the division in the usual manner and evaluate the power of 10. Change the sign(s) of the exponent(s) of 10 in the denominator and move the 10 and its exponent(s) to the numerator. Then add all the exponents of 10 together. For example,

$$\frac{10^5}{10^3} = 10^5 \times 10^{-3} = 10^{(5-3)} = 10^2$$

$$\frac{10^3 \times 10^4}{10^{-2}} = 10^3 \times 10^4 \times 10^2 = 10^{(3+4+2)} = 10^9$$

Significant Figures in Calculations   The result of a calculation based on experimental measurements cannot be more precise than the measurement that has the greatest uncertainty. (See Section 2.5 for additional discussion.)

   Addition and Subtraction The result of an addition or subtraction should contain no more digits to the right of the decimal point than are contained in the quantity that has the least number of digits to the right of the decimal point.

   Perform the operation indicated and then round off the number to the proper number of significant figures:

$$
\begin{array}{ll}
142.8 \ \ \text{g} & \\
18.843 \ \text{g} & \qquad 93.45 \ \text{mL} \\
\underline{36.42 \ \ \text{g}} & \qquad \underline{-18.0 \ \ \text{mL}} \\
198.063 \ \text{g} & \qquad 75.45 \ \text{mL} \\
198.1 \ \ \ \text{g (Answer)} & \qquad 75.5 \ \ \text{mL (Answer)}
\end{array}
$$

   Multiplication and Division In calculations involving multiplication or division, the answer should contain the same number of significant figures as the measurement that has the least number of significant figures. In multiplication or division, the position of the decimal point has nothing to do with the number of significant figures in the answer. Study the following examples:

	Round off to
$(2.05)(2.05) = 4.2025$	4.20
$(18.48)(5.2) = 96.096$	96
$(0.0126)(0.020) = 0.000252$ or	
$(1.26 \times 10^{-2})(2.0 \times 10^{-2}) = 2.520 \times 10^{-4}$	2.5×10^{-4}
$\dfrac{1369.4}{41} = 33.4$	33
$\dfrac{2268}{4.20} = 540.$	540.

Dimensional Analysis   Many problems of chemistry can be readily solved by dimensional analysis using the factor-label or conversion-factor method. Dimensional analysis involves the use of proper units of dimensions for all factors that are multiplied, divided, added, or subtracted in setting up and solving a problem. Dimensions are physical quantities such as length, mass, and time, which are expressed in such units as centimeters, grams, and seconds, respectively. In solving a problem, we treat these units mathematically just as though they were numbers, which gives us an answer that contains the correct dimensional units.

   A measurement or quantity given in one kind of unit can be converted to any other kind of unit having the same dimension. To convert from one kind of unit to another, the original quantity or measurement is multiplied or divided by a conversion factor. The key to success lies in choosing the correct conversion factor. This general method of calculation is illustrated in the following examples.

   Suppose we want to change 24 ft to inches. We need to multiply 24 ft by a conversion factor containing feet and inches. Two such conversion factors can be written relating inches to feet:

$$\frac{12 \text{ in.}}{1 \text{ ft}} \quad \text{or} \quad \frac{1 \text{ ft}}{12 \text{ in.}}$$

We choose the factor that will mathematically cancel feet and leave the answer in inches. Note that the units are treated in the same way we great numbers, multiplying or dividing as required. Two possibilities then arise to change 24 ft to inches:

$$(24 \text{ ft})\left(\frac{12 \text{ in.}}{1 \text{ ft}}\right) \quad \text{or} \quad (24 \text{ ft})\left(\frac{1 \text{ ft}}{12 \text{ in.}}\right)$$

In the first case (the correct method), feet in the numerator and the denominator cancel, giving us an answer of 288 in. In the second case, the units of the answer are $ft^2/in.$, the answer being $2.0 \ ft^2/in.$ In the first case, the answer is reasonable because it is expressed in units having the proper dimensions. That is, the dimension of length expressed in feet has been converted to length in inches according to the mathematical expression

$$ft \times \frac{in.}{ft} = in.$$

In the second case, the answer is not reasonable because the units (ft^2/in.) do not correspond to units of length. The answer is therefore incorrect. The units are the guiding factor for the proper conversion.

The reason we can multiply 24 ft times 12 in./ft and not change the value of the measurement is that the conversion factor is derived from two equivalent quantities. Therefore, the conversion factor 12 in./ft is equal to unity. When you multiply any factor by 1, it does not change the value:

$$12 \text{ in.} = 1 \text{ ft} \qquad \text{and} \qquad \frac{12 \text{ in.}}{1 \text{ ft}} = 1$$

Convert 16 kg to milligrams. In this problem it is best to proceed in this fashion:

$$kg \longrightarrow g \longrightarrow mg$$

The possible conversion factors are

$$\frac{1000 \text{ g}}{1 \text{ kg}} \quad \text{or} \quad \frac{1 \text{ kg}}{1000 \text{ g}} \qquad \frac{1000 \text{ mg}}{1 \text{ g}} \quad \text{or} \quad \frac{1 \text{ g}}{1000 \text{ mg}}$$

We use the conversion factor that leaves the proper unit at each step for the next conversion. The calculation is

$$(16 \text{ kg})\left(\frac{1000 \text{ g}}{1 \text{ kg}}\right)\left(\frac{1000 \text{ mg}}{1 \text{ g}}\right) = 1.6 \times 10^7 \text{ mg}$$

Many problems can be solved by a sequence of steps involving unit conversion factors.

Graphical Representation of Data   A graph is often the most convenient way to present or display a set of data. Various kinds of graphs have been devised, but the most common type uses a set of horizontal and vertical coordinates to show the relationship of two variables. It is called an x–y graph because the data of one variable are represented on the horizontal or x-axis (abscissa) and the data of the other variable are represented on the vertical or y-axis (ordinate). See Figure I.1.

As a specific example of a simple graph, let us graph the relationship between Celsius and Fahrenheit temperature scales. Assume that initially we have only the information in the table next to Figure I.2.

On a set of horizontal and vertical coordinates (graph paper), scale off at least 100 Celsius degrees on the x-axis and at least 212 Fahrenheit degrees on the y-axis. Locate and mark the three points corresponding to the three temperatures given and draw a line connecting these points (see Figure I.2).

Here is how a point is located on the graph: Using the 50°C–122°F data, trace a vertical line up from 50°C on the x-axis and a horizontal line across from 122°F on the y-axis and mark the point where the two lines intersect. This process is called *plotting*. The other two points are plotted on the graph in the same way. [*Note:* The number of degrees per scale division was chosen to give a graph of convenient size. In this case, there are 5 Fahrenheit degrees per scale division and 2 Celsius degrees per scale division.]

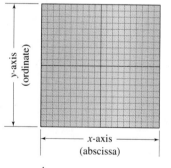

▲ **FIGURE I.1**

The graph in Figure I.2 shows that the relationship between Celsius and Fahrenheit temperature is that of a straight line. The Fahrenheit temperature corresponding to any given Celsius temperature between 0 and 100° can be determined from the graph. For example, to find the Fahrenheit temperature corresponding to 40°C, trace a perpendicular line from 40°C on the x-axis to the line plotted on the graph. Now trace a horizontal line from this point on the plotted line to the y-axis and read the corresponding Fahrenheit temperature (104°F). See the dashed lines in Figure I.2. In turn, the Celsius temperature corresponding to any Fahrenheit temperature between 32 and 212° can be determined from the graph by tracing a horizontal line from the Fahrenheit temperature to the plotted line and reading the corresponding temperature on the Celsius scale directly below the point of intersection.

The mathematical relationship of Fahrenheit and Celsius temperatures is expressed by the equation $°F = 1.8 \times °C + 32$. Figure I.2 is a graph of this equation. Because the graph is a straight line, it can be extended indefinitely at either end. Any desired Celsius temperature can be plotted against the corresponding Fahrenheit temperature by extending the scales along both axes as necessary.

Figure I.3 is a graph showing the solubility of potassium chlorate in water at various temperatures. The solubility curve on this graph was plotted from the data in the table next to the graph.

In contrast to the Celsius–Fahrenheit temperature relationship, there is no simple mathematical equation that describes the exact relationship between temperature and the solubility of potassium chlorate. The graph in Figure I.3 was constructed from experimentally determined solubilities at the six temperatures shown. These experimentally determined solubilities are all located on the smooth curve traced by the unbroken-line portion of the graph. We are therefore confident that the unbroken line represents a very good approximation of the solubility data for potassium chlorate over the temperature range from 10 to 80°C. All points on

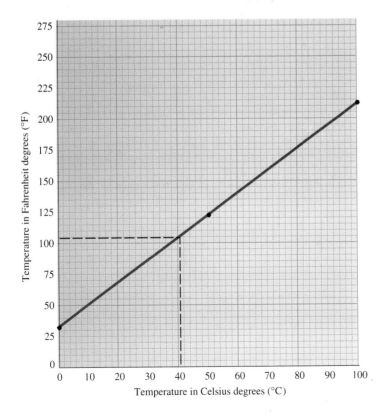

°C	°F
0	32
50	122
100	212

◀ FIGURE I.2

Temperature (°C)	Solubility (g KClO₃/ 100 g water)
10	5.0
20	7.4
30	10.5
50	19.3
60	24.5
80	38.5

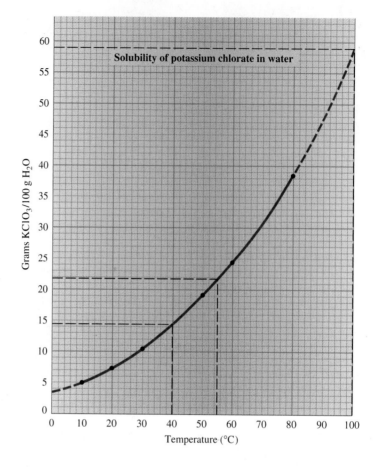

Figure 1.3 ▶

the plotted curve represent the composition of saturated solutions. Any point below the curve represents an unsaturated solution.

The dashed-line portions of the curve are *extrapolations;* that is, they extend the curve above and below the temperature range actually covered by the plotted solubility data. Curves such as this one are often extrapolated a short distance beyond the range of the known data, although the extrapolated portions may not be highly accurate. Extrapolation is justified only in the absence of more reliable information.

The graph in Figure I.3 can be used with confidence to obtain the solubility of $KClO_3$ at any temperature between 10 and 80°C, but the solubilities between 0 and 10°C and between 80 and 100°C are less reliable. For example, what is the solubility of $KClO_3$ at 55°C, at 40°C, and at 100°C?

First draw a perpendicular line from each temperature to the plotted solubility curve. Now trace a horizontal line to the solubility axis from each point on the curve and read the corresponding solubilities. The values that we read from the graph are

      40°C   14.2 g $KClO_3$/100 g water
      55°C   22.0 g $KClO_3$/100 g water
      100°C   59 g $KClO_3$/100 g water

Of these solubilities, the one at 55°C is probably the most reliable because experimental points are plotted at 50°C and at 60°C. The 40°C solubility value is a bit less reliable because the nearest plotted points are at 30°C and 50°C. The 100°C solubility is the least reliable of the three values because it was taken from the extrapolated part of the curve, and the nearest plotted point is 80°C. Actual handbook solubility values are 14.0 and 57.0 g of $KClO_3$/100 g water at 40°C and 100°C, respectively.

The graph in Figure I.3 can also be used to determine whether a solution is saturated or unsaturated. For example, a solution contains 15 g of $KClO_3$/100 g of water and is at a temperature of 55°C. Is the solution saturated or unsaturated? *Answer:* The solution is unsaturated because the point corresponding to 15 g and 55°C on the graph is below the solubility curve; all points below the curve represent unsaturated solutions.

APPENDIX
II

Using a Scientific Calculator

A calculator is useful for most calculations in this book. You should obtain a scientific calculator, that is, one that has at least the following function keys on its keyboard.

Addition $\boxed{+}$ Equals $\boxed{=}$

Subtraction $\boxed{-}$ Change sign $\boxed{+/-}$

Multiplication $\boxed{\times}$ Exponential number $\boxed{\text{Exp}}$

Division $\boxed{\div}$ Lorgarithm $\boxed{\text{Log}}$

Second function $\boxed{\text{F}}$ Antilogarithm $\boxed{10^x}$

Not all calculators use the same symbolism for these function keys, nor do all calculators work in the same way. The following discussion may not pertain to your particular calculator. Refer to your instruction manual for variations from the function symbols shown above and for the use of other function keys.

The second function key may have a different designation on your calculator.

Some keys have two functions, upper and lower. In order to use the upper (second) function, the second function key $\boxed{\text{F}}$ must be pressed in order to activate the desired upper function.

The display area of the calculator shows the numbers entered and often shows more digits in the answer than should be used. Therefore, the final answer should be rounded to reflect the proper number of significant figures of the calculations.

Addition and Subtraction   To add numbers using your calculator,

1. Enter the first number to be added followed by the plus key $\boxed{+}$.
2. Enter the second number to be added followed by the plus key $\boxed{+}$.
3. Repeat Step 2 for each additional number to be added, except the last number.
4. After the last number is entered press the equal key $\boxed{=}$. You should now have the answer in the display area.
5. When a number is to be subtracted, use the minus key $\boxed{-}$ instead of the plus key.

As an example, to add $16.0 + 1.223 + 8.45$ enter 16.0 followed by the $\boxed{+}$ key; then enter 1.223 followed by the $\boxed{+}$ key; then enter 8.45 followed by the $\boxed{=}$ key. Display shows 25.673, which is rounded to the answer 25.7.

Examples of Addition and Subtraction

Calculation	Enter in sequence	Display	Rounded answer
a. $12.0 + 16.2 + 122.3$	$12.0\boxed{+}16.2\boxed{+}122.3\boxed{=}$	150.5	150.5
b. $132 - 62 + 141$	$132\boxed{-}62\boxed{+}141\boxed{=}$	212	212
c. $46.23 + 13.2$	$46.23\boxed{+}13.2\boxed{=}$	59.43	59.4
d. $129.06 + 49.1 - 18.3$	$129.06\boxed{+}49.1\boxed{-}18.3\boxed{=}$	159.86	159.9

Multiplication   To multiply numbers using your calculator

1. Enter the first number to be multiplied followed by the multiplication key $\boxed{\times}$.
2. Enter the second number to be multiplied followed by the multiplication key $\boxed{\times}$.
3. Repeat Step 2 for all other numbers to be multiplied except the last number.
4. Enter the last number to be multiplied followed by the equal key $\boxed{=}$. You now have the answer in the display area.

Round off to the proper number of significant figures.

As an example, to calculate $(3.25)(4.184)(22.2)$ enter 3.25 followed by the $\boxed{\times}$ key; then enter 4.184 followed by the $\boxed{\times}$ key; then enter 22.2 followed by the $\boxed{=}$ key. The display shows 301.8756, which is rounded to the answer 302.

Examples of Multiplication

Calculation	Enter in sequence	Display	Rounded answer
a. $12 \times 14 \times 18$	$12\boxed{\times}14\boxed{\times}18\boxed{=}$	3024	3.0×10^3
b. $122 \times 3.4 \times 60.$	$122\boxed{\times}3.4\boxed{\times}60.\boxed{=}$	24888	2.5×10^4
c. $0.522 \times 49.4 \times 6.33$	$0.522\boxed{\times}49.4\boxed{\times}6.33\boxed{=}$	163.23044	163

Division   To divide numbers using your calculator,

1. Enter the numerator followed by the division key $\boxed{\div}$.
2. Enter the denominator followed by the equal key to give the answer.
3. If there is more than one denominator, enter each denominator followed by the division key except for the last number, which is followed by the equal key.

As an example, to calculate $\dfrac{126}{12}$ enter 126 followed by the $\boxed{\div}$ key; then enter 12 followed by the $\boxed{=}$ key. The display shows 1512, which is rounded to the answer 1.5×10^3.

Examples of Division			
Calculation	**Enter in sequence**	**Display**	**Rounded answer**
a. $\dfrac{142}{25}$	142 $\div$ 25 $=$	5.68	5.7
b. $\dfrac{0.422}{5.00}$	0.422 $\div$ 5.00 $=$	0.0844	0.0844
c. $\dfrac{124}{0.022 \times 3.00}$	124 $\div$ 0.022 $\div$ 3.00 $=$	1878.7878	1.9×10^3

Exponents In scientific measurements and calculations we often encounter very large and very small numbers. To express these large and small numbers conveniently, we use exponents or powers of 10. A number in exponential form is treated like any other number; that is, it can be added, subtracted, multiplied, or divided.

To enter an exponential number into your calculator first enter the nonexponential part of the number, then press the exponent key $\boxed{\text{Exp}}$, followed by the exponent. For example, to enter 4.9×10^3, enter 4.94, then press $\boxed{\text{Exp}}$, then press 3. When the exponent of 10 is a negative number, press the Change of Sign key $\boxed{+/-}$ after entering the exponent. For example, to enter 4.94×10^{-3}, enter in sequence 4.94 $\boxed{\text{Exp}}$ 3 $\boxed{+/-}$. In most calculators the exponent will appear in the display a couple of spaces after the nonexponent part of the number—for example, 4.94 03 or 4.94 -03.

Examples Using Exponential Numbers			
Calculation	**Enter in sequence**	**Display**	**Rounded answer**
a. $(4.94 \times 10^3)(21.4)$	4.94 $\boxed{\text{Exp}}$ 3 $\boxed{\times}$ 21.4 $\boxed{=}$	105716	1.06×10^5
b. $(1.42 \times 10^4)(2.88 \times 10^{-5})$	1.42 $\boxed{\text{Exp}}$ 4 $\boxed{\times}$ 2.88 $\boxed{\text{Exp}}$ 5 $\boxed{+/-}$ $\boxed{=}$	0.40896	0.409
c. $\dfrac{8.22 \times 10^{-5}}{5.00 \times 10^7}$	8.22 $\boxed{\text{Exp}}$ 5 $\boxed{+/-}$ $\boxed{\div}$ 5.00 $\boxed{\text{Exp}}$ 7 $\boxed{=}$	1.644 -12	1.64×10^{-12}

Logarithms The logarithm of a number to the base 10 is the power(exponent) to which 10 must be raised to give that number. For example, the log of 100 is 2.0(log $100 = 10^{2.0}$). The log of 200 is 2.3(log $200 = 10^{2.3}$). Logarithms are used in chemistry to calculate the pH of an aqueous acidic solution. The answer (log) should contain the same number of significant figures in the decimal as is in the original number. Thus, the log $100 = 2.0$ but the log 100. is 2.000.

The log key on most calculators is a function key. To determine the log using your calculator, enter the number, then press the function log key. For example, to determine the log of 125, enter 125, then the [Log] key. The answer is 2.097.

Examples Using Logarithms
Determine the log of the following:

	Enter in sequence	Display	Rounded answer
a. 42	42 [F] [Log]	1.6232492	1.62
b. 1.62×10^5	1.62 [Exp] 5 [F] [Log]	5.209515	5.210
c. 6.4×10^{-6}	6.4 [Exp] 6 [+/−] [F] [Log]	−5.19382	−5.19

Antilogarithms (Inverse Logarithms) An antilogarithm is the number from which the logarithm has been calculated. It is calculated using the [10^x] key on your calculator. For example, to determine the antilogarithm of 2.891, enter 2.891 into your calculator, then press the second function key followed by the [10^x] key, 2.891 [F] [10^x]. The display shows 8.03655, which rounds to the answer 778.

Examples Using Antilogarithms
Determine the antilogarithm of the following:

	Enter in sequence	Display	Rounded answer
a. 1.628	1.628 [F] [10^x]	42.461956	42.5
b. 7.086	7.086 [F] [10^x]	12189896	1.22×10^7
c. −6.33	6.33 [+/−] [F] [10^x]	4.6773514 −07	4.7×10^{-7}

**Additional Practice Problems. Only the problem,
the display, and the rounded answer are given.**

Problem	Display	Rounded answer
1. $143.5 + 14.02 + 1.202$	158.722	158.7
2. $72.06 - 26.92 - 49.66$	-4.42	-4.42
3. $2.168 + 4.288 - 1.62$	4.836	4.84
4. $(12.3)(22.8)(1.235)$	346.3434	346
5. $(2.42 \times 10^6)(6.08 \times 10^{-4})(0.623)$	916.65728	917
6. $\dfrac{(46.0)(82.3)}{19.2}$	197.17708	197
7. $\dfrac{0.0298}{243}$	1.2263374 -04	1.23×10^{-4}
8. $\dfrac{(5.4)(298)(760)}{(273)(1042)}$	4.299554	4.3
9. $(6.22 \times 10^6)(1.45 \times 10^3)(9.00)$	8.1171 10	8.12×10^{10}
10. $\dfrac{(1.49 \times 10^6)(1.88 \times 10^6)}{6.02 \times 10^{23}}$	4.6531561 -12	4.65×10^{-12}
11. $\log 245$	2.389166	2.389
12. $\log 6.5 \times 10^{-6}$	-5.1870866	-5.19
13. $\log 24 \times \log 34$	2.1137644	2.11
14. antilog 6.34	2187761.6	2.2×10^6
15. antilog -6.34	4.5708818 -07	4.6×10^{-7}

APPENDIX III

Vapor Pressure of Water at Various Temperatures

Temperature (°C)	Vapor pressure (torr)	Temperature (°C)	Vapor pressure (torr)
0	4.6	26	25.2
5	6.5	27	26.7
10	9.2	28	28.3
15	12.8	29	30.0
16	13.6	30	31.8
17	14.5	40	55.3
18	15.5	50	92.5
19	16.5	60	149.4
20	17.5	70	233.7
21	18.6	80	355.1
22	19.8	90	525.8
23	21.2	100	760.0
24	22.4	110	1074.6
25	23.8		

Units of Measurement

Physical Constants

Constant	Symbol	Value
Atomic mass unit	amu	1.6606×10^{-27} kg
Avogadro's number	N	6.022×10^{23}/mol
Gas constant	R (at STP)	0.08205 L atm/K mol
Mass of an electron	m_e	9.11×10^{-31} kg
		5.486×10^{-4} amu
Mass of a neutron	m_n	1.675×10^{-27} kg
		1.00866 amu
Mass of a proton	m_p	1.673×10^{-27} kg
		1.00728 amu
Speed of light	c	2.997925×10^8 m/s

SI Units and Conversion Factors

Length

SI unit: meter (m)

1 meter	= 1.0936 yards
	= 100 centimeters
	= 1000 millimeters
1 centimeter	= 0.3937 inch
1 inch	= 2.54 centimeters (exactly)
1 kilometer	= 0.62137 mile
1 mile	= 5280 feet
	= 1.609 kilometers
1 angstrom	= 10^{-10} meter

Mass

SI unit: kilogram (kg)

1 kilogram	= 1000 grams
	= 2.20 pounds
1 gram	= 1000 milligrams
1 pound	= 453.59 grams
	= 0.45359 kilogram
	= 16 ounces
1 ton	= 2000 pounds
	= 907.185 kilograms
1 ounce	= 28.3 g
1 atomic mass unit	= 1.6606×10^{-27} kilogram

Volume

SI unit: cubic meter (m^3)

1 liter	= 10^{-3} m^3
	= 1 dm^3
	= 1.0567 quarts
1 gallon	= 4 quarts
	= 8 pints
	= 3.785 liters
1 quart	= 32 fluid ounces
	= 0.946 liter
1 fluid ounce	= 29.6 milliliters

Temperature

SI unit: kelvin (K)

$$0 \text{ K} = -273.15°C$$
$$= -459.67°F$$
$$K = °C + 273.15$$
$$°C = \frac{°F - 32}{1.8}$$
$$°F = 1.8(°C) + 32$$
$$°F = 1.8(°C + 40) - 40$$

Energy

SI unit: joule (J)

1 joule	= 1 kg m^2/s^2
	= 0.23901 calorie
1 calorie	= 4.184 joules

Pressure

SI Unit: pascal (Pa)

1 pascal	= 1 kg/m s^2
1 atmosphere	= 101.325 kilopascals
	= 760 torr (mm Hg)
	= 14.70 pounds per square inch (psi)

Solubility Table

	F^-	Cl^-	Br^-	I^-	O^{2-}	S^{2-}	OH^-	NO_3^-	CO_3^{2-}	SO_4^{2-}	$C_2H_3O_2^-$
H^+	S	S	S	S	S	ss	S	S	ss	S	S
Na^+	S	S	S	S	S	S	S	S	S	S	S
K^+	S	S	S	S	S	S	S	S	S	S	S
NH_4^+	S	S	S	S	—	S	S	S	S	S	S
Ag^+	S	I	I	I	I	I	—	S	I	I	I
Mg^{2+}	I	S	S	S	I	d	I	S	I	S	S
Ca^{2+}	I	S	S	S	I	d	I	S	I	I	S
Ba^{2+}	I	S	S	S	ss	d	ss	S	I	I	S
Fe^{2+}	ss	S	S	S	I	I	I	S	ss	S	S
Fe^{3+}	I	S	S	—	I	I	I	S	I	S	I
Co^{2+}	S	S	S	S	I	I	I	S	I	S	S
Ni^{2+}	ss	S	S	S	I	I	I	S	I	S	S
Cu^{2+}	ss	S	S	—	I	I	I	S	I	S	S
Zn^{2+}	ss	S	S	S	I	I	I	S	I	S	S
Hg^{2+}	d	S	I	I	I	I	I	S	I	d	S
Cd^{2+}	ss	S	S	S	I	I	I	S	I	S	S
Sn^{2+}	S	S	S	ss	I	I	I	S	I	S	S
Pb^{2+}	I	I	I	I	I	I	I	S	I	I	S
Mn^{2+}	ss	S	S	S	I	I	I	S	I	S	S
Al^{3+}	I	S	S	S	I	d	I	S	—	S	S

Key: S = soluble in water
  ss = slightly soluble in water
  I = insoluble in water (less than 1 g/100 g H_2O)
  d = decomposes in water

Answers to Even-Numbered Questions and Exercises

Chapter 2

2. 7.6 cm

4. The most dense (mercury) at the bottom and the least dense (glycerin) at the top. In the cylinder the solid magnesium would sink in the glycerin and float on the liquid mercury.

6. 0.789 g/mL < ice < 0.91 g/mL

8. $d = m/V$ specific gravity $= \dfrac{d_{substance}}{d_{water}}$.

10. The number of degrees between the freezing and boiling points of water are

Fahrenheit	180°F
Celsius	100°C
Kelvin	100 K

12. (a) mg   (d) nm
(b) kg   (e) Å
(c) m   (f) μL

14. (a) not significant   (d) significant
(b) significant   (e) significant
(c) not significant   (f) significant

16. (a) 40.0 (3)   (c) 129,042 (6)
(b) 0.081 (2)   (d) 4.090×10^{-3} (4)

18. (a) 8.87   (c) 130. (1.30×10^2)
(b) 21.3   (d) 2.00×10^6

20. (a) 4.56×10^{-2} (c) 4.030×10^1
(b) 4.0822×10^3 (d) 1.2×10^7

22. (a) 28.1   (d) 2.010×10^3
(b) 58.5   (e) 2.49×10^{-4}
(c) 4.0×10^1 (f) 1.79×10^3

24. (a) $\frac{1}{4}$ (c) $1\frac{2}{3}$ or $\frac{5}{3}$
(b) $\frac{5}{8}$ (d) $\frac{8}{9}$

26. (a) 1.0×10^2
(b) 4.6 mL
(c) 22

28. (a) 4.5×10^8 Å   (e) 6.5×10^5 mg
(b) 1.2×10^{-6} cm   (f) 5.5×10^3 g
(c) 8.0×10^6 mm   (g) 468 mL
(d) 0.164 g   (h) 9.0×10^{-3} mL

30. (a) 117 ft   (d) 4.3×10^4 g
(b) 10.3 mi   (e) 75.7 L
(c) 7.4×10^4 mm^3 (f) 1.3 m^3

32. 50. ft/s

34. 0.102 km/s

36. 5.0×10^2 s

38. 3×10^4 mg

40. 3.0×10^3 hummingbirds to equal the mass of a condor

42. $2800

44. 57 L

46. 160 L

48. 4×10^5 m^2

50. 5 gal

52. 113°F Summer!

54. (a) 90.°F   (c) 546 K
(b) −22.6°C   (d) −300 K

56. −11.4°C = 11.4°F

58. 3.12 g/mL

60. 1.28 g/mL

62. 3.40×10^2 g

64. 7.0 lb

66. Yes, 116.5 L additional solution

68. −15°C > 4.5°F

70. *B* is 14 mL larger than *A*.

72. 76.9 g

74. 3.57×10^3 g

76. The container must hold at least 50 mL.

78. The gold bar is not pure gold.

80. 0.841 g/mL

Chapter 3

2. (a) Attractive forces among the ultimate particles of a solid (atoms, ions, or molecules) are strong enough to hold these particles in a fixed position within the solid and thus maintain the solid in a definite shape. Attractive forces among the ultimate particles of a liquid (usually molecules) are sufficiently strong to hold them together (preventing the liquid from rapidly becoming a gas) but are not strong enough to hold the particles in fixed positions (as in a solid).

(b) The ultimate particles in a liquid are quite closely packed (essentially in contact with each other) and thus the volume of the liquid is fixed at a given temperature. But the ultimate particles in a gas are relatively far apart and essentially independent of each other. Consequently, the gas does not have a definite volume.

(c) In a gas the particles are relatively far apart and are easily compressed, but in a solid the particles are closely packed together and are virtually incompressible.

4. mercury and water

6. Three phases are present.

8. A system containing only one substance is not necessarily homogeneous.

10. 30 g Si/1 g H. There are more Si atoms than H atoms.

12. P   Na
Al   N
H   Ni
K   Ag
Mg   Pu

14. sodium   silver
potassium tungsten
iron   gold
antimony   mercury
tin   lead

16. In an element all atoms are alike, while a compound contains two or more elements that are chemically combined. Compounds may be decomposed into simpler substances while elements cannot.

18. 7 metals   1 metalloid   2 nonmetals

20. aurum

22. A *compound* contains two or more elements chemically combined in a definite proportion by mass. Its properties differ from those of its components. A *mixture* is the physical combining of two or more substances (not necessarily elements). The composition may vary; the substances retain their properties and may be separated by physical means.

24. Compounds are distinguished from one another by their characteristic physical and chemical properties.

26. Cations are positively charged; anions are negatively charged.

28. Homogeneous mixtures contain only one phase; heterogeneous have two or more phases.

30. (a) H_2 **(c)** HCl **(e)** NO

32. (a) magnesium, bromine
(b) carbon, chlorine
(c) hydrogen, nitrogen, oxygen
(d) barium, sulfur, oxygen
(e) aluminum, phosphorus, oxygen

34. (a) $AlBr_3$ **(c)** $PbCrO_4$
(b) CaF_2 **(d)** C_6H_6

36. (a) 1 atom Al, 3 atoms Br
(b) 1 atom Ni, 2 atoms N, 6 atoms O
(c) 12 atoms C, 22 atoms H, 11 atoms O

38. (a) 2 atoms   **(d)** 5 atoms
(b) 2 atoms   **(e)** 17 atoms
(c) 9 atoms

40. (a) 2 atoms H   **(d)** 4 atoms H
(b) 6 atoms H   **(e)** 8 atoms H
(c) 12 atoms H

42. (a) element   **(c)** element
(b) compound **(d)** mixture

44. (a) mixture   **(c)** mixture
(b) element   **(d)** compound

46. (a) HO
(b) C_2H_6O
(c) $Na_2Cr_2O_7$

48. No. The only common liquid elements (at room temperature) are mercury and bromine.

50. 75% solids

52. 420 atoms

54. 40 atoms H

56. (a) magnesium, manganese, molybdenum, mendelevium, mercury
(b) carbon, phosphorus, sulfur, selenium, iodine, astatine, boron
(c) sodium, potassium, iron, silver, tin, antimony

58. (a) As temperature decreases, density increases.
(b) 1.28 g/L, 1.19 g/L, 1.09 g/L

Chapter 4

2. solid

4. Water disappears. Gas appears above each electrode and as bubbles in the solution.

6. A new substance is always formed during a chemical change, but never formed during physical changes.

8. (a) $118.0°C + 273 = 391.0 \text{ K}$
(b) $(118.0°C)1.8 + 32 = 244.4°F$

10. (a) chemical   **(d)** chemical
(b) physical   **(e)** chemical
(c) physical   **(f)** physical

12. The copper wire, like the platinum wire, changed to a glowing red color when heated. Upon cooling, a new substance, black copper(II) oxide, had appeared.

14. Reactant: water
Products: hydrogen, oxygen

16. the transformation of kinetic energy to thermal energy

18. (a) +   **(d)** +
(b) −   **(e)** −
(c) −

20. $2.2 \times 10^3 \text{ J}$

22. 5.03×10^{-2} J/g °C

24. 5°C

26. 29.1°C

28. 2.929×10^4 J; 44 g coal

30. 654°C

32. 16.7°C

34. 6:06 and 54 s

36. at the same rate

38. The mercury and sulfur react to form a compound since the properties of the product are different from the properties of either reactant. 1.36×10^3 g mercury. This supports the law of conservation of matter since the mass of the product is equal to the mass of the reactants.

Chapter 5

2. A neutron is about 1840 times heavier than an electron.

4. An atom is electrically neutral. An ion has a charge.

6. Isotopes have the same number of protons and electrons. Isotopes have different numbers of neutrons and thus different atomic masses.

8. (a) The nucleus of the atom contains most of the mass.
(b) The nucleus of the atom is positively charged.
(c) The atom is mostly empty space.

10. The nucleus of an atom contains nearly all of its mass.

12. Electrons:
Dalton—Electrons are not part of his model.
Thomson—Electrons are scattered throughout the positive mass of matter in the atom.
Rutherford—Electrons are located out in space away from the central positive mass.
Positive matter:
Dalton—No positive matter in his model.
Thomson—Positive matter is distributed throughout the atom.
Rutherford—Positive matter is concentrated in a small central nucleus.

14. Yes. The isotope of $^{12}_{6}C$ with a mass of 12 is an exact number. The mass of other isotopes is not an exact number.

16. Three isotopes of hydrogen have the same number of protons and electrons but differ in the number of neutrons.

18. (a) 80 protons; +80 is the electrical charge on the nucleus.
(b) Hg

20. The most abundant Ca isotope has a mass number of 40.

22. (a) $^{58}_{27}Co$; 27 protons, 32 neutrons
(b) $^{31}_{15}P$; 15 protons, 16 neutrons
(c) $^{184}_{74}W$; 74 protons 110 neutrons
(d) $^{235}_{92}U$; 92 protons, 143 neutrons

24. 24.31 amu

26. 6.716 amu

28. $1.0 \times 10^5 : 1.0$

30. (a) These atoms are isotopes.
(b) These atoms are adjacent to each other on the periodic table.

32. $1.9 \times 10^8 : 1.0$

34. ^{210}Bi has the largest number of neutrons (127).

36. 131 g (atomic mass of unknown element)

38.

Element	Protons	Neutrons	Electrons
He	2	2	2
C	6	6	6
N	7	7	7
O	8	8	8
Ne	10	10	10
Mg	12	12	12
Si	14	14	14
S	16	16	16
Ca	20	20	20

40.

Element		Symbol	Atomic No.	Protons	Neutrons	Electrons
(a)	platinum	^{195}Pt	78	78	117	78
(b)	phosphorus	^{30}P	15	15	15	15
(c)	iodine	^{127}I	53	53	74	53
(d)	krypton	^{84}Kr	36	36	48	36
(e)	selenium	^{79}Se	34	34	45	34
(f)	calcium	^{40}Ca	20	20	20	20

Chapter 6

2. Charges on their ions must be equal and opposite in sign.

4. The system for naming binary compounds composed of two nonmetals uses the stem of the second element plus *ide*. A prefix is attached to each element indicating the number of atoms of that element in the compound. Thus N_2O_5 is called dinitrogen pentoxide.

6. Magnesium forms only one series of compounds in which the cation is Mg^{2+}. Thus the name for $MgCl_2$ (magnesium chloride) does not need to be distinguished from any other compound. Copper forms two series of compounds in which the ions are Cu^+ and Cu^{2+}. Thus the name copper chloride does not indicate which compound is in question. Therefore $CuCl_2$ is called copper(II) chloride to indicate that the compound contains the Cu^{2+} ion.

8. (a) BaO   (d) $BeBr_2$
  (b) H_2S (e) Li_4Si
  (c) $AlCl_3$ (f) Mg_3P_2

10. Cl^- HSO_4^-
  Br^- HSO_3^-
  F^- CrO_4^{2-}
  I^- CO_3^{2-}
  CN^- HCO_3^-
  O^{2-} $C_2H_3O_2^-$
  OH^- ClO_3^-
  S^{2-} MnO_4^-
  SO_4^{2-} $C_2O_4^{2-}$

12. $(NH_4)_2SO_4$ NH_4Cl $(NH_4)_3AsO_4$ $NH_4C_2H_3O_2$ $(NH_4)_2CrO_4$
  $CaSO_4$ $CaCl_2$ $Ca_3(AsO_4)_2$ $Ca(C_2H_3O_2)_2$ $CaCrO_4$
  $Fe_2(SO_4)_3$ $FeCl_3$ $FeAsO_4$ $Fe(C_2H_3O_2)_3$ $Fe_2(CrO_4)_3$
  Ag_2SO_4 $AgCl$ Ag_3AsO_4 $AgC_2H_3O_2$ Ag_2CrO_4
  $CuSO_4$ $CuCl_2$ $Cu_3(AsO_4)_2$ $Cu(C_2H_3O_2)_2$ $CuCrO_4$

14. (a) carbon dioxide   (f) dinitrogen tetroxide
  (b) dinitrogen oxide   (g) diphosphorus pentoxide
  (c) phosphorus pentachloride   (h) oxygen difluoride
  (d) carbon tetrachloride   (i) nitrogen trifluoride
  (e) sulfur dioxide   (j) carbon disulfide

16. (a) potassium oxide   (e) sodium phosphate
  (b) ammonium bromide   (f) aluminum oxide
  (c) calcium iodide   (g) zinc nitrate
  (d) barium carbonate   (h) silver sulfate

18. (a) $SnBr_4$ (d) $Hg(NO_2)_2$
  (b) Cu_2SO_4 (e) TiS_2
  (c) $Fe_2(CO_3)_3$ (f) $Fe(C_2H_3O_2)_2$

20. (a) $HC_2H_3O_2$ (d) H_3BO_3
  (b) HF   (e) HNO_2
  (c) HClO   (f) H_2S

22. (a) phosphoric acid   (e) hypochlorous acid
  (b) carbonic acid   (f) nitric acid
  (c) iodic acid   (g) hydroiodic acid
  (d) hydrochloric acid   (h) perchloric acid

24. (a) Na_2CrO_4 (h) $Co(HCO_3)_2$
  (b) MgH_2 (i) $NaClO$
  (c) $Ni(C_2H_3O_2)_2$ (j) $As_2(CO_3)_5$
  (d) $Ca(ClO_3)_2$ (k) $Cr_2(SO_3)_3$
  (e) $Pb(NO_3)_2$ (l) $Sb_2(SO_4)_3$
  (f) KH_2PO_4 (m) $Na_2C_2O_4$
  (g) $Mn(OH)_2$ (n) KSCN

26. (a) calcium hydrogen sulfate
  (b) arsenic(III) sulfite
  (c) tin(II) nitrite
  (d) iron(III) bromide
  (e) potassium hydrogen carbonate
  (f) bismuth(III) arsenate
  (g) iron(II) bromate
  (h) ammonium monohydrogen phosphate
  (i) sodium hypochlorite
  (j) potassium permanganate

28. (a) FeS_2 (e) $Mg(OH)_2$
  (b) $NaNO_3$ (f) $Na_2CO_3 \cdot 10\,H_2O$
  (c) $CaCO_3$ (g) C_2H_5OH
  (d) $C_{12}H_{22}O_{11}$

30. *-ide:* Suffix is used to indicate a binary compound except for hydroxides, cyanides, and ammonium compounds.
  -ous: Used as a suffix in naming an oxy-acid that has a lower oxygen content than the *ic* acid; also used as a suffix for naming the lower ionic charge of a multivalent metal (e.g., Fe^{2+}, ferrous).
  hypo: Used as a prefix in acids or salts when the polyatomic ion contains less oxygen than that of *-ous* acid or the *-ite* salt.
  per: Used as a prefix in acids or salts when the polyatomic ion contains more oxygen than that of the *-ic* acid or the *-ate* salt.
  -ite: The suffix of a salt derived from an *-ous* acid.
  -ate: The suffix of a salt derived from an *-ic* acid.
  Roman numerals indicate the charge on the metal cation.

32. (a) $50e^-$, 50p (tin)
  (b) $48e^-$, 50p (Sn^{2+})
  (c) $46e^-$, 50p (Sn^{4+})

34. $Li_3Fe(CN)_6$
  $AlFe(CN)_6$
  $Zn_3[Fe(CN)_6]_2$

36. ammonium oxide, $(NH_4)_2O$ zinc chloride, $ZnCl_2$
ammonium carbonate,   zinc acetate, $Zn(C_2H_3O_2)_2$
   $(NH_4)_2CO_3$ carbonic acid, H_2CO_3
ammonium chloride, NH_4Cl acetic acid, $HC_2H_3O_2$
ammonium acetate,   hydrochloric acid, HCl
   $NH_4C_2H_3O_2$ water, H_2O
zinc oxide, ZnO
zinc carbonate, $ZnCO_3$

Chapter 7

2. A mole of gold has a higher mass than a mole of potassium.

4. A mole of gold atoms contains more electrons than a mole of potassium atoms.

6. 6.022×10^{23}

8. (a) 6.022×10^{23} atoms   (d) 16.00 g
   (b) 6.022×10^{23} molecules (e) 32.00 g
   (c) 1.204×10^{24} atoms

10. Choosing 100.0 g of a compound allows us to simply drop the % sign and use grams for each percent.

12. Molar masses
   (a) NaOH   40.00   (f) C_6H_5COOH 122.1
   (b) Ag_2CO_3 275.8   (g) $C_6H_{12}O_6$ 180.2
   (c) Cr_2O_3 152.0   (h) $K_4Fe(CN)_6$ 368.4
   (d) $(NH_4)_2CO_3$ 96.09 (i) $BaCl_2 \cdot 2\,H_2O$ 244.2
   (e) $Mg(HCO_3)_2$ 146.3

14. (a) 0.625 mol NaOH
   (b) 0.275 mol Br_2
   (c) 7.18×10^{-3} mol $MgCl_2$
   (d) 0.462 mol CH_3OH
   (e) 2.03×10^{-2} mol Na_2SO_4
   (f) 5.97 mol ZnI_2

16. (a) 0.0417 g H_2SO_4 (c) 0.122 g Ti
   (b) 11 g CCl_4 (d) 8.0×10^{-7} g S

18. (a) 1.05×10^{24} molecules Cl_2
   (b) 1.6×10^{23} molecules C_2H_6O
   (c) 1.64×10^{23} molecules CO_2
   (d) 3.75×10^{24} molecules CH_4

20. (a) 3.271×10^{-22} g Au
   (b) 3.952×10^{-22} g U
   (c) 2.828×10^{-23} g NH_3
   (d) 1.795×10^{-22} g $C_6H_4(NH_2)_2$

22. (a) 0.886 mol S
   (b) 42.8 mol NaCl
   (c) 1.05×10^{24} atoms Mg
   (d) 9.47 mol Br_2

24. One mole of ammonia contains
   (a) 6.022×10^{23} molecules NH_3
   (b) 6.022×10^{23} N atoms
   (c) 1.807×10^{24} H atoms
   (d) 2.409×10^{24} atoms

26. (a) 6.0×10^{24} atoms O
   (b) 5.46×10^{24} atoms O
   (c) 5.0×10^{18} atoms O

28. (a) 1.27 g Cl
   (b) 9.25×10^{-2} g H
   (c) 2.74 g H

30. (a) 47.98% Zn   (d) 21.20% N
      52.02% Cl   6.100% H
   (b) 18.17% N   24.26% S
      9.153% H   48.41% O
      31.16% C   (e) 23.09% Fe
      41.51% O   17.37% N
   (c) 12.26% Mg   59.53% O
      31.24% P (f) 54.39% I
      56.48% O   45.61% Cl

32. (a) KCl   47.55% Cl (c) $SiCl_4$ 83.46% Cl
   (b) $BaCl_2$ 34.05% Cl (d) LiCl   83.63% Cl
   highest % Cl is LiCl
   lowest % Cl is $BaCl_2$

34. 24.2% C
   4.04% H
   71.72% Cl

36. (a) $KClO_3$ lower % Cl
   (b) $KHSO_4$ higher % S
   (c) Na_2CrO_4 lower % Cr

38. Empirical formulas
   (a) CuCl   (d) K_3PO_4
   (b) $CuCl_2$ (e) $BaCr_2O_7$
   (c) Cr_2S_3 (f) PBr_8Cl_3

40. V_2O_5

42. The empirical formula is CH_2O. The molecular formula is $C_6H_{12}O_6$.

44. 5.88 g Na

46. 5.54×10^{19} m

48. (a) 8×10^{16} drops
   (b) 8×10^6 mi^3

50. 10.3 mol H_2SO_4

52. (a) H_2O contains the most molecules.
   (b) CH_3OH contains the most atoms.

54. 8.66 g Li will combine with 20.0 g S.

56. There is not sufficient S to reset with 19.5 g Zn.

58. 4.77 g O in 8.50 g $Al_2(SO_4)_3$

60. Empirical Formula   Molecular Formula
   (a) CCl_4 CCl_4
   (b) CCl_3 C_2Cl_6
   (c) CCl C_6Cl_6
   (d) C_3Cl_8 C_3Cl_8

62. 2.4×10^{22} atoms Cu

64. 8.3×10^{-15} mol people

66. 32 g Mg

68. Carbon is the mystery element.

Chapter 8

2. the number of moles of each of the chemical species in the reaction

4. A chemical change that absorbs heat energy is said to be an endothermic reaction. The products are at a higher energy level than the reactants. A chemical change that liberates energy is said to be an exothermic reaction. The products are at a lower energy level than the reactants.

6. **(a)** $H_2 + Br_2 \longrightarrow 2\ HBr$
  (b) $4\ Al + 3\ C \xrightarrow{\Delta} Al_4C_3$
  (c) $Ba(ClO_3)_2 \xrightarrow{\Delta} BaCl_2 + 3\ O_2$
  (d) $CrCl_3 + 3\ AgNO_3 \longrightarrow Cr(NO_3)_3 + 3\ AgCl$
  (e) $2\ H_2O_2 \longrightarrow 2\ H_2O + O_2$

8. **(a)** combination
  (b) combination
  (c) decomposition
  (d) double displacement
  (e) decomposition

10. **(a)** $2\ SO_2 + O_2 \longrightarrow 2\ SO_3$
  (b) $4\ Al + 3\ MnO_2 \xrightarrow{\Delta} 3\ Mn + 2\ Al_2O_3$
  (c) $2\ Na + 2\ H_2O \longrightarrow 2\ NaOH + H_2$
  (d) $2\ AgNO_3 + Ni \longrightarrow Ni(NO_3)_2 + 2\ Ag$
  (e) $Bi_2S_3 + 6\ HCl \longrightarrow 2\ BiCl_3 + 3\ H_2S$
  (f) $2\ PbO_2 \xrightarrow{\Delta} 2\ PbO + O_2$
  (g) $2\ LiAlH_4 \xrightarrow{\Delta} 2\ LiH + 2\ Al + 3\ H_2$
  (h) $2\ KI + Br_2 \longrightarrow 2\ KBr + I_2$
  (i) $2\ K_3PO_4 + 3\ BaCl_2 \longrightarrow 6\ KCl + Ba_3(PO_4)_2$

12. **(a)** $2\ Cu + S \longrightarrow Cu_2S$
  (b) $2\ H_3PO_4 + 3\ Ca(OH)_2 \xrightarrow{\Delta} Ca_3(PO_4)_2 + 6\ H_2O$
  (c) $2\ Ag_2O \xrightarrow{\Delta} 4\ Ag + O_2$
  (d) $FeCl_3 + 3\ NaOH \longrightarrow Fe(OH)_3 + 3\ NaCl$
  (e) $Ni_3(PO_4)_2 + 3\ H_2SO_4 \longrightarrow 3\ NiSO_4 + 2\ H_3PO_4$
  (f) $ZnCO_3 + 2\ HCl \longrightarrow ZnCl_2 + H_2O + CO_2$
  (g) $3\ AgNO_3 + AlCl_3 \longrightarrow 3\ AgCl + Al(NO_3)_3$

14. **(a)** $Cu(s) + FeCl_3(aq) \longrightarrow$ no reaction
  (b) $H_2(g) + Al_2O_3(s) \xrightarrow{\Delta}$ no reaction
  (c) $2\ Al(s) + 6\ HBr(aq) \longrightarrow 3\ H_2(g) + 2\ AlBr_3(aq)$
  (d) $I_2(s) + HCl(aq) \longrightarrow$ no reaction

16. **(a)** $SO_2 + H_2O \longrightarrow H_2SO_3$
  (b) $SO_3 + H_2O \longrightarrow H_2SO_4$
  (c) $Ca + 2\ H_2O \longrightarrow Ca(OH)_2 + H_2$
  (d) $2\ Bi(NO_3)_3 + 3\ H_2S \longrightarrow Bi_2S_3 + 6\ HNO_3$

18. **(a)** $C + O_2 \xrightarrow{\Delta} CO_2$
  (b) $2\ Al(ClO_3)_3 \xrightarrow{\Delta} 9\ O_2 + 2\ AlCl_3$
  (c) $CuBr_2 + Cl_2 \longrightarrow CuCl_2 + Br_2$
  (d) $2\ SbCl_3 + 3\ (NH_4)_2S \longrightarrow Sb_2S_3 + 6\ NH_4Cl$
  (e) $2\ NaNO_3 \xrightarrow{\Delta} 2\ NaNO_2 + O_2$

20. **(a)** 2 mol of Na react with 1 mol of Cl_2 to produce 2 mol of NaCl and release 822 kJ of energy. Exothermic
  (b) 1 mol of PCl_5 absorbs 92.9 kJ of energy to produce 1 mol of PCl_3 and 1 mol of Cl_2. Endothermic

22. **(a)** $2\ Al + 3\ I_2 \longrightarrow 2\ AlI_3 + $ heat
  (b) $4\ CuO + CH_4 + $ heat $\longrightarrow 4\ Cu + CO_2 + 2\ H_2O$
  (c) $Fe_2O_3 + 2\ Al \longrightarrow 2\ Fe + Al_2O_3 + $ heat

24. $P_4O_{10} + 12\ HClO_4 \longrightarrow 6\ Cl_2O_7 + 4\ H_3PO_4$
  58 atoms O on each side

26. A balanced equation tells us
  (a) the type of atoms/molecules involved in the reaction.
  (b) the relationship between quantities of the substances in the reaction.
  A balanced equation gives no information about
  (a) the time required for the reaction.
  (b) odor or colors that may result.

28. Zn metal is below Mg on the activity series.

30. **(a)** $4\ K + O_2 \longrightarrow 2\ K_2O$
  (b) $2\ Al + 3\ Cl_2 \longrightarrow 2\ AlCl_3$
  (c) $CO_2 + H_2O \longrightarrow H_2CO_3$
  (d) $CaO + H_2O \longrightarrow Ca(OH)_2$

32. **(a)** $Zn + H_2SO_4 \longrightarrow H_2 + ZnSO_4$
  (b) $2\ AlI_3 + 3\ Cl_2 \longrightarrow 2\ AlCl_3 + 3\ I_2$
  (c) $Mg + 2\ AgNO_3 \longrightarrow Mg(NO_3)_2 + 2\ Ag$
  (d) $2\ Al + 3\ CoSO_4 \longrightarrow Al_2(SO_4)_3 + 3\ Co$

34. **(a)** $AgNO_3(aq) + KCl(aq) \longrightarrow AgCl(s) + KNO_3(aq)$
  (b) $Ba(NO_3)_2(aq) + MgSO_4(aq) \longrightarrow$
    $Mg(NO_3)_2(aq) + BaSO_4(s)$
  (c) $H_2SO_4(aq) + Mg(OH)_2(aq) \longrightarrow$
    $2\ H_2O(l) + MgSO_4(aq)$
  (d) $MgO(s) + H_2SO_4(aq) \longrightarrow H_2O(l) + MgSO_4(aq)$
  (e) $Na_2CO_3(aq) + NH_4Cl(aq) \longrightarrow$ no reaction

36. **(a)** combustion of fossil fuels
  (b) destruction of the rain forests by burning
  (c) increased population

38. The effects of global warming can be reduced by
  (a) developing new energy sources (not dependent on fossil fuels)
  (b) conservation of energy resources
  (c) recycling
  (d) decreased destruction of the rain forests and other forests

Chapter 9

2. (a) (c) (e) are correct; (b) (d) (f) are incorrect
   (b) 10.7 mol HCN, (d) 36 mol H_2O, (f) 2 mol HCN

4. (a) 25.0 mol $NaHCO_3$
   (b) 3.85×10^{-3} mol $ZnCl_2$
   (c) 16 mol CO_2
   (d) 4.3 mol C_2H_5OH

6. (a) 1.31 g $NiSO_4$ (d) 1.35 g $C_6H_{12}O_6$
   (b) 3.60 g $HC_2H_3O_2$ (e) 18 g K_2CrO_4
   (c) 373 g Bi_2S_3

8. HCl

10. (a) $\dfrac{3 \text{ mol } CaCl_2}{1 \text{ mol } Ca_3(PO_4)_2}$ (d) $\dfrac{1 \text{ mol } Ca_3(PO_4)_2}{2 \text{ mol } H_3PO_4}$

   (b) $\dfrac{6 \text{ mol HCl}}{2 \text{ mol } H_3PO_4}$ (e) $\dfrac{6 \text{ mol HCl}}{1 \text{ mol } Ca_3(PO_4)_2}$

   (c) $\dfrac{3 \text{ mol } CaCl_2}{2 \text{ mol } H_3PO_4}$ (f) $\dfrac{2 \text{ mol } H_3PO_4}{6 \text{ mol HCl}}$

12. 2.80 mol Cl_2

14. 8.33 mol H_2O

16. 19.7 g $Zn_3(PO_4)_2$

18. 117 g H_2O (steam)
   271 g Fe

20. (a) 0.500 mol Fe_2O_3 (d) 65.6 g SO_2
   (b) 12.4 mol O_2 (e) 0.871 mol O_2
   (c) 6.20 mol SO_2 (f) 332 g FeS_2

22. (a) H_2S is the limiting reactant and $Bi(NO_3)_3$ is in excess.
   (b) H_2O is the limiting reactant and Fe is in excess.

24. (a) 3.0 mol CO_2
   (b) 9.0 mol CO_2
   (c) 1.8 mol CO_2

26. zinc

28. 95.0% yield of Cu

30. 77.8% CaC_2

32. A subscript is used to indicate the number of atoms in a formula. It cannot be changed without changing the identity of the substance. Coefficients are used only to balance atoms in chemical equations. They may be changed as needed to achieve a balanced equation.

34. (a) 380 g C_2H_5OH
      370 g CO_2
   (b) 480 mL C_2H_5OH

36. 65 g O_2

38. 1.0 g Ag_2S

40. (a) 2.0 mol Cu, 2.0 mol $FeSO_4$, and 1.0 mol $CuSO_4$
   (b) 15.9 g Cu, 38.1 g $FeSO_4$, 6.0 g Fe, and no $CuSO_4$

42. (a) 3.2×10^2 g C_2H_5OH
   (b) 1.10×10^3 g $C_6H_{12}O_6$

44. 3.7×10^2 kg Li_2O

46. 13 tablets

Chapter 10

2. A second electron may enter an orbital already occupied if its spin is opposite the electron already in the orbital, and if all other orbitals of the same sublevel contain an electron.

4. Both $1s$ and $2s$ orbitals are spherical in shape and located symmetrically around the nucleus. The radius of the $2s$ orbital is larger than the $1s$. The electrons in the $2s$ orbital are further from the nucleus.

6. $1s, 2s, 2p, 3s, 3p, 4s, 3d, 4p$

8. The Bohr orbit has an electron traveling a specific path, while an orbital is a region of space where the electron is most probably found.

10. s orbital

p orbitals
   p_x p_y p_z

12. Transition elements are found in the center of the periodic table. The last electrons for these elements are found in the d or f orbitals. Representative elements are located on either side of the periodic table (Groups IA–VIIA). The valence electrons for these elements are found in the s and/or p orbitals.

14.

Atomic number	Symbol
8	O
16	S
34	Se
52	Te
84	Po

All these elements have an outermost electron structure of s^2p^4.

16. 32; the 6th period has this number of elements.

18. Ar and K; Co and Ni; Te and I; Th and Pa; U and Np; Lr and Rf; Pu and Am.

20. (a) F   9 protons   (c) Br   35 protons
   (b) Ag   47 protons   (d) Sb   51 protons

22. (a) Cl $1s^22s^22p^63s^23p^5$
   (b) Ag $1s^22s^22p^63s^23p^64s^23d^{10}4p^65s^14d^{10}$
   (c) Li $1s^22s^1$
   (d) Fe $1s^22s^22p^63s^23p^64s^23d^6$
   (e) I $1s^22s^22p^63s^23p^64s^23d^{10}4p^65s^24d^{10}5p^5$

24. Bohr said that a number of orbits were available for electrons, each corresponding to an energy level. When an electron falls from a higher energy orbit to a lower orbit, energy is given off as a specific wavelength of light. Only those energies in the visible range are seen in the hydrogen spectrum. Each line corresponds to a change from one orbit to another.

26. 32 electrons in the fourth energy level

28. (a) (14p 14n) $2e^- 8e^- 4e^-$ $^{28}_{14}Si$

 (b) (16p 16n) $2e^- 8e^- 6e^-$ $^{32}_{16}S$

 (c) (18p 22n) $2e^- 8e^- 8e^-$ $^{40}_{18}Ar$

 (d) (23p 28n) $2e^- 8e^- 11e^- 2e^-$ $^{51}_{23}V$

 (e) (15p 16n) $2e^- 8e^- 5e^-$ $^{31}_{15}P$

30. (a) Sc
 (b) Zn
 (c) Sn
 (d) Cs

32. **Atomic number   Electron structure**
 (a) 9 $[He]2s^2 2p^5$
 (b) 26 $[Ar]4s^2 3d^6$
 (c) 31 $[Ar]4s^2 3d^{10} 4p^1$
 (d) 39 $[Kr]5s^2 4d^1$
 (e) 52 $[Kr]5s^2 4d^{10} 5p^4$
 (f) 10 $[He]2s^2 2p^6$

34. (a) (13p 14n) $2e^- 8e^- 3e^-$ $^{27}_{13}Al$

 (b) (22p 26n) $2e^- 8e^- 10e^- 2e^-$ $^{48}_{22}Ti$

36. K is in the fourth energy level because the $4s$ orbital is at a lower energy level than the $3d$ orbital.

38. Noble gases each have filled s and p orbitals in the outer energy level.

40. All the elements in a group have the same number of outer-shell electrons.

42. All of these elements have an $s^2 d^{10}$ electron configuration in their outermost energy levels.

44. (a) and (f)
 (e) and (h)

46. 7, 33 since they are in the same periodic group.

48. (a) nonmetal, I
 (b) metal, W
 (c) metal, Mo
 (d) metalloid, Ge

50. Period 4, Group IIIB

52. Group IIIA contains 3 valence electrons. Group IIIB contains 2 electrons in the outermost level and 1 electron in an inner d orbital. Group A elements are representative while Group B elements are transition elements.

54. Nitrogen has more valence electrons on more energy levels. More varied electron jumps are possible.

56. (a) 100% (d) 24%
 (b) 100% (e) 20%
 (c) 19%

58. $\dfrac{1.5 \times 10^8}{1}$

60. The outermost electron structure for both sulfur and oxygen is $s^2 p^4$.

62. In transition elements the last electron added is in a d or f orbital. The last electron added in a representative element is an s or p orbital.

64. The outermost energy level is the 7th. Element 87 contains one electron in the $7s$ orbital.

66. Answers will vary but should include at least a statement about:
 (a) numbering of the elements and their relationship to atomic structure
 (b) division of the elements into periods and groups
 (c) division of the elements into metals, nonmetals, and metalloids
 (d) identification and location of the representative and transition elements

68. (a) The two elements are isotopes.
 (b) The two elements are in the same period and are adjacent to each other.

Chapter 11

2. More energy is required for neon because it has a very stable outer electron structure consisting of an octet of electrons in filled orbitals (noble gas electron structure).

4. The first ionization energy decreases from top to bottom because, in successive alkali metals, the outermost electron is farther away from the nucleus and is more shielded from the positive nucleus by additional electron shells.

6. The electron to be removed from Ba is located in an energy level farther away from the nucleus than for Be. Therefore, it requires less energy to remove an electron from Ba than from Be.

8. (a) K > Na   (d) I > Br
   (b) Na > Mg   (e) Zr > Ti
   (c) O > F

10. Atomic size increases down the column since each successive element has an additional energy level that contains electrons located farther from the nucleus.

12. Cs· Ba: Tl: ·Pb: ·Po: ·At: :Rn:
    Each of these is a representative element and has the same number of electrons in its outer shell as its periodic group number.

14. Valence electrons are the electrons found in the outermost energy level of an atom.

16. An aluminum ion has a +3 charge because it has lost 3 electrons in acquiring a noble gas electron structure.

18. A bromine atom is smaller since it has one less electron than the bromine ion in its outer shell. Also, the bromine ion has 35 protons and 36 electrons resulting in a lessening of the attraction between the electrons and the nucleus.

20.   +   −
    (a) H   Cl   (d) I   Br
    (b) Li   H   (e) Mg   H
    (c) C   Cl   (f) O   F

22. (a) covalent   (c) covalent
    (b) ionic   (d) ionic

24. (a) $F + 1e^- \rightarrow F^-$ (b) $Ca \rightarrow Ca^{2+} + 2e^-$

26. (a) Ca: + :O: → CaO

    (b) Na· + ·Br: → NaBr

28. Si (4)   N (5)   P (5)   O (6)   Cl (7)

30. (a) Chloride ion, none
    (b) Nitrogen atom, gain $3e^-$ or lose $5e^-$
    (c) Potassium atom, lose $1e^-$

32. (a) The potassium atom is larger because it has one more energy level containing an electron than does a potassium ion.
    (b) A bromide ion is larger than a bromine atom because it has one more electron than a bromine atom. Both are in the same energy level.

34. (a) SbH_3, Sb_2O_3
    (b) H_2Se, SeO_3
    (c) HCl, Cl_2O_7
    (d) CCl_4, CO_2

36. $BeBr_2$, beryllium bromide
    $MgBr_2$, magnesium bromide
    $SrBr_2$, strontium bromide
    $BaBr_2$, barium bromide
    $RaBr_2$, radium bromide

38. (a) Ga: (b) $[Ga]^{3+}$ (c) $[Ca]^{2+}$

40. (a) covalent   (c) covalent
    (b) ionic   (d) covalent

42. (a) covalent   (c) ionic
    (b) covalent

44. (a) :O::O: (b) :Br:Br: (c) :I:I:

46. (a) :S:H   (c) H:N:H
        H   H

    (b) :S::C::S: (d) [H:N:H]+ [:Cl:]−
                        H

48. (a) [:I:]−   (d) [:O:Cl:O:]−
                           :O:

    (b) [:S:]2−   (e) [:O:N::O:]−
                           :O:

    (c) [:O:C::O:]2−
             :O:

50. (a) F_2, nonpolar
    (b) CO_2, nonpolar
    (c) NH_3, polar

52. (a) 2 electron pairs, linear
    (b) 4 electron pairs, tetrahedral
    (c) 4 electron pairs, tetrahedral

54. (a) tetrahedral
    (b) pyramidal
    (c) tetrahedral

56. (a) tetrahedral
    (b) bent
    (c) bent

58. potassium

60. (a) Zn
    (b) Be
    (c) N

62. Lithium has a +1 charge after the first electron is removed. It takes more energy to overcome that charge and to remove one e^- than to remove an e^- from helium

64. $SnBr_2$, $GeBr_2$

66. A covalent bond results from the sharing of a pair of electrons between two atoms, while an ionic bond involves the transfer of one or more electrons from one atom to another.

68. F, O, N, Cl.

70. It is possible for a molecule to be nonpolar even though it contains polar bonds.

72. (a) 105° (b) 107° (c) 109.5° (d) 109.5°

74. Fluorine's electronegativity is greater than any other element. Ionic bonds form between atoms of widely different electronegativities. Therefore, CsF, RbF, KF, and FrF would be ionic substances with the greatest electronegativity difference.

76. $S \dfrac{1.40\ g}{32.07\ \dfrac{g}{mol}} = 0.0437\ mol \qquad \dfrac{0.0437}{0.0437} = 1.00$

$O \dfrac{2.10\ g}{16.00\ \dfrac{g}{mol}} = 0.131\ mol \qquad \dfrac{0.131}{0.0437} = 3.00$

Empirical formula is SO_3:

$$:\!\overset{\displaystyle ..}{\underset{\displaystyle |}{O}}\!:$$
$$:\!\ddot{O}\!-\!S\!=\!\ddot{O}:$$

Chapter 12

2. The air pressure inside the balloon is greater than the air pressure outside the balloon.

4. 1 torr = 1 mm Hg.

6. 1 atm corresponds to 4 L.

8. The piston would move downward.

10. O_2, H_2S, HCl, F_2, CO_2

12. Rn, F_2, N_2, CH_4, He, H_2, velocities increase as molar masses decrease.

14. (a) pressure   (c) temperature
(b) volume   (d) number of moles

16. A gas is least likely to behave ideally at low temperatures. At very low temperatures, the velocity of the molecules decreases, and attractive forces between the molecules begin to play a significant role.

18. Equal volumes of H_2 and O_2 at the same T and P:
(a) have equal numbers of molecules (Avogadro's law)
(b) mass O_2 = 16 × mass H_2
(c) moles O_2 = moles H_2
(d) average kinetic energies are the same (T same)
(e) rate H_2 = 4 × rate O_2 (Graham's law of effusion)
(f) density O_2 = 16 × density H_2

20. $N_2(g) + O_2(g) \longrightarrow 2NO(g)$
1 vol + 1 vol $\longrightarrow$ 2 vol
Therefore, each N_2 and O_2 molecule must consist of two atoms.

22. Conversion of oxygen to ozone is an endothermic reaction. 286 kJ/3 mol O_2 is required to convert O_2 to O_3.

24. Oxygen atom, O; oxygen molecule, O_2; ozone molecule, O_3. An oxygen molecule contains 16 electrons.

26. (a) 715 torr
(b) 953 mbar
(c) 95.3 kPa

28. (a) 0.082 atm
(b) 55.92 atm
(c) 0.296 atm
(d) 0.0066 atm

30. (a) 132 mL
(b) 615 mL

32. 711 mm Hg

34. (a) 6.17 L
(b) 8.35 L

36. 7.8×10^2 mL

38. 33.4 L

40. 681 torr

42. 1.450×10^3 torr

44. 1.19 L C_3H_8

46. 28.0 L N_2

48. 1.33 g NH_3

50. (a) 11 mol H_2S
(b) 14.7 L H_2S
(c) 31.8 L H_2S

52. 2.69×10^{22} molecules CH_4

54. (a) 0.179 g/L He
(b) 2.50 g/L C_4H_8

56. (a) 3.17 g/L Cl_2
(b) 1.46 g/L Cl_2

58. 19 L Kr

60. 72.8 L

62. (a) 5.6 mol NH_3
(b) 0.640 L NO
(c) 1.1×10^2 g O_2

64. 153 L SO_2

66. The can will explode due to buildup of pressure.

68. (a) 5 L Cl_2
(b) 5.0 L NH_3
(c) 5.9 L SO_3

70. (a) CH_3 (b) C_2H_6 (c)

$$H-\underset{\underset{\textstyle H}{|}}{\overset{\overset{\textstyle H}{|}}{C}}-\underset{\underset{\textstyle H}{|}}{\overset{\overset{\textstyle H}{|}}{C}}-H$$

72. 430 mL CO_2

74. (a) 0.18 mol air
(b) 5.2 g air
(c) 0.3 g air

76. 1.0×10^2 atm

78. 1.5×10^3 torr

80. 65 atm

82. 6.1 L

84. 7.39×10^{21} molecules; 2.22×10^{22} atoms

86. (a) 34 mol H_2
(b) 121 g H_2

88. 43.2 g/mol (molar mass)

90. 39.9 g/mol (molar mass)

92. 164 g/mol

94. 0.13 mol N_2

96. 279 L H_2 at STP

98. (a) Helium effuses twice as fast as CH_4.
(b) The gases meet 66.7 cm from the helium end.

100. (a) 10.0 mol CO_2; 3.0 mol O_2; no CO_2
(b) 29 atm

102. Some ammonia gas dissolves in the water squirted into the flask, lowering the pressure inside the flask. The atmospheric pressure outside is greater than the pressure inside the flask and thus pushes water from the beaker up the tube and into the flask.

104. 0.72 g/L

106. 60.4 atm

Chapter 13

2. H_2S, H_2Se, and H_2Te are gases at 0°C

4.

6 Prefixes preceding the word *hydrate* are used to indicate the number of molecules of water present in the formulas.
mono = 1   penta = 5
di = 2   hexa = 6
tri = 3   hepta = 7
tetra = 4   octa = 8

8. about 70°C

10. case (b), the closed container

12. The vapor pressure would remain unchanged.

14. (a) 88°C
(b) 78°C
(c) 16°C

16. melting point, 0°C; boiling point, 100°C (at 1 atm pressure); colorless; odorless; tasteless; heat of fusion, 335 J/g (80 cal/g); heat of vaporization, 2.26 kJ/g (540 cal/g); density = 1.0 g/mL (at 4°C); specific heat = 4.184 J/g °C

18. If you apply heat to an ice-water mixture, the heat energy is absorbed to melt the ice, rather than to warm the water, so the temperature remains constant until all the ice has melted.

20. Ice floats in water because it is less dense than water. Ice sinks in ethyl alcohol because it is more dense than the alcohol.

22. Ethyl alcohol exhibits hydrogen bonding; ethyl ether does not. The higher heat of vaporization indicates stronger intermolecular attraction between the alcohol molecules.

24. Ammonia exhibits hydrogen bonding; methane does not.

26. $H_2NCH_2CH_2NH_2$

28. (a) mercury, acetic acid, water, toluene, benzene, carbon tetrachloride, methyl alcohol, bromine
(b) Highest boiling point is mercury; lowest is bromine.

30. In a pressure cooker, the temperature at which water boils increases above its normal boiling point, because the water vapor (steam) formed by boiling cannot escape. This results in an increased pressure over the water, and consequently, an increased boiling temperature.

32. As temperature increases, molecular velocities increase, resulting in a higher vapor pressure.

34. Ammonia

36. HF has a higher boiling point because of the strong H-bonding in HF. Neither F_2 nor Cl_2 show hydrogen bonding so F_2, with the lower molar mass, has the lower boiling point.

38. 34.6°C, the boiling point of ether

40. The expected temperature would be 4°C at the bottom of the lake.

42. (a) An anhydride is an oxide that reacts with water to form an acid or a base.
(b) An acid anhydride is an oxide of a nonmetal.
(c) A basic anhydride is an oxide of a metal.

44. SO_2, SO_3, N_2O_5

46. [KOH, K_2O] [Ba(OH)$_2$, BaO] [Ca(OH)$_2$CaO]

48. (a) $Li_2O + H_2O \longrightarrow 2\,LiOH$
(b) $2\,KOH \overset{\Delta}{\longrightarrow} K_2O + H_2O$
(c) $Ba + 2\,H_2O \longrightarrow Ba(OH)_2 + H_2$
(d) $Cl_2 + H_2O \longrightarrow HCl + HClO$
(e) $SO_3 + H_2O \longrightarrow H_2SO_4$
(f) $H_2SO_3 + 2\,KOH \longrightarrow K_2SO_3 + 2\,H_2O$

50. (a) magnesium ammonium phosphate hexahydrate
(b) iron(II) sulfate heptahydrate
(c) tin(IV) chloride pentahydrate

52. (a) Distilled water has been vaporized by boiling and re-condensed.
(b) Natural waters are generally not pure, but contain dissolved minerals and suspended matter, and can even contain harmful bacteria.

54. 0.262 mol $FeI_2 \cdot 4\,H_2O$

56. 1.05 mol H_2O

58. 48.67% H_2O

60. $FePO_4 \cdot 4\,H_2O$

62. 5.5×10^4 J

64. 42 g of steam needed; not enough steam (35 g)

66. The system will be at 0°C. It will be a mixture of ice and water.

68. (a) 18.0 g H_2O
  (b) 36.0 g H_2O
  (c) 18.0 g H_2O

70. Eventually the water will lose enough energy to change from a liquid to a solid (freeze) as the alcohol evaporates.

72. (a) From 0°C to 40°C solid X warms until at 40°C it begins to melt. The temperature remains at 40°C until all of X is melted. After that, liquid X will warm steadily to 65°C where it will begin to boil and remain at 65°C until all the liquid becomes vapor. Beyond 65°C the vapor will warm steadily until 100°C.

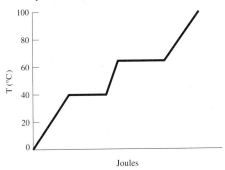

Joules

  (b) 37,000 J

74. 75°C at 270 torr pressure

76. $MgSO_4 \cdot 7 H_2O$ $Na_2HPO_4 \cdot 12 H_2O$

78. chlorine

80. When organic pollutants in water are oxidized by dissolved oxygen, there may not be sufficient dissolved oxygen to sustain marine life.

82. Na_2 zeolite(s) + $Mg^{2+}(aq) \longrightarrow$ Mg zeolite(s) + $2Na^+(aq)$

84. 1.6×10^5 cal

86. 3.1 kcal

88. 2.30×10^6 J

90. 40.2 g H_2O

92. Volume of 1.00 mol of liquid water is 18.0 mL.
Volume of 1.00 mol of water vapor at STP is 22.4 L.

94. (a) Some O_2 remains unreacted.
  (b) 20.0 mL O_2 unreacted

Chapter 14

2. 4.5 g NaF

4. For lithium and sodium halide, solubility is
  $F^- < Cl^- < Br^- < I^-$
  For potassium halides, solubility is
  $Cl^- < Br^- < F^- < I^-$

6. KNO_3

8. 6×10^2 cm^2

10. The dissolving process involves solvent molecules attaching to the solute ions or molecules. This rate decreases as more of the solvent molecules are already attached to solute molecules. As the solution becomes saturated, the number of unused solvent molecules decreases. Also, the rate of recrystallization increases as the concentration of dissolved solute increases.

12. The solution level in the thistle tube will fall.

14. It is not always apparent which component in a solution is the solute. For example, when mixing equal volumes of two liquids, designation of the solute and solvent is arbitrary.

16. Yes. Metal alloys are metals dissolved in metals.

18. Hexane and benzene are both nonpolar molecules. Therefore, hexane will dissolve in benzene. Benzene, being nonpolar, has no attraction for ionic sodium chloride.

20. Air is considered to be a solution because it is a homogeneous mixture of several gaseous substances and does not have a fixed composition.

22. The solubility of gases in liquids is greatly affected by the pressure of a gas above the liquid; little effect for solids in liquids.

24. In a saturated solution, the net rate of dissolution is zero. The rate of dissolving is equal to that of crystallization.

26. The solution contains 16 moles of HNO_3 per liter of solution.

28. The champagne would spray out of the bottle.

30. Water molecules can pass through a semipermeable membrane in both directions. The net result is that more water molecules pass from the pure water to the sugar solution.

32. A lettuce leaf immersed in salad dressing containing salt and vinegar will become limp and wilted as a result of osmosis.

34. (a) 1 M NaOH   1 L
  (b) 0.6 M Ba(OH)$_2$ 0.83 L
  (c) 2 M KOH   0.50 L
  (d) 1.5 M Ca(OH)$_2$ 0.33 L

36. The vapor pressure of the solution is lower than that of the pure solvent and intersects the vapor pressure curve of the solvent below the freezing point of the pure solvent (see Figure 14.8a).

38. The presence of the methanol lowers the freezing point of the water.

40. The molarity of a 5 molal solution is less than 5 M.

42. Reasonably soluble: (c) $CaCl_2$ (d) $Fe(NO_3)_3$
  Insoluble: (a) PbI_2 (b) $MgCO_3$ (e) $BaSO_4$

44. (a) 7.41% $Mg(NO_3)_2$
  (b) 6.53% $NaNO_3$

46. 544 g solution

48. (a) 8.815% NaOH   (b) 15% $C_6H_{12}O_6$

50. 37.5 g K_2CrO_4

52. 33.6% NaCl

54. 22% C_6H_{14}

56. (a) 2.5 M HCl
   (b) 0.59 M $BaCl_2 \cdot 2H_2O$
   (c) 2.19×10^{-3} M $Al_2(SO_4)_3$
   (d) 0.172 M $Ca(NO_3)_2$

58. (a) 3.5×10^{-4} mol NaOH
   (b) 16 mol $CoCl_2$

60. (a) 13 g HCl   (b) 8.58 g $Na_2C_2O_4$

62. (a) 3.91×10^4 mL   (b) 7.82×10^3 mL

64. (a) 6.0 M HCl   (b) 0.064 M $ZnSO_4$

66. (a) 2.0×10^2 mL 12 M HCl
   (b) 16 mL 16 M HNO_3

68. (a) 1.83 M H_2SO_4 (b) 0.30 M H_2SO_4

70. (a) 33.3 mL of 0.250 M Na_3PO_4
   (b) 1.10 g $Mg_3(PO_4)_2$

72. (a) 0.13 mol Cl_2
   (b) 16 mol HCl
   (c) 1.3×10^2 mL of 6 M HCl
   (d) 3.2 L Cl_2

74. (a) 5.5 m $C_6H_{12}O_6$ (b) 0.25 m I_2

76. (a) 10.74 m (b) −20.0°C   (c) 105.50°C

78. 163 g/mol

80. 96.5 g 10% NaOH solution

82. $C_8H_4N_2$

84. 6.4 g KNO_3; 0.14 M

86. (a) 4.5 g NaCl
   (b) 450. mL H_2O must evaporate.

88. 210. mL solution

90. 6.72 M HNO_3

92. 540. mL water to be added

94. To make 250. mL of 0.625 M KOH, take 31.3 mL of 5.00 M KOH and dilute with water to a volume of 250. mL.

96. 2.08 M HCl

98. 12.0 g $Mg(OH)_2$ is the more effective acid neutralizer.

100. 6.2 m H_2SO_4
   5.0 M H_2SO_4

102. (a) 2.9 m (b) 101.5°C

104. (a) 8.04×10^3 g $C_2H_6O_2$
   (b) 7.24×10^3 mL $C_2H_6O_2$
   (c) −4.0°F

106. 0.46 L HCl

108. (a) 7.7 L H_2O must be added
   (b) 0.0178 mol
   (c) 0.0015 mol

110. Mix together 667 mL 3.00 M HNO_3 and 333 mL 12.0 M HNO_3 to get 1000. mL of 6.00 M HNO_3.

112. 2.84 g $Ba(OH)_2$ is formed.

114. (a) 0.011 mol Li_2CO_3 (c) 3.2×10^2 mL solution
   (b) 14 g Li_2CO_3 (d) 1.5%

Chapter 15

2. An electrolyte must be present in the solution for the bulb to glow.

4. First, the orientation of the polar water molecules about the Na^+ and Cl^- ions is different. Second, more water molecules will fit around Cl^-, since it is larger than the Na^+ ion.

6. tomato juice

8. Arrhenius: HCl + NaOH $\longrightarrow$ NaCl + H_2O
   Brønsted-Lowry: HCl + KCN $\longrightarrow$ HCN + KCl
   Lewis: $AlCl_3$ + NaCl $\longrightarrow$ $AlCl_4^-$ + Na^+

10. acids, bases, salts

12. Hydrogen chloride reacts with polar water molecules to produce H_3O^+ and Cl^- ions, which conduct an electric current. HCl does not react with or ionize in hexane, hence it does not conduct an electric current.

14. CH_3OH is a nonelectrolyte; NaOH is an electrolyte. This indicates that the OH group in CH_3OH must be covalently bonded to the CH_3 group.

16. Dissociation is the separation of already existing ions in an ionic compound. The dissolving of NaCl is a dissociation process. The dissolving of HCl in water is an ionization process because ions are formed as a result of the reaction of HCl and H_2O.

18. Ions are hydrated in solution because there is an electrical attraction between the charged ions and the polar water molecules.

20. (a) $[H^+] = [OH^-]$
   (b) $[H^+] > [OH^-]$
   (c) $[OH^-] > [H^+]$

22. HCl is a polar molecule, consequently it is much more soluble in the polar solvent, water, than the nonpolar solvent, hexane.

24. The fundamental difference is in the size of the particles. The particles in true solutions are less than 1 mm in size. The particles in colloids range from 1–1000 mm in size.

26. Dialysis is the process of removing dissolved solutes from colloidal dispersion by the use of a dialyzing membrane. Dialysis is used in artificial kidneys to remove soluble waste products from the blood.

28. (a) $H_2SO_4 - HSO_4^-$; $H_2C_2H_3O_2^+ - HC_2H_3O_2$
   (b) Step 1: $H_2SO_4 - HSO_4^-$; $H_3O^+ - H_2O$
      Step 2: $HSO_4^- - SO_4^{2-}$; $H_3O^+ - H_2O$
   (c) $HClO_4 - ClO_4^-$; $H_3O^+ - H_2O$
   (d) $H_3O^+ - H_2O$; $CH_3OH - CH_3O^-$

30. (a) $NaOH(aq) + HBr(aq) \longrightarrow NaBr(aq) + H_2O(l)$
   (b) $KOH(aq) + HCl(aq) \longrightarrow KCl(aq) + H_2O(l)$
   (c) $Ca(OH)_2(aq) + 2\,HI(aq) \longrightarrow CaI_2(aq) + 2\,H_2O(l)$
   (d) $Al(OH)_3(s) + 3\,HBr(aq) \longrightarrow AlBr_3(aq) + 3\,H_2O(l)$
   (e) $Na_2O(s) + 2\,HClO_4(aq) \longrightarrow 2\,NaClO_4(aq) + H_2O(l)$
   (f) $3\,LiOH(aq) + FeCl_3(aq) \longrightarrow Fe(OH)_3(s) + 3\,LiCl(aq)$

32. (a) $NaHCO_3$—salt   **(e)** $RbOH$—base
   (c) $AgNO_3$—salt   **(f)** K_2CrO_4—salt
   (d) $HCOOH$—acid

34. (a) $0.75\ M\ Zn^{2+}$, $1.5\ M\ Br^-$
   (b) $4.95\ M\ SO_4^{2-}$, $3.30\ M\ Al^{3+}$
   (c) $0.680\ M\ NH_4^+$, $0.340\ M\ SO_4^{2-}$
   (d) $0.0628\ M\ Mg^{2+}$, $0.126\ M\ ClO_3^-$

36. (a) $4.9\ g\ Zn^{2+}$, $12\ g\ Br^-$
   (b) $8.90\ g\ Al^{3+}$, $47.6\ g\ SO_4^{2-}$
   (c) $1.23\ g\ NH_4^+$, $3.28\ g\ SO_4^{2-}$
   (d) $0.153\ g\ Mg^{2+}$, $1.05\ g\ ClO_3^-$

38. (a) $[K^+] = 1.0\ M$, $[Ca^{2+}] = 0.5\ M$, $[Cl^-] = 2.0\ M$
   (b) No ions are present in the solution.
   (c) $0.67\ M\ Na^+$, $0.67\ M\ NO_3^-$

40. (a) $0.147\ M\ NaOH$
   (b) $0.964\ M\ NaOH$
   (c) $0.4750\ M\ NaOH$

42. (a) $H_2S(g) + Cd^{2+}(aq) \longrightarrow CdS(s) + 2\,H^+(aq)$
   (b) $Zn(s) + 2H^+(aq) \longrightarrow Zn^{2+}(aq) + H_2(g)$
   (c) $Al^{3+}(aq) + PO_4^{3-}(aq) \longrightarrow AlPO_4(s)$

44. (a) $2\ M\ HCl$
   (b) $1\ M\ H_2SO_4$

46. 40.8 mL of $0.245\ M\ HCl$

48. 7.0% NaCl in the sample

50. $0.936\ L\ H_2$

52. (a) pH = 7.0   **(b)** pH = 0.30   **(c)** pH = 4.00

54. (a) pH = 4.30   **(b)** pH = 10.47

56. 3.0×10^3 mL

58. The ionization of the acetic acid solution increases the particle concentration above that of the alcohol solution. Therefore, the acetic acid solution freezes at a lower temperature.

60. A hydronium ion is a hydrated hydrogen ion. (H_3O^+)

62. (a) 100°C   pH = 6.0
      25°C   pH = 7.0
   (b) H^+ concentration is higher at 100°C.
   (c) The water is neutral at both temperatures.

64. $0.201\ M\ HCl$

66. 0.673 g KOH

68. 13.9 L of $18.0\ M\ H_2SO_4$

70. pH = 1.1, acidic

72. (a) $2\,NaOH(aq) + H_2SO_4(aq) \longrightarrow Na_2SO_4(aq) + 2\,H_2O(l)$
   (b) 1.0×10^2 mL NaOH
   (c) $0.71\ g\ Na_2SO_4$

74. $12\ M\ HNO_3$

76. Statement (c) is correct.

Chapter 16

2. The reaction is endothermic because the increased temperature increases the concentration of product (NO_2) present at equilibrium.

4. The sum of the pH and the pOH is 14. A solution whose pH is -1 would have a pOH of 15.

6. The order of molar solubilities is: $AgC_2H_3O_2$, $PbSO_4$, $BaSO_4$, $AgCl$, $BaCrO_4$, $AgBr$, AgI, PbS.

8. When 0.010 mol H^+ is added to the buffer solution, the H^+ reacts with excess $C_2H_3O_2^-$ forming un-ionized $HC_2H_3O_2$. The added H^+ is essentially removed from solution, resulting in very little change in the $[H^+]$ and the pH of the solution.

10. The rate increases because the number of collisions increases due to the added reactant.

12. Increasing the temperature causes the velocity of the molecules to increase. Faster moving molecules increase the number of collisions between molecules, resulting in an increased rate of reaction.

14. The equilibrium shifts to the right, yielding a higher percent ionization.

16. The pH of the water can be different at different temperatures, because the degree of ionization varies with temperature.

18. HNO_3 has a higher $[H^+]$ than H_2O. The HNO_3 removes $C_2H_3O_2^-$ ions from $AgC_2H_3O_2$ equilibrium forming un-ionized $HC_2H_3O_2$, which allows more $AgC_2H_3O_2$ to dissolve. Adding HCl to $AgC_2H_3O_2 \rightleftarrows Ag^+ + C_2H_3O_2^-$ forms a precipitate of AgCl resulting in increased solubility of $AgC_2H_3O_2$.

20. A buffer solution contains a weak acid or base plus a salt of that weak acid or base. When a small amount of a strong acid (H^+) is added to this buffer solution, the H^+ reacts with anions of the salt, thus neutralizing the added acid. When a strong base, OH^-, is added, it reacts with un-ionized acid to neutralize the added base. As a result, in both cases, the approximate pH of the solution is maintained.

22. (a) $H_2O(l) \overset{100°C}{\rightleftarrows} H_2O(g)$
   (b) $SO_2(l) \rightleftarrows SO_2(g)$

24. (a) $[NH_3]$, $[O_2]$, and $[N_2]$ will be increased. $[H_2O]$ will be decreased. Reaction shifts left.
(b) The addition of heat will shift the reaction to the left.

26. (a) left   I   I   D (add NH_3)
(b) left   I   I   D (increase volume)
(c) no change   N   N   N (add catalyst)
(d) ?   ?   I   I (add H_2 and N_2)

28.

Reaction	Increase temperature	Increase pressure	Add catalyst
(a)	right	left	no change
(b)	left	left	no change
(c)	left	left	no change

30. Equilibrium shift
(a) left   (b) right   (c) right

32. (a) $K_{eq} = \dfrac{[H^+][ClO_2^-]}{[HClO_2]}$

(b) $K_{eq} = \dfrac{[H^+][C_2H_3O_2^-]}{[HC_2H_3O_2]}$

(c) $K_{eq} = \dfrac{[NO]^4[H_2O]^6}{[NH_3]^4[O_2]^5}$

34. (a) $K_{sp} = [Fe^{3+}][OH^-]^3$
(b) $K_{sp} = [Sb^{5+}]^2[S^{2-}]^5$
(c) $K_{sp} = [Ca^{2+}][F^-]^2$
(d) $K_{sp} = [Ba^{2+}]^3[PO_4^{3-}]^2$

36. If H^+ is increased,
(a) pH is decreased   (c) OH^- is decreased
(b) pOH is increased   (d) K_W remains unchanged

38. (a) $Ca(CN)_2$, basic   (c) $NaNO_2$, basic
(b) $BaBr_2$, neutral   (d) NaF, basic

40. (a) $NH_4^+(aq) + H_2O(l) \rightleftharpoons H_3O^+(aq) + NH_3(aq)$
(b) $SO_3^{2-}(aq) + H_2O(l) \rightleftharpoons OH^-(aq) + HSO_3^-(aq)$

42. (a) $OCl^-(aq) + H_2O(l) \rightleftharpoons OH^-(aq) + HOCl(aq)$
(b) $ClO_2^-(aq) + H_2O(l) \rightleftharpoons OH^-(aq) + HClO_2(aq)$

44. When excess base gets into the bloodstream it reacts with H^+ to form water. Then H_2CO_3 ionizes to replace H^+, thus maintaining the approximate pH of the solution.

46. (a) $K_a = 5.7 \times 10^{-6}\,M$
(b) pH = 5.24
(c) 2.3×10^{-3}% ionized

48. $K_a = 7 \times 10^{-10}$

50. (a) pH = 3.72   (b) pH = 4.23   (c) pH = 4.72

52. $K_a = 7.3 \times 10^{-6}$

54. $[OH^-] = 1.0\,M$, pOH = 0.00, pH = 14.00, $[H^+] = 1 \times 10^{-14}\,M$

56. (a) pH = 11.4, pOH = 2.60
(b) pH = 4.73, pOH = 9.8
(c) pH = 9.1, pOH = 4.92

58. (a) $[OH^-] = 2.5 \times 10^{-6}$ (b) $[OH^-] = 1.1 \times 10^{-13}$

60. (a) $[H^+] = 2.2 \times 10^{-9}$ (b) $[H^+] = 1.4 \times 10^{-11}$

62. (a) 1.2×10^{-23} ZnS   (c) 1.81×10^{-18} Ag_3PO_4
(b) 2.6×10^{-13} $Pb(IO_3)_2$ (d) 5.13×10^{-17} $Zn(OH)_2$

64. (a) $4.5 \times 10^{-5}\,M$
(b) $7.6 \times 10^{-10}\,M$

66. (a) 8.9×10^{-4} g $BaCO_3$
(b) 9.3×10^{-9} g $AlPO_4$

68. Precipitation occurs.

70. 2.5×10^{-12} mol AgBr will dissolve.

72. pH = 4.74

74. Change in pH = $4.74 - 4.72 = 0.02$ units in the buffered solution. Initial pH = 4.74.

76. (a) 3.16 mol HI
(b) 0.57 mol I_2
(c) $K_{eq} = 57$

78. $K_{eq} = 29$

80. Hypochlorous acid: $K_a = 3.5 \times 10^{-8}$
Propanoic acid: $K_a = 1.3 \times 10^{-5}$
Hydrocyanic acid: $K_a = 4.0 \times 10^{-10}$

82. (a) Precipitation occurs.
(b) Precipitation occurs.
(c) No precipitation occurs.

84. No precipitate of $PbCl_2$

86. $K_{eq} = 1.1 \times 10^4$

88. 8.0 M NH_3

90. $K_{sp} = 4.00 \times 10^{-28}$

92. (a) The temperature could have been cooler.
(b) The humidity in the air could have been higher.
(c) The air pressure could have been greater.

94. $K_{eq} = 1$, (c) is the correct answer.

96. $K_{eq} = 3$

98. (a) After an initial increase, $[OH^-]$ will be neutralized and equilibrium shifts to the right.
(b) $[H^+]$ will be reduced (reacts with OH^-). Equilibrium shifts to the right.
(c) $[NO_2^-]$ increases as equilibrium shifts to the right.
(d) $[HNO_2]$ decreases as equilibrium shifts to the right.

100. 1.1 g $CaSO_4$

Chapter 17

2. (a) Al
(b) Ba
(c) Ni

4. (a) $2 Al + Fe_2O_3 \longrightarrow Al_2O_3 + 2 Fe + heat$
   (b) Al is more active than Fe.
   (c) No. Fe is less reactive than Al.
   (d) Yes. Al is more reactive than Cr.

6. (a) Oxidation occurs at the anode.
   $2 Cl^-(aq) \longrightarrow Cl_2(g) + 2e^-$
   (b) Reduction occurs at the cathode.
   $Ni^{2+}(aq) + 2e^- \longrightarrow Ni(s)$
   (c) The net chemical reaction is
   $Ni^{2+}(aq) + 2 Cl^-(aq) \xrightarrow[\text{energy}]{\text{electrical}} Ni(s) + Cl_2(g)$

8. (a) It would not be possible to monitor the voltage produced, but the reactions in the cell would still occur.
   (b) If the salt bridge were removed, the reaction would stop.

10. $Ca^{2+} + 2 e^- \longrightarrow Ca$ cathode reaction, reduction
    $2 Br^- \longrightarrow Br_2 + 2 e^-$ anode reaction, oxidation

12. Since lead dioxide and lead(II) sulfate are insoluble, it is unnecessary to have salt bridges in the cells of a lead storage battery.

14. Reduction occurs at the cathode.

16. A salt bridge permits movement of ions in the cell. This keeps the solution neutral with respect to the charged particles (ions) in the solution.

18. (a) $K\underline{Mn}O_4$ $+7$ **(d)** $K\underline{Cl}O_3$ $+5$
   (b) $\underline{I}_2$ 0 **(e)** $K_2\underline{Cr}O_4$ $+6$
   (c) $\underline{N}H_3$ -3 **(f)** $K_2\underline{Cr}_2O_7$ $+6$

20. (a) $\underline{O}_2$ 0 **(c)** $Fe(\underline{O}H)_3$ -2
   (b) $\underline{As}O_4^{3-}$ $+5$ **(d)** $\underline{I}O_3^-$ $+5$

22. (a) $SO_3^{2-} + H_2O \longrightarrow SO_4^{2-} + 2 H^+ + 2 e$
    S is oxidized
   (b) $NO_3^- + 4 H^+ + 3e^- \longrightarrow NO + 2 H_2O$
    N is reduced
   (c) $S_2O_4^{2-} + 2 H_2O \longrightarrow 2 SO_3^{2-} + 4 H^+ + 2 e^-$
    S is oxidized
   (d) $Fe^{2+} \longrightarrow Fe^{3+} + 1 e^-$
    Fe is oxidized

24. Equation (1):
   (a) As is oxidized, Ag^+ is reduced.
   (b) Ag^+ is the oxidizing agent, AsH_3 the reducing agent.
   Equation (2):
   (a) Br is oxidized, Cl is reduced.
   (b) Cl_2 is the oxidizing agent, NaBr the reducing agent.

26. (a) $3 Cl_2 + 6 KOH \longrightarrow KClO_3 + 5 KCl + 3 H_2O$
   (b) $3 Ag + 4 HNO_3 \longrightarrow 3 AgNO_3 + NO + 2 H_2O$
   (c) $3 CuO + 2 NH_3 \longrightarrow N_2 + 3 Cu + 3 H_2O$
   (d) $3 PbO_2 + 2 Sb + 2 NaOH \longrightarrow$
                    $3 PbO + 2 NaSbO_2 + H_2O$
   (e) $5 H_2O_2 + 2 KMnO_4 + 3 H_2SO_4 \longrightarrow$
              $5 O_2 + 2 MnSO_4 + K_2SO_4 + 8 H_2O$

28. (a) $6 H^+ + ClO_3^- + 6 I^- \longrightarrow 3 I_2 + Cl^- + 3 H_2O$
   (b) $14 H^+ + Cr_2O_7^{2-} + 6 Fe^{2+} \longrightarrow$
               $2 Cr^{3+} + 6 Fe^{3+} + 7 H_2O$
   (c) $2 H_2O + 2 MnO_4^- + 5 SO_2 \longrightarrow$
             $4 H^+ + 2 Mn^{2+} + 5 SO_4^{2-}$
   (d) $6 H^+ + 5 H_3AsO_4 + 2 MnO_4^- \longrightarrow$
             $5 H_3AsO_4 + 2 Mn^{2+} + 3 H_2O$
   (e) $8 H^+ + Cr_2O_7^{2-} + 3 H_3AsO_3 \longrightarrow$
             $2 Cr^{3+} + 3 H_3AsO_4 + 4 H_2O$

30. (a) $H_2O + 2 MnO_4^- + 3 SO_3^{2-} \longrightarrow$
             $2 MnO_2 + 3 SO_4^{2-} + 2 OH^-$
   (b) $2 H_2O + 2 ClO_2 + 2 OH^- + SbO_2^- \longrightarrow$
             $2 ClO_2^- + Sb(OH)_6$
   (c) $8 Al + 3 NO_3^- + 18 H_2O + 5 OH^- \longrightarrow$
             $3 NH_3 + 8 Al(OH)_4^-$
   (d) $4 OH^- + 2 H_2O + P_4 \longrightarrow 2 HPO_3^{2-} + 2 PH_3$
   (e) $2 Al + 6 H_2O + 2 OH^- \longrightarrow 2 Al(OH)_4^- + 3 H_2$

32. (a) The oxidizing agent is $KMnO_4$.
   (b) The reducing agent is HCl.
   (c) $3.011 \times 10^{24} \dfrac{\text{electrons}}{\text{mol } KMnO_4}$

34. $20.2 L Cl_2$

36. 66.2 mL of $0.200 M K_2Cr_2O_7$ solution

38. 91.3% KI

40. 5.560 mol H_2

42. The electrons lost by the species undergoing oxidation must be gained (or attracted) by another species, which then undergoes reduction.

44. Sn^{4+} can only be an oxidizing agent.
   Sn^0 can only be a reducing agent.
   Sn^{2+} can be both oxidizing and reducing agents.

46. Equations (a) and (b) represent oxidations.

48. (a) $F_2 + 2 Cl^- \longrightarrow 2 F^- + Cl_2$
   (b) $Br_2 + Cl^- \longrightarrow NR$
   (c) $I_2 + Cl^- \longrightarrow NR$
   (d) $Br_2 + 2 I^- \longrightarrow 2 Br^- + I_2$

50. $4 Zn + NO_3^- + 10 H^+ \longrightarrow 4 Zn^{2+} + NH_4^+ + 3 H_2O$

52. (a) Pb is the anode.
   (b) Ag is the cathode.
   (c) Pb (anode)
   (d) Ag (cathode)
   (e) Electrons flow from the lead through the wire to the silver.
   (f) Positive ions flow through the salt toward the negatively charged strip of silver; negative ions flow toward the positively charged strip of lead.

Chapter 18

2. Alpha particles are much heavier.

4. Contributions to the early history of radioactivity include
(a) Henri Becquerel—discovered radioactivity.
(b) Marie and Pierre Curie—discovered polonium and radium.
(c) Wilhelm Roentgen—discovered X rays and developed the technique for producing them.
(d) Ernest Rutherford—discovered alpha and beta particles, established the link between radioactivity and transmutation, and produced the first successful man-made transmutation.
(e) Otto Hahn and Fritz Strassmann—were first to produce nuclear fission.

6. *Isotope* is used with reference to atoms of the same element that contain different masses. *Nuclide* infers any isotope of any atom.

8.

	Charge	Mass	Nature of particles	Penetrating power
Alpha	+2	4 amu	He nucleus	low
Beta	−1	$\dfrac{1}{1837}$ amu	electron	moderate
Gamma	0	0	electro-magnetic radiation	high

10. A disintegration series is a series of α and β emissions leading to production of a stable nuclide.

12. $^{232}_{90}\text{Th} \xrightarrow{-\alpha} {}^{228}_{88}\text{Ra} \xrightarrow{-\beta} {}^{228}_{89}\text{Ac} \xrightarrow{-\beta} {}^{228}_{90}\text{Th} \xrightarrow{-\alpha}$

$^{224}_{88}\text{Ra} \xrightarrow{-\alpha} {}^{220}_{86}\text{Rn} \xrightarrow{-\alpha} {}^{216}_{84}\text{Po} \xrightarrow{-\alpha}$

$^{212}_{82}\text{Pb} \xrightarrow{-\beta} {}^{212}_{83}\text{Bi} \xrightarrow{-\beta} {}^{212}_{84}\text{Po} \xrightarrow{-\alpha} {}^{208}_{82}\text{Pb}$

14. $^{211}_{83}\text{Bi} \longrightarrow {}^{4}_{2}\text{He} + {}^{207}_{81}\text{Tl}$ $^{207}_{81}\text{Tl} \longrightarrow {}^{0}_{-1}\text{e} + {}^{207}_{82}\text{Pb}$

16. Natural uranium is 99+% U-238. Commercial nuclear reactors use U-235 enriched uranium as a fuel. Slow neutrons will cause the fission of U-235, but not U-238. Fast neutrons are capable of a nuclear reaction with U-238 to produce fissionable Pu-239. A breeder reactor converts nonfissionable U-238 to fissionable Pu-239, and in the process, manufactures more fuel than it consumes.

18. A certain amount of fissionable material (a critical mass) must be present before a self-supporting chain reaction can occur. Without a critical mass, too many neutrons from fissions will escape, and the reaction cannot reach a chain reaction status, unless at least one neutron is captured for every fission that occurs.

20. When radioactive rays pass through normal matter, they cause that matter to become ionized (usually by knocking out electrons). Therefore, the radioactive rays are classified as ionizing radiation.

22. Strontium-90 has two characteristics that create concern. Its half-life is 28 years, so it remains active for a long period of time (disintegrating by emitting β radiation). The other characteristic is that Sr-90 is chemically similar to calcium, so when it is present in milk it is deposited in bone tissue along with calcium. Red blood cells are produced in the bone marrow. If the marrow is subjected to beta radiation from strontium-90, the red blood cells will be destroyed, increasing the incidence of leukemia and bone cancer.

24. In living species, the ratio of carbon-14 to carbon-12 is constant due to the constant C-14/C-12 ratio in the atmosphere and food sources. When a species dies, life processes stop. The C-14/C-12 ratio decreases with time because C-14 is radioactive and decays according to its half-life, while the amount of C-12 in the species remains constant. Thus, the age of an archaeological artifact containing carbon can be calculated by comparing the C-14/C-12 ratio in the artifact with the C-14/C-12 ratio in the living species.

26. The half-life of carbon-14 is 5668 years.

$(4 \times 10^6 \text{ years})\left(\dfrac{1 \text{ half-life}}{5668 \text{ years}}\right) = 7 \times 10^2 \text{ half-lives}$

700 half-lives would pass in 4 million years. Not enough C-14 would remain to allow detection with any degree of reliability. C-14 dating would not prove useful in this case.

28.

	Protons	Neutrons	Nucleons
(a) $^{235}_{92}\text{U}$	92	143	235
(b) $^{82}_{35}\text{Br}$	35	47	82

30. Its atomic number increases by one, and its mass number remains unchanged.

32. Equations for alpha decays:
(a) $^{192}_{78}\text{Pt} \longrightarrow {}^{4}_{2}\text{He} + {}^{188}_{76}\text{Os}$
(b) $^{210}_{84}\text{Po} \longrightarrow {}^{4}_{2}\text{He} + {}^{206}_{82}\text{Pb}$

34. (a) $^{239}_{93}\text{Np} \longrightarrow {}^{0}_{-1}\text{e} + {}^{239}_{94}\text{Pu}$
(b) $^{90}_{38}\text{Sr} \longrightarrow {}^{0}_{-1}\text{e} + {}^{90}_{39}\text{Y}$

36. $^{30}_{15}\text{P} \longrightarrow {}^{30}_{14}\text{Si} + {}^{0}_{+1}\text{e}$

38. (a) $^{66}_{29}\text{Cu} \longrightarrow {}^{66}_{30}\text{Zn} + {}^{0}_{-1}\text{e}$
(b) $^{0}_{-1}\text{e} + {}^{7}_{4}\text{Be} \longrightarrow {}^{7}_{3}\text{Li}$
(c) $^{27}_{13}\text{Al} + {}^{4}_{2}\text{He} \longrightarrow {}^{30}_{14}\text{Si} + {}^{1}_{1}\text{H}$
(d) $^{85}_{37}\text{Rb} + {}^{1}_{0}\text{n} \longrightarrow {}^{82}_{35}\text{Br} + {}^{4}_{2}\text{He}$

40. 3 half-lives; 1/8 remaining

42. (a) 5.4×10^{11} J/mol
(b) 0.199% mass loss

44. 11,340 years old

46. (a) 0.0424 g/mol
(b) 3.8×10^{12} J/mol

48. (a) Geiger counter: Radiation passes through a thin glass window into a chamber filled with argon gas and containing two electrodes. Some of the argon ionizes, sending a momentary electrical impulse between the electrodes to the detector. This signal is amplified electronically and read out on a counter or as a series of clicks.

(b) Scintillation counter: Radiation strikes a scintillator, which is composed of molecules that emit light in the presence of ionizing radiation. A light-sensitive detector counts the flashes and converts them into a digital readout.

(c) Film badge: Radiation penetrates a film holder. The silver grains in the film darken when exposed to radiation. The film is developed at regular intervals.

50. Fission is the process of splitting a large nucleus into two roughly equal mass pieces. Fusion is the process of combining two relatively small nuclei to form a single larger nucleus.

52. (a) $^{235}_{92}U + ^{1}_{0}n \longrightarrow ^{143}_{54}Xe + 3^{1}_{0}n + ^{90}_{38}Sr$
(b) $^{235}_{92}U + ^{1}_{0}n \longrightarrow ^{102}_{39}Y + 3^{1}_{0}n + ^{131}_{53}I$
(c) $^{14}_{7}N + ^{1}_{0}n \longrightarrow ^{1}_{1}H + ^{14}_{6}C$

54. $^{236}_{92}U \longrightarrow ^{90}_{38}Sr + 3^{1}_{0}n + ^{143}_{54}Xe$

56. (a) $^{87}_{37}Rb \longrightarrow ^{0}_{-1}e + ^{87}_{38}Sr$
(b) $^{87}_{38}Sr \longrightarrow ^{0}_{+1}e + ^{87}_{37}Rb$

58. 7680 g

60. (a) 0.500 g left
(b) 0.250 g left
(c) 0.0625 g
(d) 9.77×10^{-4} g

62. 2.267×10^4 years

Chapter 19

2. tetrahedral

4. (a) $\cdot\ddot{C}\cdot$

(b)

H:C:H methane
H:C::C:H ethylene
H:C:::C:H acetylene

6. (a) CH_3-
(b) CH_3CH_2-
(c) $CH_3CH_2CH_2-$
(d) $(CH_3)_2CH-$
(e) $CH_3CH_2CH_2CH_2-$
(f) $CH_3CH_2CHCH_3$
(g) $(CH_3)_2CHCH_2-$
(h) $(CH_3)_3C-$

8. The word *ethylene* represents a compound containing a double bond. All other words represent structures having no double bonds.

10. Vinyl acetylene, C_4H_4, $CH_2=CH-CH\equiv CH$

12. Ethylene glycol is superior to methyl alcohol as an antifreeze because of its low volatility. Methyl alcohol is much more volatile than water. If the radiator leaks under pressure (normally steam), it would primarily leak methanol vapor, thus losing the antifreeze. Ethylene glycol has a lower volatility and a higher boiling point than water, so it does not present this problem.

14. The structure for 1-methylpentane would be $CH_3CH_2CH_2CH_2CH_2CH_3$. The name 1-methylpentane is not based on the longest carbon chain of 6 atoms. Therefore, the correct name is hexane.

16. Names of alkyl groups
(a) C_8H_{17} is octyl-
(b) $C_{10}H_{21}$ is decyl-

18. $CH_3CH_2CH_2CH_2CH_2CH_2CH_3$

$CH_3CH_2CH_2CH_2CHCH_3$
$\quad\quad\quad\quad\quad\quad | $
$\quad\quad\quad\quad\quad\quad CH_3$

$CH_3CH_2\underset{\underset{CH_3}{|}}{C}CH_2CH_3$

$CH_3CH_2CH_2\underset{\underset{CH_3}{|}}{C}CH_3$

$CH_3CH_2CH_2\underset{\underset{CH_3}{|}}{C}HCH_2CH_3$

$CH_3\underset{\underset{CH_3}{|}}{C}H\underset{\underset{CH_3}{|}}{C}HCH_3$

$CH_3CH_2\underset{\underset{CH_3}{|}}{C}HCHCH_3$

$CH_3\underset{\underset{CH_3}{|}}{C}H-\underset{\underset{CH_3}{|}}{C}CH_3$

$CH_3CH_2\underset{\underset{CH_2CH_3}{|}}{C}HCH_2CH_3$

20. (a) $CH_3CH_2CH_2Cl$ 1-chloropropane
(b) $(CH_3)_3CCl$ 2-chloro-2-methylpropane
(c) $CH_3CHClCH_2CH_3$ 2-chlorobutane

22. (a) 3,6-dimethyloctane
(b) 3-ethyl-2-methylhexane or (3-isopropylhexane)

24. (a) $CH_3CH_2\underset{\underset{CH_2CH_3}{|}}{C}HCH_2CH(CH_3)_2$

(b) $CH_3CH_2CH_2\underset{\underset{C(CH_3)_3}{|}}{C}HCH_2CH_2CH_3$

(c) $CH_3\underset{\underset{CH_3}{|}}{C}HCH_2\underset{\underset{CH_2CH_3}{|}}{C}CH_2CH_2\underset{\underset{CH(CH_3)_2}{|}}{C}HCHCH_2CH_3$

26. (a)

$$CH_3CCH_2CH_3$$

with CH_3 above and CH_3 below the second carbon

2,2-dimethylbutane

(b) $CH_3CHCH_2CHCH_3$

with CH_3 and CH_2CH_3 below

2,4-dimethylhexane

28. (a) (4 isomers) $CH_3CH_2CH_2CH_2I$ CH_3CHI with CH_2CH_3 below

CH_3C-I with CH_3 above and CH_3 below CH_3CHCH_2I with CH_3 below

(b) (4 isomers)
$CH_3CH_2CHCl_2$ $CH_3CCl_2CH_3$
$CH_3CHClCH_2Cl$ $CH_2ClCH_2CH_2Cl$

(c) (5 isomers)
$CH_3CH_2CHBrCl$ $CH_3CHClCH_2Br$
$CH_3CHBrCH_2Cl$ $CH_2ClCH_2CH_2Br$
$CH_3CClBrCH_3$

(d) (9 isomers)
$CH_3CH_2CH_2CHCl_2$ $CH_3CH_2CHClCH_2Cl$
$CH_3CHClCH_2CH_2Cl$ $CH_2ClCH_2CH_2CH_2Cl$
$CH_3CH_2CCl_2CH_3$ $CH_3CHClCHClCH_3$
$CH_3CHCHCl_2$ with CH_3 below CH_3CClCH_2Cl with CH_3 below
CH_3CHCH_2Cl with CH_2Cl below

30. (a) $CH_3CH=CH_2 + Br_2 \longrightarrow CH_3CHBrCH_2Br$
(b) $CH_3CH=CH_2 + HBr \longrightarrow CH_3CHBrCH_3$

32. (a) CH_3OCH_3 ether
(b) CH_3CHO aldehyde
(c) CH_3COOH carboxylic acid
(d) $HCOOCH_3$ ester

34. (a) methoxymethane   **(c)** ethanoic acid
(b) ethanal   **(d)** methyl methanoate

36. (a) hexachloroethane CCl_3CCl_3
(b) iodoethyne $CH\equiv CI$
(c) 6-bromo-3-methyl- $BrCH_2CH_2CH=CC\equiv CH$ with CH_3 below
3-hexene-1-yne
(d) 1,2-dibromoethene $CHBr=CHBr$

38. (a) 1,2-diphenylethene   **(b)** 3-pentene-1-yne
$CH\equiv CCH=CHCH_3$
(c) 3-phenyl-1-butyne
$CH\equiv CCHCH_3$

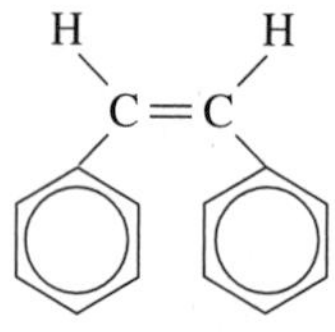

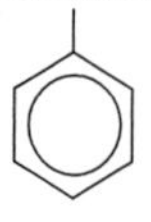

40. (a) $CH_3CH_2CHCH=CH_2$ 3-isopropyl-1-pentene
with $CH(CH_3)_2$ below
(b) $CH_3CH_2CH=CCH_2CH_3$ 3-methyl-3-hexene
with CH_3 below

42. (a) $CH_2=CH_2 + H_2O \xrightarrow{H+} CH_2CH_2OH$ (ethanol)
(b) $CH\equiv CH + 2Br_2 \longrightarrow$
$CHBr_2CHBr_2$(1,1,2,2-tetrabromoethane)
(c) $CH_2=CH_2 + H_2 \xrightarrow[Pt,\ 25°C]{1\ atm} CH_3CH_3$ (ethane)

44. (a) 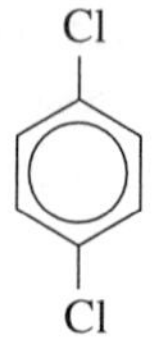
para-dichlorobenzene
(1,4-dichlorobenzene)

(b) 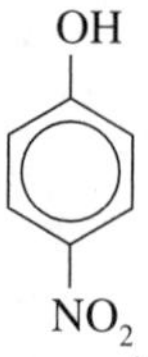
para-nitrophenol

(c) 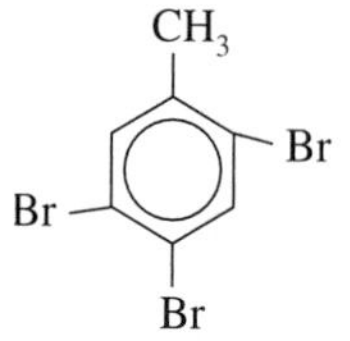
2,4,5-tribromotoluene

(d) 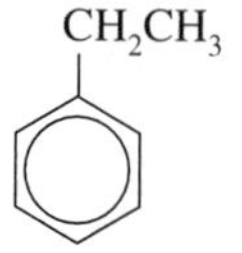
ethylbenzene

(e) 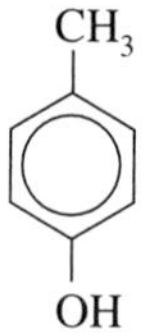
para-methylphenol

(f) 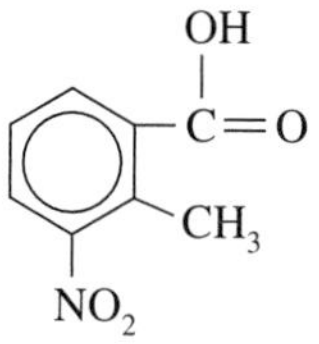
2-methyl-3-nitrobenzoic acid

46. (a) 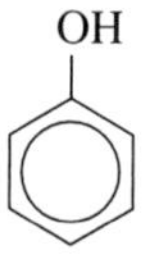
phenol

(b) 
o-bromochlorobenzene

(c) 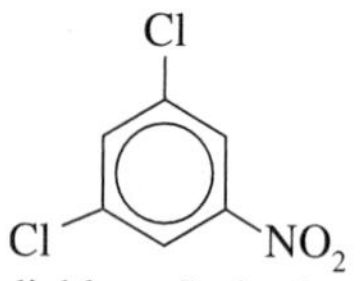
1,3-dichloro-5-nitrobenzene

(d) 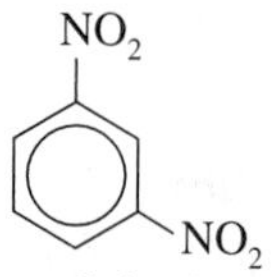
m-dinitrobenzene

48. (a) 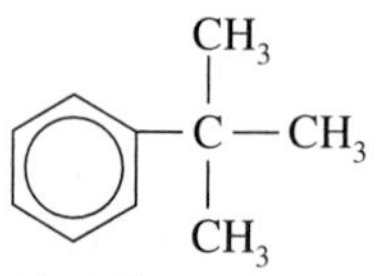
t-butylbenzene

(b) 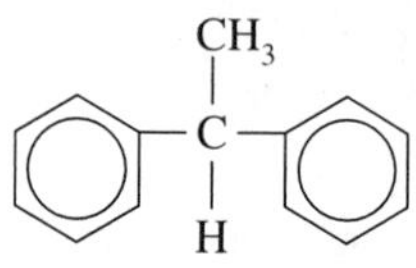
1,1-diphenylethane

50.

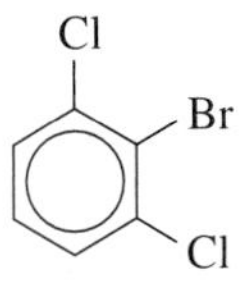

1,3-dichloro-2-bromo-
benzene

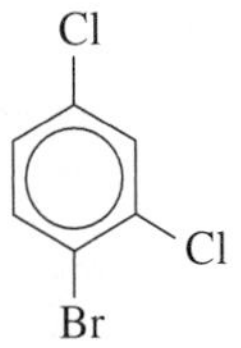

1,3-dichloro-4-bromobenzene

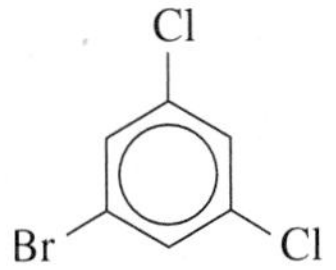

1,3-dichloro-5-bromo-
benzene

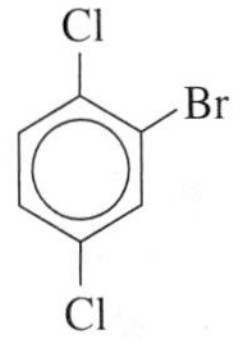

1,4-dichloro-2-bromobenzene

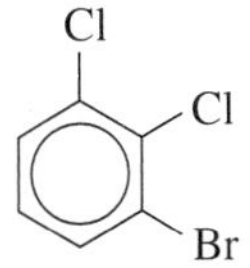

1,2-dichloro-3-bromo-
benzene

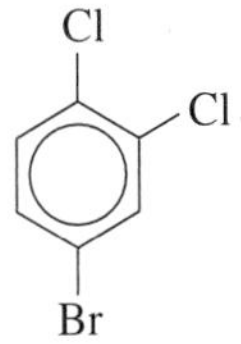

1,2-dichloro-4-bromobenzene

52. (a)

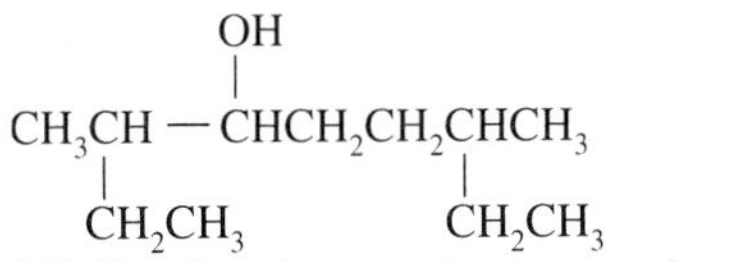

3,7-dimethyl-4-nonanol   secondary

(b) $HO-CH_2CH_2CHCH_2CH_2CH_3$

CH_3CHCH_3

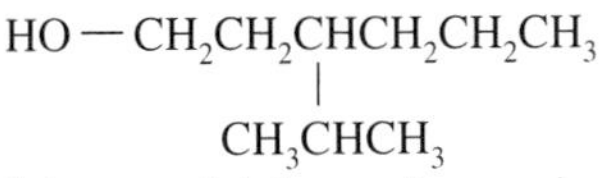

3-isopropyl-1-hexanol   primary

54. (a) 2,2-dimethyl-1-
heptanol

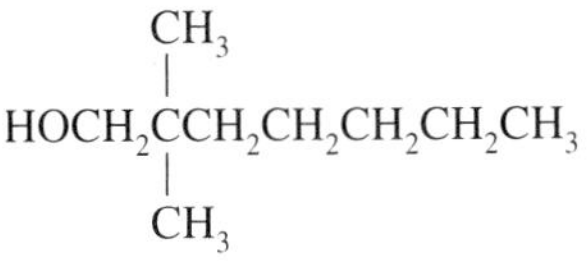

(b) 1,3-propanediol $HOCH_2CH_2CH_2OH$

56. (a)

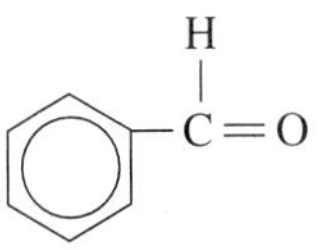

benzaldehyde

(b) $O=CCH_2CH_2C=O$ butanedial

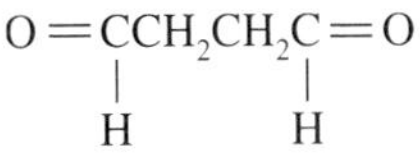

$\quad\quad H \quad\quad\quad H$

(c)

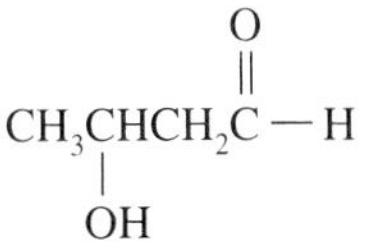

CH_3CHCH_2C-H 3-hydroxybutanal

OH

58. (a)

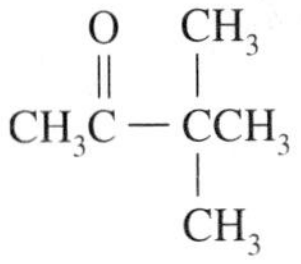

3,3-dimethyl-2-butanone
methyl t-butyl ketone

(b) $CH_3CCH_2CH_2CCH_3$ 2,5-hexanedione

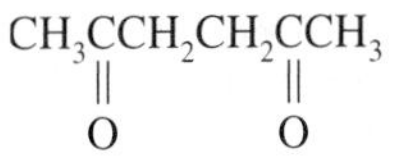

$\quad\quad\quad\quad O \quad\quad\quad O$

(c)

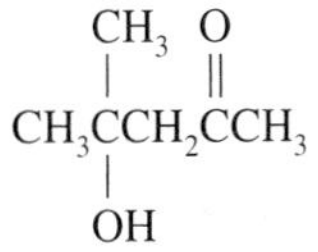

4-hydroxy-4-methyl-2-pentanone

60. (a) $CH_3CHCOOH$ 2-methylbutanoic acid

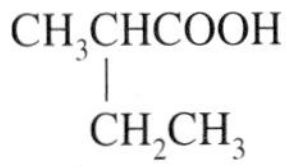

CH_2CH_3

(b)

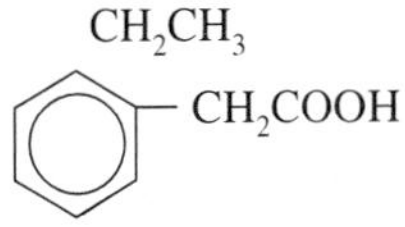

CH_2COOH phenylacetic acid

(c) $CH_3CH_2CH_2CH_2COOH$ pentanoic acid

(d)

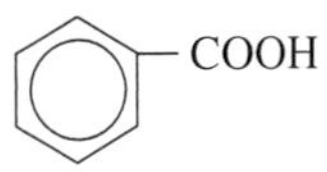

$COOH$ benzoic acid

62. Esters

(a) n-nonyl acetate
nonyl ethanoate

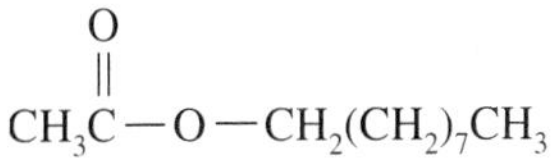

$CH_3C-O-CH_2(CH_2)_7CH_3$

(b) ethyl benzoate

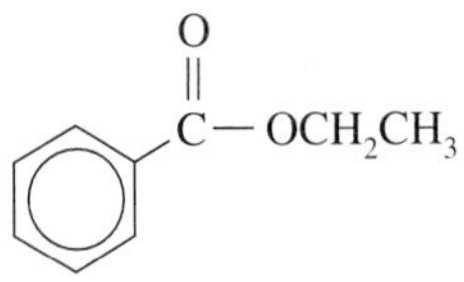

(c) methyl salicylate

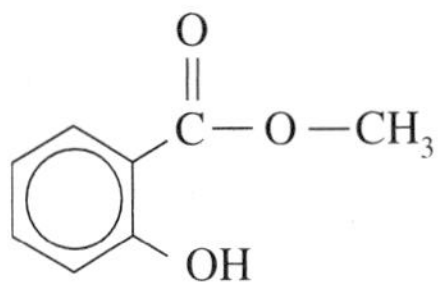

64. (a) $H-C-OCH(CH_3)_2$ isopropyl methanoate

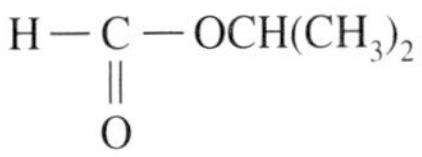

$\quad\quad\quad O$

(b)

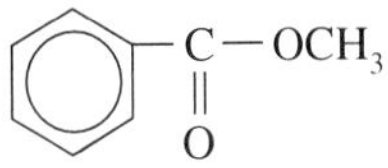

methyl benzoate

(c)

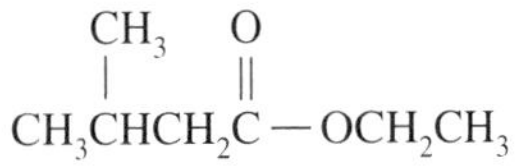

$CH_3CHCH_2C-OCH_2CH_3$ ethyl-3-methylbutanoate

66. (a) $2CH_3COOH + KOH \longrightarrow CH_3COO^-K^+ + H_2O$

(b)

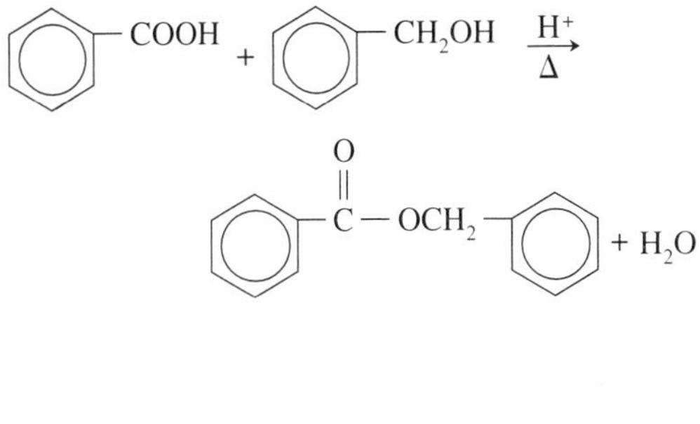

68. (a) $+CH_2-CH\xrightarrow{}_n$ polyacrylonitrile
　　　　　　　｜
　　　　　　　CN

(b) $+CF_2-CF_2\xrightarrow{}_n$ teflon

70. $CH_3CH_2CH_2CH_2CH_2OH$　　1-pentanol　　1°

$CH_3CH_2CH_2CHCH_3$　　　2-pentanol　　2°
　　　　　　　｜
　　　　　　OH

$CH_3CH_2CHCH_2CH_3$　　　3-pentanol　　2°
　　　　　｜
　　　　OH

$CH_3CH_2CHCH_2OH$　　　2-methyl-1-butanol　1°
　　　　　｜
　　　　CH_3

$CH_3CHCH_2CH_2OH$　　　3-methyl-1-butanol　1°
　　　｜
　　CH_3
　　OH
　　｜
$CH_3CCH_2CH_3$　　　　2-methyl-2-butanol　3°
　　｜
　CH_3
　OH
　｜
$CH_3CHCHCH_3$　　　　3-methyl-2-butanol　2°
　　　｜
　　CH_3

　CH_3
　｜
CH_3CCH_2OH　　　　2,2-dimethyl-1-　1°
　｜　　　　　　　　　　　propanol
　CH_3

72. (a) methyl ethyl ether　　$CH_3CH_2OCH_3$
(b) dimethyl ether　　　　CH_3OCH_3
(c) methyl ethyl ether　　$CH_3CH_2OCH_3$

74. Propanal and propanone are isomers, C_3H_6O. Butanal and butanone are isomers, C_4H_8O.

76. $CH_3(CH_2)_4COOH$　　　hexanoic acid

$CH_3CH_2CH_2CHCOOH$　　2-methylpentanoic acid
　　　　　　　｜
　　　　　　CH_3

$CH_3CHCH_2CH_2COOH$　　4-methylpentanoic acid
　　｜
　CH_3

$CH_3CH_2CHCH_2COOH$　　3-methylpentanoic acid
　　　　｜
　　　CH_3
　　　CH_3
　　　｜
CH_3CH_2CCOOH　　　2,2-dimethylbutanoic acid
　　　｜
　　CH_3

　CH_3
　｜
CH_3CCH_2COOH　　　3,3-dimethylbutanoic acid
　｜
　CH_3

　　CH_3
　　｜
$CH_3CHCHCOOH$　　　2,3-dimethylbutanoic acid
　　｜
　CH_3

$CH_3CH_2CHCOOH$　　　2-ethylbutanoic acid
　　　｜
　　CH_2CH_3

78. (a)
$+CH_2-CH-CH_2-CH-CH_2-CH-CH_2-CH\xrightarrow{}_n$
　　　　｜　　　　｜　　　　｜　　　　｜
　　　CH_3　　CH_3　　CH_3　　CH_3
　　　　　　　　polypropylene

$+CH_2-CH-CH_2-CH-CH_2-CH-CH_2-CH\xrightarrow{}_n$
　　　　｜　　　　｜　　　　｜　　　　｜
　　　CH_2　　CH_2　　CH_2　　CH_2
　　　　｜　　　　｜　　　　｜　　　　｜
　　　CH_3　　CH_3　　CH_3　　CH_3
　　　　　　　　poly-1-butene

　　CH_3　　CH_3　　CH_3　　CH_3
　　｜　　　　｜　　　　｜　　　　｜
$+CH-CH-CH-CH-CH-CH-CH-CH\xrightarrow{}_n$
　　　　｜　　　　｜　　　　｜　　　　｜
　　　CH_3　　CH_3　　CH_3　　CH_3
　　　　　　　　poly-2-butene

(b) 1.2×10^3 ethylene units

80. (a) 4 H = +4
　　　　　O = −2
　　　　　C = −2
(b) 2 O = −4
　　　　C = +4
(c) 6 H = +6
　　　　6 C = −6
　　　　C = −1

82. CH_3COOH

84. (a) $H-COOH + CH_3OH \xrightarrow{H^+} HCOOCH_3 + H_2O$
　　　　　　　　　　　　　　　　　methyl formate
(b) $CH_3CH_2CH_2COOH + CH_3CH_2CH_2CH_2OH \xrightarrow{H^+}$
　　　$CH_3CH_2CH_2COOCH_2CH_2CH_2CH_3 + H_2O$
　　　　　　　　butyl butanoate
(c) $CH_3(CH_2)_4COOH + CH_3(CH_2)_4CH_2OH \xrightarrow{H^+}$
　　　　$CH_3(CH_2)_4COOCH_2(CH_2)_4CH_3 + H_2O$
　　　　　　　　hexyl hexanoate

Chapter 20

2. Fatty acids in vegetable oils are more unsaturated than fatty acids in animal fats.

4. three disulfide linkages

6. Guanine and cytosine are mutually bonded to each other as are adenine and thymine.

8. Monosaccharides, disaccharides, and polysaccharides. Monosaccharides are the simplest.

10.

Monosaccharides	Disaccharides	Polysaccharides
glucose	sucrose	cellulose
fructose	maltose	glycogen
galactose	lactose	starch
ribose		

12.

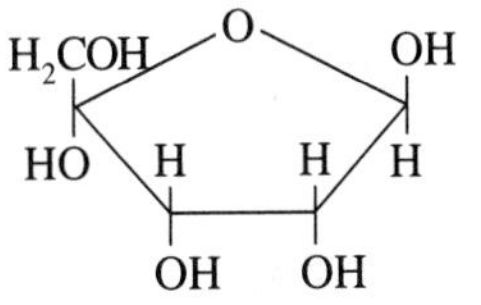

ribose

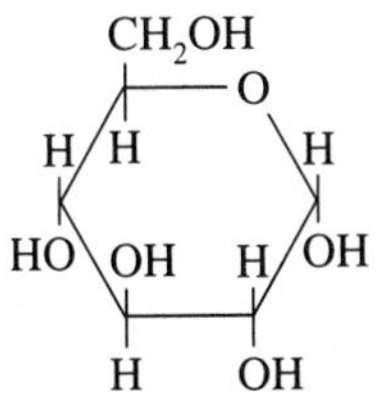

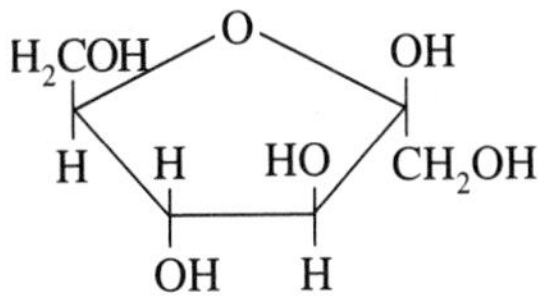

fructose

glucose

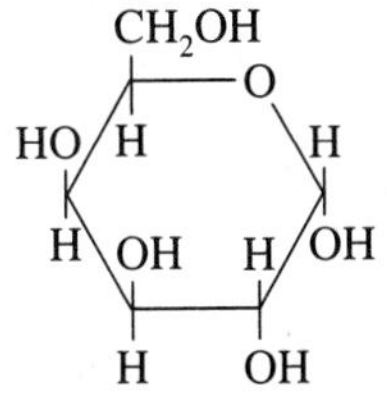

galactose

14. Lactic acid is not a carbohydrate. It does not have an aldehyde or a ketone group.

16.

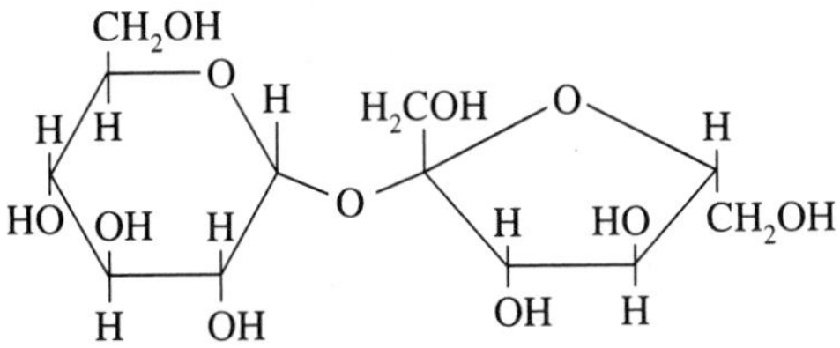

sucrose

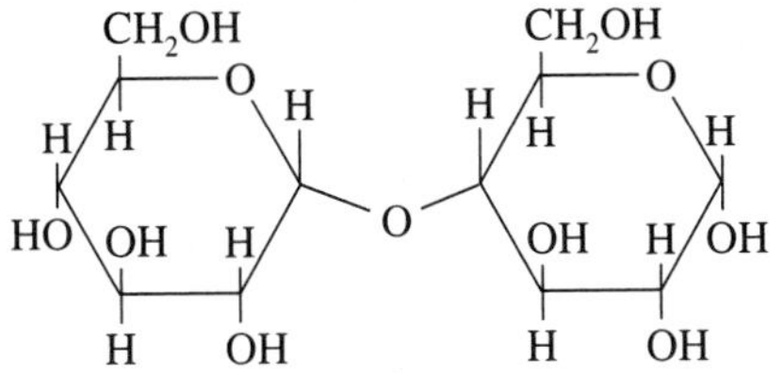

maltose

18. (a) sucrose $\xrightarrow[H_2O]{sucrase}$ glucose + fructose

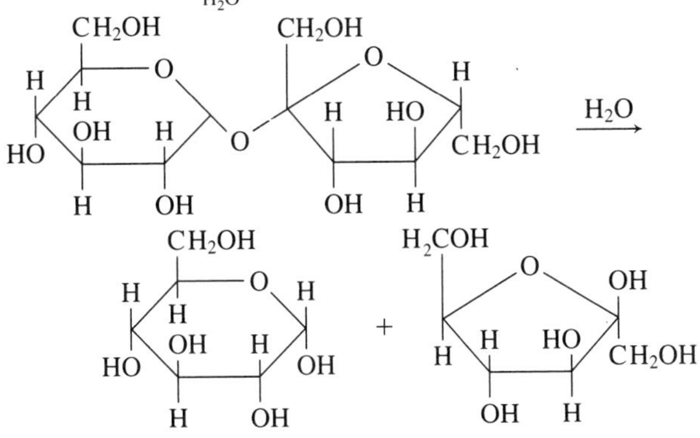

(b) maltose $\xrightarrow[H_2O]{maltase}$ glucose + glucose

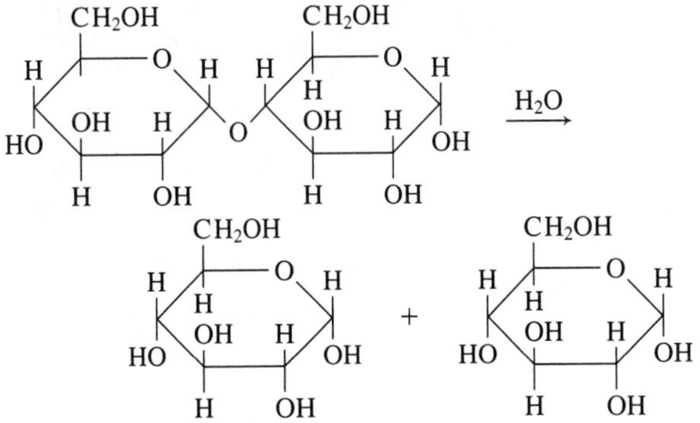

(c) Sucrase catalyzes the hydrolysis of sucrose while maltase catalyzes the hydrolysis of maltose.

20. Carbohydrates are stored in the body as glycogen.

22. Sucrose is found in the free state throughout the plant kingdom.
Maltose is found in sprouting grain.
Lactose is found free in nature mainly in the milk of mammals.
Starch is found mainly in the seeds, roots, and tubers of plants.

24. Substances are classified as lipids on the basis of their solubility in nonpolar solvents, their insolubility in water, and their greasy feeling.

26. Fats are solid and vegetable oils are liquid at room temperature. Fats contain higher amounts of saturated fatty acids and oils contain higher amounts of unsaturated fatty acids.

28.

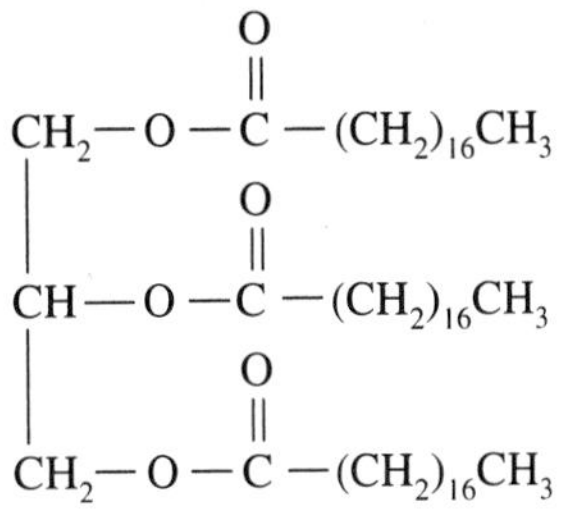

30. (a) Tripalmitin is a fat in which all the fatty acid units are palmitic acid.

(b)

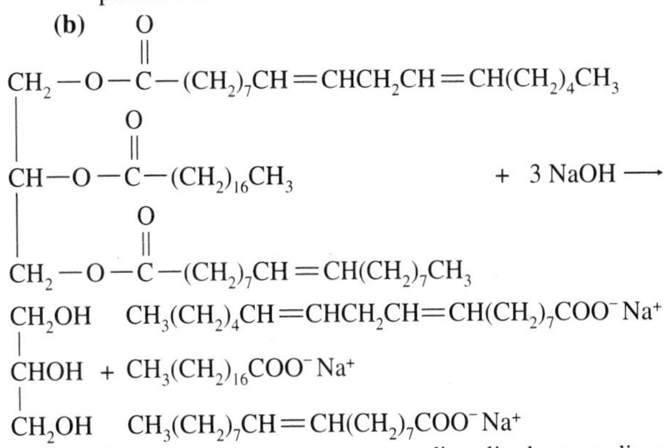

The soaps top to bottom are: sodium linoleate, sodium stearate, and sodium oleate.

32. Fats are an important food source for man. They normally account for 25 to 50% of caloric intake. Fats are the major constituent of adipose tissue. Fat deposits function to insulate the body against loss of heat and protect vital organs against mechanical injury. Source of reserve energy.

34. Cholesterol

36. Some common foods with high (over 10%) protein content are gelatin, fish, beans, nuts, cheese, eggs, poultry, and meat of all kinds.

38. The amino acids in proteins are called α-amino acids because the amine group is always attached in the α position, that is, the first carbon next to the carboxylic acid group.

40. (a)

(b)

(c)

42. Essential amino acids are those needed by the human body that cannot be synthesized by the body. They are isoleucine, leucine, lysine, methionine, phenylalanine, threonine, tryptophan, and valine.

44. Tissue proteins are continuously being broken down and resynthesized. Protein is continually needed in a balanced diet because the body does not store free amino acids.

46. (a) phosphate, ribose or deoxyribose, and one of the four nitrogen-containing bases (A, T, G, C).
　　(b) phosphate-deoxyribose-thymine
　　　　phosphate-deoxyribose-cytosine
　　　　phosphate-deoxyribose-adenine
　　　　phosphate-deoxyribose-guanine

(c)

cytosine deoxyribonucleotide

48. The two helices of the double helix are joined together by hydrogen bonds between bases.

50. G—C—T—A—A—C—G—T

52. The sequence of bases and the length of the nucleotide chains in the DNA molecules contain the coded messages that determine all the characteristics of the individual, including the reproduction of that species.

54. In mitosis, each DNA molecule forms a duplicate by uncoiling to single strands. Each strand then assembles the complementary portion from available free nucleotides to form duplicates of the original DNA molecule.

　　In meiosis, the sperm cell carries only one-half the chromosomes from its original cell, and the egg cell also carries one-half of its original chromsomes. Between them, they form a new cell that once again contains the correct number of chromosomes and all the hereditary characteristics of the species.

56. Enzymes are often specific for one particular reaction because the substrate usually fits exactly into a small part of the enzyme (known as the "active site") to form an intermediate enzyme–substrate complex.

58. Polypeptides are numbered starting with the N-terminal acid.

　　1　　2　　3　　4　　5
　　try — gly — his — phe — val

60. In the lock-and-key hypothesis, the active site of an enzyme exactly fits the complementary-shaped part of a substrate. In the induced-fit model, the enzyme changes its shape to fit the shape of the substrate.

62. Enzymes act as catalysts of biochemical reactions. Their function is to lower the activation energy of biochemical reactions.

64. CH_2O

66. The oxidation number change is from 0 to +4.

68. Cellulose: 3.7×10^3 monomer units
　　Starch: 25 monomer units
　　Cellulose: 1.9×10^{-6} m long
　　Starch: 1.3×10^{-8} m long

absolute zero $-273°C$, the zero point on the Kelvin (absolute) temperature scale. *See also* Kelvin scale [2.11, 12.6]

acid (1) A substance that produces H^+ (H_3O^+) when dissolved in water. (2) A proton donor. (3) An electron-pair acceptor. A substance that bonds to an electron pair. [15.1]

acid anhydride A nonmetal oxide that reacts with water to form an acid. [13.12]

acid ionization constant (K_a) The equilibrium constant for the ionization of a weak acid in water. [16.11]

activation energy The amount of energy needed to start a chemical reaction. [8.5, 16.8]

activity series of metals A listing of metallic elements in descending order of reactivity. [17.5]

actual yield The amount of product actually produced in a chemical reaction (as compared to the theoretical yield). [9.6]

addition reaction In organic chemistry, a reaction in which two substances join together to produce one substance. [19.9]

alcohol An organic compound consisting of an $-OH$ group bonded to a carbon atom in a nonaromatic hydrocarbon group; alcohols are classified as primary (1°), secondary (2°), or tertiary (3°), depending on whether the carbon atom to which the $-OH$ group is attached is bonded to one, two, or three other carbon atoms, respectively. [19.14]

aldehyde An organic compound that contains the $-CHO$ group. The general formula is RCHO. [19.17]

alkali metal An element (except H) from Group IA of the periodic table. [10.4]

alkaline earth metal An element from Group IIA of the periodic table. [10.4]

alkane (saturated hydrocarbon) A compound composed of carbon and hydrogen, having only single bonds between the atoms; also known as paraffin hydrocarbon. *See also* alkene and alkyne. [19.4]

alkene (unsaturated hydrocarbon) A hydrocarbon whose molecules have at least one carbon–carbon double bond. [19.7]

alkyl group An organic group derived from an alkane by removal of one H atom. The general formula is C_nH_{2n+1} (e.g., CH_3, methyl). Alkyl groups are generally indicated by the letter R. [19.6]

alkyne (unsaturated hydrocarbon) A hydrocarbon whose molecules have at least one carbon–carbon triple bond. [19.7]

allotrope A substance existing in two or more molecular or crystalline forms (example: graphite and diamond are two allotropic forms of carbon). [12.17]

alpha particle (α) A particle emitted from a nucleus of an atom during radioactive decay; it consists of two protons and two neutrons with a mass of about 4 amu and a charge of +2; it is considered to be a doubly charged helium atom. [18.3]

amino acid An organic compound containing two functional groups—an amino group (NH_2) and a carboxyl group (COOH). Amino acids are the building blocks for proteins. [20.4]

amorphous A solid without shape or form. [3.2]

amphoteric (substance) A substance having properties of both an acid and a base. [15.3]

anion A negatively charged ion. *See also* ion. [3.9, 5.5, 6.2]

anode The electrode where oxidation occurs in an electrochemical reaction. [17.6]

aqueous solution A water solution. [16.3]

aromatic compound An organic compound whose molecules contain a benzene ring, or which has properties resembling benzene. [19.10]

artificial radioactivity Radioactivity produced in nuclides during some types of transmutations. Artificial radioactive nuclides behave like natural radioactive elements in two ways: They disintegrate in a definite fashion and they have a specific half-life. Sometimes called *induced radioactivity*. [18.6]

1 atmosphere The standard atmospheric pressure; that is, the pressure exerted by a column of mercury 760 mm high at a temperature of 0°C. *See also* atmospheric pressure. [12.3]

atmospheric pressure The pressure experienced by objects on Earth as a result of the layer of air surrounding our planet. A pressure of 1 atmosphere (1 atm) is the pressure that will support a column of mercury 760 mm high at 0°C. [12.3]

atom The smallest particle of an element that can enter into a chemical reaction. [3.4]

atomic mass The average relative mass of the isotopes of an element referred to the atomic mass of carbon-12. [5.11]

atomic mass unit (amu) A unit of mass equal to one-twelfth the mass of a carbon-12 atom. [5.11]

atomic number The number of protons in the nucleus of an atom of a given element. *See also* isotopic notation. [5.9]

atomic theory The theory that substances are composed of atoms, and that chemical reactions are explained by the properties and the interactions of these atoms. [5.2, Ch. 10]

Avogadro's law Equal volumes of different gases at the same temperature and pressure contain equal numbers of molecules. [12.11]

Avogadro's number 6.022×10^{23}; the number of formula units in 1 mole. [7.1, 9.1]

balanced equation A chemical equation having the same number of each kind of atom and the same electrical charge on each side of the equation. [8.2]

barometer A device used to measure atmospheric pressure. [12.3]

base A substance whose properties are due to the liberation of hydroxide (OH^-) ions into a water solution. [15.1]

basic anhydride A metal oxide that reacts with water to form a base. [13.12]

beta particle (β) A particle identical in charge (-1) and mass to an electron. [18.3]

binary compound A compound composed of two different elements. [6.4]

biochemistry The branch of chemistry concerned with chemical reactions occurring in living organisms. [20.1]

boiling point The temperature at which the vapor pressure of a liquid is equal to the pressure above the liquid. [13.5]

bond length The distance between two nuclei that are joined by a chemical bond. [13.10]

Boyle's law At constant temperature, the volume of a fixed mass of gas is inversely proportional to the pressure ($PV = $ constant). [12.5]

Brownian movement The random motion of colloidal particles. [15.14]

buffer solution A solution that resists changes in pH when diluted or when small amounts of a strong acid or strong base are added. [16.14]

calorie (cal) A commonly used unit of heat energy; 1 calorie is a quantity of heat energy that will raise the temperature of 1 g of water 1°C (from 14.5 to 15.5°C). Also, 4.184 joules = 1 calorie exactly. *See also* joule. [4.6]

capillary action The spontaneous rising of a liquid in a narrow tube, which results from the cohesive forces within the liquid and the adhesive forces between the liquid and the walls of the container. [13.4]

carbohydrate A polyhydroxy aldehyde or polyhydroxy ketone, or a compound that upon hydrolysis yields a polyhydroxy aldehyde or ketone; sugars, starch, and cellulose are examples. [20.2]

carbonyl group The structure $\diagdown C = O$. [19.17]

carboxyl group The functional group of carboxylic acids:

$$\begin{matrix} & O \\ & \parallel \\ -& C - OH \end{matrix} \qquad [19.19]$$

carboxylic acid An organic compound having a carboxyl group. [19.19]

catalyst A substance that influences the rate of a reaction and can be recovered essentially unchanged at the end of the reaction. [16.8]

cathode The electrode where reduction occurs in an electrochemical reaction. [17.6]

cation A positively charged ion. *See also* ion. [3.9, 5.5]

Celsius temperature scale (°C) The temperature scale on which water freezes at 0°C and boils at 100°C at 1 atm pressure. [2.11]

chain reaction A self-sustaining nuclear or chemical reaction in which the products cause the reaction to continue or to increase in magnitude. [18.8]

Charles' law At constant pressure, the volume of a fixed mass of any gas is directly proportional to the absolute temperature ($V/T = $ constant). [12.6]

chemical bond The attractive force that holds atoms together in a compound. [Ch. 11]

chemical change A change producing products that differ in composition from the original substances. [4.3]

chemical equation A shorthand expression showing the reactants and the products of a chemical change (for example, $2 H_2O = 2 H_2 + O_2$). [4.3, 8.1]

chemical equilibrium The state in which the rate of the forward reaction equals the rate of the reverse reaction in a chemical change. [16.3]

chemical family *See* groups or families of elements.

chemical formula A shorthand method for showing the composition of a compound using symbols of the elements. [3.11]

chemical kinetics The study of reaction rates and reaction mechanisms. [16.2]

chemical properties The ability of a substance to form new substances either by reaction with other substances or by decomposition. [4.1]

chemistry The science of the composition, structure, properties, and reactions of matter, especially of atomic and molecular systems. [1.2]

chlorofluorocarbons (CFCs) A group of compounds made of carbon, chlorine, and fluorine. [12.17]

colligative properties Properties of a solution that depend on the number of solute particles in solution and not on the nature of the solute (examples: vapor-pressure lowering, freezing-point depression, boiling-point elevation). [14.7]

colloid A dispersion in which the dispersed particles are larger than the solute ions or molecules of a true solution and smaller than the particles of a mechanical suspension. [15.13]

combination reaction A direct union or combination of two substances to produce one new substance. [8.4]

combustion A chemical reaction in which heat and light are given off; generally, the process of burning or uniting a substance with oxygen. [19.4]

common ion effect The shift of a chemical equilibrium caused by the addition of an ion common to the ions in the equilibrium. [16.12]

common names Arbitrary names that are not based on the chemical composition of compounds (examples: quicksilver for mercury, laughing gas for nitrous oxide). [6.1]

compound A distinct substance composed of two or more elements combined in a definite proportion by mass. [3.9]

concentrated solution A solution containing a relatively large amount of dissolved solute. [14.6]

concentration of a solution A quantitative expression of the amount of dissolved solute in a certain quantity of solvent. [14.2]

condensation The process by which molecules in the gaseous state return to the liquid state. [13.3]

conjugate acid-base Two molecules or ions whose formulas differ by one H^+. (The acid is the species with the H^+, and the base is the species without the H^+.) [15.1]

copolymer A polymer containing two different kinds of monomer units. [19.21]

covalent bond A chemical bond formed between two atoms by sharing a pair of electrons. [11.5]

critical mass The minimum quantity of mass required to support a self-sustaining chain reaction. [18.8]

curie (Ci) A unit of radioactivity indicating the rate of decay of a radioactive substance: $1 \text{ Ci} = 3.7 \times 10^{10}$ disintegrations per second. [18.7]

Dalton's atomic theory The first modern atomic theory to state that elements are composed of minute individual particles called *atoms*. [5.2]

Dalton's law of partial pressures The total pressure of a mixture of gases is the sum of the partial pressures exerted by each of the gases in the mixture. [12.10]

decomposition reaction A breaking down, or decomposition, of one substance into two or more different substances. [8.4]

deliquescence The absorption of water from the atmosphere by a substance until it forms a solution. [13.14]

density The mass of an object divided by its volume. [2.12]

dialysis The process by which a parchment membrane allows the passage of true solutions but prevents the passage of colloidal dispersions. [15.15]

diatomic molecules The molecules of elements that always contain two atoms. Seven elements occur as diatomic molecules: H_2, N_2, O_2, F_2, Cl_2, Br_2, and I_2. [3.10]

diffusion The property by which gases and liquids mix spontaneously because of the random motion of their particles. [12.2]

dilute solution A solution containing a relatively small amount of dissolved solute. [14.6]

dipeptide Two α-amino acids joined by a peptide linkage. [20.4]

dipole A molecule that is electrically asymmetrical, causing it to be oppositely charged at two points [11.6]

disaccharide A carbohydrate that yields two monosaccharide units when hydrolyzed. [20.2]

disintegration series The spontaneous decay of a certain radioactive nuclide by emission of alpha and beta particles from the nucleus, finally stopping at a stable isotope of lead or bismuth. [18.4]

dissociation The process by which a salt separates into individual ions when dissolved in water. [15.6]

DNA Deoxyribonucleic acid; a high molar-mass polymer of nucleotides, present in all living matter, that contains the genetic code that transmits hereditary characteristics. [20.5]

double bond A covalent bond in which two pairs of electrons are shared. [11.5, 19.2]

double-displacement reaction A reaction of two compounds to produce two different compounds by exchanging the components of the reacting compounds. [8.4]

---

effusion The process by which gas molecules pass through a tiny orifice from a region of high pressure to a region of lower pressure. [12.2]

Einstein's mass-energy equation $E = mc^2$: the relationship between mass and energy. [18.12]

electrolysis The process whereby electrical energy is used to bring about a chemical change. [17.6]

electrolyte A substance whose aqueous solution conducts electricity. [15.5]

electrolytic cell An electrolysis apparatus in which electrical energy from an outside source is used to produce a chemical change. [17.6]

electron A subatomic particle that exists outside the nucleus and carries a negative electrical charge. [5.6]

electron configuration The orbital arrangement of electrons in an atom. [10.5]

electron-dot structure *See* Lewis structure.

electronegativity The relative attraction that an atom has for a pair of shared electrons in a covalent bond. [11.6]

electron shell *See* principal energy levels of electrons.

element A basic building block of matter that cannot be broken down into simpler substances by ordinary chemical changes; in 1994, there were 111 known elements. [3.4]

empirical formula A chemical formula that gives the smallest whole-number ratio of atoms in a compound—that is, the relative number of atoms of each element in the compound; also known as the simplest formula. [7.4]

endothermic reaction A chemical reaction that absorbs heat. [8.5]

energy The capacity of matter to do work. [4.5]

energy levels of electrons Areas in which electrons are located at various distances from the nucleus. [10.4]

energy sublevels The s, p, d, and f orbitals within a principal energy level occupied by electrons in an atom. [10.4]

enzyme A protein that catalyzes a biochemical reaction. [20.7]

equilibrium A dynamic state in which two or more opposing processes are taking place at the same time and at the same rate. [16.3]

equilibrium constant (K_{eq}) A value representing the unchanging concentrations of the reactants and the products in a chemical reaction at equilibrium. [16.9]

essential amino acid An amino acid that is not synthesized by the body and therefore must be supplied in the diet. [20.4]

ester An organic compound derived from a carboxylic acid and an alcohol. The general formula is

$$R - \underset{\underset{O}{\|}}{C} - OR'$$

[19.20]

ether An organic compound having two hydrocarbon groups attached to an oxygen atom. The general formula is $R - O - R'$. [19.16]

evaporation The escape of molecules from the liquid state to the gas or vapor state. [13.2]

exothermic reaction A chemical reaction in which heat is released as a product. [8.5]

Fahrenheit temperature scale ($°F$) The temperature scale on which water freezes at $32°F$ and boils at $212°F$ at 1 atm pressure. [2.11]

fats and oils Esters of fatty acids and glycerol. *See also* triacylglycerol. [20.3]

fatty acids Long-chain carboxylic acids present in lipids (fats and oils). [20.3]

formula unit The atom or molecule indicated by the formula of the substance under consideration (examples: Mg, O_2, H_2O). [7.1]

free radical A neutral atom or group of atoms having one or more unpaired electrons. [12.17]

freezing or melting point The temperature at which the solid and liquid states of a substance are in equilibrium. [13.6]

frequency A measurement of the number of waves that pass a particular point per second. [10.2]

functional group An atom or group of atoms that characterizes a class of organic compounds. For example, $-COOH$ is the functional group of carboxylic acids. [19.2]

---

galvanic cell *See* voltaic cell.

gamma ray (γ) High-energy photons emitted by radioactive nuclei; they have no electrical charge and no measurable mass. [18.3]

gas A state of matter that has no shape or definite volume so that the substance completely fills its container. [3.2]

Gay-Lussac's law At constant volume, the pressure of a fixed mass of gas is directly proportional to the absolute temperature (P/T = constant). [12.7]

Gay-Lussac's law of combining volumes (of gases) When measured at the same temperature and pressure, the ratios of the volumes of reacting gases are small whole numbers. [12.11]

genes Basic units of heredity that consist primarily of DNA and proteins and occur in the chromosomes. [20.6]

Graham's law of effusion The rates of effusion of two gases at the same temperature and pressure are inversely proportional to the square roots of their densities or molar masses. [12.2]

ground state The lowest available energy level within an atom. [10.3]

groups or families (of elements) Vertical groups of elements in the periodic table (IA, IIA, and so on). Families of elements that have similar outer-orbital electron structures. [10.6]

---

half-life The time required for one-half of a specific amount of a radioactive nuclide to disintegrate; half-lives of the elements range from a fraction of a second to billions of years. [18.2]

halogenation The substitution of a halogen atom for a hydrogen atom in an organic compound. [19.3]

halogens Group VIIA of the periodic table; consists of the elements fluorine, chlorine, bromine, iodine, and astatine. [10.4]

heat A form of energy associated with the motion of small particles of matter. [2.11]

heat of fusion The energy required to change 1 gram of a solid into a liquid at its melting point. [13.7]

heat of reaction The quantity of heat produced by a chemical reaction. [8.5]

heat of vaporization The amount of heat required to change 1 gram of a liquid to a vapor at its normal boiling point. [13.7]

Henry's law The amount of gas that will dissolve in a liquid varies directly with the pressure above the liquid. [CIA, p. 284]

heterogeneous Matter without uniform composition—having two or more components or phases. [3.3]

homogeneous Matter that has uniform properties throughout. [3.3]

homologous series A series of compounds in which the members differ from one another by a regular increment. For example, each successive member of the alkane series of hydrocarbons differs by a CH_2 group. [19.4]

hydrate A solid that contains water molecules as a part of its crystalline structure. [13.13]

hydrocarbon A compound composed entirely of carbon and hydrogen. [8.5, 19.3]

hydrogen bond The intermolecular force acting between molecules that contain hydrogen covalently bonded to the highly electronegative elements, F, O, and N. [13.11]

hydrolysis Chemical reaction with water in which the water molecule is split into H^+ and OH^-. [16.13]

hydronium ion The result of a protein combining with a polar water molecule to form a hydrated hydrogen ion, H_3O^+. [15.1]

hygroscopic substance A substance that readily absorbs and retains water vapor. [13.14]

hypothesis A tentative explanation of certain facts to provide a basis for further experimentation. [1.4]

ideal gas A gas that behaves precisely according to the Kinetic Molecular Theory; also called a perfect gas. [12.2]

ideal gas equation $PV = nRT$; that is, the volume of a gas varies *directly* with the number of gas molecules and the absolute temperature and *inversely* with the pressure. [12.14]

immiscible Incapable of mixing; immiscible liquids do not form a solution with one another. [14.2]

induced radioactivity *See* artificial radioactivity.

ion A positively or negatively charged atom or group of atoms. *See also* cation, anion. [3.9, 5.5]

ionic bond The chemical bond between a positively charged ion and a negatively charged ion. [11.3]

ionic compound A compound that is composed of ions (e.g., Na^+Cl^-). [3.9, 6.3]

ionization The formation of ions, which occurs as the result of a chemical reaction of certain substances with water. [15.6]

ionization energy The energy required to remove an electron from an atom, an ion, or a molecule. [11.1]

ionizing radiation Radiation with enough energy to dislocate bonding electrons and create ions when passing through matter. [18.4]

ion product constant for water (K_w) An equilibrium constant defined as the product of the H^+ ion concentration and the OH^- ion concentration, each in moles per liter. $K_w = [H^+][OH^-] = 1 \times 10^{-14}$ at 25°C. [16.10]

isomerism The phenomenon of two or more compounds having the same molecular formula but different molecular structures. [19.5]

isomers Compounds having identical molecular formulas but different structural formulas. [19.5]

isotope An atom of an element that has the same atomic number but a different atomic mass. Since their atomic numbers are identical, isotopes vary only in the number of neutrons in the nucleus. [5.10]

isotopic notation Notation for an isotope of an element where the subscript is the atomic number, the superscript is the mass number, and they are attached on the left of the symbol for the element. (For example, hydrogen-1 is notated as ^1_1H.) *See also* atomic number, mass number. [5.10, 19.1]

IUPAC International Union of Pure and Applied Chemistry, which devised (in 1921) and continually upgrades the system of nomenclature for inorganic and organic compounds. [6.1, 19.6]

joule (J) The SI unit of energy. *See also* calorie. [4.6]

Kelvin temperature scale (K) Absolute temperature scale starting at absolute zero, the lowest temperature possible. Freezing and boiling points of water on this scale are 273 K and 373 K, respectively, at 1 atm pressure. *See also* absolute zero. [2.11, 12.6]

ketone An organic compound that contains a carbonyl group between two other carbon atoms. The general formula is $R_2C = O$. [19.17]

kilocalorie (kcal) 1000 cal; the kilocalorie is also known as the nutritional or large Calorie, used for measuring the energy produced by food. [4.6]

kilogram (kg) The standard unit of mass in the metric system; 1 kilogram equals 2.205 pounds. [2.9]

kilojoule (kJ) 1000J. [4.6]

kinetic energy (KE) The energy that matter possesses due to its motion; $KE = 1/2\ mv^2$. [4.5]

kinetic-molecular theory (KMT) A group of assumptions used to explain the behavior and properties of gases. [12.2]

law A statement of the occurrence of natural phenomena that occur with unvarying uniformity under the same conditions. [1.4]

law of conservation of energy Energy can neither be created nor destroyed, but it can be transformed from one form to another. [4.8]

law of conservation of mass No change is observed in the total mass of the substances involved in a chemical reaction; that is, the mass of the products equals the mass of the reactants. [4.4]

law of definite composition A compound always contains two or more elements in a definite proportion by mass. [5.3]

law of multiple proportions Atoms of two or more elements may combine in different ratios to produce more than one compound. [5.3]

Le Chatelier's principle If a stress is applied to a system in equilibrium, the system will respond in such a way as to relieve that stress and restore equilibrium under a new set of conditions. [16.4]

Lewis structure A method of indicating the covalent bonds between atoms in a molecule or an ion such that a pair of electrons (:) represents the valence electrons forming the covalent bond. [11.2]

limiting reactant A reactant that limits the amount of product formed because it is present in insufficient amount compared to the other reactants. [9.6]

linear structure In the VSEPR model, an arrangement where the pairs of electrons are arranged 180° apart for maximum separation. [11.11]

line spectrum Colored lines generated when light emitted by a gas is passed through a spectroscope. Each element possesses a unique set of line spectrum. [10.3]

lipids Organic compounds found in living organisms that are water insoluble, but soluble in such fat solvents as diethyl ether, benzene, and carbon tetrachloride; examples are fats, oils, and steroids. [20.3]

liquid A state of matter in which the particles move about freely while the substance retains a definite volume; thus liquids flow and take the shape of their containers. [3.2]

liter (L) A unit of volume commonly used in chemistry; 1 L = 1000 mL; the volume of a kilogram of water at 4°C. [2.10]

logarithm (log) The power to which 10 must be raised to give a certain number. The log of 100 is 2.0. [15.9]

---

macromolecule *See* polymer.

mass The quantity or amount of matter that an object possesses. [2.1]

mass defect The difference between the actual mass of an atom of an isotope and the calculated mass of the protons and neutrons in the nucleus of that atom. [18.12]

mass number The sum of the protons and neutrons in the nucleus of a given isotope of an atom. *See also* isotopic notation. [5.10]

mass percent solution The grams of solute in 100 g of a solution. [14.6]

matter Anything that has mass and occupies space. [3.1]

meiosis The process of cell division to form a sperm cell and an egg cell in which each cell formed contains half of the chromosomes found in the normal single cell. [20.6]

melting or freezing point *See* freezing or melting point.

meniscus The curved upper surface of a liquid when placed in a glass cylinder. It can be concave or convex. [13.4]

metal An element that is solid at room temperature and whose properties include luster, ductility, malleability, and good conductivity of heat and electricity; metals tend to lose their valence electrons and become positive ions. [3.8]

metalloid An element having properties that are intermediate between those of metals and nonmetals (for example, silicon); these elements are useful in electronics. [3.8]

meter (m) The standard unit of length in the SI and metric systems; 1 meter equals 39.37 inches. [2.7]

metric system A decimal system of measurements. *See also* SI. [2.6]

miscible Capable of mixing and forming a solution. [14.2]

mitosis Ordinary cell division in which a DNA molecule is duplicated by uncoiling to single strands and then re-assembling with complementary nucleotides. Each new cell contains the normal number of chromosomes. [20.6]

mixture Matter containing two or more substances, which can be present in variable amounts; mixtures can be homogeneous (sugar water) or heterogeneous (sand and water). [3.3]

molality (m) An expression of the number of moles of solute dissolved in 1000 g of solvent. [14.7]

molarity (M) The number of moles of solute per liter of solution. [14.6]

molar mass The mass of Avogadro's number of atoms or molecules. [7.1, 9.1]

molar solution A solution containing 1 mole of solute per liter of solution. [14.6]

molar volume (of a gas) The volume of 1 mol of a gas at STP equals 22.4 L/mol. [12.12]

mole The amount of a substance containing the same number of formula units (6.022×10^{23}) as there are in exactly 12 g of ^{12}C. One mole is equal to the molar mass in grams of any substance. [7.1]

mole ratio A ratio between the number of moles of any two species involved in a chemical reaction; the mole ratio is used as a conversion factor in stoichiometric calculations. [9.2]

molecular equation *See* un-ionized equation.

molecular formula The total number of atoms of each element present in one molecule of a compound; also known as the true formula. *See also* empirical formula. [7.4]

molecule The smallest uncharged individual unit of a compound formed by the union of two or more atoms. [3.9]

monomer The small unit or units that undergo polymerization to form a polymer. [19.21]

monosaccharide A carbohydrate that cannot be hydrolyzed to simpler carbohydrate units (for example, simple sugars like glucose or fructose). [20.2]

---

net ionic equation A chemical equation that includes only those molecules and ions that have changed in the chemical reaction. [15.12]

neutralization The reaction of an acid and a base to form a salt plus water. [15.10]

neutron A subatomic particle that is electrically neutral and is found in the nucleus of an atom. [5.6]

noble gases A family of elements in the periodic table—helium, neon, argon, krypton, xenon, and radon—that contain a particularly stable electron structure. [10.4]

nonelectrolyte A substance whose aqueous solutions do not conduct electricity. [15.5]

nonmetal An element that has properties the opposite of metals: lack of luster, relatively low melting point and density, and generally poor conduction of heat and electricity. Nonmetals may or may not be solid at room temperature (examples: carbon, bromine, nitrogen); many are gases. They are located mainly in the upper right-hand corner of the periodic table. [3.8]

nonpolar covalent bond A covalent bond between two atoms with the same electronegativity value; thus the electrons are shared equally between the two atoms. [11.6]

normal boiling point The temperature at which the vapor pressure of a liquid equals 1 atm or 760 torr pressure. [13.5]

nuclear binding energy The energy equivalent to the mass defect; that is, the amount of energy required to break a nucleus into its individual protons and neutrons. [18.12]

nuclear fission The splitting of a heavy nuclide into two or more intermediate-sized fragments when struck in a particular way by a neutron. As the atom is split, it releases energy and two or three more neutrons that can then cause another nuclear fission. [18.8]

nuclear fusion The uniting of two light elements to form one heavier nucleus, which is accompanied by the release of energy. [18.11]

nucleic acids Complex organic acids essential to life and found in the nucleus of living cells. They consist of thousands of units called nucleotides. Includes DNA and RNA. [20.5]

nucleon A collective term for the neutrons and protons in the nucleus of an atom. [18.1]

nucleotide The building-block unit for nucleic acids. A phosphate group, a sugar residue, and a nitrogenous organic base are bonded together to form a nucleotide. [20.5]

nucleus The central part of an atom that contains all its protons and neutrons. The nucleus is very dense and has a positive electrical charge. [5.8]

nuclide A general term for any isotope of any atom. [18.1]

---

oils *See* fats and oils.

one atmosphere The standard atmospheric pressure; that is, the pressure exerted by a column of mercury 760 mm high at a temperature of 0°C. [12.3]

orbital A cloudlike region around the nucleus where electrons are located. Orbitals are considered to be energy sublevels (*s, p, d f*) within the principal energy levels. *See also* principal energy levels. [10.3, 10.4]

orbital diagram A way of showing the arrangement of electrons in an atom where boxes with small arrows indicating the electrons represent orbitals. [10.5]

organic chemistry The branch of chemistry that deals with carbon compounds but does not imply that these compounds must originate from some form of life. *See also* vital-force theory. [19.1]

osmosis The diffusion of water, either pure or from a dilute solution, through a semipermeable membrane into a solution of higher concentration. [14.8]

oxidation An increase in the oxidation number of an atom as a result of losing electrons. [17.2]

oxidation number A small number representing the state of oxidation of an atom. For an ion it is the positive or negative charge on the ion; for covalently bonded atoms it is a positive or negative number assigned to the more electronegative atom; in free elements it is zero. [17.1]

oxidation–reduction A chemical reaction wherein electrons are transferred from one element to another; also known as redox. [17.2]

oxidation state *See* oxidation number. [17.1]

oxidizing agent A substance that causes an increase in the oxidation state of another substance. The oxidizing agent is reduced during the course of the reaction. [17.2]

---

partial pressure The pressure exerted independently by each gas in a mixture of gases. [12.10]

parts per million (ppm) A measurement of the concentration of dilute solutions now commonly used by chemists in place of mass percent. [14.6]

Pauli exclusion principle An atomic orbital can hold a maximum of two electrons, which must have opposite spins. [10.4]

peptide linkage The amide bond in a protein molecule; bonds one amino acid to another. [20.4]

percent composition of a compound The mass percent represented by each element in a compound. [7.3]

percent yield The ratio of the actual yield to the theoretical yield multiplied by 100. [9.6]

perfect gas A gas that behaves precisely according to theory; also called an *ideal gas* [12.2]

periodic table An arrangement of the elements according to their atomic numbers. The table consists of horizontal rows or periods and vertical columns or families of elements. Each period ends with a noble gas. [10.6]

period of elements The horizontal groupings (rows) of elements in the periodic table. [10.6]

pH A method of expressing the H^+ concentration (acidity) of a solution; $pH = -\log [H^+]$, $pH = 7$ is a neutral solution, $pH < 7$ is acidic, and $pH > 7$ is basic. [15.9]

phase A homogeneous part of a system separated from other parts by a physical boundary. [3.3]

photon Theoretically, a tiny packet of energy that streams with others of its kind to produce a beam of light. [10.2]

photosynthesis The process by which green plants utilize light energy to synthesize carbohydrates. [CIA p. 157]

physical change A change in form (such as size, shape, physical state) without a change in composition. [4.2]

physical properties Inherent physical characteristics of a substance that can be determined without altering its composition: color, taste, odor, state of matter, density, melting point, boiling point. [4.1]

physical states of matter Solids, liquids, and gases. [3.2]

physiological saline solution A solution of 0.90% sodium chloride that is isotonic (has the same osmotic pressure) with blood plasma. [14.8]

pOH A method of expressing the basicity of a solution. $pOH = -\log [OH^-]$. $pOH = 7$ is a neutral solution, $pOH < 7$ is basic, and $pOH > 7$ is acidic. [16.11]

polar covalent bond A covalent bond between two atoms with differing electronegativity values, resulting in unequal sharing of bonding electrons. [11.5]

polyatomic ion An ion composed of more than one atom. [6.5]

polyhydroxyl alcohol An alcohol that has more than one —OH group. [19.14]

polymer (macromolecule) A natural or synthetic giant molecule formed from smaller molecules (monomers). [19.21]

polymerization The process of forming large, high-molar-mass molecules from smaller units. [19.21]

polypeptide A peptide chain containing up to 50 amino acid units. [20.4]

polysaccharide A carbohydrate that can be hydrolyzed to many monosaccharide units; cellulose, starch, and glycogen are examples. [20.2]

positron A particle with a +1 charge having the mass of an electron (a positive electron). [18.1]

potential energy (PE) Stored energy, or the energy of an object due to its relative position. [4.5]

pressure Force per unit area; expressed in many units, such as mm Hg, atm, lb/in.^{2}, torr, pascal. [12.3]

primary alcohol An alcohol in which the carbon atom bonded to the —OH group is bonded to only one other carbon atom. [19.14]

principal energy levels of electrons Existing within the atom, these energy levels contain orbitals within which electrons are found. *See also* orbital, electron. [10.4]

product A chemical substance produced from reactants by a chemical change. [4.3]

properties The characteristics, or traits, of substances that give them their unique identities. Properties are classified as physical or chemical. [4.1]

protein A polymer consisting mainly of α-amino acids linked together; occurs in all animal and vegetable matter. [20.4]

proton a subatomic particle found in the nucleus of the atom that carries a positive electrical charge and a mass of about 1 amu. An H^+ ion is a proton. [5.6]

quanta Small discrete increments of energy. From the theory proposed by physicist Max Planck that energy is emitted in energy *quanta* rather than a continuous stream. [10.3]

quantum mechanics or wave mechanics The modern theory of atomic structure based on the wave properties of matter. [10.3]

rad (radiation absorbed dose) A unit of absorbed radiation indicating the energy absorbed from any ionizing radiation; 1 rad = 0.01 J of energy absorbed per kilogram of matter. [18.7]

radioactive decay The process by which a radioactive element emits particles or rays and is transformed into another element. [18.2]

radioactivity The spontaneous emission of radiation from the nucleus of an atom. [18.1]

rate of reaction The rate at which the reactants of a chemical reaction disappear and the products form. [16.2]

reactant A chemical substance entering into a reaction. [4.3]

redox *See* oxidation–reduction. [17.2]

reducing agent A substance that causes a decrease in the oxidation state of another substance; the reducing agent is oxidized during the course of a reaction. [17.2]

reduction A decrease in the oxidation number of an element as a result of gaining electrons. [17.2]

rem (roentgen equivalent to man) A unit of radiation-dose equivalent taking into account that the energy absorbed from different sources does not produce the same degree of biological effect. [18.7]

representative element An element in one of the A groups in the periodic table. [10.6]

resonance structure A molecule or ion that has multiple Lewis structures. *See also* Lewis structure. [11.8]

reversible chemical reaction A chemical reaction in which the products formed react to produce the original reactants. A double arrow is used to indicate that a reaction is reversible. [16.1]

RNA Ribonucleic acid; a high-molar-mass polymer of nucleotides present in all living matter. Its main function is to direct the synthesis of proteins. [20.5]

roentgen (R) A unit of exposure of gamma radiation based on the quantity of ionization produced in air. [18.7]

rounding off numbers The process by which the value of the last digit retained is determined after dropping nonsignificant digits. [2.3]

salts Ionic compounds of cations and anions. [Ch. 6, 15.4]

saturated hydrocarbon A hydrocarbon that has only single bonds between carbon atoms; classified as *alkanes*. [19.3]

saturated solution A solution containing dissolved solute in equilibrium with undissolved solute. [14.3]

scientific laws Simple statements of natural phenomena to which no exceptions are known under the given conditions. [1.4]

scientific method A method of solving problems by observation; recording and evaluating data of an experiment; formulating hypotheses and theories to explain the behavior of nature; and devising additional experiments to test the hypotheses and theories to see if they are correct. [1.4]

scientific notation Writing a number as a power of 10; to do this, move the decimal point in the original number so that it is located after the first nonzero digit, follow the new number by a multiplication sign and 10 with an exponent (called its *power*) that is the number of places the decimal point was moved.
Example: $2468 = 2.468 \times 10^3$. [2.4]

secondary alcohol An alcohol in which the carbon atom bonded to the $-$OH group is bonded to two other carbon atoms. [19.14]

semipermeable membrane A membrane that allows the passage of water (solvent) molecules through it in either direction but prevents the passage of larger solute molecules or ions. [14.8]

SI An agreed-upon standard system of measurements used by scientists around the world (*Système Internationale*). *See also* metric system. [2.6]

significant figures The number of digits that are known plus one estimated digit are considered significant in a measured quantity; also called significant digits. [2.2, Appendix I]

simplest formula *See* empirical formula.

single bond A covalent bond in which one pair of electrons is shared between two atoms. [11.5, 19.2]

single-displacement reaction A reaction of an element and a compound that yields a different element and a different compound. [8.4]

soap A salt of a long-carbon-chain fatty acid. [20.3]

solid A state of matter having a definite shape and a definite volume, whose particles cohere rigidly to one another, so that a solid can be independent of its container. [3.2]

solubility An amount of solute that will dissolve in a specific amount of solvent under stated conditions. [14.2]

solubility product constant (K_{sp}) The equilibrium constant for the solubility of a slightly soluble salt. [16.12]

solute The substance that is dissolved—or the least abundant component—in a solution. [14.1]

solution A system in which one or more substances are homogeneously mixed or dissolved in another substance. [14.1]

solvent The dissolving agent or the most abundant component in a solution. [14.1]

specific gravity The ratio of the density of one substance to the density of another substance taken as a standard. Water is usually the standard for liquids and solids; air, for gases. [2.12]

specific heat The quantity of heat required to change the temperature of 1 g of any substance by 1°C. [4.6]

spectator ion An ion in solution that does not undergo chemical change during a chemical reaction. [15.10]

speed (of a wave) A measurement of how fast a wave travels through space. [10.2]

spin A property of an electron that describes its appearance of spinning on an axis like a globe; the electron can only spin in two directions and, to occupy the same orbital, two electrons must spin in opposite directions. *See also* orbital. [10.4]

standard boiling point *See* normal boiling point.

standard conditions *See* STP.

standard temperature and pressure *See* STP.

Stock (nomenclature) System A system that uses Roman numerals to name elements that form more than one type of cation. (For example: Fe^{2+}, iron(II); Fe^{3+}, iron(III).) [6.4]

stoichiometry The area of chemistry that deals with the quantitative relationships among reactants and products in a chemical reaction. [9.2]

STP (standard temperature and pressure) 0°C (273 K) and 1 atm (760 torr); also known as standard conditions. [12.8]

strong electrolyte An electrolyte that is essentially 100% ionized in aqueous solution. [15.7]

subatomic particles Particles found within the atom, mainly protons, neutrons, and electrons. [5.6]

sublimation The process of going directly from the solid-state to the vapor state without becoming a liquid. [13.2]

subscript Number that appears partially below the line and to the right of a symbol of an element (example: H_2SO_4). [3.11]

substance Matter that is homogeneous and has a definite, fixed composition; substances occur in two forms—as elements and as compounds. [3.3]

substrate In biochemical reactions the substrate is the unit acted upon by an enzyme. [20.7]

supersaturated solution A solution containing more solute than needed for a saturated solution at a particular temperature. Supersaturated solutions tend to be unstable; jarring the container or dropping in a "seed" crystal will cause crystallization of the excess solute. [14.3]

surface tension The resistance of a liquid to an increase in its surface area. [13.4]

symbol In chemistry an abbreviation for the name of an element. [3.7]

system A body of matter under consideration. [3.3]

---

temperature A measure of the intensity of heat, or of how hot or cold a system is; the SI unit is the kelvin (K). [2.11]

tertiary alcohol An alcohol in which the carbon atom bonded to the $-$OH group is bonded to three other carbon atoms. [19.14]

tetrahedral structure An arrangement of the VSEPR model where four pairs of electrons are placed 109.5° degrees apart to form a tetrahedron. [11.11, 19.2]

theoretical yield The maximum amount of product that can be produced according to a balanced equation. [9.6]

theory An explanation of the general principles of certain phenomena with considerable evidence to support it; a well-established hypothesis. [1.4]

Thomson model of the atom Thomson asserted that atoms are not indivisible but are composed of smaller parts; they contain both positively and negatively charged particles—protons as well as electrons. [5.6]

titration The process of measuring the volume of one reagent required to react with a measured mass or volume of another reagent. [15.10]

torr A unit of pressure (1 torr = 1 mm Hg). [12.3]

total ionic equation An equation that shows compounds in the form in which they actually exist. Strong electrolytes are written as ions in solution, whereas nonelectrolytes, weak electrolytes, precipitates, and gases are written in the un-ionized form. [15.12]

transcription The process of forming RNA from DNA. [20.5]

transition elements The metallic elements characterized by increasing numbers of d and f electrons. These elements are located in Groups IB–VIIB and in Group VIII of the periodic table. [10.6]

transmutation The conversion of one element into another element. [18.5]

transuranium element An element that has an atomic number higher than that of uranium (> 92). [18.13]

triacylglycerol (triglyceride) An ester of glycerol and three molecules of fatty acids. [20.3]

trigonal planar An arrangement of atoms in the VSEPR model where the three pairs of electrons are placed $120°$ apart on a flat plane. [11.11]

triple bond A covalent bond in which three pairs of electrons are shared between two atoms. [11.5, 19.2]

Tyndall effect An intense beam of light passed through a colloidal dispersion is clearly visible but is not visible when passed through a true solution. [15.14]

un-ionized equation A chemical equation in which all the reactants and products are written in their molecular, or normal, formula expression; also called a molecular equation. [15.12]

unsaturated hydrocarbon A hydrocarbon whose molecules contain one or more double or triple bonds between two carbon atoms; classified as *alkenes, alkynes,* and *aromatic* compounds. [14.3]

unsaturated solution A solution containing less solute per unit volume than its corresponding saturated solution. [14.6]

valence electron An electron in the outermost energy level of an atom; these electrons are the ones involved in bonding atoms together to form compounds. [10.5]

vapor pressure The pressure exerted by a vapor in equilibrium with its liquid. [13.3]

vaporization *See* evaporation. [13.2]

vapor-pressure curve A graph generated by plotting the temperature of a liquid on the x-axis and its vapor pressure on the y-axis. Any point on the curve represents an equilibrium between the vapor and liquid. [13.5]

vital-force theory A theory that held that organic substances could originate only from some form of living material. The theory was overthrown early in the 19th century. [19.1]

volatile (substance) A substance that evaporates readily; a liquid with a high vapor pressure and a low boiling point. [13.3]

voltaic cell A cell that produces electric current from a spontaneous chemical reaction. [17.6]

volume The amount of space occupied by matter; measured in SI units by cubic meters (m^3), but also commonly in liters and milliliters. [2.10]

volume percent (solution) The volume of solute in 100 mL of solution. [14.6]

VSEPR Valence shell electron pair repulsion; a simple model for predicting the shapes of molecules. [11.11]

water of crystallization Water molecules that are part of a crystalline structure, as in a hydrate; also called water of hydration. [13.13]

water of hydration *See* water of crystallization.

wavelength The distance between consecutive peaks and troughs in a wave; symbolized by the Greek letter lambda. [10.2]

weak electrolyte A substance that is ionized to a small extent in aqueous solution. [15.7]

weight A measure of the earth's gravitational attraction for a body (object). [2.1]

word equation A statement in words, in equation form, of the substances involved in a chemical reaction. [8.2]

INDEX

Words that appear in bold type are defined in the Glossary.

PHOTO CREDITS

Chapter 1
p. 1, Photodisc; p. 3, *Top:* J. Crawford/Image Works; *Bottom:* Mel Kindstrom/Photo Researchers; p. 4, Tom Pantages (2 photos); p. 5, *Top:* Rita Amaya; p. 7, T. McCarthy/PhotoEdit; p. 8, T. McCarthy/PhotoEdit; p. 9, B. Mahoney/Image Works.

Chapter 2
p. 10, Murray Alcosser/Image Bank; p. 11, SKA Archives; p. 12, Tom Pantages; p. 15, *Left:* SKA Archives; *Right:* David M. Phillips/Visuals Unlimited; p. 19, Ex Libris; p. 26, T. P. Dickerson/Image Works; p. 27, Rita Amaya (3 photos); p. 33, Poterfield/Chickoring/Photo Researchers; p. 35, *Top:* Richard Megna/Fundamental Photographs; *Bottom:* Tom McCarthy.

Chapter 3
p. 44, John McAnulty/Corbis; p. 45, Tom Pantages; p. 47, Edward S. Ross/Phototake; p. 49, Tom Pantages; p. 50, *Top:* Jerry Schad/Photo Researchers; *Middle:* Science Source/Photo Researchers; *Bottom:* Photo Researchers; p. 51, *Top:* Tom Pantages; *Bottom:* Richard Megna/Fundamental Photographs; p. 52, *Top:* Phil Silverman/Fundamental Photographs; *Bottom:* Rita Amaya; p. 53, *Top:* Courtesy of IBM; *Bottom:* Rita Amaya; p. 56, Damlier-Benz; p. 59, Rita Amaya; p. 60, Clockwise, starting from *Top:* Phil Hayson/Photo Researchers, Lowell Georgia/Photo Researchers. Tom Pantages (2 photos); Ken Edward/Photo Researchers, Tom Pantages.

Chapter 4
p. 64, Charles E. Rotkin/Corbis; p. 66, *Left:* Susan Van Etten/PhotoEdit; *Right:* Martin Dohrn/Photo Researchers; p. 67, Tom Pantages (3 photos); p. 68, Yoav Levy/Phototake; p. 70, B. Yaruin/Image Works; p. 72, Consumer Reports (2 photos); p. 74, Lee F. Snyder; p. 75, Ex Libris.

Chapter 5
p. 82, Stock Boston/PNI/Jeff Albertson; p. 86, *Top:* Tom Pantages; *Bottom:* Courtesy of IBM Almaden Research Center; p. 87: *Top:* Ex Libris; *Bottom: Left:* Tom Pantages; *Middle:* Richard Megna/Fundamental Photographs; *Right:* Larry Stepanowicz/Fundamental Photographs; p. 92, *Top:* Photodisc; *Bottom:* D. Young-Wolfe/PhotoEdit; p. 94, Leslye Bordon/PhotoEdit; p. 96, *Left:* Owen Franken/Corbis; *Center:* Rachael Epstein; *Right:* Ex Libris.

Chapter 6
p. 101, Antonio Rosario/Image Bank; p. 104, Clockwise, *Top Left:* Lawrence Berkeley National Laboratory; Image Works; Corbis; Achim Zschau/GSI; Science Source/Photo Researchers; Corbis; p. 107, Tom Pantages (3 photos); p. 109, Courtesy of Sargent-Welch/A VWR Corporation; p. 111, Tom Pantages; p. 114, Tom Pantages; p. 116, Tom Pantages; p. 117, Tom McCarthy; p. 120, *Left:* G. Chapman/P. Devadoss/Visuals Unlimited; *Right:* Tom Pantages.

Chapter 7
p. 126, Bill Bachmann/Image Works; p. 127, Tom McCarthy; p. 128, Tom Pantages (3 photos); p. 133, Tom Pantages; p. 134, Tom Pantages; p. 138, Tom McCarthy; p. 139, Rachael Epstein; p. 144, Tom Pantages.

Chapter 8
p. 149, Richard Megna/Fundamental Photographs; p. 157, Frederica Georgia/Photo Researchers; p. 158, Tom McCarthy; p. 159, Tom Pantages; p. 160, Richard Megna/Fundamental Photographs; p. 162, Richard Megna/Fundamental Photographs; p. 163, M. Dalton/Fundamental Photographs; p. 164, Peticolas/Megna/Fundamental Photographs; p. 165, Dave Bjorn/Tony Stone Images; p. 166, Rita Amaya; p. 167, Tom Pantages; p. 168, Jacques Jangour/Photo Researchers.

Chapter 9
p. 173, Michael Taufic/Courtesy of Merck; p. 174, Richard Megna/Fundamental Photographs; p. 179, Courtesy of NASA; p.180, Tom Pantages; p. 183, Hank Morgan/Photo Researchers.

Chapter 10
p. 198, Joe Sohm/Image Works; p. 199, Tony Freeman/PhotoEdit; p. 203, Craig Wells/Tony Stone Images; p. 205, Oregon Scientific; p. 206, Courtesy of IBM Almaden Research Center; p. 208, *Left:* Jack Bostrack/Visuals Unlimited; *Middle:* Gopel Murti/Photo Researchers; *Right:* Lawrence Livermore Laboratory/Science Photo Library/Photo Researchers.

Chapter 11
p. 219, Andrea Pistolesi/Image Bank; p. 224, Tom Pantages; p. 227, Tom Pantages; p. 228, Tom Pantages; p. 230, IBM Research/Peter Arnold; p. 235, Tom Pantages; p. 238, *Left:* IBM Almaden Research Center; *Right:* Bharat Bhushan/Ohio State University; p. 243, Tom McCarthy; p. 249, Tom McCarthy.

Chapter 12
p. 258, Joe Sohm/Image Works; p. 259, Tom Pantages; p. 271, Rita Amaya; p. 277, Rachael Epstein; p. 281, Tom Pantages; p. 284, Stephen Frink/Tony Stone Images; p. 292, SKA Archives; p. 293, Kay Pfortner/Peter Arnold.

Chapter 13
p. 300, S. R. Maglione/Photo Researchers; p. 304, *Top:* Yoav Levy/Phototake; *Bottom:* Alan Oddie/PhotoEdit; p. 307, Peter G. Aitken/Photo Researchers; p. 308, Gaillarmo Gonzalez/Visuals Unlimited; p. 310, Tom Pantages; p. 312, Ex Libris; p. 315, Richard Megna/Fundamental Photographs; p. 318, John Pinkston/Laura Stern/USGS; p. 319 *Top left and right:* Richard Megna/Fundamental Photographs; p. 320: Sara Hotchkiss/SKA Archive; p. 321, David Young-Wofl/PhotoEdit; p. 323, Mitch Kaufman/Courtesy of Southern California Edison.

Chapter 14
p. 330, Louisa Preston; p. 332, Richard Megna/Fundamental Photographs; p. 333, Kip Peticolas/Fundamental Photographs; p. 336, Ex Libris; p. 337, Richard Megna/Fundamental Photographs; p. 340, Eric Bouvet/Gamma Liaison; p. 349, Courtesy of Sargent Welch/ A VWR Corporation; pp. 352 and 353, Ex Libris; p. 354, Craig Newbauer/Peter Arnold; p. 357, Dennis Kunkel/Phototake.

Prefixes and Numerical Values for SI Units

Prefix	Symbol	Numerical value	Power of 10 equivalent
exa	E	1,000,000,000,000,000,000	10^{18}
peta	P	1,000,000,000,000,000	10^{15}
tera	T	1,000,000,000,000	10^{12}
giga	G	1,000,000,000	10^{9}
mega	M	1,000,000	10^{6}
kilo	k	1,000	10^{3}
hecto	h	100	10^{2}
deka	da	10	10^{1}
—	—	1	10^{0}
deci	d	0.1	10^{-1}
centi	c	0.01	10^{-2}
milli	m	0.001	10^{-3}
micro	μ	0.000001	10^{-6}
nano	n	0.000000001	10^{-9}
pico	p	0.000000000001	10^{-12}
femto	f	0.000000000000001	10^{-15}
atto	a	0.000000000000000001	10^{-18}

SI Units and Conversion Factors

Length

SI unit: meter (m)

1 meter	=	1.0936 yards
1 centimeter	=	0.3937 inch
1 inch	=	2.54 centimeters (exactly)
1 kilometer	=	0.62137 mile
1 mile	=	5280 feet
	=	1.609 kilometers
1 angstrom	=	10^{-10} meter

Mass

SI unit: kilogram (kg)

1 kilogram	=	1000 grams
	=	2.20 pounds
1 pound	=	453.59 grams
	=	0.45359 kilogram
	=	16 ounces
1 ton	=	2000 pounds
	=	907.185 kilograms
1 ounce	=	28.3 grams
1 atomic mass unit	=	1.6606×10^{-27} kilograms

Volume

SI unit: cubic meter (m³)

1 liter	=	10^{-3} m^3
	=	1 dm^3
	=	1.0567 quarts
1 gallon	=	4 quarts
	=	8 pints
	=	3.785 liters
1 quart	=	32 fluid ounces
	=	0.946 liter
1 fluid ounce	=	29.6 mL

Temperature

SI unit: kelvin (K)

0 K	=	−273.15°C
	=	−459.67°F
K	=	°C + 273.15

$$°C = \frac{(°F - 32)}{1.8}$$

°F = 1.8(°C) + 32

°F = 1.8(°C + 40) − 40

Energy

SI unit: joule (J)

1 joule	=	1 kg m^2/s^2
	=	0.23901 calorie
1 calorie	=	4.184 joules

Pressure

SI unit: pascal (Pa)

1 pascal	=	1 kg/m^{1}s^2
1 atmosphere	=	101.325 kilopascals
	=	760 torr
	=	760 mm Hg
	=	14.70 pounds per square inch (psi)